# Virtual Seminar Series on Pseudo-Hermitian Hamiltonians in Quantum Physics

Journal of Physics: Conference Series Volume 2038

Online
5 March 2021

ISBN: 978-1-7138-5009-0
ISSN: 1742-6588

**Printed from e-media with permission by:**

Curran Associates, Inc.
57 Morehouse Lane
Red Hook, NY 12571

CURRAN ASSOCIATES INC.
proceedings
.com

**Some format issues inherent in the e-media version may also appear in this print version.**



Institute of Physics
Dirac House, Temple Back
Bristol BS1 6BE UK

Phone: 44 1 17 929 7481
Fax: 44 1 17 920 0979

techtracking@iop.org

**Additional copies of this publication are available from:**

Curran Associates, Inc.
57 Morehouse Lane
Red Hook, NY 12571 USA
Phone: 845-758-0400
Fax: 845-758-2633
Email: curran@proceedings.com
Web: www.proceedings.com

# TABLE OF CONTENTS

Journal of Physics: Conference Series **2038** (2021) 011001 doi:10.1088/1742-6596/2038/1/011001

# Preface PT Seminars 2020

Non-Hermitian theories that are explicitly PT-symmetric and/or quasi/pseudo-Hermitian have become the subject of an active research field for the last two decades. They constitute well-defined classical and quantum theories possessing real eigenvalue spectra and unitary time-evolution. Since the underlying concepts are generic, the general principles have been applied in essentially all areas of physics and have unravelled new theoretical insights and experimental realisations especially in optical applications.

Beginning in 2003 a regular conference series entitled "Pseudo-Hermitian Hamiltonians in Quantum Physics" has taken place annually to bring together researchers in order to take stock of the current state of the field and to assist participants to collaborate in developing new ideas. Unfortunately, the 20th meeting in this series, which was scheduled originally for 2020 to be held in Los Alamos had to be postponed to 2022 due to the outbreak of the Covid 19 pandemic. To bridge the gap and to provide a platform to allow for a continued exchange of ideas and novelties a virtual seminar series via Zoom with a dedicated website (https://vphhqp.com) was organised by Andreas Fring and Francisco Correa. With currently more than 50 seminars scheduled and well over 200 registered participants, the series is still ongoing. The main motivation behind this special issue has been to allow speakers and participants in this series to present their results in a coherent written in the form of mini-reviews of certain subtopics in the field and new research results.

We hope that this special issue will become a valuable reference and inspiration for a broad scientific community, experts and newcomers alike, working in mathematical and theoretical physics.

**Carl M. Bender**, Department of Physics, Washington University, St. Louis, MO 63130, USA; cmb@wuphys.wustl.edu
**Francisco Correa**, Instituto de e Ciencias Físicas y Matemáticas, Universidad Austral de Chile, Casilla 567, Valdivia, Chile; francisco.correa@uach.cl
**Andreas Fring**, Department of Mathematics, City, University of London, Northampton Square, London EC1V 0HB, UK; a.fring@city.ac.uk

Journal of Physics: Conference Series **2038** (2021) 011002 doi:10.1088/1742-6596/2038/1/011002

# Peer review declaration

All conference organisers/editors are required to declare details about their peer review. Therefore, please provide the following information:

- **Type of peer review:**
  Single-blind (for some papers we asked for second and third opinions)
- **Conference submission management system:**
  All papers were submitted directly to the editors.
- **Number of submissions received: 29**
- **Number of submissions sent for review: 29**
- **Number of submissions accepted: 26**
- **Acceptance Rate (Number of Submissions Accepted / Number of Submissions Received X 100): 90%**
- **Average number of reviews per paper: 1.2**
- **Total number of reviewers involved: 30**
- **Any additional info on review process:**
- **Contact person for queries: Professor Andreas Fring**

Please submit this form along with the rest of your files on the submission date written in your publishing agreement.

The information you provide will be published as part of your proceedings.

# Pseudo-bosons and bi-coherent states out of $\mathcal{L}^2(\mathbb{R})$

**Fabio Bagarello**[1,2]

[1] Dipartimento di Ingegneria, Università di Palermo, I-90128 Palermo, Italy

[2] I.N.F.N., Sezione di Napoli

E-mail: `fabio.bagarello@unipa.iot`

Home page: www1.unipa.it/fabio.bagarello

**Abstract.** In this paper we continue our analysis on deformed canonical commutation relations and on their related pseudo-bosons and bi-coherent states. In particular, we show how to extend the original approach outside the Hilbert space $\mathcal{L}^2(\mathbb{R})$, leaving untouched the possibility of defining eigenstates of certain number-like operators, manifestly non self-adjoint, but opening to the possibility that these states are not square-integrable. We also extend this possibility to bi-coherent states, and we discuss in many details an example based on a couple of superpotentials first introduced in [30]. The results deduced here belong to the same distributional approach to pseudo-bosons first proposed in [29].

## 1. Introduction

It is today out of doubt that non self-adjoint Hamiltonians describe relevant physical systems: gain-loss systems, or systems having $PT$-symmetry are often connected with observables which, at least apparently, do not satisfy the *standard* self-adjointness condition: if $\mathcal{S}$ is such a system, described in a certain Hilbert space $\mathcal{H}$, with scalar product $\langle .,. \rangle$, its *natural* adjoint is defined in term of the scalar product: given $X \in B(\mathcal{H})$, the set of bounded operators on $\mathcal{H}$[1], its adjoint $X^\dagger$ is defined by $\langle X^\dagger f, g\rangle = \langle f, Xg\rangle$, $\forall f, g \in \mathcal{H}$. A self-adjoint operator $H$ satisfies the equality $H = H^\dagger$. Since the seminal paper [1] it became clear that it is possible to have a *meaningful physical models* also in presence of non self-adjoint Hamiltonians, at least if these Hamiltonians satisfy some different, and surely more physically motivated, symmetry conditions. Since then the interest in this *extended* quantum mechanics diffused among physicists, and produced several results, both from a theoretical and from an experimental point of view. Also, this interest was often shared by mathematicians, who started to contribute to the subject. What follows is a very partial (and biased) list of references, mainly focused on theoretical and mathematical aspects of this research: [2]-[16].

[1] We restrict here to bounded operators for simplicity. The adjoint can be defined also for unbounded operators, adding some conditions on the domains.

From a mathematical side, loosing self-adjointness of the Hamiltonian $H$ immediately implies that its eigenvalues could be complex and its eigenvectors can be not mutually orthogonal. While the first aspect can be easily controlled, the second is more subtle. More explicitly, simple conditions ensuring that, even if $H \neq H^\dagger$, its eigenvalues are all real, are well known. For instance, it is sufficient that $H$ and $H^\dagger$ are similar, i.e., that an invertible operator $S$ exists such that $H = SH^\dagger S^{-1}$. It is important to stress once again that we are restricting here to bounded operators. Otherwise, the situation is not that simple, but many results exist also in this case, see [9], for instance. As for the eigestates of $H$, these are no longer orthogonal, in general. However, a biorthogonal set can usually be found, made of the eigenstates of $H^\dagger$. However, even if the two sets of eigenvectors we find are usually complete (or total) in $\mathcal{H}$, they are not necessarily bases for $\mathcal{H}$. These aspects are discussed, for instance, in [10, 17].

Suppose now that $H$ is factorizable, i.e. that two operators $A$ and $B$ exist, bounded or not, such that $H = BA$. This is what happens, in particular, for the very well known quantum harmonic oscillator, for which $H_0 = c^\dagger c$, where $[c, c^\dagger] = 1\!\!1$, the identity operator on $\mathcal{H}$. It is known that, in this case, $c$, $c^\dagger$ and $H_0$ are unbounded. Still there is a common domain of functions for all these (and other) operators. $H_0$ has an orthonormal (o.n.) basis of eigenvectors, written in terms of Hermite polynomials. Not surprisingly, in the very same way in which an o.n set of eigenvectors of, say, $H_0$ is *doubled* in two biorthogonal sets of eigenstates of $H$ and $H^\dagger$ when $c$ and $c^\dagger$ are replaced by $A$ and $B$, $[A, B] = 1\!\!1$, the standard coherent state, eigenstate of $c$, is doubled into two different vectors, the so-called *bi-coherent states*, which are eigenstates of $A$ and $B^\dagger$ respectively. This was discussed first in [18], and then, with a slightly more mathematical taste, in [19]-[23].

In all what discussed in the cited references, the role of the Hilbert space $\mathcal{H}$ was essential: operators are defined on $\mathcal{H}$ (or on some subspace of $\mathcal{H}$) and the vectors all belong to $\mathcal{H}$. However, it is known that there exist alternative approaches to quantum mechanics, particularly relevant when unbounded operators are involved. This is, for instance, the approach based on *rigged Hilbert spaces*, see for instance [24]-[27]. Another possibility, which was recently introduced in a context which is relevant for this paper, makes use of distribution theory, [28] and [29]. The essential remark, in both these papers, is that in concrete and simple physical problems it can be useful to abandon Hilbert spaces and to work outside, say, some $\mathcal{L}^2(\mathbb{R})$, with or without a measure. In other words, the (analytic) problems are not really solved even changing the scalar product, and replacing $\mathcal{L}^2(\mathbb{R})$ with $\mathcal{L}^2(\mathbb{R}, p(x)\,dx)$, for some specific measure $p(x)$. This is part of what we will see also here. In fact, this aspect will be discussed here in some details in connection with a pseudo-bosonic system first introduced in [30], and then considered in [31] and, more recently, in [23]. As we will see, this system still has something to teach, both in the context of weak pseudo-bosons, and in connection with coherent states.

The paper is organized as follows: in the next section we will list few results and definitions on pseudo-bosons and bi-coherent states which are relevant for us. In Section 3 we will consider two pseudo-bosonic operators defined by means of two different functions known as *superpotentials*, and their connected number-like operators, and we will focus on their eigenstates, considering the case in which these are square-integrable and when they are not. We will try to stay as general as possible, not fixing the form of the superpotentials: in doing so, we will extend many of the results deduced in [23, 30, 31], which were found under very special choices of superpotentials. The same dual situation is also considered in Section 4: the first part of this section is devoted to bi-coherent states living in $\mathcal{L}^2(\mathbb{R})$. In the second part we will define, for the first time in our knowledge, what we call *weak* bi-coherent states using the language of distribution, i.e., as continuous functionals on some particular function space. Section 5 contains our conclusions.

## 2. Preliminaries

In this paper we will reconsider a pair of ladder operators $A$ and $B$, with $B \neq A^\dagger$, originally introduced in [30], and more recently considered in [23], to discuss how the notion of pseudo-bosons, and bi-coherent states, can be extended outside $\mathcal{L}^2(\mathbb{R})$. This analysis continues what was done in [29], where our interest was motivated by a rigorous analysis of the position and the momentum operators. To keep the paper self-contained, we devote this section to list few useful definitions and results on pseudo-bosons and on bi-coherent states.

### *2.1. $\mathcal{D}$-pseudo bosons: basic facts*

Let $\mathcal{H}$ be a given Hilbert space with scalar product $\langle .,. \rangle$ and related norm $\|.\|$. Let $a$ and $b$ be two operators on $\mathcal{H}$, with domains $D(a) \subset \mathcal{H}$ and $D(b) \subset \mathcal{H}$ respectively, $a^\dagger$ and $b^\dagger$ their adjoint, and let $\mathcal{D}$ be a dense subspace of $\mathcal{H}$ such that $a^\sharp\mathcal{D} \subseteq \mathcal{D}$ and $b^\sharp\mathcal{D} \subseteq \mathcal{D}$. Here with $x^\sharp$ we indicate $x$ or $x^\dagger$. Of course, $\mathcal{D} \subseteq D(a^\sharp)$ and $\mathcal{D} \subseteq D(b^\sharp)$.

**Definition 1** *The operators $(a, b)$ are $\mathcal{D}$-pseudo bosonic if, for all $f \in \mathcal{D}$, we have*

$$a\,b\,f - b\,a\,f = f. \tag{1}$$

When $b = a^\dagger$, this is simply the canonical commutation relation (CCR) for ordinary bosons. However, when the CCR is replaced by (1), the situation changes, becoming mathematically more interesting. In particular, it is useful to assume the following:

**Assumption $\mathcal{D}$-pb 1.**– there exists a non-zero $\varphi_0 \in \mathcal{D}$ such that $a\,\varphi_0 = 0$.

**Assumption $\mathcal{D}$-pb 2.**– there exists a non-zero $\Psi_0 \in \mathcal{D}$ such that $b^\dagger\,\Psi_0 = 0$.

It is obvious that, since $\mathcal{D}$ is stable under the action of $b$ and $a^\dagger$, then $\varphi_0 \in D^\infty(b) := \cap_{k\geq 0} D(b^k)$ and $\Psi_0 \in D^\infty(a^\dagger)$, so that the vectors

$$\varphi_n := \frac{1}{\sqrt{n!}}\, b^n \varphi_0, \qquad \Psi_n := \frac{1}{\sqrt{n!}}\, {a^\dagger}^n \Psi_0, \tag{2}$$

$n \geq 0$, can be defined and they all belong to $\mathcal{D}$. Hence, they also belong to the domains of $a^\sharp$, $b^\sharp$ and $N^\sharp$, where $N = ba$. We see that, from a practical point of view, $\mathcal{D}$ is the natural space to work with and, in this sense, it is even more relevant than $\mathcal{H}$. $\mathcal{D}$ is a sort of *physical vector space*, endowed with a scalar product and with its related norm, but $\mathcal{D}$ is not complete (in general) with respect to this norm. Let's put $\mathcal{F}_\Psi = \{\Psi_n,\, n \geq 0\}$ and $\mathcal{F}_\varphi = \{\varphi_n,\, n \geq 0\}$. It is simple to deduce the following lowering and raising relations:

$$\begin{cases} b\,\varphi_n = \sqrt{n+1}\varphi_{n+1}, & n \geq 0, \\ a\,\varphi_0 = 0, \quad a\varphi_n = \sqrt{n}\,\varphi_{n-1}, & n \geq 1, \\ a^\dagger\Psi_n = \sqrt{n+1}\Psi_{n+1}, & n \geq 0, \\ b^\dagger\Psi_0 = 0, \quad b^\dagger\Psi_n = \sqrt{n}\,\Psi_{n-1}, & n \geq 1, \end{cases} \tag{3}$$

as well as the eigenvalue equations $N\varphi_n = n\varphi_n$ and $N^\dagger\Psi_n = n\Psi_n$, $n \geq 0$. If $\langle\varphi_0, \Psi_0\rangle = 1$, then

$$\langle\varphi_n, \Psi_m\rangle = \delta_{n,m}, \tag{4}$$

for all $n, m \geq 0$. Hence $\mathcal{F}_\Psi$ and $\mathcal{F}_\varphi$ are biorthonormal. It is easy to see that, if $b = a^\dagger$, then $\varphi_n = \Psi_n$, so that biorthogonality is replaced by a simpler orthonormality. Moreover, the relations in (3) collapse, and only one number operator exists, since in this case $N = N^\dagger$.

The analogy with ordinary bosons suggests us to consider the following:

**Assumption $\mathcal{D}$-pb 3.**– $\mathcal{F}_\varphi$ is a basis for $\mathcal{H}$.

This is equivalent to requiring that $\mathcal{F}_\Psi$ is a basis for $\mathcal{H}$ as well, [32]. However, several physical models show that $\mathcal{F}_\varphi$ is **not** always a basis for $\mathcal{H}$, but it is still complete in $\mathcal{H}$: if $f \in \mathcal{H}$ is orthogonal to $\varphi_n$, for all $n$, then $f = 0$. For this reason we adopt the following weaker version of Assumption $\mathcal{D}$-pb 3, [17]:

**Assumption $\mathcal{D}$-pbw 3.**– For some subspace $\mathcal{G}$ dense in $\mathcal{H}$, $\mathcal{F}_\varphi$ and $\mathcal{F}_\Psi$ are $\mathcal{G}$-quasi bases.

This means that, for all $f$ and $g$ in $\mathcal{G}$,

$$\langle f, g\rangle = \sum_{n\geq 0}\langle f, \varphi_n\rangle\langle\Psi_n, g\rangle = \sum_{n\geq 0}\langle f, \Psi_n\rangle\langle\varphi_n, g\rangle, \tag{5}$$

which can be seen as a weak form of the resolution of the identity, restricted to $\mathcal{G}$. Of course, if $f \in \mathcal{G}$ is orthogonal to all the $\varphi_n$'s, or to all the $\Psi_n$'s, then (5) implies that $f = 0$. Hence $\mathcal{F}_\varphi$ and $\mathcal{F}_\Psi$ are complete in $\mathcal{G}$, [33].

The families $\mathcal{F}_\varphi$ and $\mathcal{F}_\Psi$ can be used to define two densely defined operators $S_\varphi$ and $S_\Psi$ via their action respectively on $\mathcal{F}_\Psi$ and $\mathcal{F}_\varphi$:

$$S_\varphi\Psi_n = \varphi_n, \qquad S_\Psi\varphi_n = \Psi_{\mathbf{n}}, \tag{6}$$

for all $n$. These operators play a very import role in the analysis of pseudo-bosons, since they map $\mathcal{F}_\varphi$ into $\mathcal{F}_\Psi$ and vice-versa, and define new scalar products in $\mathcal{H}$ is terms of which, for instance, the (new) adjoint of $b$ turns out to coincide with $a$. These, and many other aspects which are not relevant here, can be found in [22, 29, 17]. The role of operators like the $S_\varphi$ and $S_\Psi$ in (6) in different, but related situations is discussed, for instance, in [2, 5].

### *2.2. Bi-coherent states*

Concerning coherent states, rather than considering the strict $\mathcal{D}$-pseudo bosonic situation described before, where $a$, $b$, $\varphi_n$ and $\Psi_n$ satisfy the ladder equations in (3), we give here a more general result which is again based on certain biorthogonal sets, and some ladder operators, but not necessarily pseudo-bosonic. In particular, as we will show in a moment, our results here hold when replacing $\sqrt{n}$ in the second and in the fourth rows in (3) with some $\alpha_n$, see below. This approach is convenient since bi-coherent states for pseudo-bosons can be seen as a special case of a much more general situation, described by (7) below and by Theorem 2.

Let us consider two biorthogonal families of vectors, $\mathcal{F}_{\tilde{\varphi}} = \{\tilde{\varphi}_n \in \mathcal{H}, n \geq 0\}$ and $\mathcal{F}_{\tilde{\Psi}} = \{\tilde{\Psi}_n \in \mathcal{H}, n \geq 0\}$ which are $\mathcal{G}$ -quasi bases for some dense subset of $\mathcal{H}$, see (5). Consider an increasing sequence of real numbers $\alpha_n$ satisfying the inequalities $0 = \alpha_0 < \alpha_1 < \alpha_2 < \ldots$. We call $\overline{\alpha}$ the limit of $\alpha_n$ for $n$ diverging, which coincides with $\sup_n \alpha_n$. We further consider two operators, $A$ and $B^\dagger$, which act as lowering operators respectively on $\mathcal{F}_{\tilde{\varphi}}$ and $\mathcal{F}_{\tilde{\Psi}}$ in the following way:

$$A\,\tilde{\varphi}_n = \alpha_n \tilde{\varphi}_{n-1}, \qquad B^\dagger\,\tilde{\Psi}_n = \alpha_n \tilde{\Psi}_{n-1}, \tag{7}$$

for all $n \geq 1$, with $A\,\tilde{\varphi}_0 = B^\dagger\,\tilde{\Psi}_0 = 0$. These are the lowering equations which replace those in (3). Then the following theorem holds, [34]:

**Theorem 2** *Assume that four strictly positive constants $A_\varphi$, $A_\Psi$, $r_\varphi$ and $r_\Psi$ exist, together with two strictly positive sequences $M_n(\varphi)$ and $M_n(\Psi)$, for which*

$$\lim_{n\to\infty} \frac{M_n(\varphi)}{M_{n+1}(\varphi)} = M(\varphi), \qquad \lim_{n\to\infty} \frac{M_n(\Psi)}{M_{n+1}(\Psi)} = M(\Psi), \tag{8}$$

*where $M(\varphi)$ and $M(\Psi)$ could be infinity, and such that, for all $n \geq 0$,*

$$\|\tilde{\varphi}_n\| \leq A_\varphi\, r_\varphi^n M_n(\varphi), \qquad \|\tilde{\Psi}_n\| \leq A_\Psi\, r_\Psi^n M_n(\Psi). \tag{9}$$

*Then, putting $\alpha_0! = 1$ and $\alpha_k! = \alpha_1\alpha_2\cdots\alpha_k$, $k \geq 1$, the following series:*

$$N(|z|) = \left(\sum_{k=0}^{\infty} \frac{|z|^{2k}}{(\alpha_k!)^2}\right)^{-1/2}, \tag{10}$$

$$\varphi(z) = N(|z|)\sum_{k=0}^{\infty} \frac{z^k}{\alpha_k!}\tilde{\varphi}_k, \qquad \Psi(z) = N(|z|)\sum_{k=0}^{\infty} \frac{z^k}{\alpha_k!}\tilde{\Psi}_k, \tag{11}$$

*are all convergent inside the circle $C_\rho(0)$ in $\mathbb{C}$ centered in the origin of the complex plane and of radius $\rho = \overline{\alpha}\, \min\left(1, \frac{M(\varphi)}{r_\varphi}, \frac{M(\Psi)}{r_\Psi}\right)$. Moreover, for all $z \in C_\rho(0)$,*

$$A\varphi(z) = z\varphi(z), \qquad B^\dagger\Psi(z) = z\Psi(z). \tag{12}$$

*Suppose further that a measure $d\lambda(r)$ does exist such that*

$$\int_0^\rho d\lambda(r)\, r^{2k} = \frac{(\alpha_k!)^2}{2\pi}, \tag{13}$$

*for all $k \geq 0$. Then, putting $z = re^{i\theta}$ and calling $d\nu(z,\overline{z}) = N(r)^{-2}d\lambda(r)d\theta$, we have*

$$\int_{C_\rho(0)} \langle f, \Psi(z)\rangle \langle \varphi(z), g\rangle\, d\nu(z,\overline{z}) = \int_{C_\rho(0)} \langle f, \varphi(z)\rangle \langle \Psi(z), g\rangle\, d\nu(z,\overline{z}) = \langle f, g\rangle, \tag{14}$$

*for all $f, g \in \mathcal{G}$.*

Some comments are in order: first we observe that, contrarily to what happens for ordinary coherent states, [35, 36, 37], the norms of the vectors $\tilde{\varphi}_n$ and $\tilde{\Psi}_n$ need not being uniformly bounded, here. On the contrary, they can diverge rather fast with $n$. Of course, this is reflected by the fact that bi-coherent states of this kind only exist inside $C_\rho(0)$. Secondly, we are here fully working in an Hilbert space. This is clear because, for each fixed $n$, the norms of both $\tilde{\varphi}_n$

and $\tilde{\Psi}_n$ are bounded, as well as the norms in $\mathcal{H}$ of $\varphi(z)$ and $\Psi(z)$. However, and this is part of our new results in this paper, a similar strategy can be extended also to the case in which $\tilde{\varphi}_n$ or $\tilde{\Psi}_n$, or both, do not belong to $\mathcal{H}$, so that $\|\tilde{\varphi}_n\| = \|\tilde{\Psi}_n\| = \infty$, for all (or some) $n$. The analysis of this situation, motivated by what discussed in Section 3, is contained in Section 4. Finally, if $A$ and $B$ satisfy Definition 1, then $\alpha_n = \sqrt{n}$ and $\overline{\alpha} = \rho = \infty$. Hence we have convergence of the series above in all the complex plane, and the measure $d\lambda(r)$ can be easily identified. We will return on this aspect later.

## 3. A class of choices

The Hilbert space we work with, here, is $\mathcal{H} = \mathcal{L}^2(\mathbb{R})$ with the usual scalar product, and we consider, [30, 31, 23], the operators

$$A = \frac{d}{dx} + w_A(x), \qquad B = -\frac{d}{dx} + w_B(x). \tag{15}$$

where $w_A(x)$ and $w_B(x)$ are two $C^\infty$ functions such that $w_A(x) \neq \overline{w_B(x)}$, in order for $B^\dagger$ to be different from $A$. The reason for asking this regularity to $w_A(x)$ and $w_B(x)$ is because it will be used in the following, in deriving some useful result. However, most of the times, one could require much less regularity, as one can see in what follows. Adopting the terminology used in [23], we call these functions superpotentials, the reason being that they can be used, in general to construct two supersymmetric Hamiltonians $H_1 = BA$ and $H_2 = AB$ whose eigenvectors are related as in usual supersymmetric quantum mechanics, [38, 39, 23]. The fact that $H_1$ and $H_2$ are supersymmetric partners was considered in details in [23]. In this paper this aspect is not particularly relevant, since with our constraint on the superpotentials, we have $H_2 = H_1 + 1\!\!1$. This is due to the commutation rule $[A, B] = 1\!\!1$ which we are going to assume on $A$ and $B$, following Definition 1. Moreover, as we will show in a moment, these two superpotentials must be connected. Now, since

$$H_1 = BA = -\frac{d^2}{dx^2} + q_1(x)\frac{d}{dx} + V_1(x), \qquad H_2 = AB = -\frac{d^2}{dx^2} + q_1(x)\frac{d}{dx} + V_2(x), \tag{16}$$

where

$$q_1(x) = w_B(x) - w_A(x), \quad V_1(x) = w_A(x)w_B(x) - w_A'(x), \quad V_2(x) = w_A(x)w_B(x) + w_B'(x), \tag{17}$$

we find that

$$[A, B] = H_2 - H_1 = V_2(x) - V_1(x) = w_A'(x) + w_B'(x). \tag{18}$$

Hence, to have $[A, B] = 1\!\!1$, which is the case which is interesting for us here, we must have $w_A(x) + w_B(x) = x + k$, for a generic $k$ which we take to be real from now on: $k \in \mathbb{R}$. This does not prevent the superpotentials to be complex valued, in general. We will call these functions *pseudo-bosonic superpotentials* (PBSs). It is worth noticing that the commutators here must be understood in the sense of Definition 1, since they involve unbounded operators for which commutators is mathematically a *risky operation*.

It is clear that the vacua of $A$ and $B^\dagger$, see Section 2.1, are the following:

$$\varphi_0(x) = N_\varphi \exp\left\{-s_A(x)\right\}, \qquad \Psi_0(x) = N_\Psi \exp\left\{-\overline{s_B(x)}\right\}, \tag{19}$$

where $s_A(x) = \int w_A(x)dx$ and $s_B(x) = \int w_B(x)dx$, and $N_\varphi$ and $N_\Psi$ are two normalization constants which we will compute soon. Indeed a direct check shows that $A\varphi_0(x) = B^\dagger\Psi_0(x) = 0$.

It is useful to stress once more that formulas in (15) somehow extend the standard definitions in SUSY quantum mechanics, see [38, 39] for instance, where a single superpotential $w(x)$ is introduced, and found solving a Riccati equation, in the attempt to factorize a given, self-adjoint, Hamiltonian. As already observed, while this aspect can be useful in general, when $[A, B]$ has not a simple expression, it is not so relevant here exactly because this commutator is simple and, as such, informations on $H_1$ can be easily translated to analogous informations on $H_2$.

What is interesting for us is to extend now our previous analysis in [30, 31, 23]. In particular, we are not interested in requiring anymore that $\varphi_0(x)$ and $\Psi_0(x)$ are both in $\mathcal{L}^2(\mathbb{R})$ or, even more, in some dense subspace $\mathcal{D}$ of $\mathcal{L}^2(\mathbb{R})$, as Assumptions $\mathcal{D}$-pb 1 and $\mathcal{D}$-pb2 in Section 2.1. We rather want to see how much of the results in Section 2.1 can be deduced also weakening these requirements. This is exactly on the same line of our recent results on weak pseudo-bosons, [29], where Hilbert spaces were not so important and were replaced by other functional spaces, as those of distributions $\mathcal{D}'(\mathbb{R})$ or of tempered distributions $\mathcal{S}'(\mathbb{R})$. To be concrete, in this paper we will not assume, except when explicitly stated, that $\varphi_0(x), \Psi_0(x) \in \mathcal{L}^2(\mathbb{R})$. Nevertheless, we have the following result:

**Lemma 3** *If $w_A(x)$ and $w_B(x)$ are $C^\infty$ PBSs, then $\varphi_0(x)\,\overline{\Psi_0(x)} \in \mathcal{L}^1(\mathbb{R})$. Moreover, if $N_\varphi\overline{N_\Psi} = \frac{e^{-k^2/4}}{\sqrt{2\pi}}$, then $\langle\Psi_0, \varphi_0\rangle = 1$.*

**Proof** – Indeed we have, recalling that $w_A(x) + w_B(x) = x + k$,

$$\varphi_0(x)\,\overline{\Psi_0(x)} = N_\varphi\overline{N_\Psi}e^{-(s_A(x)+s_B(x))} = N_\varphi\overline{N_\Psi}e^{-\frac{x^2}{2}-kx-q},$$

where $q$ is an integration constant which, from now on, we put equal to zero, reabsorbing its effect in the normalization constants $N_\varphi$ and $N_\Psi$. Hence $\varphi_0(x)\,\overline{\Psi_0(x)}$ is a gaussian and, as such, is in $\mathcal{L}^1(\mathbb{R})$. The rest of the Lemma follows from an explicit simple computation.

□

The result of this Lemma depicts what happens, in general, for the so-called PIP-spaces[2], [40], where the scalar product in $\mathcal{L}^2(\mathbb{R})$ is extended to a sesquilinear form over *compatible spaces*: roughly speaking, $V_1$ and $V_2$ are compatible in $\mathcal{H}$ if, taken $f_1 \in V_1$ and $f_2 \in V_2$, a sesquilinear form $\langle f_1, f_2\rangle$ which extends the ordinary scalar product on $\mathcal{H}$, which we also indicate as $\langle ., .\rangle$, makes sense. Then Lemma 3 states that $\varphi_0(x)$ and $\Psi_0(x)$ belong to compatible spaces, for all possible choices of PBSs.

**Remarks:**– (1) In Lemma 3 it is not important that $w_A(x)$ and $w_B(x)$ are $C^\infty$ functions. It is sufficient that they can be integrated, so that $s_A(x)$ and $s_B(x)$ both exist. We prefer to introduce this assumption already here because it will be quite relevant in the following.

(2) According to the Lemma, from now on we will assume that

$$N_\varphi\overline{N_\Psi} = \frac{e^{-k^2/4}}{\sqrt{2\pi}}, \tag{20}$$

[2] Here PIP stands for partial inner product.

in order to have $\langle \Psi_0, \varphi_0 \rangle = 1$. Moreover, with our choice of $q$ in the proof of Lemma 3, we fix

$$s_A(x) + s_B(x) = \frac{x^2}{2} + kx, \tag{21}$$

all throughout the paper.

Let us now introduce, in analogy with (2), the following vectors:

$$\varphi_n(x) = \frac{1}{\sqrt{n!}} B^n \varphi_0(x), \qquad \psi_n(x) = \frac{1}{\sqrt{n!}} {A^\dagger}^n \psi_0(x), \tag{22}$$

$n = 1, 2, 3, \ldots$. It is clear that, since there is no guarantee that $\varphi_0(x)$ and $\Psi_0(x)$ are square-integrable, we are not in the situation described in Section 2.1: $\varphi_0(x)$ and $\Psi_0(x)$ are not, except that for special choices of PBSs, in any suitable $\mathcal{D}$ as required in Assumptions $\mathcal{D}$-pb 1. and $\mathcal{D}$-pb 2. Nevertheless, $\varphi_n(x)$ and $\Psi_n(x)$ can be computed and turn out to be $C^\infty$ functions for all possible choices of $C^\infty$ PBSs. This is the content of the following Proposition, which generalizes a similar result deduced in [23] for specific forms of $w_A(x)$ and $w_B(x)$, and in which the role of $\mathcal{L}^2(\mathbb{R})$ was essential.

**Proposition 4** *For any choice of $C^\infty$ PBSs we have*

$$\frac{\varphi_n(x)}{\varphi_0(x)} = \frac{\psi_n(x)}{\psi_0(x)} = p_n(x,k), \tag{23}$$

*for all $n \geq 0$, where $p_n(x,k)$ is independent of $w_A(x)$ and $w_B(x)$ and is defined recursively as follows:*

$$p_0(x,k) = 1, \qquad p_n(x,k) = \frac{1}{\sqrt{n}} \left( p_{n-1}(x,k)(x+k) - p'_{n-1}(x,k) \right), \tag{24}$$

*$n \geq 1$. Moreover, for all $n \geq 0$,*

$$p_n(x,k) = \frac{1}{\sqrt{2^n\, n!}} H_n\left(\frac{x+k}{\sqrt{2}}\right). \tag{25}$$

**Proof** – We start proving (23) for the $\varphi_n(x)$. The proof for the $\Psi_n(x)$ is completely analogous.

We use induction on $n$. It is clear that (23) is true if $n = 0$, since $p_0(x,k) = 1$. Suppose now that the statement is true for $n-1$, so that $\varphi_{n-1}(x) = p_{n-1}(x,k)\,\varphi_0(x)$, with $p_{n-1}(x,k)$ satisfying (24), for $n$ replaced by $n-1$. Let us then compute $\varphi_n(x)$. From (22) we have

$$\sqrt{n}\,\varphi_n(x) = B\varphi_{n-1}(x) = \left(-\frac{d}{dx} + w_B(x)\right)\left(p_{n-1}(x,k)\,\varphi_0(x)\right).$$

Now, recalling that $\frac{d}{dx}\varphi_0(x) = -w_A(x)\varphi_0(x)$ and that $w_A(x) + w_B(x) = x + k$, we get the following:

$$\sqrt{n}\,\varphi_n(x) = \left(-p'_{n-1}(x,k) + (x+k)p_{n-1}(x,k)\right)\varphi_0(x),$$

from which our first claim easily follows. Equality (25) is a consequence of the Rodriguez formula for the Hermite polynomials,

$$H_n(y) = (-1)^n e^{y^2} \frac{d^n}{dy^n} e^{-y^2},$$

and from the fact that, again using induction on $n$, we can check that

$$p_n(x,k) = (-1)^n \frac{1}{\sqrt{n!}} e^{x^2/2+kx} \frac{d^n}{dx^n} e^{-x^2/2-kx}.$$

□

Summarizing we have

$$\varphi_n(x) = \frac{N_\varphi}{\sqrt{2^n\, n!}} H_n\left(\frac{x+k}{\sqrt{2}}\right) e^{-s_A(x)}, \tag{26}$$

and

$$\Psi_n(x) = \frac{N_\Psi}{\sqrt{2^n\, n!}} H_n\left(\frac{x+k}{\sqrt{2}}\right) e^{-\overline{s_B(x)}}. \tag{27}$$

In analogy with what discussed in Section 2.1, these functions satisfy the following eigenvalue equations:

$$N\varphi_n(x) = n\varphi_n(x), \qquad N^\dagger \Psi_n(x) = n\Psi_n(x), \tag{28}$$

where $N = BA$ and $N^\dagger = A^\dagger B^\dagger$.

It is clear that, without further assumptions on the PBSs, these functions (or part of these functions) could be not square-integrable. However, Lemma 3 can be extended as follows

**Proposition 5** *If $w_A(x)$ and $w_B(x)$ are $C^\infty$ PBSs, then $\varphi_n(x)\,\overline{\Psi_m(x)} \in \mathcal{L}^1(\mathbb{R})$ and $\langle \Psi_m, \varphi_n\rangle = \delta_{n,m}$, for all $n, m \geq 0$.*

**Proof** – Indeed we have, in analogy with Lemma 3,

$$\varphi_n(x)\,\overline{\Psi_m(x)} = \frac{N_\varphi \overline{N_\Psi}}{\sqrt{2^{n+m}\, n!\, m!}} H_n\left(\frac{x+k}{\sqrt{2}}\right) H_m\left(\frac{x+k}{\sqrt{2}}\right) e^{-\frac{x^2}{2}-kx}$$

Hence, putting $y = \frac{x+k}{\sqrt{2}}$ in $\langle \Psi_m, \varphi_n\rangle$ and using (20), together with the integral

$$\int_{\mathbb{R}} H_n(x) H_m(x) e^{-x^2}\, dx = \sqrt{\pi}\, 2^n\, n! \delta_{n,m},$$

our claim follows.

□

Incidentally we observe that this approach significantly simplifies the proof given in [23], which furthermore was deduced for a very specific choice of PBSs, $w_A(x) = e^x + k$ and $w_B(x) = x - e^x$. Also, it is useful to stress that this Proposition produces, using explicit computations, the result in (4), which however was deduced under much stronger assumptions on $\mathcal{F}_\varphi = \{\varphi_n(x),\, n \geq 0\}$ and $\mathcal{F}_\Psi = \{\Psi_n(x),\, n \geq 0\}$.

Using the same language adopted after Lemma 3, Proposition 5 shows that $\varphi_n(x)$ and $\Psi_m(x)$ belong to compatible spaces so that, even if one of the two functions is not square-integrable, still the ordinary scalar product in $\mathcal{L}^2(\mathbb{R})$ can be extended to any product of these functions.

In principle, without extra assumptions on the PBSs, we don't know if the functions of $\mathcal{F}_\varphi$ and $\mathcal{F}_\Psi$ are square-integrable or not. For this reason, it makes not much sense to check if they are biorthogonal bases, or if they are complete, in $\mathcal{L}^2(\mathbb{R})$ or not, see also [29]. However, as in [29], something can still be deduced. To do so, we start rewriting $\varphi_n(x)$ and $\Psi_n(x)$ as follows:

$$\varphi_n(x) = c_n(x)\rho_A(x), \qquad \Psi_n(x) = c_n(x)\rho_B(x), \tag{29}$$

where

$$c_n(x) = \frac{1}{2^{1/4}}\, e_n\left(\frac{x+k}{\sqrt{2}}\right), \qquad \text{where} \quad e_n(x) = \frac{1}{\sqrt{2^n n! \sqrt{\pi}}} H_n(x)\, e^{-\frac{x^2}{2}}, \tag{30}$$

is the well known $n$-th eigenstate of the harmonic oscillator, and

$$\rho_A(x) = N_\varphi (2\pi)^{1/4} e^{\frac{1}{2}\left(\frac{x+k}{\sqrt{2}}\right)^2} e^{-s_A(x)}, \qquad \rho_B(x) = N_\Psi (2\pi)^{1/4} e^{\frac{1}{2}\left(\frac{x+k}{\sqrt{2}}\right)^2} e^{-\overline{s_B(x)}}. \tag{31}$$

Notice that these functions are independent of $n$. Because of the (21) $\rho_A(x)$ and $\rho_B(x)$ are not mutually independent. Indeed they satisfy

$$\rho_A(x)\,\overline{\rho_B(x)} = 1, \tag{32}$$

so that one is (a part a complex conjugation) the inverse of the other. This is in agreement, of course, with the factorization in (29) and with Proposition 5, since the set $\mathcal{F}_c = \{c_n(x),\, n \geq 0\}$ is an o.n. basis for $\mathcal{L}^2(\mathbb{R})$. Let us now define the following set:

$$\mathcal{E} = \left\{h(x) \in \mathcal{L}^2(\mathbb{R}) : \, h(x)\rho_j(x) \in \mathcal{L}^2(\mathbb{R}),\, j = A, B\right\} \tag{33}$$

This is clearly a subset of $\mathcal{L}^2(\mathbb{R})$, and is dense in it, since it contains the set $\mathcal{D}(\mathbb{R})$ of the $C^\infty$ compactly supported functions, which is dense in $\mathcal{L}^2(\mathbb{R})$. The inclusion $\mathcal{D}(\mathbb{R}) \subseteq \mathcal{E}$ can be easily checked: take $d(x) \in \mathcal{D}(\mathbb{R})$. Then, since $\rho_A(x)$ and $\rho_B(x)$ are $C^\infty$ functions, $d(x)\rho_j(x) \in \mathcal{D}(\mathbb{R})$, $j = A, B$. Hence $d(x) \in \mathcal{E}$, since $\mathcal{D}(\mathbb{R}) \subset \mathcal{L}^2(\mathbb{R})$. Now, it is possible to prove, using the same terminology as in (5), that $(\mathcal{F}_\varphi, \mathcal{F}_\Psi)$ are $\mathcal{E}$-quasi bases. More explicitly, recalling that not all the functions considered here are in $\mathcal{L}^2(\mathbb{R})$, we need to check that, for all $\epsilon(x) \in \mathcal{E}$, $\langle \epsilon, \varphi_n\rangle$ and $\langle \epsilon, \Psi_n\rangle$ are well defined for all $n \geq 0$, and that, for all $f(x), g(x) \in \mathcal{E}$,

$$\langle f, g\rangle = \sum_{n\geq 0} \langle f, \varphi_n\rangle\langle \Psi_n, g\rangle = \sum_{n \geq 0} \langle f, \Psi_n\rangle\, \langle \varphi_n, g\rangle\,, \tag{34}$$

The proof of these claims is based on the definition of $\mathcal{E}$ and on the fact that $\mathcal{F}_c$ is an o.n. basis for $\mathcal{L}^2(\mathbb{R})$. Indeed using (29) we have, for instance,

$$\langle \epsilon, \varphi_n\rangle = \int_{\mathbb{R}} \overline{\epsilon(x)}\, \varphi_n(x)\, dx = \int_{\mathbb{R}} \overline{\overline{\rho_A(x)}\,\epsilon(x)}\, c_n(x)\, dx,$$

which is well defined for all $n$ since $\overline{\rho_A(x)}\,\epsilon(x) \in \mathcal{L}^2(\mathbb{R})$. Similarly,

$$\langle \epsilon, \Psi_n\rangle = \int_{\mathbb{R}} \overline{\overline{\rho_B(x)}\,\epsilon(x)}\, c_n(x)\, dx,$$

which is also well defined for all $n$, since $\overline{\rho_B(x)}\,\epsilon(x) \in \mathcal{L}^2(\mathbb{R})$ as well. To check (34), we now use completeness of $\mathcal{F}_c$ in $\mathcal{L}^2(\mathbb{R})$:

$$\sum_{n\geq 0}\langle f, \varphi_n\rangle\langle \Psi_n, g\rangle = \sum_{n\geq 0}\langle \overline{\rho_A}\, f, c_n\rangle\langle c_n, \overline{\rho_B}\, g\rangle = \langle \overline{\rho_A}\, f, \overline{\rho_B}\, g\rangle = \int_{\mathbb{R}} \overline{f(x)}\, g(x)\, \rho_A(x)\overline{\rho_B(x)} = \langle f, g\rangle\,,$$

because of the (32). The other part of (34) can be proved similarly. We can conclude that, even if not all the functions considered in our analysis necessarily belong to $\mathcal{L}^2(\mathbb{R})$, still we can say that $(\mathcal{F}_\varphi, \mathcal{F}_\Psi)$ are $\mathcal{E}$-quasi bases, even if in a generalized sense. This is because $\mathcal{E}$ is *so well behaved* to take care of the bad asymptotic behaviour of the $\varphi_n(x)$ and $\Psi_n(x)$, when needed[3]. This is very close to what happens in distributions theory, where the *bad analytic properties* of distributions are somehow cured by very well behaved test functions. We will see something similar to this in Section 4.

Since $(\mathcal{F}_\varphi, \mathcal{F}_\Psi)$ are $\mathcal{E}$-quasi bases, it easily follows that both $\mathcal{F}_\varphi$ and $\mathcal{F}_\Psi$ are complete in $\mathcal{E}$, in the sense of [33]: if $\epsilon(x) \in \mathcal{E}$ is orthogonal to all the $\varphi_n(x)$, or to all the $\Psi_n(x)$, then $\epsilon(x) = 0$ almost everywhere (a.e.) in $\mathbb{R}$. To check this, it is enough to take $f(x) = g(x)$ in (34) and assume that, say, $\langle f, \varphi_n \rangle = 0$ for all $n$. Hence $\|f\|^2 = 0$, so that $f = 0$. We recall, see [33], that this does not imply that $\mathcal{F}_\varphi$ or $\mathcal{F}_\Psi$ are also complete in $\mathcal{L}^2(\mathbb{R})$. To check completeness of these sets in $\mathcal{L}^2(\mathbb{R})$, or in some even larger set, we recall first the following useful result, [41].

**Lemma 6** *Suppose $\eta(x)$ is a Lebesgue-measurable function which is different from zero a.e. in $\mathbb{R}$. Suppose further that there exist two positive constants $\delta, C$ such that $|\eta(x)| \leq C\, e^{-\delta|x|}$ a.e. in $\mathbb{R}$, then the set $\{x^n\, \eta(x)\}$ is complete in $\mathcal{L}^2(\mathbb{R})$.*

Of course, a similar result holds if the monomials $\{x^n\}$ are replaced by other polynomials of the $n$-th degree, like the shifted and dilated version of the Hermite polynomial $H_n\left(\frac{x+k}{\sqrt{2}}\right)$ appearing in (30).

In order to apply this Lemma, recalling (26) and (27) together with (21), we conclude that $s_A(x)$ or $s_B(x) = \frac{x^2}{2} + kx - s_A(x)$, or both, must diverge to $+\infty$ at least as fast as $|x|$, for $|x|$ diverging. Of course, as also (32) clearly shows, if $s_A(x)$ diverges too fast to $+\infty$, then $s_B(x)$ also diverges, but to $-\infty$ so that $e^{-s_B(x)}$ cannot satisfy the main assumption of the Lemma above. Stated differently, if we want both $\mathcal{F}_\varphi$ and $\mathcal{F}_\Psi$ to be complete in $\mathcal{L}^2(\mathbb{R})$, the PBSs must be chosen properly. Otherwise, only one of the two sets can be made of square-integrable functions.

Another completeness result follows from the following Theorem, [42]:

**Theorem 7** *Let $f(x)$ be a tempered distribution, $f(x) \in \mathcal{S}'(\mathbb{R})$, such that $\langle f, e_n \rangle = 0$, for all $n \geq 0$. Then $f(x) = 0$.*

Here $e_n(x)$ has been introduced in (30). This theorem can be restated by saying that the family $\{e_n(x)\}$ is complete not only in $\mathcal{L}^2(\mathbb{R})$, but also in $\mathcal{S}'(\mathbb{R})$. Using this result it is now possible to prove that the sets $\mathcal{F}_\varphi$ and $\mathcal{F}_\Psi$ are complete in $\hat{\mathcal{E}}$, defined as

$$\hat{\mathcal{E}} = \left\{h(x) \in \mathcal{L}^2(\mathbb{R}) : h(x)\rho_j(x) \in \mathcal{S}'(\mathbb{R}),\, j = A, B\right\}$$

Notice that this set is much larger than $\mathcal{E}$, since $\mathcal{S}'(\mathbb{R})$ is obviously larger than $\mathcal{L}^2(\mathbb{R})$.

We refer to [30] for more information on the operators in (15) for choices of PBSs for which $\varphi_n(x), \Psi_n(x) \in \mathcal{L}^2(\mathbb{R})$. In the next section, we will analyse bi-coherent states mostly in the *worst case*, i.e. when a fully Hilbert space treatment of the system is not possible. In doing so, we will generalize some recent results on bi-coherent states, see Section 2.2, to their distributional versions.

[3] In fact, for special choices of PBSs, it might still happen that these functions are all in $\mathcal{L}^2(\mathbb{R})$, [30, 31].In this case, $(\mathcal{F}_\varphi, \mathcal{F}_\Psi)$ are $\mathcal{E}$-quasi bases in the standard sense, [17].

## 4. Weak bi-coherent states

As we have discussed in the previous section, if we don't choose properly the PBSs $w_A(x)$ and $w_B(x)$, we likely find eigenvectors of $N = BA$ ($\varphi_n(x)$, see (26)) and of $N^\dagger = A^\dagger B^\dagger$ ($\Psi_n(x)$, see (27)) which are not necessarily in $\mathcal{L}^2(\mathbb{R})$. Therefore, even if lowering equations like those in (7) are satisfied, there is no reason why these vectors should satisfy the inequalities in (9). In fact, it may happen that $\|\varphi_n\|$ or $\|\Psi_n\|$ are infinite. Of course, this is not the case if $\rho_A(x)$ and $\rho_B(x)$ belong to $\mathcal{L}^\infty(\mathbb{R})$. In fact, when this is true, from (29) we deduce that

$$\|\varphi_n\| = \|c_n\,\rho_A\| \leq \|\rho_A\|_\infty \|c_n\| = \|\rho_A\|_\infty, \qquad \|\Psi_n\| = \|c_n\,\rho_B\| \leq \|\rho_B\|_\infty,$$

for all $n \geq 0$. Moreover, comparing (7) with (28), we deduce that $\alpha_n = \sqrt{n}$, and therefore $\overline{\alpha} = \infty$. Hence Theorem 2 holds with the following choice of the parameters involved: $A_\varphi = \|\rho_A\|_\infty$, $A_\Psi = \|\rho_B\|_\infty$, $r_\varphi = r_\Psi = M_n(\varphi) = M_n(\Psi) = 1$, for all $n \geq 0$. Hence $\rho = \infty$, and the series in (10) and (11) converge in all the complex plane. An example in which these conditions on $\rho_A(x)$ and $\rho_B(x)$ are satisfied is when we choose $s_A(x)$ and $s_B(x)$ as follows:

$$s_A(x) = \frac{x^2}{4} + \frac{kx}{2} + \Phi(x), \qquad s_B(x) = \frac{x^2}{4} + \frac{kx}{2} - \Phi(x),$$

where $\Phi(x)$ is any real $C^\infty$ function bounded from below and from above, i.e. when there exist $m, M$ such that $-\infty < m \leq \Phi(x) \leq M < \infty$, a.e. in $\mathbb{R}$. In fact, in this case we can check that

$$\|\rho_A\|_\infty = N_\varphi (2\pi)^{1/4} e^{k^2/4-m}, \qquad \|\rho_B\|_\infty = N_\Psi (2\pi)^{1/4} e^{k^2/4+M},$$

while the PBSs are $w_A(x) = \frac{x}{2} + k + \Phi'(x)$ and $w_B(x) = \frac{x}{2} + k - \Phi'(x)$. It is not difficult to find non trivial choices of $\Phi(x)$: $\cos(x)$ or $\sin(x)$, or some combinations of these, are just some examples. It may be interesting to notice that, if $\Phi(x) = 0$, then $\varphi_n(x)$ coincides with $\Psi_n(x)$, except for a normalization constant, since both $\rho_A(x)$ and $\rho_B(x)$ turn out to be constant.

### *4.1. An asymmetric example in $\mathcal{L}^2(\mathbb{R})$*

Let us take now $s_A(x) = \frac{x^2}{4}$ and $s_B(x) = \frac{x^2}{4} + kx$. Condition (21) is satisfied and we have

$$\varphi_n(x) = \frac{N_\varphi}{\sqrt{2^n\, n!}} H_n\left(\frac{x+k}{\sqrt{2}}\right) e^{-x^2/4}, \qquad \Psi_n(x) = \frac{N_\Psi}{\sqrt{2^n\, n!}} H_n\left(\frac{x+k}{\sqrt{2}}\right) e^{-x^2/4-kx}. \tag{35}$$

Both these functions are square-integrable and the norm of $\varphi_n(x)$ can be computed using the following integral, [43]:

$$\int_{\mathbb{R}} e^{-x^2} H_m(x+y)\, H_n(x+z)\, dx = 2^n\, m! \sqrt{\pi} z^{n-m} L_m^{n-m}(-2yz),$$

which holds if $m \leq n$. Here $L_m^{n-m}$ is a Laguerre polynomial. We get

$$\|\varphi_n\|^2 = |N_\varphi|^2 \sqrt{2\pi}\, L_n(-k^2), \tag{36}$$

where $L_n(x) = L_n^0(x)$. As for the norm of $\Psi_n(x)$, the same formula above produces

$$\|\Psi_n\|^2 = |N_\Psi|^2 \sqrt{2\pi}\, e^{2k^2}\, L_n(-k^2), \tag{37}$$

which differs from (36) only for a constant factor. Now, to check if this example fits the assumptions of Theorem 2, we observe that the asymptotic behaviour (in $n$) of the Laguerre polynomials for negative arguments, as deduced in [44], Theorem 8.22.3, is the following:

$$L_n(x) = \frac{e^{x/2} 2^{2(-nx)^{1/2}}}{2\sqrt{\pi}\,(-xn)^{1/4}}\left(1 + O(n^{-1/2})\right),$$

$x < 0$, so that

$$\|\varphi_n\| \simeq \frac{|N_\varphi|}{(2|k|)^{1/4}}\, e^{-k^2/4} \frac{e^{|k|\sqrt{n}}}{n^{1/8}}, \qquad \|\Psi_n\| \simeq \frac{|N_\Psi|}{(2|k|)^{1/4}}\, e^{3k^2/4} \frac{e^{|k|\sqrt{n}}}{n^{1/8}}, \tag{38}$$

where $\simeq$ stands for *except for corrections* $O(n^{-1/2})$. Now, since we have clearly $e^{|k|\sqrt{n}} < e^{|k|n}$ for all non zero $k$ and $n \geq 1$, we can identify the various ingredients of Theorem 2 as follows:

$$A_\varphi = \frac{|N_\varphi|}{(2|k|)^{1/4}}\, e^{-k^2/4}, \quad A_\Psi = \frac{|N_\Psi|}{(2|k|)^{1/4}}\, e^{3k^2/4}, \quad r_\varphi = r_\Psi = e^{|k|}, \quad M_n(\varphi) = M_n(\Psi) = \frac{1}{n^{1/8}}.$$

Of course, this is valid if $k \neq 0$, which is the only interesting case for us, since if $k = 0$ then $s_A(x) = s_B(x)$. Hence $M(\varphi) = M(\Psi) = 1$, $\rho = \infty$, and the bi-coherent states in (11) are well defined in all the complex plane, belong to $\mathcal{L}^2(\mathbb{R})$, and satisfies the eigenvalue equations (11) and the weak resolution of the identity (14), with $\mathcal{G} = \mathcal{E}$. In fact, taking

$$d\lambda(r) = \frac{1}{\pi} e^{-r^2} r\, dr, \tag{39}$$

condition (13) is satisfied:

$$\int_0^\infty d\lambda(r) r^{2k} = \frac{k!}{2\pi}.$$

Then (14) follows from the fact that, as proved in Section 3, $(\mathcal{F}_\varphi, \mathcal{F}_\Psi)$ are $\mathcal{E}$-quasi bases for all possible choices of PBSs satisfying (21). Hence, in particular, they are $\mathcal{E}$-quasi bases for our choice $s_A(x) = \frac{x^2}{4}$ and $s_B(x) = \frac{x^2}{4} + kx$, or $w_A(x) = \frac{x}{2}$ and $w_B(x) = \frac{x}{2} + k$.

### *4.2. Working outside $\mathcal{L}^2(\mathbb{R})$*

Let us now concentrate on the situation in which $\mathcal{L}^2(\mathbb{R})$ is not the natural space where to work. Of course, what we have in mind is that changing the metric does not change much, leaving most of the problems unsolved. For instance, changing the scalar product can produce a convergent series for, say, $\varphi(z)$ but does help much when convergence of $\Psi(z)$ is also considered. Or vice-versa. In other words, if we introduce a non trivial metric $\eta$ and a related scalar product $\langle .,. \rangle_\eta = \langle .,\eta . \rangle$, and the norm $\|.\|_\eta$, it may happen that $\|\varphi_n\|_\eta$ satisfies an inequality like the one in (9). But $\|\Phi_n\|_\eta$ does not, in general. For this reason, in what follows we will describe a rather general version of the problem, and propose a solution for it. Then we will adopt the strategy we are going to construct to the situation considered in Section 3.

Let $\mathcal{F}_c = \{c_n(x),\, n \geq 0\}$ be an o.n. basis in $\mathcal{L}^2(\mathbb{R})$, not necessarily coincident with the basis in (30), and let $\rho_f(x)$ and $\rho_g(x)$ be two Lebesgue-measurable functions such that, calling

$$f_n(x) = c_n(x)\, \rho_f(x), \qquad g_n(x) = c_n(x)\, \rho_g(x), \tag{40}$$

we have $f_n(x)\, g_m(x) \in \mathcal{L}^1(\mathbb{R})$, for all $n, m \geq 0$. This implies that, despite of $f_n(x)$ or $g_n(x)$ being square-integrable or not, the form $\langle f_n, g_m\rangle$ is always well defined. With a slight abuse of language, we still call $\langle f_n, g_m\rangle$ the scalar product between $f_n(x)$ and $g_m(x)$. Motivated by (32), we restrict here to the case in which $\rho_f(x) = \overline{\rho_g(x)}^{-1}$. In this way it is clear that $\mathcal{F}_f = \{f_n(x)\}$ and $\mathcal{F}_g = \{g_n(x)\}$ are biorthonormal:

$$\langle f_n, g_m\rangle = \delta_{n,m}. \tag{41}$$

We stress once more that we use this term, biorthonormal, in an extended sense, since we are not requiring here that $f_n(x)$ and $g_m(x)$ belong to $\mathcal{L}^2(\mathbb{R})$, while we are sure that $f_n(x)g_m(x) \in \mathcal{L}^1(\mathbb{R})$ anyhow. In analogy with (33) we define now the set

$$\mathcal{V} = \left\{v(x) \in \mathcal{L}^2(\mathbb{R}) : v(x)\rho_j(x) \in \mathcal{L}^2(\mathbb{R}),\ j = f, g\right\} \tag{42}$$

Repeating what we have already discussed for $\mathcal{E}$ we conclude that, if $\rho_f(x)$ and $\rho_g(x)$ are $C^\infty$ functions, then $\mathcal{D}(\mathbb{R}) \subseteq \mathcal{V}$, which is therefore dense in $\mathcal{L}^2(\mathbb{R})$. Moreover, it is easy to check that $\mathcal{V}$ is closed under linear combinations: if $v_1(x), v_2(x) \in \mathcal{V}$, then $\alpha_1 v_1(x) + \alpha_2 v_2(x) \in \mathcal{V}$ as well, for all complex $\alpha_1, \alpha_2$. In our conditions we can check that $(\mathcal{F}_f, \mathcal{F}_g)$ are $\mathcal{V}$-quasi bases,

$$\langle v, w\rangle = \sum_{n\geq 0} \langle v, f_n\rangle\langle g_n, w\rangle = \sum_{n\geq 0} \langle v, g_n\rangle\langle f_n, w\rangle, \tag{43}$$

for all $v(x), w(x) \in \mathcal{V}$, so that they are both complete in $\mathcal{V}$.

Let us now observe that, even if, say, $f_n(x) \notin \mathcal{L}^2(\mathbb{R})$, for instance because its asymptotic behaviour is not the proper one, $af_n(x)$ can make sense, $a$ being some specific operator, even if $af_n(x)$ could not be in $\mathcal{L}^2(\mathbb{R})$. Or, maybe, $b^\dagger g(x)$ can be well defined even if $g_n(x) \notin \mathcal{L}^2(\mathbb{R})$ for some operator $b^\dagger$, even if $b^\dagger g_n(x) \notin \mathcal{L}^2(\mathbb{R})$. This is exactly what happens in Section 3. For this reason, it is natural to suppose the following: there exist two linear operators $a$ and $b$, together with a strictly increasing sequence $\{\alpha_n\}$, with $\alpha_0 = 0$, such that $af_n = \alpha_n f_{n-1}$, $b^\dagger g_n = \alpha_n g_{n-1}$, $n \geq 1$, and $af_0 = b^\dagger g_0 = 0$. Calling, as before, $\overline{\alpha} = \sup_n \alpha_n$ we know that $N(|z|)$ defined as in (10) exists inside $C_{\overline{\alpha}}(0)$, a circle in the complex plane centered in the origin and of radius $\overline{\alpha}$. So far, there are not many differences with what stated by Theorem 2. But now, it is clear that it makes no sense to use conditions (9), since these would refer to the norm of $f_n(x)$ and $g_n(x)$, which can be infinite here. Hence, no bound as those in (9) is expected to hold now, and the convergence of the analogous of the $\varphi(z)$ and $\Psi(z)$ in (11) is not granted, at all. For this reason we need to adopt a different strategy which is quite close to what one does when introducing distributions as continuous functionals on some special set of functions, see [45] for instance.

Let $v \in \mathcal{V}$, and let us consider the following series

$$S_{f,v}(z) = \sum_{n\geq 0} \frac{z^n}{\alpha_n!}\langle f_n, v\rangle, \qquad S_{g,v}(z) = \sum_{n\geq 0} \frac{z^n}{\alpha_n!}\langle g_n, v\rangle. \tag{44}$$

Both these series converge, for all $v \in \mathcal{V}$, inside $C_{\overline{\alpha}}(0)$. In fact,

$$|\langle f_n, v\rangle| = |\langle c_n\rho_f, v\rangle| = |\langle c_n, \overline{\rho_f} v\rangle| \leq \|c_n\|\|\overline{\rho_f} v\| = \|\overline{\rho_f} v\| < \infty,$$

since $\overline{\rho_f(x)}v(x)$ belongs to $\mathcal{L}^2(\mathbb{R})$ by the definition of $\mathcal{V}$. Hence

$$|S_{f,v}(z)| \leq \|\overline{\rho_f}v\| \sum_{n\geq 0} \frac{|z|^n}{\alpha_n!},$$

which converges inside $C_{\overline{\alpha}}(0)$, as stated, independently of $v \in \mathcal{V}$. Similarly we find

$$|S_{g,v}(z)| \leq \|\overline{\rho_g}v\| \sum_{n\geq 0} \frac{|z|^n}{\alpha_n!},$$

which is also converging inside $C_{\overline{\alpha}}(0)$ for all $v \in \mathcal{V}$. Then we introduce two vectors, $f(z)$ and $g(z)$, $z \in C_{\overline{\alpha}}(0)$, using the following definitions:

$$\langle f(z), v\rangle = N(|z|)\, S_{f,v}(\overline{z}) = N(|z|) \sum_{n\geq 0} \frac{\overline{z}^n}{\alpha_n!}\langle f_n, v\rangle, \tag{45}$$

and

$$\langle g(z), v\rangle = N(|z|)\, S_{g,v}(\overline{z}) = N(|z|) \sum_{n\geq 0} \frac{\overline{z}^n}{\alpha_n!}\langle g_n, v\rangle. \tag{46}$$

Of course, these equations do not give us the explicit analytic expressions for $f(z)$ and $g(z)$, but they can be used to define $f(z)$ and $g(z)$ *in a weak sense.* For that, we start introducing two maps, from $\mathcal{V}$ to $\mathbb{C}$, as follows:

$$F(z)[v] = \langle f(z), v\rangle, \qquad G(z)[v] = \langle f(z), v\rangle. \tag{47}$$

It is clear that both $F(z)$ and $G(z)$ are linear functionals on $\mathcal{V}$. Next we introduce on $\mathcal{V}$ a topology $\tau_{\mathcal{V}}$ as follows: we say that a sequence $\{v_n(x)\}$ in $\mathcal{V}$ is $\tau_{\mathcal{V}}$-convergent to a certain $v(x) \in \mathcal{L}^2(\mathbb{R})$ if $\{v_n(x)\}$ converges to $v(x)$ in the norm $\|.\|$, and if $\{\rho_j(x)\, v_n(x)\}$, $j = f, g$, are Cauchy sequences in $\|.\|$ and converge to $\rho_j(x)v(x)$. It is clear that, when this is true, $v(x) \in \mathcal{V}$. Hence, $\mathcal{V}$ is closed in $\tau_{\mathcal{V}}$.

Now, if we call $\mathcal{V}'$ the set of all the continuous functionals on $\mathcal{V}$, we can prove the following result:

**Proposition 8** *$F(z)$ and $G(z)$ both belong to $\mathcal{V}'$.*

**Proof** – We only need to prove that $F(z)$ and $G(z)$ are continuous. We will give the proof for $F(z)$, since that for $G(z)$ is completely analogous.

Let $\{v_k(x)\}$ be a sequence in $\mathcal{V}$ $\tau_{\mathcal{V}}$-convergent to a certain $v(x) \in \mathcal{V}$. We need to prove that $F(z)[v_k - v] \to 0$, when $k$ diverges. Indeed, since for all $k, n \geq 0$, $\langle f_n, v_k - v\rangle = \langle c_n, \overline{\rho_f}(v_k - v)\rangle$, we have

$$|F(z)[v_k - v]| = \left| N(|z|) \sum_{n\geq 0} \frac{\overline{z}^n}{\alpha_n!}\langle f_n, v_k - v\rangle \right| \leq N(|z|) \sum_{n\geq 0} \frac{|z|^n}{\alpha_n!}|\langle f_n, v_k - v\rangle|,$$

so that

$$|F(z)[v_k - v]| \leq N(|z|) \sum_{n\geq 0} \frac{|z|^n}{\alpha_n!}|\langle c_n, \overline{\rho_f}(v_k - v)\rangle| \leq \|\overline{\rho_f}(v_k - v)\| N(|z|) \sum_{n\geq 0} \frac{|z|^n}{\alpha_n!} \to 0,$$

for $k \to \infty$, due to the definition of $\tau_{\mathcal{V}}$ and to the convergence of the series inside $C_{\overline{\alpha}}(0)$.

□

The continuous functionals $F(z)$ and $G(z)$ are what we call *weak bi-coherent states* (wbcs). The reason is contained in the following proposition, where the analogous of (12) and (14), two of the essential properties of any (bi-)coherent states, are proven. We need first to introduce an useful subspace of $\mathcal{V}$:

$$\mathcal{V}_0 = \left\{ w(x) \in \mathcal{V} :\ w(x) \in D(a^\dagger) \cap D(b), \quad a^\dagger w(x), bw(x) \in \mathcal{V} \right\} \tag{48}$$

Of course the explicit form of $\mathcal{V}_0$ depends also on $a$ and $b$. We will show later that, at least in the situation considered in Section 3, $\mathcal{V}_0$ is dense in $\mathcal{L}^2(\mathbb{R})$ since it contains $\mathcal{D}(\mathbb{R})$.

**Proposition 9** *The wbcs satisfy the following properties:*
*(i) for all $v(x) \in \mathcal{V}_0$ we have*

$$\langle v, af(z)\rangle = z\langle v, f(z)\rangle, \qquad \langle v, b^\dagger g(z)\rangle = z\langle v, g(z)\rangle, \tag{49}$$

*for all $z \in C_{\overline{\alpha}}(0)$.*
*(ii) Suppose that a measure $d\lambda(r)$ does exist such that*

$$\int_0^{\overline{\alpha}} d\lambda(r) r^{2k} = \frac{(\alpha_k!)^2}{2\pi}, \tag{50}$$

*for all $k \geq 0$. Then, putting $z = re^{i\theta}$ and calling $d\nu(z,\overline{z}) = N(r)^{-2} d\lambda(r) d\theta$, we have*

$$\int_{C_{\overline{\alpha}}(0)} \langle v, f(z)\rangle \langle g(z), w\rangle \, d\nu(z,\overline{z}) = \int_{C_{\overline{\alpha}}(0)} \langle v, g(z)\rangle \langle f(z), w\rangle \, d\nu(z,\overline{z}) = \langle v, w\rangle, \tag{51}$$

*for all $v, w \in \mathcal{V}$.*

**Proof** – We start proving (i). Since, by assumption, $v(x) \in \mathcal{V}_0$, then $v(x) \in D(a^\dagger)$ and $a^\dagger v(x) \in \mathcal{V}$. Hence we have

$$\langle v, af(z)\rangle = \langle a^\dagger v, f(z)\rangle = N(|z|) \sum_{n\geq 0} \frac{z^n}{\alpha_n!} \langle a^\dagger v, f_n\rangle = N(|z|) \sum_{n\geq 0} \frac{z^n}{\alpha_n!} \langle v, af_n\rangle.$$

using (45). Now, since $af_0 = 0$ and $af_n = \alpha_n f_{n-1}$, $n \geq 1$, the first equality in (49) easily follows. The second equality can be proved in a similar way.

The proof of (ii) does not differ significantly from that of Theorem 2, and will not be given here. We only stress that the crucial aspect of the proof is the fact that $(\mathcal{F}_f, \mathcal{F}_g)$ are $\mathcal{V}$-quasi bases.

□

This proposition shows that our wbcs produce, together, a resolution of the identity on $\mathcal{V}$. It also shows that they are *weak* eigenstates of the lowering operators $a$ and $b^\dagger$, but this is only granted on $\mathcal{V}_0$, and not on $\mathcal{V}$.

In the last part of this section we apply the general results deduced so far to what we have seen in Section 3, to check if and when wbcs can be defined. Formula (29) shows that we are indeed in the situation described in (40), with $\varphi_n(x)$, $\rho_A(x)$, $\Psi_n(x)$ and $\rho_B(x)$ identified respectively with $f_n(x)$, $\rho_f(x)$, $g_n(x)$ and $\rho_g(x)$. The o.n. basis $\mathcal{F}_c$ in (40) is constructed as in (30). Now, since $A$ and $B$ are pseudo-bosonic operators, $\alpha_n$ is easily identified: $\alpha_n = \sqrt{n}$, which implies that $\overline{\alpha} = \infty$. Condition (50) on the measure $d\lambda(r)$ reads therefore

$$\int_0^\infty d\lambda(r) r^{2k} = \frac{k!}{2\pi},$$

as in Section 4.2. Hence $d\lambda(r) = \frac{1}{\pi} e^{-r^2} r\,dr$, which is the same result we have found in (39). This means that the forms $F(z)[v]$ and $G(z)[v]$ are well defined for all $v \in \mathcal{V}$ and for all $z \in \mathbb{C}$, and that (51) is satisfied with these particular choices.

As for (49), it is useful to show that $\mathcal{V}_0$ is *large enough*. Otherwise these equalities are not particularly useful. For instance, if $\mathcal{V}_0$ would contain only the zero vector, (49) is surely true, but it becomes completely trivial. Luckily enough, this is not the case here. Indeed, we can check that $\mathcal{D}(\mathbb{R}) \subseteq \mathcal{V}_0$. In fact, we start observing that $\mathcal{V}$ coincides, in this case, with the set $\mathcal{E}$ in (33) which, as already noticed, contains $\mathcal{D}(\mathbb{R})$. Therefore $\mathcal{D}(\mathbb{R}) \subseteq \mathcal{V}$. Now, let $v_0(x) \in \mathcal{D}(\mathbb{R})$. Then $v_0(x) \in \mathcal{V}$. Moreover, $v_0(x)$ belongs to the domain of both $A^\dagger$ and $B$. This is because, for instance

$$Bv_0(x) = \left(-\frac{d}{dx} + w_B(x)\right) v_0(x) = -v_0'(x) + w_B(x)v_0(x),$$

which is again in $\mathcal{D}(\mathbb{R})$ being the difference of two functions, $v_0'(x)$ and $+w_B(x)v_0(x)$, which are both in $\mathcal{D}(\mathbb{R})$. In particular, we are using here the fact that $w_B(x)$ is $C^\infty$. So the conclusion is clear: $Bv_0(x)$ and $A^\dagger v_0(x)$ are both in $\mathcal{D}(\mathbb{R})$, so that they are also both in $\mathcal{V}$. Hence $v_0(x)$ belongs to $\mathcal{V}_0$, which for this reason is dense in $\mathcal{L}^2(\mathbb{R})$.

We can conclude that, for the operators considered in Section 3, the statements proved in Proposition 9 hold in *large sets*, so that wbcs can be defined under very general conditions on $w_A(x)$ and $w_B(x)$.

## 5. Conclusions

This paper is a continuation of our previous analysis on deformed canonical commutation relations and its related pseudo-bosons and bi-coherent states. In particular, we have shown how to extend the original approach outside $\mathcal{L}^2(\mathbb{R})$, keeping untouched the possibility of defining eigenstates of certain number-like operators, manifestly non self-adjoint, but opening to the possibility that these states are not square-integrable. The example discussed in this paper is constructed in terms of two different, but related, superpotentials. For this reason the approach described here could be relevant for supersymmetric quantum mechanics. But this is possible only when considering choices of $w_A(x)$ and $W_B(x)$ not giving rise to pseudo-bosonic operators. Otherwise, as discussed before, the super-symmetric partner of the starting Hamiltonian coincides with this, except that for an additive constant.

In the second part of the paper we have shown how bi-coherent states can be introduced for the PBSs considered here, and we have constructed a general framework to consider a situation in which $\mathcal{L}^2(\mathbb{R})$ is not sufficient, in the same line as what was done recently in [29]. In particular we have given conditions on the superpotentials for the wbcs to be well defined, and in order for them to satisfy analogous properties as those of ordinary coherent and bi-coherent states.

**Acknowledgements**

The author acknowledges partial support from Palermo University and from G.N.F.M. of the INdAM.

## References

[1] Bender C M and Boettcher S 1998 Real Spectra in Non-Hermitian Hamiltonians Having PT-Symmetry *Phys. Rev. Lett.* **80** 5243–5246

[2] Bender C M 2019 *PT Symmetry In Quantum and Classical Physics* (World Scientific Publishing Europe Ltd., London)

[3] 2012 *Special issue on quantum physics with non-Hermitian operators* ed Bender C, Fring A, Günther U, Jones H, J. Phys. A: Math. and Ther. **45**

[4] Bender C M 2007 Making Sense of Non-Hermitian Hamiltonians *Rep. Progr. Phys.* **70** 947–1018

[5] Mostafazadeh A 2010 Pseudo-Hermitian representation of Quantum Mechanics *Int. J. Geom. Meth. Mod. Phys.* **7** 1191–1306

[6] 2016 *Non-Hermitian Hamiltonians in Quantum Physics; Selected Contributions from the 15th International Conference on Non-Hermitian Hamiltonians in Quantum Physics, Palermo, Italy, 18-23 May 2015* ed Bagarello F, Passante R and Trapani C (Springer)

[7] Znojil M 2009 Three-Hilbert-space formulation of Quantum Mechanics *SIGMA* **5** 001–019

[8] Dey S, Fring A and Khantoul B 2013 Hermitian versus non-Hermitian representations for minimal length uncertainty relations *J. Phys. A: Math. and Ther.* **46** 335304

[9] 2015 *Non-selfadjoint operators in quantum physics: Mathematical aspects* ed Bagarello F, Gazeau J P, Szafraniec F H and Znojil M (John Wiley and Sons)

[10] Davies E B and Kuijlaars B J 2004 Spectral asymptotics of the non-self-adjoint harmonic oscillator *J. London Math. Soc.* **70** 420–426

[11] Davies E B 1999 Pseudospectra, the harmonic oscillator and complex resonances *Proc. Roy. Soc. London A* **455** 585–599

[12] Bagarello F 2013 Pseudo-bosons for the $D_2$ type quantum Calogero model *J. Math. Anal. Appl.* **407** 90–96

[13] Bagarello F, Gargano F, Spagnolo S and Triolo S 2017 Coordinate representation for non Hermitian position and momentum operators *Proc. Roy. Soc. A* **473** 20170434

[14] da Providencia J, Bebiano N, da Providencia J P 2016 Mathematical aspects of quantum systems with a pseudo-Hermitian Hamiltonian *Braz. J. of Phys.* **46** (2) 152–156

[15] Dorey P, Dunning C and Tateo R 2001 Spectral equivalences, Bethe ansatz equations, and reality properties in PT-symmetric quantum mechanics, *J. Phys. A: Math. Gen.* **34** 5679-5704

[16] Krejcirik D and Siegl P 2012 On the metric operator for the imaginary cubic oscillator *Phys. Rev. D* **86** 121702(R)

[17] Bagarello F 2015 Deformed canonical (anti-)commutation relations and non hermitian Hamiltonians, in [9]

[18] Trifonov D A 2008 Pseudo-boson coherent and Fock states *Proceedings of the 9th International Workshop on Complex Structures, Integrability and Vector Fields* pp 241–250 (Sofia, Bulgaria) *Preprint* arXiv: quant-ph/0902.3744

[19] Bagarello F 2010 Pseudo-bosons, Riesz bases and coherent states *J. Math. Phys.* **50** 023531

[20] Bagarello F and Triolo S 2014 Some invariant biorthogonal sets with an application to coherent states *J. Math. Anal. Appl.* **415** 462–476

[21] Bagarello F 2016 Intertwining operators for non self-adjoint Hamiltonians and bicoherent states *J. Math. Phys.* **57** 103501

[22] Bagarello F 2017 *kq*-representation for pseudo-bosons, and completeness of bi-coherent states *J. Math. Anal. Appl.* **450** 631-643

[23] Bagarello F 2020 *Susy for non-Hermitian Hamiltonians, with a view to coherent states Math. Phys. Anal. and Geom.* **23** 28

[24] Roberts J E 1966 Rigged Hilbert spaces in quantum mechanics *Commun. Math. Phys.* **3** 98–119
[25] Bohm A 1967 Rigged Hilbert Space and mathematical description of physical systems *Boulder Lectures in Theoretical Physics IX A: Mathematical Methods of Theoretical Physics* (Wiley, New York)
[26] De la Madrid R 2005 The role of the rigged Hilbert space in Quantum Mechanics *Eur. J. Phys.* **26** 287–312
[27] Bellomonte G, Di Bella S and Trapani C 2014 Operators in Rigged Hilbert spaces: some spectral properties *J. Math. Anal. Appl.* **411** 931–946
[28] Bagarello F, Gargano F, Spagnolo S and Triolo S 2017 *Coordinate representation for non Hermitian position and momentum operators Proc. Roy. Soc. A* **473** 20170434
[29] Bagarello F 2020 Weak pseudo-bosons *J. Phys. A: Math. and Ther.* **53**, 135201
[30] Bagarello F 2010 Construction of pseudo-bosons systems *J. Math. Phys.* **51** 053508
[31] Bagarello F 2011 Pseudo-bosons, so far *Rep. Math. Phys.* **68** No. 2 175–210
[32] Christensen O 2003 *An Introduction to Frames and Riesz Bases* (Birkhäuser, Boston)
[33] Bagarello F and Bellomonte G 2017 Hamiltonians defined by biorthogonal sets *J. Phys. A: Math. and Ther.* **50** N. 14 145203
[34] Bagarello F, Gargano F. and Spagnolo S 2019 Two-dimensional non commutative Swanson model and its bicoherent states *Geometric Methods in Physics, XXXVI* Kielanowski P, Odzijewicz A and Previato E eds., Trends in Mathematics, pp. 9-19 (Birkhäuser, Switzerland)
[35] Ali S T, Antoine J P and Gazeau J P 2000 *Coherent States, Wavelets and Their Generalizations* (Springer-Verlag, New York)
[36] Combescure M and Didier R 2012 *Coherent States and Applications in Mathematical Physics* (Springer)
[37] Gazeau J P 2009 *Coherent states in quantum physics* (Wiley-VCH, Berlin)
[38] Cooper F, Khare A and Sukhatme U 2001 *Supersymmetry and quantum mechanics* (World Scientific, Singapore)
[39] Junker G 1996 *Supersymmetric methods in quantum and statistical physics* (Springer-Verlag, Berlin Heidelberg)
[40] Antoine J P and Trapani C 2017 Reproducing pairs of measurable functions and partial inner product spaces *Adv. in Op. Th.* **2** 126–146
[41] Kolmogorov A and Fomine S 1973 *Eléments de la théorie des fonctions et de l'analyse fonctionelle* (MIR)
[42] Walter G G 1994 *Wavelets and other orthogonal systems with applications* (CRC Press, Boca Raton)
[43] Gradshteyn I S and Ryzhik I M 2007 *Table of integrals, series and products* Seventh Edition (Academic Press, San Diego)
[44] Szegö G 1939 *Orthogonal Polynomials* (AMS, Providence)
[45] Gelfand I M and Shilov G E 1964 *Generalized Functions* vol. I (Academic Press-New York and London)

Journal of Physics: Conference Series **2038** (2021) 012002 doi:10.1088/1742-6596/2038/1/012002

# Ermakov-Pinney equation for time-varying mass systems

**Bijan Bagchi**[1] **and Achal Vinod**[2]

[1,2] Physics Department, Shiv Nadar University, Gautam Buddha Nagar, Uttar Pradesh 201314, India

E-mail: [1]`bbagchi123@gmail.com`, [2]`av927@snu.edu.in`

**Abstract.** We extend Fring-Tenney approach of constructing invariants of constant mass time-dependent system to the case of a time-dependent mass particle. From a coupled set of equations described in terms of guiding parameter functions, we track down a modified Ermakov-Pinney equation involving a time-dependent mass function. As a concrete example we focus on an exponential choice of the mass function.

## 1. Introduction

For some time now, there have been several efforts to study time-dependent quantum systems from different perspectives. One such consists in the use of optical traps (see for some representative articles [1, 2, 3, 4, 5]). In the non-relativistic context, the governing equation of concern is the time-dependent Schrödinger equation with the explicit presence of an optical potential. It is usually taken to be in a separable form which is made up of a time-dependent modulation term together with some trapping shape described by a spatial profile [6, 7].

Interest in exploring the dynamics of time-dependent quantum systems started with the method of seeking Hermitian invariants [8, 9, 10] for them. Recently, Fring and Tenney [11, 12] made a systematic study of time-independent approximations for a class of time-dependent optical potentials. Their general strategy was to write down an approximate Lewis-Riesenfeld scheme [9] and explore, among other things, time-dependent potentials with a Stark term. While constructing time-independent invariants, even for non-Hermitian systems [12, 13, 14], one of the central issues was to derive the highly nonlinear Ermakov-Pinney ($EP$) equation [15, 16]. Interestingly, they could obtain a regular solution to it, determine the electric field and finally construct the invariants. Note that they concentrated only on a constant mass system. For other approaches towards calculating dynamical invariants we refer to [17].

Historically, derivation of the $EP$ equation (see [18] in which some symmetry aspects of this equation were explored) was first undertaken in an early work of Ermakov [15] while studying integrablity of certain classes of ordinary second-order differential equation

$$\ddot{x} + \omega^2 x = \frac{\lambda^2}{x^3}, \quad \lambda > 0 \tag{1.1}$$

We see that the $EP$ equation describes an oscillator in the presence of an inverse-cube force term.

Much later, Pinney [16] provided the solution and gave a complete representation of it in the form

$$x(t) = \left(Ar^2 + 2Brs + Cs^2\right)^{\frac{1}{2}} \tag{1.2}$$

where $r, s$ are both functions of $t$ and stand as the independent solutions of the standard harmonic oscillator $\ddot{x}+\omega^2 x = 0$ while $A, B, C$ satisfy the constraint $AC-B^2 = \frac{h^2}{W}$, $W$ being the Wronskian.

The purpose of this article is to pursue an analog procedure as that of [11] but consider a time-dependent mass instead of a constant one. The variation of the mass with time is a well studied topic in problems of unstable particles [19, 20] in the realm of particle physics. A modification to the form of the Dirac equation in seeking invariants has also been considered for time-dependent masses in Dirac motivated relativistic systems [21]. Cases of both time-dependent and position-dependent mass have also been explored in the literature (see, for instance, [22, 23]).

In the following, we basically utilize invariants to solve the time-dependent Schrödinger equation when the guiding time-dependent potential is factorizable in terms of a Gaussian profile apart from a time-dependent piece. We see that a modified EP-equation emerges as an auxiliary equation whose form differs from what is obtained in the constant mass case. However, its highly nonlinear nature precludes us from finding any exact analytical solution. We have therefore gone for a specific case-study of the mass profile as given by an exponential representation. It reduces the basic equations to a tractable form thus facilitating a simple numerical assessment.

This work is organized as follows:

In section 2, we highlight the consistency conditions that emerge when the basic equation for an invariant Hermitian operator $\hat{I}$ is subjected to the Hamiltonian obeying the time-dependent Schrödinger equation. In section 3 we derive a modified $EP$ equation by installing time-dependence in the mass. Such an equation constitutes our subsequent point of analysis which is taken up numerically. In Section 4 we present a particular class of solution corresponding to the exponential choice of the mass. Finally, in section 5, we present a summary.

## 2. Constraining equations for time-dependent coefficients

Let us begin with the time-dependent non-relativistic Schrödinger equation which in one-dimension reads

$$i\frac{\partial}{\partial t}\psi(x,t) = \left[-\frac{1}{2m}\frac{\partial^2}{\partial x^2} + V(x,t)\right]\psi(x,t) \tag{2.1}$$

where $m = m(t)$. We work with natural units $\hbar = c = 1$.

For a general quantum mechanical Hamiltonian $\hat{H}$, the time evolution operator $U(t, t_0)$ obeys [24]

$$i\dot{U}(t,t_0) = \hat{H}U(t,t_0), \quad U(t_0,t_0) = I, \quad t > t_0 \tag{2.2}$$

If the Hamiltonian does not depend on time then the solution of the above equation is simply

$$U(t,t_0) = e^{-i\hat{H}(t-t_0)} \tag{2.3}$$

where Hamiltonian is considered to be Hermitian.

Against the background of the evolution equation obeying (2.2), we note that the time-dependent invariant operator $\hat{I}(t)$ satisfies

$$\frac{d\hat{I}}{dt} = \dot{\hat{I}}(t) + \frac{1}{i}[\hat{I}, H] = 0 \tag{2.4}$$

where $H$ is time-dependent $\hat{H} = \hat{H}(x,t)$. As a concrete model of our analysis we consider the following representative Hamiltonian

$$\hat{H}(x,t) = \frac{p^2}{2m(t)} + V(x,t) + xE(t) \tag{2.5}$$

where the mass and the potential have been taken to be time-dependent. The latter is assumed to be factorizable, for a wider class of time-dependent potentials that are factorizable and also exactly solvable see [25, 26], in that $V(x,t) = \kappa(t)V(x)$, where $\kappa$ is a function of $t$. Additionally, a Stark like term is included in (2.5) which points to the presence of an electric field $E(t)$. The angular frequency $\omega$ is taken to be constant.

The choice of the representation for the invariant is flexible. In the following we adopt the form

$$\hat{I}(t) = \frac{1}{2}\left[\alpha(t)p^2 + \gamma(t)x + \delta(t)\{x,p\} + \varepsilon(t)x^2\right] \tag{2.6}$$

where $\{x,p\} = xp + px$ and the coefficient functions $\alpha, \gamma, \delta$ and $\varepsilon$ are time-dependent functions. Of course, these are distinct from $m(t)$ or $\kappa(t)$ and have to be solved for through the consistency equations.

We now address the simplest class of time-dependent Hamiltonian as induced by a varying mass harmonic oscillator potential. In such a case $\kappa$ corresponds to $m(t)$ while $V(x) = \frac{1}{2}\omega^2 x^2$. Substituting in (2.4) we obtain through the use of (2.5) the commutator

$$[\hat{I},\hat{H}] = \frac{1}{2}\Bigg(\alpha m(-\hbar^2\omega^2 - 2i\hbar\omega^2 xp) + \alpha E(-2i\hbar)p + \frac{\gamma}{2m}(2i\hbar)p + \frac{\delta}{2m}(4i\hbar)p^2 \\ + \delta m(-2i\hbar)\omega^2 x^2 + \delta E(-2i\hbar)x + \frac{\varepsilon}{2m}(4i\hbar xp + 2\hbar^2)\Bigg) \tag{2.7}$$

When substituted in (2.4) the following set of conditions is easily obtained

$$\dot{\alpha} = -\frac{2\delta}{m} \tag{2.8}$$

$$\dot{\delta} = \alpha m\omega^2 - \frac{\varepsilon}{m} \tag{2.9}$$

$$\dot{\varepsilon} = 2\delta m\omega^2 \tag{2.10}$$

$$\dot{\gamma} = 2\delta E \tag{2.11}$$

We also have an additional relation

$$\gamma = 2m\alpha E \tag{2.12}$$

along with a closed form for $E(t)$ given by

$$E(t) = \frac{1}{m}e^{\int Q(t)dt}, \quad Q(t) = \frac{3\delta}{\alpha m} \tag{2.13}$$

We should mention that in the constant mass case, the set of equations (2.8) – (2.12) reduce to those of the corresponding ones in [11].

## 3. Ermakov-Pinney equation and its modification

We aim at writing down a modified EP equation in the presence of $m(t)$. Redefining $\alpha = \sigma^2$ the remaining parameters read

$$\delta = -m\dot{\sigma}\sigma \tag{3.1}$$

$$\gamma = 2\sigma^2 mE(t) \tag{3.2}$$

$$\varepsilon = m^2\dot{\sigma}^2 + m^2\sigma\ddot{\sigma} + m\dot{m}\dot{\sigma}\sigma + m^2\sigma^2\omega^2 \tag{3.3}$$

Introducing $\tau \equiv \tau(\sigma)$ as an arbitrary function of $\sigma$ which is defined through

$$m^2\sigma\ddot{\sigma} + m\dot{m}\dot{\sigma}\sigma + m^2\sigma^2\omega^2 = \frac{\tau m^2}{\sigma^2} \tag{3.4}$$

it is clear that (3.4) stands as the signature of a dissipative EP equation for the time-varying mass case

$$\ddot{\sigma} + \omega^2\sigma + \frac{\dot{m}}{m}\dot{\sigma} = \frac{\tau}{\sigma^3} \tag{3.5}$$

We immediately see that when $m$ is stationary and $\tau$ is a constant equal to $\lambda^2$ the standard equation (1.1) is recovered. (3.3) now reads

$$\varepsilon = m^2\dot{\sigma}^2 + \frac{\tau m^2}{\sigma^2} \tag{3.6}$$

Clearly, no unique set of solutions of the parameter functions exists satisfying $\sigma$-dependent equations (3.1), (3.2) and (3.6), for a given functional choice of the mass function $m(t)$ and the electric field $E(t)$.

Identifying the EP equation in the extended form (3.5) constitutes the central result of this note. Although not exactly coincident with (3.5), a dissipative $EP$ equation was also derived by Fring and Tenney [12] but in a different context

$$\ddot{\chi} + f^2\chi - \frac{\dot{f}}{f}\dot{\chi} = \frac{f^2}{\chi^3}, \quad \chi = \chi(t) \tag{3.7}$$

where $f$ is a time-dependent parametrizing function. $f$ seems to play the analog of the mass function but notice that while the coefficient of $\sigma$ in (3.5) is a constant, the same in the corresponding (3.7) is a function of time.

Our next endeavour would be to analyse (3.6) by taking its time-derivative and comparing with (2.10). Towards this end we obtain the pair of equations

$$\dot{\varepsilon} = 2m\dot{m}\dot{\sigma}^2 + 2\dot{\sigma}\ddot{\sigma}m^2 + \frac{1}{\sigma^2}(2m\dot{m}\tau + m^2\tau'\dot{\sigma}) - \frac{2m^2\tau\dot{\sigma}}{\sigma^3} \tag{3.8}$$

$$\dot{\varepsilon} = -2m^2\omega^2\sigma\dot{\sigma} \tag{3.9}$$

where (3.1) is used. Equating them we straightforwardly obtain a differential equation for $\tau$

$$\tau' + 2\frac{\dot{m}}{m\dot{\sigma}}\tau = 0 \tag{3.10}$$

where the prime denotes a derivative with respect to $\sigma$.

## 4. Solving the modified *EP* equation

Since $\tau = \tau(\sigma)$ let us assume a monomial form for it

$$\tau = \tau_0 \sigma^4 \tag{4.1}$$

where $\tau_0$ is a constant assumed to be $> 0$. The EP equation, in principle, admits of complex solutions given by the general form

$$\sigma(t) = \xi(t) e^{i\eta(t)}, \quad \xi, \eta \in \Re \tag{4.2}$$

where $\xi$ and $\eta$ are linked by a set of coupled equations

$$\ddot{\xi} + \frac{\dot{m}}{m}\dot{\xi} + (\Omega^2 - \dot{\eta}^2) = 0 \tag{4.3}$$

$$2\dot{\xi}\dot{\eta} + \xi\left(\ddot{\eta} + \frac{\dot{m}}{m}\dot{\eta}\right) = 0 \tag{4.4}$$

where $\Omega^2 = (\omega^2 - \tau_0)$. These are very complicated equations to tackle. So we concentrate on real solutions only and in this regard consider a mass function $m(t)$ of the exponential type

$$m(t) = m_0 e^{-qt}, \quad q > 0 \tag{4.5}$$

where its damping character is in keeping with [21, 22] but with respect to the t-variable. For $\sigma$ we find the solution

$$\sigma(t) = e^{qt/2} \tag{4.6}$$

subject to the the following constraint on $\tau_0$

$$\tau_0 = \omega^2 - \frac{q^2}{4} \tag{4.7}$$

$\tau$ is insensitive to the sign of $q$. For completeness we note that $\delta$ turns out to be a constant while the coefficient function $\gamma$ and the electric field $E$ show a decaying behaviour $\sim e^{-qt/2}$.

We display the behavior of the mass variation in Figure 1 corresponding to different values of $\omega$. We took the input value of $\tau_0 = 0.01$. The results reflect a typical exponential trajectory of $m$.

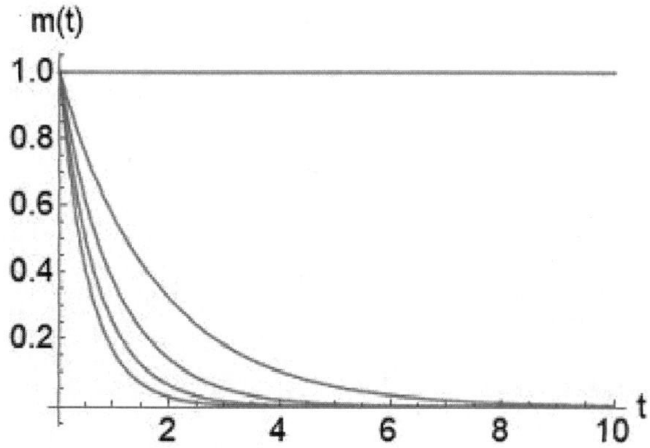


**Figure 1.** $m(t)$ variation for $\omega = 0.1, 0.3, 0.5, 0.7, 0.9$

## 5. Summary

In this article we considered the possibility of introducing a time-dependent mass while constructing an invariant for a time-dependent quantum Hamiltonian. Assuming a time-dependent harmonic oscillator with a Stark-like Hamiltonian we were able to derive the invariant conditions in their entirety. We then obtained the time-dependent mass version of the $EP$ equation that included an additional term as compared to the constant-mass case. Because of its strongly nonlinear character we inquired into simple parameter variations to assess the mass function. For the specific case of a damping mass with a certain input value of $\tau_0$, its graphical behavior was plotted over a wide range of $t$.

## Acknowledgments

We are indebted to Andreas Fring and Rebecca Tenney for pointing out errors in the previous version of the draft.

## References

[1] Ashkin A 1970 Acceleration and trapping of particles by radiation pressure *Phys. Rev. Lett.* **24** 156
[2] Neuman K C and Block S M 2004 Optical trapping Review of scientific instruments **75** 2787
[3] Ashkin A, Dziedzic J M, Bjorkholm J E and Chu S 1986 Observation of a single-beam gradient force optical trap for dielectric particles *Optics Letters* **11** 288
[4] Metcalf H J and Van der Straten P 2007 Laser cooling and trapping of neutral atoms, *The Optics Encyclopedia: Basic Foundations and Practical Applications.*
[5] Ashkin A, Dziedzic J M and Yamane T 1987 Optical trapping and manipulation of single cells using infrared laser beams Nature **330** 769
[6] Ohara K M, Gehm M E, Granade S R and Thomas J E 2001 Scaling laws for evaporative cooling in time-dependent optical traps *Phys. Rev.* A **64** 051403
[7] Fulton R, Bishop A I, Shneider M N and Barker P F 2006 Controlling the motion of cold molecules with deep periodic optical potentials *Nature Physics* **2** 465
[8] Lewis H R, Jr 1968 Class of exact invariants for classical and quantum time-dependent harmonic oscillators *J. Math. Phys.* **9** 1976
[9] Lewis H R, Jr and Riesenfeld W 1969 An exact quantum theory of the time dependent harmonic oscillator and of a charged particle time dependent electromagnetic field *J. Math. Phys.* **10** 1458
[10] Lai Y -Z, Liang J Q, Müller Kirsten H J W and Zhou J C 1996 Time dependent quantum systems and the invariant Hermitian operator *Phys. Rev.* A **53** 3691
[11] Fring A and Tenney R 2020 Time-independent approximations for time-dependent optical potentials *Eur. Phys. J. Plus* **135** 163
[12] Fring A and Tenney R 2021 Perturbative approach for strong and weakly coupled time-dependent non-Hermitian quantum systems *Phys Scrip* **96** 045211
[13] Fring A and Frith T 2017 Mending the broken PT-regime via an explicit time-dependent Dyson map *Phys. Lett.* A**381** 2318
[14] Koussa W and Maamache M 2020 Pseudo-Invariant Approach for a Particle in a Complex Time-Dependent Linear Potential *Int. J. Theor. Phys.* **59** 1490
[15] Ermakov V 1880 Second order differential equations. Conditions of complete integrability *Universita Izvestia Kiev Series III* **9** 1
[16] Pinney E 1950 The nonlinear differential equation $y'' + p(x)y + \frac{c}{y^3} = 0$ *Proc. Amer. Math. Soc.* **1** 681
[17] Bertin M C, Pimentek B M and Ramirez J A 2012 Construction of time-dependent dynamical invariants: A new approach *J. Math. Phys.* **53** 042104
[18] Morris R M and Leach P G L 2015 The Ermakov-Pinney Equation: its varied origins and the effects of the introduction of symmetry-breaking functions *Preprint* arXiv:1510.08992v1 (math.CA)
[19] Bekkar H, Benamira F and Maamache M 2003 Comment on "Solution of the Schrödinger equation for the time-dependent linear potential" *Phys. Rev.* A **68** 016101
[20] Landim R R and Guedes I 2000 Wave functions for a Dirac particle in a time-dependent potential *Phys. Rev.* A**61** 054101
[21] Zhang Z-G 2007 Complete solutions for a class of time-dependent Dirac equations *Phys. Scr.* **76** 349
[22] Pedrosa I A 1997 Exact wave functions of a harmonic oscillator with time-dependent mass and frequency *Phys. Rev.* A **55** 3219

[23] Bagchi B, Gorain P, Quesne C and Roychoudhury R 2005 New approach to (quasi)-exactly solvable Schrödinger equations with a position-dependent effective mass *Europhys. Lett.* **72** 155
[24] Sakurai J J 1994 *Modern quantum mechanics* (Addison-Wesley Publishing Company, Inc.)
[25] Rogers J and Spector D 1992 Quantum mechanics with explicit time-dependence *Phys. Lett.* A **170** 344
[26] Efthimion C J and Spector D 1994 Separation of variables and exactly soluble time-dependent potentials in quantum mechanics *Phys. Rev.* A **49** 2301

# $\mathcal{PT}$-symmetric classical mechanics

**Carl M Bender[1] and Daniel W Hook[1,2]**
[1]Physics Department, Washington University, St. Louis, MO 63130, USA
E-mail: cmb@wustl.edu
[2]Centre for Complexity Science, Imperial College London, London SW7 2AZ, UK
E-mail: d.hook@imperial.ac.uk

**Abstract.** This paper reports the results of an ongoing in-depth analysis of the classical trajectories of the class of non-Hermitian $\mathcal{PT}$-symmetric Hamiltonians $H = p^2 + x^2(ix)^\varepsilon$ ($\varepsilon \geq 0$). A variety of phenomena, heretofore overlooked, have been discovered such as the existence of infinitely many separatrix trajectories, sequences of critical initial values associated with limiting classical orbits, regions of broken $\mathcal{PT}$-symmetric classical trajectories, and a remarkable topological transition at $\varepsilon = 2$. This investigation is a work in progress and it is not complete; many features of complex trajectories are still under study.

## 1. Introduction

The family of non-Hermitian $\mathcal{PT}$-symmetric quantum-mechanical Hamiltonians

$$\hat{H} = \hat{p}^2 + \hat{x}^2(i\hat{x})^\varepsilon \quad (\varepsilon \text{ real}) \tag{1}$$

is a complex deformation in the parameter $\varepsilon$ of the quantum harmonic-oscillator Hamiltonian $\hat{p}^2 + \hat{x}^2$. At $\varepsilon = 0$ the spectrum of $\hat{H}$ consists of the harmonic-oscillator eigenvalues $E_n = 2n + 1$ ($n = 0, 1, 2, \ldots$). As $\varepsilon$ varies away from 0, the eigenvalues exhibit several characteristic behaviors: For $\varepsilon > 0$ the eigenvalues are all real, positive, and discrete, and the eigenvalues increase as $\varepsilon$ increases; for $\varepsilon < 0$ only a finite number of the eigenvalues are real and the infinite number of remaining complex eigenvalues occur as complex-conjugate pairs [1, 2, 3, 4, 5, 6].

The parametric region $\varepsilon \geq 0$ is called the *region of unbroken $\mathcal{PT}$ symmetry* because all of the eigenfunctions of $\hat{H}$ are simultaneously eigenstates of the $\mathcal{PT}$ operator; the parametric region $\varepsilon < 0$ is called the *region of broken $\mathcal{PT}$ symmetry* because the eigenfunctions of $\hat{H}$ are not all eigenstates of $\mathcal{PT}$ [6]. A transition at $\varepsilon = 0$ separates the regions of unbroken and broken $\mathcal{PT}$ symmetry. Typically, $\mathcal{PT}$-symmetric quantum Hamiltonians exhibit a transition with only real eigenvalues on one side and partly real and partly complex eigenvalues on the other. For $\varepsilon > 0$ there exists a formal similarity transformation from the non-Hermitian Hamiltonian $\hat{H}$ to a Hermitian Hamiltonian $\hat{h}$, which is *isospectral* to (has the same eigenvalues as) $\hat{H}$: $\hat{h} = S\hat{H}S^{-1}$. However, this isospectral equivalence is only formal because $S$ and $S^{-1}$ are unbounded operators, and thus the vectors in the domains of $\hat{H}$ and $\hat{h}$ are not in one-to-one correspondence [6].

The quartic case $\varepsilon = 2$ is special because only for this value of $\varepsilon$ can the transformation from $\hat{H}$ to $\hat{h}$ be carried out in closed form with $\hat{h}$ having the simple local structure $\hat{p}^2 + V(\hat{x})$, where the potential function $V$ is also quartic [7, 8, 9]. For other values of $\varepsilon$ the corresponding isospectral Hermitian Hamiltonian $\hat{h}$ contains $\hat{p}^2$, $\hat{p}^4$, $\hat{p}^6$, …, and thus is nonlocal [10].

This paper reports our study of the remarkable underlying *classical* properties of the quantum Hamiltonian (1). The classical trajectories are curves $x(t)$ that lie in the complex-$x$ plane. At the classical level there is a transition at $\varepsilon = 0$ that corresponds with the quantum transition described above. Specifically, in the region of unbroken $\mathcal{PT}$ symmetry the classical orbits tend to be closed but in the broken region the orbits are open [11]. This change of topology is shown in Fig. 1. The transition between regions of broken and unbroken $\mathcal{PT}$ symmetry has been observed repeatedly in laboratory experiments [12, 13, 14, 15, 16, 17, 18, 19, 20, 21, 22].

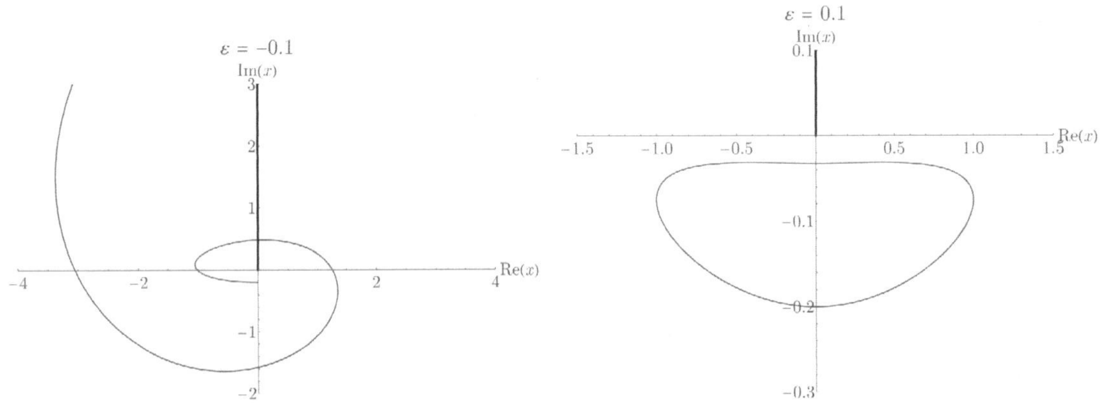


**Figure 1.** [Color online] Complex classical trajectories for the Hamiltonian $H = p^2 + x^2(ix)^\varepsilon$ for $\varepsilon = -0.1$ and $\varepsilon = 0.1$. The energy (the value of $H$) for the classical orbits is 1 and the initial value in both cases is $-0.2\,i$. For negative $\varepsilon$ the orbits are open (left panel) and for positive $\varepsilon$ the orbits are usually closed (right panel).

There have already been many studies of complex classical mechanics [2, 23, 24, 25, 26, 27, 28] and studies of complex quantum-mechanical systems at the classical level [29, 30, 31] and at the semiclassical level [32, 33, 34, 35]. Here, we have undertaken a detailed re-examination of the classical trajectories of $H$ in (1) in the unbroken region $\varepsilon \geq 0$. We report here that the classical trajectories exhibit many surprising properties that to date have not been noticed. For example, there are sequences of critical initial values that give rise to separatrix trajectories; these separatrices serve as boundaries between regions of topologically distinct classes of orbits. A particularly interesting feature is that at $\varepsilon = 2$, the unique value of $\varepsilon$ at which there exists a simple transformation from a quantum $\mathcal{PT}$-symmetric non-Hermitian quantum Hamiltonian to a local Hermitian quantum Hamiltonian, there is a corresponding dramatic transition in the topology of the classical orbits. We give a detailed description of the current status of this work in this paper.

This presentation is organized as follows. Section 2 reviews the most well known features of the complex classical trajectories for the Hamiltonian (1). We examine the generic properties of complex classical trajectories in the parametric region $0 \leq \varepsilon < 2$ in Sec. 3 with a focus on the case of irrational $\varepsilon$. (The case of rational $\varepsilon$ is more complicated because when $\varepsilon$ is rational there are new discrete symmetries in addition to $\mathcal{PT}$ symmetry that one must consider, and this complicates the discussion. We will discuss the case of rational $\varepsilon$ at length in a future paper.) We discuss the properties of complex trajectories in region $0\varepsilon \geq 2$ in Sec. 4. Some concluding remarks and conjectures about results of future investigations are given in Sec. 5.

## 2. General properties of complex classical trajectories

We begin by reviewing some of the properties of complex classical trajectories for the Hamiltonian (1). Hamilton's classical equations of motion are

$$\dot{x} = \frac{\partial}{\partial p}H = 2p, \qquad \dot{p} = -\frac{\partial}{\partial x}H = -(2+\varepsilon)x(ix)^{\varepsilon}, \tag{2}$$

where the dot indicates differentiation with respect to time $t$. (The time variable $t$ is treated as real.) As is the case with any Hamiltonian system, an immediate consequence of these equations is that the energy (that is, the value of $H$) is conserved in time. For any $\varepsilon$ we may perform a scaling of $x$ and $p$ in the Hamiltonian (1) so that without loss of generality the dimensionless total energy of a classical particle can be taken to be 1:

$$p^2 + x^2(ix)^{\varepsilon} = 1. \tag{3}$$

The system of first-order equations (2) is equivalent to Newton's equation of motion, which is a second-order differential equation, but the fact that energy is conserved allows us to use the first equation in (2) to rewrite (3) as a *first-order* differential equation for the motion of the classical particle:

$$\dot{x}^2 + 4x^2(ix)^{\varepsilon} = 4. \tag{4}$$

This equation emphasizes that if we know where the particle is, we know the velocity of the particle up to a sign, and therefore we can predict where the particle will go. (The choice in sign corresponds to whether the particle is moving forward or backward in time.) Furthermore, because this equation is complex, the path of the classical particle lies in the complex-$x$ plane.

In order to understand the trajectory of a classical particle it is necessary to find the *classical turning points* of the Hamiltonian. At these points, the velocity of the particle vanishes. From (4) we see that the turning points of $H$ in (1) are solutions to

$$x^2(ix)^{\varepsilon} = 1, \tag{5}$$

so the turning points all lie on the unit circle in the complex-$x$ plane.

By solving (4) we find the classical particle trajectories $x(t)$ and we can easily establish that the period $T$ of any closed orbit $C$ is given by [3]

$$T = \frac{1}{2}\oint_C \frac{dx}{\sqrt{1 - x^2(ix)^{\varepsilon}}}. \tag{6}$$

Typically, a closed path $C$ in the complex-$x$ plane (or, more generally, the Riemann surface) encloses the square-root branch-cut singularity joining a $\mathcal{PT}$-symmetric pair of turning points, as we can see from (6). Cauchy's theorem, which guarantees path independence of complex contour integrals, implies that for any value of $\varepsilon$ the period of all closed orbits that enclose a given pair of turning points is the same.

For the elementary case $\varepsilon = 0$, where the classical trajectories are governed by the classical harmonic oscillator Hamiltonian $H = p^2 + x^2$, there are only two turning points, which are located at $x = \pm 1$. All classical trajectories for this Hamiltonian are closed and periodic and lie in a (one-sheeted) complex plane. Five such trajectories are shown in Fig. 2.

All but one of the trajectories in Fig. 2 belong to the continuous family of nested ellipses whose foci are the turning points; these ellipses fill the entire complex-$x$ plane except for the real axis between the turning points. In addition, there is one classical trajectory for a particle that oscillates between the two turning points; this particle reaches a turning point in finite time, stops, and then retraces its path as it moves away from the turning point. We call this special

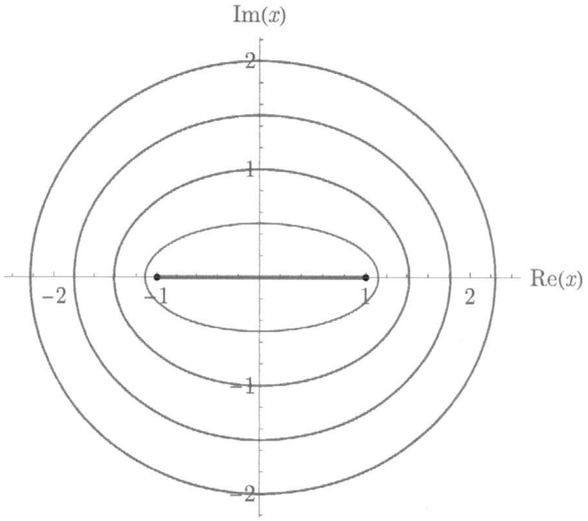


**Figure 2.** [Color online] Five closed complex trajectories of the classical harmonic oscillator $H = p^2 + x^2$, which is the $\varepsilon = 0$ case of $H$ in (1). The energy (the value of $H$) for all classical orbits is 1. Four orbits are nested ellipses whose foci are the turning points (black dots) located at $x = \pm 1$. These elliptical orbits pass through the points $-0.5\,i$, $-i$, $-1.5\,i$, and $-2.0\,i$. A special singular orbit (bold red) lies on the real axis and terminates at the turning points. The period of every orbit has the same value $T = \pi$. Although the orbits are elliptical, they are unlike Galilean planetary orbits, which are also nested ellipses, because the periods of Galilean orbits increase as the size of the ellipses increases. (Jupiter's year is longer than the Earth's year.) The Hamiltonian $H = p^2 + x^2$ is both $\mathcal{P}$ symmetric and $\mathcal{T}$ symmetric, so all trajectories are left-right and also up-down symmetric.

trajectory, which representing the motion of a particle that oscillates between a pair of turning points, a *terminating* trajectory.

The terminating trajectory is *singular* because both ends of the trajectory terminate at turning points while all other trajectories represent a particle that *never* stops or reverses its direction. Note that in the limit as the semiminor axis of an elliptical path approaches 0 we obtain a *degenerate* ellipse that represents a particle that travels parallel to and infinitesimally above (or below) the real axis, goes around a turning point infinitesimally close to the turning point, and continues on again parallel to and infinitesimally below (or above) the real axis. This limiting trajectory does not terminate. The singular terminating trajectory lies *exactly* on the real axis between the two turning points; the terminating trajectory is special because it does not encircle and enclose the turning points.

To summarize: All elliptical classical orbits enclose the branch cut on the real axis that joins the two turning points while the singular orbit that oscillates between the turning points terminates at the turning points and does not enclose the branch cut. All classical orbits, the elliptical orbits and the terminating orbit, have period $T = \pi$, which we calculate by evaluating the integral (6).

In general, the shape of a classical trajectory in the complex plane is determined by the turning points in the vicinity of the trajectory. Recall what happens in the neighborhood of a turning point on the real axis: The turning points in Fig. 2 mark the edges of the *classically allowed* region, which is the portion of the real axis between the turning points. A particle in the classically allowed region slows down as it approaches a turning point. The particle stops

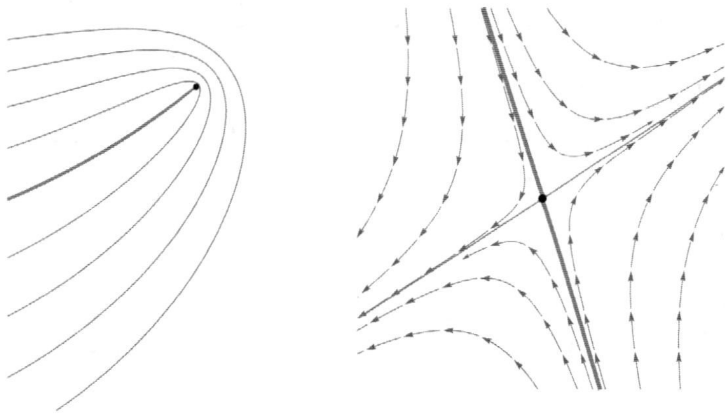

**Figure 3.** [Color online] Schematic comparison of the behavior of trajectories of a complex one-degree-of-freedom Hamiltonian near a typical turning point (black dot) in the complex plane (left panel) and trajectories of a real two-degree-of-freedom dynamical system near a typical saddle point (black dot) (right panel). For the complex Hamiltonian there is a unique terminating path (bold red) along which the trajectory approaches the turning point, stops, and then moves away from the turning point. All other trajectories (blue) exhibit a sling-shot-like behavior in which they make a 360° turn as they go around the turning point. For a saddle point there are two paths (bold red) along which particles approach this critical point, but these particles cannot reach the critical point in finite time. Along all other curves (blue) particles at first approach the saddle point and then veer away from it.

momentarily at the turning point and then speeds up as it goes back along the path and moves away from the turning point. This trajectory follows (and is parallel to) the real axis. However, in the *classically forbidden* regions on the real axis to the right of the right turning point and to the left of the left turning point, all classical trajectories are *orthogonal* to the real axis.

Let us extend this picture into the complex plane: In the neighborhood of a turning point in the complex plane there is just one direction towards and away from the turning point and there is a unique singular terminating curve along which the classical particle (i) slows down as it approaches the turning point; (ii) stops instantaneously at the turning point; and then (iii) speeds up as it retraces this path and moves away from the turning point [9]. Particles following other complex paths are flung around the turning point. The turning point acts as if it is pulling on the classical particle; paths near the terminating (in-and-out) trajectory make a sling-shot-like 360° turn about the turning point. Some typical classical trajectories are illustrated schematically in Fig. 3 (left panel). Note that the behavior illustrated in this figure is specific to theories for which the kinetic-energy term in the Hamiltonian is quadratic in $p$. For Hamiltonians whose kinetic-energy term is $p^3$ or $p^4$, for example, the behavior of classical trajectories near the turning points is significantly more complicated [9, 36]. We do not discuss this possibility here.

Note that we are discussing here a Hamiltonian dynamical system having *one complex degree of freedom*. Its classical trajectories are paths $x(t)$ in the complex-$x$ plane. This is not the same as a system having two real degrees of freedom. For the real system, the structure analogous to a complex turning point is a critical point called a *simple saddle point*. Near a typical (quadratic) saddle point [see Fig. 3 (right panel)] there are *two* terminating paths (red) directly towards the critical point. On these two special curves the particle slows down as it approaches the critical point but does not reach the critical point in finite time. On all other curves the particle first slows down as it approaches the saddle point and then speeds up as it turns and moves away from the saddle point.

As we increase $\varepsilon$ in $H$ in (1) above 0, the right and left turning points shown on Fig. 2 rotate downward into the lower-half complex plane and are located at

$$x_{\rm right} = \exp\left(-\tfrac{i\pi\varepsilon}{2\varepsilon+4}\right) \quad \text{and} \quad x_{\rm left} = \exp\left(-i\pi + \tfrac{i\pi\varepsilon}{2\varepsilon+4}\right), \tag{7}$$

which are $\mathcal{PT}$ reflections of one another. [Under a $\mathcal{PT}$ reflection a point $x$ in the complex-$x$ plane goes into $-x^*$ [6], and because $H$ in (1) is $\mathcal{PT}$ symmetric, its turning points are symmetric with respect to reflection about the imaginary axis.] Because the potential becomes multivalued when $\varepsilon > 0$, the equations of motion (2) and (4) of the classical particle are defined on a Riemann surface. Thus, the complex-$x$ plane in Fig. 2 extends onto a multisheeted Riemann surface with a branch cut emanating from the origin $x = 0$. To be consistent with $\mathcal{PT}$ symmetry we take the branch cut to lie on the positive-imaginary axis (see Fig. 4).

We evaluate the integral (6) and find that the period of the closed orbits is [3]

$$T = 2\sqrt{\pi}\,\cos\left(\tfrac{\pi\varepsilon}{4+2\varepsilon}\right)\Gamma\left(\tfrac{3+\varepsilon}{2+\varepsilon}\right)/\Gamma\left(\tfrac{4+\varepsilon}{4+2\varepsilon}\right). \tag{8}$$

The derivation of (8) assumes that the contour $C$ in (6) *does not cross the branch cut on the positive-imaginary axis.*

Figure 4 displays some classical trajectories for $\varepsilon = 1/\pi$. Plotted are the terminating trajectory for a particle that oscillates between the turning points, which lie at $t_0 = \pm 0.84496 - 0.53483\,i$, and three more trajectories that enclose this pair of turning points. These trajectories are $\varepsilon$-deformed versions of the $\varepsilon = 0$ ellipses in Fig. 2. No trajectories shown in Fig. 4 cross the branch cut. The periods for all of the closed trajectories that do not cross the branch cut, as obtained from (8) (and verified numerically), are $T = 2.93702$.

The maximal trajectory (the largest trajectory in Fig. 4 that does *not* cross the branch cut on the positive-imaginary axis) begins at the critical point $x_0 = -0.6438554\,i$. This trajectory is a separatrix curve that serves as the boundary between (i) trajectories that begin above $x_0$ and remain on the principal sheet of the Riemann surface, and (ii) trajectories that begin below $x_0$, cross the branch cut, and enter other sheets of the Riemann surface. As $\varepsilon$ increases, $x_0$ moves down the imaginary axis. As $\varepsilon$ approaches 1.0, $x_0$ approaches $-2.0\,i$ and as $\varepsilon$ approaches 2.0, $x_0$ approaches $-i\infty$ (see Fig. 5). Thus, *there is a critical topological transition in the classical theory at* $\varepsilon = 2$. This is a new discovery.

Between the origin and the critical point $x_0$, where the separatrix trajectory crosses the negative-imaginary axis, is the special point $s_0$. The point $s_0$ lies on the negative imaginary axis and is the center of the terminating trajectory connecting the turning points $t_0$. Like the plot of $x_0$ in Fig. 5, Fig. 6 shows the value of $s_0$ as a function of $\varepsilon$. Unlike $x_0$, $s_0$ remains *finite* for all $\varepsilon$. Observe that $s_0$ attains the value $-i$ when $\varepsilon = 2$. Then, as $\varepsilon$ increases further, $s_0$ reaches a minimum value of about $-1.21188\,i$ at $\varepsilon = 7.62547$ and then rises to $-i$ as $\varepsilon \to \infty$.

It is easy to explain this asymptotic limit. We see from (7) that the turning points approach $-i$ on the negative-imaginary axis as $\varepsilon \to \infty$. Furthermore, we see from (2) that a trajectory that crosses the imaginary axis at the point $-iL$, where $L > 0$, has a *real* velocity of $\pm\sqrt{1 + L^{\varepsilon+2}}$, and thus the trajectory is *perpendicular* to this axis. (The reality of the velocity is a consequence of $\mathcal{PT}$ symmetry.) We conclude that the terminating trajectory becomes infinitely short and crosses the imaginary axis at $-i$ as $\varepsilon$ approaches $\infty$. The triviality of this trajectory at $\varepsilon = \infty$ corresponds to a dramatic simplification in the corresponding quantum-mechanical theory, which becomes exactly solvable (the complex equivalent of a square-well potential) for infinite $\varepsilon$ [37].

A feature of the terminating curve that one can calculate analytically is the *slope* of the terminating curve at a turning point. To determine this slope we use Newton's law to find the

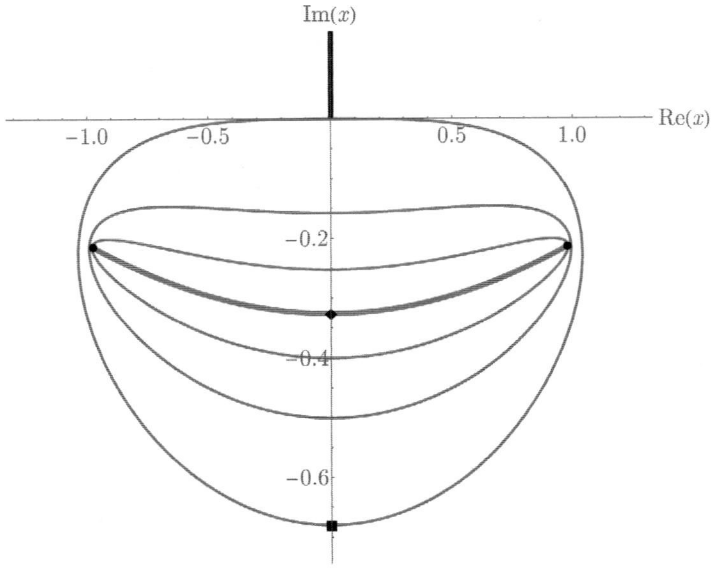


**Figure 4.** Four complex classical trajectories of $H$ in (1) for $\varepsilon = 1/\pi$. The energy is 1. A terminating singular trajectory (bold red) begins at $s_0 = -0.325235\,i$ (black diamond) and joins the two turning points (bold dots), which are located at $t_0 = \pm 0.88496 - 0.53483\,i$. Three more trajectories, which begin at $-0.4\,i$, $-0.5\,i$, and $= -0.6438554\,i$ enclose the turning points. The point $x_0 = -0.679076\,i$ (black square) is a *critical point* because the trajectory that emerges from this point is the *last trajectory that does not cross the branch cut* (black line on the positive-imaginary axis); this trajectory is a *separatrix*. The periods for all closed periodic orbits shown in this figure is $T = 2.93702$. None of the trajectories shown cross the branch cut but, as we will see, trajectories that begin below $x_0$ on the imaginary axis cross the branch cut on the positive-imaginary axis at successively higher points.

acceleration $\ddot{x}$ of a classical particle at the turning point. To derive Newton's law we take the time derivative of the first equation in (2) and use the second equation in (2) to eliminate $\dot{p}$:

$$\ddot{x} = -2(2+\varepsilon)x(ix)^{\varepsilon}. \tag{9}$$

This equation gives the (complex) acceleration $\ddot{x}$ as a function of the position of the classical particle. If we substitute the location of the right turning point $x_{\rm right}$ in (7) for $x$ in (9), we obtain

$$\ddot{x} = -(4+2\varepsilon)\exp\left(i\tfrac{\pi\varepsilon}{4+2\varepsilon}\right). \tag{10}$$

Thus, at the right turning point the particle accelerates downward and to the left and the terminating curve slopes upward at an angle of $\frac{\pi\varepsilon}{4+2\varepsilon}$. For example, in Fig. 4 the slope of the terminating curve is $\frac{\pi}{2+4\pi}$, which we have verified numerically.

## 3. Trajectories that leave the principal sheet

The class of nested trajectories shown in Fig. 4 densely fills the closed region in the complex plane that is bounded by the critical trajectory passing through the critical point $x_0$. We refer to his region as $R_0$, where the subscript 0 indicates that $R_0$ contains the 0th pair of turning points $t_0$ that the terminating trajectory $s_0$ connects.

A trajectory that begins slightly below the critical point $x_0 = -0.679076\,i$ in Fig. 4 is shown in Fig. 7. This trajectory wraps around the edge of the region of periodic trajectories in Fig. 4

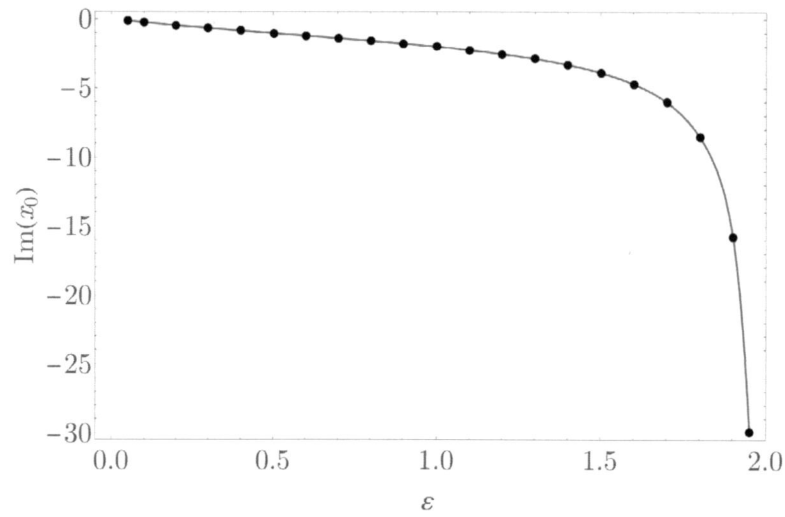


**Figure 5.** Plot of the critical point $x_0$ (the lower point at which a closed classical trajectory crosses the negative-imaginary axis for which the corresponding upper crossing point is the origin) as a function of $\varepsilon$. As $\varepsilon$ approaches 1.0, $x_0$ approaches $-2.0\,i$ and as $\varepsilon$ approaches 2.0, $x_0$ approaches $-i\infty$. The trajectory that originates from the critical point is a *separatrix* that serves as the boundary curve between orbits that remain on the principal sheet and orbits that cross the branch cut and visit other sheets of the Riemann surface.

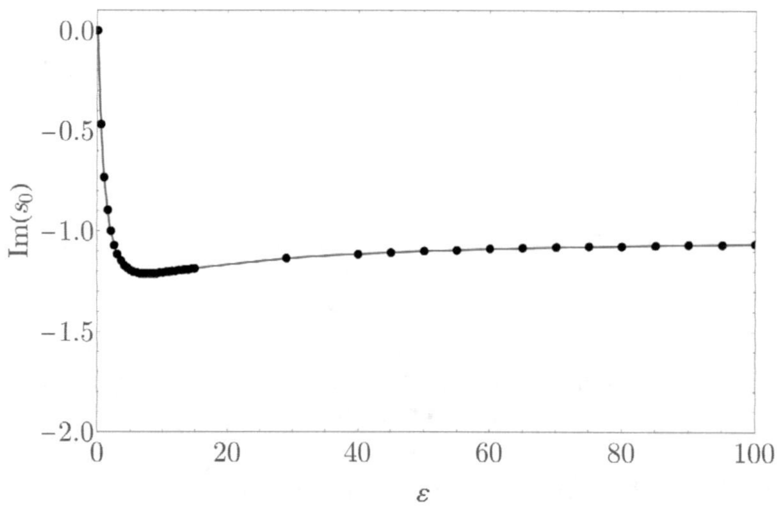


**Figure 6.** Plot of the special initial value $s_0$ that gives rise to the terminating trajectory as a function of $\varepsilon$. Note that $s_0 = -i$ at $\varepsilon = 2$. As $\varepsilon$ continues to increase, $s_0$ passes through a minimum value of $-1.21188\,i$ at $\varepsilon = 7.62547$ and then approaches the limiting value $-i$ as $\varepsilon \to \infty$.

that are confined to the principal sheet (sheet 0, where $\arg z$ ranges from $-\frac{3}{2}\pi$ to $\frac{1}{2}\pi$), and then briefly dips into the sheet above sheet 0 (sheet 1, where $\arg z$ ranges from $\frac{1}{2}\pi$ to $\frac{5}{2}\pi$) and also the sheet below sheet 0 (sheet $-1$, where $\arg z$ ranges from $-\frac{7}{2}\pi$ to $-\frac{3}{2}\pi$).

The trajectory shown in Fig. 7 is a separatrix that bounds a new region that we call $R_1$. The subscript 1 indicates that this region contains the pair of turning points $t_1$, which are connected by the terminating trajectory $s_1$. The full set of turning points are located at

$$\exp\left[\frac{(4N-\varepsilon)\pi i}{2\varepsilon+4}\right] \quad (N = 0,\ \pm1,\ \pm2,\ ...),$$

where the 0th pair of turning points $t_0$ is given by $N = 0, -1$, the 1st pair of turning points $t_1$ is given by $N = 1, -2$, the 2nd pair of turning points $t_2$ is given by $N = 2, -3$, and so on.

To compute and plot trajectories like that in Fig. 7, which visit multiple sheets of the Riemann surface, we rewrite Hamilton's equations in (2) in terms of polar variables. We introduce polar representations for $x(t)$ and $p(t)$:

$$x(t) = r(t)e^{i\theta(t)}, \qquad p(t) = a(t)e^{i\alpha(t)},$$

where $r$, $\theta$, $a$, $\alpha$ are real and $r,\ \alpha \geq 0$. We then take the real and imaginary parts of the equations of motion (2) to obtain four coupled real first-order differential equations:

$$\begin{aligned}
\dot{r}(t) &= 2a(t)\cos[\alpha(t)-\theta(t)],\\
\dot{\theta}(t) &= \tfrac{2a(t)}{r(t)}\sin[\alpha(t)-\theta(t)],\\
\dot{a}(t) &= (2+\varepsilon)r^{\varepsilon+1}(t)\cos[(1+\varepsilon)\theta(t)-\alpha(t)+(\varepsilon-2)\pi/2],\\
\dot{\alpha}(t) &= \tfrac{\varepsilon+2}{a(t)}r^{\varepsilon+1}(t)\sin[(\varepsilon+1)\theta(t)-\alpha(t)+(\varepsilon-2)\pi/2].
\end{aligned} \qquad (11)$$

The coupled system (11) is easy to solve numerically because it is first order and we can use Runge-Kutta methods to do so. An advantage of using this approach is that we can calculate the energy at each time step to verify that it is conserved. This is a crucial diagnostic that we use to verify the accuracy of our results.

The shape of a trajectory is controlled by turning points in the complex plane. In general, turning points tend to *pull* on trajectories [36]. We see in Fig. 4 that there are two turning points $t_0$ on the principal sheet, and from (7) we know that they are located in the lower-half plane at the angles $-\frac{\pi}{2+4\pi}$ and at $-\pi+\frac{\pi}{2+4\pi}$. The angular distance between these turning points is $\pi-\frac{\pi}{1+2\pi}$, and if we rotate clockwise or anticlockwise by this angular amount from the turning points $t_0$ on the principal sheet we find the next pair of turning points $t_1$, one at $\frac{2\pi-1}{2+4\pi}\pi$ on sheet 1 and one at $-\frac{1+8\pi}{2+4\pi}\pi$ on sheet $-1$.

In general, even though there are many turning points that influence the shape of a classical trajectory, a closed trajectory encloses (encircles) exactly two turning points; that is, if we think of the trajectory as a directed closed curve, then exactly two turning points lie in the *interior* of the region delimited by the trajectory. This is a key property of a classical Hamiltonian whose kinetic term is $p^2$, regardless of the number of Riemann sheets visited by the trajectory. (Closed trajectories for a $p^3$ or $p^4$ Hamiltonian may enclose more than two turning points.)

One surprising and nonintuitive result of the numerical study in Fig. 7 is that the trajectory that lies at the border but immediately *outside* the gray Region $R_0$ in Fig. 7 does *not* enclose the turning points $t_0$ in $R_0$. It is important to understand how a trajectory that follows the border of a region does not enclose that region. To assist in visualizing why this is so, we provide the schematic diagram in Fig. 8. This figure shows that a trajectory just outside the border of the gray region snaps and heads off in one direction, turns around, heads off in another direction, turns around, and then returns to its starting point without enclosing the gray region.

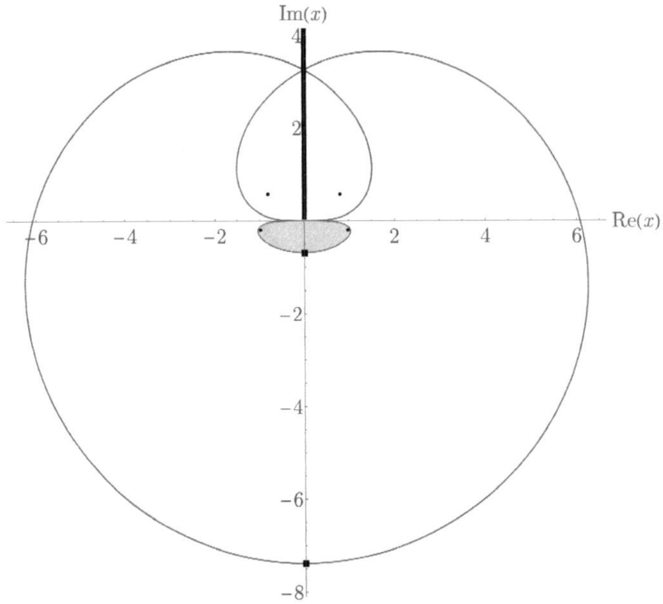


**Figure 7.** [Color online] A trajectory that begins at $-0.68\,i$, which is just below the critical point at $x_0 = -0.679076\,i$ (black square) for $\varepsilon = 1/\pi$. The trajectory can go right or left, but if it goes to the right, it follows the edge of Region $R_0$ (gray) containing trajectories that lie entirely on the principal sheet (sheet 0) of the Riemann surface. It then crosses the branch cut (black line) on the positive-imaginary axis in the positive sense and enters the sheet above the principal sheet (sheet 1, where the trajectory is colored red). The trajectory then curves to the right and crosses back onto the principal sheet, where it is colored purple. After making a full 360° clockwise turn in the principal sheet, it crosses the branch cut in the negative direction and enters the sheet below the principal sheet (sheet $-1$, where the trajectory is colored blue). It then turns around, crosses the branch cut in the positive direction, returns to the principal sheet, and follows the edge of the gray Region $R_0$ back to the starting point $-0.68\,i$ on the negative-imaginary axis. This trajectory is $\mathcal{PT}$ symmetric; that is, it is invariant under reflection through the imaginary axis on sheet 0. The trajectory also crosses the negative-imaginary axis just above the critical point $x_1 = -7.389098\,i$ (black square), which marks the boundary between curves in Region $R_1$ and curves in Region $R_2$. This closed trajectory encloses the *next* pair of turning points (black dots), which are located at $t_1 = \pm 0.81697 + 0.57668, i$ in the upper-half complex planes on sheets 1 and $-1$, but does *not* enclose the turning points at $t_0 = \pm 0.97908 - 0.20346\,i$, which are also shown on Fig. 4.

Having seen the schematic diagram in Fig. 8, we display in Fig. 9 a three-dimensional view of the trajectory in Fig. 7. This figure emphasizes that the trajectory crosses from sheet 0 (the principal sheet) up onto sheet $+1$ and then down through sheet 0 and onto sheet $-1$ before returning back to sheet 0 of the Riemann surface.

Like the special trajectory in Fig. 4, which connects the two turning points on the principal sheet, there is also a special terminating trajectory that connects the two new turning points $t_1$ indicated in Fig. 7, which are located on sheets $+1$ and $-1$. This special trajectory is the complex generalization of real oscillatory classical motion: Rather than going around the turning points, the classical particle that follows this trajectory reaches a turning point, stops, and then retraces its path as it moves away from the turning point. This terminating trajectory, which is

Journal of Physics: Conference Series **2038** (2021) 012003 doi:10.1088/1742-6596/2038/1/012003

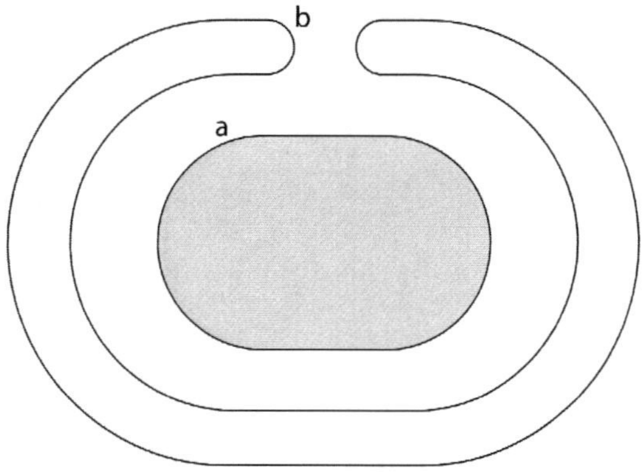


**Figure 8.** Schematic diagram showing how a closed trajectory like that in Fig. 7, which follows the border of the gray region, does not enclose that region. The biggest trajectory inside the gray region is labeled **a**; a trajectory just outside the border of the gray region is labeled **b**. Note that **a** and **b** are different. Outside the the gray region the **b** trajectory snaps and turns around. It snakes in one direction on a new sheet of the Riemann surface and then snakes in the opposite direction on another sheet, and thus does not enclose any part of the gray region.

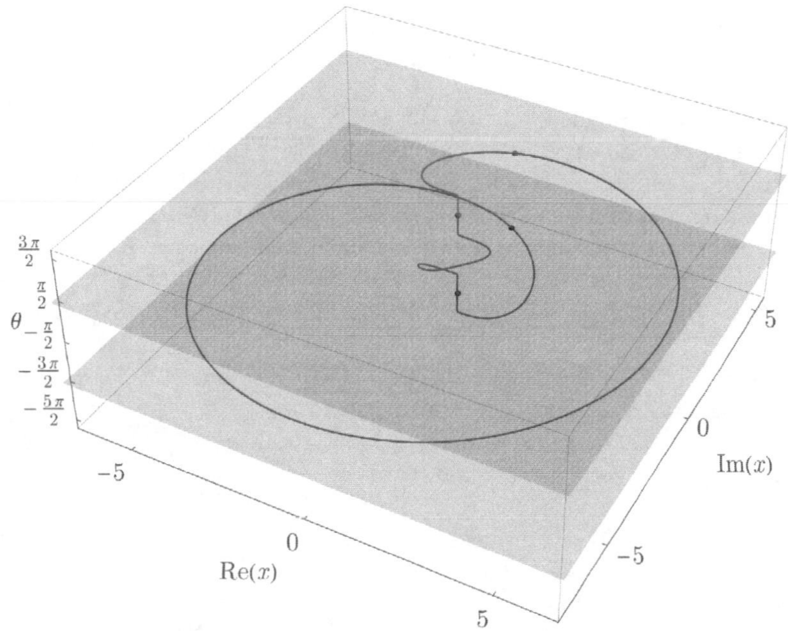


**Figure 9.** [Color online] The trajectory in Fig. 7 plotted in three-dimensional perspective to make precise the generic features of the curve **b** in the schematic diagram in Fig. 8. At the red dots the trajectory enters and leaves sheet 1 and at the blue dots the trajectory enters and leaves sheet $-1$ on the Riemann surface. The trajectory begins at $-0.68\,i$.

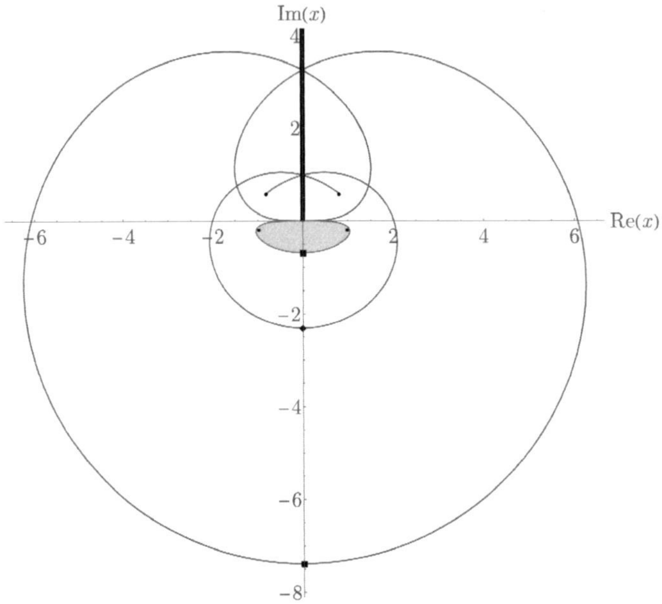


**Figure 10.** [Color online] Same as Fig. 7 but with one additional trajectory that terminates at the two new turning points at $t_1 = \pm 0.81697 + 0.57668, i$ in the upper-half complex planes on sheets $-1$ and 1. This new trajectory begins lower down the imaginary-$x$ axis at $s_1 = -2.31061\, i$. As in Fig. 7, trajectories are colored purple on sheet 0, red on sheet 1, and blue on sheet $-1$.

shown in Fig. 10, begins at $s_1 = -2.31061\, i$.

All of the trajectories in Region $R_0$ are $\mathcal{PT}$ symmetric and cross the negative-imaginary axis on the principal sheet of the Riemann surface between the origin and $x_0$. Similarly, all of the trajectories in Region $R_1$ are $\mathcal{PT}$ symmetric and cross the negative-imaginary axis on the principal sheet of the Riemann surface between $x_0$ and $x_1$. This might lead us to conjecture that there is an infinite tower of regions $R_n$ ($n = 0, 1, 2, ...,$ where $R_n$ contains the pair of turning points $t_n$ joined by the terminating trajectory $s_n$, and one might guess that these regions cover the negative-imaginary axis and that they are bounded by separatrices passing through the critical points $x_n$. While this is a very neat and simple picture, this hypothesis is completely wrong!

The actual situation is much more complicated. Only a *finite* number of regions cover the entire negative-imaginary axis, and this number decreases to 1 as $\varepsilon$ approaches 2 from below. The lower end of the lowest region extends to $-i\infty$. Indeed, we see in Fig. 5 that as $\varepsilon \to 2^-$, the critical point $x_0$ reaches $-i\infty$. For the case were are considering, namely, $\varepsilon = \frac{1}{\pi}$, the negative axis is completely covered by just four Regions: $R_0$, $R_1$, $R_2$, and $R_5$, where the lower end of Region $R_5$ is at $-i\infty$. Because the negative-imaginary axis is covered by these four regions, it is not possible for trajectories associated with any other regions to cross this axis, and thus all other trajectories are not $\mathcal{PT}$ symmetric.

It is possible that some of these broken $\mathcal{PT}$-symmetric paths might approach limit cycles in the complex plane, but we have not yet been found such a trajectory. It is much more likely that non-$\mathcal{PT}$-symmetric trajectories spiral outward towards infinity as we have seen in Fig. 1 (left panel).

We have found a particularly useful class of trajectories that help us to understand the topology of complex classical orbits, and these are asymptotically radial trajectories that terminate at infinity. To find such trajectories we begin with (4), and study the asymptotic behavior of $x(t)$ for large $|x|$. We introduce radial coordinates via

$$x(t) = r(t)e^{i\theta_0}, \tag{12}$$

where $r(t)$ and $\theta_0$ are real and $r(t) > 0$. In general, the polar angle is a function of $t$, as we see in Fig. 1 (left panel), but here we seek *asymptotically radial* paths so we are interested in the simple case for which $\theta_0$ is a constant. Substituting (12) into (4), we obtain an ordinary differential equation for $r(t)$:

$$\dot{r}^2 e^{2i\theta_0} + 4r^{2+\varepsilon}e^{i\pi\varepsilon/2}e^{(2+\varepsilon)i\theta_0} = 4. \tag{13}$$

If there is a solution for which $r \gg 1$, we may neglect the 4 on the right side of (13). We therefore have a two-term differential equation,

$$\dot{r}^2 \sim -4r^{2+\varepsilon}e^{i\pi\varepsilon/2}e^{i\varepsilon\theta_0} \quad (r \to \infty), \tag{14}$$

which we can solve. Since the left side of (14) is real and positive, the right side must also be real and positive, so we obtain an algebraic equation for the phase

$$1 = -e^{i\pi\varepsilon/2}e^{i\varepsilon\theta_0}, \tag{15}$$

and a separable asymptotic differential equation for the radial variable $r(t)$:

$$\dot{r}^2 \sim 4r^{2+\varepsilon} \quad (r \to \infty). \tag{16}$$

First, we solve (16). Remembering that both sides are positive, we take the square root and get $\dot{r} \sim \pm 2r^{1+\varepsilon/2}$ $(r \to \infty)$. We then *separate* the equation $\dot{r}r^{-1-\varepsilon/2} \sim \pm 2$ $(r \to \infty)$ and integrate both sides with respect to $t$: $r^{-\varepsilon/2} \sim \pm\varepsilon(C-t)$ $(r \to \infty)$, where we have divided by $-2$ and $C$ is a constant of integration. Finally, we square this equation, invert it, and raise it to the power $1/\varepsilon$:

$$r^{\varepsilon} \sim [\varepsilon^2(C-t)^2]^{-1/\varepsilon} \quad (r \to \infty). \tag{17}$$

This shows that for all $\varepsilon > 0$ the particle reaches $r = \infty$ in **finite time**. Also we verify that the asymptotic approximation of neglecting the 4 on the right side of (13) is consistent. Finally, we solve (15) to find the angular locations of the asymptotic rays. The result is

$$\theta_0 = -\frac{\pi}{2} + (2N-1)\frac{\pi}{\varepsilon}, \tag{18}$$

where $N = 0, \pm1, \pm2, \ldots$ is an integer.

Here are a few simple cases: If $\varepsilon = 1$, then for $N = 1$ the escape path (the radial trajectory to infinity) is exactly at $\pi/2$. (See Fig. 11.) This is a particularly interesting trajectory that terminates at *one* end. The trajectory runs straight up the positive-imaginary axis from the turning point at $i$ and bounces off the point at $i\infty$. If $\varepsilon = 2$, then the escape path is located at $(N-1)\pi$. Thus, for $N = 1$ and $N = 0$ the escape paths are at 0 and $-\pi$ (that is, the positive-real and negative-real axes).

A more interesting and nontrivial asymptotically radial escape path for $\varepsilon = \frac{1}{\pi}$ is shown in Fig. 12. This path, which corresponds to $N = 0$ and $N = 1$ in (18), lies on five sheets of the Riemann surface. The path is $\mathcal{PT}$ symmetric and approaches $\infty$ on sheets $\pm2$ of the Riemann surface. This path is a separatrix trajectory between Regions $R_2$ and $R_5$.

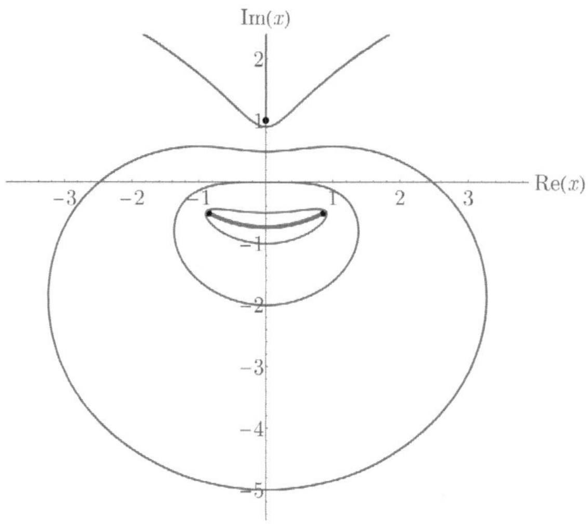


**Figure 11.** [Color online] Complex trajectories for the simple case $\varepsilon = 1$. There is only one sheet in the Riemann surface and three turning points (black dots). Five trajectories are shown: (i) A two-ended terminating trajectory connecting the $t_0$ pair of turning points in the lower-half plane; (ii) A single-ended (semi-terminating) trajectory that begins at the turning point at $i$ and runs off to $i\infty$ along the positive-imaginary axis; (iii) Three closed trajectories, one beginning at $-i$, a second beginning at $-2i$ and passing through the origin (as predicted in Fig. 5), and a third beginning at $-5i$ and dipping to avoid the semi-terminating trajectory that begins at $i$.

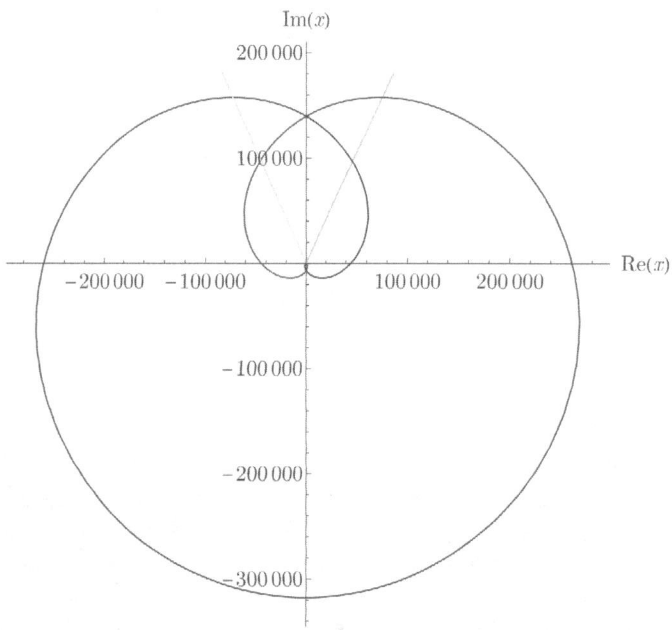


**Figure 12.** [Color online] A $\mathcal{PT}$-symmetric five-sheeted separatrix path for $\varepsilon = \frac{1}{\pi}$ corresponding the $N = 0$ and $N = 1$ cases in (18). The path becomes asymptotically straight on sheets $\pm 2$. The color scheme for the trajectory is as follows: lime green on sheet $+2$, red on sheet $+1$, purple on sheet 0, blue on sheet $-1$, and mint green on sheet $-2$.

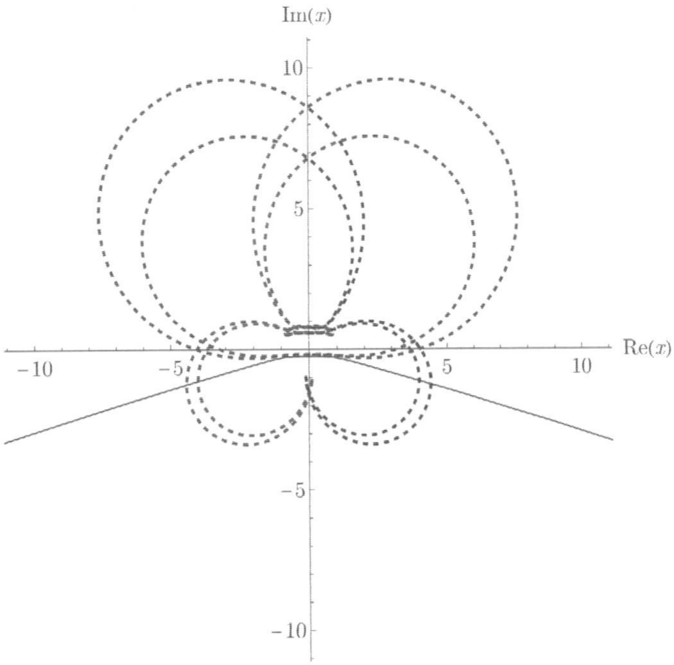


**Figure 13.** [Color online] Two complex trajectories for $\varepsilon = 1 + \sqrt{2}$. One closed trajectory (solid line) begins at $x_0 = -0.25i$, which lies just below the upper edge of Region $R_0$, and lies just inside the boundary Region $R_0$. This trajectory crosses the negative-imaginary axis on the principal sheet again (not shown). A second closed trajectory (dashed line) begins at $-0.2i$, which lies in the gap on the negative-imaginary axis below the origin and above the upper edge of Region $R_0$. This trajectory is $\mathcal{PT}$ symmetric and lies inside Region $R_1$. Unlike the situation for which $0 < \varepsilon < 2$, Region $R_1$ lies *above* Region $R_0$. The color scheme is the same as that used in Fig. 12.

## 4. Behavior of complex trajectories for $\varepsilon > 2$

A strange remarkable change in the topology of the trajectories on Riemann surface occurs as $\varepsilon$ increases above 2. When $\varepsilon > 2$, the upper end of Region $R_0$ on the negative-imaginary axis lies *below* the origin; a narrow *gap* on the negative-imaginary axis opens up between the upper end of Region $R_0$ and the origin. This gap is illustrated in Fig. 13, where we have taken $\varepsilon = 1 + \sqrt{2}$. A trajectory lying inside and very near the upper edge of Region $R_0$ and below the gap begins at $-0.25i$ (solid line). A second trajectory (dashed line) begins at $-0.2i$, which lies inside the gap. This $\mathcal{PT}$-symmetric trajectory lies inside Region $R_1$, which now lies *above, and not below* Region $R_0$, unlike the configuration of regions for the case $0 < \varepsilon < 2$. Both trajectories are closed.

The gap on the negative-imaginary axis has a truly remarkable substructure. The gap consists of many, and perhaps infinitely many, subintervals through which trajectories that originate on sheets $+1, +2, +3, \ldots$ can pass in order to enter sheets $-1, -2, -3, \ldots$ and thereby become $\mathcal{PT}$ symmetric. This means that the higher $\mathcal{PT}$-symmetric regions develop extremely narrow throats. To illustrate this substructure we have investigated the terminating trajectories for $\varepsilon = 1 + \sqrt{2}$ belonging to Regions $R_0, \ldots, R_8$ to determine the points at which these paths pass through the gap. These crossing points are listed in Table 1 and the terminating trajectories are plotted in Figs. 14 – 22. The trajectories that originate at the $n = 2$ and $n = 8$ turning points

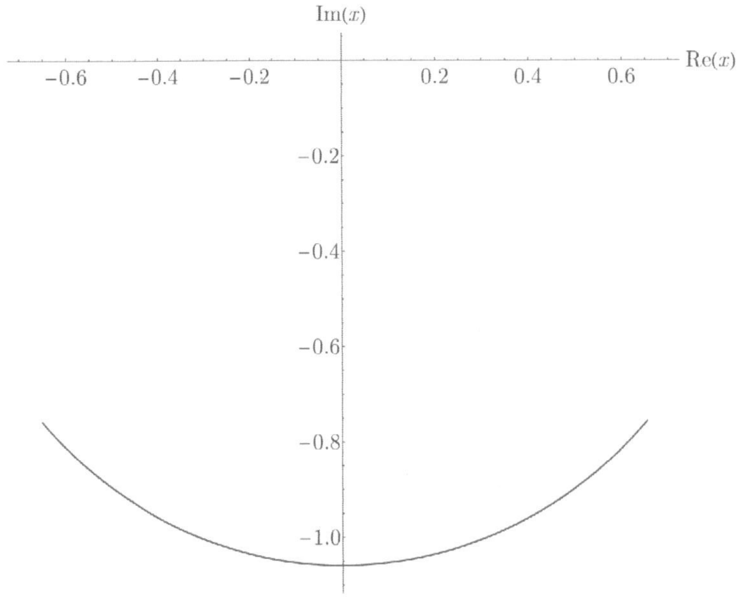


**Figure 14.** [Color online] Terminating path on the principal sheet for $\varepsilon = \sqrt{2}+1$. This path connects the $t_0$ pair of turning points and crosses the negative-imaginary axis at $-1.05872i$, which is consistent with Fig. 6. The color scheme is the same as that used in Fig. 12.

do not cross the negative imaginary axis on the principal sheet and instead spiral out to infinity. These trajectories are shown in Figs. 21 and 22. Note that the trajectories become increasingly complicated as the crossing point gets closer to the top of the gap (at the origin).

| $n$ | Crossing points $s_n$ on negative-imaginary axis |
|---|---|
| 0 | $-1.05872i$ |
| 4 | $-0.191947i$ |
| 1 | $-0.0958837i$ |
| 7 | $-0.0469295i$ |
| 5 | $-0.0312037i$ |
| 3 | $-0.0167618i$ |
| 6 | $-0.00378715i$ |

**Table 1.** Points $s_n$ on the negative-imaginary axis on the principal sheet at which terminating trajectories (that is, trajectories that connect the $t_n$ pair of turning points $t_n$) for $\sqrt{2}+1$ go from positive sheets to negative sheets on the Riemann surface. The terminating trajectory in Region $R_0$ passes through the point $s_0$, which is near $-i$, and this is consistent with the graph in Fig. 6. The remaining points $s_n$ lie inside the narrow gap that lies between the origin and $-0.25i$, but these points come in an unexpected order: 4, 1, 7, 5, 3, 6. There are *no* terminating trajectories for $n=2$ and $n=8$. For these cases the paths that begin at one of the turning points $t_2$ and $t_8$ are unable to pass through the gap; these paths are not $\mathcal{PT}$ symmetric and instead spiral outward to infinity.

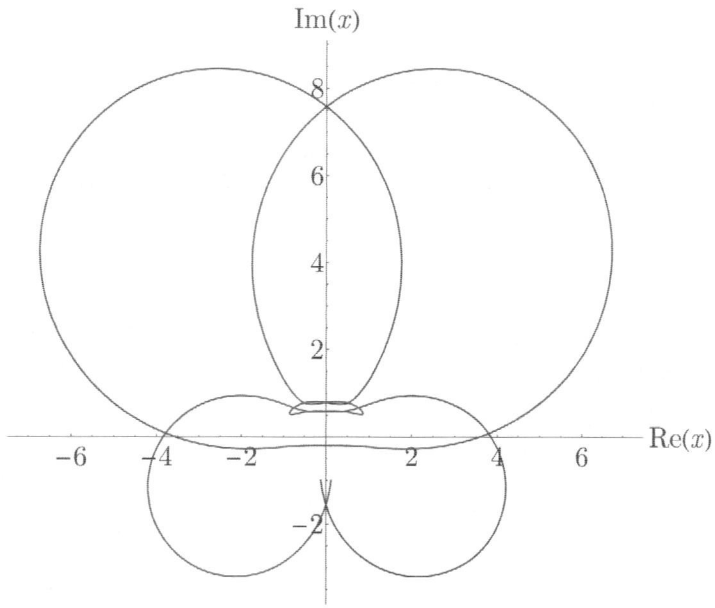


**Figure 15.** [Color online] Terminating path for $\varepsilon = \sqrt{2}+1$ that connects the $t_4$ pair of turning points and crosses the negative-imaginary axis on the principal sheet at $-0.191947i$, which lies inside the gap. The color scheme is the same as that used in Fig. 12.

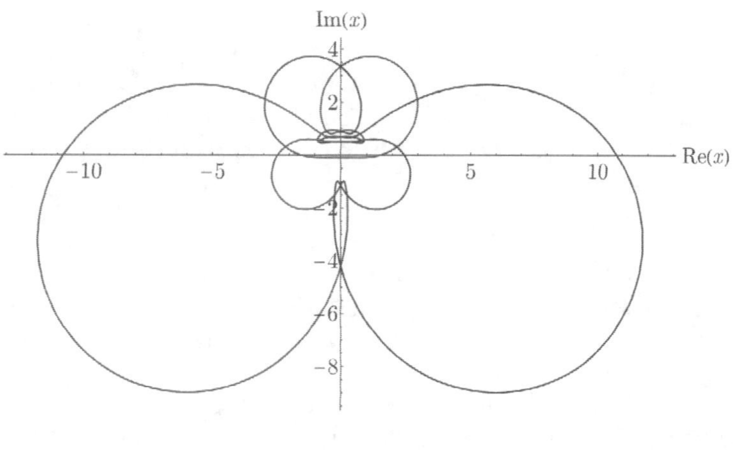


**Figure 16.** [Color online] Terminating path for $\varepsilon = \sqrt{2}+1$ that connects the $t_1$ pair of turning points and crosses the negative-imaginary axis on the principal sheet at $-0.0958837i$ inside the gap. The color scheme is the same as that used in Fig. 12. This trajectory is slightly more complicated than that in Fig. 15.

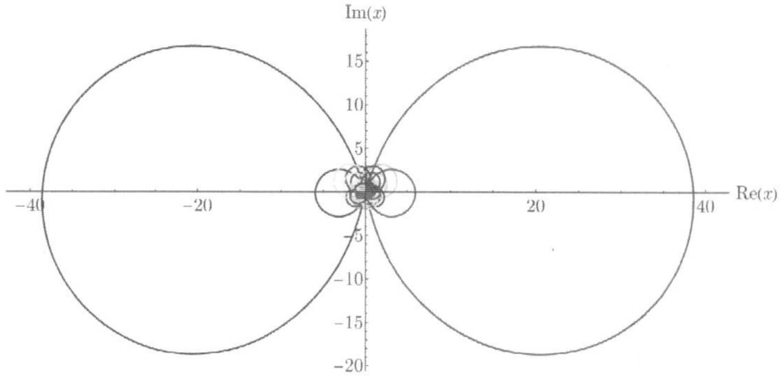


**Figure 17.** [Color online] Terminating path for $\varepsilon = \sqrt{2}+1$ that connects the $t_7$ pair of turning points and crosses the negative-imaginary axis on the principal sheet at $-0.0469295i$ in the gap. The color scheme is the same as that used in Fig. 12. This trajectory is much more complicated than that in Fig. 16 and visits 5 sheets in the Riemann surface.

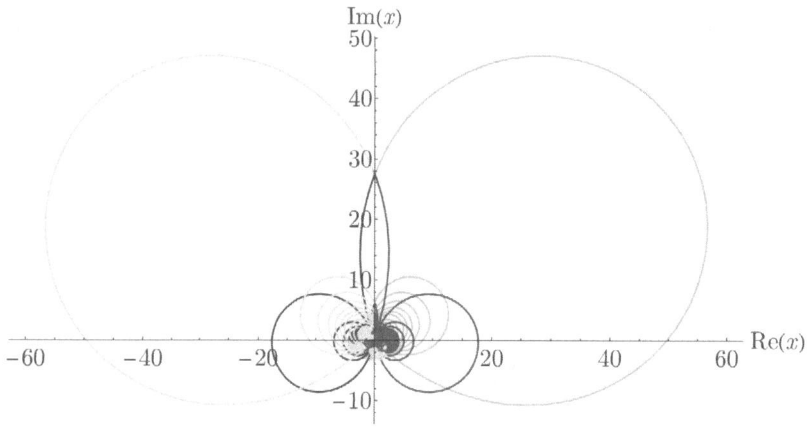


**Figure 18.** [Color online] Terminating path for $\varepsilon = \sqrt{2}+1$ that connects the $t_5$ pair of turning points and crosses the negative-imaginary axis on the principal sheet at $-0.0312037i$ in the gap. The color scheme is the same as that used in Fig. 12. This trajectory is much more complicated than that in Fig. 17. The trajectory winds back and forth repeatedly as it searches for the very tiny subgap through with it must pass in order to cross through the negative-imaginary axis on the principal sheet.

## 5. Conclusions and conjectures regarding future studies

We have opened a Pandora's box; this study reveals the immense complexity of the classical paths for the quantum-mechanical Hamiltonian in (1), much of which had not been discovered in previous studies. Perhaps, the most surprising discovery is the narrow gap on the negative-imaginary axis on the principal sheet that opens up as $\varepsilon$ increases above 2. Evidently, the point $\varepsilon = 2$ is an elaborate and unexpected singularity in the classical-mechanical theory.

The presence of the narrow gap on the negative-imaginary axis of the principal sheet of the Riemann surface of complex trajectories suggests two possibilities. For those regions $R_n$ containing $\mathcal{PT}$-symmetric classical trajectories, all of these trajectories must pass through an extremely narrow subgap. Therefore, these regions must be barbell-shaped with an extremely

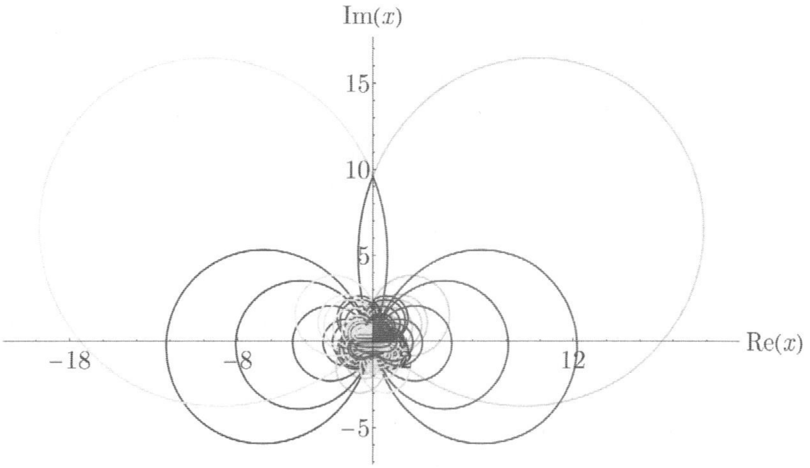


**Figure 19.** [Color online] Terminating path for $\varepsilon = \sqrt{2}+1$ that connects the $t_3$ pair of turning points and crosses the negative-imaginary axis on the principal sheet at $-0.0167618i$ in the gap. The color scheme is the same as that used in Fig. 12. This trajectory is even more complicated than that in Fig. 18.

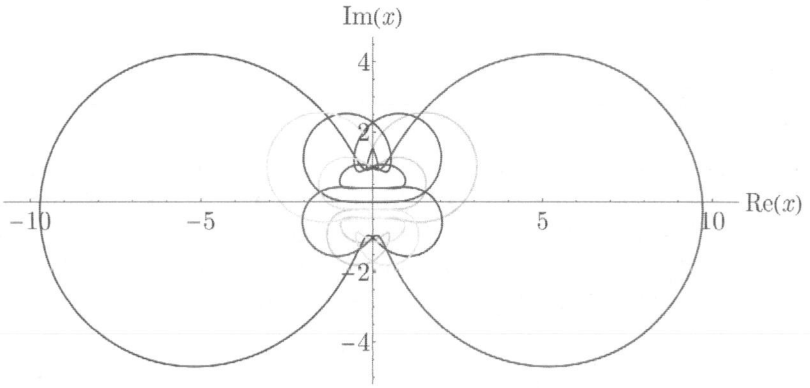


**Figure 20.** [Color online] Terminating path for $\varepsilon = \sqrt{2}+1$ that connects the $t_6$ pair of turning points and crosses the negative-imaginary axis on the principal sheet at $-0.00378715i$ in the gap. The color scheme is the same as that used in Fig. 12. It is interesting that this terminating trajectory has a simpler structure than those in the previous figures. We do not have an explanation for this.

narrow throat. Mapping the boundaries of these regions is not an easy numerical project but it is one of our objectives and it is currently under way. All other regions contain trajectories that are not $\mathcal{PT}$ symmetric and these regions must come in $\mathcal{PT}$-symmetric pairs. One of our long-term objectives is to construct a clear map of the Riemann surface. We expect to discover further structures. For example, we may find that some of the region boundaries have fractal structure [39].

Most importantly, we believe that there may be more singular values of $\varepsilon$ at which additional gaps on the negative-imaginary axis appear. We conjecture that a new gap appears at $\varepsilon = 6$, $\varepsilon = 10$, $\varepsilon = 14$, and so on, and that these gaps have quantum-mechanical analogs that were

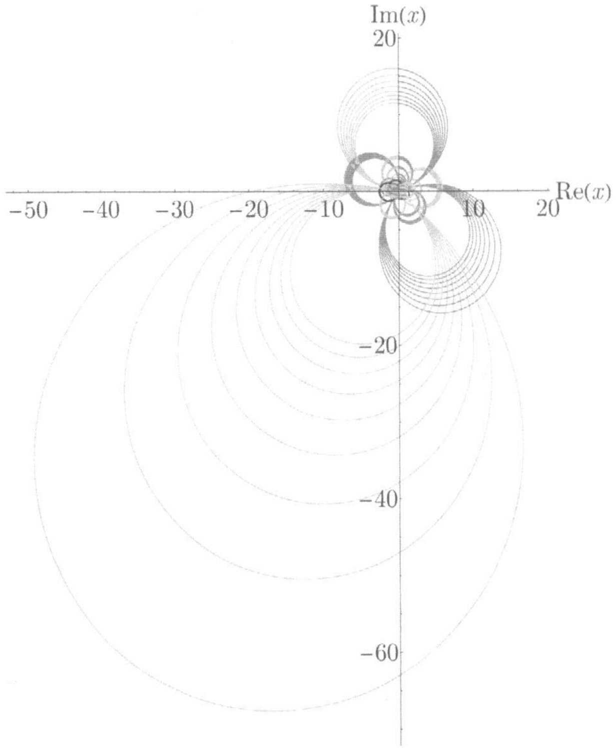


**Figure 21.** [Color online] Complex trajectory for $\varepsilon = \sqrt{2}+1$ that begins at one of the $t_2$ turning points, namely, the turning point on sheet −1. The path is unable to find a way to cross through the gap on the negative-imaginary axis on the principal sheet and thus it is not $\mathcal{PT}$ symmetric. The path has a remarkable structure in which it dips briefly onto sheet 0 and then oscillates between sheets −2 and −3 as it gradually spirals outward to complex infinity. The color scheme is the same as that used in Fig. 12 except that the trajectory is colored brown on sheet −3.

discovered in Ref. [38]. In the quantum theory these are the points at which new families of entirely real spectra appear as a consequence of the emergence of new $\mathcal{PT}$-symmetric pairs of Stokes sectors.

CMB is supported by the Alexander von Humboldt Foundation, the Simons Foundation, and the UK Engineering and Physical Sciences Research Council. Mathematica was used to perform the numerical calculations in this paper.

[1] Bender C M and Boettcher S 1998 *Phys. Rev. Lett.* **80**, 5243
[2] Bender C M, Boettcher S and Meisinger P N 1999 *J. Math. Phys.* **40**, 2201
[3] Bender C M 2007 *Reps. Prog. Phys.* **70**, 947
[4] Dorey P E, Dunning C and Tateo R 2001 *J. Phys. A: Math. Gen.* **34**, L391 (2001) and **34**, 5679
[5] Dorey P E, Dunning C and Tateo R 2007 *J. Phys. A: Math. Gen.* **40**, R205
[6] Bender C M *et al.* 2019 *PT symmetry: In quantum and classical systems* (Singapore: World Scientific)
[7] Jones H F and Mateo J 2006 *Phys. Rev. D* **73**, 085002
[8] Bender C M, Brody D C, Chen J-H, Jones H F, Milton K A and Ogilvie M C 2006 *Phys. Rev. D* **74**, 025016
[9] Bender C M and Hook D W 2008 *J. Phys. A: Math. Theor.* **41**, 244005

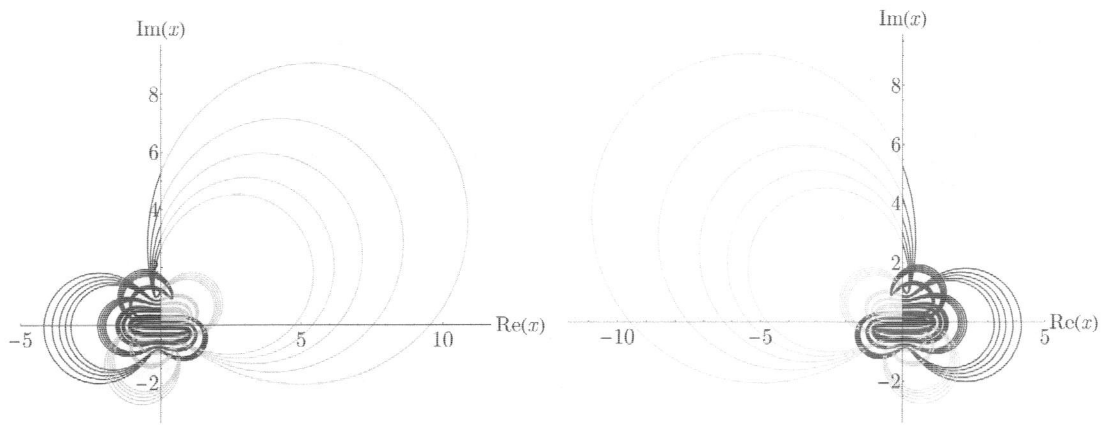


**Figure 22.** [Color online] Complex trajectories for $\varepsilon = \sqrt{2}+1$ that begin at each of the $t_8$ turning points. Neither path is unable to cross through the gap on the negative-imaginary axis on the principal sheet and thus they are not $\mathcal{PT}$ symmetric. However, they are $\mathcal{PT}$ reflections of one another. The color scheme is the same as that used in Fig. 12.

[10] Bender C M, Brody D C and Jones H F 2004 *Phys. Rev. Lett.* **93**, 251601
[11] Bender C M, Brody D C and Hook D W 2008 *J. Phys. A: Math. Theor.* **41**, 352003
[12] Rubinstein J, Sternberg P and Ma Q 2007 *Phys. Rev. Lett.* **99**, 167003
[13] Guo A, Salamo G J, Duchesne D, Morandotti R, Volatier-Ravat M, Aimez V, Siviloglou G A and Christodoulides D N 2009 *Phys. Rev. Lett.* **103**, 093902
[14] Rüter C E, Makris K G, El-Ganainy R, Christodoulides D N Segev M and Kip D 2010 *Nat. Phys.* **6**, 192
[15] Zhao K F, Schaden M and Wu Z 2004 *Phys. Rev.* A **81**, 042903
[16] Chong Y D, Ge L and Stone A D 2011 *Phys. Rev. Lett.* **106**, 093902
[17] Lin Z, Ramezani H, Eichelkraut T, Kottos T, Cao H and Christodoulides D N 2011 *Phys. Rev. Lett.* **106**, 213901
[18] Schindler J, Li A, Zheng M C, FEllis F M and Kottos T 2011*Phys. Rev.* A **84**, 040101
[19] Bittner S, Dietz B, Guenther U, Harney H L, Miski-Oglu M, Richter A and Schaefer F 2012 *Phys. Rev. Lett.* **108**, 024101
[20] Chtchelkatchev N M, Golubov A A, Baturina T I and Vinokur V M 2012 *Phys. Rev. Lett.* **109**, 150405
[21] Peng B, Ozdemir S K, Lei F, Monifi F, Gianfreda M, Long G L, Fan S, Nori F, Bender, C M and Yang L 2014 *Nat. Phys.* **10**, 394
[22] Assawaworrarit S, Yu X and Fan S, 2017*Nat. 546*, 387
[23] Bender C M, Holm D D and Hook D W 2006 *J. Phys. A: Math. Theor.* **40**, F81
[24] Bender C M, Chen J-H, Darg D W and Milton K A 2006 *J. Phys. A: Math. Theor.* **39**, 4219
[25] Bender C M and Darg D W 2007*J. Math. Phys.* **48**, 042703
[26] Arpornthip T and Bender C M 2009 *Pramana* **73**, 259
[27] Bender C M, Hook D W and Kooner K S 2010 *J. Phys. A: Math. Theor.* **43**, 165201
[28] Anderson A G and Bender C M 2012 *J. Phys. A: Math. Theor.* **45**, 455101
[29] Bender C M and Hook D W 2011 *J. Phys. A: Math. Theor.* **44**, 372001
[30] Bender C M and Hook D W 2012 *Phys. Rev. A* **86**, 022113
[31] Bender C M, Hook D W, Mavromatos N E and Sarker S 2016 *J. Phys. A: Math. and Theor.* **49**, 45lt01
[32] Miller W H and George T F 1972 *J. Chem. Phys.* **56**, 5668 (1972).
[33] Xavier A L Jr. and de Aguiar A M 1996 *Ann. Phys.* **252**, 458
[34] Takatsuka K, Ushiyama H and Inoue-Ushiyama A 1999 *Phys. Rep.* **322**, 347
[35] Turok N 2014 *New J. Phys.* **16**, 063006
[36] Bender C M and Hook D W 2014 *Stud. Appl. Math.* **133**, 318
[37] Bender C M, Boettcher S, Jones H F and Savage V M 1999 *J. Phys. A: Math. Gen.* **32**, 6771 (1999).
[38] Bender C M and Klevansky S P 2010 *Phys. Rev. Lett.* **105**, 031601
[39] Bender C M and Vinson J P 1996 *J. Math. Phys.* **37**, 4103

# $\mathcal{PT}$ symmetry and renormalisation in quantum field theory

**Carl M Bender[1], Alexander Felski[2], Sandra P Klevansky[3] and Sarben Sarkar[4]**

[1]Department of Physics, Washington University, St Louis, Missouri 63130, USA

[2]Institut für Theoretische Physik, Universität Heidelberg, 69120 Heidelberg, Germany

[3]Institut für Theoretische Physik, Universität Heidelberg, 69120 Heidelberg, Germany

[4]Department of Physics, King's College London, London WC2R 2LS, UK

E-mail: cmb@wustl.edu

E-mail: felski@thphys.uni-heidelberg.de

E-mail: spk@physik.uni-heidelberg.de

E-mail: sarben.sarkar@kcl.ac.uk

**Abstract.**

Quantum systems governed by non-Hermitian Hamiltonians with $\mathcal{PT}$ symmetry are special in having real energy eigenvalues bounded below and unitary time evolution. We argue that $\mathcal{PT}$ symmetry may also be important and present at the level of Hermitian quantum field theories because of the process of renormalisation. In some quantum field theories renormalisation leads to $\mathcal{PT}$-symmetric effective Lagrangians. We show how $\mathcal{PT}$ symmetry may allow interpretations that evade ghosts and instabilities present in an interpretation of the theory within a Hermitian framework. From the study of examples $\mathcal{PT}$-symmetric interpretation is naturally built into a path integral formulation of quantum field theory; there is no requirement to calculate explicitly the $\mathcal{PT}$ norm that occurs in Hamiltonian quantum theory. We discuss examples where $\mathcal{PT}$-symmetric field theories emerge from Hermitian field theories due to effects of renormalisation. We also consider the effects of renormalisation on field theories that are non-Hermitian but $\mathcal{PT}$-symmetric from the start.

## 1. Introduction

Non-Hermitian Hamiltonians govern systems that in general receive energy from and/or dissipate energy into their environment and so they are typically not in equilibrium. Their energy is not conserved and their energy levels are complex. However, in 1998 [1] non-Hermitian $\mathcal{PT}$ systems were shown to offer new possibilities for unitary time evolution in quantum mechanics. Since then there has been extensive research activity [2], particularly in material science and optics, which has implemented the ideas of $\mathcal{PT}$-symmetric quantum mechanics.

$\mathcal{PT}$-symmetric quantum mechanics[1] can be considered to be a one-dimensional quantum field theory. To date there have been no clear examples of $\mathcal{PT}$-symmetric quantum field theory in

[1] The term quantum mechanics is used to refer to quantum systems with a finite number of degrees of freedom; the term quantum field theory (QFT) is used to refer to systems with an infinite number of degrees of freedom.

higher dimensions. We examine the possibility that $\mathcal{PT}$ symmetry may emerge when considering the effect of quantum fluctuations in a higher-dimensional field theory. We also examine the converse process where we consider the effect of quantum fluctuations on an initially $\mathcal{PT}$-symmetric quantum field theory.

Using numerical techniques the work of Bender and Boettcher [1] showed that quantum-mechanical Hamiltonians of the form

$$H = \frac{1}{2}p^2 + x^2 (ix)^\epsilon \tag{1}$$

have a real positive spectrum. Using a correspondence between ordinary differential equations and integrable field theory models, Dorey *et al.* [3] provided an analytical proof of the result of Bender and Boettcher.

Mostafazadeh [4] considered the framework of pseudo-Hermiticity [5] which contained $\mathcal{PT}$ symmetry as a special case. A Hamiltonian $H$ is pseudo-Hermitian if $H$ is not selfadjoint but

$$H^\dagger = \eta H \eta^{-1}, \tag{2}$$

where $\eta$ is a positive-definite Hermitian operator. In terms of $\eta$ the Hilbert space of states has a positive-definite inner product given by

$$\langle \varphi, \eta\chi \rangle \tag{3}$$

where $\langle , \rangle$ is a conventional inner product on the Hilbert space of states for the Hermitian part of the Hamiltonian. There is not a universal expression for $\eta$, which is difficult to calculate in general [2]. This would seem to imply an impediment to calculating correlation functions in $\mathcal{PT}$-symmetric field theories. Within the context of quantum mechanics, formulated in terms of functional integrals, Jones and Rivers [6] showed through examples that it was not necessary to calculate $\eta$. They also gave a general field theoretic argument [7] (valid for the case of unbroken $\mathcal{PT}$-symmetry and symmetric Hamiltonians) that the calculation of vacuum expectation values of field operators did not require the explicit calculation of $\eta$ or equivalently the related $C$-operator[2]. This latter argument relied on the relation $A_H = \exp(iHt)A\exp(-iHt)$, where $A_H$, $A$ and $H$ were a general Heisenberg boson field operator, the corresponding Schrödinger operator and a Hamiltonian operator which can be quasi-Hermitian. The versatility of the path-integral calculation of Green's functions is further confirmed in [8]; there it is shown that in quantum mechanics a given Hamiltonian can have different spectra depending on boundary conditions, which are imposed in the complex plane in Stokes sectors. For a left-right ($\mathcal{PT}$) symmetric choice of two such sectors the spectra can be real. In QFT the boundary conditions are imposed on the path of integration of the functional integral. The Schwinger-Dyson equations for the Greens functions, which are obtained formally from the path integral, do not depend on such details of the path-integration measure[2]. For the $g\phi^4$ scalar field theory there are two solutions one for $g > 0$ and the other for $g < 0$. From the two-point function we can calculate the pole. For $D = 1$ we can also calculate the difference between the energy for the first excited state and the ground state. This difference is found by solving the Schrödinger equation using $\mathcal{PT}$-symmetric Stokes wedges and the Hermitian Stokes wedges. The path-integral computation of the mass and this excitation energy agree. Hence, we shall take the path-integral representation as the definition of a quantum field theory and, except for the Lee model, not consider $\eta$ explicitly. The path integral is a convenient formulation for both perturbative and nonperturbative calculations in our context.

We consider four examples of field theories to illustrate three different aspects of $\mathcal{PT}$ symmetry:

[2] We should stress that in other applications of path-integration ( such as the computation of S-matrix elements) the $C$-operator cannot be neglected.

(i) the Lee model and the elimination of ghosts [9];
(ii) the role of the top quark and the Higgs instability [10] in QFTs which go beyond the Standard Model (SM) of particle physics [12, 13];
(iii) the new epsilon expansion for $\mathcal{PT}$-symmetric scalar field theory [14, 15];
(iv) the functional renormalisation group and the preservation of $\mathcal{PT}$ symmetry in the infrared after renormalisation [16].

## 2. Role of $\mathcal{PT}$ symmetry in the Lee model

From the study of dispersion relations in QFT [17, 18] it has been shown that it is possible that a formally Hermitian QFT may contain ghosts. The Lee model [19], which we discuss next, is just such an example where the vertex function and, in particular, coupling-constant renormalisation is related to the emergence of a non-Hermitian Hamiltonian. The role of $\mathcal{PT}$ symmetry is crucial in banishing the ghosts and making the field theory viable [9].

### *2.1. Lee model*

The Lee model is a field theory which is very different from the Standard Model of particle physics but it has the advantage of being soluble; so it provides a convenient framework for the emergence of non-Hermiticity in the process of renormalisation. Indeed, the model was devised as a field theory whose coupling-constant and wave-function renormalisation could be computed exactly in principle. There are variants of the model. One variant of the model contains fields for infinitely heavy spinless fermions, $V$ and $N$, as well as a neutral scalar field $\theta$. Antifermions are not in the theory and so there is no crossing symmetry. The permitted reaction channel is

$$V \rightleftharpoons N + \theta, \tag{4}$$

but the channel

$$N \rightleftharpoons V + \theta \tag{5}$$

is not allowed.

In momentum space the Hamiltonian of the Lee model (in a space of finite volume $\mathcal{V}$) is

$$H = H_0 + H_{\text{int}}, \tag{6}$$

where

$$H_0 = m_V \overline{\psi}_V \psi_V + m_N \overline{\psi}_N \psi_N + \sum_{\mathbf{k}} \omega_{\mathbf{k}} a^\dagger_{\mathbf{k}} a_{\mathbf{k}} \tag{7}$$

and

$$H_{\text{int}} = \delta m_V \overline{\psi}_V \psi_V - g_0 \mathcal{V}^{-\frac{1}{2}} \sum_{\mathbf{k}} \frac{u(\omega_{\mathbf{k}})}{(2\omega_{\mathbf{k}})^{\frac{1}{2}}} (\overline{\psi}_V \psi_N a_{\mathbf{k}} + \overline{\psi}_N \psi_V a^\dagger_{\mathbf{k}}). \tag{8}$$

$u(\omega_{\mathbf{k}})$ is a dimensional cut-off function which is chosen to tend to 0 for large $\omega_{\mathbf{k}}$ where $\omega_{\mathbf{k}} = \sqrt{\mathbf{k}^2 + \mu^2}$. Standard commutation and anti-commutation rules are assumed:

$$\{\overline{\psi}_V, \psi_V\} = \{\overline{\psi}_N, \psi_N\} = 1 \tag{9}$$

$$\left[a_{\mathbf{k}}, a^\dagger_{\mathbf{k}'}\right] = \delta_{\mathbf{k},\mathbf{k}'}. \tag{10}$$

All other commutators and anticommutators vanish. Bare operators and coupling constant appear in $H$. However $m_V$, $m_N$, and, $\mu$ are renormalised parameters and so are determined by experiment; $\delta m_V$ is the mass counterterm for the $V$ particle and is a function of $g_0$. No further mass renormalisation is actually needed. The coupling $g_0$ in principle is determined from the scattering cross-section of $N$ and $\theta$ although there are interesting complications of non-Hermiticity if the coupling is not small.

However, let us for the moment persist in the view that $g_0$ is real and so $H$ is Hermitian. Let us choose as a basis for the Hilbert space states of the form

$$|\rangle = |n_V, n_N, \{n_{\mathbf{k}}\}\rangle \,. \tag{11}$$

From $H$ it is clear that there are two conserved quantities $B$ and $Q$:

$$B = n_V + n_N, \quad Q = -N_\theta + n_N, \tag{12}$$

where $N_\theta = \sum_{\mathbf{k}} n_{\mathbf{k}}$. Hence the Hilbert space is partitioned into an infinite number of independent (superselection) sectors with fixed $B$ and $Q$. Although this makes the model soluble, the analysis of sectors with large $B$ and $Q$ is complicated. Since our main point is to show how renormalisation can lead to a non-Hermitian Hamiltonian, we simplify our analysis further by considering (i) a nontrivial sector $B = 1$ and $Q = 0$ and (ii) modifying the model so that there is no $\mathbf{k}$ dependence. We choose to work with this simplified model since it suffices to illustrate the emergence of non-Hermiticity from renormalisation.(The more complicated field-theoretic model can also be analysed leading to similar results.) The model now is a *quantum-mechanical* model since the quantum fields in (7) and (8) are replaced by quantum operators. Hence, infinities that arise from summing over an infinite set of modes in quantum field theory are absent; the model remains interesting since some features of coupling-constant renormalisation persist. The resulting Hamiltonian is $\mathcal{H} = \mathcal{H}_0 + \mathcal{H}_1$, where

$$\mathcal{H}_0 = m_V V^\dagger V + m_N N^\dagger N + \mu a^\dagger a \tag{13}$$

and

$$\mathcal{H}_1 = g_0(V^\dagger N a + a^\dagger N^\dagger V) + \delta m_V V^\dagger V. \tag{14}$$

We will look at the 2-dimensional Hilbert space associated with this sector and consider the eigenstates, which we will denote by

$$|V\rangle = c_{11}\,|1,0,0\rangle + c_{12}\,|0,1,1\rangle \,, \tag{15}$$

$$|N\theta\rangle = c_{21}\,|1,0,0\rangle + c_{22}\,|0,1,1\rangle \,, \tag{16}$$

with eigenvalues $m_V$ and $E_{N\theta}$. Let us denote $(m_V + \delta m_V)$ by $m_{V_0}$, the bare mass of $V$. The eigenvalues can be shown to satisfy the equations

$$m_V = \frac{1}{2}(m_N + m_\theta + m_{V_0} - \sqrt{M_0^2 + 4g_0^2}), \tag{17}$$

$$E_{N\theta} = \frac{1}{2}(m_N + m_\theta + m_{V_0} + \sqrt{M_0^2 + 4g_0^2}), \tag{18}$$

where $M_0 \equiv m_N + m_\theta - m_{V_0}$. Following field theory, we define the wave-function renormalisation constant $Z_V$ through the relation

$$1 = \langle 0|\, \frac{1}{\sqrt{Z_V}} V\,|V\rangle \,, \tag{19}$$

which gives

$$Z_V = \frac{2g_0^2}{\sqrt{M_0^2 + 4g_0^2}\,(\sqrt{M_0^2 + 4g_0^2} - M_0)}, \tag{20}$$

and the coupling constant renormalisation through

$$\frac{g^2}{g_0^2} = Z_V. \tag{21}$$

We deduce that

$$g_0^2 = \frac{g^2}{1 - \frac{g^2}{M^2}}, \tag{22}$$

where $M \equiv m_N + m_\theta - m_V$. If $g$, the experimentally determined value, exceeds $M$ then $g_0$ becomes pure imaginary and the Hamiltonian becomes non-Hermitian. Although the Hamiltonian has become non-Hermitian, it is $\mathcal{PT}$-symmetric. Explicitly, the transformations due to $\mathcal{P}$ are [9]

$$\mathcal{P}V\mathcal{P} = -V, \qquad \mathcal{P}N\mathcal{P} = -N, \qquad \mathcal{P}a\mathcal{P} = -a, \tag{23}$$

$$\mathcal{P}V^\dagger\mathcal{P} = -V^\dagger, \qquad \mathcal{P}N^\dagger\mathcal{P} = -N^\dagger, \qquad \mathcal{P}a^\dagger\mathcal{P} = -a^\dagger. \tag{24}$$

The transformations due to $\mathcal{T}$ are

$$\mathcal{T}V\mathcal{T} = V, \qquad \mathcal{T}N\mathcal{T} = N, \qquad \mathcal{T}a\mathcal{T} = a \tag{25}$$

$$\mathcal{T}V^\dagger\mathcal{T} = V^\dagger \qquad \mathcal{T}N^\dagger\mathcal{T} = N^\dagger \qquad \mathcal{T}a^\dagger\mathcal{T} = a^\dagger. \tag{26}$$

It is now straightforward to check that $i|g_0|(V^\dagger Na + a^\dagger N^\dagger V)$ is $\mathcal{PT}$-symmetric. This is our first example of a $\mathcal{PT}$-symmetric field theory that arises from a Hermitian field theory after renormalisation. On introducing a $\mathcal{PT}$-symmetric inner product in the ghost regime of the Lee model, the so-called ghost state (identified within the Hermitian frame work) turns out to have a positive norm[9]. The Lee model then can be interpreted as an acceptable quantum field theory.

## 3. Higgs instability and $\mathcal{PT}$ symmetry in beyond-the-SM physics

The conventional approach to renormalisation involves the regularisation of loop integrals in Feynman diagrams. This led to scale dependence of the parameters of the theory which can be understood in a different way using the approach [20] of Wilson to renormalisation. Wilson's approach leads naturally to the concept of effective actions, and the form of the effective actions for some theories in high-energy physics is non-Hermitian and $\mathcal{PT}$-symmetric. Low energy and long wavelength approximations of quantum evolution equations obtained from the effective actions can be studied using the methods of $\mathcal{PT}$-symmetric quantum mechanics.

The change from the conventional to the Wilson approach can be seen by considering a scalar field $\Phi$ whose (conventional) action is given by

$$S_\Lambda[\Phi] = \int d^Dx(\frac{1}{2}\partial_\mu\Phi\partial^\mu\Phi + U_\Lambda(\Phi)) \tag{27}$$

where the UV cut-off is $\Lambda$. If we integrate over Fourier modes with momenta of magnitude $p$ in the shell $k \le p \le \Lambda$, we arrive at an action which we will denote by $S_k[\phi]$ where $\phi$ has only Fourier modes with $p \le k$ (a sharp infrared cut-off).The associated partition function is

$$Z_k[j] \equiv \exp(-W_k[j]) \equiv \int D\phi \exp\Big(-S_k[\phi] - \int_p j_p\phi_{-p}\Big) \tag{28}$$

and a Legendre transformation on $W_k[j]$ leads to the effective action $\Gamma_k[\phi_c]$

$$\Gamma_k[\phi_c] = W_k[j] - \int d^Dx j\phi_c - \Delta_k[\phi_c]. \tag{29}$$

The renormalisation group flow equation[3] (reviewed in [23]) is

$$\partial_k \Gamma_k = \frac{1}{2} \mathrm{Tr} \left\{ \partial_k R_k \left( \frac{\delta^2 \Gamma_k}{\delta \phi_c(p) \delta \phi_c(q)} + R_k(p) \delta(p+q) \right)^{-1} \right\}. \tag{30}$$

There is a systematic (derivative expansion) approximation schemes [21, 24] for solving this integro-differential equation for the effective action.

$$\Gamma_k [\phi_c] = \int d^D x \left( U_k (\phi_c) + \frac{Z_k (\phi_c)}{2} \partial_\mu \phi_c \partial^\mu \phi_c + Y_k (\phi_c) (\partial_\mu \phi_c \partial^\mu \phi_c)^2 + W_k (\phi_c) (\partial_\mu \partial^\mu \phi_c)^2 + \cdots \right). \tag{31}$$

The effective action can be represented (at the lowest level of approximation) by the *ansatz*

$$\Gamma_k[\phi_c] = \int d^D x (\frac{1}{2} \partial_\mu \phi_c \partial^\mu \phi_c + U_k(\phi_c)). \tag{32}$$

The counterpart of the classical equation of motion is

$$\frac{\delta \Gamma_k[\phi_c]}{\delta \phi_c(x)} = 0. \tag{33}$$

The effective potential, $U_k(\phi_c)$, which results from renormalisation, shows the emergence of $\mathcal{PT}$ symmetry in the next models that we will discuss. For the effective action of (32), in the low energy approximation and small $k$, equation (33) leads to a Schrödinger equation which we will analyse now. The flow of the effective potential with $k$ will be analysed in section 5. We consider two theories of current interest using this formalism:

(i) a theory of dynamical breaking of supergravity via a gravitino condensate field [25] $\varphi_c$ with $U(\varphi_c) = -{\varphi_c}^4 \log(i\varphi_c)$ for large $\varphi_c$,

and

(ii) Beyond the Standard Model of particle physics for which $\varphi_c$ is a pseudoscalar field and $U(\varphi_c) = -{\varphi_c}^4 \log(\varphi_c^2)$ for large $\varphi_c$ [10, 11].

We now give brief motivations for the study of the above two models.

- Motivation for the model underlying (i): This model was studied in [25] in order to understand the possible role of a supersymmetric (SUSY) phase transition occurring in the early universe in the study of inflation. SUSY is at best a broken symmetry. The breaking of SUSY within the context of a local symmetry, supergravity, can lead to a Starobinsky type model [26] for inflation.
- Motivation for the model underlying (ii): Although a scalar Higgs particle has been discovered at the LHC, there is still interest in looking for other Higgs bosons which can exist in models which go *beyond the Standard Model* (BSM) of particle physics.The interest in BSM comes from trying to understand the origin of Dark Matter and neutrino masses. One such model is the Minimal Supersymmetric Standard Model (MSSM) [13]. The MSSM has two Higgs doublets which leads to five physical Higgs bosons, both $CP$ odd and even. Our aim is not phenomenological but to show that the considerations of $\mathcal{PT}$ symmetry may have a role to play in understanding the scalar sector of extensions of the Standard Model (SM).

[3] Wetterich [21] introduced a smooth infrared cut-off function $R_k(p^2)$ rather than a sharp cut-off so that $S_k[\phi] = \int d^D x[\frac{1}{2}\partial_\mu \phi \partial^\mu + V_k(\phi)] + \Delta_k[\phi]$, where $\Delta_k[\phi] = \frac{1}{2}\int d^D p \phi_p R_k(p^2)\phi_{-p}$. A common choice for $R_k(p^2)$ is $R_k(p^2) = (k^2 - p^2)\Theta(k^2 - p^2)$ [22] .

Both examples, in conventional quantum mechanics, would show unstable behaviour for large $\varphi_c$. Under $\mathcal{P}$ we have $\varphi_c \to -\varphi_c$ and under $\mathcal{T}$ we have $i \to -i$. So $U(\varphi_c)$ is $\mathcal{PT}$ symmetric in both cases.

We study such effective potentials by considering three related quantum-mechanical Hamiltonians $H_i$, $i = 1, 2, 3$:

$$H_1 = p^2 + x^4 \log(ix), \tag{34}$$
$$H_2 = p^2 - x^4 \log(ix), \tag{35}$$
$$H_3 = p^2 - x^4 \log(x^2), \tag{36}$$

which are non-Hermitian but $\mathcal{PT}$-symmetric. We will find that the first two Hamiltonians will show unbroken $\mathcal{PT}$ symmetry and the Hamiltonian $H_3$ will show broken $\mathcal{PT}$ symmetry.

*3.1. The spectra for $H_1$*

We use the WKB method of semiclassical quantum mechanics to determine the energy spectrum of $H_1$:

- Locate turning points.
- Examine the complex classical trajectories on an infinite-sheeted Riemann surface.
- Determine the open or closed nature of the trajectories.

The turning points satisfy the equation

$$E = x^4 \log(ix). \tag{37}$$

We take $E = 1.24909$ because this is the numerical value of the ground-state energy obtained separately by solving the Schrödinger equation. (See the table below.)

One turning point lies on the negative imaginary-$x$ axis. To find this point we set $x = -ir$ ($r > 0$) and obtain the algebraic equation $E = r^4 \log r$. Solving this equation by using Newton's method, we find that the turning point lies at $x = -1.39316i$. To find the other turning points we seek solutions to (37) in polar form $x = re^{i\theta}$ ($r > 0$, $\theta$ real). Substituting for $x$ in (37) and taking the imaginary part, we obtain

$$\log r = -(2k\pi + \theta + \pi/2)\cos(4\theta)/\sin(4\theta), \tag{38}$$

where $k$ is the sheet number in the Riemann surface of the logarithm. (We choose the branch cut to lie on the positive-imaginary axis.) Using (38), we simplify the real part of (37) to

$$E = -r^4(2k\pi + \theta + \pi/2)/\sin(4\theta). \tag{39}$$

We then use (38) to eliminate $r$ from (39) and use Newton's method to determine $\theta$. For $k = 0$ and $E = 1.24909$, two $\mathcal{PT}$-symmetric (left-right symmetric) pairs of turning points lie at $\pm 0.93803 - 0.38530i$ and at $\pm 0.32807 + 0.75353i$. For $k = 1$ and $E = 1.24909$ there is a turning point at $-0.53838 + 0.23100i$; the $\mathcal{PT}$-symmetric image of this turning point lies on sheet $k = -1$ at $0.53838 + 0.23100i$.

The turning points determine the shape of the classical trajectories. Two topologically different kinds of classical paths are shown in Figs. 1 and 2. All classical trajectories are *closed* and *left-right symmetric*, and this implies that the quantum energies are all real [27].

The WKB quantization condition is a complex path integral on the principal sheet of the logarithm ($k = 0$). On this sheet a branch cut runs from the origin to $+i\infty$ on the imaginary

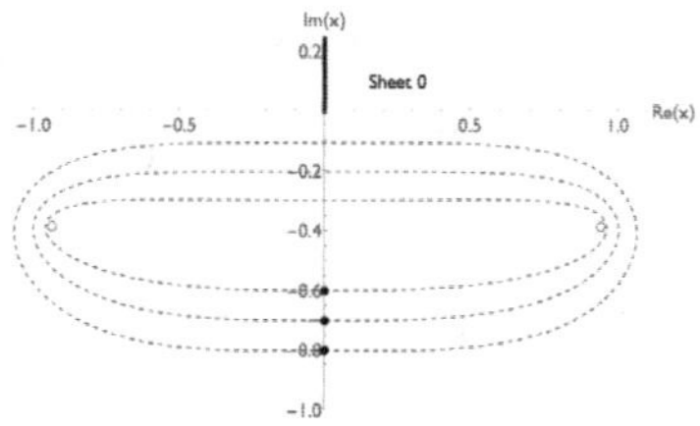


**Figure 1.** Three nested closed classical paths.

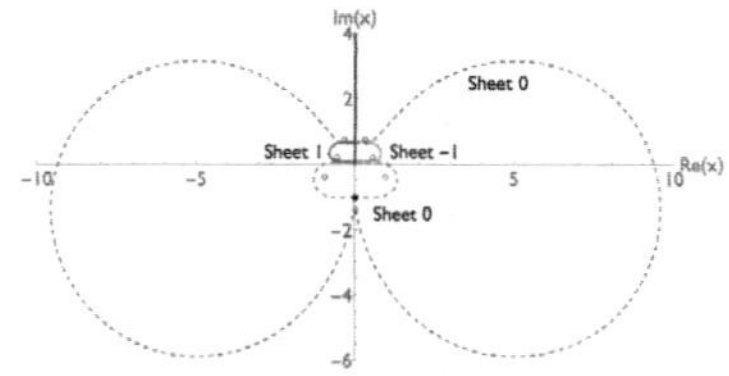


**Figure 2.** *Closed* classical path for energy $E = 1.24909$.

axis; this choice of branch cut respects the $\mathcal{PT}$ symmetry of the configuration. The integration path goes from the left turning point $x_{\mathrm{L}}$ to the right turning point $x_{\mathrm{R}}$ [28]:

$$\left(n+\frac{1}{2}\right)\pi \sim \int_{x_{\mathrm{L}}}^{x_{\mathrm{R}}} dx\sqrt{E-V(x)} \quad (n >> 1). \tag{40}$$

If the energy is large ($E_n \gg 1$), then from (38) with $k = 0$ we find that the turning points lie slightly below the real axis at $x_{\mathrm{R}} = re^{i\theta}$ and at $x_{\mathrm{L}} = re^{-\pi i - i\theta}$ with

$$\theta \sim -\pi/(8\log r) \quad \text{and} \quad r^4 \log r \sim E. \tag{41}$$

We choose the path of integration in (40) to have a constant imaginary part so that the path is a horizontal line from $x_{\mathrm{L}}$ to $x_{\mathrm{R}}$. Since $E$ is large, $r$ is large and thus $\theta$ is small. We obtain the simplified approximate quantization condition

$$\left(n+\frac{1}{2}\right)\pi \sim r^3 \log r \int_{-1}^{1} dt\sqrt{1-t^4}, \tag{42}$$

which leads to the WKB approximation for $n \gg 1$:

$$\frac{E_n}{[\log(E_n)]^{1/3}} \sim \left[\frac{\Gamma(7/4)(n+1/2)\sqrt{\pi}}{\Gamma(5/4)\sqrt{2}}\right]^{4/3}. \tag{43}$$

| Calculation of values of energy | | | | |
|---|---|---|---|---|
| Energy level $n$ | $E_n$ | $\frac{E_n}{[\log(E_n)]^{1/3}}$ | WKB | % error |
| 0 | 1.24909 | 2.06161 | 0.54627 | 73.5028 |
| 3 | 13.7383 | 9.96525 | 7.31480 | 26.5969 |
| 6 | 31.6658 | 20.9458 | 16.6979 | 20.2804 |
| 9 | 52.9939 | 33.4674 | 27.6956 | 17.2463 |
| 12 | 76.9748 | 47.1776 | 39.9324 | 15.3573 |

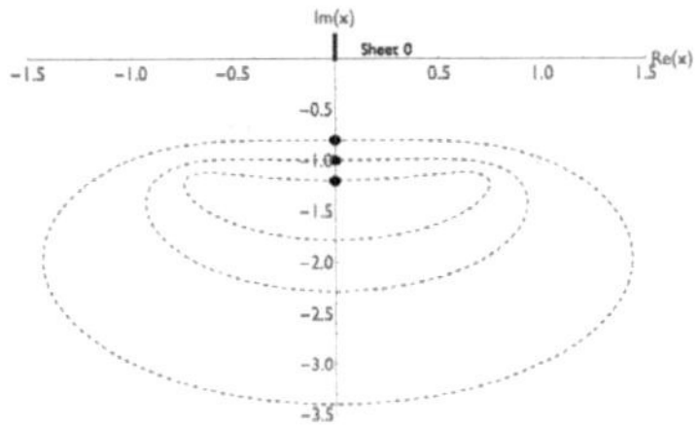


**Figure 3.** Three nested classical trajectories for $H_2$ with $E = 2.07734$.

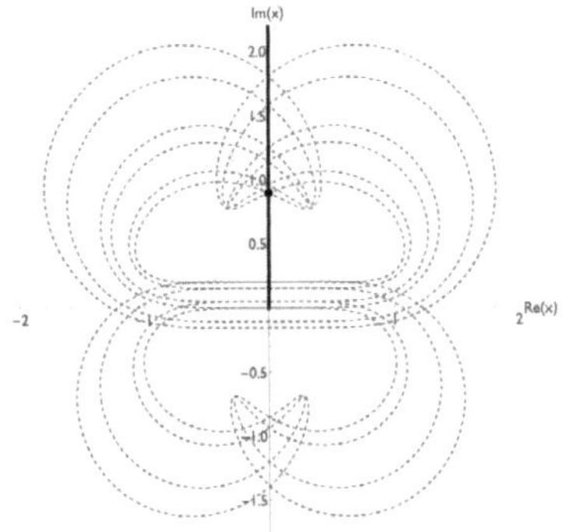


**Figure 4.** Complex classical trajectory for $H_2$ with $E = 2.07734$.

**Analysis of the supergravity model Hamiltonian** $H_2$: The classical trajectories for the Hamiltonian $H_2$ are plotted in Figs. 3 and 4. Like the classical trajectories for the Hamiltonian $H_1$), these trajectories are closed, which implies that all the eigenvalues for $H_2$ are real.

**Analysis of the Higgs model Hamiltonian** $H_3$**:** To make sense of $H_3$ we again introduce a parameter $\epsilon$ and we define $H_3$ as the limit of $H = p^2 + x^2(ix)^\epsilon \log\left(x^2\right)$ as $\epsilon : 0 \to 2$. This case is distinctly different from that for $H_2$. Figure 5 shows that the $\mathcal{PT}$ symmetry is broken for all $\epsilon \neq 0$. When $\epsilon = 2$, there are only four real eigenvalues: $E_0 = 1.1054311$, $E_1 = 4.577736$, $E_2 = 10.318036$, and $E_3 = 16.06707$. To confirm this result we plot a classical trajectory for $\epsilon = 2$ in Fig. 6. In contrast with Fig. 4, the trajectory is open and not left-right symmetric.

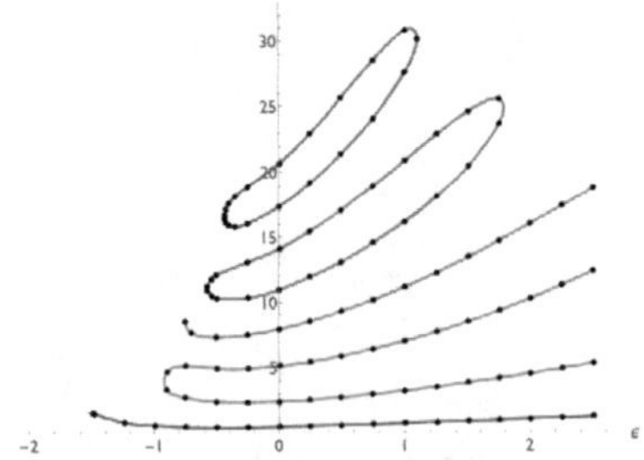


**Figure 5.** *Energies of the Hamiltonian* $H = p^2 + x^2(ix)^\epsilon \log(ix)$ plotted versus $\epsilon$.

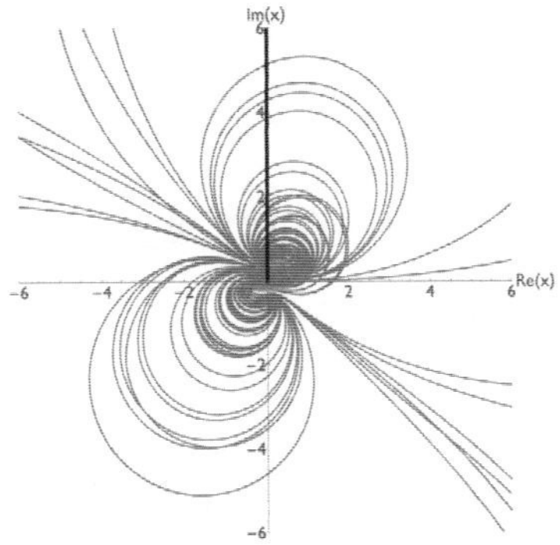


**Figure 6.** *Classical path for the Hamiltonian* $H = p^2 - x^4 \log\left(x^2\right)$.

## 4. Renormalisation and canonical scalar field theory

We have shown examples of $\mathcal{PT}$-symmetric quantum field theories arising from the process of renormalisation in field theories. The properties of $\mathcal{PT}$-symmetric field theories have yet to be established. In quantum mechanics much progress in understanding $\mathcal{PT}$ symmetry has been made through the study of the Hamiltonian in (1). A natural generalisation to an Euclidean field theory Hamiltonian in $D$-dimensions is to consider the Lagrangian

$$L = \frac{1}{2}(\nabla\phi)^2 + \frac{1}{2}\mu^2\varphi^2 + \frac{1}{2}g\mu_0^2\phi^2(i\mu_0^{\frac{\delta}{2}}\phi)^\epsilon \tag{44}$$

where $\phi$ is a dimensional pseudoscalar field, $\delta = 2 - D$ and $\epsilon \geq 0$. It is natural to study this class of field theories since we can then compare our findings in certain limits with those found in quantum mechanical systems. For noninteger $\epsilon$ the interaction is nonpolynomial and the methods that are normally used for calculating Greens functions in quantum field theory do not apply. The method we will use to set up a quantum field theory involves

(i) rewriting of the interaction in terms of a formal series of polynomial interactions;

(ii) creating modified Feynman rules in a nonperturbative expansion;

(iii) consideration of infinite contributions and renormalisation.

The first part of the method, used in (i), was introduced decades ago and has been recently developed in (ii) and (iii) within the context of $\mathcal{PT}$-symmetric field theory. The procedure of (i) allows us to use Wick's theorem and the linked-cluster theorem in conjunction with (ii).

In a conventional field theory described by a Lagrangian $\mathcal{L}$ within polynomial interactions in a field $\varphi$, we calculate connected Green's functions using a partition functional $Z(J)$ defined through a path integral as

$$Z(J) = \int \mathcal{D}\varphi \exp\Big(-\int d^Dx(\mathcal{L} + J\varphi)\Big), \tag{45}$$

where $J$ is a source. It is well known that the normalised partition functional $\frac{Z(J)}{Z(0)}$ satisfies

$$\ln\frac{Z(J)}{Z(0)} = \exp\Big(\sum \text{Connected source} - \text{to} - \text{source diagrams}\Big). \tag{46}$$

This result simplifies our calculations in (ii) and (iii). First, we note that formally, on writing $\psi = \mu_0^{\frac{\delta}{2}}\varphi$ (and suppressing the argument of $\psi$)

$$\psi^2(i\psi)^\epsilon = \sum_{n=0}^{\infty}\frac{\epsilon^n}{n!}\sum_{r=0}^{n}\left(\begin{array}{c} n \\ r \end{array}\right)\left(\frac{i\pi|\psi|}{\psi}\right)^r\psi^2\left(\frac{1}{2}\ln\psi^2\right)^{n-r} = \sum_{n=1}^{\infty}\epsilon^n\mathcal{L}_n + \psi^2 \tag{47}$$

because

$$\ln(i\psi) = \frac{1}{2}i\pi|\psi|/\psi + \frac{1}{2}\ln\psi^2. \tag{48}$$

The series in (47) can be expressed in terms of powers of $\psi$ on noting that

(i)
$$\left(\frac{|\psi|}{\psi}\right)^r = \sum_{w=0}^{\infty}\mathcal{I}_w\psi^{(2w+1)r}, \tag{49}$$

where $\mathcal{I}_w = \frac{2}{\pi}\int_0^\infty dt\frac{(-t^2)^w}{(2w+1)!}$.

(ii)
$$(\ln\psi^2)^s = \mathcal{F}^0_{N,s}\psi^{2N}, \tag{50}$$

where $\mathcal{F}^0_{N,s} = \lim_{N\to 0}(\frac{d}{dN})^s$.

Thus, we can rewrite the partition function in (45) as

$$Z[J] = \int\mathcal{D}\varphi\exp\left(-\int d^Dx(\sum_{n=0}^{\infty}\epsilon^n\mathcal{L}_n + J\varphi)\right), \tag{51}$$

where $\mathcal{L}_0$ is the free (that is, $\epsilon$-independent) part of the Lagrangian. All $n$-point Green's functions can be evaluated as

$$\frac{\delta^n Z[J]}{\delta J(x_1)\delta J(x_2)\cdots\delta J(x_n)}\Big|_{J=0} = \int\mathcal{D}\varphi\exp\left(-\int d^Dx\sum_{n=0}^{\infty}\epsilon^n\mathcal{L}_n\right)\varphi(x_1)\cdots\varphi(x_n).$$

We are only interested in the connected Green's functions $G_c(x_1, x_2, \cdots, x_n)$. Expanding the integrand in the path integral in powers of $\epsilon$ we obtain products of $\mathcal{L}_n$ which consist of integrals of powers of $\phi$ and the operations denoted by $\mathcal{I}_w$ and $\mathcal{F}^0_{N,s}$. On passing the operators $\mathcal{I}_w$ and $\mathcal{F}^0_{N,s}$ from inside the path integral to outside the path integral, we are left with a path integral which can be evaluated using Wick's theorem. This is a well-defined procedure (explored in two recent papers [14, 15]) which has the advantage that the path integral is performed along the real $\phi$-axis. Thus, we need not be concerned with the hopelessly complicated (infinite-dimensional) integration paths that terminate in complex Stokes sectors.

### *4.1. Results to* O($\epsilon$)

In general $Z = \exp(-E_0V)$ where $E_0$ is the ground-state energy density and $V$ is the spacetime volume. Using the above method to $O(\epsilon)$ we find for general $D$ that

$$\Delta E = \frac{1}{4}\epsilon(4\pi)^{-D/2}\Gamma(1-\frac{1}{2}D)\Big\{\ln[2(4\pi)^{-D/2}\Gamma(1-\frac{1}{2}D)] + \psi(3/2)\Big\}. \tag{52}$$

As a check, for $D = 0$ and first order in $\epsilon$, we have

$$\Delta E = \frac{\epsilon}{2\sqrt{2\pi}}\int_{-\infty}^{\infty}d\phi\exp\left(-\frac{1}{2}\phi^2\right)\phi^2\ln(i\phi) = \frac{\epsilon}{4}[\psi(3/2) + \ln(2)], \tag{53}$$

which agrees with (52). For $D = 1$, $\Delta E$ is the expectation of the interaction Hamiltonian to $O(\epsilon)$ in the unperturbed (Gaussian) ground state

$$\Delta E = \frac{\epsilon}{2}\int_{-\infty}^{\infty} dx \exp(-x^2)x^2 \ln(ix) \Big/ \int_{-\infty}^{\infty} dx \exp(-x^2) = \frac{\epsilon}{8}\psi(3/2). \tag{54}$$

The calculated higher order connected Green's function to $O(\epsilon)$ is

$$G_c(y_1, \cdots, y_n) = -\frac{1}{2}\epsilon(-i)^n \Gamma(\frac{n}{2} - 1)[\frac{1}{2}(4\pi)^{-D/2}\Gamma(1 - \frac{1}{2}D)]^{1-n/2} \int d^D x \prod_{k=1}^{n} \triangle_1(y_k - x), \tag{55}$$

where the free propagator $\triangle_\lambda(x)$ is associated with $L_0 = \frac{1}{2}(\nabla\phi)^2 + \frac{1}{2}\lambda^2\phi^2$ and obeys the equation

$$(-\nabla^2 + \lambda^2)\triangle_\lambda(x) = \delta^{(D)}(x). \tag{56}$$

The solution of (56) is

$$\triangle_\lambda(x) = \lambda^{D/2-1}|x|^{1-D/2}(2\pi)^{-D/2}K_{1-D/2}(\lambda|x|) \tag{57}$$

and

$$\triangle_\lambda(0) = \lambda^{D-2}(4\pi)^{-D/2}\Gamma(1 - \frac{D}{2}) \sim \frac{1}{2\pi\delta} \quad (\delta \to 0). \tag{58}$$

From (55) it is clear that as $\delta \to 0+$, the connected Green's functions $G_c(y_1, \cdots, y_n) \to 0$ for $n \geq 3$. This indicates that at least to $O(\epsilon)$ the theory is noninteracting at $D = 2$. When we consider $O(\epsilon^2)$ contributions, we will reexamine this issue.

Turning to $n = 1$ and $n = 2$ we have

$$G_1 = -i\epsilon\sqrt{\frac{1}{2}\pi(4\pi)^{-D/2}\Gamma(1 - \frac{D}{2})} \tag{59}$$

and the two-point function $\widetilde{G}_2(p)$ in momentum space is

$$\widetilde{G}_2(p) = 1/[p^2 + 1 + \epsilon K + O(\epsilon^2)] \tag{60}$$

where $K = \frac{3}{2} - \frac{1}{2}\gamma + \frac{1}{2}\ln[\frac{1}{2}(4\pi)^{-D/2}\Gamma(1 - \frac{D}{2})]$. Thus the renormalised mass to $O(\epsilon)$ is

$$M_R^2 = 1 + K\epsilon + O(\epsilon^2). \tag{61}$$

So near $D = 2$

$$G_1 \sim -i\epsilon\frac{1}{2\sqrt{\delta}} \tag{62}$$

and

$$M_R^2 \sim -\frac{1}{2}\epsilon \ln \delta + A, \tag{63}$$

where $A = 1 + \epsilon[\frac{3}{2} - \frac{\gamma}{2} - \frac{1}{2}\ln(4\pi)]$. Because of the singularities in $G_1$ and $M_R^2$ as $\delta \to 0$ some renormalisation is needed to remove these infinities. The question is whether perturbative renormalisation can be performed in the context of the novel $\epsilon$-expansion method. We can remove the divergence in $G_1$ by introducing in the Lagrangian a linear counter term $iv\phi$ where $v$ has dimension $(\text{mass})^{1+D/2}$ and $v = v_1\epsilon + v_2\epsilon^2 + v_3\epsilon^3 + \cdots$. Since $v$ is real such a term is compatible with $\mathcal{PT}$-symmetry. Adding also a mass counter term $\mu$ we can consider the Lagrangian density

$$L = \frac{1}{2}(\nabla\phi)^2 + \frac{1}{2}\mu^2\phi^2 + \frac{1}{2}g\mu_0^2\phi^2(i\mu_0^{1-D/2}\phi)^\epsilon, \tag{64}$$

where $\phi$ is dimensionful. Using this Lagrangian we obtain

$$G_1 = -\frac{i\epsilon g}{m^2}\mu_0^{D/2-1}\sqrt{\frac{1}{2}\pi m^{D-2}\triangle_1(0)} + \frac{i\epsilon v_1}{\mu_0^2 m^2} \tag{65}$$

and for the renormalised mass

$$M_R^2 = (m\mu_0)^2 + \frac{1}{2}\epsilon g\mu_0^2\{3 - \gamma + \ln[\frac{1}{2}m^{D-2}\triangle_1(0)]\}. \tag{66}$$

In both expressions we have introduced the dimensionless quantity $m^2 = g + \mu^2/\mu_0^2$ and as $\delta \to 0$

$$G_1 \sim \frac{i\epsilon}{g\mu_0^2 + \mu^2}(v_1 - \frac{g\mu_0^2}{2\sqrt{\delta}}) \tag{67}$$

and

$$M_R^2 \sim \mu^2 - \frac{1}{2}\epsilon g\mu_0^2 \ln\delta + A, \tag{68}$$

where $A = g\mu_0^2\{1 + \epsilon[\frac{3}{2} - \frac{\gamma}{2} - \frac{1}{2}\ln(4\pi)]\}$ is a finite quantity. By setting $v_1 = \frac{g\mu_0^2}{2\sqrt{\delta}}$ we have a finite $G_1 = 0$. $M_R$ is logarithmically divergent in $\delta$. We absorb this divergence into $\mu$ by setting

$$\mu^2 = B + \frac{\epsilon}{2}g\mu_0^2\ln\delta, \tag{69}$$

so that $M_R^2 = A + B$ and $B$ is a finite quantity determined in principle from experiment.

### *4.2. Calculations to second order in $\epsilon$*

In second order the calculations become much more involved. The $O(\epsilon^2)$ contribution to the connected part of $G_1$, which we denote by $G_{1,2}$, can be shown to be

$$G_{1,2} = -\frac{1}{2}igm^{-2}\sqrt{\frac{1}{2}\pi\triangle(0)}\Big\{[\ln[2\mu_0^{2-D}\triangle(0)] + \psi(2)\Big\} + \frac{1}{8}ig^2m^{-4}\sqrt{\frac{1}{2}\pi\triangle(0)}\Big\{(\ln[2\mu_0^{2-D}\triangle(0)] + \psi(\frac{3}{2}))(6-D) + 2D - 4 + 4\triangle(0)\mu_0^2m^2\int d^Dx(1 + \frac{\triangle(x)}{\triangle(0)})^2\ln[1 + \frac{\triangle(x)}{\triangle(0)}]\Big\}, \tag{70}$$

where $\triangle(0)$ denotes for brevity $\triangle_{\mu_0 m}(0)$. As $\delta \to 0$,

$$G_{1,2} \sim -\frac{i}{4}gm^{-2}\delta^{-1/2}[\psi(2) - \ln(\pi\delta)] + \frac{i}{4}g^2m^{-4}\delta^{-1/2}[1 + \psi(3/2) - \ln\pi - \ln\delta] + O(\delta). \tag{71}$$

So the algebraic divergence $\delta^{-1/2}$ persists to $O(\epsilon^2)$. The divergence can be removed through $v_2$. Similarly the $\ln\delta$ divergence persists for $G_2$ at second order. Interestingly, the higher-order Green's functions continue to vanish for $D = 2$. Avoidance of a noninteracting theory remains an open question in this approach [15].

## 5. Renormalisation group flows of $\mathcal{PT}$-symmetric theories

So far we have considered whether a Hermitian theory can lead to a non-Hermitian $\mathcal{PT}$-symmetric theory due to the effect of renormalisation. We ask the opposite question in this section: Can a $\mathcal{PT}$-symmetric field theory retain its $\mathcal{PT}$ symmetry as the Lagrangian flows due to renormalisation?

We review some preliminary work on this question. In its full generality this is an intractable problem. We turn to the framework of the *functional renormalisation group* [20], which combines the functional formulation of quantum field theory with the Wilsonian renormalisation group. As mentioned in Sec. 3, it is possible to make some progress in solving the functional equations using the simpler approach developed by Wetterich [21] and Morris [24] which, in the local potential approximation (32), leads to a nonlinear *partial differential equation* (PDE) rather than a functional equation. This simplification enables substantial progress in understanding the effective potential in arbitrary spacetime dimensions. In $D$ dimensions this PDE is

$$\partial_k U_k(\phi_c) = \frac{1}{\pi_D} \frac{k^{D+1}}{k^2 + U_k''(\phi_c)}, \tag{72}$$

where $\pi_D = \frac{D(2\pi)^D}{S_{D-1}}$ and $S_{D-1} = \frac{2\pi^{D/2}}{\Gamma(D/2)}$ is the surface area of a unit $D$-dimensional sphere. We may assume that the equations for $U_k(\phi_c)$ involve dimensionless quantities. If this were not so, we could achieve dimensionlessness by introducing a mass scale $M$. For example, for $D = 1$ the dimensionless variables, denoted by a tilde, are $\tilde{\phi} = M^{1/2}\phi$, $\tilde{g} = M^{-3}g$, $\tilde{k} = M^{-1}k$, and $\tilde{\mu} = M^{-1}\mu$. The Wetterich equation (72) can be thought of as being in terms of such dimensionless variables.

To avoid difficulties in numerical analysis associated with boundary conditions, in the past the PDE (72) has been analyzed by approximating $U_k(\phi_c)$ as a finite series in powers of the field $\phi_c$. This *ansatz* leads to a sequence of coupled nonlinear *ordinary* differential equations. The consistency of such a procedure has not been established.

For $D = 1$ (the quantum-mechanical case), (72) becomes

$$\partial_k U_k(\phi_c) = \frac{1}{32\pi^2} \frac{k^2}{k^2 + U_k''(\phi_c)}. \tag{73}$$

Even in this one-dimensional setting, no exact solution to this nonlinear PDE is known and only numerical solutions have been discussed.

We depart from the above treatment by performing an asymptotic analysis for large values of the cut-off $k$ [16]. The novelty of the approach used here is that it avoids the appearance of coupled nonlinear ordinary differential equations. To leading order the results of this analysis are qualitatively different depending on whether the space-time dimension $D$ is greater or less than 2. Of course, we are interested in the $k \to 0$ limit which we will, as we shall see, approach through the method of Padé approximants.

Letting $z = k^{2+D}$ and $U_k(\phi_c) = U(z, \phi)$, we can rewrite (72) as

$$U_z(z, \phi) = \frac{1}{(2+D)\pi_D} \frac{1}{z^{\frac{2}{2+D}} + U_{\phi\phi}(z, \phi)}, \tag{74}$$

where the subscripts on $U$ indicate partial derivatives. We now assume that for large $z$ we can neglect the $U_{\phi\phi}$ term in the denominator. (The consistency of this assumption is easy to verify when $D < 2$.) Then, for large $z$ we have to leading order in our approximation scheme

$$U_z(z, \phi) \sim \frac{1}{(2+D)\pi_D} z^{-\frac{2}{2+D}} \quad (z \gg 1). \tag{75}$$

On incorporating a correction $\epsilon$ to this leading behavior

$$U(z,\phi) = \frac{1}{D\pi_D} z^{\frac{D}{2+D}} + \epsilon(z,\phi), \tag{76}$$

we get to order $O(\epsilon^2)$

$$\epsilon_z(z,\phi) = -\frac{1}{(D+2)\pi_D} z^{-\frac{4}{2+D}} \epsilon_{\phi\phi}(z,\phi). \tag{77}$$

On making the further change of variable $t = \frac{D+2}{2-D} z^{\frac{D-2}{D+2}}$, (77) becomes

$$\epsilon_t(t,\phi) = \frac{1}{(D+2)\pi_D} \epsilon_{\phi\phi}(t,\phi). \tag{78}$$

The variable $t$ is positive for $D < 2$ and negative for $D > 2$ and is not defined at $D = 2$. Thus, (78) is a conventional diffusion equation for $D < 2$ but is a *backward* diffusion equation for $D > 2$. The backward diffusion equation is an inverse problem that is ill-posed. The problems associated with this ill-posedness may be connected with difficulties in solving (72) numerically when $D = 4$.

In the preliminary study these issues were not addressed further. The case $D = 1$ was considered and some simple $\mathcal{PT}$-symmetric theories were studied [16]. From (72), on defining $\widehat{U}_k \equiv U_k - \frac{k^D}{D\pi_D}$, we can deduce that

$$\partial_k \widehat{U}_k = -\frac{k^{D-1}\widehat{U}''_k}{\pi_D\left(k^2 + \widehat{U}''_k\right)}. \tag{79}$$

From (79) it is consistent to assume that $\widehat{U} \to V(\phi)$ and $\frac{\widehat{U}''}{k^2} \to 0$ as $k \to \infty$. (For simplicity of notation we have dropped the suffix $k$ in $\widehat{U}_k$.) Let us write the correction as $V(\phi) + \frac{1}{k}U_1(\phi)$. On substituting in (79), we obtain

$$U_1(\phi) = \frac{1}{\pi} V''(\phi). \tag{80}$$

We can proceed in this way and write the next correction as $\widehat{U}(\phi) = V(\phi) + \frac{1}{k\pi}V''(\phi) + \frac{1}{k^2}U_2(\phi)$. This leads to $U_2(\phi) = \frac{1}{2\pi^2}V^{(4)}$ where $V^{(4)}(\phi) \equiv \frac{d^4}{d\phi^4}V(\phi)$. Repeating this procedure, we get $U_3(\phi) = \frac{1}{3}\left(\frac{1}{2\pi^3}V^{(6)}(\phi) - V^{(2)}(\phi)^2\right)$. This procedure can be formalized: On writing $\delta = \pi/k$ (not to be confused with $\delta$ in the last section) and $x = \pi\phi$, (79) becomes

$$\frac{\partial}{\partial\delta}\widehat{U} = \frac{\frac{\partial^2}{\partial x^2}\widehat{U}}{1 + \delta^2 \frac{\partial^2}{\partial x^2}\widehat{U}} \tag{81}$$

and $\widehat{U} = \sum_{n=0}^{\infty} \delta^n U_n(\phi)$.

When $\delta$ is small, the scale $k$ is large and we are probing the microscopic potential. As $\delta \to \infty$ we probe the infrared limit of the effective potential. However, the analysis outlined above was based on a perturbation theory in $\delta$. There is a parallel to the theory of phase transitions, where a physical entity is expanded in inverse powers of the temperature $T$, and then an extrapolation procedure is used to find the critical behavior at lower $T$. This often-used extrapolation technique is based on Padé approximation.

So far, our treatment has been for a general potential; we will now specialize $V(x)$ to two cases $V(x) = gx^3$ and $V(x) = gx^4$, which are $\mathcal{PT}$-symmetric for $g = i$ and $g$ real. We also include a mass term. Our purpose is to consider the series $\sum_{n=0}^{N} \delta^n U_n(x)$ generated from the series relevant to these two cases and to consider the $\delta \to 0$ limit.

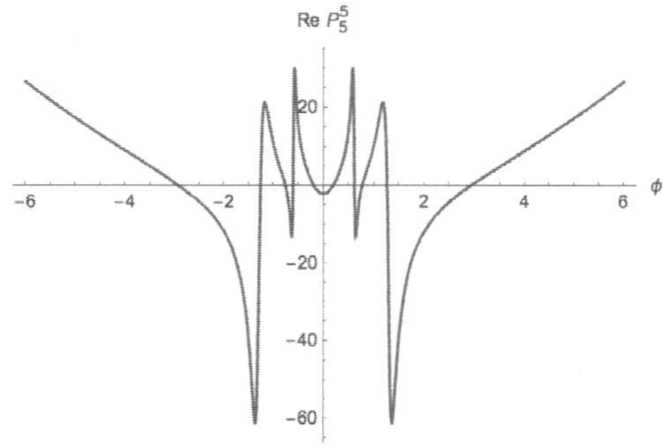


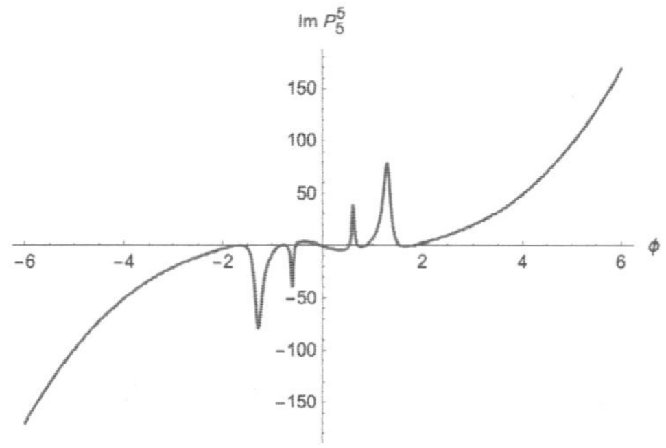


**Figure 7.** Effective potential flow for massive $i\phi^3$ with the mass parameter $\mu = 1$. Shown is a plot of the real and imaginary parts of the $P_5^5$ approximant plotted as functions of real $\phi$. Observe that there are no poles. The real part of the potential is right-side-up, so there is no instability. Furthermore, apart from small fluctuations near the origin, the imaginary part of the potential behaves like $i\phi^3$ for large $|\phi|$.

### *5.1. Cubic potentials*

We consider both massless and massive cubic potentials. In the former case $U_0(\phi) = g\phi^3$ and in the latter case $U_0(\phi) = \mu^2\phi^2 + g\phi^3$. For the massless case the solution to (81) is

$$\begin{aligned}
U_1(\phi) &= 6g\phi, & U_{10}(\phi) &= \tfrac{1621728}{175} g^5\phi^3, \\
U_2(\phi) &= 0, & U_{11}(\phi) &= \tfrac{17882208}{1925} g^5\phi - \tfrac{46656}{11} g^6\phi^6, \\
U_3(\phi) &= -12g^2\phi^2, & U_{12}(\phi) &= -\tfrac{30608064}{385} g^6\phi^4, \\
U_4(\phi) &= -6g^2, & U_{13}(\phi) &= \tfrac{279936}{13} g^7\phi^7 - \tfrac{4537967328}{25025} g^6\phi^2, \\
U_5(\phi) &= \tfrac{216}{5} g^3\phi^3, & U_{14}(\phi) &= \tfrac{4512754944}{7007} g^7\phi^5 - \tfrac{5900745888}{175175} g^6, \\
U_6(\phi) &= \tfrac{456}{5} g^3\phi, & U_{15}(\phi) &= \tfrac{1439488599552}{525525} g^7\phi^3 - \tfrac{1679616}{15} g^8\phi^8, \\
U_7(\phi) &= -\tfrac{1296}{7} g^4\phi^4, & U_{16}(\phi) &= \tfrac{2953243339824}{175175} g^7\phi - \tfrac{25063743168}{5005} g^8\phi^6, \\
U_8(\phi) &= -\tfrac{34668}{35} g^4\phi^2, & U_{17}(\phi) &= \tfrac{10077696}{17} g^9\phi^9 - \tfrac{105103706989824}{2977975} g^8\phi^4, \\
U_9(\phi) &= \tfrac{7776}{9} g^5\phi^5 - \tfrac{89496}{315} g^4, & U_{18}(\phi) &= \tfrac{3212310887424}{85085} g^9\phi^7 - \tfrac{101014620538752}{2127125} g^8\phi^2.
\end{aligned} \tag{82}$$

These expressions, which are valid for pure imaginary $g$ as well as for real $g$, are cumbersome but manageable. We stress that there has been no truncation of the function space on which $\widehat{U}(\phi)$ has support. This contrasts with the usual approach which requires a truncation at the onset of the calculation of the renormalisation group flow. This absence of truncation continues to be a feature for the massive case.

The iterative solution to (81) for the massive case is similar to that for the massless case, but the expressions for $U_n(\phi)$ have many more terms. For example, the coefficient of $\delta^9$ is

$$U_9(\phi) = \tfrac{8}{315}(34020g^5\phi^5 + 56700g^4\mu^2\phi^4 - 11187g^4 + 37800g^3\mu^4\phi^3 + 12600g^2\mu^6\phi^2 + 2100g\mu^8\phi + 140\mu^{10}),$$

which in the massless limit $\mu \to 0$ reduces to the two-term expression in (82). We refrain from listing the coefficients explicitly and instead proceed directly to the large-$\delta$ behavior of the diagonal Padé approximants. We denote the diagonal Padé approximants in the massless case by $P_N^{0,N}(\delta)$ and in the massive case by $P_N^N(\delta)$. Let us examine some low-order Padé approximants in the limit $\delta \to \infty$. For example,

$$\lim_{\delta\to\infty} P_2^2(\delta) = \frac{18g^3\phi^5 + 3g^2\left(10\mu^2\phi^4 - 9\right) + 14g\mu^4\phi^3 + 2\mu^6\phi^2}{2\left(3g\phi + \mu^2\right)^2}.$$

[As a check, when $\mu = 0$ this expression agrees with the corresponding expression $g\phi^3 - \frac{3}{2\phi^2}$ for $P_2^{0,2}(\delta)$.] For $N = 3$ we obtain

$$\lim_{\delta\to\infty} P_3^{0,3}(\phi) = \frac{g\phi^3\left(736g\phi^5 + 1705\right)}{25 - 800g\phi^5} \qquad \text{and} \qquad P_3^3(\phi) = -u_3/d_3,$$

where

$$u_3 = 1609632g^8\phi^8 + 729g^7\phi^3\left(5888\mu^2\phi^4 + 5115\right) + 243g^6\mu^2\phi^2\left(23008\mu^2\phi^4 + 15345\right) + 1296g^5\mu^4\phi\left(3376\mu^2\phi^4 + 945\right) + 15120g^4\mu^6\left(142\mu^2\phi^4 + 9\right) + 658944g^3\mu^{10}\phi^3 + 121824g^2\mu^{12}\phi^2 + 12288g\mu^{14}\phi + 512\mu^{16},$$

$$d_3 = 225g^2[7776g^5\phi^5 + 81g^4(160\mu^2\phi^4 - 3) + 8640g^3\mu^4\phi^3 + 2880g^2\mu^6\phi^2 + 480g\mu^8\phi + 32\mu^{10}].$$

[As a check, we have $\lim_{\mu\to 0} P_3^3(\phi) = P_3^{0,3}(\phi)$.]

Note that for the massless case $P_2^{0,2}$ has a double pole for both real and imaginary $g$. In fact, the same is true for $P_n^{0,n}$ for even $n$. This pole is an artifact of the Padé approximation and is not present when $n$ is odd, so we will only consider the behavior of these approximants for odd $n$. In general, the odd-$n$ diagonal Padé approximants have no singularities at all on the real-$\phi$ axis when $g$ is imaginary but singularities occur for the case of real $g$ . These findings indicate that the $\mathcal{PT}$-symmetric effective potential is well behaved in the infrared limit . From the expressions for the diagonal Padé approximants we see that for large $|\phi|$, the leading behavior of the imaginary part of the effective potential is exactly $i\phi^3$. Consequently the $\mathcal{PT}$ nature of the interaction is preserved under renormalisation.

A similar investigation of quartic potentials also shows that $\mathcal{PT}$-symmetry is maintained under renormalisation [16].

## 6. Conclusions

The study of $\mathcal{PT}$-symmetric quantum field theory is still in its infancy. We have shown that there is a link between renormalisation of some Hermitian field theories and $\mathcal{PT}$-symmetric field theories. This provides a motivation for the study of $\mathcal{PT}$-symmetric field theories. Often, the usual tools of quantum field theory cannot be applied directly. We have reviewed some interesting and promising approaches to studying these new field theories. There are many opportunities for future work in these approaches. The models we chose were based on simplicity but, the SM Lagrangian has many types of fields and interactions which can lead to the emergence of other $\mathcal{PT}$-symmetric QFTs.

*Acknowledgments*
CMB thanks the Alexander von Humboldt and Simons Foundations for financial support. SS and CMB thank the UK Engineering and Physical Sciences Research Council for financial support.

## References

[1] Bender C M and Boettcher S 1998 *Phys. Rev. Lett.* **80** 5243
[2] Bender C M 2019 *$\mathcal{PT}$ Symmetry in Quantum and Classical Physics* (Singapore: World Scientific)
[3] Dorey P E, Dunning C and Tateo R 2001 *J. Phys.* A **34** 5679
[4] Mostafazadeh A 2002 *J. Math. Phys.* **43**, 205 ; 2003 *J. Phys.* A **36** 7081
[5] Scholtz F G, Greyer H B and Hahne F J W 1992 *Ann. Phys.* **213** 74
[6] Jones H F and Rivers R J 2007 *Phys. Rev.* D **75** 025023
[7] Jones H F and Rivers R J 2009 *Phys. Lett.* A **373** 3304
[8] Bender C M and Klevansky S P 2009 *Phys. Rev. Lett.* **105** 3304
[9] Bender C M, Brandt S F, Chen J-H and Wang Q 2005 *Phys. Rev.* D **71** 025014
[10] Sher M 2010 *Phys. Rept.* **179** 031601
[11] Coleman S and Weinberg E J 1973 *Phys. Rev.* D **8** 1888
[12] Bender C M, Hook D , Mavromatos N E and Sarkar S 2016 *J. Phys.* A **49** 45LT01
[13] Haber H E and Kane G L 1985 *Phys. Rept.* **117** 75
[14] Bender C M, Hassanpour N, Klevansky S P and Sarkar S 2018 *Phys. Rev.* D **98** 125003
[15] Felski A, Bender C M, Klevansky S P and Sarkar S 2021 Towards perturbative renormalization of $\phi^2(i\phi)^\epsilon$ quantum field theory *Preprint* (*Preprint* hep-th/2103.07577)
[16] Bender C M and Sarkar S 2018 *J. Phys.* A **51** 225202
[17] Lehmann H, Symanzik K and Zimmermann W 1955 *Nuovo Cimento* **1** 205
[18] Barton G 1963 *Introduction to Advanced Field Theory* (New York: Wiley & Sons)
[19] Lee T D 1954 *Phys. Rev.* **95** 1329
[20] Gies H 2012 *Lect. Notes Phys.* **852** 287
[21] Wetterich C 2001 *Int. J. Mod. Phys.* A **16** 1951
[22] Litim D 2001 *Phys. Rev.* D **64** 105007
[23] Wipf A 2013 *Statistical Approach to Quantum Field Theory* (Lect. Notes Phys. 864) (Berlin : Springer)
[24] Morris T R 1994 *Phys. Lett.* B **329** 241
[25] Alexandre J , Houston N and Mavromatos N E 2013 *Phys. Rev.* D **88** 125017; 2015 *Int. J. Mod. Phys.* D **24** 1541004
[26] Starobinsky A A 1980, *Phys. Lett.* B **91**, 99
[27] Bender C M, Holm D D and Hook D W 2007 *J. Phys.* A **40** F81; Brody D C and Hook D W 2008 *J. Phys.* A **41** 352003; Fring A and Bagchi B 2011 *J. Phys.* A **44** 325201
[28] Bender C M and Orszag S A 1978 *Advanced Mathematical Methods for Scientists and Engineers* (New York: McGraw Hill)

# PT symmetry and the evolution speed in open quantum systems [1]

**Dorje C Brody**[1,2]

[1]Department of Mathematics, University of Surrey, Guildford GU2 7XH, UK

[2]St Petersburg National Research University of Information Technologies, Mechanics and Optics, St Petersburg 197101, Russia

E-mail: d.brody@surrey.ac.uk

**Abstract.** The dynamics of an open quantum system with balanced gain and loss is not described by a PT-symmetric Hamiltonian but rather by Lindblad operators. Nevertheless the phenomenon of PT-symmetry breaking and the impact of exceptional points can be observed in the Lindbladean dynamics. Here we briefly review the development of PT symmetry in quantum mechanics, and the characterisation of PT-symmetry breaking in open quantum systems in terms of the behaviour of the speed of evolution of the state.

## 1. PT symmetry and quantum physics

We begin with a brief account of the development of PT symmetry in quantum mechanics. In 1998 Bender and Boettcher found that there is a family of Hamiltonians of the form $\hat{H} = \frac{1}{2}\hat{p}^2 + \hat{x}^2(\mathrm{i}\hat{x})^\epsilon$ that possesses real, positive, and discrete eigenvalues for all $\epsilon \geq 0$ [1]. These Hamiltonians, while not Hermitian, are nevertheless invariant under the action of parity-and-time (PT) reversal: $\hat{p} \to \hat{p}$, $\hat{x} \to -\hat{x}$, and $\mathrm{i} \to -\mathrm{i}$. This observation led them to conjecture that generic PT-symmetric Hamiltonians may possess entirely real eigenvalues. Subsequently, eigenvalues of many other PT-symmetric Hamiltonians were investigated by numerous authors.

A natural question thus arising is whether a consistent quantum theory can be developed by use of a PT-symmetric Hamiltonian. In standard quantum mechanics the reality of observables are ensured by the Hermiticity condition on the Hilbert space endowed with a Hermitian inner product. While Hermiticity may be replaced by the condition of PT symmetry to enforce the reality of observables, a Hilbert space endowed with a PT-conjugation inner product is problematic, for, the parity operator is trace free and squares to the identity so that half of its eigenvalues are negative. In other words, such Hilbert spaces are of the Pontryagin-type that come equipped with indefinite metrics [2]. Because Hilbert-space norms are related to probabilities in quantum mechanics, negative norms would imply negative probabilities, which makes the theory unphysical.

In 2002, however, Mostafazadeh [3], and independently Bender *at al.* [4] observed that for a PT-symmetric Hamiltonian $\hat{H}$ there exists another PT-symmetric observable that commutes with $\hat{H}$ such that its eigenvalues are either plus or minus one, depending on the parity type of the eigenfunctions of $\hat{H}$. Such an operator naturally has the interpretation analogous to a charge

[1] This paper is based on the talk presented at the PHHQP Seminar Series.

(C) operator. This means that a PT-symmetric Hamiltonian is automatically CPT-symmetric, so in fact we can speak of CPT-symmetric quantum theory for which the Hilbert-space inner product is defined with respect to a positive-definite CPT conjugation. In particular, writing $\hat{g}$ for the operator associated with the CP conjugation, the inner product of a pair of elements $|\psi\rangle$ and $|\varphi\rangle$ is given by $\langle\psi,\varphi\rangle = \langle\psi|\hat{g}|\varphi\rangle$. It can then be shown that the operator $\hat{g}$ is Hermitian and positive, so that it admits the interpretation of defining a metric on Hilbert space, and thus is sometimes referred to as the metric operator.

Because $\hat{g}$ is Hermitian and positive, there is an operator $\hat{u}$, not unique, such that we can write $\hat{g} = \hat{u}\hat{u}^\dagger$. Then the operator $\hat{u}$ can be used to map between the Hilbert space endowed with a standard Hermitian inner product and that endowed with a CPT inner product, because $\langle\psi,\varphi\rangle = \langle\hat{u}^\dagger\psi|\hat{u}^\dagger\varphi\rangle$. Further, the operator $\hat{u}$ can be used to define a similarity transformation between a PT-symmetric observable and a Hermitian observable; whereas $\hat{g}$ can be used to determine the Hermitian adjoint of an observable $\hat{F}$ in the sense that $\hat{F}^\dagger = \hat{g}^{-1}\hat{F}\hat{g}$. An operator satisfying such a condition is sometimes referred to as being 'pseudo-Hermitian'. Once these structures associated with PT symmetry were uncovered, it became appreciated that the notion of pseudo-Hermiticity in itself has been around for some time [5], and likewise the idea of defining Hilbert space inner product in terms of a nontrivial metric operator for characterising quantum theory has been proposed previously [6], albeit in a different context.

The fact that there is a similarity transformation relating a PT-symmetric Hamiltonian to a Hermitian one naturally raises the question whether a quantum system described by a PT-symmetric Hamiltonian is in fact equivalent to standard Hermitian quantum mechanics. That they are indeed equivalent on account of the existence of a similarity transformation is the point that has been argued for long by Mostafazadeh [7], although in infinite dimensions one has to be cautious in making such an assertion.

To discuss the question of equivalence it will be useful first to consider quantum systems modelled on finite-dimensional Hilbert spaces. In this case, all operators are bounded, so mathematical subtleties associated with unbounded operators do not enter the discussion. It is then tempting to conclude that a system described by a PT-symmetric Hamiltonian is just an alternative representation of the same system described by a Hermitian Hamiltonian, i.e. they are equivalent. This 'equivalence' argument by itself does not quite account for exactly what goes on for two reasons: First, a PT-symmetric Hamiltonian admits more parametric degrees of freedom than a Hermitian Hamiltonian. For instance, in the case of a two-level system, a Hermitian Hamiltonian has four independent parameters, while a PT-symmetric Hamiltonian can have six independent parameters. More generally, for an $n$-level system a PT-symmetric Hamiltonian can have $n(2n-1)$ independent parameters, as opposed to $n^2$ for a Hermitian Hamiltonian. Because parameters in a Hamiltonian represent experimental setup, a blind statement of equivalence does not explain this discrepancy. Second, for a PT-symmetric Hamiltonian it is possible to vary parameters in the Hamiltonian to effect a phase transition for which in one phase the eigenstates of the Hamiltonian are PT symmetric (the unbroken phase), while in the other, broken phase they are not. In the broken phase, the eigenvalues of the Hamiltonian are no longer real—some, or all of them, come in complex conjugate pairs. No such phenomenon is seen in a Hermitian Hamiltonian in finite dimensions.

This apparent contradiction was further clarified in 2016 when it was shown, using a biorthogonal formalism of [8], that there is a canonical separation of the degrees of freedom in a PT-symmetric Hamiltonian into two sets: one corresponding to the degrees of freedom associated with its Hermitian counterpart Hamiltonian, and one corresponding to the degrees of freedom associated with the specification of the metric operator $\hat{g}$ [9]. Further, it was shown that for closed quantum systems there are no laboratory experiments to adjust or determine the values of the latter degrees of freedom. That is, the expectation values of physical observables are independent of $\hat{g}$. Thus, perhaps a better characterisation is that for a finite-dimensional system the two descriptions are *indistinguishable* rather than equivalent. In particular, without the ability to adjust the parameters in the operator $\hat{g}$ it is not possible to experimentally realise the phase transition discussed above.

In the infinite-dimensional case, the situation is more complicated even in the countable case, because the equivalence of eigenvalues between a PT-symmetric Hamiltonian and its Hermitian 'counterpart' Hamiltonian related by a similarity transformation involving the operator $\hat{g}$ is ensured only if both operators $\hat{g}$ and $\hat{g}^{-1}$ are bounded. Otherwise, the image of an eigenstate of one of the Hamiltonians under the map, or more generally the image of any state from the domain of this Hamiltonian, may not belong to the other Hilbert space, or to the domain of the other Hamiltonian. For many model Hamiltonians studied in the literature, either $\hat{g}$ or $\hat{g}^{-1}$ (or both) are unbounded, and in these cases it is not simple to determine whether the two descriptions are equivalent: It may be the case that there are genuinely new physics of closed quantum systems modelled on infinite-dimensional Hilbert spaces, possibly with a partially broken PT symmetry, although this remains to be further investigated.

The above discussion is concerned with closed systems. Around 2006, however, it was observed that PT symmetry can be realised in a laboratory for open systems by balancing gain and loss [10, 11, 12]. For example, suppose that with respect to a choice of reference frame the left side of the system absorbs energy, while the right side emits energy, in a symmetric configuration, by the same amount. Then under parity (left-right) reversal, gain turns into loss and loss turns into gain; but time reversal has the same effect so that the system is PT symmetric. Intuitively, although the system is open, it can behave in a manner similar to a closed one because the overall energy is conserved. Such an open system can thus be described by a PT-symmetric Hamiltonian, where now the degrees freedom associated with the operator $\hat{g}$ can be controlled in a laboratory by adjusting, for instance, the gain and loss strengths. It should be evident, however, that for such a system to remain in a quasi-equilibrium state of energy, the internal energy transport, i.e. the coupling strength of the left and the right side of the system, has to be sufficiently strong, for otherwise the system cannot redistribute the energy to maintain its quasi-equilibrium state. Thus by increasing gain and loss strengths while keeping the internal interaction strength fixed, the system undergoes a phase transition such that although the Hamiltonian remains PT symmetric, its eigenstates are no longer PT symmetric and consequently the energy eigenvalues become complex. For PT-symmetric systems, such transitions are accompanied by the degeneracies of the energy eigenstates, and these critical points are often referred to as exceptional points in the literature [13, 14, 15, 16, 17].

The facts that PT symmetry can be implemented relatively straightforwardly for open systems in a laboratory, and that such systems typically exhibit nontrivial phase transitions with interesting and sometimes counterintuitive features, led many researchers to explore properties of PT-symmetric systems both theoretically and experimentally—perhaps most notably in the context of optical systems such as optical waveguides. In the context of open systems, however, those whose dynamics are described by PT-symmetric Hamiltonians are classical rather than quantum mechanical. Likewise, many of the experiments that were implemented concern classical systems. The reason is because a perfect balancing of gain and loss of energy (or of particle number, or of volume) is not feasible in quantum mechanics.

To see this, it suffices to consider the simple system described above where energy is absorbed on the left side and emitted on the right side. Suppose that $\epsilon$ joules of energy is absorbed on the left side between time $t$ and time $t + dt$ where $dt \ll 1$. Then to perfectly balance this, it is necessary that exactly $\epsilon$ joules of energy is emitted from the right side precisely over the same time interval. Such a process can be achieved classically. Quantum mechanically, to realise a perfect PT symmetry one requires a sharp precision in both energy and time, but such a precision amounts to violating the energy-time uncertainty relation. Another way of seeing the issue is to consider a quantum system having a discrete set of energy eigenvalues. Then unless the eigenvalues are equally spaced (such as an oscillator) and the value of $\epsilon$ coinciding with the energy gap, the system will always remain in a state of indefinite energy (i.e. not an energy eigenstate) and one can only consider the expectation value of the Hamiltonian in order to speak about the energy of the system.

The impossibility of maintaining a perfect PT symmetry, however, does not mean that a PT-symmetric open environment cannot be realised in quantum mechanics, because it remains meaningful to speak about balancing gain and loss *on average*. A PT-symmetric configuration of the absorption and emission *rates* does not violate Heisenberg constraints. However, the fact that one cannot determine whether the system has, say, absorbed or emitted energy, implies that as time passes, the observer gradually loses information about the state of the system. The theoretical implication is that it is not possible to characterise the dynamics of a PT-symmetric open quantum system in terms of a PT-symmetric Hamiltonian. Instead, time evolution is governed by an equation of the Lindblad type, or more generally of the Krauss type.

Focusing in particular on systems described by the Lindblad equation, it would be of interest then to enquire what are the characteristic features of PT-symmetric open quantum systems and the associated symmetry breaking. Indeed, there are growing interests in studying the effects of exceptional points in genuinely open quantum systems described by Lindblad equations [18, 19]. With this in mind, the purpose here is to explore this question by focusing specifically on the behaviour of the speed of the evolution of the state that has been examined in [20].

## 2. Open quantum system dynamics

Intuitively, by an open system $S$ we have in mind one that is coupled to another system $B$. The latter may be large enough to act as a bath, or may be another small quantum system. Suppose that the total system $S + B$ is described by a pure-state wave function $|\Psi_{S+B}\rangle$. The state of an open quantum system $S$ is then determined by tracing out the degrees of freedom associated with $B$:

$$\hat{\rho}_S = \mathrm{tr}_B \left( \frac{|\Psi_{S+B}\rangle\langle\Psi_{S+B}|}{\langle\Psi_{S+B}|\Psi_{S+B}\rangle} \right).$$

Provided that the state $|\Psi_{S+B}\rangle$ is not a product state of the form $|\Psi_{S+B}\rangle = |\Psi_S\rangle \otimes |\Psi_B\rangle$, or equivalently stated if the two systems $S$ and $B$ are entangled, the result of the tracing is necessarily a mixed-state density matrix so that we have $\hat{\rho}_S^2 \neq \hat{\rho}_S$. As an example, suppose that $S$ and $B$ are both spin-$\frac{1}{2}$ particles, and $|\Psi_{S+B}\rangle$ is the pure spin-0 singlet state $\frac{1}{\sqrt{2}}(|\uparrow\downarrow\rangle - |\downarrow\uparrow\rangle)$. Then tracing over $B$ we obtain

$$\hat{\rho}_S = \tfrac{1}{2}\big(|\uparrow\rangle\langle\uparrow| + |\downarrow\rangle\langle\downarrow|\big),$$

which is a state distinct from the coherent superposition $\frac{1}{\sqrt{2}}(|\uparrow\rangle + |\downarrow\rangle)$.

We can think of a density matrix as representing a (classical) statistical average over pure states, in the sense that it can be written as the expectation of a random pure-state projection operator:

$$\hat{\rho} = \mathbb{E}\left[\frac{|\psi\rangle\langle\psi|}{\langle\psi|\psi\rangle}\right].$$

For instance, suppose that an experimentalist is aware that the state of the system will be $|\psi_1\rangle$ if the system absorbs energy, but otherwise will be $|\psi_0\rangle$, and that the probability of absorption is $p$. Then without a measurement to acquire more information (and thus altering the state) one can only assert that the system is in the averaged state

$$\hat{\rho} = p\frac{|\psi_1\rangle\langle\psi_1|}{\langle\psi_1|\psi_1\rangle} + (1-p)\frac{|\psi_0\rangle\langle\psi_0|}{\langle\psi_0|\psi_0\rangle},$$

which again is not the same state as the pure state $|\psi\rangle = \sqrt{p}\,|\psi_1\rangle + \sqrt{1-p}\,|\psi_0\rangle$. Thus the density matrix $\hat{\rho}$ has the interpretation of 'either $|\psi_0\rangle$ or $|\psi_1\rangle$' in the sense of classical probability, whereas the pure state $|\psi\rangle$ has the interpretation of 'neither $|\psi_0\rangle$ nor $|\psi_1\rangle$' in the sense of quantum probability; and the two states are experimentally distinguishable. A mixed state density matrix can always be expressed in such an ensemble average of pure states. However, given a state $\hat{\rho}$ there are uncountably many ways of expressing it in terms of averages of pure states, so it is not possible to determine which experimental setup has resulted in that state.

The time evolution of the density matrix that preserves its positivity and the trace condition can typically be described by the dynamical equation of the Lindblad type:

$$\partial_t \hat{\rho} = -\mathrm{i}[\hat{H}, \hat{\rho}] + \sum_k \left[ \hat{L}_k \hat{\rho} \hat{L}_k^\dagger - \frac{1}{2}\left( \hat{L}_k^\dagger \hat{L}_k \rho + \hat{\rho} \hat{L}_k^\dagger \hat{L}_k \right) \right].$$

This is the so-called GKLS equation of Gorini, Kossakowski, Sudarshan [21] and Lindblad [22], or just the Lindblad equation for short. Interestingly, in the context of a proposal of Hawking [23] that the dynamical equation for the density matrix will not preserve unitarity, Banks, Susskind and Peskin [24] explored a general dynamics for the state that preserves positivity and trace condition, and evidently unaware of the GKLS equation arrived at essentially the same conclusion.

The Lindblad operators $\hat{L}_k$ characterise the way in which the system $S$ interacts with the environment $B$. So for example if there is a particle gain at the left side and particle loss at the right side, then the dynamics of this system can be characterised by a pair of Lindblad operators with $\hat{L}_1 = \gamma_L \hat{a}_L^\dagger$ and $\hat{L}_2 = \gamma_R \hat{a}_R$, expressed in terms of the particle creation and annihilation operators, where the parameters $\gamma_L, \gamma_R$ determine the gain/loss strengths. That an averaged balancing of gain and loss can be realised in this way to describe a PT-symmetric quantum system in the context of Bose-Einstein condensates has been known for long to the Stuttgart group [25, 26]. In the case of a many body system, such a system can be treated effectively in terms of a mean-field approximation, which in turn may be described by an effective PT-symmetric Hamiltonian [27].

From the linearity of the dynamics we can think of the right side of the Lindblad equation as representing the action of a Liouville operator $\mathcal{L}$ on $\hat{\rho}$, and write

$$\partial_t \hat{\rho} = \mathcal{L}\hat{\rho}.$$

Here the Liouville operator $\mathcal{L}$ can be interpreted as a superoperator, which becomes evident if we express operators using the index notation by writing $\partial_t \rho^\alpha_\beta = \mathcal{L}^{\alpha\alpha'\beta'}_{\beta\beta'} \rho^{\beta'}_{\alpha'}$ with repeated indices summed over. It turns out that the eigenvalues of $\mathcal{L}$ are real or else come in complex conjugate pairs, just like any PT-symmetric operator. Thus, from a certain point of view one could argue that every open system dynamics is PT symmetric, although exactly what that might mean physically is not well understood.

## 3. Unitary evolution speed for quantum states

There are various features of the dynamics of an open system that one can investigate, but following [20], here we shall be focused on one particular aspect concerning the speed of evolution of the state. The evolution speed of a quantum system in itself is of interest for several reasons. For example, to implement a quantum operation such as a quantum algorithm in a quantum computer, it is useful to have an idea on how fast a given task can be realised. The evolution speed is also linked to the sensitivity or stability of the state against changes in time, which in turn is useful in arriving at error lower bounds in estimating quantum states [28].

For pure states, the evolution speed under a unitary motion is given by the Anandan-Aharonov relation [29]. The idea can be described as follows. We start with the Schrödinger-Kibble equation

$$\frac{\mathrm{d}}{\mathrm{d}t}|\psi(t)\rangle = -\mathrm{i}\frac{1}{\hbar}\left(\hat{H} - \langle\psi(t)|\hat{H}|\psi(t)\rangle\right)|\psi(t)\rangle$$

for the dynamics of the state. Note that the Schrödinger-Kibble equation is a nonlinear differential equation defined on Hilbert space. However, because the energy expectation $E = \langle\psi(t)|\hat{H}|\psi(t)\rangle$ is a constant of motion, the equation is indeed linear along each unitary orbit. The effect of removing the energy expectation from the Hamiltonian is to eliminate dynamical phase under the evolution, and the equation has the advantage that for a stationary state we have $|\dot{\psi}(t)\rangle = 0$ and hence we automatically recover the time-independent Schrödinger equation $\hat{H}|\psi\rangle = E|\psi\rangle$, which in undergraduate quantum mechanics texts can only be derived by evoking the somewhat mysterious correspondence between energy and time derivative. Mathematically, the Schrödinger-Kibble equation gives what is called a horizontal lift of the Schrödinger equation in Hilbert space to the projective Hilbert space where the tangent vector $|\dot{\psi}\rangle$ is everywhere orthogonal to $|\psi\rangle$.

If the state $|\psi\rangle$ of a system satisfies the Schrödinger-Kibble equation, then the squared speed of evolution is determined by

$$v^2(t) = 4\langle\dot{\psi}(t)|\dot{\psi}(t)\rangle,$$

where the factor of 4 here is conventional in fixing the scale of the metric, and is related to the fact that the Bloch sphere of spin-$\frac{1}{2}$ systems has radius $\frac{1}{2}$. Alternatively, if we work with the solutions to the conventional Schrödinger equation $|\dot{\psi}(t)\rangle = -\mathrm{i}\hbar^{-1}\hat{H}|\psi(t)\rangle$, then we can define the proper 'velocity' vector by

$$|v(t)\rangle = |\dot{\psi}(t)\rangle - \langle\psi(t)|\dot{\psi}(t)\rangle\,|\psi(t)\rangle,$$

and calculate $v^2(t) = 4\langle v(t)|v(t)\rangle$. Either way, we deduce as a result the Anandan-Aharanov relation:

$$v^2(t) = \frac{4\Delta H^2}{\hbar^2},$$

which asserts that the speed of the evolution is determined by the energy uncertainty in the state. But because energy uncertainty is constant of motion under unitarity, we see that the speed $v(t)$ only depends on the initial state, and is independent of $t$.

For mixed states the situation is a little more complicated. The reason is because the notion of speed, which is distance divided by time, depends on the choices of distance and time. As for time, this is measured by the parameter $t$ in the evolution equation, but we are still left with the choice of distance. For pure states we are able to circumvent the discussion because there is only one unitary invariance notion of distance, given by the Fubini-Study metric on the state space [30]. For mixed states, there are choices in the space to represent the density matrix (for example, a Hilbert space or a Euclidean space), and these different choices result in different distance measures.

To make a direct comparison with the speed for pure states evolution in Hilbert space, we consider here embedding of density matrices in a Hilbert space. This is because while the density matrix itself has trace unity, in infinite dimensions its square need not have finite trace, if it has continuous spectrum. So instead of working with $\hat{\rho}$ we consider a Hermitian square-root of the density matrix given by

$$\hat{\rho} \to \hat{\xi} = \sqrt{\hat{\rho}}$$

so that $\hat{\rho} = \hat{\xi}^2$. The choice of $\hat{\xi}$ is not unique, but any such square-root density matrix would suffice. Then integrability of $\hat{\rho}$ makes $\hat{\xi}$ square-integrable, i.e. it belongs to a real Hilbert space endowed with the trace norm. Under a unitary time evolution, the dynamical equation satisfied by $\hat{\xi}$ is

$$\partial_t\hat{\xi} = -\mathrm{i}[\hat{H},\hat{\xi}].$$

Then the squared speed of evolution is defined by

$$v^2(t) = \mathrm{tr}(\partial_t\hat{\xi}\,\partial_t\hat{\xi}),$$

and a short calculation shows that

$$v^2(t) = 2\left[\mathrm{tr}(\hat{H}^2\hat{\rho}) - \mathrm{tr}(\hat{H}\sqrt{\hat{\rho}}\hat{H}\sqrt{\hat{\rho}})\right].$$

The right side is twice the Wigner-Yanase skew information measure [31, 32]. It is not difficult to show that

$$\Delta H^2 \geq \mathrm{tr}(\hat{H}^2\hat{\rho}) - \mathrm{tr}(\hat{H}\sqrt{\hat{\rho}}\hat{H}\sqrt{\hat{\rho}}) \geq 0,$$

where the upper bound $\Delta H^2$ is attained for a pure state, because if $\hat{\rho} = |\phi\rangle\langle\phi|$ for some $|\phi\rangle$ then $\sqrt{\hat{\rho}} = |\phi\rangle\langle\phi|$ so $\mathrm{tr}(\hat{H}\sqrt{\hat{\rho}}\hat{H}\sqrt{\hat{\rho}}) = \langle\phi|\hat{H}|\phi\rangle^2$; whereas the lower bound 0 is attained if $\hat{\rho}$ is a stationary state so that $[\hat{\rho},\hat{H}] = 0$. It follows that the speed of evolution under a unitary motion is somewhat slowed down for mixed states, as compared to their pure-state counterparts. It is worth noting that the energy variance $\Delta H^2$ is nonzero for all impure stationary states, whereas the Wigner-Yanase skew information vanishes for all stationary states. It follows that the Anandan-Aharonov relation for the speed of unitary evolution is valid only for pure states; for general states – pure or mixed – the speed is given by the Wigner-Yanase skew information, and this is related to the fact that the Fisher information associated with time estimation for mixed states is not given by the energy uncertainty [32].

## 4. Embedding mixed states in Euclidean space

In the case of an open system dynamics, to make a direct comparison to the foregoing analysis on unitary motions it would be desirable to first formulate the evolution equation satisfied by the square-root density matrix $\hat{\xi}$ and then work out the squared-speed by the formula $v^2(t) = 2\,\mathrm{tr}(\partial_t\hat{\xi}\,\partial_t\hat{\xi})$. However, the evolution equation for $\hat{\xi}$ associated with a Lindblad dynamics for $\hat{\rho} = \hat{\xi}^2$ is not known. Therefore, rather than working in a Hilbert space and useing a Hilbert-space norm to determine the speed, we shall instead regard the space of density matrices as a subspace of a Euclidean space and use the Euclidean norm to evaluate the speed [20, 33, 34]. It follows that in the unitary limit we will not recover the Wigner-Yanase skew information because a different distance measure is used. Indeed there are various different measures being considered in the literature to analyse the speed of evolution [35, 36, 37, 38, 39, 40, 41, 42].

The space of density matrices in a Hilbert space $\mathcal{H}^n$ of dimension $n$ forms a subset of the interior of a sphere $S^{n^2-2}$ in a Euclidean space $\mathbb{R}^{n^2-1}$. To see this, recall that the outer boundary of the space of density matrices consists of pure states. Then writing $|\phi\rangle = (\phi^1,\phi^2,\ldots,\phi^n)$ for a normalised pure state vector we can define coordinates of pure states in $\mathbb{R}^{n^2}$ by $(x^k, x^{hk}, y^{hk})$, where $h,k = 1,2,\ldots,n$, $k \neq h$, if we set

$$x^k = \sqrt{2}\,\phi^k\bar{\phi}^k, \quad x^{hk} = \phi^h\bar{\phi}^k + \phi^k\bar{\phi}^h, \quad y^{hk} = \mathrm{i}(\phi^h\bar{\phi}^k - \phi^k\bar{\phi}^h).$$

Then the trace condition on $|\phi\rangle\langle\phi|$ implies that we have a linear constraint $x^1+x^2+\cdots+x^n = \sqrt{2}$ corresponding to a hyperplane $\mathbb{R}^{n^2-1}$, and hence the pure states lie on a sphere $S^{n^2-2}$, forming the boundary for general mixed states [43]. Accordingly, every density matrix $\hat{\rho}$ can be thought of as being represented by a vector $\boldsymbol{r} \in \mathbb{R}^{n^2-1}$.

We let $\{\hat{\sigma}_j\}_{j=0,\ldots,n^2-1}$ be an orthonormal basis for the linear space of bounded operators on $\mathcal{H}^n$ with the Hilbert-Schmidt inner product

$$\langle\hat{\sigma},\hat{\tau}\rangle = \mathrm{tr}\left(\hat{\sigma}^\dagger\hat{\tau}\right),$$

and set $\hat{\sigma}_0 = n^{-1/2}\mathbb{1}$. Hence the operators $\{\hat{\sigma}_j\}_{j=1,\ldots,n^2-1}$ are trace free, and together they satisfy the orthonormality condition

$$\langle\hat{\sigma}_i,\hat{\sigma}_j\rangle = \delta_{ij}.$$

An arbitrary density matrix $\hat{\rho}$ can then be expressed in the form

$$\hat{\rho} = \frac{1}{\sqrt{n}}\,\hat{\sigma}_0 + \sum_{j=1}^{n^2-1} r_j\,\hat{\sigma}_j,$$

where $r_j = \mathrm{tr}(\hat{\rho}\hat{\sigma}_j)$. For a pure state we have

$$\mathrm{tr}(\hat{\rho}^2) = \frac{1}{n} + \sum_{j=1}^{n^2-1} r_j^2 = 1,$$

from which it follows that the squared radius of the sphere $S^{n^2-2}$ in $\mathbb{R}^{n^2-1}$ is given by $1-n^{-1}$. For $n=2$ we then get the usual Bloch sphere with radius $\frac{1}{2}$.

Consider now a one-parameter family of density matrices $\hat{\rho}(t)$ satisfying the Lindblad equation

$$\partial_t\hat{\rho} = -\mathrm{i}[\hat{H},\hat{\rho}] + \sum_k\left[\hat{L}_k\hat{\rho}\hat{L}_k^\dagger - \frac{1}{2}\left(\hat{L}_k^\dagger\hat{L}_k\rho + \hat{\rho}\hat{L}_k^\dagger\hat{L}_k\right)\right],$$

along with an initial condition $\hat{\rho}(0)$. Substituting

$$\hat{\rho} = \frac{1}{\sqrt{n}}\,\hat{\sigma}_0 + \sum_{j=1}^{n^2-1} r_j\,\hat{\sigma}_j$$

in here, we find that the dynamical equation in $\mathbb{R}^{n^2-1}$ is given by

$$\dot{r}_i = \sum_{j=1}^{n^2-1}\Lambda_{ij}r_j + b_i,$$

where

$$\Lambda_{ij} = \mathrm{tr}\left[-\mathrm{i}\,[\hat{\sigma}_j,\hat{\sigma}_i]\hat{H} + \sum_k\hat{L}_k\hat{\sigma}_j\hat{L}_k^\dagger\hat{\sigma}_i - \tfrac{1}{2}\sum_k\left(\hat{L}_k^\dagger\hat{L}_k\hat{\sigma}_j\hat{\sigma}_i + \hat{L}_k^\dagger\hat{L}_k\hat{\sigma}_i\hat{\sigma}_j\right)\right]$$

is a real matrix, and

$$b_i = \frac{1}{n}\sum_k\mathrm{tr}\left([\hat{L}_k,\hat{L}_k^\dagger]\hat{\sigma}_i\right)$$

is a real vector. Writing the dynamical equation in the form $\partial_t\hat{\rho} = \mathcal{L}\hat{\rho}$, the components $\mathcal{L}_{ij}$ of $\mathcal{L}$ in the basis $\{\hat{\sigma}_j\}$ are

$$\mathcal{L}_{ij} = \langle\hat{\sigma}_i,\mathcal{L}\hat{\sigma}_j\rangle = \mathrm{tr}\left(\hat{\sigma}_i\mathcal{L}\hat{\sigma}_j\right),$$

and we have $(\mathcal{L}\hat{\rho})_j = \sum_i \mathrm{tr}(\hat{\sigma}_j\mathcal{L}\hat{\sigma}_i)\,\mathrm{tr}(\hat{\sigma}_i\hat{\rho})$. Therefore, writing

$$\dot{r}_j = \sum_{i=1}^{n^2-1} \mathcal{L}_{ji} r_i + \frac{1}{\sqrt{n}} \mathcal{L}_{j0} \, ,$$

we deduce that $\mathcal{L}_{ji} = \Lambda_{ji}$ for $i, j \neq 0$ and $\mathcal{L}_{j0} = \sqrt{n}\, b_j$. Because the real matrix $\mathcal{L}_{ji}$ is in general not symmetric, its eigenvalues are either real or else come in complex conjugate pairs. If the Lindblad operators are normal, then we have $b_j = 0$ and the eigenvalues of the Liouville operator are thus determined by the matrix $\Lambda_{ji}$.

## 5. Evolution speed for mixed states

In terms of the Euclidean norm, the squared speed of evolution of the state in $\boldsymbol{r} \in \mathbb{R}^{n^2-1}$ is given by

$$v^2(t) = \sum_{j=1}^{n^2-1} \dot{r}_j^2 = \mathrm{tr}\left[(\mathcal{L}\hat{\rho})^2\right] .$$

Writing

$$\mathcal{L}\hat{\rho} = -\mathrm{i}[\hat{H}, \hat{\rho}] + \mathcal{D}\hat{\rho}$$

for the Lindblad equation, where

$$\mathcal{D}\hat{\rho} = \sum_k \left[ \hat{L}_k \hat{\rho} \hat{L}_k^\dagger - \frac{1}{2} \left( \hat{L}_k^\dagger \hat{L}_k \rho + \rho \hat{L}_k^\dagger \hat{L}_k \right) \right] ,$$

we obtain:

$$v^2(t) = 2\left[ \mathrm{tr}\left(\hat{H}^2\hat{\rho}^2\right) - \mathrm{tr}\left(\hat{H}\hat{\rho}\hat{H}\hat{\rho}\right)\right] - 2\mathrm{i}\,\mathrm{tr}\left(\hat{\rho}\,[\mathcal{D}\hat{\rho}, \hat{H}]\right) + \mathrm{tr}\left[(\mathcal{D}\hat{\rho})^2\right] .$$

We see that there are three terms contributing to the speed of evolution. The contribution to $v^2$ from the unitary evolution resembles, but is different from, the Wigner-Yanase skew information $I = \mathrm{tr}(\hat{H}^2\hat{\rho}) - \mathrm{tr}(\hat{H}\sqrt{\hat{\rho}}\hat{H}\sqrt{\hat{\rho}})$. Let us call

$$S(X) = \mathrm{tr}(\hat{X}^\dagger\hat{X}\hat{\rho}^2) - \mathrm{tr}(\hat{X}\hat{\rho}\hat{X}^\dagger\hat{\rho})$$

the 'modified skew information' for $\hat{X}$, which reduces to $\Delta X^2 = \langle \hat{X}^\dagger\hat{X}\rangle - \langle \hat{X}^\dagger\rangle\langle\hat{X}\rangle$ for a pure state [20]. In particular, if $\hat{X}$ is Hermitian then $S(X)$ is just its variance. Thus for pure states we recover the Anandan-Aharonov relation. This follows on account of the fact that the metric on the space of pure states induced by the ambient Euclidean metric on $\mathbb{R}^{n^2-1}$ is the Fubini-Study metric [43].

In contrast to unitary time evolution, in an open system the velocity will in general obtain a 'radial' component so that the purity $\mathrm{tr}(\hat{\rho}^2)$ changes. The measure of purity-change rate is then given by the squared magnitude of the radial velocity

$$v_R^2(t) = \frac{(\boldsymbol{r}\cdot\dot{\boldsymbol{r}})^2}{\boldsymbol{r}\cdot\boldsymbol{r}} = \frac{[\mathrm{tr}(\hat{\rho}\mathcal{L}\hat{\rho})]^2}{\mathrm{tr}[(\hat{\rho} - n^{-1}\mathbb{1})^2]},$$

which vanishes for unitary dynamics. A short calculation shows remarkably that

$$v_R(t) = \sum_k \frac{S(L_k)}{\sqrt{\mathrm{tr}\left[(\hat{\rho} - n^{-1}\mathbb{1})^2\right]}}.$$

In other words, the speed of the change of the purity is determined by the modified skew information associated with the Lindblad operators. Note that if for some $t$ we have $\hat{\rho}(t) = n^{-1}\mathbb{1}$, the state of total ignorance, the denominator for $v_R(t)$ vanishes but in this state we also have $S(L) = 0$ for all $\hat{L}$ so that $v_R(t)$ remains finite.

Let us examine the behaviour of the evolution speed via illustrative examples of $2 \times 2$ models [20]. (i) For the first example we take the Hamiltonian to be $\hat{H} = \frac{1}{2}g\hat{\sigma}_z$ and the Lindblad operator to be $\hat{L} = \sqrt{\gamma}\hat{\sigma}_z$. Because the Hamiltonian and the Lindblad operator commute here, we get a pure decoherence dynamics in which the off-diagonal elements of the density matrix decay exponentially, while the diagonal elements remain constants of motion. In this example it is easy to deduce that

$$v^2(t) = \mathrm{e}^{-4\gamma t}(4\gamma^2 + g^2)[r_x^2(0) + r_y^2(0)],$$

and hence that $v(t)$ decreases exponentially in time. The motion of the state under the dynamics can easily be visualised. Suppose that the initial state is a pure state on the surface of the Bloch sphere in $\mathbb{R}^3$. Then under the Lindblad dynamics, the state spirals inwards in the $x - y$ plane while keeping the value of $z = \mathrm{tr}(\hat{\sigma}_z\hat{\rho})$ constant, and is eventually absorbed to the $z$-axis.

(ii) In the next example we let the Hamiltonian be $\hat{H} = \frac{1}{2}g\hat{\sigma}_x$, and we choose the Lindblad operator to be $\hat{L} = \sqrt{\gamma}\hat{\sigma}_z$ so that it does not commute with $\hat{H}$. One can think of this example as the simplest nontrivial model of a PT-symmetric open quantum system. It is important to keep in mind, however, that the notion of PT-symmetry in quantum mechanics for open systems is distinct from that in classical systems. Suppose that one starts with a pure initial state. Then the effect of the Lindblad operator $\hat{L} = \sqrt{\gamma}\hat{\sigma}_z$ on the space of pure states is to turn the state into an eigenstate of $\hat{\sigma}_z$. That is, an initial pure state $|\psi\rangle$ turns into either $|\uparrow_z\rangle$ or $|\downarrow_z\rangle$, in a manner similar to the effect of a measurement being performed on the observable $\hat{\sigma}_z$. Thinking of $\hat{\sigma}_z$ as representing 'energy' we see that sometimes the system will gain energy so that the state of the system converges to the eigenstate of $\hat{\sigma}_z$ with larger eigenvalue; sometimes it will lose energy so that the state of the system converges to the other eigenstate of $\hat{\sigma}_z$ with smaller eigenvalue—but which way it might go is random and is governed by the Born probability rule. This picture is often referred to as the stochastic unravelling of the Lindblad equation: The pure state evolves randomly, but the expectation of the random pure-state projector gives the state (density matrix) of the system that obeys a deterministic equation of the Lindblad type. The important point is that while each realisation of such a process is random, on average the energy is conserved in the sense that $\mathrm{tr}(\hat{\sigma}_z\hat{\rho})$ is constant of motion under the pure Lindblad dynamics (when $\hat{H} = 0$). Hence on average the gain and loss are balanced, although in each realisation the system either gains energy or loses energy. This is in contrast to classical situations whereby (in the unbroken phase) gain and loss are exactly balanced; not merely on average.

The addition of the unitary term generated by a Hamiltonian that does not commute with the Lindblad operator, however, breaks this balance and the system no longer has conserved (on average) observables. Nevertheless, the characteristic features of a PT-symmetric system described by the Hamiltonian $\hat{H} = g\hat{\sigma}_x - \mathrm{i}\gamma\hat{\sigma}_z$, and in particular the impact of the exceptional points $g = \pm\gamma$ for this Hamiltonian, manifests itself. To see this we note that the matrix elements $\Lambda_{ij}$ of the Liouville operator can easily be worked out to give:

$$\Lambda = \begin{pmatrix} -2\gamma & 0 & 0 \\ 0 & -2\gamma & -g \\ 0 & g & 0 \end{pmatrix}$$

along with $b_i = 0$. Hence the four eigenvalues of the Liouville operator are $0$, $-2\gamma$, and $-\gamma \pm \sqrt{\gamma^2 - g^2}$. It follows that in the region of unbroken PT-symmetry $g > \gamma$ the eigenvalues are

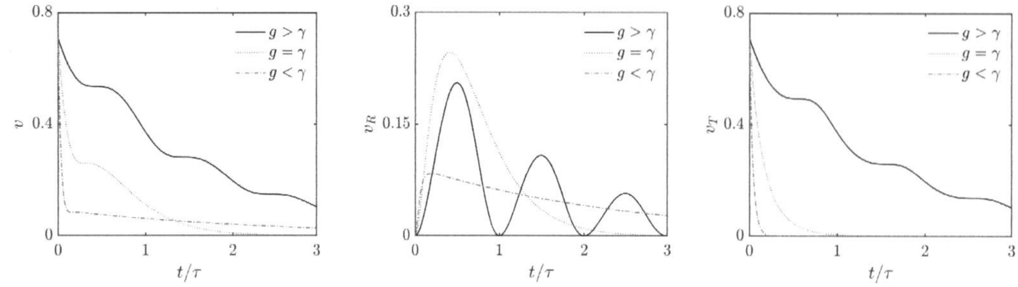


**Figure 1.** The behaviour of the evolution speed $v(t)$, its radial component $v_R(t)$, and its tangential component $v_T(t)$ are plotted here in the PT-symmetric model for which $\hat{H} = \frac{1}{2}g\hat{\sigma}_x$ and $\hat{L} = \sqrt{\gamma}\hat{\sigma}_z$. Each plot shows the speed as a function of time in three phases: the unbroken phase $g > \gamma$, the critical point $g = \gamma$, and the broken phase $g < \gamma$.

either real or come in complex conjugate pairs; at the exceptional point $g = \gamma$ the PT-symmetry gets broken; and in the broken phase $g < \gamma$ all the eigenvalues are real. At first this may appear to be the reversal of the usual situation with PT symmetry, but recall that we have defined the Liouville operator according to the equation $\partial_t\hat{\rho} = \mathcal{L}\hat{\rho}$ without the complex number.

We remark that the lack of real eigenvalues for i$\mathcal{L}$ is due to the fact that there is always loss of information in an open quantum dynamics, leading to an overall decay; but the signature of PT symmetry is seen in the remaining details of the dynamics. Thus, from the existence of PT-symmetry breaking in this model, we expect the system to exhibit different behaviours in each of these phases. Indeed, the solutions to the dynamical equations are:

$$\begin{aligned}
\mathrm{e}^{\gamma t}\, r_x(t) &= \mathrm{e}^{-\gamma t}\, r_x(0), \\
\mathrm{e}^{\gamma t}\, r_y(t) &= [\cos\omega t - (\gamma/\omega)\sin\omega t]\, r_y(0) - [(g/\omega)\sin\omega t]\, r_z(0), \\
\mathrm{e}^{\gamma t}\, r_z(t) &= [(g/\omega)\sin\omega t]\, r_y(0) + [\cos\omega t + (\gamma/\omega)\sin\omega t]\, r_z(0),
\end{aligned}$$

with $\omega = \sqrt{g^2 - \gamma^2}$. We see therefore that the solutions are oscillatory in the unbroken phase $g > \gamma$, whereas in the broken phase $g < \gamma$ the oscillations are completely suppressed. Notice that by writing the solutions in the form $\mathrm{e}^{\gamma t}\, r_\alpha(t)$, $\alpha = x, y, z$, we remove the background damping effect so that the effect of PT symmetry becomes transparent.

The squared speed of evolution $v^2(t)$, its radial component $v_R^2(t)$, and its tangential component $v_T^2(t) = v^2(t) - v_R^2(t)$, are shown below for a system prepared initially in the spin-$z$ up state $|\psi(0)\rangle = |{\uparrow}\rangle$. Because the initial state is chosen to be an eigenstate of $\hat{L}$, we have $v_R(t) = 0$ at $t = 0$. In the unbroken phase, the speed exhibits a decay superimposed with oscillations. Here $v_R(t)$ oscillates periodically with the period $\tau = \pi/\sqrt{g^2 - \gamma^2}$, where the minima correspond to the times at which the Bloch vector is aligned with the $z$-axis, i.e. when $\mathcal{D}\hat{\rho} \propto \hat{\rho}$. Moving into the broken phase, the speed decays rapidly at short times and the oscillation in $v_T(t)$ is completely damped out. However, in this phase the velocity remains nonzero for a longer duration.

**Acknowledgements.** The author acknowledges support from the Russian Science Foundation, grant 20-11-20226, and is grateful to B. Longstaff for discussion.

## References

[1] Bender C M and Boettcher S 1998 Real spectra in non-Hermitian Hamiltonians having $\mathcal{PT}$-symmetry *Phys Rev Lett* **80** 5243–5246

[2] Pontryagin L S 1944 Hermitian operators in spaces with indefinite metrics *Bull Acad Sci URSS Ser Math [Izvestiya Akad Nauk SSSR]* **8** 243–280

[3] Mostafazadeh A 2002 Pseudo-Hermiticity versus PT-symmetry III *J Math Phys* **43** 3944–3951
[4] Bender C M, Brody D C and Jones H F 2002 Complex extension of quantum mechanics *Phys Rev Lett* **89** 27040
[5] Mackey G W 1959 *Commutative Banach Algebras* (Rio De Janeiro: Instituto de Matematica pura e Aplicada do Conselho Nacional de Pesquisa)
[6] Scholtz F G, Geyer H B and Hahne F J W 1992 Quasi-Hermitian operators in quantum mechanics and the variational principle *Ann Phys* **213** 74–101
[7] Mostafazadeh, A. 2003 Exact PT-symmetry Is equivalent to Hermiticity *J Phys* A**36** 7081–7091
[8] Brody D C 2014 Biorthogonal quantum mechanics *J Phys* A**47** 035305
[9] Brody D C 2016 Consistency of PT-symmetric quantum mechanics *J Phys* A**49**, 10LT03
[10] Ruschhaupt A, Delgado F and Muga J G 2005 Physical realization of PT-symmetric potential scattering in a planar slab waveguide *J Phys* A**38** L171–L176
[11] Makris K G, El-Ganainy R, Christodoulides D N and Musslimani Z H 2008 Beam dynamics in PT-symmetric optical lattices *Phys Rev Lett* **100** 103904
[12] Klaiman S, Günther U and Moiseyev N 2008 Visualization of branch points in PT-symmetric waveguides *Phys Rev Lett* **101** 080402
[13] Kato T 1966 *Perturbation Theory of Linear Operators* (Berlin: Springer)
[14] Berry M V 2004 Physics of nonhermitian degeneracies *Czech J Phys* **54** 1039–1047
[15] Heiss W D 2012 The physics of exceptional points *J Phys* A**45** 444016
[16] Brody D C and Graefe E M 2013 Information geometry of complex Hamiltonians and exceptional points *Entropy* **15** 3361–3378
[17] Miri M. A. and Alu A 2019 Exceptional points in optics and photonics *Science* **363** eaar7709
[18] Hatano N 2019 Exceptional points of the Lindblad operator of a two-level system *Mol Phys* **117** 2121–2127
[19] Minganti F, Miranowicz A, Chhajlany R W, and Nori F 2019 Quantum exceptional points of non-Hermitian Hamiltonians and Liouvillians: The effects of quantum jumps *Phys Rev* A**100** 062131
[20] Brody D C and Longstaff B 2019 Evolution speed of open quantum dynamics *Phys Rev Res* **1** 033127
[21] Gorini V, Kossakowski A and Sudarshan E C G 1976 Completely positive dynamical symmetry groups of $N$-level systems *J Math Phys* **17** 821–825
[22] Lindblad G 1976 On the generators of quantum dynamical semigroups. *Commun Math Phys* **48** 119–130
[23] Hawking S W 1982 The unpredictability of quantum gravity *Commun Math Phys* **87** 395–415
[24] Banks T, Susskind L and Peskin M 1984 Difficulties for the evolution of pure states into mixed states *Nucl Phys* B**244** 125–134
[25] Dast D, Haag D, Cartarius H and Wunner G 2014 Quantum master equation with balanced gain and loss *Phys Rev* A**90** 052120
[26] Dast D, Haag D, Cartarius H, Main J and Wunner G 2016 Bose Einstein condensates with balanced gain and loss beyond mean-field theory *Phys Rev* A**94** 053601
[27] Graefe E M, Korsch H J and Niederle A E 2008 Mean-field dynamics of a non-Hermitian Bose-Hubbard dimer *Phys Rev Lett* **101** 150408
[28] Brody D C and Hughston L P 1996 Geometry of quantum statistical inference *Phys Rev Lett* **77** 2851–2854
[29] Anandan J and Aharonov Y 1990 Geometry of quantum evolution *Phys Rev Lett* **65** 1697–1700
[30] Brody D C and Hughston L P 2001 Geometric quantum mechanics *J Geo Phys* **38** 19–53
[31] Luo S 2003 Wigner-Yanase skew information and uncertainty relations *Phys Rev Lett* **91** 180403
[32] Brody D C 2011 Information geometry of density matrices and state estimation. *J Phy* A**44** 252002
[33] Campaioli F, Pollock F A, Binder F C and Modi K 2018 Tightening quantum speed limits for almost all states *Phys Rev Lett* **120** 060409
[34] Campaioli F, Pollock F A and Modi K 2019 Tight, robust, and feasible quantum speed limits for open dynamics *Quantum* **3** 168
[35] Mondal D and Pati A K 2016 Quantum speed limit for mixed states using an experimentally realizable metric *Phys Lett* A**380** 1395–1400
[36] Deffner S and Campbell S 2017 Quantum speed limits: from Heisenberg's uncertainty principle to optimal quantum control *J Phys* A**50** 453001
[37] Alipour S, Mehboudi M and Rezakhani A T 2014 Quantum metrology in open systems: Dissipative Cramér-Rao bound *Phy Rev Lett* **112** 120405
[38] Funo K, Shiraishi N and Saito K 2019 Speed limit for open quantum systems *New J Phys* **21** 013006
[39] Uzdin R and Kosloff R 2016 Speed limits in Liouville space for open quantum systems *Europhys Lett* **115** 40003
[40] Deffner S and Lutz E 2013 Quantum speed limit for non-Markovian dynamics *Phys Rev Lett* **111** 010402
[41] Taddei M M, Escher B M, Davidovich L and de Matos Filho R L 2013 Quantum speed limit for physical processes *Phys Rev Lett* **110** 050402

[42] del Campo A, Egusquiza I L, Plenio M B and Huelga S F 2013 Quantum speed limit in open system dynamics *Phys Rev Lett* **110** 050403
[43] Brody D C 2013 Geometry of the complex extension of Wigner's theorem *J Phy* A**46** 395301

# On some mathematical models tracking epidemics spreading

**Jose M Cerveró**

Departamento de Física Fundamental. Universidad de Salamanca, 37008 Salamanca, SPAIN

E-mail: `cervero@usal.es`

**Abstract.** A complete integrability of one SEIRD-like dynamical system is presented. Many models like this have been used nowadays in epidemiology and several other descriptions of virological spreading. In this paper we show that one of them is exactly solvable. Only one almost trivial condition for integrability is needed. The statistical perspective is not considered. Our solution is an exact one and the result hereby presented cast some doubts on the interest in this class of deterministic models. One entirely new avenue for tackling the problem of spreading diseases is then proposed. Curiously enough is surprisingly related to **Quantum Mechanics in its non-hermitic version also called PT-Quantum Mechanics**.

## 1. The SEIRD-1 system

There is a growing interest nowadays in describing the spread of a pandemia, for obvious reasons. Many models of dynamical systems have been presented but until now, no one of them seems to account for a very rapidly growing expansion of obvious examples of these terrible scourges of the human kind. One of this model has been used frequently: **We shall call SEIRD-1 Dynamical System** (See SIR-Dynamical Systems and generalizations in the web, and also [1] (specially equation (1) on it), and [2]). Many more have appeared recently ([3], [4] and [5]). Also interest on modelling dynamical systems to account for the present pandemic has fostered an immense body of research as it is easy to understand [6]. In the dynamical system we would like to concentrate, we can easily identify the **S** with the **Susceptible population**, **E** with the **Infected Asymptomatic population**, **I** with the **Infected Symptomatic population**, **R** with the **Recovered population** and finally **D** with the **Dead population**. The sum of these quantities should be a constant integer number $\mathcal{N}$: **The total population.**

$$\mathcal{N} = \mathbf{S}(\xi) + \mathbf{E}(\xi) + \mathbf{I}(\xi) + \mathbf{R}(\xi) + \mathbf{D}(\xi) \tag{1}$$

The system under scrutiny has the form:

$$\frac{d\mathbf{S}(\xi)}{d\xi} = -r_1\mathbf{S}(\xi)\mathbf{E}(\xi) - r_2\mathbf{S}(\xi)\mathbf{I}(\xi) \tag{2}$$

$$\frac{d\mathbf{E}(\xi)}{d\xi} = r_1\mathbf{S}(\xi)\mathbf{E}(\xi) + r_2\mathbf{S}(\xi)\mathbf{I}(\xi) - (a_1 + c_1)\mathbf{E}(\xi) \tag{3}$$

$$\frac{d\mathbf{I}(\xi)}{d\xi} = c_1\mathbf{E}(\xi) - (a_2 + c_2)\mathbf{I}(\xi) \tag{4}$$

$$\frac{d\mathbf{R}(\xi)}{d\xi} = a_1\mathbf{E}(\xi) + a_2\mathbf{I}(\xi) \tag{5}$$

$$\frac{d\mathbf{D}(\xi)}{d\xi} = c_2\mathbf{I}(\xi) \tag{6}$$

It is important to remark the fact that this system is consistent with the conservation equation (1) At any rate many virologist [2] have made use of these approaches to analyze their data. Let us now proceed to manipulate these equations. From (6) it is trivial to see that:

$$\mathbf{I}(\xi) = \frac{1}{c_2}\frac{d\mathbf{D}(\xi)}{d\xi} \tag{7}$$

From (4) for the **Infected population** easily follows that:

$$\mathbf{E}(\xi) = \frac{1}{c_1c_2}\left[\frac{d^2\mathbf{D}(\xi)}{d\xi^2} + (a_2 + c_2)\frac{d\mathbf{D}(\xi)}{d\xi}\right] \tag{8}$$

Substitution of (7) and (8) in (5) yields for $\mathbf{R}(\xi)$ the equation

$$\frac{d\mathbf{R}(\xi)}{d\xi} = \frac{1}{c_1c_2}\left[a_1\frac{d^2\mathbf{D}(\xi)}{d\xi^2} + (a_1a_2 + a_1c_2 + a_2c_1)\frac{d\mathbf{D}(\xi)}{d\xi}\right] \tag{9}$$

which easily be integrated as

$$\mathbf{R}(\xi) = \frac{1}{c_1c_2}\left[a_1\frac{d\mathbf{D}(\xi)}{d\xi} + (a_1a_2 + a_1c_2 + a_2c_1)\mathbf{D}(\xi)\right] + R_0 \tag{10}$$

A similar substitution performed in (2) provides for $\mathbf{S}(\xi)$ the following differential equation:

$$\frac{d\mathbf{S}(\xi)}{d\xi} = -\mathbf{S}(\xi)\frac{1}{c_1c_2}\left[r_1\frac{d^2\mathbf{D}(\xi)}{d\xi^2} + (r_1a_2 + r_1c_2 + c_1r_2)\frac{d\mathbf{D}(\xi)}{d\xi}\right] \tag{11}$$

whose trivial integration yields for $\mathbf{S}(\xi)$ the expression:

$$\mathbf{S}(\xi) = S_0 \exp\left\{-\frac{1}{c_1c_2}\left[r_1\frac{d\mathbf{D}(\xi)}{d\xi} + (r_1a_2 + r_1c_2 + c_1r_2)\mathbf{D}(\xi)\right]\right\} \tag{12}$$

and the substitution of (7) and (8) in (2) provides:

$$\mathbf{S}(\xi)\left[r_1\frac{d^2\mathbf{D}(\xi)}{d\xi^2} + (r_1a_2 + r_1c_2 + c_1r_2)\frac{d\mathbf{D}(\xi)}{d\xi}\right] =$$

$$= \left[\frac{d^3\mathbf{D}(\xi)}{d^3\xi} + (a_1 + c_1 + a_2 + c_2)\frac{d^2\mathbf{D}(\xi)}{d\xi^2} + (a_1 + c_1)(a_2 + c_2)\frac{d\mathbf{D}(\xi)}{d\xi}\right] \tag{13}$$

Therefore, the compatibility between (12) and (13) implies

$$S_0 \exp\left\{-\frac{1}{c_1c_2}\left[r_1\frac{d\mathbf{D}(\xi)}{d\xi} + (r_1a_2 + r_1c_2 + c_1r_2)\mathbf{D}(\xi)\right]\right\} = \tag{14}$$

$$= \frac{\left[\frac{d^3\mathbf{D}(\xi)}{d^3\xi} + (a_1 + c_1 + a_2 + c_2)\frac{d^2\mathbf{D}(\xi)}{d\xi^2} + (a_1 + c_1)(a_2 + c_2)\frac{d\mathbf{D}(\xi)}{d\xi}\right]}{\left[r_1\frac{d^2\mathbf{D}(\xi)}{d\xi^2} + (r_1a_2 + r_1c_2 + c_1r_2)\frac{d\mathbf{D}(\xi)}{d\xi}\right]} \tag{15}$$

whose logarithmic derivative yields the following fourth order differential equation for just $\mathbf{D}(\xi)$:

$$\left[\frac{d^4\mathbf{D}(\xi)}{d\xi^4} - \frac{d^3\mathbf{D}(\xi)}{d\xi^3}\right]\left[\frac{d^2\mathbf{D}(\xi)}{d\xi^2} + k_3\frac{d\mathbf{D}(\xi)}{d\xi}\right] + \tag{16}$$

$$+\frac{k_0}{c_1c_2}\left[\frac{d^2\mathbf{D}(\xi)}{d\xi^2} + k_3\frac{d\mathbf{D}(\xi)}{d\xi}\right]^2\left[\frac{d^3\mathbf{D}(\xi)}{d\xi^3} + (k_1+k_2)\frac{d^2\mathbf{D}(\xi)}{d\xi^2} + k_1k_2\frac{d\mathbf{D}(\xi)}{d\xi}\right] + \tag{17}$$

$$+(k_3(k_1+k_2) - k_1k_2)\left[\frac{d\mathbf{D}(\xi)}{d\xi}\frac{d^3\mathbf{D}(\xi)}{d\xi^3} - \left[\frac{d^2\mathbf{D}(\xi)}{d\xi^2}\right]^2\right] \tag{18}$$

and the following redefinition of constant has been made:

$$r_1 = k_0 \tag{19}$$

$$a_1 + c_1 = k_1 \tag{20}$$

$$a_2 + c_2 = k_2 \tag{21}$$

$$\frac{1}{r_1}(r_1a_2 + r_1c_2 + c_1r_2) = k_3 \tag{22}$$

With this change of variables (7), (8) and (10) read:

$$\mathbf{I}(\xi) = \frac{1}{c_2}\frac{d\mathbf{D}(\xi)}{d\xi} \tag{23}$$

$$\mathbf{E}(\xi) = \frac{1}{c_1c_2}\left[\frac{d^2\mathbf{D}(\xi)}{d\xi^2} + k_2\frac{d\mathbf{D}(\xi)}{d\xi}\right] \tag{24}$$

$$\mathbf{R}(\xi) = \frac{1}{c_1c_2}\left[a_1\frac{d\mathbf{D}(\xi)}{d\xi} + (a_1c_2 + a_2c_1)\mathbf{D}(\xi)\right] + \mathcal{M} \tag{25}$$

$$\mathbf{S}(\xi) = \frac{1}{k_0}\frac{\left[\frac{d^3\mathbf{D}(\xi)}{d^3\xi} + (k_1+k_2)\frac{d^2\mathbf{D}(\xi)}{d\xi^2} + k_1k_2\frac{d\mathbf{D}(\xi)}{d\xi}\right]}{\left[\frac{d^2\mathbf{D}(\xi)}{d\xi^2} + k_3\frac{d\mathbf{D}(\xi)}{d\xi}\right]} \tag{26}$$

where $\mathcal{M}$ is a constant proportional to the total initial population. Furthermore, the conservation condition (1), can be also used to obtain another equation for $\mathbf{D}(\xi)$. The substitution of equations (26) in (2) with the help of (7) give us the following third order differential equation for $\mathbf{D}(\xi)$:

$$\frac{c_1c_2}{k_0}\left[\frac{d^3\mathbf{D}(\xi)}{d\xi^3} + (k_1+k_2)\frac{d^2\mathbf{D}(\xi)}{d\xi^2} + k_1k_2\frac{d\mathbf{D}(\xi)}{d\xi}\right] +$$
$$+\left[\frac{d^2\mathbf{D}(\xi)}{d\xi^2} + k_3\frac{d\mathbf{D}(\xi)}{d\xi}\right]\left[\frac{d^2\mathbf{D}(\xi)}{d\xi^2} + (k_1+k_2)\frac{d\mathbf{D}(\xi)}{d\xi} + k_1k_2\mathbf{D}(\xi) + \mathcal{M}c_1c_2\right] = 0 \tag{27}$$

Actually, (27) is a first integral for (18). This property can easily be proved by solving (27) for the third derivative of $\mathbf{D}(\xi)$. The substitution of this third order derivative in (18) satisfies identically this equation. Now one can easily see that (27) can also be written as:

$$\frac{c_1 c_2}{k_0}\frac{d}{d\xi}\left[\frac{d^2\mathbf{D}(\xi)}{d\xi^2}+(k_1+k_2)\frac{d\mathbf{D}(\xi)}{d\xi}+k_1k_2\mathbf{D}(\xi)+\mathcal{M}c_1c_2\right]+$$
$$+\left[\frac{d^2\mathbf{D}(\xi)}{d\xi^2}+k_3\frac{d\mathbf{D}(\xi)}{d\xi}\right]\left[\frac{d^2\mathbf{D}(\xi)}{d\xi^2}+(k_1+k_2)\frac{d\mathbf{D}(\xi)}{d\xi}+k_1k_2\mathbf{D}(\xi)+\mathcal{M}c_1c_2\right]=0 \quad (28)$$

The equation is identically satisfied if and only if:

$$\left[\frac{d^2\mathbf{D}(\xi)}{d\xi^2}+(k_1+k_2)\frac{d\mathbf{D}(\xi)}{d\xi}+k_1k_2\mathbf{D}(\xi)+\mathcal{M}c_1c_2\right]=0 \quad (29)$$

This equation is linear and easy to integrate. The general solution with two initial conditions dictated by $\mathbf{D}(0)=\mathbf{K}=\mathcal{M}c_1c_2$ and $\frac{d\mathbf{D}}{d\xi}\mid_{\xi=0}=0$:

$$\mathbf{D}(\xi)=\frac{\mathbf{K}}{k_2-k_1}\{k_2e^{-k_1\xi}-k_1e^{-k_2\xi}\}-\mathbf{K} \quad (30)$$

Now in (22) the numerator vanishes. In order to set to zero the denominator as well we insert $\mathbf{D}(\xi)$ in the expression:

$$\left[\frac{d^2\mathbf{D}(\xi)}{d\xi^2}+k_3\frac{d\mathbf{D}(\xi)}{d\xi}\right]=\frac{r_1\mathbf{K}}{k_2-k_1}(k_1k_2-k_3)\{k_1e^{-k_1\xi}-k_2e^{-k_2\xi}\} \quad (31)$$

The only way for this expression to be zero is:

$$k_1k_2=k_3 \Longrightarrow or \Longrightarrow r_1(a_1+c_1)(a_2+c_2)=r_1a_2+r_1c_2+r_2c_1 \quad (32)$$

The last relationship among all these constants has to be seen as the **Fundamental Condition for the Integrability of the SEIRD**. The relationship (32) is extremely easy to fulfill and we do not consider such a condition as a very stringent condition for integrability. Note however that the parameters of a real pandemic may or may not be arranged to fulfill equation (32) .

As we have shown the SEIRD systems are of very limited use when we try to apply them to the real world. This is the main reason why we should look for other methods (probably not of a purely deterministic nature) that may account for the spreading of epidemics. These methods may combine statistical background with a well defined mathematical foundations. We would like to present here one of these alternatives which combines statistics with a set of mathematical rules.

The main advantage of this method is that there exist as many forms of curves as one wish, given the couple of parameters $\{A, B\}$. The pair of constants can be adjusted at will. In fact we have generated dozens of curves of different shapes just by changing judiciously the pair$\{A, B\}$. But all this method and its intrinsic features will be the subject of the next section.

## 2. Another quite different approach

**Brief Introduction to PT-Quantum Mechanics** Non Hermitian Potentials have been extensively studied, scrutinized and exploited with different physical applications from the beginning of the second millenium. For a recent account see the fundamental reference [7].

Actually the present authors have applied some features of this Non Hermitian Hamiltonians to a set of various kinds of complex potentials and more precisely **the band structure of periodic potentials** [8]. The diagrams we shall discuss below hasn't been found by statistical procedures but rather by exactly solving a non Hermitian potential with the conventional Schrödinger equation ([9] and [10]). In this paper we would like to call reader's attention to one of the most striking properties on one of the Pöschl-Teller potential with complex extension: the Property of Absorption ([11] and [12])

There is no direct relationship between SEIRDS and the PT-symmetric method that will be described below. We have only tried to present an alternative method to systems of deterministic differential equations, using quantum mechanics.

The enormous amount of shapes arising from the solutions provides a manifold of curves that is given by the fact that almost any form of the different rates can be adjusted to empirical evidence by using just two positive constants. The explicit form of the non Hermitic Potential that we shall use, in non-dimensional units is:

$$V(x) = \frac{V_1}{\cosh^2 x} + iV_2 \frac{\sinh x}{\cosh^2[x]} \tag{33}$$

Let us define:

$$\mathbf{A} = \sqrt{V_1 + V_2 - \frac{1}{4}} \quad ; \quad \mathbf{B} = \sqrt{V_1 - V_2 - \frac{1}{4}} \tag{34}$$

As the values of $V_1$ and $V_2$, are always positive but not necesarily real, the constant **A** must be real and positive as well but **B** can be real positive and also pure imaginary. Let us now define other set of constants as functions of **A** and **B**:

$$\mathbf{C}_+ = \frac{1}{2}\{\cosh[2\pi\mathbf{A}] + \cosh[2\pi\mathbf{B}]\} \tag{35}$$

$$\mathbf{C}_- = \{\cosh[\pi\mathbf{A}] - \cosh[\pi\mathbf{B}]\} \tag{36}$$

$$\mathbf{C}_\otimes = \{\cosh[\pi\mathbf{A}]\cosh[\pi\mathbf{B}]\} \tag{37}$$

These constants $\mathbf{C}_+$, $\mathbf{C}_-$ and $\mathbf{C}_\otimes$ enter into the expressions for the **Absorption**, the **Transmission** and the **Reflection** which are given by [11]:

$$\mathcal{A} = \frac{\mathbf{C}_-\{\exp[2\pi x]\cosh[\pi\mathbf{B}] - \exp[-2\pi x]\cosh[\pi\mathbf{A}]\} + \mathbf{C}_-{}^2}{\cosh^2[2\pi x] + 2\mathbf{C}_\otimes\cosh[\mathbf{2\pi x}] + \mathbf{C}_+} \tag{38}$$

$$\mathcal{T} = \frac{\{\sinh[2\pi x]\}^2}{\cosh^2[2\pi x] + 2\mathbf{C}_\otimes\cosh[\mathbf{2\pi x}] + \mathbf{C}_+} \tag{39}$$

$$\mathcal{R} = \frac{\{\exp[\pi x]\cosh[\pi\mathbf{B}] + \exp[-\pi x]\cosh[\pi\mathbf{A}]\}^2}{\cosh^2[2\pi x] + 2\mathbf{C}_\otimes\cosh[2\pi x] + \mathbf{C}_+} \tag{40}$$

Notice that as required for conservation of probability:

$$\mathcal{R} + \mathcal{T} + \mathcal{A} = 1 \tag{41}$$

The above condition is the counterpart of (1) in **SEIRD-like** systems.

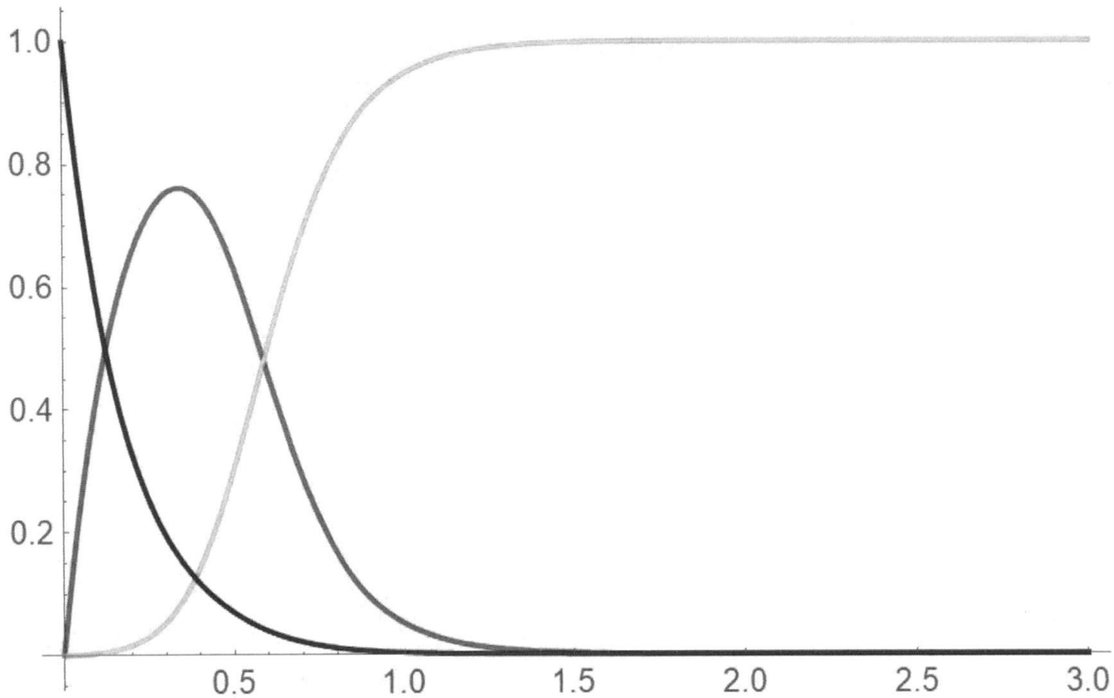


**Figure 1.** Red: **Deaths** , Blue: **Susceptible Population** , Green: **Recovered Population**; Value of the Parameters $\mathbf{A} = \mathbf{1}$ ; $\mathbf{B} = \mathbf{0.28i}$

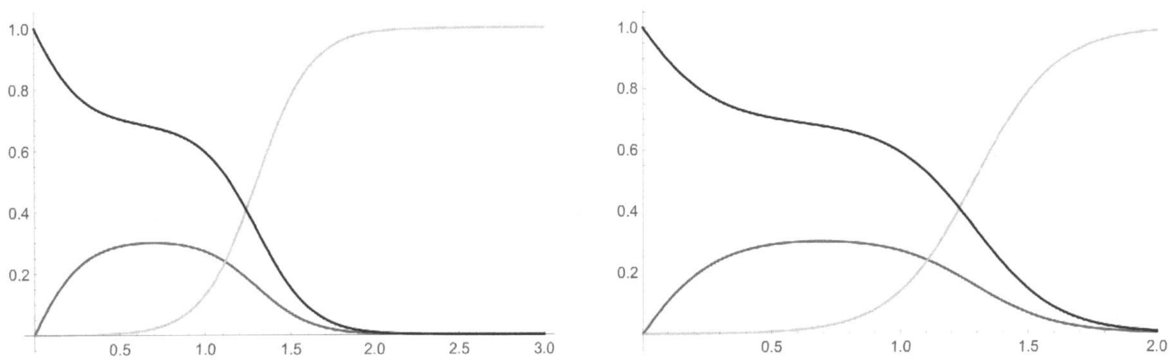


**Figure 2.** Parameters; Left: $\mathbf{A} = \mathbf{1.36}$ , $\mathbf{B} = \mathbf{1.24}$ ; Right: $\mathbf{A} = \mathbf{1.36}$ , $\mathbf{B} = \mathbf{1.18}$

**Figure 1** attach colours to kinds of potential patients. One can easily fix the average statistical data to practically any form of the curves by adequately changing the parameters. The blue curve there shows a deviation from the initial pattern in the direction of increasing the **Susceptible** people in the middle of the spreading which corresponds to non-immunology effect. In the last pair of drawings one can see a sustained number of **Deaths** over a long time due to lacking of control measures or useless medication. Actually any intermediate effect in the epidemics can be reproduced by a judicious change of only two constants.

**Acknowledgments.** This research has been supported in part by MICINN (Grant PID2019-106820RBC22) and Junta de Castilla y León (Grant SA256P18).

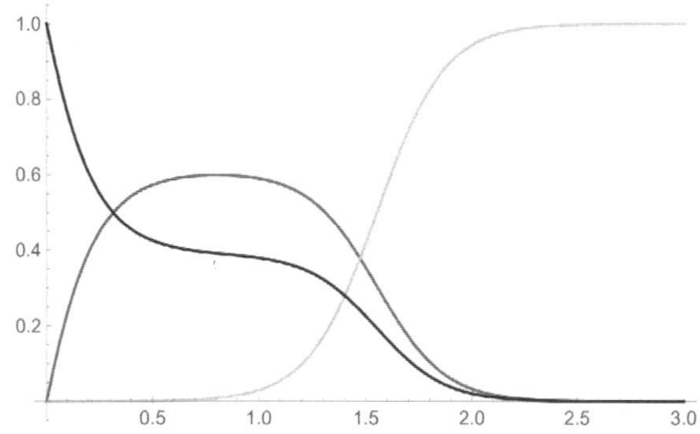


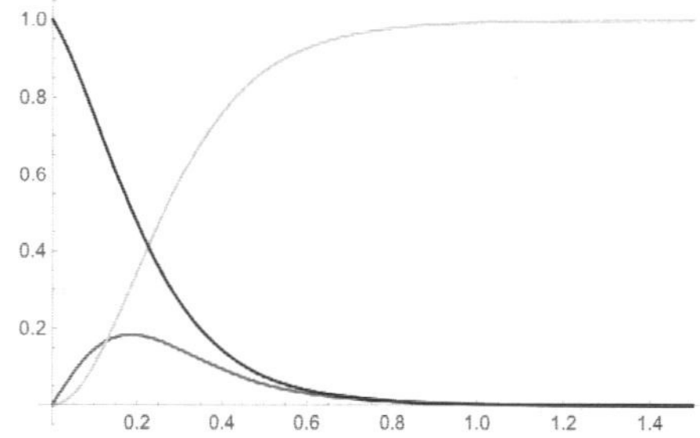


**Figure 3.** Parameters; Left: $\mathbf{A = 7.7}$ , $\mathbf{B = 1.4}$ ; Right: $\mathbf{A = 6.1}$ , $\mathbf{B = 0.27i}$

## References

[1] Karthikeyan R. et al 2020 A fractional model for the novel coronavirus (COVID- 19) outbreak *Nonlinear Dynamics* **101** 711-718
[2] Chadi M, Saad-Roy et al 2020 Dynamics in a simple evolutionary-epidemiological model for the evolution of an initial asymptomatic infection stage *Proc. Natl. Acad. Sci. USA* **117** 11541-11550
[3] Saha S, Samanta G P and Nieto J 2020 Epidemic model of COVID-19 outbreak by inducing behavioural response in population *Nonlinear Dynamics* **102** 455–487
[4] Khyar O and Allali K 2020 Global dynamics of a multi-strain SEIR epidemic model with general incidence rates: application to COVID-19 pandemic *Nonlinear Dynamics* **102** 489-509
[5] Kumar Biswas S, Kumar Ghosh J, Susmita Sarkar S and Ghosh U 2020 COVID-19 pandemic in India: a mathematical model study *Nonlinear Dynamics* **102** 537–553
[6] Freeman T 2020 COVID Research and Resources Group brings physicists together *Physics World*
[7] Bender C M 2019 *PT-Symmetry In Quantum And Classical Physics* (World Scientific Publishing Ltd)
[8] Mateo J and Cerveró J M 2003 PT-Symmetry in quantum periodic potentials *Phys. Lett.* A **317** 26-31
[9] Dabrowska J W, Khare A and Sukhatme U P 1988 Explicit wavefunctions for shape-invariant potentials by operator techniques *J. Phys.* A **21** L195-L200
[10] Khare A and Sukhatme U P 1988 Scattering amplitudes for supersymmetric shape-in variant potentials by operator methods *J. Phys.* A **21** L501-L508
[11] Cerveró J M and Rodriguez A 2004 Absorption in atomic wires *Phys. Rev.* A **70** 52705
[12] Rodriguez A and Cerveró J M 2006 One dimensional disordered wires with Poschl-Teller potentials *Phys. Rev.* B **74** 104201

Journal of Physics: Conference Series **2038** (2021) 012007 doi:10.1088/1742-6596/2038/1/012007

# Algebraic integrability of $\mathcal{PT}$-deformed Calogero models

**Francisco Correa[1] and Olaf Lechtenfeld[2]**

[1]Instituto de Ciencias Físicas y Matemáticas,
Universidad Austral de Chile, Casilla 567, Valdivia, Chile

[2]Institut für Theoretische Physik and Riemann Center for Geometry and Physics
Leibniz Universität Hannover
Appelstraße 2, 30167 Hannover, Germany

E-mail: francisco.correa@uach.cl, olaf.lechtenfeld@itp.uni-hannover.de

**Abstract.** We review some recents developments of the algebraic structures and spectral properties of non-Hermitian deformations of Calogero models. The behavior of such extensions is illustrated by the $A_2$ trigonometric and the $D_3$ angular Calogero models. Features like intertwining operators and conserved charges are discussed in terms of Dunkl operators. Hidden symmetries coming from the so-called algebraic integrability for integral values of the coupling are addressed together with a physical regularization of their action on the states by virtue of a $\mathcal{PT}$-symmetry deformation.

## 1. Introduction

Calogero models, also known as Calogero–Moser–Sutherland models, represent one of the best examples of many-particle integrable models and find applications in a wide range of areas in physics and mathematics. Introduced first by Calogero for pairwise inverse-square interactions with three and $n$ particles [1], it was then generalized to different type of potentials. The rational potential can be extended to a trigonometric, a hyperbolic [2] or an elliptic one. Moreover, all mentioned cases can be formulated for any finite Coxeter group [3, 4], enabling a large class of many-particle integrable models. There is a vast literature on this topic; for an overview in the subject and many of the applications, see for instance [3, 4, 5, 6, 7]. In recent decades, the familiy of Calogero systems has been studied under the light of non-Hermitian Hamiltonians [8, 9, 10]. Non-Hermitian extensions include a wide range of integrable systems, see for instance [11], and realizations in nature due to the application of integrable non-linear equations in optics [12]. In this work, we focus in two particular features of Calogero models: $\mathcal{PT}$-symmetric deformations and the algebraic structure related with conserved quantities and intertwining operators. Both topics are briefly reviewed here.

The first non-Hermitian extensions of Calogero models were done back in 2000 by Znojil and Tater, performing imaginary shifts on the coordinates in two- and three-particle systems [13]. In the same year Basu–Mallick and Kundu proposed an extension of the rational $A_{n-1}$ Calogero model inspired by long-range interaction with explicit momentum dependence [14]. Despite their model not being Hermitian, the energy eigenvalues are real and bounded from below. This idea was extended later in the $\mathcal{PT}$-symmetric regime to other models [15], see also [16, 17, 18]. A next step further was done by Fring in [19], studying the extensions in [14, 15, 17, 18] from a generic perspective including all Coxeter groups beyond the rational case, towards the trigonometric, hyperbolic and elliptic models. As a result, the rational non-Hermitian deformations turned out to remain integrable, but all other cases require compensating terms to keep integrability. Until now, the most elegant way to introduce non-hermiticity is enlarging the Coxeter root systems [20, 21]. Ways to generically construct complex root systems were developed in a series of papers by Fring and Smith [22, 23, 24]. Other results for Calogero models in the non-Hermitian realm include, analysis of complex domains [25], quasi exactly solvable approaches [26], complex extensions of the coupling constants [27], random matrix theories [28, 29], spectral singularities [30], isospectral and supersymmetric deformations [31, 32]. More recently, $\mathcal{PT}$-symmetric deformations of Calogero models have been used to construct invisible and reflectionless potentials by means of complex Darboux transformations related with the Korteweg–de Vries integrable hierarchy [33] playing a role in conformal and supersymmetric theories [34, 35, 36]. The quantum behavior of Calogero systems from a Hamiltonian formulation considering balanced gain and loss was studied in [37, 38], for a recent review see [39]. The idea of introducing non-Hermiticity was also studied from the point of view of spectral degeneracies, conserved quantities and intertwining operators [40, 41]. The objective of this brief review is to discuss the main results of these works and the future prospects for the topic.

Intertwining operators for Calogero models were introduced in the 1990s [42, 43] connecting the Liouville integrals at different coupling values. They play a crucial role when the couplings take integer values, allowing one to obtain Liouville eigenstates from the free theory but also to build up algebraically independent conserved quantities, on top of the Liouville integrals and beyond superintegrability. In this regime the models are known to be algebraically or analytically integrable. All those features can be adressed by means of Dunkl operators [44], see also [45, 46]. These integrals were treated as conserved quantities in formal sense since they commute with the Liouville integrals. In the rational case, they generate supersymmetric algebras [47, 48]. However, the action of the additional charges is not well defined, mapping physical states to non-physical ones. We show that this can be remedied by means of a $\mathcal{PT}$-regularization. In fact, the idea of healing the action of the additional conserved charges is not new and has been studied in one dimensional cases [49, 50] but also in regularizing degenerate soliton solutions of the Korteweg–de Vries equation [51].

This paper is organized as follows. In Section 2 we summarize the main features of the trigonometric Calogero–Sutherland model for the $A_2$ root system. Both conserved quantities and intertwining operators are presented together with their algebra and their action on the energy eigenstates. Then we introduce $\mathcal{PT}$-symmetry in a simple way in order to discuss the spectral degeneration and the physical restoration of a nonlinear conserved charge. A similar approach is given in Section 3 but for the angular Calogero model associated with the $D_3$ root system. The last section is devoted to conclusions and open problems.

## 2. $\mathcal{PT}$-symmetry in Calogero–Sutherland models

The quantum Calogero–Sutherland Hamiltonian [2] was introduced as a toy model in nuclear physics due to the type of short-range interaction in comparison with the rational version. We consider three interacting particles with coordinates $x_i \in \mathbb{R}/2\pi\mathbb{Z}$, $i = 1, 2, 3$, on a circle governed by the $A_2$ Coxeter root system and the Hamiltonian

$$H_{CS}(g) = -\frac{1}{2}\sum_{i=1}^{3}\partial_i^2 + \sum_{i<j}^{3}\frac{g(g-1)}{\sin^2(x_i-x_j)}\ , \tag{1}$$

where $g$ is a coupling parameter. As we shall see below this coupling parameter plays an important role in relation with degeneracy and conserved quantities in the non-Hermitian case. The spectral and algebraic properties of the system (1) can be studied by different methods, in the present discussion we focus on the approaches given in [52, 53, 54, 55]. In particular, we will use the Dunkl operator approach, which in this model takes the form

$$D_i(g) \;=\; \partial_i - g\sum_{j(\neq i)}\cot(x_i-x_j)\, s_{ij} \tag{2}$$

where $s_{ij}$ permutes the coordinates $x_i$ and $x_j$. This method is particularly useful for a number of reasons, including the construction of

- all conserved quantities by means of Weyl-*invariant* polynomials in the $D_i$ operators,
- intertwining operators by means of Weyl-*anti-invariant* polynomials in the $D_i$ operators,
- the energy eigenstates in terms of the Jack polynomials in an algebraic manner.

These features will be briefly revisited below. For the $A_2$ root system described by (1), the conserved quantities are constructed by means of the Newton sums

$$I_m(g) = \text{res}\,[D_1^m(g) + D_2^m(g) + D_3^m(g)]\ , \quad m = 1, 2, 3\ , \tag{3}$$

but we have only three independent integrals. The notation "res" stands for the restriction to completely symmetric functions, which removes all permutation operators. The selection the charges is not unique, any other permutation-invariant polynomial in the Dunkl operators will also provide an integral of motion. Instead, the following basis is considered,

$$C_1(g) = I_1(g)\ , \quad C_2 = I_2(g) - 8g^2 = -2\,H_{CS}(g), \quad C_3(g) = I_3(g) - I_1(g)I_2(g)\ , \tag{4}$$

which besides satisfying $[C_i(g), C_j(g)] = 0$ provide the simplest intertwining relations. The intertwining operators are constructed by the restriction of any permutation anti-symmetric polynomial. The simplest one has differential order three,

$$M(g) \;=\; \tfrac{1}{3}\,\text{res}\,\big(D_{12}(g)D_{23}(g)D_{31}(g) + D_{23}(g)D_{31}(g)D_{12}(g) + D_{31}(g)D_{12}(g)D_{23}(g)\big)\ , \tag{5}$$

where we denote $D_{ij}(g) = D_i(g) - D_j(g)$. Further higher-order intertwining operators from anti-symmetric polynomials may be constructed following the same recipe. In the current case only one intertwiner is needed for a complete algebraic description. However, as we shall see in the next section, sometimes higher-order ones are required. The explicit expressions of conserved quantities (4) and intertwiners (5) are given in [41]. With the basis (4), the intertwining relations take the standard form

$$M(g)\,C_\ell(g) \;=\; C_\ell(g{+}1)\,M(g)\ . \tag{6}$$

As a consequence of the above relation, the action of intertwining operators will not change the energy on the wavefunctions. In fact, (6) is nothing else than the shape-invariant feature studied in the context of supersymmetric quantum mechanics. The presence of shape-invariance and the construction of the spectrum have been studied in Calogero models [56, 57] but using first order interwiners of a different nature. The energy spectrum of the stationary Schrödinger equation

$$H_{CS}(g)\Psi^{(g)}_{n_1,n_2} = E_{n_1,n_2}\Psi^{(g)}_{n_1,n_2}\,, \tag{7}$$

depends quadratically on two quantum numbers, $n_1$ and $n_2$,

$$E_{n_1,n_2}(g) = (n_1 + 2g)^2 + \tfrac{1}{3}(n_1 - 2n_2)^2, \tag{8}$$

obeying the relation $n_1 \geq n_2 \geq 0$ [41, 53]. The wavefunctions

$$\Psi^{(g)}_{n_1,n_2} \;=\; e^{-\frac{2\mathrm{i}}{3}(n_1+n_2)(x_1+x_2+x_3)}\,\Delta^g P^{(g)}_{n_1,n_2}(x_1,x_2,x_3) \tag{9}$$

are given in terms of the Vandermonde determinant $\Delta = \prod_{i<j}\sin(x_i - x_j)$ and the so-called Jack polynomials $P^{(g)}_{n_1,n_2}$, which are homogeneous polynomials of degree $n_1 + n_2$ in the $x_i$ coordinates and symmetric under permutations. They can be constructed analytically in terms of deformed Dunkl operators. For more details of their construction and properties, see [53, 41] and references therein. The action of the intertwining operators on the wavefunctions reads

$$M(g)\,\Psi^{(g)}_{n_1,n_2} \;=\; n_2(n_1+g)(n_1-n_2)\Psi^{(g+1)}_{n_1-2,n_2-1}\,, \tag{10}$$

$$M^\dagger(g)\,\Psi^{(g)}_{n_1,n_2} \;=\; (n_1+3g-1)(n_1-n_2+2g-1)(n_2+2g-1)\Psi^{(g-1)}_{n_1+2,n_2+1}\,, \tag{11}$$

where $M(1-g) = M^\dagger(g)$. As the action on the intertwining operators on the wavefunctions does not change the energy values (6), the shifting on the $g$ parameter in (8) is compensated with modifications in the quantum numbers $n_1$ and $n_2$. The conserved quantities (4) action on the states take the form

$$C_1(g)\Psi^{(g)}_{n_1,n_2} = 0\,, \tag{12}$$

$$C_2(g)\Psi^{(g)}_{n_1,n_2} = 2\big[(n_1+2g)^2 + \tfrac{1}{3}(n_1-2n_2)^2\big]\Psi^{(g)}_{n_1,n_2}\,, \tag{13}$$

$$C_3(g)\Psi^{(g)}_{n_1,n_2} = -\tfrac{8}{9}\mathrm{i}\,(n_1-2n_2)(2n_1-n_2+3g)(n_1+n_2+3g)\Psi^{(g)}_{n_1,n_2}\,. \tag{14}$$

In the second relation, (8) is used into the definition of $C_2$ in coherence with the changes in (10). The degeneration due to $n_2 \mapsto n_1 - n_2$ flips the overall sign of the action of $C_3$ in (14), and therefore these two degenerate energy states can be distinguished by this integral of motion.

*Issues on the algebraic integrability and the symmetry restoration by $\mathcal{PT}$ deformations*

The idea of revisiting the non-Hermiticity in the Calogero–Sutherland model is inspired mainly by three problems.

(i) In the Hermitian case discussed above, the ground state of (1) is given by

$$\Psi^{(g)}_{0,0} = \prod_{i<j}^{3}\sin^g(x_i - x_j)\,, \quad \text{with} \quad E_{0,0}(g) = 4g^2\,. \tag{15}$$

It is clear what $\Psi^{(g)}_{0,0}$ vanishes when the coordinate values coincide, those regions correspond to the Weyl-alcove walls defining also the singularities of the interacting potential (1) given by the $A_2$ structure. Because of the power dependence, when $g < 0$ the ground state and more generically the wave-functions $\Psi^{(g)}_{n_1,n_2}$ become non-physical due to the non-normalizability resulting from such singularities.

(ii) The Calogero–Sutherland Hamiltonian (1) displays a naive but relevant symmetry changing the coupling constant parameter according to $g \leftrightarrow 1-g$. This symmetry suggests that states of two different values can be considered simultaneously within a single unique Hilbert space. However, because of the previous point, states containing negative powers of $g$ will be non-physical making this symmetry meaningless.

(iii) The cases when $g \in \mathbb{N}$ are special. In this situation the Calogero–Sutherland model is called "algebraically integrable". This feature appears by a combination of the symmetry $g \leftrightarrow 1-g$ and the intertwining operator $M(g)$. As the action of a single interwiner shifts the coupling constant by unity, for integer values of the coupling, we can step from $1-g$ to $g$ and vice versa by iteration of the process. In other words, we can build a chain of $2g-1$ consecutive interwiners in the form

$$Q(g) \;=\; M(g-1)M(g-2)\cdots M(2-g)M(1-g) \tag{16}$$

which acting on the Hamiltonian and using the invariance under $g \leftrightarrow 1-g$, gives

$$Q(g)H(g) = H(1-g)Q(g) = H(g)Q(g)\ . \tag{17}$$

In this way the operator $Q(g)$ turns out to be an extra conserved quantity. However, because of the previous arguments for $g > 1$, the action of $Q(g)$ will transform physical states into non-physical ones, and the reverse for $g < 0$.

We can tackle all these points at the same time by introducing $\mathcal{PT}$-symmetry into the system. Among the different approaches one may follow [13, 19, 20, 22, 23, 24, 41], here we will use the simplest one by shifting the coordinates by an imaginary amount,

$$x_\ell \to x_\ell + \mathrm{i}\epsilon_\ell, \quad \ell = 1, 2, 3. \tag{18}$$

In this way the consider both $\mathcal{P}$ and $\mathcal{T}$ operators in a standard way,

$$\mathcal{P}:\ (x_1, x_2, x_3) \mapsto (-x_1, -x_2, -x_3), \quad \text{and} \quad \mathcal{T}: \mathrm{i} \mapsto -\mathrm{i}\ . \tag{19}$$

In order to find a complete regularization of the system we must turn on all three parameters $\epsilon_\ell$. Figure 1 shows how the absolute value of the potential looks when the complex regularization (18) is introduced.

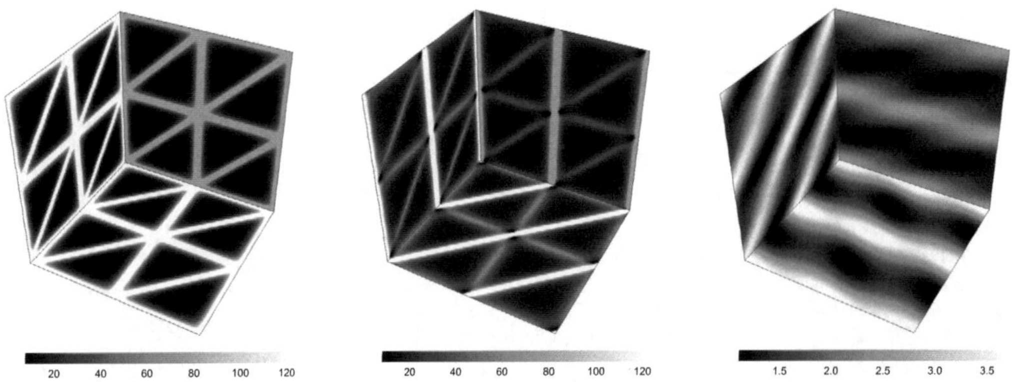


**Figure 1.** 3D sliced density plots the absolute value of the potential term in (1). (Left) The pure real potential with all $\epsilon_\ell = 0$. The boundaries of the Weyl alcoves appear as white lines. (Center) The same plot but with $\epsilon_3 = 0.2$ turned on. The potential still displays some singular regions. (Right) When all parameters are turned on $\epsilon_1 = 2, \epsilon_2 = -0.5$ and $\epsilon_3 = 0.2$, the potential is regularized. Note the change of the scale in comparison with other plots.

Once the potential does no longer display singularities we are able to use the symmetry $g \leftrightarrow 1-g$ and join the states from both sides considering $g > \frac{1}{2}$. The enhancement of energy degeneracy becomes apparent when we write the energy (8) in the weight space notation

$$E_{n_1,n_2}(g) = (\lambda_1 - 2g)^2 + \lambda_2^2 \ . \tag{20}$$

Here we identify

$$(\lambda_1, \lambda_2) = \left(-n_1, \tfrac{1}{\sqrt{3}}(n_1-2n_2)\right) , \tag{21}$$

and the condition $n_1 \geq n_2 \geq 0$ is translated into $\lambda_1 \leq -\sqrt{3}\,|\lambda_2|$. Thus, in the $\lambda$-space the set of all allowed states form a $\frac{\pi}{3}$ wedge, as can be seen in Figure 2. Considering a circle centered at $(2g, 0)$ of radius $R_g = \sqrt{E_{n_1,n_2}(g)}$, all the states lying on the circle will share the same energies. For instance, the states with $\pm\lambda_2$, i.e. $\Psi^{(g)}_{n_1,n_2}$ and $\Psi^{(g)}_{n_1,n_1-n_2}$, belong to those cases. On top of that, after $\mathcal{PT}$ regularization and because of the symmetry $g \leftrightarrow 1-g$, we can also take into account the states on the circle centered at $2(1-g, 0)$ of same radius. Albeit the cases with $g \geq 0$ are rarely degenerated, the cases when $g < 0$ display a high degeneracy, up to order 12.

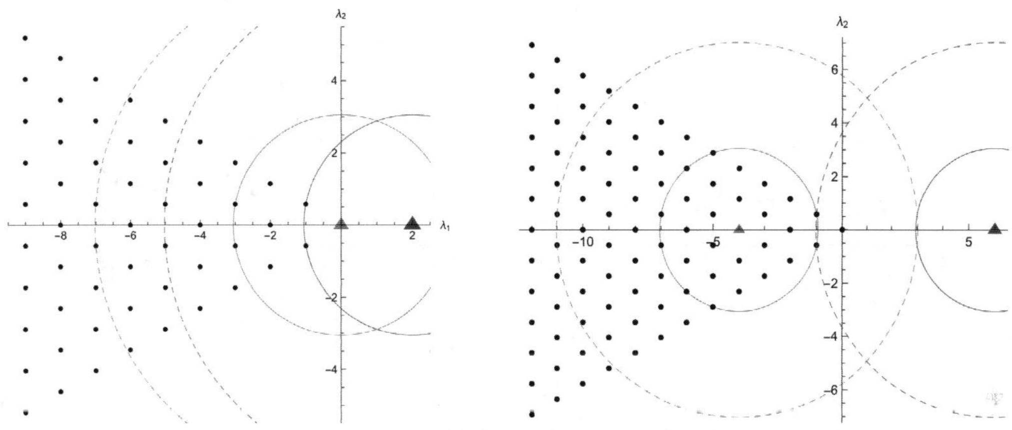


**Figure 2.** In the weight space, the states are represented as black dots and circles correspond to fixed "energy shells". (Left) Solid and dashed lines stand for energies $E = \frac{28}{3}$ and $E = \frac{148}{3}$ while blue and red colors stand for $g$=1 and $g$=0 respectively. (Right) Same energy shells as before but blue and red colors correspond to $g$=3 and $g$=−2 respectively.

As we mentioned above, the case $g \in \mathbb{Z}$ is peculiar, and the degeneracy related with $g \leftrightarrow 1-g$ is also reflected by an extra conserved charge $Q(g)$, which has always odd differential order $3(2g-1)$. If we define that states with $g>0$ and $g\leq 0$ have certain parity, for instance even and odd respectively, then the operator $Q(g)$ is of odd nature, in the sense that it maps those sets of states into each other. The explicit action on the states (9) takes the form

$$Q(g)\Psi^{(g)}_{n_1,n_2} \propto \Psi^{(1-g)}_{n_1-4g+2,n_2-2g+1} , \tag{22}$$

and we note that both states belong to the same Hilbert space due to $g \leftrightarrow 1-g$. In fact, $Q(g)$ does not change the Liouville eigenvalues because it commutes with all charges,

$$[Q(g), C_\ell(g)] = 0, \quad \ell = 1, 2, 3. \tag{23}$$

The idea of making well-defined the action of an odd type of conserved charge is not new and has been studied in one-dimensional cases in the past [49, 50]. The notion of $Q(g)$ as an odd integral, allows one to build different types of hidden supersymmetry structures without fermion degrees of freedom [58, 59]. In Calogero models we can choose as a grading operator any of the permutations $s_{ij}$ due to $\{Q(g), s_{ij}\} = 0$, and therefore $Q(g)$ may be treated as a supercharge. This notion of algebraic structures was studied in detail for the rational Hermitian Calogero model [47], see also [48]. It is natural to wonder whether the operators $Q(g)$ are completely independent of the Liouville integrals. After shifting the coupling from $g$ to $g+1$ by the action of $M(g)$, one may shift it back to $g$ by applying $M(1-g) = M^\dagger(g)$. Therefore the combination $M^\dagger(g)M(g)$ should commutes with the Hamiltonian. We can verify this by virtue of

$$M^\dagger(g)M(g) = R(g) = 18C_3^2+8C_3C_1^3-3C_2^3+3C_2^2C_1^2-C_2C_1^4+C_1^6-6g^2(3C_2-C_1^2+8g^2)^2\,, \quad (24)$$

which is nothing else than a polynomial in the conserved charges. We can elucidate the meaning of $Q(g)$ by taking its square,

$$\begin{aligned} Q^2(g) &= M(g-1)\cdots M(3-g)M(2-g)M(1-g)M(g-1)M(g-2)M(g-3)\cdots M(1-g) \\ &= M(g-1)\cdots M(3-g)M(2-g)M^\dagger(g-1)M(g-1)M(g-2)M(g-3)\cdots M(1-g) \\ &= M(g-1)\cdots M(3-g)M(2-g)R(g-1)M(g-2)M(g-3)\cdots M(1-g) \\ &= M(g-1)\cdots M(3-g)M(2-g)M(g-2)R(g-2)M(g-3)\cdots M(1-g) \\ &= M(g-1)\cdots M(3-g)\,(R(g-2))^2\,M(g-3)\cdots M(1-g) \\ &\;\vdots \\ &= M(g-1)M(1-g)\,(R(1-g))^{2g-2} = (R(1-g))^{2g-1} = (M^\dagger(g)M(g))^{2g-1}\,. \end{aligned} \quad (25)$$

So using the relation (24) we identify $Q^2(g)$ as a higher-order polynomial in the conserved charges. Thus $Q(g)$ is not a standard supercharge but a nonlinear supercharge in a wider sense, see [47]. We finish this section presenting in Fig. 3 a plot of the different degeneracies for the low values of $g$.

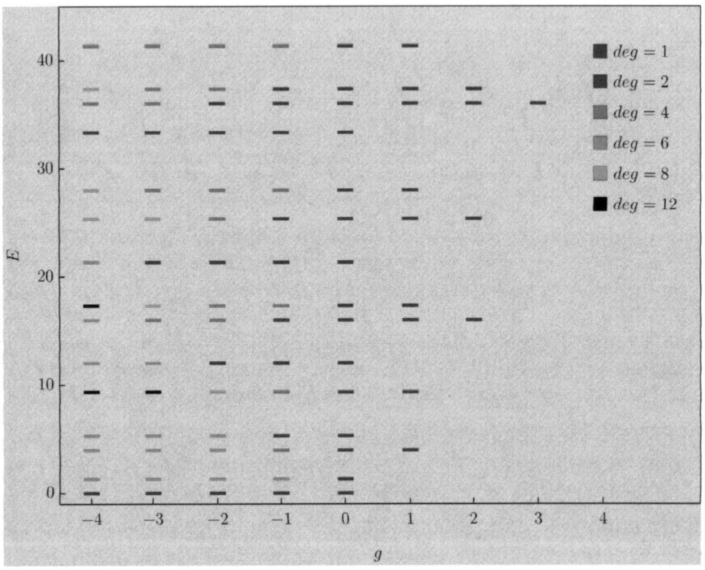


**Figure 3.** Energy spectrum and degeneracies for the $A_2$ model at integer coupling $g \in \mathbb{Z}$.

## 3. $\mathcal{PT}$-symmetry in angular Calogero models

In this section we review a different kind of many-particle system and its non-Hermitian extension. The $n$-particle Calogero model with rational interaction potential displays a conformal symmetry, which enables the superintegrability of the system [60, 61]. Alternatively, the system may be interpreted as a conformal particle living in $\mathbb{R}^n$ and being subject to an external potential. It is possible to separate the radial and the angular part of the Hamiltonian which defines an angular Calogero model living on the hypersphere $S^{n-1}$. One may naively guess that the angular system is simpler than the original one but, despite the fact the angular model is still superintegrable, the converse is true. This is why the angular Calogero models have been studied recently at the classical and quantum level [40, 62, 63, 64, 65, 66, 67], but also regarding some of their algebraic structures [68, 69, 70]. In order to illustrate the main features of these models, we focus on the $D_3$ angular version. A more detailed discussion of the following ideas is given in [40, 67]. The Hamiltonian in this case includes the angular momentum as a kinetic term and a tetrahexahedral potential,

$$H(g) = -\frac{1}{2}\sum_{i<j}^{3}(x_i\partial_j - x_j\partial_i)^2 + 2\,g(g-1)\,(x_1^2+x_2^2+x_3^2)\sum_{i<j}^{3}\frac{x_i^2+x_j^2}{(x_i^2-x_j^2)^2}\,, \tag{26}$$

and is also invariant under $g \leftrightarrow 1-g$. Figure 4 shows the potential and the tessellation of the sphere in 24 isosceles triangles defined by the Weyl chambers.

**Figure 4.** Density plots of the potential term in (26) before and after the $\mathcal{PT}$ regularization, see (45) below. (Left) In the Hermitian case, $\epsilon_1 = \epsilon_2 = 0$, the plot is scaled by a composition of a term $\log\circ\log\circ\log$. The Weyl walls are represented by the white lines. (Center) Absolute value of the potential in the non-Hermitian case with $\epsilon_1 = 2.5$ and $\epsilon_2 = 0$, where some singular lines remain present. (Right) Absolute value of the complete regularized potential for $\epsilon_1 = 2.1$ and $\epsilon_2 = 1.5$.

In the same spirit of the previous section, we focus the discussion on the Dunkl operator approach. The analogues of (2) involve the angular momenta instead of the linear ones and take the form

$$\mathcal{L}_1 = x_2\partial_3 - x_3\partial_2 + g\left(\tfrac{x_3}{x_1-x_2}s_{12} - \tfrac{x_3}{x_1+x_2}\widetilde{s}_{12} - \tfrac{x_2+x_3}{x_2-x_3}s_{23} + \tfrac{x_2-x_3}{x_2+x_3}\widetilde{s}_{23} + \tfrac{x_2}{x_3-x_1}s_{31} + \tfrac{x_2}{x_3+x_1}\widetilde{s}_{31}\right)\,, \tag{27}$$

$$\mathcal{L}_2 = x_3\partial_1 - x_1\partial_3 + g\left(\tfrac{x_1}{x_2-x_3}s_{23} - \tfrac{x_1}{x_2+x_3}\widetilde{s}_{23} - \tfrac{x_3+x_1}{x_3-x_1}s_{31} + \tfrac{x_3-x_1}{x_3+x_1}\widetilde{s}_{31} + \tfrac{x_3}{x_1-x_2}s_{12} + \tfrac{x_3}{x_1+x_2}\widetilde{s}_{12}\right)\,, \tag{28}$$

$$\mathcal{L}_3 = x_1\partial_2 - x_2\partial_1 + g\left(\tfrac{x_2}{x_3-x_1}s_{31} - \tfrac{x_2}{x_3+x_1}\widetilde{s}_{31} - \tfrac{x_1+x_2}{x_1-x_2}s_{12} + \tfrac{x_1-x_2}{x_1+x_2}\widetilde{s}_{12} + \tfrac{x_1}{x_2-x_3}s_{23} + \tfrac{x_1}{x_2+x_3}\widetilde{s}_{23}\right)\,, \tag{29}$$

where from now on we omit the explicit dependence on $g$ in the operators unless necessary. As the dynamics is governed here by the $D_3$ Coxeter group, besides the permutations $s_{ij}$ we introduce the reflections $\widetilde{s}_{ij}$ given by

$$\widetilde{s}_{12} : (x_1, x_2, x_3) \mapsto (-x_2, -x_1, +x_3) \ , \tag{30}$$
$$\widetilde{s}_{31} : (x_1, x_2, x_3) \mapsto (-x_3, +x_2, -x_1) \ , \tag{31}$$
$$\widetilde{s}_{23} : (x_1, x_2, x_3) \mapsto (+x_1, -x_3, -x_2) \ . \tag{32}$$

which together generate Weyl group $S_4$. Now, with the angular Dunkl operators we are able to construct

- all conserved quantities by means of Weyl-*invariant* polynomials in the $\mathcal{L}_i$ operators.
- intertwining operators by means of Weyl-*anti-invariant* polynomials in the $\mathcal{L}_i$ operators.
- the energy eigenstates in terms of harmonic polynomials.

One possible choice to build up the conserved charges is

$$J_k = \ \mathrm{res}\big(\mathcal{L}_1^k + \mathcal{L}_2^k + \mathcal{L}_3^k\big) \qquad \text{for} \quad k = 2, 4, 6 \ , \tag{33}$$

where $J_2 = -2\,H(g) + 6g(6g{+}1)$ is the shifted Hamiltonian. The higher-order integrals commute with the Hamiltonian, $[J_2, J_\ell] = 0$ for $\ell = 4, 6$, but, in contrast with the previous case $[J_4, J_6]$ is different from zero so it is not a Liouville system. For the sake of simplicity, we are using Cartesian coordinates to describe the Dunkl operators and the wavefunctions. Nevertheless, the Hamiltonian (26) is two-dimensional and can be expressed completely in terms of a polar and an azimuthal angle [40, 67]. As we have three $(2 \times 2{-}1)$ integrals of motion, the system is superintegrable. The conserved quantities $J_4$ and $J_6$ have differential order greater than two, hence the two-dimensional system is not separable [71]. The peculiarity of the angular model is revealed by the specific form of the $J_k$ algebra. The non-vanishing commutator reads

$$[J_6, J_4] = \ 12 M_3^\dagger M_6 + 24(3{+}4g) J_6 J_2 - 12(3{+}2g) J_4^2 - 48(1{+}2g) J_4 J_2^2 + 12(1{+}2g) J_2^4 \tag{34}$$
$$+ \text{ lower-order terms} \ . \tag{35}$$

It cannot be expressed only in terms of the $J_k$ basis integrals and depends explicitly on *two* interwining operators, defined next. In the trigonometric case, only one intertwining operator was required to completely describe the algebraic structure. Here we need two intertwiners of differential order three and six respectively,

$$M_3 = \tfrac{1}{6}\mathrm{res}\big(\mathcal{L}_1\mathcal{L}_2\mathcal{L}_3 + \mathcal{L}_1\mathcal{L}_3\mathcal{L}_2 + \mathcal{L}_2\mathcal{L}_3\mathcal{L}_1 + \mathcal{L}_2\mathcal{L}_1\mathcal{L}_3 + \mathcal{L}_3\mathcal{L}_1\mathcal{L}_2 + \mathcal{L}_3\mathcal{L}_2\mathcal{L}_1\big) \ , \tag{36}$$
$$M_6 = \mathrm{res}\big(\{\mathcal{L}_1^4, \mathcal{L}_2^2\} - \{\mathcal{L}_2^4, \mathcal{L}_1^2\} + \{\mathcal{L}_2^4, \mathcal{L}_3^2\} - \{\mathcal{L}_3^4, \mathcal{L}_2^2\} + \{\mathcal{L}_3^4, \mathcal{L}_1^2\} - \{\mathcal{L}_1^4, \mathcal{L}_3^2\}\big) \ . \tag{37}$$

They intertwine the Hamiltonian in the standard way,

$$M_s(g) H(g) = H(g{+}1) M_s(g) \ , \tag{38}$$

but the generic intertwining relations for the two charges take a more complicated form in comparison to (6),

$$M_s(g) J_\ell(g) \ = \ \sum_{s',\ell'} \gamma_{s\ell}^{s'\ell'}(g)\, J_{\ell'}(g{+}1)\, M_{s'}(g) \ , \tag{39}$$

where in the sum of the right-hand side could appear more than one interwiner. The functions $\gamma_{s\ell}^{s'\ell'}(g)$ are polynomials in $g$, see [67]. We briefly review the energy spectrum for the angular model (26),

$$H(g)\ \Psi_{\ell_3,\ell_4}^{(g)} = E_\ell\ \Psi_{\ell_3,\ell_4}^{(g)} \ . \tag{40}$$

The energy depends on a combination of two quantum numbers $\ell_3$ and $\ell_4$,

$$E_\ell = \tfrac{1}{2}q\,(q{+}1) \qquad \text{and} \qquad q \;=\; 6g+\ell \;=\; 6g+3\ell_3{+}4\ell_4\;. \tag{41}$$

The allowed values $\ell_3, \ell_4 = 0, 1, 2...$ lead to degeneracies firstly for $\ell$ and secondly due to the fact $E_\ell$ is quadratic in $\ell$. The energy eigenfunctions can be written as

$$\Psi^{(g)}_{\ell_3,\ell_4} = (x_1{+}x_2{+}x_3)^{-q/2}\Delta^g h^{(g)}_{\ell_3,\ell_4}(x)\;, \tag{42}$$

where the Vandermonde determinant takes the form $\Delta = \prod_{i<j}(x_i^2{-}x_j^2)$. The $h^{(g)}_{\ell_3,\ell_4}(x)$ are homogenous polynomials of degree $\ell = 3\ell_3 + 4\ell_4$ in the $x_i$ coordinates. They can be constructed in terms of Dunkl operators and are invariant under the action of the $S_4$ group. For more details of their construction and specific examples, see [40, 67]. So far, there are no closed formulas for the action of the conserved quantities $J_4$, $J_6$ or the intertwiners $M_3$ and $M_6$. Still, the latter act on the wavefunctions (42) according to

$$M_s(g)\Psi^{(g)}_{\ell_3,\ell_4} \propto \sum_{\ell'=\ell-6} \mu^{s,\ell_3,\ell_4}_{\ell'_3,\ell'_4}(g)\Psi^{(g+1)}_{\ell'_3,\ell'_4}\;, \qquad s=3,6 \tag{43}$$

where the $\mu$'s are some polynomials in $g$. As the wavefunctions (42) contain $\Delta^g$, for $g<0$ we find singularities at the vanishing locus of the Vandermonde determinant, forcing $g \geq 0$ for a physical spectrum. The degeneracy for the allowed energy levels in this case can be computed exactly and reads

$$\deg(E_\ell) \;=\; \left\lfloor\frac{\ell}{12}\right\rfloor + \begin{cases} 0 & \text{for} \quad \ell = 1,2,5 \mod 12 \\ 1 & \text{for} \quad \ell = \text{else} \mod 12 \end{cases}. \tag{44}$$

*$\mathcal{PT}$-symmetry regularization, once again.*

The set of ideas coming from points (i), (ii) and (iii) in Section 2 also applies to the angular model (26). It is possible to remove all singularities by a $\mathcal{PT}$-symmetric deformation. This is achieved by introducing spherical coordinates as follows,

$$\begin{pmatrix} x_1 \\ x_2 \\ x_3 \end{pmatrix} = r \begin{pmatrix} \sin(\theta{+}\mathrm{i}\epsilon_1)\cos(\phi{+}\mathrm{i}\epsilon_2) \\ \sin(\theta{+}\mathrm{i}\epsilon_1)\sin(\phi{+}\mathrm{i}\epsilon_2) \\ \cos(\theta{+}\mathrm{i}\epsilon_1) \end{pmatrix}. \tag{45}$$

The $\mathcal{PT}$-operator can be chosen as $\mathcal{P}:\ (\theta,\phi) \mapsto (-\theta,-\phi)$, which means

$$\mathcal{P}:\ (x_1,x_2,x_3) \mapsto (-x_1,x_2,x_3), \quad \text{and} \quad \mathcal{T}: \mathrm{i} \mapsto -\mathrm{i}\;. \tag{46}$$

The Hamiltonian (26) clearly is invariant under the combined action. In order to remove both potential and wave-function singularities both parameters $\epsilon_1$ and $\epsilon_2$ must be turned on, see Figure 4. Because of the regularization there now exist physical states for $g<0$, and we must combine them with the tower of states at $1-g>0$. In this way the degeneracy (44) heavily increases (for large values of the energy) giving as a result

$$\deg(E_\ell) \;=\; \begin{cases} g{-}1 \;+ \begin{Bmatrix} 0 & \text{for} & q+6g = 0,3,4,7,8,11 \bmod 12 \\ 1 & \text{for} & q+6g = 1,2,5,6,9,10 \bmod 12 \end{Bmatrix} & \text{if} \quad q<6g{-}6 \\ \left\lfloor\frac{q}{6}\right\rfloor \;+ \begin{Bmatrix} 0 & \text{for} & q = 1,2,5 \bmod 6 \\ 1 & \text{for} & q = 0,3,4 \bmod 6 \end{Bmatrix} & \text{if} \quad q \geq 6g{-}6\,. \end{cases} \tag{47}$$

In Fig. 5 we present the distribution of allowed states and degeneracies for low values of the energy.

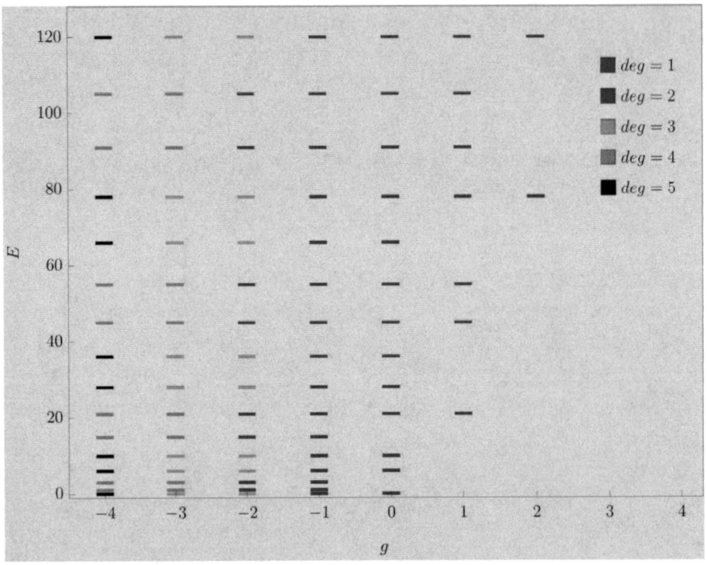


**Figure 5.** Energy spectrum and degeneracies for the $D_3$ model at integer coupling $g \in \mathbb{Z}$.

Like the Calogero–Sutherland model, the angular one becomes analytically integrable for integer values of the coupling constant $g$. However, because we have two different intertwining operators we have more ways to construct the additional odd charges

$$Q(g) = M_*(g-1)M_*(g-2)\cdots M_*(2-g)M_*(1-g) \tag{48}$$

where $M_*$ stands for using in every step either $M_3$ or $M_6$. The odd nature of these conserved charges can understood from the relations

$$\begin{aligned} &M_3^\dagger M_3 \propto 2J_6-3J_4J_2+J_2^3 + \text{ lower-order terms },\\ &M_6^\dagger M_6 \propto -12J_6^2+12\{J_6,J_4\}J_2-\tfrac{16}{3}J_6J_2^3+2J_4^3-14J_4^2J_2^2+6J_4J_2^4-\tfrac{2}{3}J_2^6\\ &\qquad +\text{lower-order terms },\end{aligned} \tag{49}$$

which tell us that $Q^2(g)$ is a polynomial in the conserved even charges. For example, in the case of $g=2$ we have

$$\left(Q_{333}^{(2)}\right)^2 = (2J_6-3J_4J_2+J_2^3)^3 + \text{ lower-order terms }. \tag{50}$$

## 4. Outlook and open problems

In this review we addressed non-Hermitian extensions of the trigonometric and angular Calogero models under the scope of integrability. Both systems exhibit a set of conserved charges, intertwining operators and -for integer couplings- a higher-order aditional odd integral of motion $Q(g)$. The latter flips the coupling $g \leftrightarrow 1-g$ of the states, which means to transform physical states into singular ones. Introducing a $\mathcal{PT}$-symmetric deformation as a regularization removes all singularities of potentials and wavefunctions. In this way the conserved charges $Q(g)$ acquire a physical nature. Taking into account the symmetry $g \leftrightarrow 1-g$, the spectral degeneracy is radically increased by the deformation. There are further results we have not presented here which are more involved but not less interesting. These features have been studied for the Calogero–Sutherland model $G_2$ model describing the so-called Calogero–Marchioro–Wolfes problem [72].

This is a non-simply-laced case, so there are two couplings associated to the short and long roots of the corresponding Coxeter group, which translates to a richer structure with different types of conserved quantities [41]. Regarding the angular model, the $BC_3$, $A_1^{\oplus 3}$ and $H_3$ systems have also been studied in a similar way [40]. Analogous investigations for the hyperbolic or elliptic Calogero interactions are still missing. Further deformations of Calogero models may also be considered [73]. For a Hamiltonian discussed there, can be arranged to

$$H_D = -\frac{1}{2}\sum_{i=1}^{3}\partial_i^2 + \frac{m(m-1)}{\sin^2(x_1-x_2)} + \frac{1-m}{\sin^2(x_1-i\sqrt{m}x_3)} + \frac{1-m}{\sin^2(x_2-i\sqrt{m}x_3)}, \tag{51}$$

which displays two extra conserved charges. Trigonometric and elliptic deformations of such systems were also studied, see [74] and references therein.

*Acknowledgments*

FC was partially supported by Fondecyt grant 1211356. OL has been supported by the Deutsche Forschungsgemeinschaft under grant LE 838/12 and by the COST Action MP1405 QSPACE.

## References

[1] Calogero F 1969 Solution of a three-body problem in one dimension *J. Math. Phys.* **10** 2191–2196 Calogero F 1971 Solution of the one-dimensional N-body problem with quadratic and/or inversely quadratic pair potentials *J. Math. Phys.* **12** 419–436; Erratum, *ibidem* 1996 **37** 3646
[2] Sutherland B 1971 Exact results for a quantum many-body problem in one dimension I & II *Phys. Rev. A* **4**, 2019-2021 & 1972 *Phys. Rev. A* **5** 1372–1376
[3] Olshanetsky M A and Perelomov A M 1981 Classical integrable finite-dimensional systems related to Lie algebras *Phys. Rept.* **71** 313–400
[4] Olshanetsky M A and Perelomov A M 1983 Quantum integrable systems related to Lie algebras *Phys. Rept.* **94** 313–404
[5] Polychronakos A P 2006 Physics and mathematics of Calogero particles *J. Phys. A: Math. Gen.* **39** 12793 *Preprint* arXiv:hep-th/0607033
[6] Sutherland B 2004 *Beautiful Models: 70 Years of Exactly Solved Quantum Many-Body Problems* (World Scientific, Singapore)
[7] 2000 *Calogero–Moser–Sutherland Models* CRM Series in Mathematical Physics ed van Diejen J F and Vinet L (Springer, New York)
[8] Mostafazadeh A 2002 PseudoHermiticity versus $\mathcal{PT}$-symmetry. The necessary condition for the reality of the spectrum *J. Math. Phys.* **43** 205-214 *Preprint* arXiv:math-ph/0107001
[9] Bender C M 2007 Making sense of non-Hermitian Hamiltonians *Rept. Prog. Phys.* **70** 947 *Preprint* arXiv:hep-th/0703096
[10] Bender C M, Dorey P E, Dunning C, Fring A, Hook D W, Jones H F, Kuzhel S, Levai G and Tateo R 2019 *$\mathcal{PT}$ Symmetry: In Quantum and Classical Physics* (World Scientific, Singapore)
[11] Fring A $\mathcal{PT}$-symmetric deformations of integrable models 2013 *Phil. Trans. Roy. Soc. Lond. A* **371** 20120046 *Preprint* arXiv:1204.2291
[12] Christodoulides Dand Yang J 2018 *Parity-time symmetry and its applications* (Springer, Berlin)
[13] Znojil M and Tater M 2001 Complex Calogero model with real energies *J. Phys. A* **34** 1793-1803 *Preprint* arXiv:quant-ph/0010087
[14] Basu–Mallick B and Kundu A 2000 Exact solution of Calogero model with competing long range interactions *Phys. Rev. B* **62** 9927 *Preprint* arXiv:cond-mat/0003425
[15] Basu–Mallick B and Mandal B P 2001 On an exactly solvable B(N) type Calogero model with nonHermitian $\mathcal{PT}$ invariant interaction *Phys. Lett. A* **284** 231-237 *Preprint* arXiv:cond-mat/0101349
[16] Basu–Mallick B 2002 Fractional statistics in some exactly solvable Calogero like models with $\mathcal{PT}$ invariant interactions *Int. J. Mod. Phys. B* **16** 1875-1882 *Preprint* arXiv:cond-mat/0201074
[17] Basu–Mallick B, Bhattacharyya T, Kundu A and Mandal B P 2004 Bound and scattering states of extended Calogero model with an additional PT invariant interaction *Czech. J. Phys.* **54** 5-12 *Preprint* arXiv:hep-th/0309136

[18] Basu–Mallick B, Bhattacharyya T and Mandal B P 2005 Phase shift analysis of PT-symmetric nonhermitian extension of $A_{N-1}$ Calogero model without confining interaction *Mod. Phys. Lett. A* **20** 543-552 *Preprint* arXiv:nlin/0405068
[19] Fring A 2006 A note on the integrability of non-Hermitian extensions of Calogero–Moser–Sutherland models *Mod. Phys. Lett. A* **21** 691-699 *Preprint* arXiv:hep-th/0511097
[20] Fring A and Znojil M 2008 $\mathcal{PT}$-symmetric deformations of Calogero models *J. Phys. A* **41** 194010 *Preprint* arXiv:0802.0624.
[21] Assis P E G and Fring A 2009 From real fields to complex Calogero particles *J. Phys. A* **42** 425206 *Preprint* arXiv:0907.1079 [hep-th]
[22] Fring A and Smith M 2010 Antilinear deformations of Coxeter groups, an application to Calogero models *J. Phys. A* **43** 325201 *Preprint* arXiv:1004.0916 [hep-th]
[23] Fring A and Smith M 2011 $\mathcal{PT}$ invariant complex $E_8$ root spaces *Int. J. Theor. Phys.* **50** 974 *Preprint* arXiv:1010.2218 [math-ph]
[24] Fring A and Smith M 2012 Non-Hermitian multi-particle systems from complex root spaces *J. Phys. A* **45** 085203 *Preprint* arXiv:1108.1719 [hep-th]
[25] Jakubsky V 2004 $\mathcal{PT}$-symmetric Calogero-type model *Czechoslovak Journal of Physics* **54** 67-69
[26] Brihaye Y and Nininahazwe A 2004 On $\mathcal{PT}$ symmetric extensions of the Calogero model *Int. J. Mod. Phys. A* **19** 4391-4400 *Preprint* arXiv:hep-th/0311081
[27] Ghosh P K and Gupta K S 2004 On the real spectra of Calogero model with complex coupling *Phys. Lett. A* **323** 29-33 *Preprint* arXiv:hep-th/0310276
[28] Shukla P 2001 Non-Hermitian random matrices and the Calogero–Sutherland model *Phys. Rev. Lett.* **87** 194102
[29] Jain S R 2006 Random matrix theories and exactly solvable models *Czech. J. Phys.* **56** 1021-1032
[30] Mandal B P and Ghatak A 2012 Spectral singularity and non-Hermitian $\mathcal{PT}$-symmetric extension of $A_{N-1}$ type Calogero model without confining potential *J. Phys. A* **45** 444022 *Preprint* arXiv:1209.0535 [math-ph]
[31] Ghosh P K 2011 Constructing Exactly Solvable Pseudo-hermitian Many-particle Quantum Systems by Isospectral Deformation Int. J. Theor. Phys. **50** 1143-1151 *Preprint* arXiv:1012.0907 [quant-ph]
[32] Ghosh P K 2012 Supersymmetric many-particle quantum systems with inverse-square interactions *J. Phys. A* **45** 183001 *Preprint* arXiv:1111.6255 [hep-th]
[33] Correa F, Jakubsky V and Plyushchay M S 2015 $\mathcal{PT}$-symmetric invisible defects and confluent Darboux-Crum transformations *Phys. Rev. A* **92** no.2 023839 *Preprint* arXiv:1506.00991 [hep-th]
[34] Mateos Guilarte J and Plyushchay M S 2017 Perfectly invisible $\mathcal{PT}$-symmetric zero-gap systems, conformal field theoretical kinks, and exotic nonlinear supersymmetry *JHEP* **12** 061 *Preprint* arXiv:1710.00356 [hep-th]
[35] Mateos Guilarte J and Plyushchay M S 2019 Nonlinear symmetries of perfectly invisible $\mathcal{PT}$-regularized conformal and superconformal mechanics systems *JHEP* **01** 194 *Preprint* arXiv:1806.08740 [hep-th]
[36] Inzunza L and Plyushchay M S 2021 Conformal bridge transformation and $\mathcal{PT}$-symmetry (*Preprint* arXiv:2104.08351 [hep-th])
[37] Ghosh P K and Sinha D 2018 Hamiltonian formulation of systems with balanced loss-gain and exactly solvable models *Annals Phys.* **388** 276-304 *Preprint* arXiv:1707.01122 [hep-th]
[38] Sinha D and Ghosh P K 2019 On the bound states and correlation functions of a class of Calogero-type quantum many-body problems with balanced loss and gain *J. Phys. A* **52** no.50, 505203 *Preprint* arXiv:1709.09648 [hep-th]
[39] Ghosh P K 2021 Classical hamiltonian systems with balanced loss and gain (*Preprint* arXiv:2104.03745 [math-ph])
[40] Correa F and Lechtenfeld O 2017 $\mathcal{PT}$ deformation of angular Calogero models *JHEP* **1711** 122 *Preprint* arXiv:1705.05425 [hep-th]
[41] Correa F and Lechtenfeld O 2019 $\mathcal{PT}$ deformation of Calogero–Sutherland models *JHEP* **2019** 166 *Preprint* arXiv:1903.06481 [hep-th]
[42] Chalykh O A and Veselov A P 1990 Commutative rings of partial differential operators and Lie algebras *Commun. Math. Phys.* **126** 597–611
[43] Chalykh O A 1996 Additional integrals of the generalized quantum Calogero–Moser system *Theor. Math. Phys.* **109** 1269–1273
[44] Dunkl C F 1989 Differential-difference operators associated to reflection groups *Trans. Amer. Math. Soc.* **311** 167–183
[45] Opdam E M 1988 Root systems and hypergeometric functions III, IV *Comp. Math.* **67** 21–49, 191–209
[46] Heckman G J 1991 A remark on the Dunkl differential-difference operators *Harmonic analysis on reductive groups* ed Barker W and Sally Progr. Math. **101** 181–191 (Birkhäuser, Switzerland)

[47] Correa F, Lechtenfeld O and Plyushchay M S 2014 Nonlinear supersymmetry in the quantum Calogero model *JHEP* **1404** 151 *Preprint* arXiv:1312.5749 [hep-th]
[48] Carrillo–Morales F, Correa F and Lechtenfeld O 2021 Integrability, intertwiners and non-linear algebras in Calogero models JHEP **2021** 163 *Preprint* arXiv:2101.07274 [hep-th]
[49] Correa F and Plyushchay M S 2012 Self-isospectral tri-supersymmetry in $\mathcal{PT}$-symmetric quantum systems with pure imaginary periodicity *Annals Phys.* **327** 1761–1783 *Preprint* arXiv:1201.2750 [hep-th]
[50] Correa F and Plyushchay M S 2012 Spectral singularities in $\mathcal{PT}$-symmetric periodic finite-gap systems *Phys. Rev. D* **86** 085028 *Preprint* arXiv:1208.4448 [hep-th]
[51] Correa F and Fring A 2016 Regularized degenerate multi-solitons *JHEP* **09** 008 *Preprint* arXiv:1605.06371 [nlin.SI]
[52] Polychronakos A P 1992 Exchange operator formalism for integrable systems of particles, *Phys. Rev. Lett.* **69** 703–705 arXiv:hep-th/9202057
[53] Lapointe L and Vinet L 1996 Exact operator solution of the Calogero–Sutherland model *Commun. Math. Phys.* **178** 425 *Preprint* arXiv:q-alg/9509003
[54] Perelomov A M, Ragoucy E and Zaugg Ph 1998 Explicit solution of the quantum three-body Calogero–Sutherland model *J. Phys. A: Math. Gen.* **31** L559 *Preprint* arXiv:hep-th/9805149
[55] García Fuertes W, Lorente M and Perelomov A M 2001 An elementary construction of lowering and raising operators for the trigonometric Calogero–Sutherland model *J. Phys. A: Math. Gen.* **34** 10963 *Preprint* arXiv:math-ph/0110038
[56] Efthimiou C J and Spector D 1997 Shape invariance in the Calogero and Calogero-Sutherland models *Phys. Rev. A* **56** 208 *Preprint* arXiv:quant-ph/9702017
[57] Ghosh P K, Khare A and Sivakumar M 1998 Supersymmetry, shape invariance and solvability of $A_{N-1}$ and $BC_N$ Calogero-Sutherland model *Phys. Rev. A* **58** 821 *Preprint* arXiv:cond-mat/9710206
[58] Plyushchay M S 1996 Deformed Heisenberg algebra, fractional spin fields and supersymmetry without fermions *Annals Phys.* **245** 339–360 *Preprint* arXiv:hep-th/9601116
[59] Plyushchay M S 2000 Hidden nonlinear supersymmetries in pure parabosonic systems *Int. J. Mod. Phys. A* **15** 3679–3698 *Preprint* arXiv:hep-th/9903130
[60] Wojciechowski S 1983 Superintegrability of the Calogero–Moser system *Phys. Lett.* **95A** 279–281.
[61] Kuznetsov V 1996 Hidden symmetry of the quantum Calogero–Moser system *Phys. Lett. A* **218** 212–222 *Preprint* arXiv:solv-int/9509001
[62] Hakobyan T, Nersessian A, Yeghikyan V 2009 The cuboctahedric Higgs oscillator from the rational Calogero model *J. Phys. A: Math. Theor.* **42** 205206 *Preprint* arXiv:0808.0430 [hep-th]
[63] Hakobyan T, Krivonos S, Lechtenfeld O and Nersessian A 2010 Hidden symmetries of integrable conformal mechanical systems *Phys. Lett. A* **374** 801–806 *Preprint* arXiv:0908.3290 [hep-th]
[64] Hakobyan T, Lechtenfeld O, Nersessian A and Saghatelian A 2011 Invariants of the spherical sector in conformal mechanics *J. Phys. A: Math. Theor.* **44** 055205 *Preprint* arXiv:1008.2912 [hep-th]
[65] Hakobyan T, Lechtenfeld O and Nersessian A 2012 The spherical sector of the Calogero model as a reduced matrix model *Nucl. Phys. B* **858** 250–266 *Preprint* arXiv:1110.5352 [hep-th]
[66] Feigin M, Lechtenfeld O and Polychronakos A 2013 The quantum angular Calogero-Moser model *JHEP* **1307** 162 *Preprint* arXiv:1305.5841[math-ph]
[67] Correa F and Lechtenfeld O 2015 The tetrahexahedric angular Calogero model *JHEP* **1510** 191 *Preprint* arXiv:1508.04925 [hep-th]]
[68] Feigin M V 2003 Intertwining relations for the spherical parts of generalized Calogero operators *Theor. Math. Phys.* **135** 497–509
[69] Feigin M and Hakobyan T 2015 On Dunkl angular momenta algebra *JHEP* **1511** 107 *Preprint* arXiv:1409.2480 [math-ph]
[70] Feigin M and Hakobyan T 2019 Algebra of Dunkl Laplace–Runge–Lenz vector (*Preprint* arXiv:1907.06706 [math-ph])
[71] Miller Jr W, Post S and Winternitz P 2013 Classical and quantum superintegrability with applications *J. Phys. A* **46** 423001 *Preprint* arXiv:1309.2694 [math-ph]
[72] Quesne C 1995 Exchange operators and extended Heisenberg algebra for the three-body Calogero–Marchioro–Wolfes problem *Mod. Phys. Lett. A* **10** 1323 *Preprint* arXiv:hep-th/9505071
[73] Chalykh O A, Feigin M V and Veselov A P 1998 New integrable generalizations of Calogero–Moser quantum problem *Journal of Math. Physics* **39** (2) 5341–5355
[74] Khodarinova L A 2005 Quantum integrability of the deformed elliptic Calogero–Moser problem *Journal of Math. Physics* **46** 033506

# An introductory review on resource theories of generalized nonclassical light

**Sanjib Dey**

Department of Physical Sciences, Indian Institute of Science Education and Research Mohali, Sector 81, SAS Nagar, Manauli 140306, India

E-mail: `sanjibdey4@gmail.com`

**Abstract.** Quantum resource theory is perhaps the most revolutionary framework that quantum physics has ever experienced. It plays vigorous roles in unifying the quantification methods of a requisite quantum effect as wells as in identifying protocols that optimize its usefulness in a given application in areas ranging from quantum information to computation. Moreover, the resource theories have transmuted radical quantum phenomena like coherence, nonclassicality and entanglement from being just intriguing to being helpful in executing realistic thoughts. A general quantum resource theoretical framework relies on the method of categorization of all possible quantum states into two sets, namely, the free set and the resource set. Associated with the set of free states there is a number of free quantum operations emerging from the natural constraints attributed to the corresponding physical system. Then, the task of quantum resource theory is to discover possible aspects arising from the restricted set of operations as resources. Along with the rapid growth of various resource theories corresponding to standard harmonic oscillator quantum optical states, significant advancement has been expedited along the same direction for generalized quantum optical states. Generalized quantum optical framework strives to bring in several prosperous contemporary ideas including nonlinearity, $\mathcal{PT}$-symmetric non-Hermitian theories, $q$-deformed bosonic systems, etc., to accomplish similar but elevated objectives of the standard quantum optics and information theories. In this article, we review the developments of nonclassical resource theories of different generalized quantum optical states and their usefulness in the context of quantum information theories.

## 1. Introduction

The value of a substance is a relative property that is dictated by its inadequacy, which means that the objects that are easily available in nature are not valuable. A resource theory is nothing but a systematic extension of this fundamental principle of economics. The basic idea of a resource theory is to classify the allowed (also called free) and prohibited set of actions for a given purpose, and then to figure out what can be achieved out of the allowed operations. Certain objects that are not possible to generate via the free operations are considered to be resources in the given framework. For example, consider a person is going out for a nearby day-trip. Certain things like nonperishable foods, sun-glass, money, etc., are allowed to carry. Suppose that the person does not own a car and, therefore, accompanying a personal car is prohibited in the given conjecture. Since carrying the money is allowed, the person is free to hire a cab or to catch a mode of public transport. Thus, a hired car or public transport is a resource of the day-trip. Likewise, fuel is the resource for an automobile driver and a cooking ingredient requiring refrigeration is a resource for a group of campers.

Quantum resource theory [1, 2] is a generic framework to study the resources of the atomic and sub-atomic phenomena that are governed by quantum mechanical laws. Such frameworks systematically unify different aspects of a particular quantum phenomenon by removing the associated arbitrariness and ambiguities. Technically, the fundamental goal of a quantum resource theory is to establish a framework for quantifying a quantum notion via a restricted set of operations, while a full set of operations being implied to all the physically allowed processes in quantum mechanics. If one defines a permissible set of free quantum operations through which only some of the physically comprehensible states can be prepared, the complement of the free set is coined as the resource set. A state that neither belongs to the free set nor can be generated via the associated free quantum operations is called a resource state.

In recent days, quantum resource theory has become an immense area of interest to people working in quantum information theories. Currently, several quantum resource theories including the resource theories of coherence [3, 4, 5, 6, 7], nonclassicality [8, 9, 1, 10, 11, 12, 13], quantum memory [14], nonlocality [15, 16], EPR steering [17], quantum network [18], imaginarity of quantum mechanics [19], quantum hypothesis testing [20], optimization of programmable quantum computers [21], etc., are under intensive scrutiny. Whereas the same for quantum entanglement [22, 23, 24, 25, 26] has become the most acclaimed example. Entanglement is a resource for long-distance communication between two quantum labs with respect to the free operations being local quantum operations and classical communication (LOCC). When Alice and Bob are separated by a long distance, as far as our current technology is concerned, the only mode of communication is the classical channel, like a telephone. Since Alice and Bob each own a quantum lab, they are allowed to prepare different quantum states locally via quantum operations. However, since the communication channel is restricted to be classical, the local quantum states cannot be shared with each other to prepare a joint quantum state that is entangled. But, it is known that a pair of quantum states having opposite spins is inherently entangled even if they are located far apart and, thus, entanglement is a resource in this scheme.

While the resource theories of various quantum phenomena have experienced rapid progress in recent years [7, 1, 11], a profound advancement for the same for generalized nonclassical states has also been noticed [27, 28, 29, 30, 31, 32]. This review article seeks to summarize the nonclassical resource theories of the generalized nonlinear and $q$-deformed quantum optical states and their versatility in the context of quantum information and computation.

## 2. Resource theories of nonclassicality: Preliminaries

Classical physics facilitates a complete description of an object by a point in the phase space, however, the uncertainty principle strongly resists such a point-like characterization of quantum mechanical particles. The Glauber coherent state being a minimum uncertainty state without any preferred axial direction in the optical phase space, its dynamics resemble most closely to a classical system [33]. This is the reason why the coherent state is usually agreed to be the most accurate classical state of light [34, 35]. Any state having a quantum-like behavior with respect to coherent states is called nonclassical. The higher the deviation of the quantum-like behavior, the higher the nonclassicality of the state. In reality, nonclassical states fill most of the space of the Hilbert space, whereas the classical-like states occupy only a corner of it. Despite the fact that the coherent states have ample usage in modern physics, it is well-established that the nonclassical states manifest genuine supremacy over coherent states in many applications. Furthermore, often it turns out that the states with a higher degree of nonclassicality are more useful. Therefore, the quantification of nonclassicality is treated as an important objective, which can be achieved within the resource theoretical framework with ease.

### *2.1. The general scheme*

In order to find resources of nonclassicality, the first task is to define a set of classical-like states, say $\mathbb{S}$, which belongs to the Hilbert space $\mathcal{H}$. Naturally, a nonclassical state resides in the complement set of $\mathbb{S}$. Let us also consider that associated with the set of classical state $\mathbb{S}$, there exists a set of permissible operations $\mathbb{O}$, which is a strict subset of the set of all possible allowed quantum operations. The permissible operations $\mathbb{O}$ are restricted in such a way that if $\mathbb{F} \in \mathbb{O}$ as well as $\psi \in \mathbb{S}$ then $\mathbb{F}(\psi) \in \mathbb{S}$, where $\psi$ represents an arbitrary quantum state. We are now ready to devise a resource theory for nonclassicality. We may propose that, for a particular resource theory, there exists a non-negative measure $M(\psi)$ of nonclassicality, which satisfies the following three fundamental properties:

$$M(\psi) = 0 \quad \text{if} \quad \psi \in \mathbb{S}, \tag{1}$$

$$M(\psi) \geq M(\mathbb{F}(\psi)) \quad \text{if} \quad \mathbb{F} \in \mathbb{O}, \tag{2}$$

$$M\left(\textstyle\sum_i q_i \psi_i\right) \leq \textstyle\sum_i q_i M(\psi_i). \tag{3}$$

The property (1) is obviously the first and foremost requirement that the measure $M(\psi)$ becomes positive if and only if the state $\psi$ turns out to be nonclassical. Property (2) implies that, given the set of operations $\mathbb{O}$, the measure $M(\psi)$ is a monotonic function of nonclassicality. It means that if $\mathbb{F} \in \mathbb{O}$, the nonclassicality measure $M(\psi)$ must decrease monotonically under the operation $\mathbb{F}$. This property, in turn, ensures that the nonclassical states can neither be generated from classical ones, nor the nonclassicality of the state can be enhanced via the set of operations $\mathbb{O}$. The only possible way to increase the nonclassicality is by replacing the state $\psi$ with another nonclassical state, say $\chi$, so that $M(\psi) < M(\chi)$. Therefore, the sources of nonclassicality can be characterized as supplementary resources which are needed to overcome the restrictions inherent to the classical states $\mathbb{S}$ and operations $\mathbb{O}$. Property (3) is a convexity property, which reinforces the concept that the nonclassicality can never be increased by mixing two states $\psi$ and $\chi$ statistically. It should be noted that a statistical mixture of type $q\psi + (1-q)\chi$ is intrinsically a classical procedure and, therefore, such a process can never be allowed to generate extra nonclassicality.

The framework discussed above is a general procedure to formulate a resource theory of nonclassicality. For further understanding of the mathematical treatment of the resource theory, one may refer, for instance [36]. It is worth mentioning that so far it has not been possible to unify all types of existing nonclassicalities and provide a single nonclassicality measure for all of them. But, one is compelled to find the resources of each type of nonclassicality separately by associating a corresponding measure $M(\psi)$ that fits the framework mentioned above. In the rest of the section, we shall discuss the resource theory for some types of nonclassicalities.

### *2.2. Resource theory of quasi-probability distribution functions*

Since the Glauber coherent state $|\alpha\rangle$ is a predominant example of a classical-like pure state, a classical statistical mixture of them will obviously remain classical. The corresponding density matrix is given by

$$\rho_{\mathrm{c}} = \int P_{\mathrm{c}}(\alpha)\, |\alpha\rangle\langle\alpha|\, \mathrm{d}^2\alpha, \tag{4}$$

with $P_{\mathrm{c}}(\alpha)$ being a probability density, such that $\int P_{\mathrm{c}}(\alpha)\, \mathrm{d}^2\alpha = 1$. A similar type of representation is also possible for general mixed quantum states. In fact, Glauber [34] and Sudarshan [35] showed that any quantum optical state can be expressed in the overcomplete coherent state basis and, thus, the density matrix of the corresponding states can be represented in the following form

$$\rho = \int P(\alpha)\, |\alpha\rangle\langle\alpha|\, \mathrm{d}^2\alpha, \tag{5}$$

where $P(\alpha)$ is familiar as the Glauber-Sudarshan's $P$-function. Notice the structural similarity of (5) with (4). The only dissimilarity lies in the interpretation of the two functions $P_{\rm c}(\alpha)$ and $P(\alpha)$. Since $P(\alpha)$ is applicable to general quantum states, it is interpreted as a quasi-probability distribution function rather than the usual positive probability density. In other words, although $P(\alpha)$ is normalized, i.e. $\int P(\alpha)\,{\rm d}^2\alpha = 1$, it may not be positive always. The reason is that when an arbitrary quantum optical state is expressed in the overcomplete coherent state basis, one obtains extra weight functions that ensure the resolution of identity. These extra weight functions are solely responsible for the negativity of $P(\alpha)$ for some quantum optical states. It becomes a positive probability density only when the corresponding state is classical. The states for which $P(\alpha)$ is negative are termed nonclassical.

There are some other quasi-probability distribution functions dictating the nonclassical properties of quantum optical states; such as Husimi $Q$-function [37], Wigner distribution function [38], etc., which are discussed in most of the standard textbooks of quantum optics [39, 40, 41, 42, 43, 44]. In a recent work [45], the authors have introduced $s$-parameterized quasi-probabilities $P_{\rm s}(\alpha)$ with which they have unified the notion of nonclassicality for all the quasi-probability distribution functions in the framework of resource theory. As per their formalism, the s-parameterized measure of nonclassicality is given by

$$M_{\rm s}(\rho) = \frac{1}{2}\left(\int |P_{\rm s}(\alpha)|\,{\rm d}^2\alpha - 1\right), \quad s \in [-1, 1], \tag{6}$$

whose value is entirely controlled by the negativity of the associated quasi-probability distribution functions. Here, $s = 1, 0, -1$ correspond to the cases of the $P$-function, Wigner function and the $Q$-function, respectively. It was shown that with the increase of the parameter $s$, $M_{\rm s}(\rho)$ (6) becomes an increasingly stronger measure of nonclassicality. Thus, it clearly indicates that the negativity of the $P$-function is the strongest and the same for the $Q$-function is the weakest measure of nonclassicality among the three quasi-probability distribution functions.

A similar type of resource theory, albeit, from a different angle has been explored in [6, 9], where the set of free quantum operations $\mathbb{O}$ has been restricted to the operations that allow forward feeding of measurement outcomes and linear optical operations. Thus, such a resource theory is more useful to the scenarios where the measurement outcomes are utilized to determine the degree of nonclassicality. It was also shown that a nonclassicality measure of such type of theories obeys the three necessary properties (1), (2) and (3) as discussed in the earlier subsection. For further details on nonclassical resource theories of quasi-probability distributions, we refer the readers to a recent review article [11].

### *2.3. Resource theory of quadrature squeezing*

In quantum optics, quadratures $\hat{x}$ and $\hat{p}$ are the field amplitudes of the quantized electromagnetic field, whose expressions are incidentally similar to the dimensionless position and momentum operators of the linear harmonic oscillator

$$\hat{x} = \frac{\hat{a} + \hat{a}^\dagger}{2}, \quad \hat{p} = \frac{\hat{a} - \hat{a}^\dagger}{2i}. \tag{7}$$

Quadrature squeezing refers to a situation when the variance of at least one of the quadratures is less than the right-hand side of the type of uncertainty relation that the state obeys. Further interesting phenomena of quadrature squeezing in the cases of Heisenberg's and Robertson's uncertainty relations can be found in [46]. While the resource theory of the quasi-probability distribution function as discussed in the previous subsection may be interesting and helpful, often it turns out that the computation of the distribution function corresponding to some states becomes laborious. In this subsection, we shall explore a relatively simpler but equivalent resource. Before delving into the details, let us start with an example of a Gaussian wave-packet, whose position and momentum quadrature variances, $\sigma_x$ and $\sigma_p$, are not equal to each other. Given that the quadratures are statistically uncorrelated, the $P$-function of the wave-packet can be written as [47]

$$P_{\rm G}(\alpha) = \frac{1}{\mathcal{N}}\left(\frac{({\rm Re}\alpha - A)^2}{\sigma_x - 1/2} + \frac{({\rm Im}\alpha - B)^2}{\sigma_p - 1/2}\right). \tag{8}$$

Here $\mathcal{N}$ represents the normalization constant, and $A, B$ denote the central position of the distribution in the real and imaginary axes of the complex $\alpha$-plane. It is obvious that the function (8) cannot exist as a normalizable distribution if the variances of any of the two quadratures become less than 1/2. Therefore, if at least one of the quadrature variances of a quantum state becomes less than 1/2, the state does not correspond to a Gaussian distribution and, therefore, it is nonclassical. The statement not only applies to the Gaussian state, but to any arbitrary quantum state. In fact, one can write a general relation between the $x$-quadrature variance and the $P$-function of an arbitrary state [47] as

$$\sigma_x = \frac{1}{2}\int P(\alpha)\left[1 + \left(\alpha^* + \alpha - \langle \hat{a}^\dagger + \hat{a}\rangle\right)^2\right] {\rm d}^2\alpha. \tag{9}$$

The above relation tells that the $P$-function can be a positive probability density provided that $\sigma_x > 1/2$, otherwise $P(\alpha)$ will always be negative. A similar relation can also be derived for the $\hat{p}$-quadrature variance, which is not shown here. However, what we realize is that if either $\sigma_x < 1/2$ or $\sigma_p < 1/2$ holds for a quantum state, the corresponding $P$-function fails to be positive and the associated state will be nonclassical. Therefore, it is no longer essential to study the $P$-function and its corresponding resources if one gains the knowledge of quadrature squeezing. The equivalences of quadrature squeezing with the other two distribution functions, namely the $Q$-function and the Wigner function are unfamiliar. However, as we discussed in the previous subsection, the $P$-function is the strongest measure of nonclassicality and, therefore, as far as the resource is concerned, the study of Wigner and $Q$-functions are no longer essential. But, an analysis of quadrature squeezing will suffice.

Let us now discuss a formulation of the resource theory of quadrature squeezing [48, 49, 50]. Let us consider an arbitrary but fixed operator $\hat{g}$ associated with an experimental setup, with $\hat{g} \equiv \hat{g}(\hat{a}, \hat{a}^\dagger)$ being an operator-valued function of the ladder operators. Let us also propose that a quantum state is said to be nonclassical if for an observable $\hat{g}^\dagger\hat{g}$ we have

$$\langle : \hat{g}^\dagger\hat{g} : \rangle < 0, \tag{10}$$

where $: \cdot\cdot :$ denotes the normal ordering of the operators. Then, the Hermitian operators, $: \hat{g}^\dagger\hat{g} :$ and $\hat{g}^\dagger\hat{g}$, are observables of the corresponding setup. Clearly, the second observable is positive semi-definite, but the spectrum of the first one may be negative. The negativity of the difference of the corresponding expectation values can be measure with $\delta$, such that

$$\delta = \langle : \hat{g}^\dagger\hat{g} : \rangle - \langle \hat{g}^\dagger\hat{g}\rangle, \tag{11}$$

which can be derived easily by following the standard methods of normal ordering. However, since $\langle \hat{g}^\dagger\hat{g}\rangle \geq 0$, the following relation is guaranteed

$$\delta \leq \langle : \hat{g}^\dagger\hat{g} : \rangle < 0. \tag{12}$$

Therefore, the relative nonclassicality of a quantum state for a designated measurement setup can be defined by a non-negative measure $M$ as follows

$$M = \begin{cases} \langle : \hat{g}^\dagger\hat{g} : \rangle/\delta & \text{if } \langle : \hat{g}^\dagger\hat{g} : \rangle < 0 \\ 0 & \text{otherwise .} \end{cases} \tag{13}$$

In other words, the measure $M$ defined in (13) quantifies the negativity of the LHS of (10) relative to the lower bound determined by the relation (12). The relation (13) also suggests that if the negativity of $\langle : \hat{g}^\dagger \hat{g} : \rangle$ approaches the lower bound, the corresponding quantum state exhibits a maximum non-negativity of the relative measure $M$, i.e., $M = 1$. Hence, the maximal nonclassicality is achieved when $\delta = \langle : \hat{g}^\dagger \hat{g} : \rangle$, i.e., $\langle \hat{g}^\dagger \hat{g} \rangle = 0$.

In order to apply this formalism to find resources of quadrature squeezing, let us consider a perfect homodyne detection setup. The principle of homodyne detection technique relies on the measurement of the probability distribution of the phase-induced $\hat{x}$-quadrature operator

$$\hat{x}_\phi = \frac{1}{2}\left(\hat{a}e^{i\phi} + \hat{a}^\dagger e^{-i\phi}\right), \tag{14}$$

of a given radiation mode [51, 52]. One can choose the operator $\hat{g}$ as the standard deviation of the quadrature, i.e., $\hat{g} \equiv \Delta\hat{x}_\phi = \hat{x}_\phi - \langle \hat{x}_\phi \rangle$, so that the quadrature squeezing effect can be quantified with the measure given in (13). The value of $\delta$ follows from the fact that the variance of the quadrature differs from its normal-ordered value by that of the vacuum state quadrature, i.e., $\delta = \langle : (\Delta\hat{x}_\phi)^2 : \rangle - \langle (\Delta\hat{x}_\phi)^2 \rangle = -\langle (\Delta\hat{x}_\phi)^2 \rangle_{\rm vac}$. As discussed earlier, the maximum robustness of the measure is obtained when $\langle (\Delta\hat{x}_\phi)^2 \rangle$ vanishes. However, in reality, it is hardly possible to face a circumstance when the value of $\langle (\Delta\hat{x}_\phi)^2 \rangle$ is exactly zero. Rather, a realistic situation should refer to a scenario when the value is very small, say $\langle (\Delta\hat{x}_\phi)^2 \rangle_{\rm min}$. In short, when $\langle (\Delta\hat{x}_\phi)^2 \rangle = \langle (\Delta\hat{x}_\phi)^2 \rangle_{\rm min}$, we obtain $M_{\rm qs} = M_{\rm qs}^{\rm max}$, and, therefore, the maximum robustness in the measure can be expressed as

$$M_{\rm qs}^{\rm max} = 1 - \frac{\langle (\Delta\hat{x}_\phi)^2 \rangle_{\rm min}}{\langle (\Delta\hat{x}_\phi)^2 \rangle_{\rm vac}}. \tag{15}$$

Thus, it is obvious that the state will become maximally nonclassical when the quadrature variance of the corresponding state $\langle (\Delta\hat{x}_\phi)^2 \rangle$ becomes minimum with respect to the same for the vacuum state. Therefore, one may conclude that the maximal quadrature squeezing implies maximal nonclassicality.

### *2.4. Resource theory of sub-Poissonian photon statistics*

The phenomenon of the photon number squeezing is characterized by the nature of the sub-Poissonian photon distribution, where the photon number distribution of a state is narrower than the average number of photons, i.e., $\langle (\Delta\hat{n})^2 \rangle < \langle \hat{n} \rangle$, with $\hat{n} = \hat{a}^\dagger \hat{a}$ being the number operator of the radiation field. The optical noise of a sub-Poissonian light can go well-below the classical shot-noise limit (the lowest detectable noise level with the classical light) and, thus, sub-Poissonian light often turns out to be extremely useful, especially, for the purpose of quantum information processing. Experimental detection of sub-Poissonian light is carried out with photodetectors and the resulting beam of photoelectrons also show sub-Poissonian nature [53, 54]. In the framework of resource theory, the nonclassicality of a sub-Poissonian light can be characterized easily by associating the operator $\hat{g}$ with the standard deviation of the photon number

$$\hat{g} \equiv \Delta\hat{n} = \hat{n} - \langle \hat{n} \rangle. \tag{16}$$

By following a similar logic of quadrature squeezing, we can write

$$\delta = \langle : (\Delta\hat{n})^2 : \rangle - \langle (\Delta\hat{n})^2 \rangle = -\langle (\Delta\hat{n})^2 \rangle_{\rm Pois.} = \langle \hat{n} \rangle. \tag{17}$$

The resulting nonclassicality measure reads as follows

$$M_{\rm ns} = 1 - \frac{\langle (\Delta\hat{n})^2 \rangle}{\langle (\hat{n}) \rangle} = -\mathcal{Q}, \tag{18}$$

Journal of Physics: Conference Series **2038** (2021) 012008 doi:10.1088/1742-6596/2038/1/012008

where $\mathcal{Q}$ denotes the Mandel parameter. The result is unambiguous in the sense that the nonclassical measure $M_{\rm ns}$ attains a maximal value of 1 when the photon number variance $\langle(\Delta\hat{n})^2\rangle$ vanishes, i.e., when the number squeezing is maximized. One of the interesting notions of the resource (13) utilized in this and the previous subsection is that the measures are directly related to the mean values of the observables, which can be measured experimentally. Thus, it becomes easier for an experimenter to detect robust nonclassical states through an analysis of quadrature and photon number squeezing.

## 3. Generalizations of quantum optics

Before moving to the discussion of generalizations of nonclassical states, let us highlight the basic notions of non-Hermitian theories which will be utilized for the generalizations.

### *3.1. Role of non-Hermitian theories*

Observables in quantum mechanics demand Hermiticity of the corresponding operators, which assures reality of spectrum and unitary time-evolution. However, quantum physics has evolved a lot over the centuries and standing at the 21st century of physics, it is evident that Hermitian operators are only a subset of physical observables. The journey of non-Hermitian operators as observables started almost at the time of the inception of quantum physics. In the year 1929, von Neumann and Wigner [55] indicated the possibility that the non-Hermitian Hamiltonians can acquire real positive eigenspectrum and discrete eigenstates. Later, such systems went through a more intense investigation, and, nowadays, the notion of the so-called bound states in the continuum (BICs) is well-realized via different examples [56] along with their bi-orthonormal eigenfunctions [57]. Towards the end of the last century and at the beginning of this century, we have noticed a rapid growth in this direction, particularly in the $\mathcal{PT}$-symmetric theories [58] and the formulation of pseudo-Hermitian Hamiltonians [59]. Increased interest in these two theories has been dispersed in almost every branch of physics, likewise in quantum optics and information theories with no exception. While the role of non-Hermiticity in optical waveguide theories has been crucial, almost inevitably it appears in the process of the generalization of quantum optical models also. This review being focused on generalized quantum optics, we shall mainly discuss the role of non-Hermiticity in such theories. Before that, let us briefly summarize the basic notions of the two main directions of non-Hermitian theories in the rest of this subsection.

*3.1.1. $\mathcal{PT}$-symmetry* It was Wigner [60] who pointed out that any operator which is invariant under an anti-linear involution whose eigenfunctions also respect the symmetry possesses real eigenvalues. Bender and Boettcher, in a seminal paper [58], have shown that the Hamiltonians which are symmetric under the combined operation of parity $\mathcal{P}$ and time-reversal $\mathcal{T}$, i.e., which are invariant under an anti-linear operation, like $x \to -x$, $p \to p$, $i \to -i$, admit real spectrum. The result is clearly in agreement with Wigner's theorem [60] without any loss of generality. Note that there exist subtle issues for the non-existence of the real spectrum even when the non-Hermitian Hamiltonian is $\mathcal{PT}$-symmetric, i.e., $[H, \mathcal{PT}] = 0$. Since $\mathcal{PT}$ is an anti-linear operator, it is not guaranteed that all of its eigenstates will be the eigenstates of the Hamiltonian. The case when the wavefunctions are simultaneous eigenstates of the Hamiltonian and the $\mathcal{PT}$-operator, is referred to as $\mathcal{PT}$ unbroken scenario and the eigenvalues are always real. Broken $\mathcal{PT}$ cases (i.e., when $[H, \mathcal{PT}] = 0$ but $\mathcal{PT}\psi' \neq \psi'$) may exhibit conjugate pairs of complex eigenvalues and the reality of the spectrum is not guaranteed. Thus, the proper way to interpret the theory is that the non-Hermitian Hamiltonians with unbroken $\mathcal{PT}$-symmetry always possess real eigenspectrum [61]. Thus, a non-Hermitian generalization of the standard quantum theory demands a replacement of the Hermitian adjoint ($H = H^\dagger$) with the $\mathcal{PT}$-adjoint ($H = H^{\mathcal{PT}}$), such that the inner product $\langle\psi'|\phi'\rangle^{\mathcal{PT}}$ of the corresponding Hilbert space takes the form

$$\int \left[\psi'(x)\right]^{\mathcal{PT}} \phi'(x)\, \mathrm{d}x = \int \left[\psi'(-x)\right]^{*} \phi'(x)\, \mathrm{d}x. \tag{19}$$

Due to the fact that $\mathcal{PT}$ is an anti-linear operator, wevafunctions may not be simultaneous eigenfunctions of $H$ and $\mathcal{PT}$, which implies that the norm may not be positive definite always. However, the problem is resolved by introducing a charge conjugation operator $\mathcal{C}$ and replacing the $\mathcal{PT}$-norm with the $\mathcal{CPT}$-norm [62]. Since then it is widely accepted that within the regime of unbroken $\mathcal{PT}$-symmetry a $\mathcal{CPT}$-symmetric non-Hermitian Hamiltonian describes a complete self-consistent physical system with real eigenvalues and unitary time-evolution.

Rigorous developments of $\mathcal{PT}$-symmetric theories have been carried out over the years in various mathematical [63, 64, 65, 66] and physical contexts [67]. Plenty of experimental confirmation of $\mathcal{PT}$-symmetric effects as well as their applications in various branches of physics have enriched the field. Early experiments have revealed materials with controllable loss and gain by drawing an analogy between the stationary Schrödinger equation and the Helmholtz equation describing monochromatic linearly polarized light [68, 69, 70]. This has given rise to meta-materials [71] exhibiting the phenomena of unidirectional invisibility [72] and coherent perfect absorption [73]. $PT$-symmetry has also been applied in microwave cavities physics [74], nuclear magnetic resonant quantum systems [75], finding topological edge states [76, 77], stimulating the superconductivity [78], etc. It is hardly possible to provide a complete list of articles devoted to the development of the theory, however, for more information and references, one may follow some of the recent review articles [61, 79, 46] and books on the filed [80, 81, 82].

*3.1.2. Pseudo-Hermiticity* A second approach to studying non-Hermitian physics is popular as pseudo-Hermiticity [59]. The initial idea came from the work of Dirac and Pauli [83] in 1943, which was followed up by Sudarshan [84], Lee and Wick [85] to overcome the renormalization induced negative norm states during the process of quantizing the electromagnetic and other fields. It was even before the appearance of $\mathcal{PT}$-symmetry and pseudo-Hermiticity in the literature when people tried to provide a consistent quantum mechanical framework of non-Hermitian Hamiltonians using the concept of quasi-Hermiticity [86, 87]. The concept was resurfaced by Mostafazadeh in a more popular form as pseudo-Hermiticity [59], where a non-Hermitian Hamiltonian $H_{\mathrm{nh}}$ and a Hermitian Hamiltonian $H_{\mathrm{h}}$ are considered to be related by a similarity transformation

$$H_{\mathrm{h}} = \eta H_{\mathrm{nh}} \eta^{-1} = H_{\mathrm{h}}^{\dagger} = \eta^{-1} H_{\mathrm{nh}}^{\dagger} \eta, \quad \Leftrightarrow \quad H_{\mathrm{nh}}^{\dagger} = \rho H_{\mathrm{nh}} \rho^{-1}, \quad \text{with } \rho = \eta^{\dagger}\eta. \tag{20}$$

Thus, the non-Hermitian Hamiltonian $H_{\mathrm{nh}}$ becomes Hermitian with respect to a linear, non-unique, invertible and self-adjoint operator $\rho$ playing the role of a metric. Consequently, the eigenstates of the non-Hermitian Hamiltonian $H_{\mathrm{nh}}$ turns out to be orthogonal with respect to the so-called pseudo-inner product provided that the eigenstates of $H_{\mathrm{h}}$ and $H_{\mathrm{nh}}$ ($|\Phi_{\mathrm{h}}\rangle$ and $|\Phi_{\mathrm{nh}}\rangle$, respectively) are constrained by the relation $|\Phi_{\mathrm{nh}}\rangle = \eta^{-1}|\Phi_{\mathrm{h}}\rangle$. Within the formalism, all crucial properties of a physical system can be proven rigorously [59, 88, 89] and, therefore, with respect to a metric $\rho$, a non-Hermitian Hamiltonian can represent a complete physical system. It is worth mentioning that pseudo-Hermiticity does not require to assume the operator $\rho$ to be positive. Nevertheless, it is not so straightforward to obtain an exact expression of the metric, in fact, there are only a limited number of articles producing exact forms, see; for instance [90, 91, 92, 93]. However, there are various methods to obtain the same, for example, using the perturbation theory, spectral theory [94, 95], Moyal product [96], etc.

Pseudo-Hermiticity is equally important and significant as the $\mathcal{PT}$-symmetric theory. People argue that the scope of $PT$-symmetric theories is limited in comparison to that of pseudo-Hermiticity in the sense that $\mathcal{PT}$-symmetry does not always guarantee a real spectrum.

Despite the fact that such issues are debated [97, 98, 99], it is conceivable that both of the theories have made a significant development of the field of non-Hermitian physics [100, 101, 102, 103, 104, 105].

### *3.2. Different methods of generalization*

In the last few decades, the study on the generalizations of quantum optical states has been one of the foremost priorities of researchers because of their widespread applications in almost every branch of physics. Though the progress is slowed down significantly due to many mathematical hardships, the outcomes are often rewarding. The initial efforts came from the generalization of coherent states in the group theoretical framework [106]. Later they were extended to the wavelet theories, which have been found to be very useful in many areas including signal analysis and image processing in a joint time-frequency transform. It is out of our scope to discuss such developments in detail, for further interest in this area, one may refer to some existing reviews and books [107, 108, 109]. There exist several other methods of generalization, however, we shall focus on the nonlinear and $q$-deformed construction of generalized nonclassical states assisted by the non-Hermitian theories; like $\mathcal{PT}$-symmetry and pseudo-Hermiticity. This will lead us to the main goal of this review towards the resource theoretical analysis of the generalized nonclassical light which will be expounded in Sec. 4.

*3.2.1. Nonlinear generalization* The starting point of the nonlinear generalization is to gain insights on the generalizations of the ladder operators $\hat{A}^\dagger \equiv f(\hat{n})\hat{a}^\dagger = \hat{a}^\dagger f(\hat{n}+1)$, $\hat{A} \equiv \hat{a}f(\hat{n}) = f(\hat{n}+1)\hat{a}$, obeying the following nonlinear commutator algebras

$$[\hat{A}, \hat{A}^\dagger] = (\hat{n}+1)f^2(\hat{n}+1) - \hat{n}f^2(\hat{n}), \quad [\hat{n}, \hat{A}] = -\hat{A}, \quad [\hat{n}, \hat{A}^\dagger] = \hat{A}^\dagger. \tag{21}$$

Usually, $f(\hat{n})$ is a general operator-valued real nonlinear function of the standard number operator $\hat{n} = \hat{a}^\dagger\hat{a}$. The algebras (21) naturally reduce to the Heisenberg algebra, $[\hat{a}, \hat{a}^\dagger] = 1, [\hat{n}, \hat{a}] = -\hat{a}, [\hat{n}, \hat{a}^\dagger] = \hat{a}^\dagger$, when $f(\hat{n})$ is chosen to be 1. The aim is to find out the function $f(\hat{n})$ corresponding to a general physical system so that the action of the generalized operators $\hat{A}$ and $\hat{A}^\dagger$ on the Fock states can be realized as

$$\hat{A}^\dagger|n\rangle = \sqrt{n+1}f(n+1)|n+1\rangle, \quad \hat{A}|n\rangle = \sqrt{n}f(n)|n-1\rangle. \tag{22}$$

If we assume that the Hamiltonian of the general quantum mechanical model can be factorized in terms of the generalized number operator $\hat{A}^\dagger\hat{A}$, the exact form of $f(\hat{n})$ can be obtained from the eigenvalues of the associated physical system. The appearance of any constant terms in the Hamiltonian can always be eliminated by simply rescaling the eigenvalues accordingly. Therefore, the first requirement is to solve the eigenvalues of the general model which can be utilized to build the ladder operators of the corresponding model. Most of the quantum optical states (coherent and nonclassical) arise from different quantum optical relations composed of the ladder operators. Thus, on many occasions, direct use of the ladder operators $\hat{A}, \hat{A}^\dagger$ produces the generalized quantum optical states, however, it is not always the case. We shall explain such scenarios while discussing some particular models in Sec. 4. Also, it is worth mentioning that in order to claim a well-defined quantum optical state, one has to study all the necessary properties of the states individually. For example, the construction of coherent states requires the resolution of identity; see, for instance [110], for further details. Nevertheless, the method of generalization emerging from a nonlinear function $f(n)$ consisting of the creation and annihilation operators is widely known as the nonlinear generalization [111, 112, 113, 114] and, nowadays, the endeavor is widely established; see, for example [115, 116, 28, 117, 30]. The formulation has given rise to some similar type of generalizations; such as the Gazeau-Klauder generalization [118, 119, 120], generalization to supersymmetric quantum mechanics [121, 122], etc.

*3.2.2. Generalization from q-deformation* The sole purpose of $q$-deformed generalization is to provide a systematic framework for the construction of quantum optics for various deformed quantum mechanical models. $q$-deformation is a unified framework that can be related with the quantum groups and, thus, a deformed model can always be associated with a deformed canonical commutation relation of type

$$\hat{A}_q\hat{A}_q^\dagger - q^2\hat{A}_q^\dagger\hat{A}_q = q^{g(\hat{n})}, \quad |q| < 1, \tag{23}$$

with $g(\hat{n})$ being an arbitrary function of the number operator. The ladder operators $\hat{A}_q$ and $\hat{A}_q^\dagger$ are constructed in a $q$-deformed Fock space

$$|n\rangle_q := \frac{\hat{A}_q^{\dagger n}}{\sqrt{[n]_q!}}|0\rangle_q, \quad [n]_q! := \prod_{k=1}^{n}[k]_q, \quad [0]_q! := 1, \quad \hat{A}_q|0\rangle_q = 0, \quad {}_q\langle 0|0\rangle_q = 1, \tag{24}$$

such that the following operations are satisfied

$$\hat{A}_q|n\rangle_q = \sqrt{[n]_q}\,|n-1\rangle_q, \quad \hat{A}_q^\dagger|n\rangle_q = \sqrt{[n+1]_q}\,|n+1\rangle_q. \tag{25}$$

Here, the deformed number operators $[\hat{n}]_q \equiv \hat{A}_q^\dagger\hat{A}_q$ are explicitly obtained as a solution of (23) with the help of (24) and (25). This implies that one obtains a well-defined quantum mechanical framework in the sense that the states $|n\rangle_q$ form an orthonormal basis in a $q$-deformed Hilbert space $\mathcal{H}_q$ spanned by the vectors $|\psi\rangle_q := \sum_{n=0}^{\infty} c_n|n\rangle_q$ with $c_n \in \mathbb{C}$, such that ${}_q\langle\psi|\psi\rangle_q = \sum_{n=0}^{\infty}|c_n|^2 < \infty$. The commutation relation between $\hat{A}_q$ and $\hat{A}_q^\dagger$ takes the form

$$[\hat{A}_q, \hat{A}_q^\dagger] = 1 + (q^2-1)\hat{A}_q^\dagger\hat{A}_q = 1 + (q^2-1)[\hat{n}]_q, \tag{26}$$

which reduces to the canonical commutation relation in the limit $q \to 1$. The $q$-deformed oscillator algebras of type (23) have been shown to be correlated with various physical systems [123, 124, 125, 126, 127] and, thus, they have been utilized for the generalizations of various quantum optical states [128, 129, 27, 29, 130, 131]. Note that, both the nonlinear and $q$-deformed generalizations can be applied to the non-Hermitian Hamiltonian systems also. It is obvious that the non-Hermitian generalization to quantum optics creates additional complications, however, they can be resolved using the formalism discussed in Sec. 3.1. For further information in this regard, one may refer to a review article dedicated to this scenario [46].

## 4. Resources theories of nonclassicality with generalized quantum optical states

Here, we shall discuss some important generalized quantum optical states with a focus on their resources of nonclassicality.

### *4.1. Generalized nonclassical states*

*4.1.1. Squeezed states* Squeezed states are known to be one of the most useful nonclassical states. The optical noise produced by a squeezed state is less than a coherent state [132, 133] and, this is why squeezed states are used extensively in many areas [134, 135, 136, 137] including gravitational wave detection [138, 139, 140, 141]. For further interest in the applications of squeezed states, one may refer to some review articles; see, for instance [142, 143, 46]. The generalization of squeezed states starts with the replacement of the harmonic oscillator ladder operators $\hat{a}, \hat{a}^\dagger$ with the generalized ones $\hat{A}, \hat{A}^\dagger$ in any of the following two definitions [28, 31]

$$(\hat{a} + \xi\hat{a}^\dagger)|\alpha, \xi\rangle = \alpha|\alpha, \xi\rangle, \tag{27}$$

$$|\alpha, \xi\rangle = \hat{D}(\alpha)\hat{S}(\xi)|0\rangle, \quad \alpha, \xi \in \mathbb{C}. \tag{28}$$

Here, $\hat{D}(\alpha) = \exp(\alpha\hat{a}^\dagger + \alpha^*\hat{a})$ and $\hat{S}(\xi) = \exp\{(\xi\hat{a}^{\dagger 2} - \xi^*\hat{a}^2)/2\}$ are the displacement and squeezing operator, respectively. By following the definition (27), the construction of the generalized squeezed state in Fock space follows from the assumption that it can be expanded in the Fock basis as

$$|\alpha,\xi\rangle_{\text{gen}} = \frac{1}{\mathcal{N}(\alpha,\xi)} \sum_{n=0}^{\infty} \frac{\mathcal{I}(\alpha,\xi,n)}{\sqrt{n!}f(n)!}|n\rangle\ , \tag{29}$$

and then solving the unknown function $\mathcal{I}(\alpha,\xi,n)$ from the corresponding definition of the squeezed states (27). By replacing (29) in (27) along with the help of (22), one ends up with a recurrence relation

$$\mathcal{I}(\alpha,\xi,n+1) = \alpha\ \mathcal{I}(\alpha,\xi,n) - \xi n f^2(n)\mathcal{I}(\alpha,\xi,n-1), \tag{30}$$

whose solution will constitute the required state $|\alpha,\xi\rangle_{\text{gen}}$. However, the difficulty in finding a general solution of (30) for an arbitrary function $f(n)$ has barred the community for a long time to find an exact expression of the generalized squeezed states $|\alpha,\xi\rangle_{\text{gen}}$. In some of the recent studies, the squeezed states for particular quantum systems having a particular form of the function $f(n)$ have been constructed [28], however, a general solution of the squeezed state has come out recently in [31]. The usage of non-Hermitian theories becomes inevitable since, on many occasions, the Hamiltonians turn out to be non-Hermitian; see, for instance [28, 46], for more details. Nevertheless, the computation of the general solution of (30) involves certain mathematical challenges, which is, however, resolved in [31] and the solution is given by

$$\mathcal{I}(\alpha,\xi,n) = \sum_{m=0}^{[n/2]} (-\xi)^m \alpha^{n-2m} g(n,m), \tag{31}$$

with

$$g(n,m) = \sum_{j_1=1}^{(n-2m+1)} \sum_{j_2=j_1+2}^{(n-2m+3)} \sum_{j_3=j_2+2}^{(n-2m+5)} \cdots \sum_{j_m=j_{m-1}+2}^{(n-1)} \mu. \tag{32}$$

Here, $\mu = k(j_1)k(j_2)\cdots k(j_m)$, $g(n,0) = 1$ and $k(j) := jf^2(j)$. The symbol $[n]$ in (31) denotes the Floor function, which results in an integer value less than or equal to the corresponding real number $n$. It is now obvious that the generalized squeezed state comprised of (29) and (31) can be applied to any quantum mechanical system to find the corresponding squeezed state. In [31], the result has been analyzed with an example of a Rosen-Morse potential where it was shown that the general solution (31) can be utilized without any further difficulties.

*4.1.2. Schrödinger cat states* Generalization of cat states is straightforward once the same is realized for coherent states. The generalized coherent states in the $q$-deformed formulation arises from the definition $\hat{A}_q|\alpha\rangle_q^{\text{gen}} = \alpha|\alpha\rangle_q^{\text{gen}}$ with the use of (25) as follows

$$|\alpha\rangle_q^{\text{gen}} = \frac{1}{\mathcal{N}_q(\alpha,q)} \sum_{n=0}^{\infty} \frac{\alpha^n}{\sqrt{[n]_q!}}|n\rangle_q \quad \alpha \in \mathbb{C}. \tag{33}$$

The Schrödinger cat states are formed as even and odd superposition of coherent states [27]

$$|\alpha\rangle_{q,\pm}^{\text{gen}} = \frac{1}{\mathcal{N}_{q,\pm}(\alpha,q)}\Big(|\alpha\rangle_q^{\text{gen}} \pm |-\alpha\rangle_q^{\text{gen}}\Big), \tag{34}$$

with the normalization constant being

$$\mathcal{N}_{q,\pm}^2(\alpha,q) = 2 \pm \frac{2}{\mathcal{N}_q^2(\alpha,q)} \sum_{n=0}^{\infty} \frac{(-1)^n|\alpha|^{2n}}{[n]_q!}. \tag{35}$$

For further details on the construction, we refer the readers to [27].

*4.1.3. Photon-added coherent states* Photon-added coherent states are yet another interesting class of nonclassical states, the $q$-deformed generalization of which can be performed by operating the deformed photon creation operator $\hat{A}_q^\dagger$ successively on the deformed coherent states

$$|\alpha, m\rangle_q^{\text{gen}} = \frac{1}{\mathcal{N}_q(\alpha, q, m)} \hat{A}_q^{\dagger m} |\alpha\rangle_q^{\text{gen}} = \frac{1}{\mathcal{N}_q(\alpha, q, m)\mathcal{N}_q(\alpha, q)} \sum_{n=0}^{\infty} \frac{\alpha^n}{[n]_q!} \sqrt{[n+m]_q!}\, |n+m\rangle_q, \quad (36)$$

with the normalization constant being

$$\mathcal{N}_q^2(\alpha, q, m) = {}^{\text{gen}}{}_q\langle \alpha, m | \hat{A}_q^m \hat{A}_q^{\dagger m} | \alpha, m\rangle_q^{\text{gen}} = \frac{1}{\mathcal{N}_q^2(\alpha, q)} \sum_{n=0}^{\infty} \frac{|\alpha|^{2n}}{[n]_q!^2} [n+m]_q!. \quad (37)$$

A more detailed analysis of the state can be found in [29].

*4.1.4. Photon-subtracted squeezed vacuum states* Squeezed vacuum states (SVS) are nonclassical. Photon subtraction from the SVS makes the state even more nonclassical. The harmonic oscillator SVS $|\xi\rangle$ are built from the solution of the following equation

$$(\hat{a}\cosh r + \hat{a}^\dagger e^{i\theta} \sinh r)|\xi\rangle = 0, \quad (38)$$

which originally follows the definition of SVS, $|\xi\rangle = \hat{S}(\xi)|0\rangle$. The generalization of SVS does not arise from the direct replacement of $\hat{a}, \hat{a}^\dagger$ with $\hat{A}, \hat{A}^\dagger$ because the generalized squeezing operator $\hat{S}_{\text{gen}}(\xi) = \exp\{(\xi \hat{A}^{\dagger 2} - \xi^* \hat{A}^2)/2\}$ is no longer possible to disentangle and, therefore, the relations $(\hat{A}\cosh r + \hat{A}^\dagger e^{i\theta}\sinh r)|\xi\rangle_{\text{gen}} = 0$ and $|\xi\rangle_{\text{gen}} = \hat{S}_{\text{gen}}(\xi)|0\rangle$ are no longer equivalent to each other. The problem can be overcome by introducing a set of generalized auxiliary ladder operators $\hat{B} = \hat{a}[1/f(\hat{n})] = [1/f(\hat{n}+1)]\hat{a}$ and $\hat{B}^\dagger = [1/f(\hat{n})]\hat{a}^\dagger = \hat{a}^\dagger[1/f(\hat{n}+1)]$ giving rise to a new set of commutation relations $[\hat{A}, \hat{B}^\dagger] = [\hat{B}, \hat{A}^\dagger] = 1$. Therefore, the new set of conjugate ladder operators $\hat{A}, \hat{B}^\dagger$ or $\hat{B}, \hat{A}^\dagger$ will permit the disentanglement of the squeezing operator [32]. The definition of generalized SVS is, thus, altered as

$$(\hat{A}\cosh r + \hat{B}^\dagger e^{i\theta} \sinh r)|\xi\rangle_{\text{gen}} = 0, \quad (39)$$

leading to the recurrence relation

$$\mathcal{J}_{m+1} = -\frac{e^{i\theta}\tanh r}{f(m)f(m+1)} \sqrt{\frac{m}{m+1}} \mathcal{J}_{m-1}, \quad (40)$$

which when solved in an even basis, the required state comes out as follows

$$|\xi\rangle_{\text{gen}} = \frac{1}{\mathcal{N}(r)} \sum_{n=0}^{\infty} (-1)^n \frac{e^{in\theta}(\tanh r)^n \sqrt{(2n)!}}{2^n n! f(2n)!} |2n\rangle. \quad (41)$$

The odd basis solution of (40) gives rise to the generalized squeezed first excited state that may also be interesting, however, it is not our focus here. Photon-subtracted SVS $|\xi, m\rangle_{\text{gen}}$ are then prepared by removing photons from the SVS, as given by [32]

$$|\xi, m\rangle_{\text{gen}}^{\text{e}} = \hat{A}^{2m}|\xi\rangle_{\text{gen}} = \frac{1}{\mathcal{N}^{\text{e}}(r,m)} \sum_{n=0}^{\infty} \frac{(-\tanh r)^k e^{ik\theta}(2k)!}{2^k k! \sqrt{(2n)!} f(2n)!} |2n\rangle, \quad (42)$$

and

$$|\xi, m\rangle_{\text{gen}}^{\text{o}} = \hat{A}^{2m+1}|\xi\rangle_{\text{gen}} = \frac{1}{\mathcal{N}^{\text{o}}(r,m)} \sum_{n=0}^{\infty} \frac{(-\tanh r)^{k'} e^{ik'\theta}(2k')!}{2^{k'} k'! \sqrt{(2n+1)!} f(2n+1)!} |2n+1\rangle, \quad (43)$$

in even and odd basis, respectively. Here, $k = m + n$ and $k' = k + 1$.

### *4.2. Resource theories*

The sole intention of building all of the above generalized nonclassical states is to study their usefulness as resources. The three resource theories discussed in subsections 2.2, 2.3 and 2.4 have been analyzed for all of the four generalized states along with their harmonic oscillator counterparts in [27, 29, 31, 32]. Quadrature and number squeezing are relatively straightforward to compute, but due to mathematical complexity, sometimes the computations of quasi-probability distribution functions become troublesome for generalized states. However, as discussed earlier in subsection 2.2 that the phenomenon of quadrature squeezing is equivalent to the negativity of the $P$-function, which is the strongest measure of nonclassicality among the three quasi-probability distribution functions. Therefore, even if someone is not able to compute the quasi-probability distribution functions in some cases, the results will not be weakened provided that one is able to compute the quadrature squeezing. In the cases of squeezed states [31] and photon-subtracted squeezed vacuum states [32], all of the three resource theories have been discussed in this article, whereas in the case of cat states [27] and photon-added coherent states [29], only two of them, namely the quadrature and number squeezing, have been analyzed. In fact, for photon-added coherent states [29] higher order squeezing effects have been studied which are known to be a more robust measure of nonclassicality compared to the first order squeezing effects.

It is known that most of the harmonic oscillator nonclassical states are good resources of nonclassicality in the sense that they exhibit non-negativity of the quasi-probability distributions, quadrature squeezing and number squeezing with respect to the measures (6), (15) and (18), respectively. However, the novelty of the results in [27, 29, 31, 32] lies in the fact that the generalized nonclassical states behave as a richer source of nonclassicality in comparison to those of the harmonic oscillator with respect to the same measures of nonclassicality given in (6), (15) and (18). Moreover, one of the striking features of the $q$-deformed generalizations is that one obtains an additional degree of freedom in such systems with which the degree of nonclassicality can be enhanced further by controlling the deformation parameter. It was shown that in some cases, the nonclassicality measure attains a value very close to the maximal possible value, which has never been possible earlier for harmonic oscillator nonclassical states. Overall, it is indicative that the generalized nonclassical states may be utilized as enriched resources of nonclassicality and, thus, they legitimately claim further attention and exploration.

## 5. Realization of generalized nonclassical light in real life

Real-life realization of squeezed states and other nonclassical states of the harmonic oscillator is a long-term goal of researchers. The aspiration has started in the late twentieth century and, nowadays, they are fairly well-realized via numerous sophisticated methods, like; wave mixing in optical fibers [144, 145], parametric down-conversion [146], normal modes of the coupled atom-field system [147], optical [148, 149] and nanomechanical resonators [150], microwave radiation in superconducting resonant circuit [151], homodyne detection technique [152, 153], optomechanical coupling [154, 155], many-body systems [156], superconducting qubit[157], etc. Nonlinear light is nothing but a nonlinear dependence in the polarization of the light. Therefore, the simplest realization of nonlinear nonclassical lights can be brought in by sending the harmonic oscillator nonclassical lights through a nonlinear crystal or any other nonlinear mediums. Indeed, such a concept has been utilized in [158] where a Kerr-type nonlinearity has been introduced in one of the arms of a Mach-Zehnder interferometer to analyze the entanglement properties of the output. Though, at that time, the output states were not interpreted as nonlinear quantum light and the states emerging from the setup may not be identical with the generalized nonclassical states that are studied here. However, a similar concept can be considered as the starting point of the journey of nonlinear quantum optical states. Generation of the nonlinear quantum optical states in real life are underway. More sophisticated concepts of nonlinearity have been introduced over the years, however, the primitive idea remains somewhat similar. Today there exist plenty of studies exhibiting Kerr-type nonlinear effects through atomic coherence in multilevel atomic systems [159], carbon nanotubes [160], optical cavities with interacting photons [161], optomechanical cavities [162, 163], topological nanostructure [164], nanoimaging in graphene [165], etc. Several other types of nonlinearities have also been realized primarily via quantum optomechanical interaction [166, 167, 168] and their usefulness in practical scenarios has been indicated [169]. The realization of generalized nonclassical lights that we have reviewed in this article will surely require further sophistication, however, we believe that the clue to the realization of such states lies in the realization of Kerr and other types of nonlinearities, which are underway. We hope the experimental realization of generalized nonclassical light will be possible in near future.

## 6. Conclusions

This review article began with a broader understanding of the operational approach of quantum resource theory followed by a focused discussion on various nonclassical resources of different quantum optical states. We have summarized some of the most common nonclassical measures as resources, such as; quasi-probability distribution functions, quadrature squeezing and sub-Poissonian photon statistics. One of the important objectives of this review is to scrutinize the usefulness of generalized nonlinear and $q$-deformed quantum optical states in different frameworks of quantum optics and information theories, and to quantify their nonclassicality measure within the unified roof of nonclassical resource theories. There are several other interesting observations in Sec. 4 during the study of $q$-deformed nonclassical states, especially the cat states and the photon-added coherent states. A common feature that is attributed to both of the states is that when the deformation is increased, the corresponding nonclassical states demonstrate an increased degree of nonclassicality. The deformation parameter being the characteristics of the models themselves, it is conceivable that the elevated degree of nonclassicality is an inherent property of the corresponding nonclassical states, which cannot be generated from the standard quantum optical states via quantum operations. Thereby exploring a new possibility of obtaining better resources of nonclassicality that can be utilized more efficiently towards the different purposes of quantum information and computation.

## Acknowledgments

The author acknowledges the support of the research grant (DST/INSPIRE/04/2016/001391) by DST-INSPIRE, Govt. of India.

## References

[1] Chitambar E and Gour G 2019 *Rev. Mod. Phys.* **91** 025001
[2] Liu Z W, Bu K and Takagi R 2019 *Phys. Rev. Lett.* **123** 020401
[3] Baumgratz T, Cramer M and Plenio M B 2014 *Phys. Rev. Lett.* **113** 140401
[4] Winter A and Yang D 2016 *Phys. Rev. Lett.* **116** 120404
[5] Streltsov A, Rana S, Bera M N and Lewenstein M 2017 *Phys. Rev. X* **7** 011024
[6] Tan K C, Volkoff T, Kwon H and Jeong H 2017 *Phys. Rev. Lett.* **119** 190405
[7] Streltsov A, Adesso G and Plenio M B 2017 *Rev. Mod. Phys.* **89** 041003
[8] Asbóth J K, Calsamiglia J and Ritsch H 2005 *Phys. Rev. Lett.* **94** 173602
[9] Yadin B, Binder F C, Thompson J, Narasimhachar V, Gu M and Kim M S 2018 *Phys. Rev. X* **8** 041038
[10] Kwon H, Tan K C, Volkoff T and Jeong H 2019 *Phys. Rev. Lett.* **122** 040503
[11] Tan K C and Jeong H 2019 *Quantum Rep.* **1** 151–161
[12] Ge W, Jacobs K, Asiri S, Foss-Feig M and Zubairy M S 2020 *Phys. Rev. Research* **2** 023400
[13] Ferrari G, Lami L, Theurer T and Plenio M B 2020 *arXiv:2010.00044*
[14] Rosset D, Buscemi F and Liang Y C 2018 *Phys. Rev. X* **8** 021033

[15] Wolfe E, Schmid D, Sainz A, Kunjwal R and Spekkens R W 2020 *Quantum* **4** 280
[16] Rosset D, Schmid D and Buscemi F 2020 *Phys. Rev. Lett.* **125** 210402
[17] Gallego R and Aolita L 2015 *Phys. Rev. X* **5** 041008
[18] Pirandola S 2019 *Commun. Phys.* **2** 51
[19] Wu K D, Kondra T V, Rana S, Scandolo C M, Xiang G Y, Li C F, Guo G C and Streltsov A 2021 *Phys. Rev. Lett.* **126** 090401
[20] Pirandola S, Laurenza R, Lupo C and Pereira J L 2019 *npj Quantum. Inf.* **5** 50
[21] Banchi L, Pereira J, Lloyd S and Pirandola S 2020 *npj Quantum. Inf.* **6** 42
[22] Horodecki R, Horodecki P, Horodecki M and Horodecki K 2009 *Rev. Mod. Phys.* **81** 865–942
[23] Eltschka C and Siewert J 2014 *J. Phys. A: Math. Theor.* **47** 424005
[24] Killoran N, Steinhoff F E S and Plenio M B 2016 *Phys. Rev. Lett.* **116** 080402
[25] Pirandola S, Laurenza R, Ottaviani C and Banchi L 2017 *Nature Commun.* **8** 15043
[26] Contreras-Tejada P, Palazuelos C and de Vicente J I 2019 *Phys. Rev. Lett.* **122** 120503
[27] Dey S 2015 *Phys. Rev. D* **91** 044024
[28] Dey S and Hussin V 2015 *Phys. Rev. D* **91** 124017
[29] Dey S and Hussin V 2016 *Phys. Rev. A* **93** 053824
[30] Dey S, Fring A and Hussin V 2017 *Int. J. Mod. Phys. B* **31** 1650248
[31] Zelaya K, Dey S and Hussin V 2018 *Phys. Lett. A* **382** 3369–3375
[32] Dey S and Nair S S 2020 *J. Phys. A: Math. Theor.* **53** 385305
[33] Dey S and Fring A 2013 *Phys. Rev. A* **88** 022116
[34] Glauber R J 1963 *Phys. Rev.* **131** 2766–2788
[35] Sudarshan E C G 1963 *Phys. Rev. Lett.* **10** 277–279
[36] Coecke B, Fritz T and Spekkens R W 2016 *Inf. Comput.* **250** 59–86
[37] Husimi K 1940 *Proc. Phys. Math. Soc. Jpn.* **22** 264–314
[38] Wigner E 1932 *Phys. Rev.* **40** 749–759
[39] Scully M O and Zubairy M S 1997 *Quantum optics* (Cambridge, England: Cambridge Univ. Press)
[40] Dodonov V V and Man'ko V I 2003 *Theory of nonclassical states of light* (London: Taylor & Francis Group)
[41] Gerry C and Knight P 2005 *Introductory quantum optics* (Cambridge, England: Cambridge Univ. Press)
[42] Klauder J R and Sudarshan E C G 2006 *Fundamentals of quantum optics* (New York: Dover Pub.)
[43] Walls D F and Milburn G J 2007 *Quantum optics* (Berlin Heidelberg: Springer-Verlag)
[44] Agarwal G S 2013 *Quantum optics* (New York: Cambridge Univ. Press)
[45] Tan K C, Choi S and Jeong H 2020 *Phys. Rev. Lett.* **124** 110404
[46] Dey S, Fring A and Hussin V 2018 *Coherent states and their applications* vol 205 ed Antoine J P, Bagarello F and Gazeau J P (Cham: Springer) pp 209–242
[47] Mollow B R and Glauber R J 1967 *Phys. Rev.* **160** 1076–1096
[48] Gehrke C, Sperling J and Vogel W 2012 *Phys. Rev. A* **86** 052118
[49] Vogel W and Sperling J 2014 *Phys. Rev. A* **89** 052302
[50] Sperling J and Vogel W 2015 *Phys. Scr.* **90** 074024
[51] Mandel L and Wolf E 1995 *Optical coherence and quantum optics* (New York: Cambridge Univ. Press)
[52] Vogel W and Welsch D G 2006 *Quantum optics* (Weinheim: Wiley-VCH)
[53] Short R and Mandel L 1983 *Phys. Rev. Lett.* **51** 384–387
[54] Teich M C and Saleh B E A 1985 *J. Opt. Soc. Am. B* **2** 275–282
[55] Neumann J V and Wigner E 1929 *Zhurnal Physik* **30** 467–470
[56] Friedrich H and Wintgen D 1985 *Phys. Rev. A* **32** 3231–3242
[57] Persson E, Gorin T and Rotter I 1996 *Phys. Rev. E* **54** 3339–3351
[58] Bender C M and Boettcher S 1998 *Phys. Rev. Lett.* **80** 5243–5246
[59] Mostafazadeh A 2002 *J. Math. Phys.* **43** 205
[60] Wigner E P 1960 *J. Math. Phys.* **1** 409
[61] Bender C M 2007 *Rep. Prog. Phys.* **70** 947–1018
[62] Bender C M, Brody D C and Jones H F 2002 *Phys. Rev. Lett.* **89** 270401
[63] Dorey P, Dunning C and Tateo R 2001 *J. Phys. A: Math. Gen.* **34** 5679–5704
[64] Shin K C 2001 *J. Math. Phys.* **42** 2513
[65] Weigert S 2003 *J. Opt. B: Quant. Semiclas. Opt.* **5** S416–S419
[66] Dey S, Fring A and Gouba L 2012 *J. Phys. A: Math. Theor.* **45** 385302
[67] Dey S, Raj A and Goyal S K 2019 *Phys. Lett. A* **383** 125931
[68] Guo A, Salamo G J, Duchesne D, Morandotti R, Volatier-Ravat M, Aimez V, Siviloglou G A and Christodoulides D N 2009 *Phys. Rev. Lett.* **103** 093902
[69] Rüter C E, Makris K G, El-Ganainy R, Christodoulides D N, Segev M and Kip D 2010 *Nature Phys.* **6** 192–195

[70] Regensburger A, Bersch C, Miri M A, Onishchukov G, Christodoulides D N and Peschel U 2012 *Nature* **488** 167–171
[71] Feng L, Xu Y L, Fegadolli W S, Lu M H, Oliveira J E B, Almeida V R, Chen Y F and Scherer A 2013 *Nature Materials* **12** 108–113
[72] Lin Z, Ramezani H, Eichelkraut T, Kottos T, Cao H and Christodoulides D N 2011 *Phys. Rev. Lett.* **106** 213901
[73] Chong Y D, Ge L, Cao H and Stone A D 2010 *Phys. Rev. Lett.* **105** 053901
[74] Bittner S, Dietz B, Günther U, Harney H L, Miski-Oglu M, Richter A and Schäfer F 2012 *Phys. Rev. Lett.* **108** 024101
[75] Zheng C, Hao L and Long G L 2013 *Phil. Trans. R. Soc. A* **371** 20120053
[76] Xiao L, Zhan X, Bian Z H, Wang K K, Zhang X, Wang X P, Li J, Mochizuki K, Kim D, Kawakami N, Yi W, Obuse H, Sanders B C and Xue P 2017 *Nature Phys.* **13** 1117–1123
[77] Mittal V, Raj A, Dey S and Goyal S K 2021 *Scientific Reports* **11** 10262
[78] Chtchelkatchev N M, Golubov A A, Baturina T I and Vinokur V M 2012 *Phys. Rev. Lett.* **109** 150405
[79] El-Ganainy R, Makris K G, Khajavikhan M, Musslimani Z H, Rotter S and Christodoulides D N 2018 *Nature Phys.* **14** 11–19
[80] Moiseyev N 2011 *Non-Hermitian quantum mechanics* (Cambridge, England: Cambridge Univ. Press)
[81] Bagarello F, Gazeau J P, Szafraniec F H and Znojil M (eds) 2015 *Non-selfadjoint operators in quantum physics: Mathematical aspects* (New Jersey: John Wiley & Sons)
[82] Bender C M 2018 *PT symmetry: in quantum and classical physics* (Singapore: World Scientific)
[83] Pauli W 1943 *Rev. Mod. Phys.* **15** 175–207
[84] Sudarshan E C G 1961 *Phys. Rev.* **123** 2183–2193
[85] Lee T and Wick G 1969 *Nucl. Phys. B* **9** 209–243
[86] Dieudonné J 1961 *Proc. Int. Sympos. Linear Spaces (Jerusalem, 1960), Pergamon, Oxford* 115–122
[87] Scholtz F G, Geyer H B and Hahne F J W 1992 *Ann. Phys.* **213** 74–101
[88] Mostafazadeh A 2010 *Int. J. Geom. Methods Mod. Phys.* **07** 1191–1306
[89] Brody D C 2014 *J. Phys. A: Math. Theor.* **47** 035305
[90] Bagchi B, Quesne C and Roychoudhury R 2005 *J. Phys. A: Math. Gen.* **38** L647–L652
[91] Znojil M and Geyer H B 2006 *Phys. Lett. B* **640** 52–56
[92] Dey S, Fring A and Mathanaranjan T 2014 *Ann. Phys.* **346** 28–41
[93] Dey S, Fring A and Mathanaranjan T 2015 *Int. J. Theor. Phys.* **54** 4027–4033
[94] Bender C M, Brody D C and Jones H F 2004 *Phys. Rev. D* **70** 025001
[95] Ghatak A and Mandal B P 2013 *Commun. Theor. Phys.* **59** 533–539
[96] Figueira de Morisson Faria C and Fring A 2006 *Czech. J. Phys.* **56** 899–908
[97] Mostafazadeh A 2003 *arXiv:quant-ph/0310164*
[98] Bender C M, Chen J H and Milton K A 2006 *J. Phys. A: Math. Gen.* **39** 1657–1668
[99] Dey S 2014 *Ph.D. Thesis, arXiv:1410.3193*
[100] Graefe E M, Korsch H J and Niederle A E 2008 *Phys. Rev. Lett.* **101** 150408
[101] Dey S, Fring A and Khantoul B 2013 *J. Phys. A: Math. Theor.* **46** 335304
[102] Dey S and Fring A 2013 *Acta Polytechnica* **53** 268–270
[103] Dey S and Fring A 2014 *Phys. Rev. D* **90** 084005
[104] Dey S, Fring A and Gouba L 2015 *J. Phys. A: Math. Theor.* **48** 40FT01
[105] Zelaya K, Dey S, Hussin V and Rosas-Ortiz O 2020 *Quantum Rep.* **2** 12–38
[106] Perelomov A M 1972 *Commun. Math. Phys.* **26** 222–236
[107] Perelomov A 1986 *Generalized coherent states and their applications* (Berlin Heidelberg: Springer-Verlag)
[108] Zhang W M, Feng D H and Gilmore R 1990 *Rev. Mod. Phys.* **62** 867–927
[109] Ali S T, Antoine J P and Gazeau J P 2014 *Coherent states, wavelets, and their generalizations* (New York: Springer)
[110] Dey S 2017 *Physical and mathematical aspects of symmetries* ed Duarte S, Gazeau J P, Faci S, Micklitz T, Scherer R and Toppan F (Cham: Springer) pp 145–152
[111] de Matos Filho R L and Vogel W 1996 *Phys. Rev. A* **54** 4560–4563
[112] Man'ko V I, Marmo G, Sudarshan E C G and Zaccaria F 1997 *Phys. Scr.* **55** 528–541
[113] Mancini S 1997 *Phys. Lett. A* **233** 291–296
[114] Sivakumar S 1999 *J. Phys. A: Math. Gen.* **32** 3441–3447
[115] Trifonov D A 2000 *J. Opt. Soc. Am. A* **17** 2486–2495
[116] Kwek L C and Kiang D 2003 *J. Opt. B: Quantum Semiclas. Opt.* **5** 383–386
[117] Hertz A, Dey S, Hussin V and Eleuch H 2016 *Symmetry* **8** 36
[118] Gazeau J P and Klauder J R 1999 *J. Phys. A: Math. Gen.* **32** 123–132
[119] Dey S and Fring A 2012 *Phys. Rev. D* **86** 064038

[120] Dey S, Fring A, Gouba L and Castro P G 2013 *Phys. Rev. D* **87** 084033
[121] Benedict M G and Molnár B 1999 *Phys. Rev. A* **60** R1737–R1740
[122] Fernández D J, Hussin V and Rosas-Ortiz O 2007 *J. Phys. A: Math. Theor.* **40** 6491–6511
[123] Arik M and Coon D D 1976 *J. Math. Phys.* **17** 524
[124] Biedenharn L C 1989 *J. Phys. A: Math. Gen.* **22** L873–L878
[125] Macfarlane A J 1989 *J. Phys. A: Math. Gen.* **22** 4581–4588
[126] Sun C P and Fu H C 1989 *J. Phys. A: Math. Gen.* **22** L983–L986
[127] Kulish P P and Damaskinsky E V 1990 *J. Phys. A: Math. Gen.* **23** L415–L419
[128] Quesne C, Penson K and Tkachuk V 2003 *Phys. Lett. A* **313** 29–36
[129] Naderi M H, Soltanolkotabi M and Roknizadeh R 2004 *J. Phys. A: Math. Gen.* **37** 3225–3240
[130] Fakhri H and Hashemi A 2016 *Phys. Rev. A* **93** 013802
[131] Jayakrishnan M P, Dey S, Faizal M and Sudheesh C 2017 *Ann. Phys.* **385** 584–590
[132] Walls D F 1983 *Nature* **306** 141–146
[133] Loudon R and Knight P L 1987 *J. Mod. Opt.* **34** 709–759
[134] Hillery M 2000 *Phys. Rev. A* **61** 022309
[135] Menicucci N C, van Loock P, Gu M, Weedbrook C, Ralph T C and Nielsen M A 2006 *Phys. Rev. Lett.* **97** 110501
[136] Riedel M F, Böhi P, Li Y, Hänsch T W, Sinatra A and Treutlein P 2010 *Nature* **464** 1170–1173
[137] Giovannetti V, Lloyd S and Maccone L 2011 *Nature Photon.* **5** 222–229
[138] Vahlbruch H, Chelkowski S, Hage B, Franzen A, Danzmann K and Schnabel R 2006 *Phys. Rev. Lett.* **97** 011101
[139] Schnabel R, Mavalvala N, McClelland D E and Lam P K 2010 *Nature Commun.* **1** 121
[140] Aasi J, Abadie J, Abbott B, Abbott R, Abbott T, Abernathy M, Adams C, Adams T, Addesso P, Adhikari R *et al.* 2013 *Nature Photon.* **7** 613–619
[141] Chua S S Y, Slagmolen B J J, Shaddock D A and McClelland D E 2014 *Class. Quantum Grav.* **31** 183001
[142] Dodonov V V 2002 *J. Opt. B: Quantum Semiclas. Opt.* **4** R1–R33
[143] Braunstein S L and van Loock P 2005 *Rev. Mod. Phys.* **77** 513–577
[144] Reid M D and Walls D F 1985 *Phys. Rev. A* **31** 1622–1635
[145] Slusher R E, Hollberg L W, Yurke B, Mertz J C and Valley J F 1985 *Phys. Rev. Lett.* **55** 2409–2412
[146] Wu L A, Kimble H J, Hall J L and Wu H 1986 *Phys. Rev. Lett.* **57** 2520–2523
[147] Raizen M G, Orozco L A, Xiao M, Boyd T L and Kimble H J 1987 *Phys. Rev. Lett.* **59** 198–201
[148] Dodonov V V, Klimov A B and Man'ko V I 1990 *Phys. Lett. A* **149** 225–228
[149] Bose S, Jacobs K and Knight P L 1997 *Phys. Rev. A* **56** 4175–4186
[150] Rabl P, Shnirman A and Zoller P 2004 *Phys. Rev. B* **70** 205304
[151] Zagoskin A M, Il'ichev E, McCutcheon M W, Young J F and Nori F 2008 *Phys. Rev. Lett.* **101** 253602
[152] Ourjoumtsev A, Jeong H, Tualle-Brouri R and Grangier P 2007 *Nature* **448** 784–786
[153] Sychev D V, Ulanov A E, Pushkina A A, Richards M W, Fedorov I A and Lvovsky A I 2017 *Nature Photon.* **11** 379–382
[154] Gu W J, Li G X and Yang Y P 2013 *Phys. Rev. A* **88** 013835
[155] Rashid M, Tufarelli T, Bateman J, Vovrosh J, Hempston D, Kim M S and Ulbricht H 2016 *Phys. Rev. Lett.* **117** 273601
[156] Omran A, Levine H, Keesling A, Semeghini G, Wang T T, Ebadi S, Bernien H, Zibrov A S, Pichler H, Choi S *et al.* 2019 *Science* **365** 570–574
[157] Song C, Xu K, Li H, Zhang Y R, Zhang X, Liu W, Guo Q, Wang Z, Ren W, Hao J *et al.* 2019 *Science* **365** 574–577
[158] Sanders B C 1992 *Phys. Rev. A* **45** 6811–6815
[159] Wang H, Goorskey D and Xiao M 2001 *Phys. Rev. Lett.* **87** 073601
[160] Gambetta A, Manzoni C, Menna E, Meneghetti M, Cerullo G, Lanzani G, Tretiak S, Piryatinski A, Saxena A, Martin R L and Bishop A R 2006 *Nature Phys.* **2** 515–520
[161] Chang D E, Vuletić V and Lukin M D 2014 *Nature Photon.* **8** 685–694
[162] Lü X Y, Zhang W M, Ashhab S, Wu Y and Nori F 2013 *Scientific Rep.* **3** 2943
[163] Yan Y, Zhu J P and Li G X 2016 *Opt. Express* **24** 13590–13609
[164] Kruk S, Poddubny A, Smirnova D, Wang L, Slobozhanyuk A, Shorokhov A, Kravchenko I, Luther-Davies B and Kivshar Y 2019 *Nature Nanotech.* **14** 126–130
[165] Jiang T, Kravtsov V, Tokman M, Belyanin A and Raschke M B 2019 *Nature Nanotech.* **14** 838–843
[166] Pikovski I, Vanner M R, Aspelmeyer M, Kim M S and Brukner Č 2012 *Nature Phys.* **8** 393–397
[167] Dey S, Bhat A, Momeni D, Faizal M, Ali A F, Dey T K and Rehman A 2017 *Nucl. Phys. B* **924** 578–587
[168] Khodadi M, Nozari K, Dey S, Bhat A and Faizal M 2018 *Scientific Reports* **8** 1659
[169] Dey S and Hussin V 2019 *Int. J. Theor. Phys.* **58** 3138–3148

# Pseudo-hermitian random matrix theory: a review

**Joshua Feinberg**[1,2,3] **and Roman Riser**[1,2]

[1] Department of Mathematics and
[2] Haifa Research Center for Theoretical Physics and Astrophysics, University of Haifa, Haifa 31905, Israel
[3] https://orcid.org/0000-0002-2869-0010

**Abstract.** We review our recent results on pseudo-hermitian random matrix theory which were hitherto presented in various conferences and talks. (Detailed accounts of our work will appear soon in separate publications.) Following an introduction of this new type of random matrices, we focus on two specific models of matrices which are pseudo-hermitian with respect to a given indefinite metric **B**. Eigenvalues of pseudo-hermitian matrices are either real, or come in complex-conjugate pairs. The diagrammatic method is applied to deriving explicit analytical expressions for the density of eigenvalues in the complex plane and on the real axis, in the large-$N$, planar limit. In one of the models we discuss, the metric **B** depends on a certain real parameter $t$. As $t$ varies, the model exhibits various 'phase transitions' associated with eigenvalues flowing from the complex plane onto the real axis, causing disjoint eigenvalue support intervals to merge. Our analytical results agree well with presented numerical simulations.

## 1. Introduction

PT-symmetric quantum mechanics (PTQM) [1] and its broader applications (see [2] for recent reviews) have been at the focus of intensive and prolific research activity during the past quarter of century. Broadly speaking, the Hilbert space of a PT-symmetric quantum mechanical model is endowed with a non-trivial metric operator, with respect to which the hamiltonian is hermitian. Hamiltonians of PTQM models with proper metrics are sometimes referred to as *quasi-hermitian* [3, 4]. They are diagonalizable, and their spectrum is essentially real, because they are similar to a conventionally hermitian hamiltonian. In contrast, hamiltonians of PTQM models with *indefinite* metrics are referred to as *pseudo-hermitian* [5, 6], and their eigenvalues are either real or come in complex-conjugate pairs. For a mathematically precise summary of the nomenclature of quasi-hermiticity and pseudo-hermiticity see [7].

Quasi-hermitian matrices can be thought of as truncated quasi-hermitian linear operators. In [8, 9] we found it useful to further distinguish *strictly*-quasi-hermitian matrices within the broader class of quasi-hermitian ones: Strictly-quasi-hermitian matrices are hermitian with respect to positive definite (and therefore invertible) metrics. In contrast, merely quasi-hermitian matrices are associated with non-negative non-invertible metrics.

An interesting strictly-quasi-hermitian random matrix model was introduced in [10]. These authors fixed a metric, and took the hamiltonian as random, with the aim of studying numerically the dependence of the average density of eigenvalues and level spacing statistics on the metric. Yet another interesting example of a strictly-quasi-hermitian random matrix model, akin to the Dicke model of superradiance, was provided by [11], in which a numerical study of the level spacing distribution was carried out.

Upon truncation to finite vector spaces, pseudo-hermitian operators turn into pseudo-hermitian matrices. See [12] for a recent discussion of (real asymmetric) pseudo-hermitian random matrices.

In this paper we introduce a family of pseudo-hermitian random matrix models and review our analytical and numerical results for the distribution of real and complex eigenvalues. A detailed account of these investigations will appear soon separately [13].

Let us start with the definitions of strict quasi-hermiticity and pseudo-hermiticity:

**Definition 1** (strict quasi-hermiticity). *Let* $\mathbf{B}$ *and* $\phi$ *be* $N \times N$ *matrices where* $\phi$ *is a complex matrix and* $\mathbf{B}$ *a hermitian, positive definite* metric. *We call* $\phi$ *strictly-quasi-hermitian with respect to the metric* $\mathbf{B}$ *if it fulfills the* intertwining relation

$$\phi^\dagger \mathbf{B} = \mathbf{B}\phi. \tag{1}$$

The definition means that $\phi$ is hermitian with respect to the metric $\mathbf{B}$. For $\mathbf{B} = \mathbb{1}$ it reduces to ordinary hermiticity.

Given the metric $\mathbf{B}$, the general solution of the intertwining relation (1) is

$$\phi = \mathbf{AB}, \tag{2}$$

where $A = A^\dagger$ is an arbitrary hermitian matrix [10]. Thus, given $\mathbf{B}$, there are $N^2$ independent quasi-hermitian matrices with respect to $\mathbf{B}$.

An important corollary is that for a (positive definite) metric, the spectrum of $\phi$ is always real. This can be seen easily when we write

$$\phi = \sqrt{\mathbf{B}}^{-1} \underbrace{\sqrt{\mathbf{B}}\mathbf{A}\sqrt{\mathbf{B}}}_{\text{hermitian}} \sqrt{\mathbf{B}} \tag{3}$$

where it is obvious that $\phi$ is similar to a hermitan matrix, provided $\sqrt{\mathbf{B}}$ is hermitian, which requires that $\mathbf{B}$ be positive. Spectra of strictly-quasi-hermitian matrices were studied in [8, 9, 10, 11].

When we relax the condition that $\mathbf{B}$ has to be positive definite, we come to the definition of pseudo-hermiticity:

**Definition 2** (pseudo-hermiticity). *Let* $\mathbf{B}$ *and* $\phi$ *be* $N \times N$ *matrices where* $\phi$ *is a complex matrix and* $\mathbf{B}$ *an* indefinite *metric. We call* $\phi$ *pseudo-hermitian with respect to the indefinite metric* $\mathbf{B}$ *if it fulfills the* intertwining relation

$$\phi^\dagger \mathbf{B} = \mathbf{B}\phi. \tag{4}$$

The general solution of the intertwining relation (4) is still given by (2) where $\mathbf{A}$ is an arbitrary hermitian matrix [13]. However, contrary to the strictly-quasi-hermitian case, the spectrum of pseudo-hermitian matrices need no longer be purely real, and may contain pairs of complex-conjugated eigenvalues.

As a motivation, let us first give a simple physical example where strictly-quasi-hermitian matrices appear [8, 9]. Consider a *highly-connected* mechanical system with $N$ degrees of freedom with generalized coordinates $q$, executing small oscillations about a stable equilibrium state $q_0$. Its equation of motion is

$$\mathbf{M}\ddot{x} = -\mathbf{K}x, \tag{5}$$

where $\mathbf{M}$ is the hermitian and strictly positive mass matrix, $\mathbf{K}$ the hermitian and positive definite matrix of spring constants and $x = q - q_0$ denotes the deviation from the equilibrium point $q_0$. The Lagrangian of the system is given by

$$L = \dot{q}^T \mathbf{a}(q)\dot{q} - U(q), \tag{6}$$

Journal of Physics: Conference Series **2038** (2021) 012009 doi:10.1088/1742-6596/2038/1/012009

where $\mathbf{a}(q)$ is the strictly positive definite metric of the configuration space, with $\mathbf{M} = \mathbf{a}(q_0)$, $U(q)$ is the potential energy, and

$$K_{ij} = \left.\frac{\partial^2 U}{\partial q_i \partial q_j}\right|_{q_0}, \tag{7}$$

the Hessian at the minimum.

The solution of the equation of motion (5) can be found in terms of harmonic eigenmodes

$$x = Ae^{i\omega t}, \tag{8}$$

where $A$ is the amplitude vector and $\omega$ the corresponding oscillation frequency. The eigenmode equation reads

$$(-\omega^2\mathbf{M} + \mathbf{K})A = 0. \tag{9}$$

All frequencies must be real, of course, since we are dealing with small oscillations around a stable minimum.

Alternatively, we can write the Eq. 9 as

$$\phi A = \omega^2 A, \tag{10}$$

where $\phi = \mathbf{M}^{-1}\mathbf{K}$ and $\phi^\dagger = \mathbf{K}\mathbf{M}^{-1}$. Since in general, $\mathbf{M}$ and $\mathbf{K}$ do not commute, $\phi$ is clearly not hermitian, but it is strictly-quasi-hermitian in respect to the metric $\mathbf{M}$ as it fulfills the intertwining relation $\phi^\dagger\mathbf{M} = \mathbf{M}\phi$. Therefore it is an example where strictly-quasi-hermitian matrices appear.

For very large mechanical systems with high connectivity (all particles are coupled to each other) this problem naturally lends itself to analysis in terms of random matrices. The matrices $M$ and $K$ have to be drawn from sensible uncorrelated probability distributions of positive definite matrices.

In [8, 9], $\mathbf{K}$ was drawn from the Wishart ensemble with variance $\sigma^2$ and $\mathbf{M}$ from the shifted Wishart ensemble with variance $\sigma'^2$ with a shift $m_0$, i.e.

$$\mathbf{M} = \mathbf{C}^\dagger\mathbf{C} + m_0, \qquad \mathbf{K} = \tilde{\mathbf{C}}^\dagger\tilde{\mathbf{C}} \tag{11}$$

where $\mathbf{C}$, $\tilde{\mathbf{C}}$ are complex $N \times N$ matrices with probability distribution functions (PDF)

$$P(\mathbf{C}) = Z^{-1}\mathrm{e}^{-\frac{N}{\sigma^2}\operatorname{tr}\mathbf{C}^\dagger\mathbf{C}}, \qquad P(\tilde{\mathbf{C}}) = Z^{-1}\mathrm{e}^{-\frac{N}{\sigma'^2}\operatorname{tr}\tilde{\mathbf{C}}^\dagger\tilde{\mathbf{C}}}. \tag{12}$$

Using free probability theory, one can obtain the eigenvalue statistics of $\phi = \mathbf{M}^{-1}\mathbf{K}$. [8, 9]

In this review, we want to study what happens when the metric is no longer positive, i.e. we will concentrate on the case of indefinite metric $\mathbf{B}$, and therefore pseudo-hermitian matrix $\phi$. Such a scenario may arise as a modification of the mechanical model discussed above, in which the system undergoes a structural transition, rendering the mass matrix $\mathbf{M}$ "tachyonic", that is, possess negative eigenvalues. Alternatively, some springs may lose their elasticity, making the equilibrium point $q_0$ unstable (a saddle point of the potential $U(q)$), which in turn renders the matrix $\mathbf{K}$ non-positive definite. Yet another important motivation for indefinite metrics originates from quantum mechanical systems with broken $PT$ symmetry [1, 2]. The hamiltonian $H$ of a $PT$-symmetric system satisfies the intertwining relation $H^\dagger\mathcal{P}\mathcal{C} = \mathcal{P}\mathcal{C}H$, where the metric is the product of the parity operator $\mathcal{P}$ and the so-called $\mathcal{C}$ operator. As explained in Chapter 3 of [2], the operator $\mathcal{C}$ is hermitian, commutes with both $H$ and $\mathcal{PT}$, and like parity, squares to the identity. The strictly quasi-hermitian case (positive definite $\mathcal{PC}$) corresponds to unbroken $PT$ symmetry. Here each eigenstate of $H$ is also an eigenstate of the $\mathcal{PT}$ operator and the corresponding eigenenergy is real. The pseudo-hermitian case (indefinite $\mathcal{PC}$), on the other hand, corresponds to broken $PT$ symmetry. In this case, eigenstates of $H$ are either eigenstates of $\mathcal{PT}$ as well, with real eigenenergies, or come in doublets with complex-conjugate eigenenergies, and the member eigenstates of each doublet are mapped onto each other by the $\mathcal{PT}$ operator. (In the broken phase, there should be at least one such pair of complex conjugate eigenvalues.) We refer the interested reader to Sections 1.1 and 1.2 in [2] for an elementary lucid explanation of $PT$ symmetry breaking in qualitative physical terms. Now, imagine a highly structured complicated $PT$-symmetric system. Such a system, like the complex mechanical system discussed above, naturally lends itself to analysis in terms of random matrix theory.

The structure of the rest of this paper is as follows: In Section 2.1 we will define our pseudo-hermitian random matrix model regarding a given metric $\mathbf{B}$. In Section 2.2 we will briefly review the methods we have used to find an explicit analytic result of the eigenvalues in the limit when $N \to \infty$. More details will appear in [13]. In Section 2.3 we will discuss the results when the metric is of the form $\mathbf{B} = \mathrm{diag}(1, \ldots, 1, -1, \ldots, -1)$. The main results of this section are the density of eigenvalues on the real axis given by (33) and the parametrization (see Eq. (39)) of the complex domain where the density is uniform. The analytical formulas are confronted with various results from numerical simulations. Finally, in Section 2.4 we will generalize the metric to the form $\mathbf{B} = \mathrm{diag}(1, \ldots, 1, t, \ldots, t)$ where we concentrate on the case when $t < 0$. This model will show various phase transition in the density of real eigenvalues as well as in the domain of complex eigenvalues. We will summarize the results for the criticality of the parameters in a phase diagram. A brief discussion section concludes this review.

## 2. Indefinite Metric: Model of Pseudo-Hermitian Random Matrices

### *2.1. Definition of the Model*

We will study a random matrix ensemble which is pseudo-hermitian (see Definition 2) regarding a deterministic, indefinite metric $\mathbf{B}$. If $\phi$ is an element of the set of matrices which are pseudo-hermitian regarding the metric $\mathbf{B}$, it fulfills the intertwining relation (4). As we have mentioned in the introduction, the general solution is given by $\phi = \mathbf{AB}$ where $\mathbf{A} = \mathbf{A}^\dagger$ is an arbitrary hermitian matrix (see Eq. (2)).

To define an ensemble of pseudo-hermitian random matrices regarding the given metric $\mathbf{B}$, we are free to choose a probability distribution on $\mathbf{A}$ in (2). Since $\mathbf{A}$ has to be hermitian, a natural choice is to draw $\mathbf{A}$ from the Gaussian Unitary Ensemble (GUE), i.e. we put on $\mathbf{A}$ the PDF

$$P(\mathbf{A}) = \frac{1}{Z_N} \mathrm{e}^{-\frac{Nm^2}{2}\,\mathrm{tr}\,\mathbf{A}^2}, \tag{13}$$

where $m > 0$ is a parameter and $Z_N$ is the normalization constant. The PDF (13) induces the following PDF on $\phi$,

$$P(\phi) = \frac{1}{\tilde{Z}_N} \mathrm{e}^{-\frac{Nm^2}{2}\,\mathrm{tr}(\mathbf{B}^{-2}\phi^\dagger\phi)}, \tag{14}$$

where $\tilde{Z}_N$ is another normalization constant.

We will see that $\phi$ can contain complex and real eigenvalues. The main goal of this paper is to analyze the density of eigenvalues of $\phi$ on the real axis and to find the domain of complex eigenvalues in the limit when $N \to \infty$.

### *2.2. Methods for analyzing the large N limit*

The averaged density of eigenvalues of the pseudo-hermitian matrix $\phi$ we can calculate from the the resolvent of $\phi = \mathbf{AB}$ (Green's function),

$$G(w) = \left\langle \frac{1}{N}\,\mathrm{tr}\,\frac{1}{w - \mathbf{AB}} \right\rangle, \tag{15}$$

where the brackets denote averaging of $\mathbf{A}$ over the GUE ensemble. We will see that in the large $N$ limit we can obtain $G(w)$ explicitly in closed form.

Averaging over $\mathbf{A}$ becomes simpler if one can avoid the product of matrices $\mathbf{AB}$ by the method introduced in [14], where the authors define a $2N \times 2N$ block matrix $\mathbf{H}$,

$$\mathbf{H} = \begin{pmatrix} 0 & \mathbf{A} \\ \mathbf{B} & 0 \end{pmatrix}, \tag{16}$$

with its resolvent

$$\frac{1}{z - \mathbf{H}} = \begin{pmatrix} \frac{z}{z^2 - \mathbf{AB}} & \frac{1}{z^2 - \mathbf{AB}}\mathbf{A} \\ \frac{1}{z^2 - \mathbf{BA}}\mathbf{B} & \frac{z}{z^2 - \mathbf{BA}} \end{pmatrix}, \tag{17}$$

and its Green's function

$$\widetilde{G}(z) = \left\langle \frac{1}{2N} \operatorname{tr} \frac{1}{z - \mathbf{H}} \right\rangle. \tag{18}$$

From the upper left block of the resolvent of $\mathbf{H}$ we can get the desired Green's function (15) when we divide by z and substitute $w = z^2$. Helpful will be the fact, that the block of the matrix $\mathbf{H}$, which includes the random matrix $\mathbf{A}$, is decoupled from the block with the deterministic metric $\mathbf{B}$.

Since $\mathbf{H}$ is typically non-hermitian, it might have complex eigenvalues. In the large $N$ limit, they can be dense in a two-dimensional domain in $\mathbb{C}$. Therefore $\widetilde{G}$ is not an analytic function of $w$: It is not just a function of $w$ but also of $w^*$.[1]

We will overcome the difficulty mentioned above by the method of *Hermitization* (see [15, 16, 17, 18]). For this we have to double the size of the matrix again, so now we have to deal with the resolvent of a $4N \times 4N$ block matrix

$$\hat{\mathcal{G}} = \left[ \begin{pmatrix} \eta & z \\ z^* & \eta \end{pmatrix} - \begin{pmatrix} 0 & \mathbf{H} \\ \mathbf{H}^\dagger & 0 \end{pmatrix} \right]^{-1} = \begin{pmatrix} \eta & 0 & z & -\mathbf{A} \\ 0 & \eta & -\mathbf{B} & z \\ z^* & -\mathbf{B} & \eta & 0 \\ -\mathbf{A} & z^* & 0 & \eta \end{pmatrix}^{-1}. \tag{19}$$

Notice that all the elements of the matrix on the right-hand side of Eq. (19) are $N \times N$ blocks. $\hat{\mathcal{G}}$ is the resolvent of the $4N \times 4N$ matrix

$$\begin{pmatrix} 0 & z - \mathbf{H} \\ z^* - \mathbf{H}^\dagger & 0 \end{pmatrix}. \tag{20}$$

Important is the fact that the matrix given by (20) is *hermitian*. Therefore, off the real axis in the complex $\eta$-plane, $\hat{\mathcal{G}}$ is a holomorphic function of the complex spectral parameter $\eta$. This will allow us to make a series expansion in $1/\eta$, which we will do in a diagrammatic way.

[1] This can be seen for example in the complex Ginibre ensemble, where the eigenvalue density in the large $N$ limit is uniformly supported on the unit disk. Its Green's function is given by

$$G_{\text{Gin}}(z) = \begin{cases} z^* & \text{if } |z| < 1, \\ \frac{1}{z} & \text{if } |z| \geq 1. \end{cases}$$

The density of eigenvalues can be obtained from $G_{\text{Gin}}(z)$ by the anti-holomorphic derivative,

$$\varrho_{\text{Gin}}(z) = \frac{1}{\pi} \frac{\partial}{\partial z^*} G_{\text{Gin}}(z).$$

Following 't Hooft, we will use Feynman diagrams with double lines [19] in the planar limit[20], when $N \to \infty$. We will calculate the planar limit by the following expansion in Feynman graphs. We expand the resolvent $\hat{\mathcal{G}}$ in powers of bare propagators

$$\hat{\mathcal{G}}_0 = \hat{\mathcal{G}}_{|A=0} = \tag{21}$$

and rearrange the perturbative expansion in terms of the self-energy

$$\hat{\Sigma} = \tag{22}$$

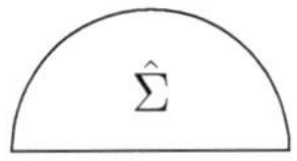


in the following way:

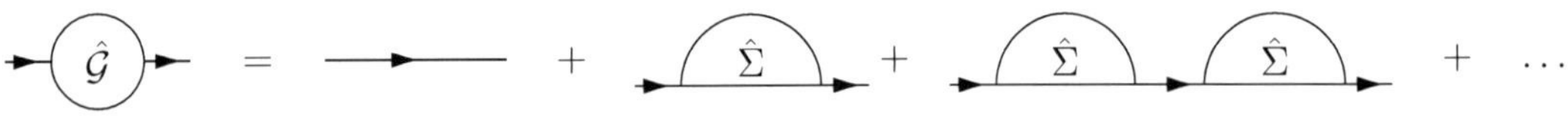


This rearrangement is allowed because $\hat{\mathcal{G}}$ is an analytic function of $\eta$, and therefore this Born series is convergent. This would not have been possible before the procedure of Hermitization.

$\hat{\Sigma}$ is the sum over all 1-quark irreducible diagrams[20]. We can express $\hat{\Sigma}$ in terms of the connected cumulants of the distribution of $A$ and the full propagator $\langle\hat{\mathcal{G}}\rangle$. Since $A$ is from the GUE ensemble, there is only one connected cumulant.

In the planar limit we arrive at following expression for the self energy, the so called *gap equation*,

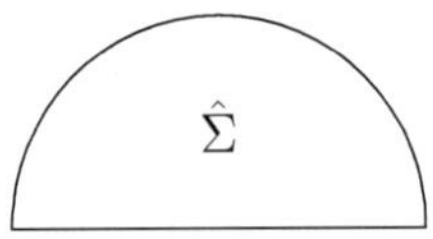


=

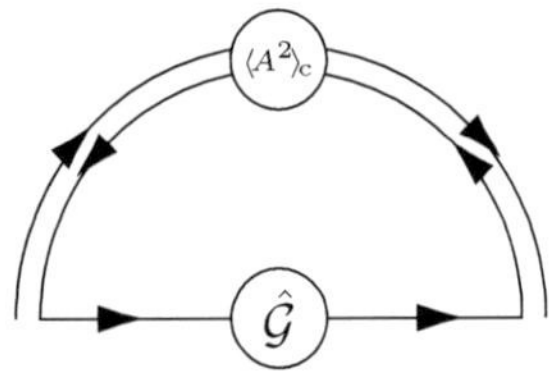


$$= \begin{pmatrix} \overline{44} & 0 & 0 & \overline{41} \\ 0 & 0 & 0 & 0 \\ 0 & 0 & 0 & 0 \\ \overline{14} & 0 & 0 & \overline{11} \end{pmatrix}, \tag{23}$$

where $\overline{\alpha\beta} = N^{-1}\langle \mathrm{tr}\, \hat{\mathcal{G}}_{\alpha\beta}\rangle$ are block traces with $\alpha = 1, 2, 3, 4$ and $\beta = 1, 2, 3, 4$. Notice that the block traces are numbers, therefore the right-hand side of Eq. (23) is a $4 \times 4$ matrix. The appearance of many zeros in this matrix is due to the decoupling of the random blocks from the deterministic blocks, which we have mentioned in the end of the paragraph below (18).

We are left to find out the four quantities $\overline{44}$, $\overline{41}$, $\overline{14}$ and $\overline{11}$. After taking $\eta \to 0$, the block traces are uniquely determined by a self-consistent equation, the *Schwinger-Dyson equation*,

$$\langle\hat{\mathcal{G}}\rangle = \frac{1}{\hat{\mathcal{G}}_0^{-1} - \hat{\Sigma}}. \tag{24}$$

All other blocks of $\langle\hat{\mathcal{G}}\rangle$ can be found in terms of the four quantities $\overline{11}, \overline{14}, \overline{41}, \overline{44}$. These four quantities are determined from their self-consistent equations (see e.g. Eq. (30) for $\overline{41}$). In particular, we find for the $\overline{31}$ block,

$$\begin{aligned} \left\langle \frac{1}{N} \mathrm{tr}\, \hat{\mathcal{G}}_{31} \right\rangle &= \left\langle \frac{1}{N} \mathrm{tr} \frac{z}{z^2 - \mathbf{AB}} \right\rangle \\ &= \frac{m^2 z}{N} \mathrm{tr} \frac{\mathbf{B}^{-1}(\overline{14} - m^2 w^* \mathbf{B}^{-1})}{\overline{11} \cdot \overline{44} - (\overline{41} - m^2 w \mathbf{B}^{-1})(\overline{14} - m^2 w^* \mathbf{B}^{-1})}, \end{aligned} \tag{25}$$

where we remind the reader that $w = z^2$.

It admits two different kind of solutions for the Green's function, a *holomorphic* one and a *non-holomorphic* solution. From the holomorphic solution we will be able to obtain the eigenvalue density on the real axis, while the non-holomorphic solution will give the complex eigenvalues.

The block traces $\overline{11}$ and $\overline{44}$ are either both zero, or non-zero. If $\overline{11} = \overline{44} = 0$, then $\overline{41}$ is a holomorphic function of $w$ away from the real axis, as well as the Green's function

$$G(w) = \left\langle \frac{1}{N} \operatorname{tr} \frac{1}{w - \mathbf{AB}} \right\rangle = -\frac{m^2}{N} \operatorname{tr} \frac{\mathbf{B}^{-1}}{\overline{41} - m^2 w \mathbf{B}^{-1}}. \tag{26}$$

It accounts for the real part of the spectrum of $\phi = \mathbf{AB}$.

On the other-hand, if $\overline{11}$ and $\overline{44}$ are non-zero, all quantities are non-holomorphic functions of $w$. This solution accounts for eigenvalues of $\phi$ in the complex plane. The boundary of the two-dimensional domain occupied by the complex eigenvalues is obtained by setting $\overline{11} = \overline{44} = 0$ in the non-holomorphic gap equations.

For any positive definite metric $\mathbf{B}$, one can consistently show that only the holomorphic solution exists, while for an indefinite metric , the holomorphic and non-holomorphic solution co-exist. So far we have not specified the deterministic metric $\mathbf{B}$. To proceed further, we will choose a specific form of metric which we will do in the following sections. Then we can find an explicit solution for the density of real eigenvalues and the domain of complex eigenvalues.

*2.3. Density of Eigenvalues if* $\mathbf{B} = \operatorname{diag}(1, \ldots, 1, -1, \ldots, -1)$

In this section we will make a particular choice of the indefinite metric $\mathbf{B}$. For simplicity we will choose $\mathbf{B}$ to be diagonal and of the form

$$\mathbf{B} = \operatorname{diag}(\underbrace{1, \ldots, 1}_{k}, \underbrace{-1, \ldots, -1}_{N-k}), \tag{27}$$

with $1 < k < N$. We will define the fraction of ones on the diagonal of $\mathbf{B}$,

$$\lambda = \frac{k}{N}, \tag{28}$$

which will be especially convenient since we usually consider the limit $N \to \infty$. Notice that from the general solution (2) we can see that the model has the obvious symmetry

$$\lambda \mapsto 1 - \lambda. \tag{29}$$

We will now come back to the holomorphic solution which we have mentioned at the end of Section (2.2). For the particular choice of metric $\mathbf{B}$ given in (27), the holomorphic gap equation reduces to the cubic equation

$$m^2 b + \frac{\lambda}{b + w} + \frac{1 - \lambda}{b - w} = 0, \tag{30}$$

where

$$b = -m^{-2}\overline{41}. \tag{31}$$

Taking the solution of the cubic equation (30) for $b$ with the asymptotic behavior

$$b(w) \sim \frac{1 - 2\lambda}{m^2 w}, \qquad (w \to \infty), \tag{32}$$

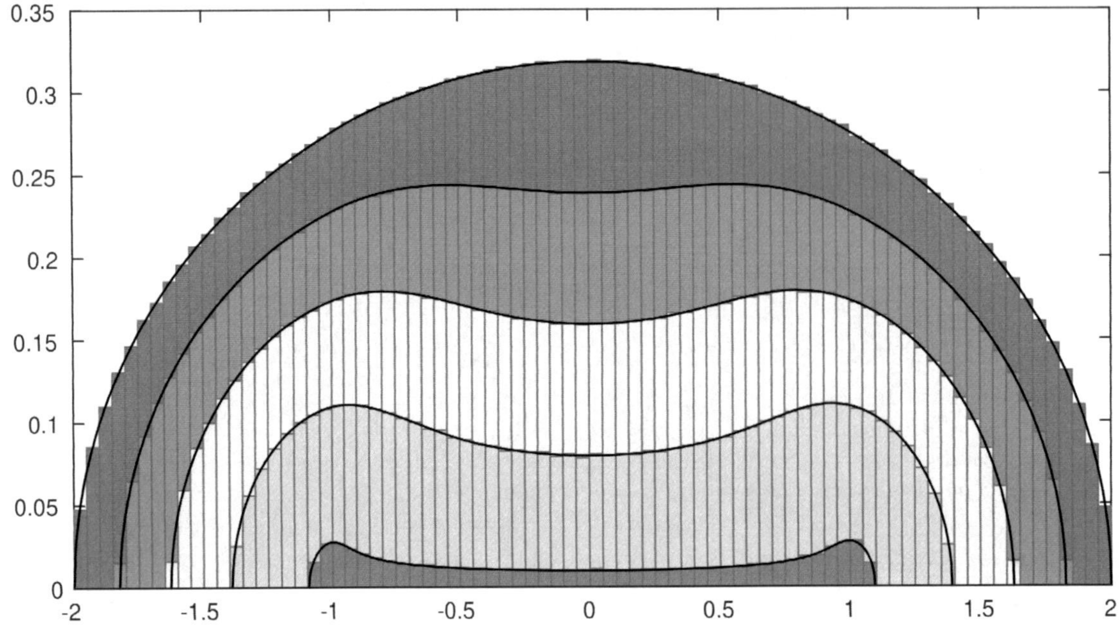


**Figure 1.** Normalized histograms of real eigenvalues from a single sample using a numerical simulation with $N = 32768$. Different color represent different values of $\lambda$: 31/64, 3/8, 1/4, 1/8, 1/32768 (ordered as density increases). The solid black line shows the corresponding large $N$ theoretical prediction $\rho^{(1)}(x)$ given by (33).

and substitute it in Eq. (26) using (31), we can get the density of real eigenvalues from the discontinuities of the Green's function $G(w)$ across the real axis. We find that in the large $N$ limit the density of real eigenvalues of $\phi$ is supported on the interval $[-a, a]$ where it is given by

$$\rho^{(1)}(x) = \frac{1}{2\pi} \lim_{\epsilon \searrow 0} \mathrm{Im} \left[ G(x - \mathrm{i}\epsilon) - G(x + \mathrm{i}\epsilon) \right]$$
$$= \mathrm{sign}(1 - 2\lambda) \frac{\left|\xi - \sqrt{\Delta}\right|^{2/3} - \left|\xi + \sqrt{\Delta}\right|^{2/3}}{\sqrt{3} \cdot 2^{2/3} \cdot 6\pi m^2 x}, \qquad |x| < a, \tag{33}$$

where

$$a = \left( \frac{3|1-2\lambda|^{2/3} \left[ \left(1-2\sqrt{\lambda(1-\lambda)}\right)^{1/3} + \left(1+2\sqrt{\lambda(1-\lambda)}\right)^{1/3} \right] + 2}{2m^2} \right)^{1/2}, \tag{34}$$

$$\xi = \xi(x) = -27m^4(1 - 2\lambda)x, \tag{35}$$

and the discriminant

$$\Delta = \Delta(x) = \xi^2 + 4 \cdot 27m^6(1 - m^2x^2)^3. \tag{36}$$

Note that for $x$ going to zero, the density behaves like

$$\lim_{x \to 0} \rho^{(1)}(x) = \frac{m|1 - 2\lambda|}{\pi}. \tag{37}$$

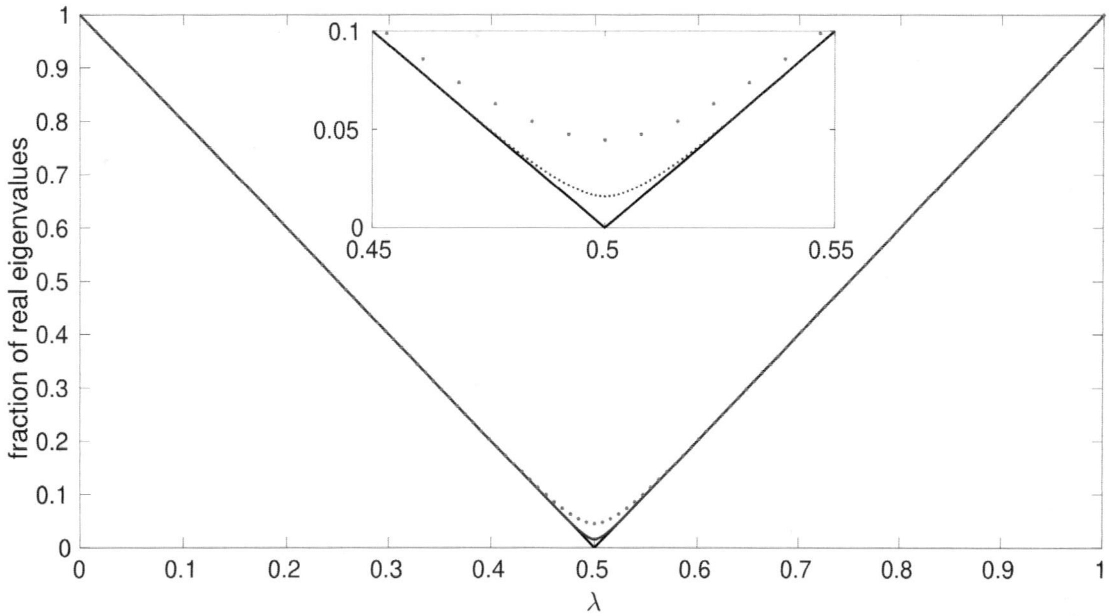


**Figure 2.** Fraction of real eigenvalues as a function of $\lambda$. Solid black line shows the theoretical large $N$ prediction given by (38), while the dots represent numerical simulations with $N = 128$ averaged over 500000 samples (red) and $N = 1024$ using 2000 samples (blue). Inset: Magnification near $\lambda = 1/2$.

In Fig. 1 we show normalized histograms obtained from numerical generated samples for various $\lambda$. We have used a single sample of large matrix size with $N = 32768$. The histograms fit well with the solid black lines from the corresponding theoretical limits $N \to \infty$ given by Eq. (33). That a single sample is suitable to represent the averaged density, comes from the self-averaging effect. This is typical for the eigenvalue density of large random matrices.

The histogram in magenta shows the result for $\lambda = 31/64$ which is close to the degenerated case where the theoretical density (33) predicts that there is no density of real eigenvalues. While lambda decreases from 1/2, the density of real eigenvalues increases. The blueish histogram shows the case where $\lambda$ is close to zero. For $\lambda = 0$ it is obvious from the model, that $\phi = \mathbf{A}$ which has been drawn from GUE, so the semi-circle law is no surprise.

Since we have normalized the density such that all real and all complex eigenvalues together sum up to one, integrating $\rho^{(1)}(x)$ gives the fraction of real eigenvalues as a function of $\lambda$. We find from (33) that the fraction of real eigenvalues is

$$\int_{-a}^{a} \rho^{(1)}(x)\mathrm{d}x = |1 - 2\lambda|, \tag{38}$$

which is consistent with the symmetry of the model (29). In Fig. 2 this result has been confronted with numerical simulations for matrix sizes $N = 128$ and $N = 1024$ averaged over many samples. For $\lambda$ away from 1/2 the convergence is exponentially fast. Near 1/2, we can see a small deviation for finite $N$. The theoretical large $N$ prediction, actually is a lower bound for the number of real eigenvalues, even for finite $N$ and for each sample. This follows from a special case of a purely algebraic theorem [21].

Next we want to discuss what we find for the choice of metric $\mathbf{B}$ given by (27) for the non-holomorphic solution mentioned at the end of section 2.2 which accounts for the complex

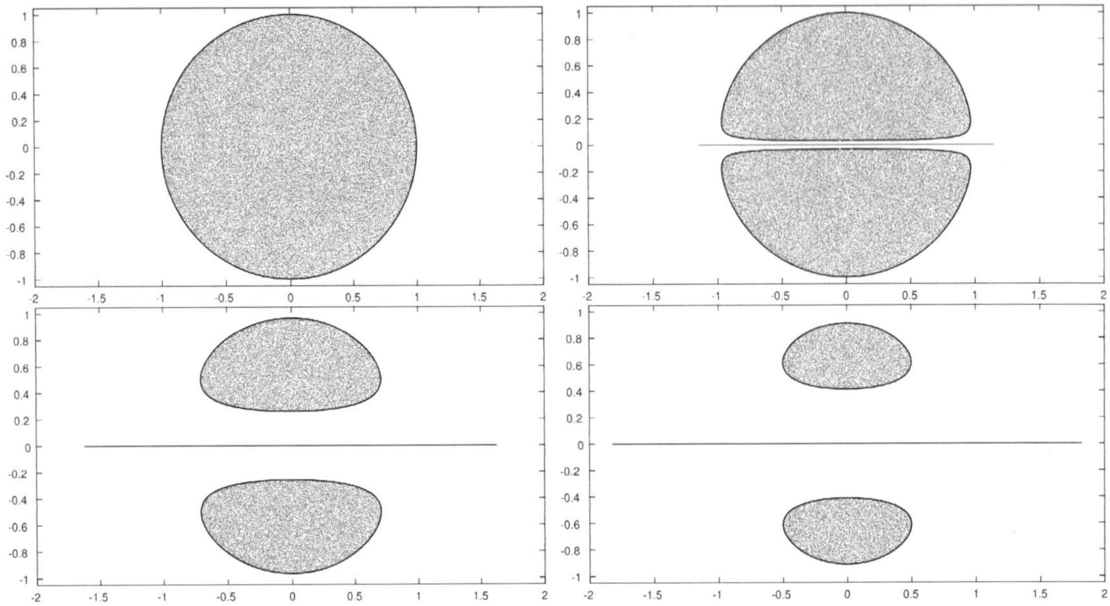

**Figure 3.** The blue dots represent the eigenvalues of $\phi = \mathbf{AB}$ obtained from numerical simulations using a single realization with $N = 32768$ for various values of $\lambda$: 1/2, 15/32, 1/4, 1/8 (from upper left to lower right). The solid black line shows the theoretical boundary when $N \to \infty$ given by (39).

eigenvalues. As we have discussed before, in this case, both $\overline{11}$ and $\overline{44}$ are non-zero and the Green's function $G(w)$ is non-holomorphic. More precise, when $N \to \infty$, $G(w)$ is a non-holomorphic in the domain which is densely filled by complex eigenvalues and holomorphic outside. At the boundary of this domain, $G(w)$ has to transform continuously to the Green's function we have obtained from the holomorphic solution. Therefore we can find the boundary of the complex domain by setting $\overline{11} = \overline{44} = 0$ in the non-holomorphic gap equation. In the large $N$ limit, we find for our particular choice of metric the following parametrization in polar coordinates,

$$r_\pm(\theta) = \frac{1}{\sqrt{2}m}\left(1 \pm \sqrt{1 - \left(\frac{\sin\theta_0}{\sin\theta}\right)^2}\right)^{1/2}, \tag{39}$$

when $\sin^2\theta \geq \sin^2\theta_0$ with $\sin\theta_0 = |2\lambda - 1|$. $r_+$ is the part of the boundary farther from the origin, and $r_-$ the closer one.

In Fig. 3 we illustrate the result for various values of $\lambda$ by scatter plots. The blue dots correspond to the eigenvalues of $\phi$ obtained from single samples of matrices of size $N = 32768$. Since the scatter plots have been produced by a single sample, it demonstrates well how uniformly the eigenvalues are distributed which is due to the repulsive force among them. The solid black lines represent the theoretical boundaries given by Eq. (39).

On the upper left plot, we see the degenerated case $\lambda = 1/2$ when the theoretical prediction no density of real eigenvalues. On the remaining scatter plots we see also eigenvalues on the real axis where the blue dots appear as a line since they are very dense as there are $|N - 2k| = N|1 - 2\lambda|$ real eigenvalues. For fixed $\lambda$, the typical distance between eigenvalues on the real lines is of order $N^{-1}$ while the mean distance between nearest neighbors of complex eigenvalues in the bulk is of order $N^{-1/2}$.

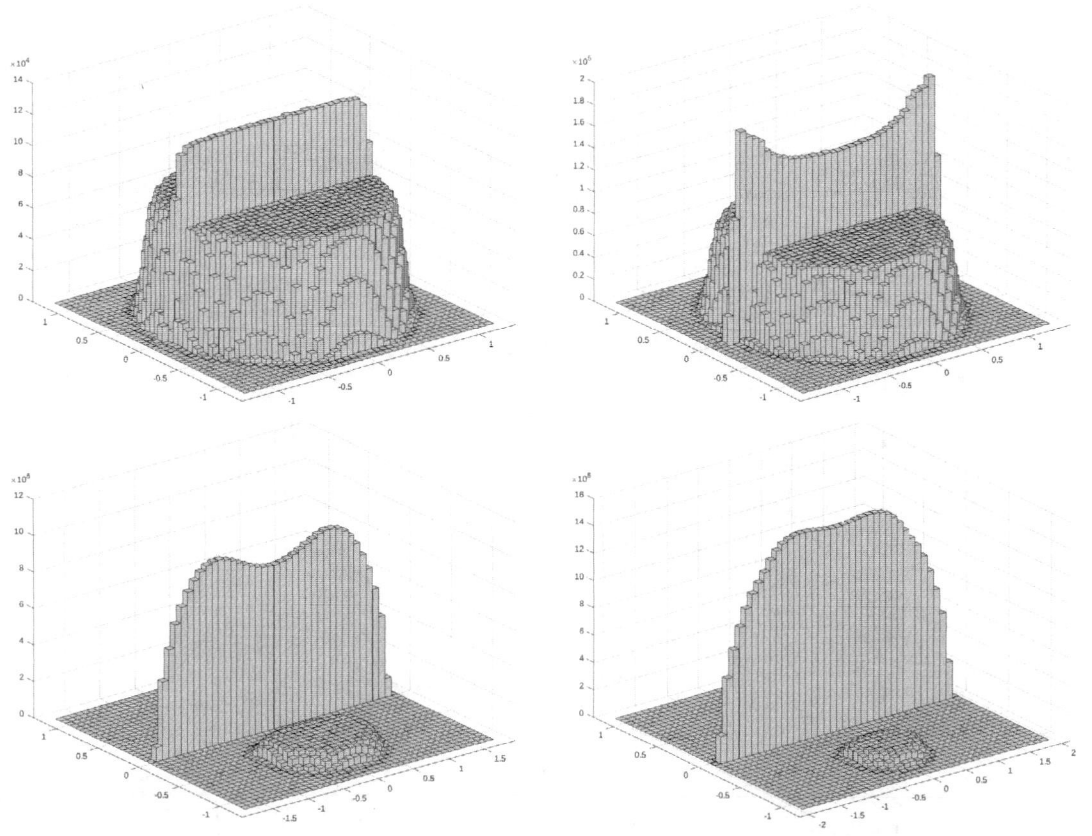

**Figure 4.** Histograms representing counts of eigenvalues of $\phi$ for two-dimensional bins in the complex plane showing the eigenvalue distribution obtained by numerical simulations of 500000 samples with $N = 128$ for various values of $\lambda$: 1/2, 15/32, 1/4, 1/8 (from upper left to lower right).

For each sample, the eigenvalue distribution is symmetric regarding the real axis. This is due to the fact that the characteristic polynomial of $\det(z - \phi)$ has real coefficients and the complex eigenvalues appear in conjugated pairs. It is a manifestation of PT symmetry. The eigenvalue distribution has an additional symmetry regarding the imaginary axis, but only after averaging over all samples.

We can calculate the area of the complex domain, the two blobs parameterized by (39), by the following integral

$$2\int_{r_-}^{r_+}\int_{\theta_0-\pi}^{2\pi-\theta_0} r dr d\theta = \int_{\theta_0-\pi}^{2\pi-\theta_0} \left(r_+^2(\theta) - r_-^2(\theta)\right) d\theta = \frac{(1-|1-2\lambda|)\pi}{m^2}. \quad (40)$$

By Gauss law we can get from the Green's function also the density of complex eigenvalues,

$$\rho^{(2)}(w) = \frac{1}{\pi}\frac{\partial}{\partial w^*}G(w) = \frac{m^2}{\pi}, \quad (41)$$

uniformly on the complex domain and independent of $\lambda$. By the product of the area and the density given by (40) and (41) respectively, we find that the fraction of complex eigenvalue is given by $1 - |1 - 2\lambda|$, consistent with what we have found on the real axis in Eq. (38).

In Fig. 4 we show two dimensional histograms obtained from eigenvalues of many samples of size N=128, for the same lambdas we have used in the scatter plots in Fig. 3. In the histograms we counted both real and imaginary eigenvalues, though the two dimensional bins are not well suited to measure the one-dimensional density and therefore the ratio of the height of the real eigenvalues and the complex eigenvalues is depending on the size of the bins. Even though the degenerated case $\lambda = 1/2$ (upper left histogram) the theoretical prediction tells us that all eigenvalues are complex, for finite $N$ there are still some real eigenvalues as we have seen in Fig. 2. We can well see that inside the bulk the density is very flat. Only at the edge one notice some deviation from the constant density which is a typical finite $N$ size effect that one can find similarly in the Ginibre ensemble.

From (39) we find that the complex domain has a distance from the real axis given by

$$m^{-1}\sin(\theta_0/2), \tag{42}$$

and it touches the axis when $\lambda$ approaches 1/2. The latter is the degenerated case which we have mentioned before. Then the complex eigenvalues uniformly fill a disk, like in the (real) Ginibre ensemble.

### *2.4. Density of Eigenvalues for* $\mathbf{B} = \mathrm{diag}(1,\ldots,1,t,\ldots,t)$

In the remaining section of this paper, we want to discuss what happens if we generalize the deterministic metric $\mathbf{B}$ from the last section to the form

$$\mathbf{B} = \mathrm{diag}(\underbrace{1,\ldots,1}_{k},\underbrace{t,\ldots,t}_{N-k}). \tag{43}$$

The fraction of ones on the diagonal of $\mathbf{B}$ we still denote by $\lambda$, like in (28). The main motivation for this generalization is that it allows us to study what happens when $t$ approaches zero and the metric becomes singular. (The case of singular metric will not be discussed in this paper.) We will concentrate on the case when $t < 0$. We shall see that this model exhibits various phases of the density of real eigenvalues and of the complex one. In the two-dimensional space spanned by the parameters $t$ and $\lambda$, phase changes in the system happens on several one-dimensional curves, where we can either describe the critical value of $\lambda$ as a function of $t$, or the critical value of $t$ as a function of $\lambda$. In this section we will variate $t$ while keeping $\lambda$ fixed. Therefore it is more convenient to introduce the critical $t$ as function of $\lambda$. We will denote the different critical values of $t$ by $t_{\mathbb{C}} = t_{\mathbb{C}}(\lambda)$ for a phase change which is only present in the density of complex eigenvalues, $t_{\mathbb{R}} = t_{\mathbb{R}}(\lambda)$ when the phase change appears in the density on the real axis, and $t_{\mathbb{CR}} = t_{\mathbb{CR}}(\lambda)$ for a phase change that happens in the density of real and complex eigenvalues at the same time.

Analogously to (30) we get the cubic equation

$$v^3 + v^2\left(u + \frac{u}{t}\right) + v\left(\frac{u^2}{t} + 1\right) + u\left(1 - \lambda + \frac{\lambda}{t}\right) = 0 \tag{44}$$

where we have introduced the rescaled variables $u = mw$ and $v = mb$. We can substitute its solution into Eq, (26) and find in a similar way as discussed in section (2.3), the density on the real axis and the domain of complex eigenvalues.

We are now going to study the change of the complex domain as we keep $\lambda = 3/4$ fixed while varrying $t$, starting at the known case $t = -1$ discussed in the previous section. The blue dots in Fig. 5 are again obtained from a scatter plot of a single sample of large size while the black solid lines show the theoretical boundary of the complex domain. We show the effect when we slightly change $t = -1$ corresponding to the model in Section 2.3 to the new model with $t = -1.2$. The boundary of the domain deforms continuously with $t$. The density is still uniform in the complex

Journal of Physics: Conference Series **2038** (2021) 012009 doi:10.1088/1742-6596/2038/1/012009

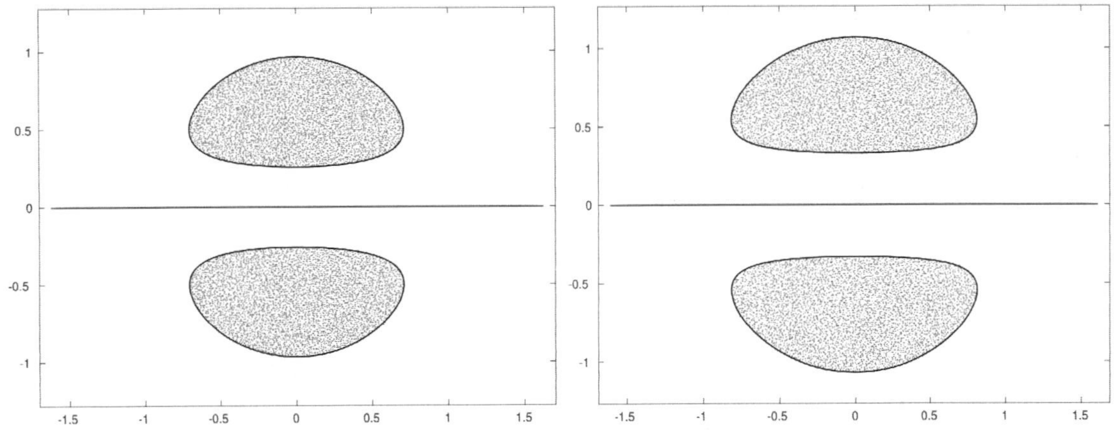

**Figure 5.** The blue dots represent the eigenvalues of $\phi = \mathbf{AB}$ obtained from numerical simulations using a single realization with $N = 16384$ and $\lambda = 3/4$ for the values of $t = -1$ (left) and $t = -1.2$ (right). The solid black line shows the theoretical boundary when $N \to \infty$. (The left plot we have put here for comparison. It has already appeared on the left bottom of Fig. 3 for a different value of $N$.)

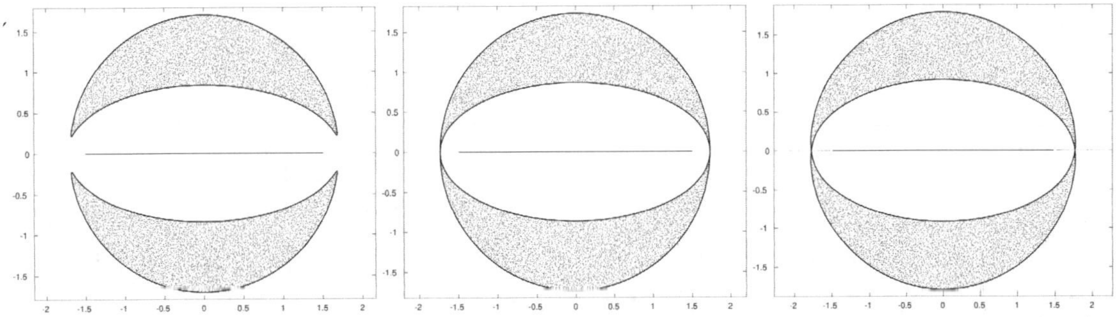

**Figure 6.** The blue dots represent the eigenvalues of $\phi = \mathbf{AB}$ obtained from numerical simulations using a single realization with $N = 16384$ and $\lambda = 3/4$ for various values of $t$: -2.9, -3, -3.2 (from left to right) which represents the pre-critical, critical and post-critical case respectively. The solid black line shows the theoretical boundary when $N \to \infty$.

domain, but it changes with t. Since there is no continuous way for a complex eigenvalue to move onto the real axis, the number of complex eigenvalues does not change, i.e. the fraction of real eigenvalues is given by $|1 - 2\lambda|$ as we have found in Eq. (38).

We are now decreasing $t$ further as presented in Fig. 6, which shows the pre-critical, critical and post-critical case where $t = -2.9$, $-3$ and $-3.2$, respectively. In the first and second case (left and center plot), the fraction of complex eigenvalues is still given by $1 - |1 - 2\lambda|$. In the critical case, when $t = t_{\mathbb{CR}}(\lambda)$, the complex phase changes from a disconnected domain to a connected one as the two crescents touch on the real axis. In the post-critical case (right plot) there is a flow of complex eigenvalues to the real axis, and the fraction of complex eigenvalues becomes less than $1 - |1 - 2\lambda|$. Since for $t = t_{\mathbb{CR}}(\lambda)$, the crescents touch the real axis outside of the interval of real eigenvalues, on each side a new interval of support of real eigenvalue density emerges in the post-critical phase. The fraction of real eigenvalues in the central interval is still given by $|1 - 2\lambda|$ since at this moment there is no continuous way for the eigenvalues from the

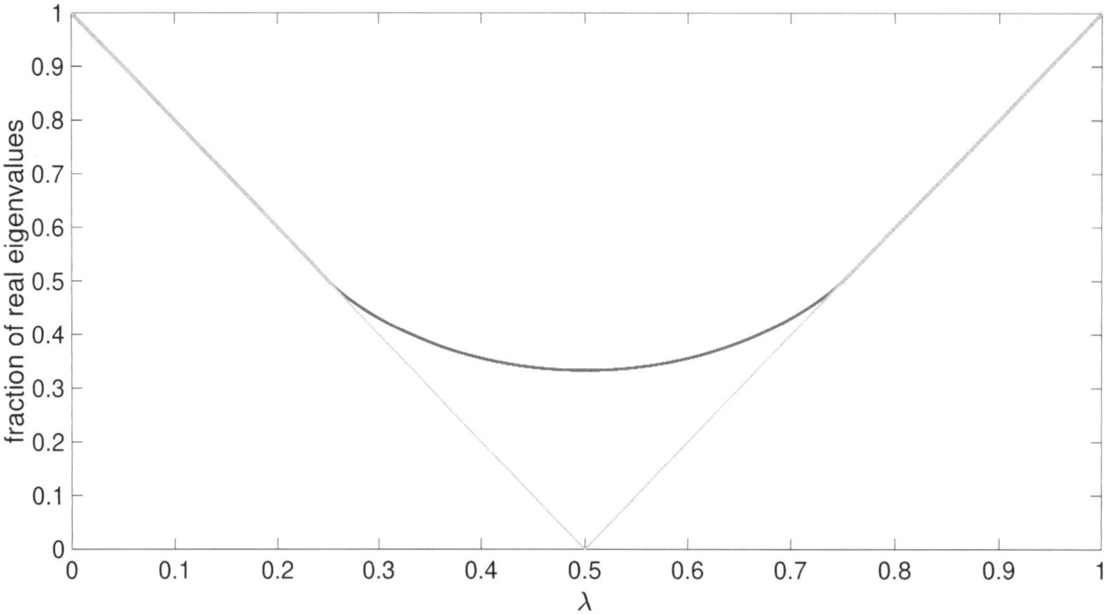


**Figure 7.** Theoretical large $N$ prediction for the fraction of real eigenvalues as a function of $\lambda$ for $t = -3$ (red) and $t = -1$ (green).

outer interval to flow into the central one.

While the total amount of density of real eigenvalues is fixed by $|1-2\lambda|$ when $t \geq t_{\mathbb{CR}}$ (or the amount of density in the central interval if $t_{\mathbb{CR}} > t > t_{\mathbb{R}}$), the endpoints of the interval and the distribution of the density on its support will still depend on the parameters $t$ and $\lambda$. Similarly as in Eq. (33) and (34) for the model where $t = -1$, one can find analytic expressions for the density of real eigenvalues and the endpoints of its support, but they look quite cumbersome. We will state them in an upcoming paper [13] where we will give more details about the calculation.

The function that gives the fraction of all real eigenvalues is no longer that simple if $t < t_{\mathbb{CR}}$ and depends on both $\lambda$ and $t$. It is plotted in Fig. 7 as a function of $\lambda$, where we compare the large-$N$ limit for the simpler case where $t = -1$ (green) given by (38) and $t = -3$ (red). The critical curves $t_{\mathbb{CR}}(\lambda)$ and $t_{\mathbb{C}}(\lambda)$ can be also inverted to obtain $\lambda_{\mathbb{CR}} = \lambda_{\mathbb{CR}}(t)$ and $\lambda_{\mathbb{C}} = \lambda_{\mathbb{C}}(t)$. For the value $t = -3$ as shown in Fig. 7, then as long as $\lambda$ is larger than the critical $\lambda_{\mathbb{CR}} = 3/4$ (or smaller than $\lambda_{\mathbb{C}} = 1 - \lambda_{\mathbb{CR}} = 1/4$) the red curve coincides with the green line. Consistent with the previous discussion, the fraction of real eigenvalues only differs from the simpler case $t = -1$, once we have reached the post-critical phase, where some complex eigenvalues have flown to the real axis. Obviously the fraction is larger than in the simpler case. As was mentioned at the end of the paragraph below (38), it follows from [21] that the simpler case gives an upper bound for the fraction of complex eigenvalues which again holds for finite $N$ and for each sample.

In Fig. 8 we show histograms of the density of real eigenvalues in the post-critical case with $t = -3.2$, obtained from 1000 samples. The red solid line shows the theoretical large-$N$ prediction. It shows a small gap between the outer intervals and the central one. Since the gap is very small, it is hard to see it in the histogram from the numerical simulations, especially since at the edge of the support, the finite-$N$ effects are typically more pronounced than in the bulk. But in the bulk we can see that the red curve fits perfectly with the histogram. When there are multiple intervals, one can obtain an expression for the size of the gap by the difference of their endpoints [13].

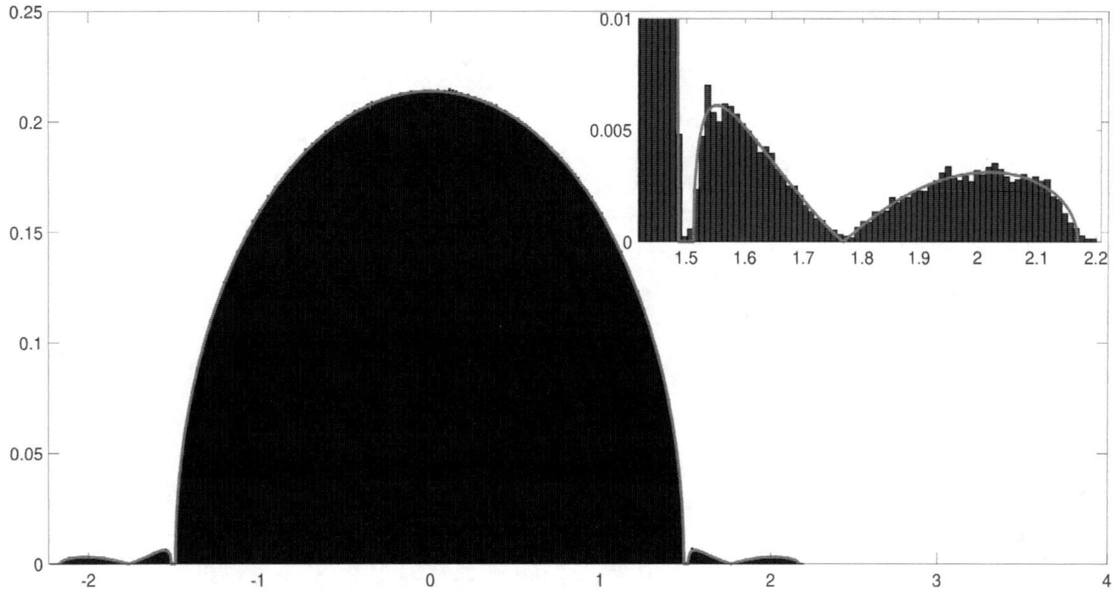


**Figure 8.** Histogram of real eigenvalues in the post-critical case obtained from 1000 samples with parameters $N = 4096$, $\lambda = 3/4$, $t = -3.2$. Red solid line shows the theoretical large $N$ result. Inset: Magnification of the right interval.

As can be also seen in Fig. 8, the density on the outer interval also vanishes (on both sides) at a point, without forming a gap in the spectrum. This is the point where the two crescents touch the real line. As this plots show, in the post-critical case, after the complex domain has started to touch the real line, the density of real eigenvalues in the outer intervals, accounts for all the complex eigenvalues which have flown to the real axis. When we decrease $t$ even further, away from $t_{\mathbb{CR}}(\lambda)$, it can happen that the disjoint intervals on the real axis merge again to a single interval. If $1/9 < \lambda < 8/9$, this merging happens at $t = t_{\mathbb{R}}(\lambda)$. Otherwise, the endpoints of the intervals will evade towards $\pm\infty$ as $t \to -\infty$ without merging of the intervals.

Finally, in Fig. 9, we summarise the various phases of real and complex eigenvalues of $\phi$ in a phase diagram of the two-dimensional parameter space. The horizontal axis represents $\lambda$ and the vertical axis $t$. Even though in this section we have concentrated on the case where the metric $\mathbf{B}$ is indefinite, i.e. $t < 0$, the discussion below (restricted to the density of real eigenvalues) will still hold in the strictly-quasi-hermitian case when $t$ is positive and there are no complex eigenvalues.

In all scatter plots in this section we have studied the case for fixed $\lambda = 3/4$. Notice that the critical values of $t$ for complex or real eigenvalues depend on $\lambda$. Below we will discuss the various critical curves which can be found in the phase diagram.

We will start with the criticality which we have already discussed and shown in Fig. 6, when the domain of complex eigenvalues consist of two crescents which touch on the real line. The critical value of $t$ as a function of $\lambda$ is given by

$$t_{\mathbb{CR}}(\lambda) = \frac{\lambda}{\lambda - 1}, \tag{45}$$

and is plotted in the phase diagram Fig. 9 by the red curve. It is both a critical phase for the complex as well as the real eigenvalues since the touching of the crescents happens outside of the interval of real eigenvalues and therefore gives birth to two new disconnected intervals.

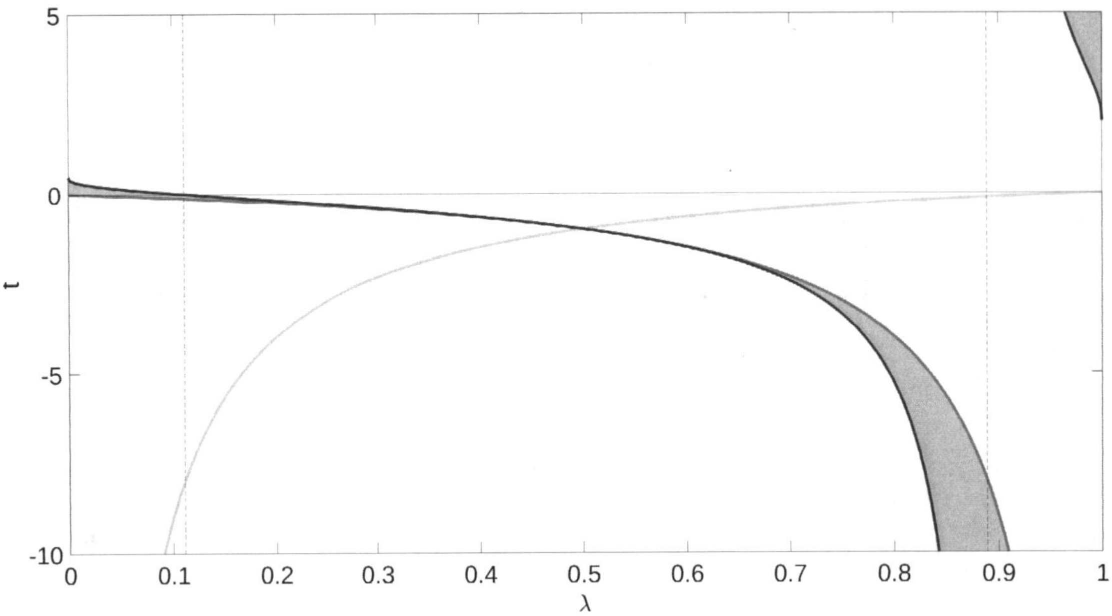


**Figure 9.** Phase diagram of the two-dimensional parameter space of the model where $\lambda$ is on the horizontal and $t$ on the vertical axis. The colored curves represent critical values of the parameters. $t_{\mathbb{C}}(\lambda)$ (green) given by (46) shows the phase transition for complex eigenvalues, $t_{\mathbb{R}}(\lambda)$ (blue) given by (47) for the real eigenvalues when disjoint intervals merge and $t_{\mathbb{CR}}(\lambda)$ (red) given by (45) is a criticality present in real and complex eigenvalues at the same time. For $t < 0$ the curve $t_{\mathbb{R}}(\lambda)$ only exists for $1/9 < \lambda < 8/9$ which is marked by the vertical dashed lines. The yellow shaded area represents the phase where the complex domain is disconnected. The magenta shaded area denotes the phase where the line density of real eigenvalues consists of multiple intervals.

There is another criticality in the complex phase given by

$$t_{\mathbb{C}}(\lambda) = \frac{\lambda - 1}{\lambda} = \frac{1}{t_{\mathbb{CR}}(\lambda)}, \tag{46}$$

represented by the green curve in the phase diagram. In contrast to the previous case it is not a criticality for the real eigenvalues. In this phase the domain of complex eigenvalues is convex and the touching point with the real axis happens at the origin. Since in this moment, there is some density of real eigenvalues in the neighborhood of the origin, the complex eigenvalues can flow in a continuous way to the existing interval of real eigenvalues.

Finally there is the criticality which only concerns the real eigenvalue density, given by

$$t_{\mathbb{R}}(\lambda) = \frac{1}{3(8 - 9\lambda)} \left( 6(2 - 3\lambda) - \frac{|\xi_\lambda + \sqrt{\Delta_\lambda}|^{1/3} + |\xi_\lambda - \sqrt{\Delta_\lambda}|^{1/3}}{2^{1/3}} \right), \tag{47}$$

where

$$\Delta_\lambda = 531441(8 - 9\lambda)^2(1 - \lambda)^2\lambda^2 \geq 0, \tag{48}$$

and

$$\xi_\lambda = -729\lambda(\lambda - 1)(7\lambda - 8). \tag{49}$$

In the phase diagram, $t_{\mathbb{R}}(\lambda)$ is plotted in blue. This critical curve is associated with the merging of three disjoint intervals of real eigenvalues into a single one. If we restrict $t$ to be negative, it only exists when $1/9 < \lambda < 8/9$ which is marked by the vertical dashed lines in Fig. 9. The right vertical line is an asymptote of the blue curve.

If for fixed $\lambda$, $t$ lies between $t_{\mathbb{C}}(\lambda)$ and $t_{\mathbb{CR}}(\lambda)$, i.e. in the yellow shaded area in the phase diagram, the domain of complex eigenvalues is disconnected. Else, i.e. when $t$ lies in the white or magenta shaded area, the complex domain always touches the real axis. When $t$ lies between $t_{\mathbb{R}}(\lambda)$ and $t_{\mathbb{CR}}(\lambda)$, i.e. in the magenta shaded area, the density of real eigenvalues consist of three disjoint intervals, while else it is supported on a single interval.

The simple law where the fraction of real eigenvalues as a function of $\lambda$ is given by $|1 - 2\lambda|$ is only valid in the yellow shaded area of the phase diagram, while it still holds for the fraction of real eigenvalues in the center interval in the magenta shaded phase.

All the critical lines, in red, blue, and green, intersect at $\lambda = 1/2$ and $t = -1$. Because of this, we have seen only one critical case in the simpler model in Section 2.3 where $t = -1$. There is corresponds to the degenerated case where the upper half disk touches the lower one.

## 3. Discussion

In this work we reviewed our recent results on the spectral properties of large random matrices which model systems whose hamiltonians are either quasi-hermitian or pseudo-hermitian. Large quasi-hermitian or pseudo-hermitian matrices can be thought of truncations of the corresponding operators to finite dimensional vector spaces. Thus, in the limit of large matrix size, one may gain useful knowledge on the corresponding operators themselves. In the usual spirit of Random Matrix Theory [22], randomness of these matrices can model complicated or chaotic systems, or systems which are disordered to begin with. As explained in detail in [8, 9], in the quasi-hermitian case, the powerful methods of free probability theory can be brought to bear to deriving analytically an explicit expression for the average density of vibrational modes of a highly connected large disordered mechanical system. An important physical feature of such systems is a universal behaviour of the density of modes at low frequencies, namely, that the density of low frequency phonons tends to a non-vanishing constant. This is also the typical behaviour of the phonon spectrum of a regular (periodic) one-dimensional chain of atoms. Thus, in more formal words, one may say that such highly connected mechanical systems universally have spectral dimension one. The bulk of this review was dedicated to the pseudo-hermitian case [13]. We have applied the diagrammatic method to obtain an explicit analytical expression for the average density of eigenvalues of such matrices both on the real axis and in the complex plane and discovered a rich phase structure of this object. We have also carried meticulous numerical analysis of these matrices. The numerical results agree very well with our analytical predictions. The results presented here are just the tip of the iceberg. Quasi- and pseudo-hermitian random matrices are clearly a promising new direction in the vast ocean of Random Matrix Theory.

## Acknowledgments

This research was supported by the Israel Science Foundation (ISF) under grant No. 2040/17. Computations presented in this work were performed on the Hive computer cluster at the University of Haifa, which is partly funded by ISF grant 2155/15. Finally, we thank T. Can for turning our attention to [21].

## References

[1] Bender C M and Boettcher S 1998 Real spectra in non-Hermitian Hamiltonians having PT symmetry *Phys. Rev. Lett.* **80** 5243

[2] Bender C M et al. 2019 *PT Symmetry in Quantum and Classical Physics* (London: World Scientific Publishing Europe)

[3] Dieudonné J 1961 *Proc. Int. Symp. on Linear Spaces (1960 Jerusalem)* (Oxford: Pergamon Press) pp 115-122
[4] Scholtz F G, Geyer H B and Hahne F J W 1992 Quasi-Hermitian operators in quantum mechanics and the variational principle *Ann. Phys.* **213** 74
[5] Froissart M 1959 Covariant formalism of a field with indefinite metric *Il Nuovo Cimento* **14** 197
[6] Mostafazadeh A 2002 Pseudo-Hermiticity versus PT symmetry: the necessary condition for the reality of the spectrum of a non-Hermitian Hamiltonian *J. Math. Phys.* **43** 205
[7] Assis P E G and Fring A 2008 Non-Hermitian Hamiltonians of Lie algebraic type *J. Phys. A: Math. Theor.* **42** 015203 (*Preprint* arXiv:0804.4677), section 3 (and references therein)
[8] Feinberg J and Riser R 2020 Universal collective behavior of the vibration spectrum of highly connected disordered systems, *Preprint* arXiv:2012.05964 [math-ph]
[9] Feinberg J and Riser R 2021 Dynamics of disordered mechanical systems with large connectivity, free probability theory, and quasi-Hermitian random matrices *Ann. Phys.* Available online 168456 (April 1) (*Preprint* arXiv:2106.01117 [math-ph])
[10] Joglekar Y N and Karr W A 2011 Level density and level-spacing distributions of random, self-adjoint, non-hermitian matrices *Phys. Rev.* E **83** 031122
[11] Deguchi T, Ghosh P K and Kudo K 2009 Level statistics of a pseudo-hermitian Dicke model *Phys. Rev.* E **80** 026213
[12] Kumar S and Ahmed Z 2017 Pseudosymmetric random matrices: Semi-Poisson and sub-Wigner statistics *Phys. Rev.* E **96** 022157
[13] Feinberg J and Riser R, in preparation
[14] Burda Z, Janik R A and Nowak M A 2011 Multiplication law and S transform for non-hermitian random matrices *Phys. Rev.* E **84** 061125
[15] Feinberg J and Zee A 1997 Non-hermitian random matrix theory: Method of hermitian reduction *Nuclear Physics* B **504** 579-608
[16] Janik R A, Nowak M A, Papp G and Zahed I 1997 Non-hermitian random matrix models *Nuclear Physics* B **501** 603-642
[17] Chalker J and Wang Z J 1997 Diffusion in a random velocity field: Spectral properties of a non-hermitian Fokker-Planck operator *Phys. Rev. Lett.* **79** 1797
[18] Efetov K B 1997 Directed quantum chaos *Phys. Rev. Lett.* **79** 491
[19] 't Hooft G 1974 A planar diagram theory for strong interactions *Nuclear Physics* B **72** 461-473; 1994 *Under the Spell of the Gauge Principle* (Singapore: World Scientific) p 378
[20] Zee A 2010 *Quantum Field Theory in a Nutshell* 2nd edition (Princeton: Princeton University Press) Chapter VII.4, p 396
[21] Carlson D H 1965 On real eigenvalues of complex matrices *Pacific J. Math.* **15** 1119-1129
[22] Mehta M L 2004 *Random Matrices* 3rd Edition (Amsterdam: Elsevier)

# Non-Hermitian gauge field theories and BPS limits

**Andreas Fring and Takanobu Taira**

Department of Mathematics, City, University of London, Northampton Square, London EC1V 0HB, UK

E-mail: a.fring@city.ac.uk, takanobu.taira@city.ac.uk

**Abstract.** We present an overview of some key results obtained in a recent series devoted to non-Hermitian quantum field theories for which we systematically modify the underlying symmetries. Particular attention is placed on the interplay between the continuous symmetry group that we alter from global to local, from Abelian to non-Abelian, from rank one to generic rank N, and the discrete anti-linear modified CPT-symmetries. The presence of the latter guarantees the reality of the mass spectrum in a certain parameter regime. We investigate the extension of Goldstone's theorem and the Higgs mechanism, which we demonstrate to work in the conventional fashion in the CPT-symmetric regime, but which needs to be modified technically at the standard exceptional points of the mass spectrum and entirely fails at what we refer to as zero exceptional points as well as in the broken CPT-symmetric regime. In the full non-Hermitian non-Abelian gauge theory we identify the t'Hooft-Polyakov monopoles by means of a fourfold Bogomol'nyi-Prasad-Sommerfield (BPS) limit. We investigate this limit further for other types of non-Hermitian field theories in 1+1 dimensions that possess complex super-exponential and inverse hyperbolic kink/anti-kink solutions and for 3+1 dimensional Skyrme models for which we find new types of complex solutions, that all have real energies due to the presence of different types of CPT-symmetries.

## 1. Introduction

Abelian and non-Abelian gauge field theories as manifested in quantum electrodynamics and the standard model, respectively, play a central role in particle physics and quantum field theory. While the general principle of gauge theories is a powerful universal concept, the concrete models have their well-known limitations see e.g. [1, 2] and the very recent evidence for the breaking of lepton universality in beauty-quark decays [3]. Many extensions of the standard model by new types of ideas such as supersymmetry or the extension to string/M-theory have and are being explored. Here we report on a series of studies [4, 5, 6, 7, 8, 9] that explore certain sectors of the standard model to allow for the possibility of non-Hermitian gauge field theories.

This approach in field theory is inspired by the success of non-Hermitian extensions to quantum mechanics. In the conventional approach to quantum mechanics one usually demands Hamiltonians to be Hermitian, as this property guarantees the reality of the energy spectrum and unitary time-evolution. However, in a seminal paper Bender and Boettcher [10] realised that Hermiticity is only a sufficient but not a necessary condition to ensure these properties. The reality of the spectrum may still be guaranteed when the Hamiltonian and the corresponding wave functions respect an anti-linear symmetry [11] of which a simultaneous parity and time-reversal ($\mathcal{PT}$)-symmetry is an example. When modifying the inner product by a suitable new metric, also the unitary time-evolution, i.e. the preservation of probability densities, can be ensured. Meanwhile these idea have found their way into a wide range of areas in physics with classical optics being especially successful as in that context many experiments and applications can be realised, see e.g. [12, 13, 14, 15, 16], thus not only confirming the theoretical formulation but also supporting the manifestation of $\mathcal{PT}$ symmetric systems in nature.

The natural extension of these ideas to the quantum field theories is still far less developed, but is currently an active field of research. In this review we shall mainly focus on the study of various classical solutions to field theories. We start by investigating how the central concepts in standard Hermitian quantum field theory, such as how the Higgs mechanism builds on the Goldstone theorem, are extended to non-Hermitian theories. We recall that the former predicts the number of massless Nambu-Goldstone fields [17, 18] as the number of global continuous symmetry generators of the theory that are broken by the vacuum around which the theory is expanded. For a concrete theory with global $U(1)$ and $SU(N)$-symmetry with two or three fields in the adjoint and fundamental representation we discuss how the discrete anti-linear symmetries determine a parameter regime in a non-Hermitian theory in which the theorem still holds, needs to be technically modified and eventually completely breaks down. When changing the global to a local symmetry the Higgs mechanism [19, 20, 21, 22] couples these fields to gauge fields in such a way that the combined field acquires a mass, such as for instance the W and Z gauge bosons in the standard model.

The non-Hermitian gauge theories obtained in this manner are then further probed for their existence of magnetic monopoles that constitute another important type of classical solutions of the quantum field theory. For our non-Hermitian toy model these solutions are obtained in a fourfold scaling limit of the corresponding equation of motion usually referred to as the Bogomol'nyi-Prasad-Sommerfield (BPS) limit [23, 24]. These type of solutions are special as they saturate the lowest energy or Bogomol'nyi bound. Here we pay particular attention to whether these energies are still real despite solving a complex equation. We provide a reality condition whose key ingredient is a modified $\mathcal{CPT}$-symmetry of the Hamiltonian and the solutions.

Subsequently we adopt the viewpoint to treat these type of non-Hermitian BPS theories in their own right and investigate them in 1+1 and also 3+1 dimensions in form of reduced versions of the Skyrme model [25] that was designed to model the nuclei.

Our manuscript is organised as follows. In section 2 we elaborate on the spontaneous symmetry breaking in non-Hermitian field theories with a discussion of Goldstone's theorem in section 2.1 and the Higgs mechanism in section 2.2. Complex solutions to BPS equations are discussed in section 3, with t'Hooft-Polyakov magnetic monopoles in section 3.1, dual BPS theories in 1+1 dimensions and BPS Skyrmion solutions in section 3.3. Our conclusions are stated in section 4.

## 2. Spontaneous symmetry breaking in non-Hermitian field theory

In the context of non-Hermitian $\mathcal{PT}$ symmetric quantum mechanics [26] it is well known that one needs to find a well-defined metric to ensure a unitary time-evolution of a Hamiltonian system and also to define meaningful observable quantities. A direct investigation of a non-Hermitian system therefore often leads to bizarre conclusions and apparent inconsistencies, as for instance when studying properties of non-observable quantities in a conventional framework with an inconsistent inner product. These issues persist and carry over to a quantum field theoretic setting, where a similar problem arises for instance when one studies the variational principle for an action of a complex scalar field theory $I = \int \mathcal{L}(\phi,\phi^*)$. The equations of motion obtained from $\delta I/\delta\phi = 0$ and $\delta I/\delta\phi^* = 0$ appear to be incompatible under complex conjugation due to the fact that $I \neq I^*$. One may attempt to remedy these issues by all kind of mechanism, e.g. [27, 28], but here we adopt the well-established and successful viewpoint relying on the construction of what is often referred to as the pseudo-Hermitian approach [29, 26] by finding a Dyson map $\eta$ [30], whose adjoint action map a non-Hermitian Hamiltonian $H$ to a Hermitian

Hamiltonian $h$ as $\eta H \eta^{-1} = h$.

Technically these transformations are usually difficult to find and are also known to be not unique [31]. Here we encounter two versions of these transformations, as a map acting on the field theoretic Hamiltonian involving non-commutative equal time commutation relations between canonical fields and also as matrix transformations on the squared-mass matrix. We will see that for some examples the action of the field of these maps is indeed identical. For references purposes and in order to establish our notation we briefly recall the standard technique used for finite dimensional Hamiltonians $H$, where in the low order field theoretic context the non-Hermitian squared-mass matrix $M^2$ is the analogue to $H$.

As a starting point the Hamiltonian is assumed to be $\mathcal{PT}$-symmetric, that is a simultaneous parity $\mathcal{P}$ and time transformation $\mathcal{T}$,

$$[H, \mathcal{PT}] = H\mathcal{P} - \mathcal{P}H^* = 0 , \quad \mathcal{P}^T\mathcal{P} = \mathbb{I}. \tag{1}$$

A bi-orthonormal basis can then be constructed from the right and left eigenvector, $\{v_n\}$ and $\{u_n\}$, respectively, of $H$

$$Hv_n = \epsilon_n v_n , \quad H^\dagger u_n = \epsilon u_n, \tag{2}$$

where the dagger denotes the usual complex conjugate transpose. This basis satisfies the following orthonormality and completeness relations

$$\langle u_n | v_m \rangle = \delta_{nm} , \quad \sum_n |u_n\rangle \langle v_n| = \sum_n |v_n\rangle \langle u_n| = \mathbb{I}, \tag{3}$$

with the left and right eigenbasis related to each other by the parity operator

$$|u_n\rangle = s_n \mathcal{P} |v_n\rangle , \tag{4}$$

with $s_n = \pm 1$ defining the signature. Combining the relations (3), (4) and the second relation in (1) we can express the parity operator as

$$\mathcal{P} = \sum_n s_n |u_n\rangle \langle u_n| , \quad \mathcal{P}^T = \sum_n s_n |v_n\rangle \langle v_n| . \tag{5}$$

The relation $\mathcal{P}^T\mathcal{P} = 1$ then automatically holds, following from $\mathcal{P}^T\mathcal{P}|v\rangle = |v\rangle$ and $\langle u| \mathcal{P}^T\mathcal{P} = \langle u|$.

Closely related, one defines an operator $\mathcal{C}$ as

$$\mathcal{C} = \sum_n s_n |v_n\rangle \langle u_n| , \tag{6}$$

that is easily seen to be related to the metric operator $\rho := \eta^\dagger \eta$ as $\mathcal{C} = \rho^{-1}\mathcal{P}$. The operator $\mathcal{C}$ satisfies the following algebraic relations

$$[\mathcal{C}, H] = 0 , \quad [\mathcal{C}, \mathcal{PT}] = 0 , \quad \mathcal{C}^2 = \mathbb{I}. \tag{7}$$

Naturally when identifying the $\mathcal{C}$ and $\mathcal{P}$ operators at the level of the squared-mass matrix we have ignored all interaction terms, which might not be left invariant. We will see below that requiring $\mathcal{PT}$ to be a symmetry of the entire Lagrangian will select out a particular subset, which for our solutions restricts the possible choices of the signatures $\{s_n\}$.

### *2.1. Nambu-Goldstone bosons in models with global $U(1)$ and $SU(2)$ symmetry*

We begin by considering a theory that respects a continuous global $SU(N)$ symmetry and a discrete modified anti-linear $\mathcal{CPT}$-symmetry. The following model is a direct generalization of a $U(1)$ symmetric model first studied in the context non-Hermitian field theories in [32]

$$\mathcal{L}_n^{SU(N)} = \sum_{i=1}^{n} \partial_\mu \phi_i^\dagger \partial^\mu \phi_i + c_i m_i^2 \phi_i^\dagger \phi_i + \sum_{i=1}^{n-1} \kappa_i \mu_i^2 \left( \phi_i^\dagger \phi_{i+1} - \phi_{i+1}^\dagger \phi_i \right) - \frac{g_i}{4} \left( \phi_1^\dagger \phi_1 \right)^2 , \tag{8}$$

containing $n$ complex scalar fields $\phi_i = (\phi_i^1, \ldots, \phi_i^N)$, $i, \ldots n$, with $N$-components, where each field $\phi_i$ is taken to be in the fundamental representation of $SU(N)$. The constant parameters in the model are $m_i, \mu_i, g_i \in \mathbb{R}$ and $\kappa_i, c_i \in \{-1, 1\}$. The respective continuous $SU(N)$ and discrete anti-linear $\mathcal{CPT}$ symmetries are realised as [4]

$$SU(N) \ : \ \phi_j \to e^{i\alpha T^a} \phi_j, \tag{9}$$

$$\mathcal{CPT}_{1/2} \ : \ \phi_i(x_\mu) \to \mp \phi_i^*(-x_\mu) \ \text{ for } \frac{i}{2} \in \mathbb{Z}, \ \ \phi_j(x_\mu) \to \pm \phi_j^*(-x_\mu) \ \text{ for } \frac{j+1}{2} \in \mathbb{Z}, \tag{10}$$

with real parameters $\alpha$ and $SU(N)$ generators $T^a$. Notice that the $\mathcal{CPT}$ symmetry is not the standard one commonly assumed in Hermitian theories.

A key ingredient in the pseudo-Hermitian approach is to find an equivalent Hermitian Lagrangian to the original non-Hermitian one by acting on it adjointly with the Dyson map. In our case we can use the map

$$\eta = \exp \left[ \frac{\pi}{2} \sum_{i=2,4,6,\ldots} \sum_{\alpha=1}^{N} \int d^3x \Pi^{\varphi_i^\alpha}(t, \vec{x}) \varphi_i^\alpha(t, \vec{x}) + \Pi^{\chi_i^\alpha}(t, \vec{x}) \chi_i^\alpha(t, \vec{x}) \right] , \tag{11}$$

expressed in terms of the real components $\varphi_i^\alpha$, $\chi_i^\alpha$ of the complex fields $\phi_i^\alpha = (\varphi_i^\alpha + i\chi_i^\alpha)/\sqrt{2}$ to transform , $\mathcal{L}_n^{SU(N)}$. This Dyson map is a generalisation of the one used in [33, 34], see $S$ after equations (37) and (40), respectively, and is, as is well-known, not unique. It does not exhibit the standard features of becoming ill-defined at the exceptional points of the theory and in the broken $\mathcal{PT}$-regime in an obvious manner. Furthermore, in [35] it has been reported that issues with ghost fields arise possibly due to the application of this map. Here we will proceed and use (11) to construct an equivalent Hermitian system. For this purpose we resort temporarily to the quantum theory and when assuming the standard equal-time commutation relation between the fields and their corresponding conjugate momentum operators, each of the real component fields transforms as

$$\varphi_{2i}^\alpha \to -i\varphi_{2i}^\alpha \, , \quad \chi_{2i}^\alpha \to -i\chi_{2i}^\alpha . \tag{12}$$

When recasting the transformed equivalent Hermitian Lagrangian into a simpler form by defining the $N \times n$-component vector field $\Phi = (\varphi_1^1, \chi_2^1, \ldots, )$ we obtain

$$\mathcal{L} = \frac{1}{2} \partial_\mu \Phi \mathcal{I} \partial^\mu \Phi + \frac{1}{2} \Phi^T H \Phi + \mathcal{O}(\Phi^4). \tag{13}$$

Notice that as a result of the transformation we have introduced the metric operator $\mathcal{I} = \text{diag}(1, -1, 1, \ldots)$ in the kinetic term. Crucially the Hessian matrix $H$ is now Hermitian. The squared-mass matrices of the above Lagrangian are then found by calculating the expression $M^2 := \mathcal{I}H$.

Depending on which vacuum we choose to expand around, we expect to find different types of spectra containing the number of massless Goldstone bosons as predicted by Goldstone's theorem

[17, 18], i.e. when expanding around the $SU(N)$-symmetric vacua we expect to find no massless Goldstone boson, whereas when expanding around the vacua that break the $SU(N)$-symmetry we anticipate to have a massless Goldstone boson in the spectrum for each broken generator of the symmetry group. In [4] we confirmed that these standard predictions from Hermitian theories also hold in the regions of the parameter space where the $\mathcal{CPT}$ is unbroken and at the boundaries of those domains. However, besides predicting the right number of zero eigenvalues for the theory to be meaningful, one also needs to be able to identify the fields corresponding to the Goldstone bosons. It is this latter property that breaks down at the boundaries of the $\mathcal{CPT}$-symmetric regime. Let us see in some more detail what kind of boundaries and physical regions we may encounter.

For this purpose we need to compute the squared-mass matrices by expanding around specific vacua. For instance, for the Abelian $U(1)$-symmetric model with $n = 3$ fields and the non-Abelian $SU(2)$-model with $n = 2$ fields expanded around the symmetry broken vacua we obtain the squared-mass matrices

$$M^2_{U(1),3} = \begin{pmatrix} \frac{3c_3 m_3^2 \mu^4}{c_2 c_3 m_2^2 m_3^2+\nu^4} + 2c_1 m_1^2 & -c_\mu \mu^2 & 0 & 0 & 0 & 0 \\ c_\mu \mu^2 & -c_2 m_2^2 & c_\nu \nu^2 & 0 & 0 & 0 \\ 0 & -c_\nu \nu^2 & -c_3 m_3^2 & 0 & 0 & 0 \\ 0 & 0 & 0 & \frac{c_3 m_3^2 \mu^4}{c_2 c_3 m_2^2 m_3^2+\nu^4} & c_\mu \mu^2 & 0 \\ 0 & 0 & 0 & -c_\mu \mu^2 & -c_2 m_2^2 & -c_\nu \nu^2 \\ 0 & 0 & 0 & 0 & c_\nu \nu^2 & -c_3 m_3^2 \end{pmatrix}, \quad (14)$$

$$M^2_{SU(2),2} = \begin{pmatrix} -\frac{\mu^4}{m_2^2} & \mu^2 & 0 & 0 & 0 & 0 & 0 & 0 \\ \mu^2 & -m_2^2 & 0 & 0 & 0 & 0 & 0 & 0 \\ 0 & 0 & \frac{\mu^4}{m_2^2} & -\mu^2 & 0 & 0 & 0 & 0 \\ 0 & 0 & -\mu^2 & m_2^2 & 0 & 0 & 0 & 0 \\ 0 & 0 & 0 & 0 & -\frac{\mu^4}{m_2^2} & -\mu^2 & 0 & 0 \\ 0 & 0 & 0 & 0 & -\mu^2 & -m_2^2 & 0 & 0 \\ 0 & 0 & 0 & 0 & 0 & 0 & \frac{3\mu^4}{m_2^2} - 2m_1^2 & \mu^2 \\ 0 & 0 & 0 & 0 & 0 & 0 & \mu^2 & m_2^2 \end{pmatrix}. \quad (15)$$

Key features of the model can be understood from the eigenvalue spectrum of $M^2_{U(1),3}$ plotted in figure 1 as a function of $\nu$ with all other parameters fixed as indicated. Crucially we recognise the predicted number of Goldstone bosons to emerge, i.e. one corresponding to the eigenvalue $\lambda_0 = 0$.

Moreover we observe in figure 1 panel (a) and (b) a standard feature in the eigenvalue spectra for complex matrices, namely that points in the parameter space at which two eigenvalues coincide and become complex conjugate pairs beyond this point. At these points also the eigenvectors coalesce and therefore the matrices are no longer diagonalisable. For energy spectra these points are commonly referred to as *exceptional points* [36]. However, unlike as for energy spectra, we have to exclude here the negative values in the $M^2$-spectrum to obtain physical, that is real, masses. Thus for instance the exceptional points $EP_1$ and $EP_2$ in figure 1 panel (a) are in the non-physical and physical regime, respectively.

In addition we identify another special point in figure 1 at which the forbidden level crossing in Hermitian theories [37] is circumvented. Our analysis shows that at the point marked as 0EP the non-zero eigenvalue coalesces with the one for the zero eigenvalue. Moreover, just as for the standard exceptional point the two eigenvalues coalesce so that the matrix is non-diagonalisable. Thus in principle this point has all the trademarks of an exceptional point except for the fact that

Journal of Physics: Conference Series **2038** (2021) 012010 doi:10.1088/1742-6596/2038/1/012010

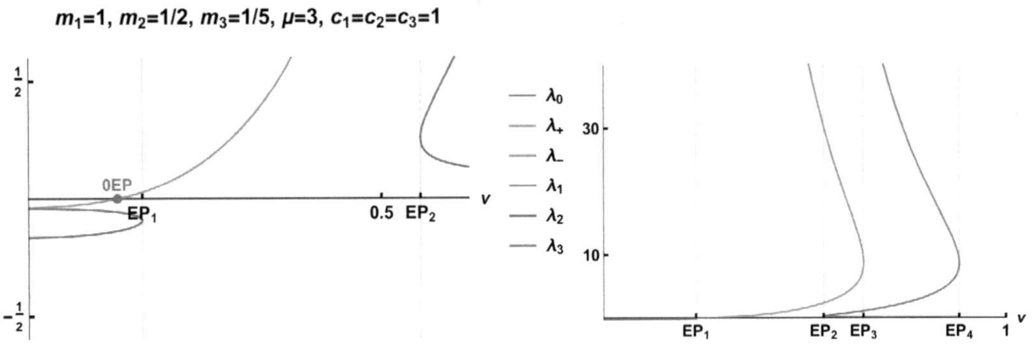


Figure 1: Real parts of the six eigenvalues $\lambda_0$, $\lambda_1$, $\lambda_2$, $\lambda_3$ and $\lambda_\pm$ of $M^2_{U(1),3}$ as functions of $\nu$, with all other parameters fixed as indicated, in the $U(1)$-broken regime. $\lambda_0 = 0$ corresponds to the Goldstone boson. Panel (b) is a zoomed out version of Panel (a).

beyond this point, in both directions, the two eigenvalues do not become complex conjugate pairs but remain real as can be seen in figure 1 panel (a). Thus in order to distinguish the behaviour at these type of points from the one of standard exceptional points, we have termed this point as *zero exceptional point.*

Next we address the question of identifying whether the Lagrangians $\mathcal{L}_n^{SU(N)}$ are indeed physically meaningful. As a first exclusion principle we identify regions in the parameter space for which the model has a well-defined classical mass spectrum. Interestingly for $M^2_{SU(2),2}$ with the two choices $c_1 = c_2 = \pm 1$ no such region exists and these versions of the model must therefore be discarded as non-physical. However, for $c_1 = -c_2 = \pm 1$ we can identify physical regions in the parameter space as seen in figure 2 where these regions are plotted with respects to the parameter combinations $\mu^4/m_1^4$ and $m_2^2/m_1^2$. Of course there might be other arguments, also resulting at the quantum level, that force these models to be non-physical, but they pass the basic test of possessing a well defined classical mass spectrum.

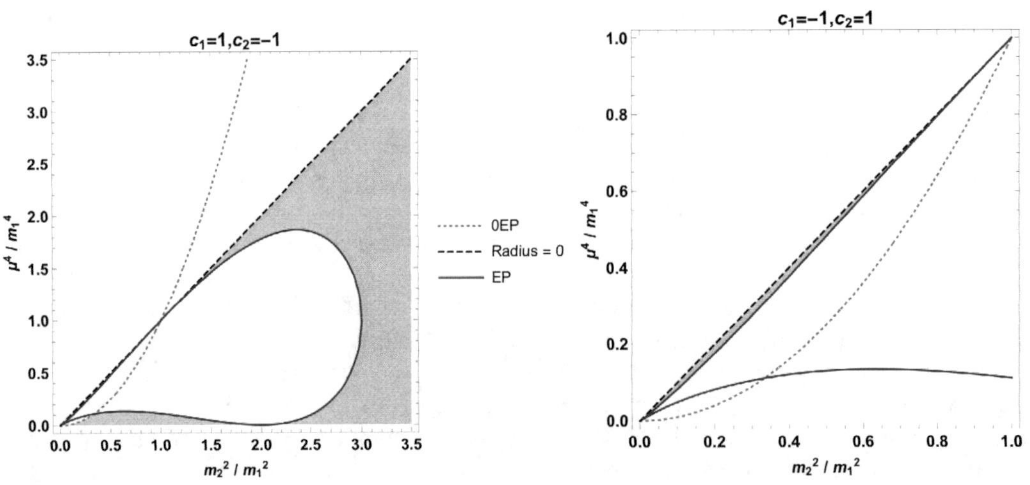


Figure 2: Physical regions (in orange), resulting from $M^2_{SU(2),2}$ in the $SU(2)$-broken regime, at which the masses are positive definite bounded by exceptional points, zero exceptional points and trivial vacua with radius $R = 0$ for different choices of the constants $c_i$.

Having discussed the explicit forms of the physical regions and their boundaries, let us return to the question of whether the Goldstone bosons may be identified in the non-Hermitian theories considered. As shown in detail in [4], in the $\mathcal{CPT}$-symmetric regime the Goldstone theorem holds in the usual sense and the fields corresponding to the Goldstone bosons can be identified explicitly. Remarkably this is still possible at the standard exceptional point, despite the fact that mass squared matrix can not be diagonalised. Technically this needs to be done differently by exploiting the fact that the mass-squared matrix is in a block diagonal form, so that the block matrix with null eigenvector can still be diagonalisable at the exceptional point of other block matrices. Hence we can still identify the Goldstone fields at standard exceptional point. However, this is no longer possible at the zero exceptional point where the mechanism breaks down and the fields corresponding to the Goldstone boson can not be constructed.

Finally let us identify the $\mathcal{CPT}$-operators that ensure the reality of the mass-squared matrix and see whether they leave the complex Lagrangian $\mathcal{L}_n^{SU(N)}$ invariant. Using the definitions (5) and (6) for these operators, we solve the equations they must obey with the non-Hermitian Hamiltonian replaced by the non-Hermitian mass-squared matrix. As a concrete example we take the $3 \times 3$ block matrix of the mass-squared matrix in the $n = 3, U(1)$ model before the similarity transformation has been carried out, and solve all the property equations (1) and (7) for $\mathcal{P}$ and $\mathcal{C}$. Given the eigenvalues of the block matrix $\{\lambda_j\} = \{0, \lambda_-, \lambda_+\}$, we define the quantities$\Lambda_j := \lambda_j + c_2 m_2^2 + c_3 m_3^2$ and$\Lambda_j^k := \lambda_j + c_k m_k^2$. Then we obtain the solutions for $\mathcal{P}$ and $\mathcal{C}$ that can be written as

$$\mathcal{P}' = \sum_{j=0,\pm} \frac{s_j}{N_j^2} \begin{pmatrix} \left(\Lambda_j^2\Lambda_j^3+\nu^4\right)^2 & i\mu^2\Lambda_j^3\left(\Lambda_j^2\Lambda_j^3+\nu^4\right) & \mu^2\nu^2\left(\Lambda_j^2\Lambda_j^3+\nu^4\right) \\ -i\mu^2\Lambda_j^3\left(\Lambda_j^2\Lambda_j^3+\nu^4\right) & \mu^4\left(\Lambda_j^3\right)^2 & -i\nu^2\mu^4\Lambda_j^3 \\ \mu^2\nu^2\left(\Lambda_j^2\Lambda_j^3+\nu^4\right) & i\nu^2\mu^4\Lambda_j^3 & \mu^4\nu^4 \end{pmatrix}, \tag{16}$$

$$\mathcal{C}' = \sum_{j=0,\pm} \frac{(-1)^{\delta_{-,j}} s_j}{N_j^2} \begin{pmatrix} \left(\Lambda_j^2\Lambda_j^3+\nu^4\right)^2 & i\mu^2\Lambda_j^3\left(\Lambda_j^2\Lambda_j^3+\nu^4\right) & \mu^2\nu^2\left(\Lambda_j^2\Lambda_j^3+\nu^4\right) \\ i\mu^2\Lambda_j^3\left(\Lambda_j^2\Lambda_j^3+\nu^4\right) & -\mu^4\left(\Lambda_j^3\right)^2 & i\nu^2\mu^4\Lambda_j^3 \\ \mu^2\nu^2\left(\Lambda_j^2\Lambda_j^3+\nu^4\right) & i\nu^2\mu^4\Lambda_j^3 & \mu^4\nu^4 \end{pmatrix} \tag{17}$$

The normalisation constants are $N_0^2 = \kappa\lambda_-\lambda_+$, $N_\pm^2 = (\kappa + \lambda_\pm\Lambda_\pm)\lambda_\pm(\lambda_+ - \lambda_-)$. As the solutions depend on the choices for the set of signatures $s_0, s_\pm = \pm 1$, we have found eight different possible solutions for these operators. The two particular parity operators $\mathcal{P}(s_0 = \pm, s_- = \mp, s_+ = \pm)$ are rather special because they can be identified as the operators involved in the $\mathcal{CPT}_{1/2}$ transformations in (9) that leave also the entire Lagrangian invariant when switching on the interaction terms. We still have an ambiguity of three operators left, which reflects the aforementioned non-uniqueness of the metric and the Dyson map in the $\mathcal{C}'$ and $\mathcal{P}'$-operators.

### *2.2. Higgs mechanism in a model with local SU(2) symmetry*

While in the previous section the continuous symmetry has been global, in the sense of being the same at all points in spacetime, we will now allow these to depend on points in spacetime, that is to be local. Via the standard mechanism of minimal coupling this will introduce local gauge fields and in turn is also expected to remove the degrees of freedom of the Goldstone bosons. The latter phenomena is known as the celebrated Higgs mechanism [19, 20, 21, 22] in the standard Hermitian setting. In [7] we showed that the mechanism is also extendable to non-Hermitian theories, although differently in different regimes characterized by the $\mathcal{CPT}$-symmetry as we shall discuss. We illustrate this for a model with local $SU(2)$ symmetry of the form

$$\mathcal{L}_l = \sum_{i=1}^{2} |D_\mu \phi_i|^2 + m_i^2 |\phi_i|^2 - \mu^2 \left(\phi_1^\dagger \phi_2 - \phi_2^\dagger \phi_1\right) - \frac{g}{4}\left(|\phi_1|^2\right)^2 - \frac{1}{4} F_{\mu\nu} F^{\mu\nu}. \tag{18}$$

This Lagrangian is identical to $\mathcal{L}^{n=2}_{SU(2)}$ introduced in the previous section, but minimally coupled with local gauge fields $A_\mu$, so that partial derivatives are replaced by covariant derivatives $D_\mu = \partial_\mu - ieA_\mu$ with charge $e \in \mathbb{R}$. A standard Yang-Mills term comprised of the Lie algebra valued field strength $F_{\mu\nu} = \partial_\mu A_\nu - \partial_\nu A_\mu - ie[A_\mu, A_\nu]$ has also been added. The two anti-linear symmetries in (9) are still respected by $\mathcal{L}_l$ as in the previous section.

We may also still map $\mathcal{L}_l$ to an equivalent Hermitian Lagrangian by means of an adjoint action of the Dyson map (11), albeit with changed conjugate momenta due to the presence of the gauge field. With this modification the field transformations due to the action of the Dyson map are unchanged. The potential term of the resulting real Lagrangian is equivalent to the one in $\mathcal{L}^{n=2}_{SU(2)}$, so when expanding the potential around the non-trivial symmetry breaking vacuum we find same mass-squared matrix as given in (15).

Expanding also the kinetic term around the non-trivial symmetry breaking vacuum, we [7] found the following mass for the gauge vector boson

$$m_g = \frac{eR_f}{m_2^2}\sqrt{m_2^4 - \mu^4}, \tag{19}$$

where $R_f = \sqrt{4(\mu^4 + c_1 c_2 m_1^2 m_2^2)/g m_2^2}$ is the radius of the spherical vacuum solution in the fundamental representation. This gauge mass is real and positive when $R_f$ and $\sqrt{m_2^4 - \mu^4}$ are simultaneously positive definite. Thus in these cases the Higgs mechanism works in the same fashion as in the Hermitian case and the gauge vector boson becomes massive. However, there are now several interesting scenarios where this might be violated. First we note that the vacuum radius can never be negative for $c_1 = c_2 = \pm 1$, but these two cases have been excluded previously as being non-physical. Therefore the vacuum radius always takes the form $R_f = 4(\mu^4 - m_1^2 m_2^2)/g m_2^2$ for the two remaining physical cases. Next we notice that the scenario with vanishing $R_f$ for $\mu^4 = m_1^2 m_2^2$ corresponds to the black dotted line in both panels of figure 2. At these lines the theory is expanded around the trivial symmetric vacuum so that no Goldstone bosons emerge that may be combined with the gauge field. The square root factor in (19) is more interesting as it may be tuned to give rise to novel phenomena only present in non-Hermitian theories. When $m_2^4 = \mu^4$ the mass of the gauge field also vanishes. This scenario corresponds to the red dotted line in both panels in figure 2, that is to the zero exceptional points. As we discussed in the previous section, at these points the Goldstone boson field can not be identified even though the zero mass eigenvalue is correctly predicted by Goldstone's theorem and found in the eigenvalue spectrum of the mass-squared matrix. This means there is no well-defined Goldstone boson that can be combined with the gauge field to ensure that it acquires a non-vanishing mass. Thus *the Higgs mechanism breaks down at the zero exceptional point with the Goldstone boson being unidentifiable and the gauge particle unable to acquire a mass.* Interestingly, with regard to the Higgs mechanism nothing special happens at the exceptional points which constitute the remaining boundary of the physical region indicated in figure 2 by solid blue lines. Despite the fact that the Goldstone boson field has to be constructed technically in a different fashion, it can actually be identified and hence can be combined with the gauge field to give the latter a non-vanishing mass.

Having discussed the breakdown of the Higgs mechanism at two types of boundaries of the physical regions as indicated by the orange region in figure 2, let us now briefly explain how the Higgs mechanism works inside the physical region of our non-Hermitian model. Rewriting for this purpose the Lagrangian in terms of the Goldstone fields $\{G^a\}$, we found in [7] the following expression for the kinetic term

$$\begin{aligned}\mathcal{L} &= \sum_{a=1}^{3} \partial_\mu G^a \partial^\mu G^a - m_g A^1_\mu \partial^\mu G^1 + m_g A^2_\mu \partial^\mu G^1 + m_g A^3_\mu \partial^\mu G^3 + \frac{1}{2} m_g^2 A^a_\mu A^{a\mu} + \ldots \\ &= \frac{1}{2} m_g^2 \left( A^1_\mu - \frac{1}{m_g} \partial_\mu G^1 \right)^2 + \frac{1}{2} m_g^2 \left( A^2_\mu + \frac{1}{m_g} \partial_\mu G^2 \right)^2 + \frac{1}{2} m_g^2 \left( A^3_\mu + \frac{1}{m_g} \partial_\mu G^3 \right)^2 + \ldots \\ &= \frac{1}{2} m_g^2 \sum_{a=1}^{3} B^a_\mu B^{a\mu} + \ldots . \end{aligned} \tag{20}$$

Thus by introducing the new gauge field $B^a_\mu = A^a_\mu \pm \frac{1}{m_g} \partial_\mu G^a$ we have removed the degrees of freedom of the Goldstone bosons and introduced a mass term for these combined gauge field-Goldstone boson fields. Evidently when the Goldstone boson fields can not be found, as at the zero exceptional points, or do not exist at all, as in a theory expanded around a trivial symmetry preserving vacuum, and the new $B_\mu$ fields can not be defined. Remarkably at the standard exceptional points this is still possible, when adapting the mechanism to identify the Goldstone bosons appropriately as explained in the previous section.

We note that the analysis presented in this section can be extended to the $SU(N)$-symmetric model in a straightforward fashion for which we found $N^2 - 1$ gauge fields with the same masses as given in (19) and the same behaviour in all regions of the parameter space.

## 3. Complex BPS solitons and magnetic monopoles with real energy

Non-Abelian gauge theories are known to possess an intriguing variety of different types of solutions to their equations of motion, such as solitons and almost unavoidably also magnetic monopole solutions [38, 39, 40]. More than fourty years ago Olive and Montonen [41] noticed that different types of solutions may be related to each other and that the soliton solutions in non-Abelian gauge theories become equivalent to massive gauge fields in a dual theory. Here we wish to address the question of whether these type of features will survive, or in which way they need to be altered, in a non-Hermitian theory [6] and pay special attention to whether the energies of the solutions might still be real.

Extremely instrumental in finding explicit analytic solutions is the Bogomol'nyi-Prasad-Sommerfield (BPS) [23, 24] limit, consisting of taking a multi-scaling limit in a physically motivated fashion by acknowledging that certain mass ratios are very small. While most of the equations in this context are very complicated coupled differential equations, this approach allows to convert to a solvable system with explicit solutions that, however, still contains the key features of the physics involved. In the next subsection we will first demonstrate how these limits may be taken to obtain t'Hooft-Polyakov monopole solutions and subsequently consider a general setting of theories for which the limit is assumed to have been already carried out in 1+1 and also 3+1 dimensions.

For our solutions to be physically meaningful we require a reality condition on them as well as on the Hamiltonian so that the corresponding energies become real. A simple argument that adapts the reality condition from the quantum mechanical to the field theory [42] is simply demanding the Hamiltonian and field solution to be $\mathcal{PT}$-symmetric. When

(a) the Hamiltonian satisfies $\mathcal{H}[\phi(x_\mu)] = \mathcal{H}^\dagger[\phi(-x_\mu)]$, and

(b) the solutions to the equations of motion are $\mathcal{PT}$ symmetric, $\mathcal{PT}\phi(t, \vec{x}) = \phi(-t, -\vec{x}) = \phi(t, \vec{x})$, then the energy $E$ of the solutions $\phi$ is real. This follows from the simple argument

$$E = \int_{-\infty}^{\infty} dx \mathcal{H}[\phi(x)] = -\int_{\infty}^{-\infty} dx \mathcal{H}[\phi(-x)] = \int_{-\infty}^{\infty} dx \mathcal{H}^{\dagger}[\phi(x)] = E^{*}. \tag{21}$$

In [8, 9] we found that these requirements need to be extended to include more possibilities when there are non-trivial anti-linear symmetry relating two degenerate solutions. In such a scenario we replace the $\mathcal{PT}$ symmetry of the solution with a generic anti-linear symmetry that we refer to as modified $\mathcal{CPT}$, which is not to be confused with the standard $\mathcal{CPT}$ in quantum field theory. Then the reality of the energy is guaranteed if the following three conditions hold:

**(i)** The Hamiltonian transforms under the modified $\mathcal{CPT}$-symmetry as

$$\mathcal{CPT}: \mathcal{H}[\phi(x_\mu)] \rightarrow \mathcal{H}^{\dagger}[\phi(-x_\mu)]. \tag{22}$$

**(ii)** Two solutions to the equations of motion $\phi_1$ and $\phi_2$, not necessarily distinct, are related to each other by the $\mathcal{CPT}$-symmetry as

$$\mathcal{CPT}: \phi_1(x_\mu) \rightarrow \phi_2(-x_\mu). \tag{23}$$

The $\mathcal{CPT}$ symmetry is here a generic anti-linear symmetry which may or may not coincide with the $\mathcal{PT}$ symmetry.

**(iii)** The energies $E[\phi]$ of the two solutions are degenerate

$$E[\phi_1] = E[\phi_2]. \tag{24}$$

When $\phi_1 = \phi_2$ and $\mathcal{CPT} = \mathcal{PT}$, the condition (i), (ii) and (iii) coincide with (a) and (b). While for a quantum mechanical systems the anti-linear symmetries can usually be read off trivially from the Hamiltonian and are then simply verified for the wave functions in the $\mathcal{PT}$-symmetric regime, here the starting point is mostly reversed and it appears more practical to identify the relevant symmetry from the explicit solutions first.

For N-soliton solutions these reality conditions have to be enlarged by employing also the integrability of the model and using the fact that asymptotically the N-soliton solutions separate into N one-soliton solutions [43, 44].

Let us now construct concrete solutions in some explicit non-Hermitian systems and study their properties.

### *3.1. t'Hooft-Polyakov magnetic monopoles*

In order to study how complex monopole solutions might arise we consider a complex extended version of a Lagrangian [6] for which real t'Hooft-Polyakov monopole solutions were shown to exist [38, 39] by adding a complex part in a similar fashion as in our model used to investigate the non-Hermitian version of the Higgs mechanism in (18)

$$\begin{aligned} \mathcal{L}_{cm} &= \frac{1}{2}Tr\left(D\phi_1\right)^2 + \frac{1}{2}Tr\left(D\phi_2\right)^2 - c_1 m_1^2 Tr\left(\phi_1^2\right) + c_2 m_2^2 Tr\left(\phi_2^2\right) \\ &\quad -i\mu^2 Tr\left(\phi_1\phi_2\right) - \frac{g}{4}Tr\left(\phi_1^2\right)^2 - \frac{1}{4}Tr\left(F_{\mu\nu}F^{\mu\nu}\right). \end{aligned} \tag{25}$$

Here we take $\phi_1$ and $\phi_2$ to belong to the adjoint representation of $SU(2)$, which means we can factorise the fields as $\phi_i(x) = \phi_i^a(x)\tau^a$, where $\phi_i^a(x)$ are a real fields and $\{\tau^a\}$ denote the generators of $SU(2)$. Noting that the adjoint action of the Dyson map

$$\eta_{\pm} = \prod_{a=1}^{3} \exp\left(\pm\frac{\pi}{2}\int d^3\Pi^a\phi_2^a\right) \tag{26}$$

transforms the fields as $\phi_1 \to \phi_1$, $\phi_2 \to c_3 i\phi_2$ with $c_3 = \pm 1$, we can map the complex monopole Lagrangian $\mathcal{L}_{cm}$ to two equivalent Hermitian ones. The equations of motion and the Higgs vacuum may then be found in a standard fashion as described above. Next we assume the parametrization

$$(\phi^{cl}_\alpha)^a = h_\alpha(r)\hat{r}^a_{n_\alpha}\ , \quad (A^{cl}_i)^a = \epsilon^{iaj}\hat{r}^j_n\left(\frac{u(r)-1}{er}\right), \quad \hat{r}^a_n = \begin{pmatrix} \sin(\theta)\cos(n\varphi) \\ \sin(\theta)\sin(n\varphi) \\ \cos(\theta) \end{pmatrix} \tag{27}$$

for the scalar and the gauge fields, respectively, with $n$ denoting the winding number. This Ansatz is motivated by the fact that the static solutions have to converge to the vacuum solution in order to have finite energy [40, 45]

$$\lim_{r\to\infty} h_1(r) = \pm R_a\ , \qquad \lim_{r\to\infty} h_2(r) = \mp\frac{c_2 c_3 \mu^2}{m_2^2}R_a, \tag{28}$$

with $R_a = \sqrt{(m_1^2 m_2^2 - \mu^4)/2gm_2^2}$ denoting the radius of the spherical vacuum solution in the adjoint representation. The equations of motion then acquire the form

$$u''(r) + \frac{u(r)\left[1-u^2(r)\right]}{r^2} + \frac{e^2 u(r)}{2}\left\{h_2^2(r) - h_1^2(r)\right\} = 0, \tag{29}$$

$$h_1''(r) + \frac{2h_1'(r)}{r} - \frac{2h_1(r)u^2(r)}{r^2} + g\left\{c_1\frac{m_1^2}{g}h_1(r) + c_3\frac{\mu^2}{g}h_2(r) + 2h_1^3(r)\right\} = 0, \tag{30}$$

$$h_2''(r) + \frac{2h_2'(r)}{r} - \frac{2h_2(r)u^2(r)}{r^2} + c_2 m_2^2\left\{h_2(r) + c_3\frac{\mu^2}{m_2^2}h_1(r)\right\} = 0, \tag{31}$$

extending the standard set of equations previously obtained in [38, 39] by a new field $h_2$ and the additional equation (31). Evidently these equations are difficult to solve. In order to facilitate the construction of explicit analytic solutions we derive the BPS equations by carrying out the fourfold limit

$$\lim_{g,m_1,m_2,\mu\to 0} (29),(30),(31) \quad \text{with} \quad X := \frac{m_1^2}{g} < \infty, Y := \frac{\mu^2}{g} < \infty, Z := \frac{\mu^2}{m_2^2} < \infty. \tag{32}$$

Remarkably the equations obtained in this manner may be solved explicitly. Adjusting the boundary conditions appropriately, as indicated in (28), we obtain two distinct sets of solutions, which connect two different vacua

$$u(r) = \pm\frac{erlR_a}{\sinh(erlR_a)}, \tag{33}$$

$$h_1^\pm(r) = \pm\text{Sign}(n)\frac{1}{l}\left\{|lR_a|\coth\left(e|lR_a|r\right) - \frac{1}{er}\right\}, \tag{34}$$

$$h_2^\pm = \mp\text{Sign}(n)\frac{c_2 c_3 Z}{l}\left\{|lR_a|\coth\left(e|lR_a|r\right) - \frac{1}{er}\right\}, \tag{35}$$

with $l := \sqrt{1-Z^2}$ and $R_a = \sqrt{(X-YZ)/2}$.

Let us now consider whether the energies for these solutions are real despite the fact that the Hamiltonian of the system is non-Hermitian. After the similarity transformation has been carried out, the energy for a static solution to our system $\mathcal{L}_{cm}$ is given by

$$E = \int d^3x Tr\left(B^2\right) + Tr\left\{(D_i\phi_1)^2\right\} - Tr\left\{(D_i\phi_2)^2\right\} + V, \tag{36}$$

which when evaluated for the two solutions in (33)-(35) leads to the same energies for both

$$E = \frac{8|n|\pi R_a}{e}\left(\frac{1-Z^2}{\sqrt{1-Z^2}}\right) = \frac{8|n|\pi R_a}{e} l. \tag{37}$$

Next we compare the behaviour of the gauge mass, given in (19) with $R_f \to R_a$ as derived in [6], versus the monopole mass taken to be the rest mass, i.e. it is equal to its energy E in (37). In figure 3 we notice that in the weak coupling regime for $e = 2$ we have $m_g < m_m$, with exchanged ordering in the strong regime for $e = 10$. Most notably, however, is the behaviour seen in panels (a) and (b) showing that both masses vanish simultaneously at the zero exceptional point, where they both become purely imaginary and hence non-physical. As functions of $Z$ they revive as real valued beyond the trivial vacuum for which $R_a = 0$. In panels (c) and (d) we show that the non-physical region can be made to vanish when the zero-exceptional point and the trivial vacuum coincide, which is the case for the choice $Z = 1$.

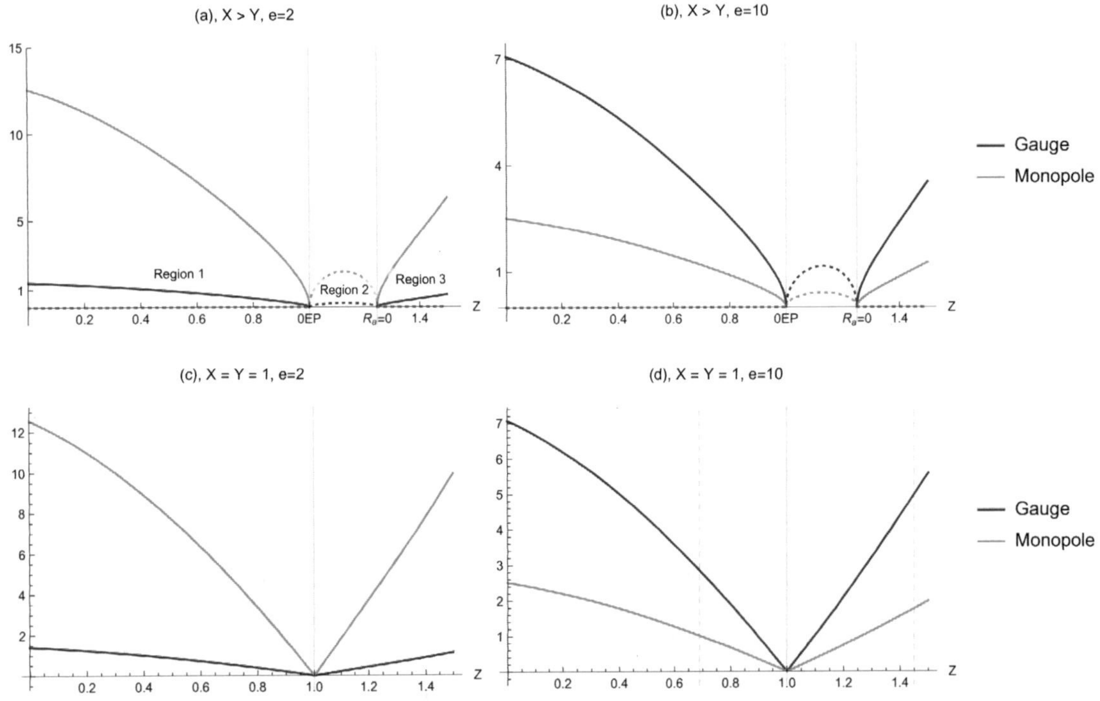


Figure 3: Revival of the gauge and magnetic monopole masses in the weak and strong regime, panels (a), (c) and (b), (d), respectively. In panel (a) and (b) we have $X = 1$, $Y = 4/5$ and $n = 1$. Solid lines correspond to real and dotted lines to imaginary parts.

We conclude this section by explaining the reality of the energies in different regions employing the conditions (i)-(iii) in (22)-(24). First, we identify the $\mathcal{CPT}$-symmetry from the Lagrangian by noticing that the Hamiltonian resulting from $\mathcal{L}_{cm}$ is mapped to it conjugate under the transformation

$$\mathcal{CPT}: \phi_i(x) \to (-1)^{\delta_{i2}} \left[\phi_i(-x)\right]^\dagger . \tag{38}$$

This constitutes the condition (i) in (22). Acting with this symmetry on the two scalar fields in

(27) we find condition (ii) realised as

$$\mathcal{CPT}: \quad \phi_1^{\pm}(x) \to \left[\phi_1^{\pm}(-x)\right]^{\dagger} = \begin{cases} \phi_1^{\mp}(x) & \text{in region 1} \\ \phi_1^{\pm}(x) & \text{in region 3} \end{cases},$$

$$\phi_2^{\pm}(x) \to -\left[\phi_2^{\pm}(-x)\right]^{\dagger} = \begin{cases} \phi_2^{\pm}(x) & \text{in region 1} \\ \phi_2^{\mp}(x) & \text{in region 3} \end{cases}.$$

We used here the properties $\left(h_1^{\pm}\right)^{\dagger} = h_1^{\pm}$, $\left(h_2^{\pm}\right)^{\dagger} = h_2^{\mp}$ in region 1 whereas in region 3 we have $\left(h_1^{\pm}\right)^{\dagger} = h_1^{\mp}$, $\left(h_2^{\pm}\right)^{\dagger} = h_2^{\pm}$. In region 2 the conjugate fields $\left(h_1^{\pm}\right)^{\dagger}$ and $\left(h_2^{\pm}\right)^{\dagger}$ are not solving the BPS equations. Hence in region 2 the $\mathcal{CPT}$-symmetry is broken and we therefore expect complex energies. In regions 1 and 3 we have always one of the fields mapped to itself, whereas the mapping of the other field exchanges dual and self-dual fields. Since the energies of the dual and self-dual fields are the same, also condition (iii) is satisfied so that the energies of the solutions must be real.

### *3.2. Dual BPS theories in 1+1 dimensions*

Motivated by the success of the theories obtained in the BPS limit, as exemplified in the previous subsection, one may take a different standpoint and regard the BPS theories as the starting point of one's considerations in their own right. For this purpose one may formulate a generic theory that possesses the general properties observed in BPS theories based on the original procedure [23] by completing the square or other methods such employing the concept of strong necessary conditions [46]. In [47] the authors build on these observations by pointing out that the energy functional and topological charges of a BPS theory can be cast into the generic forms

$$E = \int d^d x \left(A^2 + \tilde{A}^2\right) = \int d^d x \left[(A \mp \tilde{A})^2 \pm 2A\tilde{A}\right], \quad Q = \int d^d x A\tilde{A}, \tag{39}$$

where the quantities $A$ and $\tilde{A}$ are functions of the fields and their derivatives appearing in the defining Lagrangian of the theory. The energy in (36) can for instance be shown to be of that precise form too when the BPS limit is carried out. Using the definitions in (39) one may then show [47] that the compatibility condition between the Euler-Lagrange equation resulting from functionally varying the energy $E$ and the topological charge conservation $\delta Q = 0$ when varied with respect to a field change, leads to the anti/self-dual equations $A = \pm\tilde{A}$. The latter equation is interpreted as the BPS equation, since it achieves at the same time that the charge $Q$ saturates the Bogomolny bound $E = |Q|$, as seen from the relations in (39).

Specifying the expression for $E$ and $Q$ further by introducing scalar fields $\phi$, a non-trivial target space metric $\eta$ and a pre-potential $U$, one may take them to be of the still fairly general form [48]

$$E = \frac{1}{2}\int \partial_\mu \phi_a \eta_{ab} \partial^\mu \phi_b + \eta_{ab}^{-1} \frac{\delta U}{\delta \phi_a} \frac{\delta U}{\delta \phi_b}, \tag{40}$$

$$Q = \int_{\infty}^{\infty} dx \frac{\partial U}{\partial x} = \lim_{x \to \infty} U[\phi(x)] - \lim_{x \to -\infty} U[\phi(x)]. \tag{41}$$

Comparing with the solutions in (39) one identifies

$$A = \rho_{ab} \partial_x \phi_b, \quad \tilde{A} = \frac{\delta U}{\delta \phi_b} \rho_{ba}^{-1}, \tag{42}$$

where the metric is factorised as $\eta = \rho^T \rho$. Several examples for this setting that lead to real solutions were considered in [48]. Here we recall one of many examples of complex extensions studied in [8] that led to complex solutions with real energies.

We make a complex choice for the target space metric and keep the pre-potential real

$$\eta = \begin{pmatrix} 1 & -i\lambda \\ -i\lambda & 1 \end{pmatrix}, \qquad \text{and} \qquad U = -(\cos\phi_1 + \mu\phi_1 + \cos\phi_2), \tag{43}$$

with $\lambda, \mu \in \mathbb{R}$ and $\phi_1(t,x), \phi_2(t,x)$ being real scalar fields. The potential, being the last term in (40), resulting from these choices is a complex coupled sine-Gordon potential

$$V = \frac{1}{2(1+\lambda^2)} \left[ (\sin\phi_1 - \mu)^2 + 2i\lambda(\sin\phi_1 - \mu)\sin\phi_2 + \sin^2\phi_2 \right]. \tag{44}$$

In the Hermitian limit $\lambda = 0$ we can solve the corresponding BPS equations analytically to

$$\phi_1^{\pm(n)} = 2\text{arctan}\left\{ \frac{1}{\mu} + \frac{\sqrt{1-\mu^2}}{\mu} \tanh\left[ \frac{1}{2}\sqrt{1-\mu^2}(\pm x + \kappa_1) \right] \right\} + 2n\pi, \tag{45}$$

$$\phi_2^{\pm(n)} = 2\text{arctan}\left( e^{\pm x + \kappa_2} \right) + 2n\pi. \tag{46}$$

In the non-Hermitian case, $\lambda \neq 0$, we have to resort to solving the equations numerically. A set of sample kink and anti-kink solutions are depicted in figure 4. Crucially we observe that the asymptotic limits for the real parts do not depend on $\lambda$ and those for the imaginary parts are always vanishing.

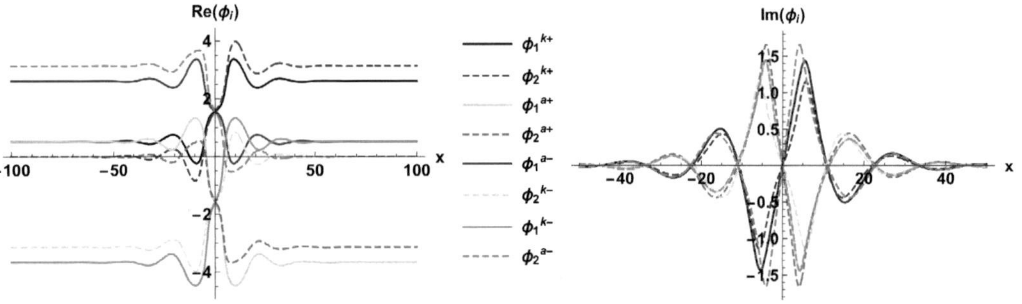


Figure 4: Complex kink $\phi^k$ and anti-kink $\phi^a$ solutions to the self-dual and anti-self-dual BPS equations for the complex coupled sine-Gordon potential (44). As initial conditions we have taken either $\phi(0) = \pi/2$ or $\phi(0) = -\pi/2$, as can be identified in panel (a), and the coupling constants are chosen as $\lambda = 3$ and $\mu = 1/2$.

These solutions connect different types of vacua, which are easily identified as the fixed points of the dynamical system described by the self-dual and anti-self-dual BPS equations. We find the infinite set of vacua

$$v_1^{(n,m)} = (\text{arcsin}\mu + 2n\pi,\ m\pi)\ , \quad v_2^{(n,m)} = (\pi - \text{arcsin}\mu + 2n\pi,\ m\pi), \tag{47}$$

with $n, m \in \mathbb{Z}$. Classifying the nature of the fixed points using standard techniques for dynamical systems by applying the linearization theorem, we can identify the nature of the fixed points as stable or unstable and conclude [8] that the solutions only interpolate between the unstable vacua $v_1^{(n,2m)}$ and stable vacua $v_2^{(n,2m+1)}$ as indicated in figure 5.

More specifically, the solutions depicted in figure 4 interpolate the following vacua

$$v_1^{(0,0)} \xrightarrow{\phi_1^{k+}\phi_2^{k+}} v_2^{(0,1)},\ v_1^{(0,0)} \xrightarrow{\phi_1^{a+}\phi_2^{a+}} v_2^{(-1,1)},\ v_1^{(0,0)} \xleftarrow{\phi_1^{a-}\phi_2^{k-}} v_2^{(0,-1)},\ v_1^{(0,0)} \xleftarrow{\phi_1^{k-}\phi_2^{a-}} v_2^{(-1,1)}. \tag{48}$$

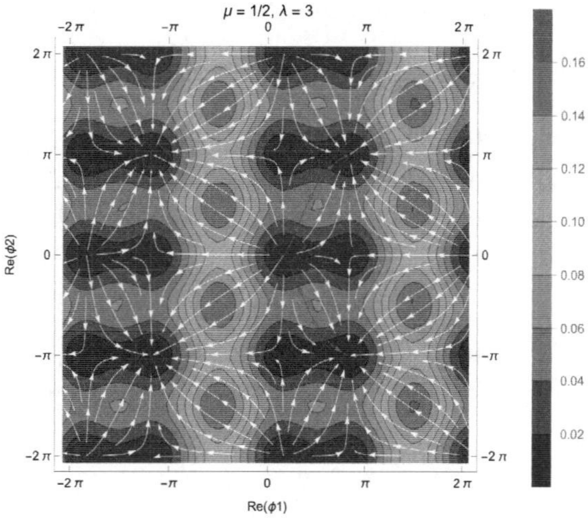


Figure 5: Real part of the complex coupled sine-Gordon potential (44) as a functions of the real parts of the fields $\phi_1$ and $\phi_2$ with real parts of the gradient flow superimposed.

The interpolation for other vacua with higher values for $n, m$ can be obtained by including the $n$ dependencies into the solutions.

Having characterized the behaviour of these solutions, we evaluate their energies by means of (40). In the Hermitian case we can use the analytic expressions from (46) obtaining

$$E = 2\left[1 + \sqrt{1-\mu^2} - \mu \arctan\left(\frac{\sqrt{1-\mu^2}}{\mu}\right)\right]. \tag{49}$$

Since the energy is of a topological nature and the asymptotic behaviour is independent of the coupling constant, we obtain for $|\mu| < 1$ the same real energy (49) also in the non-Hermitian case for all values of $\lambda$.

Finally let us use the symmetry argument (i)-(iii) to explain why the energies for these solutions must be real. The symmetry condition on the Hamiltonian (22) is satisfied by the modified $\mathcal{CPT}$ symmetry

$$\mathcal{CPT}: \phi_i(x) \to (-1)^{\delta_{i2}} \left[\phi_i(-x)\right]^\dagger. \tag{50}$$

On the level of the equations of motion this symmetry maps the self-dual BPS equations to the conjugate of the anti-self-dual BPS equations. Thus the symmetry inevitably relates two different types of solutions. Inspecting the numerical solutions, we can indeed identify a set of self-dual and anti-self-dual solutions that are related by this $\mathcal{CPT}$-symmetry

$$\phi_1^{k\pm}(x) = \left[\phi_1^{a\mp}(-x)\right]^\dagger, \quad \phi_2^{k+} = -\left[\phi_2^{k-}(-x)\right]^\dagger, \quad \phi_2^{a+} = -\left[\phi_2^{a-}(-x)\right]^\dagger. \tag{51}$$

Using now the fact that the energies are independent of the $\pm$-signs that relate the self-dual and the anti-self-dual equation and moreover that kinks and anti-kinks have identical energies we have guaranteed that the energies of the field related by the $\mathcal{CPT}$-symmetry are degenerate. Hence condition (24) is also satisfied, and therefore the energies related to these solutions must be real.

More examples involving different types of complex potentials, such as a complex extended sine-Gordon model or a logarithmic potential that admits super-exponential kink solutions can be found in [8].

### 3.3. Dual BPS theories in 3+1 dimensions, the BPS Skyrme model

In [9] we extended the study of complex BPS solutions to $3+1$ dimensions for several variants the original Skyrme model [49] that was originally designed as a theory of meson fields that also contains fermionic states interacting with those mesons. Nowadays it is interpreted as a low-energy effective field theory for quantum chromodynamics. Here we only present a summary of the main results obtained for a particular BPS version proposed in [50]. Similarly as demonstrated in the previous two sections, also this version allows for elegant analytic solutions, but in addition it resolved various discrepancies between the experimental results and the theoretical description by means of the original non-BPS full Skyrme model. The solutions to the latter are known to give binding energies that disagrees with the experiment [51].

The model we consider here is a complex BPS Skyrme Lagrangian of the form

$$\mathcal{L}_{\text{cBPSS}} = -\frac{\lambda^2}{4}(\sin\zeta - i\epsilon\cos\zeta)^4 \sin^2\Theta \mathcal{B}_\mu \mathcal{B}^\mu - \mu^2\left(\sqrt{1-\epsilon^2} - \cos\zeta - i\epsilon\sin\zeta\right), \tag{52}$$

where $\mathcal{B}^\mu := \varepsilon^{\mu\nu\rho\tau}\zeta_\nu\Theta_\rho\Phi_\tau$. Redefining the coupling constants as $\lambda \to \tilde{\lambda} = \lambda(1-\epsilon^2)$, $\mu \to \tilde{\mu} = \mu(1-\epsilon^2)^{1/4}$, the model can be thought of as a complex boosted version of the BPS Skyrme model $\mathcal{L}_{\text{BPSS}} := -\tilde{\lambda}^2 N_0^2 B_\mu B^\mu - \tilde{\mu}^2 V$ with potential $V = \frac{1}{2}Tr\,(\mathbb{I}-U)$ originally proposed in [50]. The Skyrme fields $\zeta, \Theta, \Phi$ enter through a parametrization of a $SU(2)$-group valued elements $U$, which can be used to define a Lie algebraic current $L_\mu$ in form of a right Maurer Cartan form, which in turn may be used to define the topological current $B_\mu$ as

$$U := e^{i\zeta(\sigma\cdot\vec{n})}, \quad L_\mu := U^\dagger\partial_\mu U, \quad B^\mu := \frac{1}{N_0}\epsilon^{\mu\nu\rho\tau}Tr\,(L_\nu L_\rho L_\tau) = \frac{1}{2N_0}\sin^2\zeta\sin\Theta\,\mathcal{B}^\mu. \tag{53}$$

Here $\sigma$ denotes standard Pauli matrices, we take the three component unit vector as $\vec{n} = (\sin\Theta\cos\Phi, \sin\Theta\sin\Phi, \cos\Theta)$ and $\lambda, \mu, \epsilon \in \mathbb{R}$ are constant parameters. $N_0$ is a normalization constant which is usually taken to be $N_0 = 24\pi^2$ for static solutions in order to produce integer Baryon numbers.

We notice that both Lagrangians $\mathcal{L}_{\text{cBPSS}}$ and $\mathcal{L}_{\text{BPSS}}$ are invariant under the anti-linear $\mathcal{CPT}$ transformation $\zeta \to -\zeta, i \to -i$. Moreover, their associated Hamiltonians are related by the adjoint action of a modified version of a Dyson map previously used in [52, 8]

$$\eta = \exp\left[-\text{arctanh}\,\epsilon \int d^3x \Pi^\zeta(t,r)\right], \tag{54}$$

as $\mathcal{H}_{\text{cBPSS}} = \eta^{-1}\mathcal{H}_{\text{BPSS}}\eta$. The effect of this map is to shift the field $\zeta$ by $-i\text{arctanh}\epsilon$.

The real Lagrangian is known to posses a topological static solution referred to as a BPS Skyrmion found in [50] with the following compacton structure

$$\zeta(r) = \begin{cases} 2\arccos\left(\frac{1}{\sqrt{2}}\left|\frac{\tilde{\mu}}{n\tilde{\mu}}\right|^{1/3} r\right) & \text{for } r \in \left[0, r_c = \sqrt{2}\left|\frac{\tilde{\mu}}{n\tilde{\mu}}\right|^{1/3} r\right] \\ 0 & \text{otherwise} \end{cases}, \tag{55}$$

where the Skyrmion fields are related to the spherical space-time coordinates $(r, \theta, \phi)$, $r \in [0, \infty)$, $\theta \in [0, \pi)$, $\phi \in [0, 2\pi)$ as $\Theta = \theta$, $\Phi = n\phi$ with $n \in \mathbb{Z}$.

The energy for these solutions was evaluated to be real and finite

$$E = \frac{64}{15}\sqrt{2}|n|\tilde{\mu}\tilde{\lambda}\pi(1-\epsilon^2)^{5/4}. \tag{56}$$

Considering [9] instead the complex Lagrangian $\mathcal{L}_{\text{cBPSS}}$ in (52 allows for a plethora of solutions, as we now briefly recall. We start by simplifying the calculation by re-casting the Lagrangian into the generic form (39) with

$$A = \frac{\lambda}{2}(\sin\zeta - i\epsilon\cos\zeta)^2 \sin\Theta\,\mathcal{B}_0\ , \quad \tilde{A} = \mu\left(\sqrt{1-\epsilon^2} - \cos\zeta - i\epsilon\sin\zeta\right)^{1/2}. \tag{57}$$

Keeping the same identification between the Skyrme fields $\zeta, \Theta, \Phi$ and the spherical coordinates the BPS equations acquires the form

$$\frac{n\tilde{\lambda}}{2r^2}\sin^2\left(\zeta - i\text{arctanh}\epsilon\right)\frac{d\zeta}{dr} = \pm\tilde{\mu}\sqrt{1-\cos\left(\zeta - i\text{arctanh}\epsilon\right)}. \tag{58}$$

We found an infinite set of solutions to this equation

$$\zeta^{\pm}_{\alpha,m}(r) = \tilde{\zeta}^{\pm}_{\alpha,m}(r) + i\text{arctanh}\epsilon = 2\text{arccos}\left[\omega^{\alpha}\frac{(n\tilde{\lambda}c \mp \tilde{\mu}r^3)^{1/3}}{\sqrt{2}n^{1/3}\tilde{\lambda}^{1/3}}\right] + i\text{arctanh}\epsilon, \tag{59}$$

with $\omega = \exp(i2\pi/3)$, $\alpha \in \{0,1,2\}$, $m \in \mathbb{Z}$ and $c$ denoting an integration constant. We may view this equation as a complex shifted version of the solution to the real BPS equation with different boundary conditions, thus allowing for more possibilities of solutions not considered before. In particular, we would like to identify different types of compacton solutions. For this purpose we identify the critical values $r_0$ where $\tilde{\zeta}^{\pm}_m(r_0) = 0$ and $r_\pi$ where $\tilde{\zeta}^{\pm}_m(r_\pi) = \pi$ as

$$r^{\pm}_{0,\alpha} := \omega^{\alpha}\left(\frac{\pm n\tilde{\lambda}(c-2\sqrt{2})}{\tilde{\mu}}\right)^{1/3}, \quad r^{\pm}_{\pi,\alpha} := \omega^{\alpha}\left(\frac{\pm n\tilde{\lambda}}{\tilde{\mu}}\right)^{1/3}. \tag{60}$$

Besides recovering the standard BPS Skyrmion, we also find new types of solutions by pasting different branches together including a purely imaginary solution

$$\zeta_{\text{BPS}} := \begin{cases} \tilde{\zeta}^{-}_{0,0} & \text{for} \quad 0 \le r \le r_0^- \\ 0 & \text{for} \quad r_0^- < r \end{cases}, \qquad \zeta_{\text{St}} := \begin{cases} \tilde{\zeta}^{-}_{1,0} & \text{for} \quad 0 \le r \le r_\pi^- \\ \tilde{\zeta}^{-}_{0,0} & \text{for} \quad r_\pi^- \le r \le r_0^- \\ 0 & \text{for} \quad r_0^- < r \end{cases}, \tag{61}$$

$$\zeta_{i\text{BPS}} := \begin{cases} \tilde{\zeta}^{+}_{0,0} & \text{for} \quad 0 \le r \le r_0^+ \\ 0 & \text{for} \quad r_0^+ < r \end{cases}. \tag{62}$$

Noting further that $r^+_\pi(c) = r^-_\pi(-c)$ we may even glue self-dual and anti-self dual solutions at this critical value obtaining

$$\zeta_{\text{Cusp}} := \begin{cases} \tilde{\zeta}^{+}_{0,0} & \text{for} \quad 0 \le r \le r_\pi^+ = r_\pi^- \\ \tilde{\zeta}^{-}_{0,0} & \text{for} \quad r_\pi^- \le r \le r_0^- \\ 0 & \text{for} \quad r_0^- < r \end{cases}, \quad \zeta_{\text{Shell}} := \begin{cases} 0 & \text{for} \quad 0 \le r \le r_0^+ \\ \tilde{\zeta}^{+}_{0,0} & \text{for} \quad r_0^+ \le r \le r_\pi^+ = r_\pi^- \\ \tilde{\zeta}^{-}_{0,0} & \text{for} \quad r_\pi^- \le r \le r_0^- \\ 0 & \text{for} \quad r_0^- < r \end{cases}. \tag{63}$$

We illustrate the behaviour of these solutions in figure 6.

The apparent issue that the first order derivative $d\zeta/dr$ is discontinues at the critical values is well known to occur also in the real case for the compacton solutions. We may argue therefore in a similar way as in [50] and conclude that the solutions are in fact well-defined. This is because the derivative always occurs with a multiple of $\sin^2\zeta$ in the BPS equation so that the left and right limit of $\sin^2\zeta d\zeta/dr$ have always a finite value. We note that this finite value differ in sign

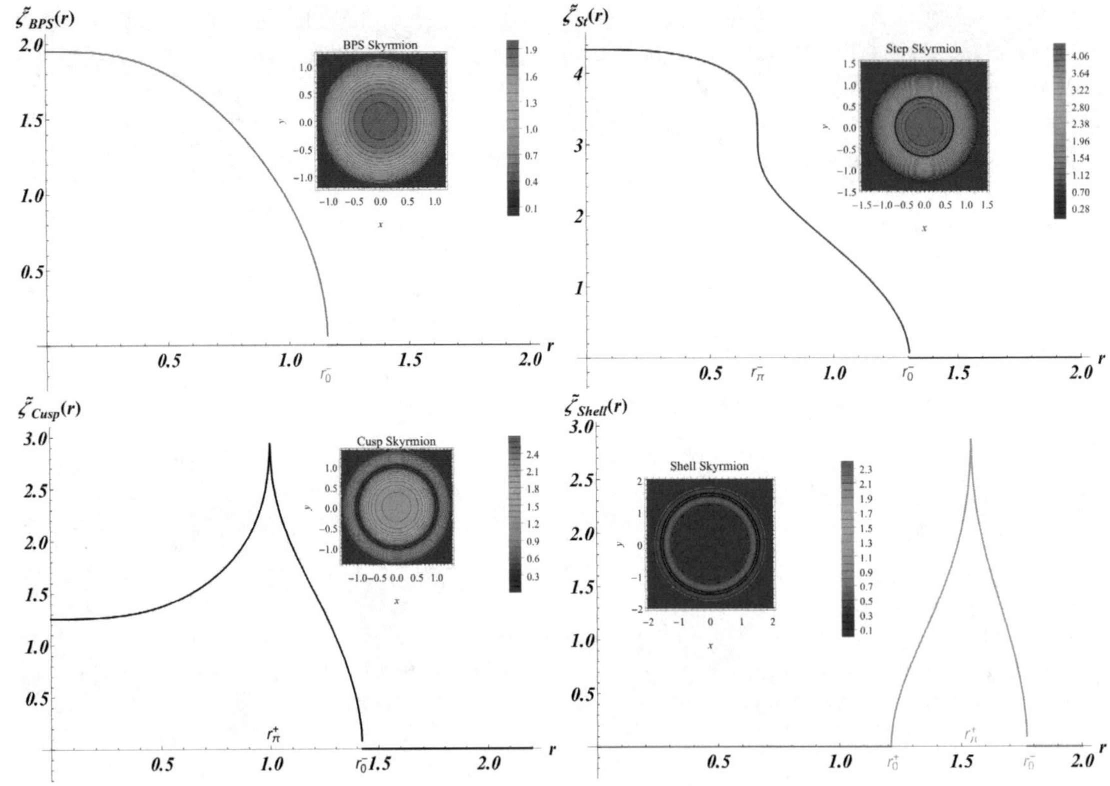


Figure 6: Differen types of Skyrmion solutions (62) and (63) of the equation of motion (58)[2]. We have have taken $n = 1$, $\tilde{\lambda} = 3/2$, $\tilde{\mu} = 2$, $s = 1$ and $c = 1/2$ (panel a), $c = -1/2$ (panel b), $c = -1$ (panel c), $c = 1$ (panel d).

for Cusp and Shell solutions. This is because we have pasted the solutions from self and anti-self dual equations. However, this sign ambiguity disappears when one considers the equations of motion instead of the BPS equations. Hence, we adopt here the view that the equation of motion is more fundamental than the BPS equations and that all the solutions presented are therefore well-defined.

The energies are computed to be finite and real, despite the solutions being complex

$$E_{\text{BPS/St, Cusp}} = \frac{8}{15} n\tilde{\mu}\tilde{\lambda}\pi \left(8\sqrt{2} \mp 10c \pm 3c^{5/3}\right), \quad E_{\text{Shell}} = \frac{128}{15}\sqrt{2}n\tilde{\mu}\tilde{\lambda}\pi, \quad E_{i\text{BPS}} = -E_{\text{BPS}}. \tag{64}$$

We conclude this section by verifying the reality conditions (i)-(iii). First we notice that condition (22) is satisfied for our non-Hermitian Hamiltonian following from $\mathcal{L}_{c\text{BPSS}}$ by the anti-linear symmetry transformation

$$\mathcal{CPT}' : \zeta(x_\mu) \to \zeta^*(-x_\mu) + 2i\text{arctanh}\epsilon. \tag{65}$$

Performing this transformation on our solutions, we find

$$\zeta^{\pm}_{\alpha,m}(r) \to \left[\zeta^{\pm}_{\alpha,m}(r)\right]^* + 2i\text{arctanh}\epsilon = \zeta^{\pm}_{\alpha,m}(r). \tag{66}$$

Thus in this case the solutions are mapped to itself so the condition (24) is automatically satisfied and the energy must therefore be real.

In [9] we gradually make the above system more complicated and consider examples for different types of potentials and also different terms in the Lagrangians. In there we consider a model that gives rise to semi-kink and massless solutions, Bender-Boettcher type potentials, a complex trigonometric model that breaks the $\mathcal{CPT}$ symmetry and a new submodel of the BPS system. In all these examples the reality argument is shown to hold, albeit in different ways.

## 4. Conclusion

In a strictly pseudo-Hermitian approach, by utilizing Dyson maps acting adjointly on a non-Hermitian field theoretic Hamiltonian, we have demonstrated that the Goldstone theorem holds and the Higgs mechanism are realised in the parameter regime in which the theory admits a modified $\mathcal{CPT}$-symmetry. This parameter regime is bounded by three different types of boundaries: (i) the standard exceptional points, at which we have to identify the Goldstone bosons technically in a modified way due the Jordan structure of the squared-mass matrix so that the Higgs mechanism is still in tact. (ii) the zero exceptional points at which the Goldstone boson can not be identified and can therefore also not be coupled to the gauge particle to give them mass and (iii) the trivial vacua for which the continuous symmetry is not broken and we do not expect Goldstone bosons to emerge in the first place that may be utilized in the Higgs mechanism. In the $\mathcal{CPT}$-broken regime not only these mechanism break down, but the non-Hermitian theory becomes entirely ill-defined.

In a non-Hermitian $SU(2)$-invariant local gauge theory we have identified complex variants of the t'Hooft-Polyakov monopoles, which have real energies in the $\mathcal{CPT}$-symmetric parameter regime. Interestingly in two disconnected regions the solutions do not respect this symmetry themselves, but instead the symmetry exchanges the solutions to the dual and the self-dual solutions to the BPS equations. As the energies of these solutions are degenerate, our reality conditions holds and the energies are guaranteed to be real. In the intermediate region the solutions do not respect the $\mathcal{CPT}$-symmetry and are therefore also not expected to have real energies. We computed the energies explicitly confirming the general assertions.

For BPS theories in 1+1 dimensions, a complex coupled sine-Gordon model, and in 3+1 dimensions, BPS versions of the Skyrme model, we have found solitonic/kink/anti-kink type soltions with real energies governed by the anti-linear $\mathcal{CPT}$-symmetry in a similar fashion as the monopole solutions. The broadening of the solutions to the larger complex regime allowed for new types of configurations we refer to as step, cusp and shell Skyrmion solutions with further examples to be found in [9].

## References

[1] Ellis J 2002 *Lectures notes PSI Summer School, Zuoz 2002, preprint hep-ph/0211168*
[2] Ellis J 2012 *Phil. Trans. Royal Soc. A: Math., Phys. and Eng. Sciences* **370** 818–830
[3] LHCb collaboration 2021 *Test of lepton universality in beauty-quark decays, preprint CERN-EP-2021-042, arXiv:2103.11769*
[4] Fring A, Taira T 2020 *Nuclear Physics B***950** 114834
[5] Fring A, Taira T 2020 *Phys. Rev. D* **101** 045014
[6] Fring A, Taira T 2020 *Phys. Lett. B* **807**, 135583
[7] Fring A, Taira T 2020 *Massive gauge particles versus Goldstone bosons in non-Hermitian non-Abelian gauge theory, preprint arXiv:2004.00723*
[8] Fring A, Taira T 2020 *Journal of Physics A: Math. and Theor.* **53** 455701
[9] Correa F, Fring A, Taira T 2021 *Complex BPS Skyrmions with real energy, preprint arXiv:2102.05781*
[10] Bender C M, Boettcher S 1998 *Phys. Rev. Lett.* **80** 5243–5246
[11] Wigner E 1960 *J. Math. Phys.* **1** 409–413
[12] Regensburger A, Bersch C, Miri M A, Onishchukov G, Christodoulides D N, Peschel U 2012 *Nature* **488** 167–171
[13] Peng B, Özdemir S K, Lei F, Monifi F, Gianfreda M, Long G L, Fan S, Nori F, Bender C M, Yang L 2014 *Nature Physics* **10**(5) 394–398
[14] Makris K G, El-Ganainy R, Christodoulides D N, Musslimani Z H 2008 *Phys. Rev. Lett.* **100** 103904

[15] Guo A, Salamo G J, Duchesne D, Morandotti R, Volatier-Ravat M, Aimez V, Siviloglou G A, Christodoulides D N 2009 *Phys. Rev. Lett.* **103** 093902
[16] Rüter C E, Makris K G, El-Ganainy R, Christodoulides D N, Segev M, Kip D 2010 *Nature Physics* **6** 192–195
[17] Goldstone J 1961 *Il Nuovo Cimento* **19** 154–164
[18] Nambu Y, Jona-Lasinio G 1961 *Phys. Rev.* **124** 246
[19] Guralnik G S, Hagen C R, Kibble T W 1964 *Phys. Rev. Lett.* **13** 585–587
[20] Higgs P W 1964 *Phys. Rev. Lett.* **13** 508–509
[21] Higgs P W 1964 *Phys. Lett.* **12** 132–133
[22] Englert F, Brout R 1964 *Phys. Rev. Lett.* **13** 321–323
[23] Bogomolny E B 1976 *Sov. J. Nucl. Phys.(Engl. Transl.)* **24** 4
[24] Prasad M K, Sommerfield C M 1975 *Phys. Rev. Lett.* **35**(12), 760–762
[25] Skyrme T H 1962 *Nuclear Physics* **31** 556–569
[26] Bender C M, Dorey P E, Dunning C, Fring A, Hook D W, Jones H F, Kuzhel S, Levai G, Tateo R 2019 *PT Symmetry: In Quantum and Classical Physics* (World Scientific, Singapore)
[27] Alexandre J, Millington P, Seynaeve D 2017 *Phys. Rev. D* **96** 065027
[28] Alexandre J, Millington P, Seynaeve D 2018 *Journal of Physics Conference Series* **952** 012012
[29] Mostafazadeh A 2010 *Int. J. Geom. Meth. Mod. Phys.* **7** 1191–1306
[30] Dyson F J 1956 *Phys. Rev.* **102** 1230–1244
[31] Scholtz F G, Geyer H B, Hahne F 1992 *Ann. Phys.* **213** 74–101
[32] Alexandre J, Ellis J, Millington P, Seynaeve D 2018 *Phys. Rev. D* **98** 045001
[33] Mannheim P D 2018 *Journal of Physics A: Math. and Theor.* **51** 315302
[34] Mannheim P D 2019 *Phys. Rev. D* **99** 045006
[35] Alexandre J, Ellis J, Millington P 2020 *Phys. Rev. D* **102** 125030
[36] Kato T 1995 *Perturbation Theory for Linear Operators*, volume 132 of *Classics in Mathematics*, (Springer Berlin Heidelberg)
[37] Von Neumann J, Wigner E 1929 *Phys. Zeit.* **30** 467–470
[38] t'Hooft G 1974 *Nuclear Physics B* **79** 276–284
[39] Polyakov A M 1996 *JETP Lett.* **20** 540–541
[40] Goddard P, Olive D 1978 *Reports on Progress in Physics* **41** 1357
[41] Montonen C, Olive D 1977 *Phys. Lett. B* **72** 117–120
[42] Fring A 2007 *Journal of Physics A: Math. and Theor.* **40** 4215–4224
[43] Cen J, Fring A 2016 *Journal of Physics A: Math. and Theor.* **49** 365202
[44] Cen J, Correa F, Fring A 2017 *J. of Math. Phys.* **58** 032901
[45] Derrick G H 1964 *J. of Math. Phys.* **5** 1252–1254
[46] Stepień L 2019 *Acta Physica Polonica B* **50** 65
[47] Adam C, Ferreira L A, Da Hora E, Wereszczynski A, Zakrzewski W J 2013 *JHEP* **2013** 62
[48] Ferreira L A, Klimas P, Wereszczyński A, Zakrzewski W J 2019 *Journal of Physics A: Math. and Theor.* **52** 315201
[49] Skyrme T H R 1962 *Nucl. Phys.* **31**, 556–569
[50] Adam C, Sanchez-Guillen J, Wereszczyński A 2010 *Phys. Lett. B* **691** 105–110
[51] Adkins G S, Nappi C R, Witten E 1983 *Nuclear Physics B* **228** 552–566
[52] Bender C M, Jones H F, Rivers R J 2005 *Phys. Lett.* **B625** 333–340

Journal of Physics: Conference Series **2038** (2021) 012011 doi:10.1088/1742-6596/2038/1/012011

# Characteristic influence of exceptional points in quantum dynamics

**Savannah Garmon[1,2], Takafumi Sawada[1], Kenichi Noba[1], Gonzalo Ordonez[3]**

[1] Department of Physical Science, Osaka Prefecture University, Gakuen-cho 1-1, Sakai 599-8531, Japan

[2] Institute of Industrial Science, University of Tokyo, Kashiwa 277-8574, Japan

[3] Department of Physics and Astronomy, Butler University, Gallahue Hall, 4600 Sunset Avenue, Indianapolis, Indiana 46208, USA

E-mail: `sgarmon@p.s.osakafu-u.ac.jp`

**Abstract.** We review some recent work on the occurrence of coalescing eigenstates at exceptional points in non-Hermitian systems and their influence on physical quantities. We particularly focus on quantum dynamics near exceptional points in open quantum systems, which are described by an outwardly Hermitian Hamiltonian that gives rise to a non-Hermitian effective description after one projects out the environmental component of the system. We classify the exceptional points into two categories: those at which two or more resonance states coalesce and those at which at least one resonance and the partnering anti-resonance coalesce (possibly including virtual states as well), and we introduce several simple models to explore the dynamics for both of these types. In the latter case of coalescing resonance and anti-resonance states, we show that the presence of the continuum threshold plays a strong role in shaping the dynamics, in addition to the exceptional point itself. We also briefly discuss the special case in which the exceptional point appears directly at the threshold.

## 1. Introduction

The traditional assumption of Hermiticity in quantum theories has enabled remarkable success by providing a key element in a formalism that is mathematically elegant and which yields excellent agreement with experiment. However, in recent decades, many researchers have found it useful to introduce non-Hermitian elements into physical theories for a variety of reasons [1–14]. Most commonly, this has been done to describe energy lost to or acquired from the surrounding environment, which has often been described in terms of coupled-mode theory. Under this kind of formalism, one assumes the most important physical properties of a given system can be described in terms of a few principal modes, one or more of which may have complex energy. The imaginary parts of the complex energies can then be associated with energy sinks or sources from the surrounding environment. This approach has been very successful in describing the physics of *exceptional points* (EPs) [15–17], at which two or more eigenstates coalesce while their respective eigenvalues become degenerate, particularly in optics [18]. Numerous interesting features of the exceptional points have been revealed both theoretically and in experiments, such as intrinsic chirality [19, 20] and modifications of the usual adiabatic properties when they are encircled [21–27]. Enhanced parametric sensitivity in the vicinity of the EPs may also lead to interesting applications [28–32]. From this point of view, some researchers have questioned if environmental influences and dissipation might better be viewed as a resource than a limitation.

However, in a deeper sense, there has always existed a tension between the strict requirement of Hermiticity and the actual implementation of the theory since the earliest days of quantum mechanics. This is evident from the fact that quantum mechanics demands that energy eigenvalues must be real, yet the exponential decay width in well-established processes such as atomic relaxation is obtained from the negative imaginary part of a complex eigenvalue, which is the so-called *resonance* eigenvalue. This 'quirk' of the theory has generally been dealt with quite differently by different researchers. Since the resonance eigenvalue is usually obtained by an analytic continuation of the original Hermitian theory into the complex energy plane, some have waved the issue away as being little more than a mathematical trick. But others have elected to embrace the role of the resonance in the theory, for example by viewing it as a generalized eigenstate that resides outside of the usual Hilbert space (that only admits $L^2$-normalizable solutions with real eigenvalues) [9,33–41]. Or, a more recent work [42] has proposed a formalism in which the resonance state and bound state are treated on precisely equal footing (albeit with the inclusion of two other generalized eigenstates: the partner anti-resonance, with positive imaginary part of the eigenvalue, as well as anti-bound or virtual states that have real eigenvalue but are non-normalizable in the usual sense) [43–46].

A clear advantage to this approach is that the resonance appears spontaneously due to the interaction between the system and the surrounding environment, rather than simply being presupposed as in the coupled mode theory. More precisely, under this approach, one describes the environment surrounding the primary system of interest in terms of an energy continuum. The details of the energy continuum are determined by the microscopic degrees of freedom associated with a particular environment. This includes the existence of a lower threshold (equivalent to a band edge or waveguide cutoff mode) below which the continuum ends and the usual bound state eigenvalues (if any) would appear (the continuum might or might not also have an upper threshold). The continuum also has a built-in density of states, describing how many modes a given level within the continuum can accommodate. Under a number of situations, either the threshold or the density of states (or both) might strongly influence the dynamical properties of a given system. Indeed, since the resonance emerges in this picture precisely due to the interaction of the discrete system modes with the continuum, the resonance generally first appears in the vicinity of the threshold. The point in the parameter space at which the resonance appears is precisely an exceptional point, which some authors have therefore interpreted as a kind of spectral phase transition at which the dynamics are abruptly modified [36, 40, 45–48] Further, since the exceptional point at which the resonance arises lies close to the threshold, one would expect the threshold itself could strongly influence the dynamics in this case, which has been confirmed by two of the present authors in Ref. [44]. We review this work below.

In this paper, we refer to models that incorporate both discrete and continuous spectral elements as examples of *open quantum systems*. We note that in the context of cavity quantum electrodynamics, this microscopic description of the environment in terms of a continuum incorporating a well-defined density of states is often referred to as a *structured reservoir* [49–58]. We also comment that the coupled-mode theory described above can be viewed as a kind of macroscopic approximation over the microscopic degrees of freedom written into the structured reservoir, which explains why the coupled-mode theory has been particularly successful in the classical context.

In the following, we review some recent work on the dynamics near exceptional points from the microscopic perspective, particularly Ref. [44] (see also [59]). We note a disadvantage to the microscopic framework is that deriving the dynamics in the context of the structured reservoir requires significantly more work than in the coupled-mode theory. However, an obvious advantage to this approach is that it should provide a more accurate description of the dynamics in certain truly quantum settings. A further advantage we will find is that the dynamics are richer in the case that we preserve the microscopic degrees of freedom in the system.

It is worth mentioning that in this work, we only deal with the problem at the Hamiltonian level. Some recent works have emphasized that a more complete description can be obtained by considering the Liouvillian formalism, which can describe quantum jumps and other effects [60–63] We comment on this point further in the conclusion.

In the next section, we introduce a quick categorization scheme for two types of exceptional points in open quantum system, before outlining the remainder of the paper.

## 2. Categorization scheme for two-level exceptional points

Here we briefly outline a convenient categorization scheme for the exceptional points as originally introduced in Refs [11, 44].

Let us begin by denoting an exceptional point at which $N$ eigenstates coalesce as an EPN. Then in the simplest case of the EP2, there are two further natural subcategories. First, an exceptional point at which two virtual states with real eigenvalues coalesce before forming a resonance/anti-resonance pair is defined as an EP2A. As mentioned in the Introduction, this usually occurs in the vicinity of a continuum threshold and, since the resonance spontaneously appears here, can be viewed as a spectral phase transition or dynamical phase transition [36, 40, 45–48]. In this picture, when the resonance is absent from the system, the system dynamics would either be oscillatory (stable) or dissipative, but non-Markovian (i.e., governed by reversible dynamics). But when the resonance appears, one would generally expect the dominant dynamics to be Markovian, with the exponential decay width determined by the imaginary part of the complex eigenvalue. However, it should be emphasized that, somewhat similar to the picture in the quantum phase transition, the dynamics in the *immediate* vicinity of the EP2A are not necessarily exponential, even after the resonance has appeared. We will show this explicitly in what follows. (As a brief aside, we note that in the context of parity-time (PT) symmetric systems, the EP2A is associated with PT-symmetry breaking [11, 14].)

Next, we denote an exceptional point at which two resonance eigenstates coalesce before forming two different resonances as an EP2B. Because the resonance exists on either side of the EP2B, this case is not associated with the same shift from non-Markovian to Markovian dynamics as in the case of the EP2A. However, the EP2B is associated with the phenomenon of resonance trapping [8, 10, 64, 65] and qualitative features of the exponential decay can change on either side of the transition [65]. Further, the dynamics in the immediate vicinity of the EP2B quite generally appears as power law-exponential decay [66–73], as demonstrated in a microwave billiard experiment [74, 75]. This occurs as a result of the double pole at the EP2B.

The generalization of this categorization scheme to higher-order EPs involving three or more coalescing states [76, 77] is fairly natural and we will comment on the higher-order cases of both A-type and B-type exceptional points in what follows.

In Sec. 3 we look at a simple model for an EP2B in an open quantum system and demonstrate the appearance of power law-exponential dynamics on the primary timescale, which is replaced by inverse power law decay at very long timescales. Then, in Sec. 4 we discuss the generalization to higher-order B-type EPs and also briefly review their influence on some other physical quantities that have been studied in the literature, including scattering processes. We then turn to the A-type exceptional point and show in Sec. 5 an example in which an EP2A near the band edge gives rise to decay of the characteristic form $1 - C_1 t^{1/2} + C_2 t$. A quick numerical simulation of the dynamics in the vicinity of an EP3A near the band edge is then presented in Sec. 6. Finally, we summarize our main results in Sec. 7 and give some comments on the literature and directions for future work, including a comment on potential experimental realization of these results.

Journal of Physics: Conference Series **2038** (2021) 012011 doi:10.1088/1742-6596/2038/1/012011

## 3. Quantum dynamics near an EP2B

To study the dynamics near an EP2B, we introduce a simple model for a qubit coupled to a semi-infinite hopping chain, described by the Hamiltonian

$$H_q = -V\left(d_1^\dagger d_2 + d_2^\dagger d_1\right) - g\left(c_1^\dagger d_2 + d_2^\dagger c_1\right) - J\sum_{j=1}^{\infty}\left(c_j^\dagger c_{j+1} + c_{j+1}^\dagger c_j\right). \tag{1}$$

Here, the intra-qubit coupling is given by $-V$ in the first term, while the chemical potential of the two qubit sites (with creation operators $d_1^\dagger, d_2^\dagger$) are both taken as zero. The second site of the qubit $d_2$ is then coupled to the endpoint of the chain $c_1^\dagger$ with coupling parameter $-g$, while the hopping parameter along the chain is given by $-J$ in the third term. We will measure energy in this system according to the unit $J = 1$. Notice that no interaction terms appear in (1), so that we can work within the single-particle framework below.

This model was originally introduced by A. D. Dente and collaborators in Ref. [47] to analyze a variety of parameter regimes in which the dynamics exceeds description by the usual Fermi golden rule. For example, the authors show that a regime of anomalous diffusion exhibiting pure inverse power law decay can be associated with virtual states (solutions with real energy eigenvalues appearing on the second Riemann sheet) [47] (see also [78–81]). However, the dynamics near the exceptional point for this model are first considered in detail in Ref. [44] We outline the main points of that calculation below.

### *3.1. Model $H_q$: effective Hamiltonian, spectrum and exceptional points*

As the first step in our analysis, we assume our boundary condition as an outgoing wave from the qubit region (the Siegert boundary condition) [39,82]

$$\psi(x) = \begin{cases} \psi_1 & \text{for } x = d_1 \\ \psi_2 & \text{for } x = d_2 \\ Ce^{ikx} & \text{for } x \geq 1 \end{cases}. \tag{2}$$

This represents the initial condition we will later consider, in which the $d_1$ site of the qubit is initially occupied at $t = 0$. We next introduce the projection operator for the qubit sector of the model as

$$P = |d_1\rangle\langle d_1| + |d_2\rangle\langle d_2|, \tag{3}$$

and the complementary projection operator for the chain (environment) portion of the model

$$Q = 1 - P = \sum_{j=1}^{\infty} |j\rangle\langle j|. \tag{4}$$

We can immediately obtain the energy eigenvalue associated with the $Q$ sector by solving the Schrödinger equation for any $x \geq 2$ as $\langle x|H|\psi\rangle = E\langle x|\psi\rangle$, which gives

$$-\psi(x-1) - \psi(x+1) = E\psi(x). \tag{5}$$

Applying the plane wave solution $\psi(x) = Ce^{ikx}$ from Eq. (2) in the above equation yields

$$E(k) = -2\cos k, \tag{6}$$

after setting the energy units as $J = 1$.

To obtain the discrete spectrum associated with the qubit, we next project out the $Q$ sector of the model by applying the Feshbach method [8,39,83–85]. Doing so, we obtain an equivalent form of the Schrödinger equation written in the projected $P$ sector as

$$H_{\text{eff}}(E_j)P|\psi_j\rangle = E_j\left(P|\psi_j\rangle\right) \tag{7}$$

in which the effective Hamiltonian is obtained according to

$$\begin{aligned} H_{\text{eff}}(E) &= PH_qP + PH_qQ\frac{1}{E - QH_qQ}QH_qP \\ &= \begin{pmatrix} 0 & -V \\ -V & -g^2e^{ik} \end{pmatrix}. \end{aligned} \tag{8}$$

We emphasize two points about the effective Hamiltonian. First, note that $H_{\text{eff}}$ has dependence on its own eigenvalue through the lower-right entry $-g^2e^{ik}$ and Eq. (6). Hence, Eq. (7) is a non-linear eigenvalue problem. Second, the factor $e^{ik}$ in this entry clearly represents the residual influence of the chain onto the qubit following the projection. Notice this factor also renders $H_{\text{eff}}$ non-Hermitian. Hence, we see that projecting out the environmental degrees of freedom has revealed the implicit non-Hermitian character of our model.

Taking the determinant of Eq. (7) with (8) and applying Eq. (6) we obtain a polynomial equation $p_B(E_j) = 0$ for the discrete eigenvalues of $H_q$. Here $p_B(E)$ is quartic in $E$ and given by

$$p_B(E) = \left(1 - g^2\right)E^4 + \left[g^4 + \left(g^2 - 2\right)V^2\right]E^2 + V^4. \tag{9}$$

As the quartic takes double quadratic form, we can immediately solve $p_B(E_j) = 0$ for the four discrete eigenvalues, given by all four sign combinations in

$$E_j = \pm\sqrt{\frac{\left(2 - g^2\right)V^2 - g^4 \pm g^2\sqrt{g^4 + 2\left(g^2 - 2\right)V^2 + V^4}}{2\left(1 - g^2\right)}}. \tag{10}$$

The real and imaginary parts of these solutions are plotted as the full blue lines for the case $g = 0.75$ in Figs. 1 and 2, respectively. From Eq. (10) we can now easily find the location of the exceptional points in the model. Let $g < 1$ be fixed while $V$ is a controllable parameter. Then there is an EP2A that appears at $V = \bar{V}_A$ with

$$\bar{V}_A = 1 + \sqrt{1 - g^2}, \tag{11}$$

which appears at about $\bar{V}_A \approx 1.66144$ in Figs. 1 and 2. However, since it easier to analyze the properties near the EP2A in the simpler model studied later in Sec. 5, we will not focus on this here. Instead, we primarily devote our attention to the EP2B located at

$$\bar{V}_B = 1 - \sqrt{1 - g^2}, \tag{12}$$

which corresponds to the coalesced eigenvalues

$$\mp\bar{E}_B = \mp i\sqrt{\frac{2 - g^2}{\sqrt{1 - g^2}} - 2} \equiv \mp i\frac{\bar{\Gamma}_B}{2}. \tag{13}$$

Here, the upper sign is for the two coalesced resonance states and the lower sign is for the coalescence of the associated anti-resonance states. Note we have also defined the decay width for the coalesced resonance at the EP2B as $\bar{\Gamma}_B$. These two EP2Bs can be seen at $\bar{V}_B \approx 0.33856$ in Figs. 1 and 2.

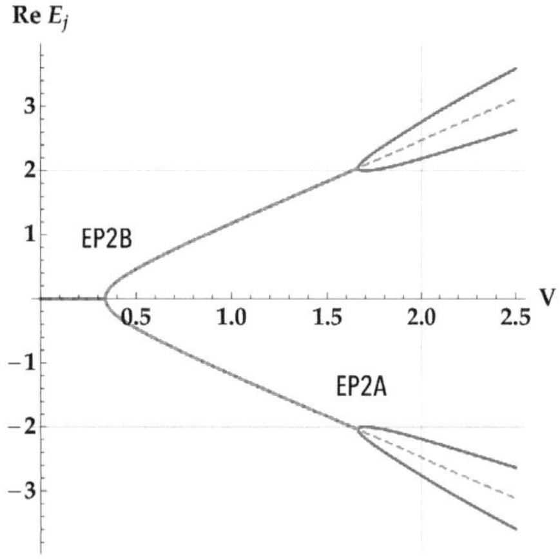


**Figure 1.** The real part of the discrete solutions for model $H_q$ from Eq. (10) are shown as the blue, thick lines for $g = 0.75$. The location of the band edges $E = \pm 2$ are indicated by lines on both axes. The EP2As occur at $\bar{V}_A \approx 1.66144$ and the EP2Bs occur at $\bar{V}_B \approx 0.33856$. The Puiseux expansions centered on the EP2B from Eq. (14) are shown with the dashed green and orange lines.

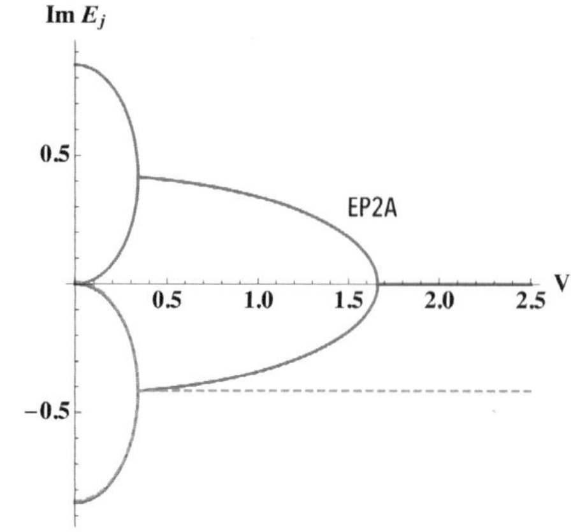


**Figure 2.** The imaginary part of the discrete solutions for model $H_q$ from Eq. (10) are shown as the blue, thick lines for $g = 0.75$. The EP2As occur at $\bar{V}_A \approx 1.66144$ and the EP2Bs occur at $\bar{V}_B \approx 0.33856$. The Puiseux expansions centered on the EP2B from Eq. (14) are shown with the dashed green and orange lines.

As mentioned in the Introduction, even when the system is detuned from the precise location of the exceptional points, their existence still strongly influences the spectral features of the model. This influence can be expressed in terms of the Puiseux expansion that gives an accurate approximation of the coalescing eigenvalues in the near vicinity of the EPs. Following methods similar to those outlined in Refs. [11,48], we obtain the Puiseux expansion for the two resonances in the present model near the EP2B as

$$E = -i\frac{\bar{\Gamma}_B}{2} \pm \frac{g^2}{2\sqrt{(1-g^2)\left(2-g^2-2\sqrt{1-g^2}\right)}}\left(V^2-\bar{V}_B^2\right)^{1/2} + \mathcal{O}\left(V^2-\bar{V}_B^2\right). \quad (14)$$

These are shown by the green and orange dashed lines in Figs. 1 and 2. Note that it is this Puiseux expansion that gives rise to enhanced parametric sensitivity near the exceptional points, as described in Refs. [28–32].

### *3.2. Linearized eigenvalue problem for $H_q$*

The detailed calculation for the dynamics near the EP2B in the present model is slightly tedious but largely straightforward. Here we just present a quick outline, emphasizing one point about the divergence of the norm that we think provides context about the properties of the exceptional points. The details of the calculation can be found in Ref. [44] (By comparison, in Sec. 5 we will give a somewhat more detailed recounting of the calculation for the dynamics near the EP2A, as that situation is a bit more subtle.)

Following Ref. [44], as a first step in analyzing the dynamics, we transform the non-linear eigenvalue problem in Eq. (7) into a generalized linear eigenvalue problem, according to the formalism of Ref. [42]. To do this we introduce the variable

$$\lambda \equiv e^{ik}, \tag{15}$$

which is convenient to work with below. Our objective now is to transform the original non-standard eigenvalue problem Eq. (7) in terms of $E$ into an equivalent linear eigenvalue problem in $\lambda$.

The continuum eigenvalue in Eq. (6) can be immediately rewritten in terms of $\lambda$ as the simple expression

$$E = -\lambda - 1/\lambda \tag{16}$$

Applying this in Eq. (7), the eigenvalue equation for the effective Hamiltonian can be shown to be a $2 \times 2$ quadratic eigenvalue problem in $\lambda$. Following the standard method to linearize the quadratic eigenvalue problem [86, 87], this in turn can be transformed into a $4 \times 4$ generalized linear eigenvalue problem of the form

$$(F - \lambda G)\,|\Psi\rangle = 0 \tag{17}$$

in which

$$F = \begin{bmatrix} 0 & 0 & 1 & 0 \\ 0 & 0 & 0 & 1 \\ 1 & 0 & 0 & -V \\ 0 & 1 & -V & 0 \end{bmatrix} \qquad G = \begin{bmatrix} 1 & 0 & 0 & 0 \\ 0 & 1 & 0 & 0 \\ 0 & 0 & -1 & 0 \\ 0 & 0 & 0 & g^2 - 1 \end{bmatrix}. \tag{18}$$

and

$$|\Psi\rangle = \begin{bmatrix} \langle d_1|\psi\rangle \\ \langle d_2|\psi\rangle \\ \lambda\langle d_1|\psi\rangle \\ \lambda\langle d_2|\psi\rangle \end{bmatrix}. \tag{19}$$

Notice the key point that the dimension of the eigenvalue problem in Eq. (17) is four, which is in one-to-one correspondence with the number of discrete eigenvalues of the original Hamiltonian $H_q$. We also emphasize that although the $F$ and $G$ matrices are Hermitian, the *generalized* eigenvalue problem in Eq. (17) is non-Hermitian because $G$ is not the identity matrix. To see this another way, one could rewrite Eq. (17) in the equivalent form $\left(G^{-1}F - \lambda I_4\right)|\Psi\rangle = 0$, in which $G^{-1}F$ is explicitly non-Hermitian.

Setting the determinant of Eq. (17) equal to zero we obtain the four $\lambda$ eigenvalues as

$$\lambda_j = \pm\sqrt{\frac{V^2 + g^2 - 2 \pm \sqrt{g^4 + 2\left(g^2 - 2\right)V^2 + V^4}}{2\left(1 - g^2\right)}}. \tag{20}$$

Plugging these into Eq. (16) verifies that they are equivalent to the original $E$ eigenvalues in Eq. (10). We can also now obtain the coalesced eigenvalues $\pm\bar{\lambda}_B$ at the EP2B as well as a Puiseux expansion, similar to Eq. (14); we leave the details to Ref. [44].

### *3.3. Dynamics near the EP2B: intermediate timescale*

We now evaluate the influence of the EP2B on the dynamics in the qubit sector of the model $H_q$. To simplify our analysis we choose as an initial state $|d_1\rangle$; although we emphasize the results would be qualitatively similar for the $|d_2\rangle$ state or any linear combination within the qubit sector. The survival probability for this state is given by $P(t) = |A(t)|^2$ in which

$$A(t) = \langle d_1|e^{-iH_q t}|d_1\rangle. \tag{21}$$

For very short evolutions, it is known that quantum systems universally exhibit parabolic dynamics; the exceptional point has no particular influence on the dynamics on this timescale. For the present model, Eq. (21) can easily be evaluated to show that $P(t) \approx 1 - V^2t^2$ for $t \ll 1/\sqrt{2}V$. This short-lived effect has been detected in a handful of experiments [88, 89] and gives rise to the so-called quantum Zeno effect, with inhibited decay upon repeated, consecutive measurements [90–95].

Let us turn now to the intermediate timescale dynamics, on which the influence of the EP2B is directly pronounced. Following the formalism of Ref. [42], in the context of the generalized eigenvalue problem Eq. (17), we can write the survival amplitude as

$$\begin{aligned} A(t) &= \langle d_1 | e^{-iH_q t} | d_1 \rangle \\ &= \frac{1}{2\pi i} \sum_{j=\{s,\pm\}} \int_C d\lambda \left(-\lambda + \frac{1}{\lambda}\right) \exp\left[i\left(\lambda + \frac{1}{\lambda}\right)t\right] \langle d_1|\psi_j\rangle \frac{\lambda_j}{\lambda - \lambda_j} \langle\tilde{\psi}_j|d_1\rangle, \end{aligned} \tag{22}$$

which is equivalent to the usual integral over the Green's function, but here the integration is rewritten in terms of the $\lambda$ variable and further decomposed into contributions from each of the four $|\psi_j\rangle$ eigenstates. We note the presence of the norm of the four eigenstates $\langle\tilde{\psi}_j|d_1\rangle\langle d_1|\psi_j\rangle$ in this expression, in which $\langle\tilde{\Psi}_j|$ denotes the left eigenstate for the corresponding left eigenvalue equation $\langle\tilde{\Psi}|\,(F - \lambda G) = 0$.

The norm can be explicitly obtained from the condition $\langle\tilde{\Psi}_j|G|\Psi_j\rangle = 1$ and the eigenvalue equation (17) as

$$\langle\tilde{\psi}_j|d_1\rangle\langle d_1|\psi_j\rangle = \langle d_1|\psi_j\rangle^2 = \frac{V^2\lambda_j^2}{1 + (1 + g^2 + V^2)\,\lambda_j^2 - (1 - 2g^2 + V^2)\,\lambda_j^4 - (1 - g^2)\,\lambda_j^6}. \tag{23}$$

Similar to the Puiseux expansion for $E_j$, we can reparameterize the $V$ parameter near the EP2B as $V = \bar{V}_B + \delta$ and then expand the eigenstate norm near the EP2B as

$$\langle d_1|\psi_{\sigma,\pm}\rangle^2 = \frac{1}{4}\left[\mp\frac{1}{\sqrt{2\delta}}\sqrt{1 - g^2 - \sqrt{1 - g^2}} + 1 + O(\delta^{1/2})\right]. \tag{24}$$

Note this expression is divergent as $\delta \to 0$, which is a general property of the EP. However, also note that the sign of the divergent term is opposite for the two lines of this equation, corresponding to the two coalescing states. Indeed, in physical quantities, the divergent contributions must always cancel at the EP.

Applying the expansion Eq. (24) as well as the Puiseux expansion for the $\lambda_j$ eigenvalues, the summed terms under the integral in Eq. (22) can in turn be expanded as

$$\sum_{j=\{s,\pm\}} \frac{\lambda_j}{\lambda - \lambda_j}\langle d_1|\psi_j\rangle^2 = \frac{\bar{\lambda}_B^4 + \lambda^2}{\left(\bar{\lambda}_B^2 - \lambda^2\right)^2} + O(\delta), \tag{25}$$

in which the double-pole is the direct result of the EP2B. Now evaluating the residue for the double pole, we find the survival probability near the EP2B takes the characteristic form

$$P(t)\left[1 + D_1 t + D_2 t^2\right] e^{-\bar{\Gamma}t}. \tag{26}$$

in which

$$D_1 = \frac{g^2\left(1 + \sqrt{1 - g^2}\right)}{2\,(1 - g^2)^{3/4}} \qquad D_2 = \frac{g^4\left(2 - g^2 + 2\sqrt{1 - g^2}\right)}{16\,(1 - g^2)^{3/2}}. \tag{27}$$

We show this prediction for the dynamics matches very well with a numerical integration for the case $g = 0.75$ in Fig. 3.

This result for the dynamics near an EP2B has been confirmed in the microwave cavity experiment in Refs. [74, 75].

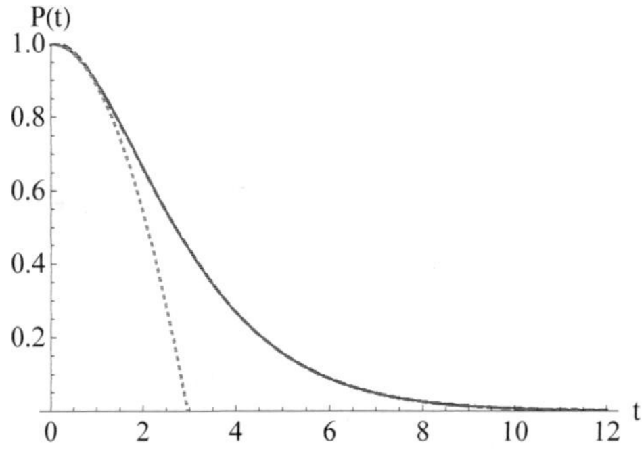


**Figure 3.** Survival probability $P(t)$ for $g = 0.75$ and $V = 0.3385622$, near the EP2B at $\bar{V}_B \approx 0.3385620$ . The purple dashed curve is the approximation from Eq. (26), which agrees well with the numerical integration (solid blue curve). The beige dotted curve shows the parabolic early-time dynamics.

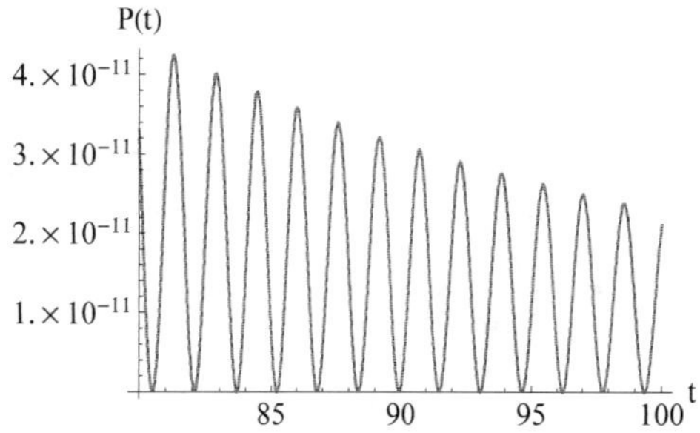


**Figure 4.** Long-time survival probability $P(t)$ exactly at the EP2B $\bar{V}_B \approx 0.3385620$ for $g = 0.75$. The approximation corresponding to Eq. (28) is shown as the green dash-dotted curve and again agrees well the numerical integration, shown as the solid blue curve.

### *3.4. Dynamics near the EP2B: asymptotic timescale*

Similar to the short-time parabolic dynamics mentioned earlier, it can be shown that quantum systems quite universally exhibit non-exponential decay on very large timescales [78, 96–102] However, under most circumstances this effect would be extremely difficult to detect because it does not appear until after many lifetimes of the exponential decay have passed, by which time almost nothing of the initial state remains to be measured in experiment [98]

For the present model, it is shown in Ref. [44] that at the EP2B the long-time dynamics exhibit the behavior

$$P(t) \sim t^{-3} \cos^2 (2t + \pi/4) \tag{28}$$

as shown for the case $g = 0.75$ in Fig. 4. We note here that the inverse power law decay $1/t^3$ is rather typical for long-time dynamics in 1-D systems [57, 78, 100, 102] and the oscillations in Eq. (28) emerge here due to the equal influence from the two band edge contributions at $E = \pm 2$. Hence, unlike the intermediate time dynamics, the EP2B doesn't seem to have any pronounced influence on the long-time dynamics; at least, not in the present circumstances. However, for the case of the EP2A in Sec. 5, we will find the long-time dynamics can be significantly enhanced compared to the usual case.

## 4. Intermediate Discussion: Dynamics, scattering and other signatures of EPNBs

In Sec. 3, we saw that near the EP2B the usual exponential dynamics were replaced with power law-exponential decay on the primary timescale during which most of the dissipation occurs. While from an engineering perspective we can view this as a potential tool for designing microscopic systems with desired dynamical properties, from a physicist's perspective we could view the power-law-exponential dynamics in Eq. (26) as a means to identify (prove) the presence of an EP2B. Further, it is immediately clear how this result should generalize for the case of $N$ coalescing resonances at an EPNB. This would lead to an $N$th order pole in the integrand for the survival amplitude, yielding a dynamical signature of the form $(1+C_1t+\ldots+C_{2N-2}t^{2(N-1)})e^{-\Gamma t}$ in the resulting survival probability [66]. Hence, in principle, the existence of the EP and its order could both be determined from the power law-exponential decay, without the need to tune the parameters to the exact location of the EPNB, which might be challenging.

A few other methods have been proposed to determine the presence of an EP2B without needing to tune to its precise location. The first would be to dynamically encircle the EP in the parameter space, and observe the resulting modification of the usual adiabatic evolution scheme [21–27]. In this case, it can be shown that the final state at the end of the adiabatic evolution depends only on the direction of encirclement, not the initial state one populates at the beginning of the cycle. This has been confirmed in several laboratory experiments [25–27]. See also the discussion in App. A of [44].

A second method to infer the existence of a nearby EP2B would be to measure its signature influence on the scattering cross section. As first reported by E. Hernández, et al, in Ref [103] in the context of scattering from a spherical double barrier potential, a double pole in the scattering matrix results in a cross section with a characteristic split-peak profile. As argued by W. D. Heiss and G. Wunner in Ref. [104], this split-peak can be viewed in relation to the well-known asymmetric Fano-Friedrichs profile, which appears due to quantum interference between two different states interacting with the same decay channel, or between two different decay paths for the same state into a single channel. Finally, A. Ben-Asher, et al, propose to observe the split-peak profile in resonance tunnelling experiments by inducing an EP2B through laser action in diodes and similar systems [105].

It has also been shown that B-type exceptional points can induce significant enhancement of spontaneous emission in their vicinity [106, 107]. Related to this, it is recently shown in Ref. [108] that an EP ring appears in the spectrum of a magnetically-sensitive doping impurity in a polyacetylene molecule, which results in a giant response in the single-spin electron resonance as measured by STM probe. This result could potentially enable single-spin detection in such experiments [108].

## 5. Quantum dynamics near an EP2A

To study the dynamics near an EP2A, we now turn to a model in which a simple quantum emitter (or quantum dot) is coupled to the endpoint of a semi-infinite chain. Our Hamiltonian in this case is given by

$$H_d = \epsilon_d d^\dagger d - J \sum_{j=1}^{\infty} \left( c_j^\dagger c_{j+1} + c_{j+1}^\dagger c_j \right) - g \left( c_1^\dagger d + d^\dagger c_1 \right). \tag{29}$$

Here the quantum dot with variable potential $\epsilon_d$ is described with the creation operator $d^\dagger$ and is coupled to the endpoint of the chain with coupling strength $g$ in the third term. The hopping strength along the semi-infinite chain is given by $J$ in the second term, just the same as the previous model studied in Sec. 3. Again, we measure energy in units of $J = 1$.

### 5.1. *Eigenvalues and effective Hamiltonian*

Similar to our initial steps for the previous model, we first write our outgoing wave boundary condition for model $H_d$ as

$$\psi(x) \equiv \langle x|\psi\rangle = \begin{cases} \psi_d & \text{for } x = d \\ Ce^{ikx} & \text{for } x \geq 1 \end{cases}. \tag{30}$$

Evaluating the Schrödinger equation along the chain for any site $x \geq 2$ yields the same continuum dispersion equation $E(k) = -2\cos k$ as we obtained for the previous model in Eq. (6).

In the next step, we write the projection operators for the present model. The projection into the discrete sector associated with the quantum dot is written simply

$$P = |d\rangle\langle d| \tag{31}$$

Journal of Physics: Conference Series **2038** (2021) 012011 doi:10.1088/1742-6596/2038/1/012011

while again the projection associated with the continuum is given by

$$Q = 1 - P = \sum_{j=1}^{\infty} |j\rangle\langle j|. \tag{32}$$

We can again obtain an effective Hamiltonian in the $Q$ subspace after projecting out the $P$ sector according to the Feshbach method by writing

$$\begin{aligned} H_{\text{eff}}(E) &= PH_dP + PH_dQ\frac{1}{E - QH_dQ}QH_dP \\ &= \epsilon_d - g^2 e^{ik}. \end{aligned} \tag{33}$$

Applying this in the Schrödinger equation (7) and applying Eq. (6), we obtain the polynomial dispersion equation, which is quadratic for the present model. The two resulting eigenvalues are given by

$$E_{\pm} = \frac{\epsilon_d\left(2 - g^2\right) \pm g^2\sqrt{\epsilon_d^2 - 4\left(1 - g^2\right)}}{2\left(1 - g^2\right)}. \tag{34}$$

Along with these, we can write the associated wave vectors $k_{\pm}$, which are obtained through $E_{\pm} = -2\cos k_{\pm}$.

The location of the exceptional points can be immediately obtained from Eq. (34) as

$$\epsilon_d = \pm 2\sqrt{1 - g^2}, \tag{35}$$

in which we assume that $\epsilon_d$ is an experimentally accessible parameter while $g$ is fixed. We immediately observe that for $g < 1$, these EPs are real-valued, while for $g > 1$ they escape into the complex plane (the parameter choice $g = 1$ gives a special case for which the quadratic dispersion reduces to a linear polynomial). For our purposes, we will assume the condition $g < 1$ is always satisfied. Under this assumption, both exceptional points in Eq. (35) denote EP2As at which two virtual states coalesce before forming a resonance/anti-resonance pair, or vice-versa.

In what follows, we will evaluate the characteristic dynamics near the exceptional point, focusing particularly on the case $g \ll 1$ for which the EP2As in turn lie near their respective thresholds $\epsilon_d \approx \pm 2$. In this case, as shown in Ref. [44] both the exceptional point itself and the threshold play key roles in shaping the dynamics. For definiteness, from this point we will focus on the lower exceptional point, which we denote $\epsilon_d = \bar{\epsilon}_A$ with

$$\bar{\epsilon}_A \equiv -2\sqrt{1 - g^2}. \tag{36}$$

(Of course, the dynamics near the other EP2A would be similar.) In the case $\epsilon_d = \bar{\epsilon}_A$, the two eigenvalues coincide at

$$E_+(\bar{\epsilon}_A) = E_-(\bar{\epsilon}_A) = \bar{E}_A \equiv -\frac{2 - g^2}{\sqrt{1 - g^2}}. \tag{37}$$

### 5.2. *Linearized eigenvalue problem for $H_d$*

From this point, if we were to attempt to evaluate the dynamics near the EP2A just applying the usual Green's function technique, we would encounter difficulties due to the exceptional point appearing near the branch cut. To circumvent these, we will analyze the problem in terms of the linearized eigenvalue problem similar to our work on the EP2B problem in Sec. 3. The fact that the contributions from the two eigenstates come pre-decomposed in the integrand enables us to rewrite the integration so that we can deal with the double pole and the branch point at separate steps in the analysis.

Just as we did in Sec. 3, we follow Ref. [44] by introducing the variable $\lambda \equiv e^{ik}$. As before, the continuum eigenvalue in Eq. (6) can then be immediately rewritten as

$$E = -\lambda - 1/\lambda. \tag{38}$$

The Schrödinger equation (7) with the effective Hamiltonian (33) corresponding to $H_d$ can be rewritten into a quadratic eigenvalue problem in terms of $\lambda$ in the form

$$\left[\left(1-g^2\right)\lambda^2 + \epsilon_d\lambda + 1\right] P|\psi\rangle = 0. \tag{39}$$

Again following the standard method to linearize the quadratic eigenvalue problem [87], we rewrite Eq. (39) in the form

$$\left(F - \lambda G\right)|\Psi\rangle = 0 \tag{40}$$

in which the $2 \times 2$ $F$ and $G$ matrices are given by

$$F = \begin{bmatrix} 0 & 1 \\ 1 & \epsilon_d \end{bmatrix} \qquad G = \begin{bmatrix} 1 & 0 \\ 0 & -1+g^2 \end{bmatrix} \tag{41}$$

and the generalized eigenstate $|\Psi\rangle$ takes the form

$$|\Psi\rangle \equiv \begin{bmatrix} \langle d|\psi\rangle \\ \lambda\langle d|\psi\rangle \end{bmatrix}. \tag{42}$$

Setting the determinant of (40) to zero yields the $\lambda$ eigenvalues as

$$\lambda_\pm = \frac{-\epsilon_d \mp \sqrt{\epsilon_d^2 - 4\left(1-g^2\right)}}{2\left(1-g^2\right)} \tag{43}$$

Finally, we can obtain the norm of the generalized eigenstates $|\Psi\rangle$ from the natural orthonormalization condition

$$\langle\tilde{\Psi}_i\,|G|\,\Psi_j\rangle = \delta_{i,j}, \tag{44}$$

in which $\langle\tilde{\Psi}_i|$ is again the left-eigenstate obtained from the left-eigenvalue problem $\langle\tilde{\Psi}|\left(F - \lambda G\right) = 0$. It is not difficult to show that $\langle\tilde{\Psi}| = |\Psi\rangle^T$ [42, 44]. Evaluating the norm from Eq. (44), we find

$$\langle d|\psi_\pm\rangle\langle\tilde{\psi}_\pm|d\rangle = \langle d|\psi_\pm\rangle^2 = \frac{1}{1-\left(1-g^2\right)\lambda_\pm^2}. \tag{45}$$

*5.3. Exceptional point properties in the $\lambda$ notation*
Here we derive a few expressions needed for the survival probability calculation.

Just as the energy eigenvalues coincided at the exceptional point $\epsilon_d = \bar{\epsilon}_A$, the two $\lambda$ eigenvalues also coincide, giving

$$\lambda_+(\bar{\epsilon}_A) = \lambda_-(\bar{\epsilon}_A) = \bar{\lambda}_A \equiv -\frac{2}{\bar{\epsilon}_A} = \frac{1}{\sqrt{1-g^2}}. \tag{46}$$

Next, to evaluate the dynamics near the EP we parameterize the $\epsilon_d$ variable by writing

$$\epsilon_d = \bar{\epsilon}_A + \delta \tag{47}$$

in which $\delta$ is taken to be a small, externally controllable parameter. We then apply this reparameterization and expand Eq. (43) to obtain the Puiseux series for the $\lambda$ eigenvalues as

$$\lambda_\pm = \bar{\lambda}_A \left(1 \pm i\delta^{1/2}\bar{\lambda}_A^{1/2} - \frac{\bar{\lambda}_A\delta}{2}\right) + O(\delta^{3/2}). \tag{48}$$

Similarly, we expand the norm from Eq. (45) to find

$$\langle d|\psi_\pm\rangle^2 = \frac{1}{2}\left(1 \pm \frac{i}{(\bar{\lambda}_A\delta)^{1/2}}\right) + O(\delta^{1/2}). \tag{49}$$

We emphasize that when we calculate the survival probability below, the divergent contributions in this expression from the two states will cancel.

### 5.4. *Pure non-Markovian dynamics near the EP2A: threshold influence*

Now we move on to the actual evaluation of the survival probability, which is again written $P(t) = |A(t)|^2$, in which

$$A(t) = \langle d|e^{-iH_d t}|d\rangle. \tag{50}$$

is the survival amplitude of the initially-prepared $|d\rangle$ state.

Similar to our analysis in Sec. 3, we can easily perform an operator expansion of the exponential in Eq. (50) to find that for early times, the system follows the usual parabolic decay of the form

$$P_Z(t) \approx 1 - g^2 t^2, \tag{51}$$

for $t \lesssim T_Z$, in which

$$T_Z = \frac{1}{\epsilon_d}. \tag{52}$$

Again, this particular result for the early timescale dynamics has no particular relationship to the EP2A and it would be difficult to detect in many circumstances. However, we will discover a situation below in which the Zeno timescale dynamics are significantly enhanced near the EP2A, although that situation is not our primary focus here.

Similar to our development in Sec. 3, to evaluate the influence of the EP2A on the dynamics on the intermediate timescale we find it useful to write the survival amplitude as a sum of the individual contributions from the two discrete eigenstates according to the method developed in Ref. [42] . Following this method as originally worked out in Ref. [44] , we can write the survival amplitude for the initial state $|d\rangle$ as

$$\begin{aligned} A(t) &\equiv \langle d|e^{-iH_I t}|d\rangle \\ &= \frac{1}{2\pi i}\sum_{j=\pm}\int_C d\lambda \left(-\lambda + \frac{1}{\lambda}\right) \exp\left[i\left(\lambda + \frac{1}{\lambda}\right)t\right] \langle d|\psi_j\rangle \frac{\lambda_j}{\lambda - \lambda_j}\langle\tilde{\psi}_j|d\rangle. \end{aligned} \tag{53}$$

in which the contour $C$ in the complex $\lambda$ plane is indicated by the brown, clockwise-oriented circle just inside the unit circle in Fig. 5. Further, the summation over $j$ indicates the two discrete eigenstates of the Hamiltonian; these consist of two virtual bound states when the condition $\epsilon_d \lesssim \bar{\epsilon}_A$ holds, or a resonance/anti-resonance pair when $\epsilon_d \gtrsim \bar{\epsilon}_A$ holds. The dynamics are qualitatively similar in these two cases very near to the exceptional point.

Applying the expansions Eq. (48) [for $\lambda_j$] and Eq. (49) [for the norm] to the integrand in Eq. (53), we can expand the survival amplitude in terms of the small parameter $\delta$ to find

$$A(t) \approx \frac{1}{2\pi i}\int_C d\lambda \left(-\lambda + \frac{1}{\lambda}\right) \exp\left[i\left(\lambda + \frac{1}{\lambda}\right)t\right] \frac{-\bar{\lambda}_A^2}{(\lambda - \bar{\lambda}_A)^2} + O(\delta^{1/2}), \tag{54}$$

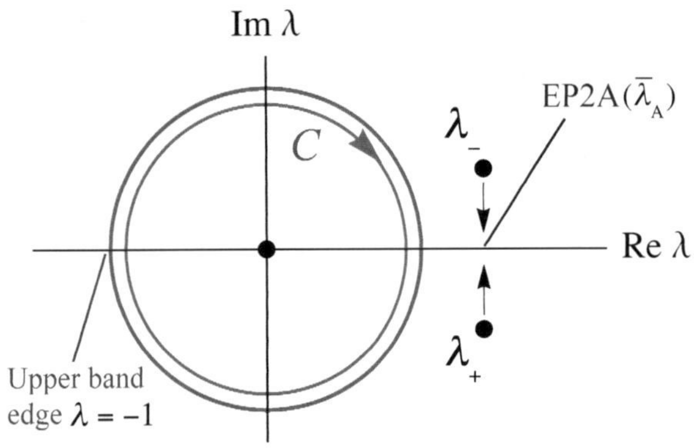


**Figure 5.** Integration contour $C$ in the complex $\lambda$ plane. The continuum coincides with the unit circle, indicated in green. The lower (upper) band edge has been mapped from $E = -2$ ($E = 2$) to $\lambda = 1$ ($\lambda = -1$). An essential singularity appears at the origin, while another occurs at infinity. The $\lambda_\pm$ eigenvalues approach the coalescent eigenvalue $\bar{\lambda}_A$ on the real axis as $\epsilon_d \to \bar{\epsilon}_A$.

in which we remind the reader that $\delta$ is defined according to our reparameterization $\epsilon_d = \bar{\epsilon}_A + \delta$ from Eq. (47). Notice in the first term of the above result that the divergent terms from the norm of the eigenstates have cancelled; further, the lowest-order term is independent of $\delta$.

At this point, if the EP2A wasn't close to the band edge, we could just follow the usual analysis and deform the contour to take the residue at the double pole, similar to our calculation for the EP2B in Sec. 3. However, as shown in Ref. [44], not only is the result of such a calculation inaccurate, it is actually non-unitary. This is because, in the present situation, the influence on the dynamics from the continuum threshold (branch-point effect) is about as strong as the influence from the EP2A itself.

We aim to disentangle these two influences on the dynamics in the following calculation. As the first step, we rewrite the survival amplitude in Eq. (54) as

$$A(t) = -\bar{\lambda}_A^2 \frac{\partial I(\bar{\lambda}_A, t)}{\partial \bar{\lambda}_A} \tag{55}$$

in which

$$I(\bar{\lambda}_A, t) = \int_C \frac{d\lambda}{2\pi i} \left(-\lambda + \frac{1}{\lambda}\right) \exp\left[i\left(\lambda + \frac{1}{\lambda}\right)t\right] \frac{1}{\lambda - \bar{\lambda}_A}. \tag{56}$$

In this step, we have reduced the order of the pole by introducing differentiation with respect to $\bar{\lambda}_A$. In the next step, we rewrite the integration into the complex energy plane through the change of integration variable $\lambda = -E/2 + i\sqrt{1 - E^2/4}$, which yields

$$I(\bar{\lambda}_A, t) = -\frac{1}{2\pi} \int_{C_E} dE\, e^{-iEt} \frac{\sqrt{1 - E^2/4}}{E - \bar{E}_A}, \tag{57}$$

in which $C_E$ is the counter-clockwise contour surrounding the branch cut in the complex energy plane and $\bar{E}_A = -\bar{\lambda}_A - \bar{\lambda}_A^{-1}$. Now, although the order of the pole has been effectively reduced, it still appears in the integration simultaneously with the branch cut $\sqrt{1 - E^2/4}$. However, following the development from App. C of Ref. [44], we can remove the pole entirely by rewriting it in terms of an additional integration, such that Eq. (57) now takes the form

$$I(\bar{\lambda}_A, t) = -\frac{i}{\pi} \int_0^\infty d\tau \int_{-2}^{2} dE\, e^{-iEt} e^{i(E - \bar{E}_A)\tau} \sqrt{1 - E^2/4}. \tag{58}$$

The integration in $E$ can now be rewritten in terms of a Bessel function, so that Eq. (58) takes the form

$$I(\bar{\lambda}_A, t) = e^{-i\bar{E}_A t} \left[1/\bar{\lambda}_A - i \int_0^t dt'\, e^{i\bar{E}_A t'} \frac{J_1(2t')}{t'}\right]. \tag{59}$$

Under the integral, the information about the branch cut has now been encoded in the Bessel function $J_1(t)$.

As detailed in Ref. [44], after re-inserting Eq. (59) back into Eq. (55), the remaining integration can be expanded in terms of the energy gap

$$\Delta_{EP} \equiv -2 - \bar{E}_A. \tag{60}$$

between the EP2A and the band edge (lower threshold). This expansion introduces a factor $t^{-1/2}$, which as detailed in Ref. [78] (see also [81]) is a direct result of the proximity of the eigenvalues to the band edge. Under the assumption $|\Delta_{EP} t| \ll 1$, or $t \ll T_{EP}$ with

$$T_{EP} \equiv \frac{1}{\Delta_{EP}}, \tag{61}$$

we find the approximation for the survival amplitude as

$$A(t) \approx e^{-i\bar{E}_A t}\left[1 - 4\sqrt{\frac{it\Delta_{EP}}{\pi}} + 2it\Delta_{EP}\right] + O(t\Delta_{EP})^{3/2}. \tag{62}$$

The second term here is our key result, which can be understood as arising due to the combination of the factor $t^{-1/2}$ coming from the band edge and the factor $te^{-i\bar{E}_A t}$ coming from the double pole at the EP2A. Out of caution, we also keep the third term to maintain consistency in keeping up with one correction term when we take the modulus squared for the survival probability.

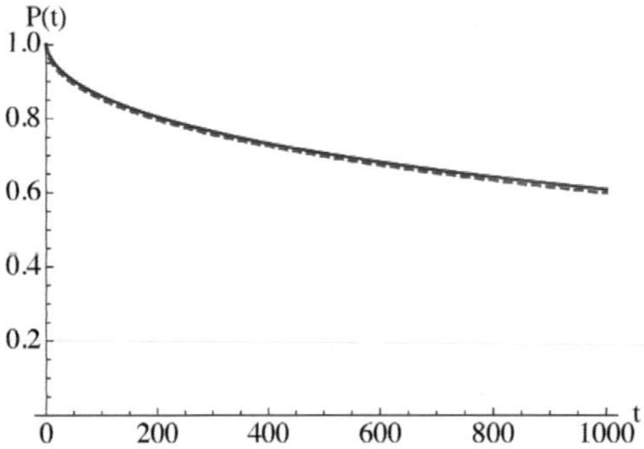


**Figure 6.** Survival probability $P(t)$ near the EP2A, which in turn is near the band edge. The solid blue curve gives the numerical integration while the purple dashed curve represents the approximation for the intermediate timescale dynamics reported in Eq. (63). The parameters used are $g = 0.1$ and $\epsilon_d = -1.989974$, very near the EP2A at $\bar{\epsilon}_A \approx -1.989975$.

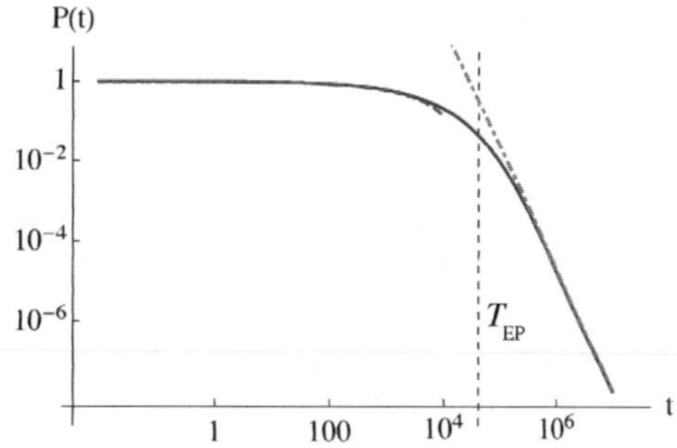


**Figure 7.** Log-log plot of the survival probability near the EP2A that is in turn near the band edge. The numerical integration (solid blue curve) is compared with the Eq. (63) approximation (purple dashed curve) for $t \ll T_{EP}$ and with the long-time approximation Eq. (64) (green chained line) for $T_{EP} \ll t$. The parameters used are the same as Fig. 6, with $T_{EP} \sim 10^5$.

Taking the square modulus of Eq. (62) for the physical quantity, we obtain the expression for the survival probability on the intermediate timescale as

$$P(t) \approx 1 - 4\sqrt{\frac{2t\Delta_{EP}}{\pi}} + \frac{16t\Delta_{EP}}{\pi}. \tag{63}$$

Journal of Physics: Conference Series **2038** (2021) 012011 doi:10.1088/1742-6596/2038/1/012011

This result is plotted (purple dashed curve) against the numerical integration (blue solid curve) in Fig. 6 (linear scale) and Fig. 7 (log scale) for the case $g = 0.1$ and $\epsilon_d = -1.989974$. For this value of $g$, the EP is located $\bar{\epsilon}_A \approx -1.989975$ and the coalesced eigenvalue is located in the energy plane at $\bar{E}_A \approx -2.0000253$, very near the band edge at $E = -2$. Fig. 7 demonstrates that Eq. (63) describes the evolution quite accurately up until about $t \sim T_{EP}$, where $T_{EP} = 1/\Delta_{EP} \sim 10^5$ as defined in Eq. (61).

For $t \gg T_{EP}$, the usual branch-point dynamics take over [78, 98, 100], yielding inverse power law decay. For the present model, it is shown in Ref. [44] that the long-time dynamics at the EP2A are given by

$$P(t) \approx \frac{g^4}{4\pi \left(1 - \sqrt{1-g^2}\right)^8 t^3}. \tag{64}$$

This result agrees well with the numerical integration for the dynamics for $t \gg T_{EP}$ as shown in Fig. 7 (green chained line).

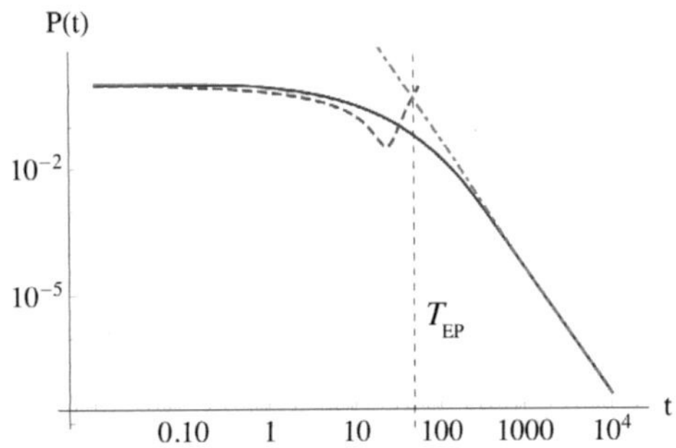


**Figure 8.** Log-log plot of the survival probability $P(t)$ near the EP2A for $g = 0.5$, giving $\bar{E}_A \approx -2.02073$ and $\bar{\epsilon}_A \approx -1.73205$. The nearby value of $\epsilon_d$ has been chosen as $\epsilon_d = -1.7321$. The solid blue curve gives the numerical integration, the dashed purple curve represents the expression reported in Eq. (63) and the chained green line gives the long time evolution $\sim t^{-3}$ from Eq. (64).

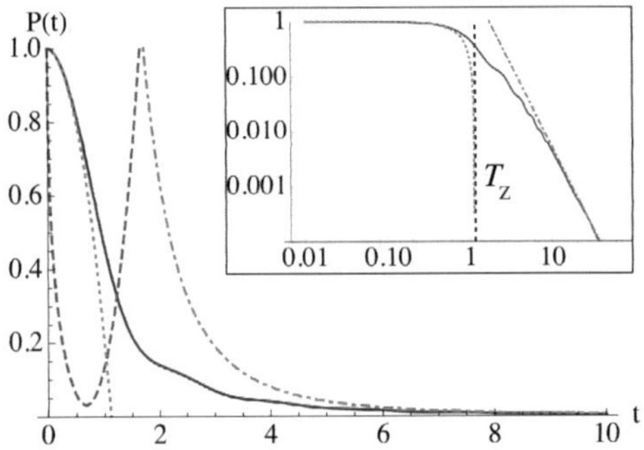


**Figure 9.** Survival probability near the EP2A when it is farther from the band edge [inset: log-log plot]. Here $g = 0.9$, which places the EP2A in the energy plane at $\bar{E}_A \approx -2.73005$ and in the parameter space at $\bar{\epsilon}_A \approx -0.87178$. We have set $\epsilon_d = -0.87$, nearby $\bar{\epsilon}_A$. In this case, the short-time parabolic dynamics $P_Z(t) \approx 1 - g^2t^2$ in Eq. (51) (beige dotted curve) is strongly pronounced, while the prediction from Eq. (63) [purple dashed curve] fails completely.

Next, we briefly comment on how the dynamics are modified as the parameters are changed so that the exceptional point is moved away from the band edge. For the case $g = 0.5$ shown in Fig. 8(a), the EP2A appears at $\bar{E}_A \approx -2.02073$, so that the transition from the $P(t) \sim 1 - Ct^{1/2}$ dynamics to the long-time dynamics $P(t) \sim 1/t^3$ occurs earlier, around $T_{EP} \sim 50$. Notice that the survival probability is also significantly less depleted when the long-time dynamics kick in, as compared to Fig. 7.

Finally, in Fig. 9 we consider the case in which $g = 0.9$ and $\epsilon_d = -0.87$ so that the EP is even farther from the band edge and the timescale $T_{EP}$ now actually falls before the timescale $T_Z \approx 1.15$ for the short-time dynamics [from Eq. (52)]. Because $T_{EP} < T_Z$, the fractional power law dynamics on the intermediate time scale are now squeezed out of the picture entirely, in favor of the short-time parabolic dynamics that are greatly enhanced from the usual picture. Further, in this interesting case, the short-time dynamics transition directly to the long-time inverse power law dynamics, which are also significantly enhanced compared to the usual picture. This case demonstrates that the dynamics associated with an EP2A *far* from the band edge could offer an opportunity to study both the short-time Zeno dynamics and the long-time inverse power law dynamics that are usually quite challenging to observe in experiment.

## 6. Quantum dynamics at an EP3A near the threshold

In the previous section, we saw that the dynamics associated with an EP2A were fully non-Markovian, and that in the case that the exceptional point was located near the band edge the survival probability took the characteristic form $P(t) \sim 1 - C_1\sqrt{t} + C_2 t$. In this section, we present an example of an EP3A near the band edge and give a quick numerical example confirming the dynamics are again non-Markovian. We discuss a possible fitting for our simulation as well, but working out the details of the evolution at an analytic level is left for future work.

Let us consider the Hamiltonian

$$H_n = \epsilon_d d^\dagger d - J \sum_{j=1}^{\infty} \left( c_j^\dagger c_{j+1} + c_{j+1}^\dagger c_j \right) - g \left( c_n^\dagger d + d^\dagger c_n \right), \tag{65}$$

which consists of a discrete, semi-infinite array with a quantum emitter or dot side-coupled to the $n$th element of the array. At least for the choices $n = 2, 4$ and 6, the model can be shown to contain an EP3A that moves nearer to the threshold with increasing $n$ (it is likely this pattern would continue for even $n > 6$). For the choice $n = 2$ the model is more analytically tractable, but the EP3A is unfortunately not so close to the band edge so that an expansion like what we employed in Sec. 5 seems inapplicable anyways. So instead we will mainly focus on the $n = 4$ model to show the properties of the exceptional point and a numerical simulation confirming the non-Markovian dynamics. We will then briefly discuss the $n = 6$ model and give a simulation for the non-Markovian dynamics in that case as well.

The variable parameters in the model in Eq. (65) are again the potential on the dot $\epsilon_d$ and the dot-chain coupling $g$. As usual, we measure the energy in units of $J = 1$.

### *6.1. The $n = 4$ model: spectrum and EP3A near the band edge*

Here we consider the $n = 4$ case of the model in Eq. (65). Since the preceding analysis based on the effective Hamiltonian was primarily introduced for performing analytic calculations in the previous EP2 cases, we will here instead rely directly on the Green's function formalism, which is a bit quicker to dive into for present purposes.

As a first step, we can partially diagonalize $H_n$ by applying the Fourier transform on the half-chain in the form

$$c_n^\dagger = \sqrt{\frac{2}{\pi}} \int_0^\pi dk \, \sin k \, \tilde{c}_k^\dagger. \tag{66}$$

This allows us to write Eq. (65) for $n = 4$ in the form

$$H_{n=4} = \epsilon_d d^\dagger d + \int_0^\pi dk \, E_k \tilde{c}_k^\dagger \tilde{c}_k + g \int_0^\pi V_k \left( \tilde{c}_k^\dagger d + d^\dagger \tilde{c}_k \right), \tag{67}$$

in which $E_k = -2\cos k$ is the usual tight-binding dispersion and $V_k = -\sqrt{2/\pi} \sin 4k$. The Green's function at the side-coupled dot impurity site can be now obtained at the single-particle level, following resummation, in the form

$$\langle d | \frac{1}{E - H_{n=4}} | d \rangle = \frac{1}{E - \epsilon_d - \Sigma_{n=4}(E)}, \tag{68}$$

in which the self-energy function $\Sigma_{n=4}(E)$ is given by

$$\Sigma_{n=4}(E) \equiv g^2 \int_0^{\pi} \frac{|V_k|^2}{E - E_k} = \frac{g^2}{\sqrt{E^2-4}} \left[ 1 - \left( \frac{-E + \sqrt{E^2-4}}{2} \right)^8 \right]. \tag{69}$$

The discrete eigenvalue spectrum is then determined by the poles of Eq. (68), or $E - \epsilon_d - \Sigma_{n=4}(E) = 0$. Applying Eq. (69) in this relation yields an octic polynomial equation $p_{n=4}(E_j) = 0$ in which

$$\begin{aligned} p_{n=4}(E) &= g^2 E^8 - \epsilon_d\, g^2 E^7 - g^2(6+g^2)E^6 + 6\epsilon_d\, g^2 E^5 + 2g^2(5+2g^2)E^4 \\ &\quad -10\epsilon_d\, g^2 E^3 - (1+2g^2)^2 E^2 + 2\epsilon_d(1+2g^2)z - \epsilon_d^2. \end{aligned} \tag{70}$$

Assuming $g \ll 1$, we find that for most values of $\epsilon_d$ the spectrum consists of three resonance/antiresonance pairs and two real-valued solutions. Approximately speaking, for values $|\epsilon_d| \lesssim 2$ inside the band, the two real solutions are virtual states, while for $|\epsilon_d| \gtrsim 2$ one is usually a bound state and the other is virtual.

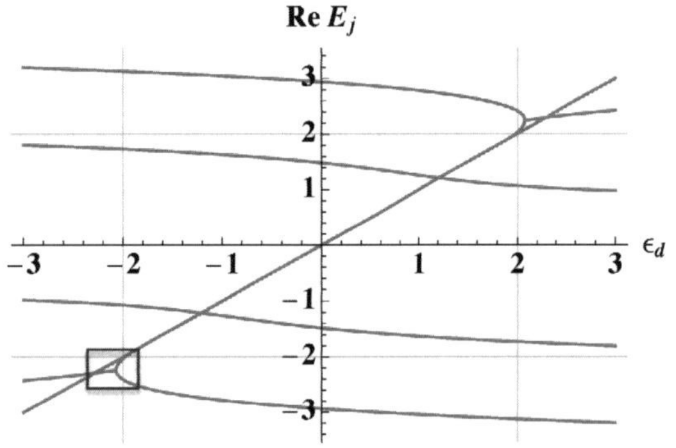


**Figure 10.** Real parts of the discrete eigenvalues for model $H_{n=4}$ as a function of $\epsilon_d$ in the case $g = 0.06$. The upper and lower band edges $\pm 2$ are indicated by grey lines on both axes. The area highlighted by a purple box around the lower band edge is given a closer view in Fig 11.

In Fig. 10, we plot the real part of the eigenvalues for the representative case $g = 0.06$. Notice that near the upper and lower band edges, a region occurs in which several eigenvalue crossings appear. The region near the lower band edge, which is highlighted with a purple box in Fig. 10, can be seen in a zoomed-in view in Fig. 11. In this region, there are two EP2As and the spectral properties depend sensitively on the precise value of $\epsilon_d$. Around $\epsilon_d \approx -2.3$ there occurs a crossing of the real part of the eigenvalues between a bound state and a resonance/anti-resonance pair. We emphasize this point is not an exceptional point. However, as we sweep the value of $\epsilon_d$ leftward in the figure, we encounter an EP2A around $\epsilon_d \approx -2.07$ at which the previously-mentioned resonance/anti-resonance pair coalesces before forming two virtual states (or two anti-bound states [42]). The virtual state with the more negative eigenvalue very slowly moves off to negative infinity for increasing $\epsilon_d$, while the other instead approaches a third virtual state very near the lower band edge $-2$. These latter two virtual states coalesce to form a new resonance/anti-resonance pair around $\epsilon_d \approx -1.985$.

Next, in Fig. 12, we show the imaginary part of the complex eigenvalues, zooming-in a bit more tightly on the region in which the two EP2As occur. The location of the EP2As at which the negative and positive parts of the imaginary eigenvalues for the resonance and anti-resonance collapse is clearly visible. We emphasize there are three virtual states (all three with real eigenvalue) in the region between the EP2As, which appear along the horizontal axis in Fig. 12.

From this point, if we slowly increase the value of $g$ we find that the two EP2As from Figs. 11 and 12 gradually approach one another along the $\epsilon_d$ axis. When $g$ reaches the value

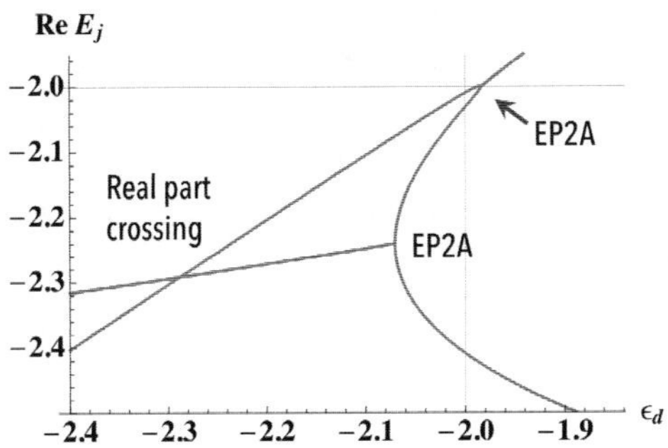


**Figure 11.** Zoomed-in view of the real part of the spectrum for $H_{n=4}$ corresponding to the purple box from Fig. 10 (for $g = 0.06$). The two EP2As near the lower band edge are indicated, as well as a crossing for the real parts of the eigenvalue.

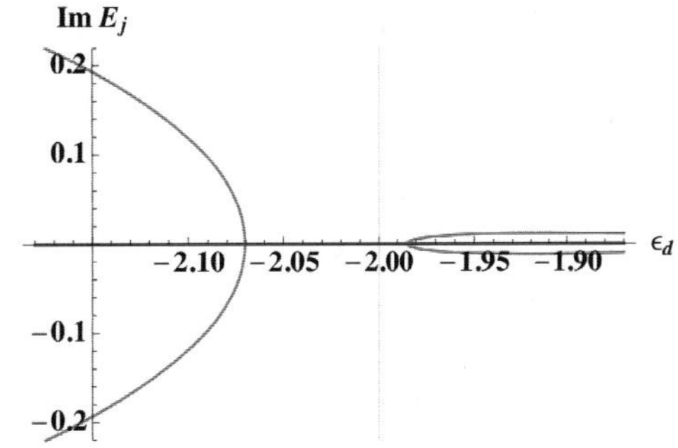


**Figure 12.** Imaginary part of the spectrum for $H_{n=4}$ showing the meeting points for two resonance/anti-resonance pairs, corresponding to the two EP2As from Fig. 11 (for $g = 0.06$). Note that the purely real eigenvalues also appear along the horizontal axis.

$g = \bar{g}_{EP3} \approx 0.0914264$ the two EP2As meet at the $\epsilon_d$ value $\epsilon_d = \bar{\epsilon}_{EP3} \approx -1.958109$ and form an EP3. If we fix $g$ at exactly $g = \bar{g}_{EP3}$ and vary $\epsilon_d$ in the vicinity of the EP3, we find that a virtual state, a resonance, and an anti-resonance all coalesce at $\epsilon_d = \bar{\epsilon}_{EP3}$, before forming another virtual state, resonance, anti-resonance trio on the other side of the EP3, as shown in Figs. 13 and 14. We refer to this type of exceptional point as an EP3A. The coalesced eigenvalue at the EP3A is given by $\bar{E}_{EP3} \approx -2.030646$.

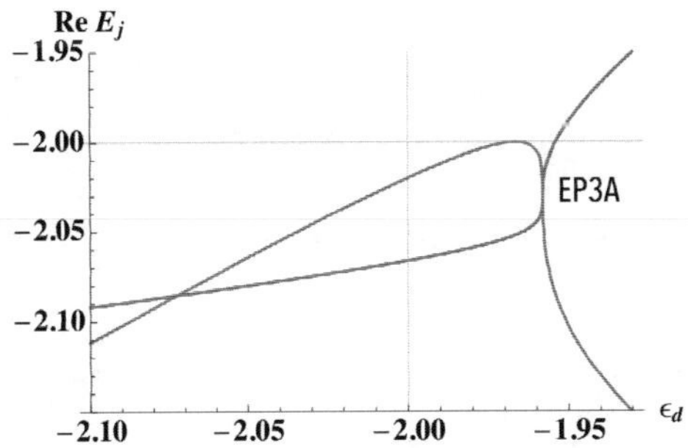


**Figure 13.** Real part of the eigenvalues for model $H_{n=4}$ near the lower band edge for $g = \bar{g}_{EP3} \approx 0.0914264$. The two EP2As from Fig. 11 and Fig. 12 have combined to form an EP3A at $\epsilon_d = \bar{\epsilon}_{EP3} \approx -1.958109$.

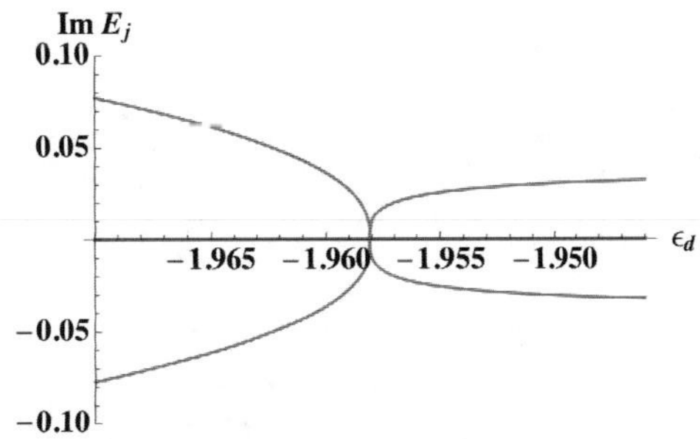


**Figure 14.** Imaginary part of the eigenvalues for model $H_{n=4}$ near the lower band edge for $g = \bar{g}_{EP3} \approx 0.0914264$. The two EP2As from Fig. 11 and Fig. 12 have combined to form an EP3A at $\epsilon_d = \bar{\epsilon}_{EP3} \approx -1.958109$.

We note that the above picture in which the two EP2As form an arc in the $\epsilon_d, g$ parameter space of $H_{n=4}$ before colliding to form a higher-order exceptional point can be viewed as an *exceptional nexus* in the language of Ref. [109].

*6.2. The $n = 4$ model: EP3A dynamics*

We note that the exceptional point eigenvalue $\bar{E}_{EP3} \approx -2.030646$ appears somewhat close to the lower band edge (threshold) at $E = -2$, so that the characteristic gap $\Delta_{EP3}$ takes the value $\Delta_{EP3} = -2 - \bar{E}_{EP3} \approx 0.030646$. Hence we expect that the decay will be fully non-Markovian and further that the threshold should have some significant influence on the dynamics in combination with the exceptional point itself.

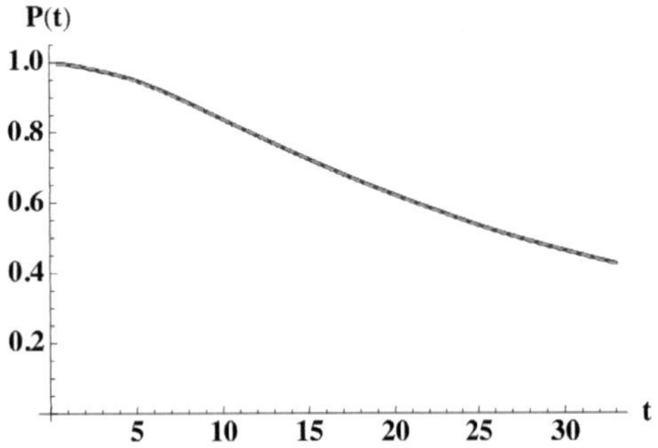


**Figure 15.** Numerical simulation (blue data points) for the survival probability $P(t)$ near the EP3A in model $H_{n=4}$. The parameters used are $\epsilon_d = -2.030647$ and $g = 0.0914264$. The data fit for Eq. (73) is shown as the orange dashed line with the parameters reported in Table 1 for $n = 4$.

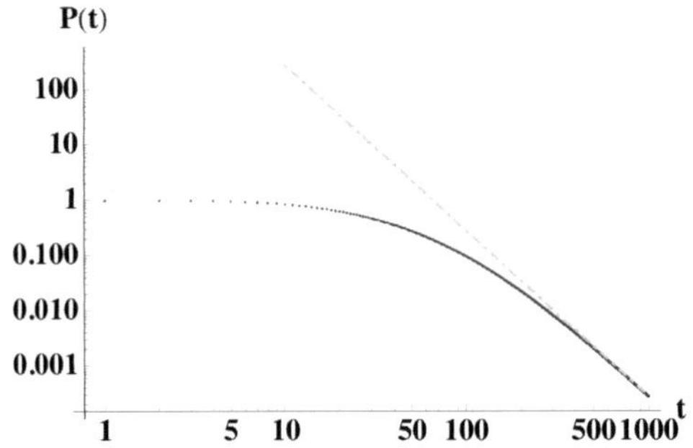


**Figure 16.** Log-log plot of the numerical simulation (blue data points) for the survival probability $P(t)$ near the EP3A in model $H_{n=4}$. The parameters used are $\epsilon_d = -2.030647$ and $g = 0.0914264$. The analytic expression for the long-time dynamics from Eq. (74) is shown with the green dot-dashed line.

We plot a numerical simulation for the dynamics in the case $\epsilon_d = -2.030647$ and $g = 0.0914264$ (very close to the EP3A) as the blue data points in Fig. 15. This numerical result was obtained by direct solution of Schrödinger's equation for a model with 1050 lattice sites beyond the $n$th site (with $n = 4$ in the present case). Visually, it is easy to ascertain that the decay is non-Markovian in this figure. To be certain, we applied an exponential data fitting (and a few variations on exponential decay), which fit the data poorly.

We attempt to better understand the non-Markovian dynamics as follows. Working by analogy with the results for the EP2A near the band edge in Sec. 5, one might expect the survival amplitude in the present case to take the form

$$A(t) \sim \left(1 + B_1 t^{1/2} + B_2 t + B_3 t^{3/2} + B_4 t^2 + B_5 t^{5/2} + B_6 t^3\right) e^{-i\bar{E}_{EP3} t}, \tag{71}$$

in which the $t^{3/2}$-dependent term is expected to appear due to the multiplication of a factor $t^2 e^{-i\bar{E}_{EP3} t}$ from the triple-pole at the EP3A and a factor $1/\sqrt{t}$ coming from the nearby band edge. This would correspond to Eq. (62) for the EP2A case in Sec. 5.4, and just as in that case, we would expect it might be valid on the scale $t \lesssim T_{EP3}$, in which the timescale $T_{EP3}$ is given by

$$T_{EP3} = \frac{1}{\left|-2 - \bar{E}_{EP3}\right|} \approx 32.6306. \tag{72}$$

Also notice that just as in the EP2A case, we have kept the higher-order $t^m$ terms with $m > 3/2$ in Eq. (71) to ensure consistency when we take the square modulus to obtain the probability as

$$P(t) \sim 1 + C_1 t^{1/2} + C_2 t + C_3 t^{3/2} + C_4 t^2 + C_5 t^{5/2} + C_6 t^3. \tag{73}$$

Let us emphasize that we would expect the higher-order terms to be correction terms with the $C_4$, $C_5$ and $C_6$ coefficients all being relatively small.

Assuming this picture is correct, a question immediately arises: would we expect the first ($t^{1/2}$) or second ($t^{3/2}$) fractional power term to be the dominant term in the evolution? If the coefficients took the comparable form as Eq. (63) for the EP2A case, then we would have $C_n t^{n/2} = A_n(\Delta_{EP3}t)^{n/2}$ appearing as powers of the small quantity $\sqrt{\Delta_{EP3}t}$. Then $C_3 t^{3/2}$ could only be expected to be the dominant term with a much larger coefficient $A_3$ than $A_1$.

**Table 1.** Fitting parameters for the hypothesized power-law decay near the EP3A [Eq. (73)], as determined in Mathematica

| Model | $C_1$ | $C_2$ | $C_3$ | $C_4$ | $C_5$ | $C_6$ |
|---|---|---|---|---|---|---|
| $n = 4$ | $-0.0217969$ | 0.0286801 | $-0.0122409$ | $-0.00225201$ | 0.000980366 | $-0.000076716$ |
| $n = 6$ | $-0.011098$ | 0.0101236 | $-0.0029150$ | $-0.00050665$ | 0.000128822 | $-0.000006557$ |

We performed several data fittings based on Eq. (73) in Mathematica, one of which is shown by the orange dashed curve in Fig. 15. The coefficients obtained for this data fitting are shown in Table 1. The fitting agrees with the numerical simulation well and the numbers reported for the coefficients seem intuitively reasonable, with the first three coefficients $C_{1,2,3}$ all being about the order of the gap $\Delta_{EP3} \approx 0.030646$, and the coefficients for the three higher-order terms decreasing in magnitude from there. The resulting RMS value for this fitting was also reasonable with RMS $\approx 0.00105$.

However, the fitting coefficients Mathematica chose were somewhat sensitive to the time length of the simulation. Reducing the simulation time length eventually results in the sign designations of the $C_{1,2,3}$ coefficients being flipped, which to us seemed qualitatively a little different than Eq. (63). And in any case, we found that using a polynomial fit with only integer powers of $t$ with the same number of fitting parameters as Eq. (73) yielded a comparable RMS value. Hence, we only claim that the hypothesis in Eq. (73) *might* be correct and should be investigated further.

Regardless of the applicability of Eq. (73), the primary point is that the dynamics are fully non-Markovian. And for the longest timescale in the system the dynamics near the EP3A can be shown to demonstrate the usual inverse power law decay. In particular, we derived the approximation for the evolution during this time domain as

$$P(t) \approx \frac{64\bar{g}_{EP3}^4}{\pi(2 + \bar{\epsilon}_{EP3} - 4\bar{g}_{EP3}^2)^4 t^3}, \tag{74}$$

which is shown as the green dot-dashed line in the log-log plot in Fig. 16.

### 6.3. *The $n = 6$ model: EP3A dynamics closer to the threshold*

Finally, we briefly consider the EP3A in the $n = 6$ model from Eq. (65). Following analysis similar to that which was applied for the $n = 4$ model, we can show that the $n = 6$ case gives rise to a dispersion polynomial $p_{n=6}(E_j) = 0$ in which $p_{n=6}(E)$ is 12th-order in $E$. The spectrum for $g \ll 1$ is qualitatively similar to that for $n = 4$, except that here it usually consists of five resonance/anti-resonance pairs and two real eigenvalues. Again, the exception to this statement is near the two band edges where a picture very similar to that shown in Figs. 11 and 12 occurs. And similar to that case, we find that two EP2As meet to form an EP3A, in this case occurring at $\bar{g}_{EP3} \approx 0.04946448$ and $\epsilon_d = \bar{\epsilon}_{EP3} \approx -1.9816623$. However, the coalesced eigenvalue in this case occurs a bit closer to the threshold, at $\bar{E}_{EP3} \approx -2.0131867$. With this value, we would expect the timescale for the validity of the proposed approximation in Eq. (73) to be about

$$T_{EP3} = \frac{1}{|-2 - \bar{E}_{EP3}|} \approx 75.8. \tag{75}$$

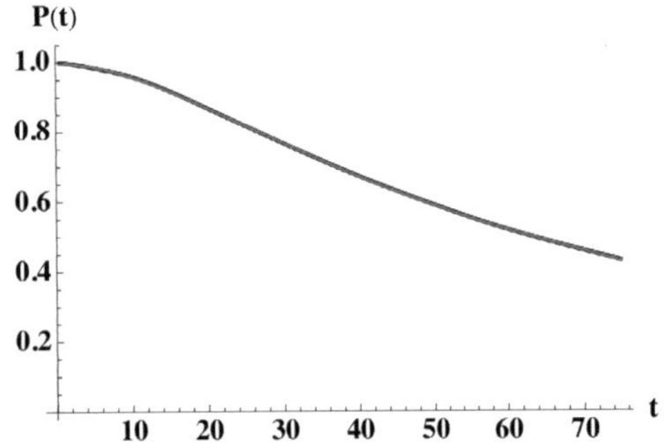


**Figure 17.** Numerical simulation (blue data points) for the survival probability $P(t)$ near the EP3A in model $H_{n=6}$. The parameters used are $\epsilon_d = -1.981662$ and $g = 0.0494644$. The data fit for Eq. (73) is shown as the orange line with the parameters reported in Table 1 for $n = 6$. The RMS value for the fitting was 0.000996.

We show a numerical simulation (blue data points) in Fig. 17 for the case $\epsilon_d = -1.981662$ and $g = 0.0494644$, very close to the EP3A. The fitted curve corresponding to Eq. (73) is shown in orange, with the fitting parameters reported again in the $n = 6$ line of Table 1. The estimated coefficients $C_i$ again are decreasing for the higher-order terms starting from $i = 4$, although here, closer to the threshold, they are notably a bit smaller than the $n = 4$ case. It is notable that in both the $n = 4$ and $n = 6$ cases, the $C_1$ and $C_2$ terms seem to be a bit more dominant than the $C_3$ term.

## 7. Conclusion

In this work, we have reviewed coalescing eigenstates and their influence on certain physical quantities in the near-vicinity of the exceptional point at which the coalescence occurs. In particular, we have focused on their signature influence on dynamical processes in quantum systems, taking into account that the presence of the continuum threshold can equally play a role in shaping these effects in some circumstances.

Near the B-type exceptional points, at which two or more resonances coalesce, we saw that the usual exponential decay was modified with power law factors determined by the number of coalescing resonances. Due to the influence of the threshold, the power law-exponential decay can be expected to be replaced by some form of non-exponential decay on very long timescales; for the EP2B considered in Sec. 3.4, we showed this took the form of inverse power-law decay, as plotted in Fig. 4. We emphasize that in most typical circumstances this long-time effect would be very difficult to detect, just as in the ordinary case involving a lone resonance. However, just as the non-exponential dynamics can be enhanced when a lone resonance approaches the threshold [110, 111], it seems likely that a B-type exceptional point near the threshold would also lead to some kind of enhanced non-exponential evolution, which could prove interesting. However, we don't know of any specific physical model in which this scenario actually occurs.

Since a coalescing resonance/anti-resonance pair at an EP2A naturally occurs in the vicinity of the continuum threshold, we found that the threshold has a much stronger influence on the dynamics near the exceptional point in this case. As originally shown in Ref. [44], this results in a significantly more complex picture for the quantum dynamics. In the case that the EP2A occurs close to the threshold, we found that the survival probability exhibited a decay of the form $P(t) \sim 1 - C_1\sqrt{t} + C_2 t$, in which the $\sim \sqrt{t}$ term results from one factor of $t$ coming from the pole at the exceptional point and a factor $1/\sqrt{t}$ coming from the threshold. Further, the decay was fully non-Markovian on all timescales. In the case that the EP2A is located further away from the threshold the characteristic $1 - C\sqrt{t}$ effect becomes less pronounced, but the decay is still fully non-Markovian as the short time parabolic decay and the long-time inverse power law decay become more strongly enhanced. This indicates that exceptional points could play a useful role in non-Markovian dynamical engineering.

We then gave an example of a model with an EP3A at which a resonance, anti-resonance and virtual state all coalesce somewhat close to the threshold. A quick numerical simulation revealed that the dynamics were non-Markovian in this case as well.

Generalizing from these observations, another natural question to consider would be the possibility of the A-type EP appearing directly at the threshold. Indeed, this scenario is possible in open quantum systems, and some examples have appeared in the literature [40, 59, 112]. However, the detailed analysis of the time evolution in these scenarios seems relatively unexplored, with the exception of a work-in-progress by two of the present authors.

In Ref. [59], which was presented in the present virtual conference series on Pseudo-Hermitian Hamiltonians in Quantum Physics, we have demonstrated the occurrence of an exceptional point directly at the threshold in generic 1-D continuum systems that is anomalous in the sense that the branch cut appearing in the Puiseux expansion is higher-order than the number of levels that actually coalesce at the EP. In particular, for the model studied in Ref. [59], there are three levels with energy eigenvalues that converge on the continuum threshold as the system coupling $g$ is shut off: a resonance, an anti-resonance and a bound state. This results in a Puiseux expansion for the energy eigenvalues of the form $E_j \approx E_{th} - C_j g^{4/3}$ in which $E_{th}$ is the energy corresponding to the threshold and $C_j$ is a constant, as usual. From our previous experience, this seems to strongly suggest the presence of a third-order exceptional point at the threshold. However, a careful analysis reveals the order of the EP is actually only second-order, as only two linear combinations of the three discrete states can be shown to coalesce, while a third linear combination instead fuses to the continuum threshold itself. Meanwhile, the dynamics associated with this anomalous-order exceptional point takes the form $1 - Ct^{3/2}$, which seems more similar to what one might expect for a third-order EP rather than a second-order one. This result suggests that at least for the EPs appearing directly at the threshold, the time evolution probably has to be worked out on a case-by-case basis. We also note the dynamics in this case are related to a well-known problem in spontaneous emission near a photonic band gap [50, 113, 114].

We conclude with a final comment noting that in this work we have confined our analysis of the role of exceptional points in open quantum systems to the level of the Hamiltonian. However, it has been emphasized in a number of recent works that a deeper understanding of the role of exceptional points in quantum systems can be obtained by evaluating the problem at the level of the density matrix under the Liouvillian master equation formalism, including Lindblad terms that can account for decoherence and quantum jump processes [60–63]. In Ref. [62] it is has been shown by F. Minganti, et al that exceptional points appearing at the level of the Hamiltonian can have their locations shifted under the Liouvillian formalism, and further that new exceptional points may appear that cannot be seen in the Hamiltonian picture at all.

Given the possibility for parameter shifting of EPs under the Liouvillian framework, several interesting questions immediately arise: would the aforementioned exceptional points occurring directly at the threshold also appear shifted? And further, for exceptional points either at or near the threshold, how would the dynamics at the Hamiltonian level be modified under the influence of quantum jumps?

However, this does not necessarily imply that it would be impossible to observe the dynamics associated with the bare Hamiltonian itself. We note two recent experiments [115,116] conducted at Washington University by the group of K. Murch, in which data post-selection is used to eliminate experimental trials in which quantum jumps occur. This approach has enabled studies of quantum state tomography and dynamics near an exceptional point in a superconducting qubit embedded in a cavity [115].

**Acknowledgments**
We thank Kater Murch for helpful discussions related to this work. S. G. acknowledges support from the Japan Society for the Promotion of Science through KAKENHI Grant No. JP18K03466.

## References

[1] Hatano N and Nelson D R 1996 *Phys. Rev. Lett.* **77** 570
[2] Hatano N and Nelson D R 1997 *Phys. Rev.* B **56** 8651
[3] Bender C M and Boettcher S 1998 *Phys. Rev. Lett.* **80** 5243
[4] Mostafazadeh A 2002 *J. Math. Phys.* **43** 205
[5] Bender C M, Brody D C, and Jones H F 2002 *Phys. Rev. Lett.* **89** 270401
[6] Ruschhaupt A, Delgado F and Muga J G 2005 *J. Phys. A: Math. Theor.* **38** L171–L176
[7] Makris K G, El-Ganainy R, Christodoulides D N and Musslimani Z H 2008 *Phys. Rev. Lett.* **100** 103904
[8] Rotter I 2009 *J. Phys. A: Math. Theor.* **42** 153001
[9] Moiseyev N 2011 *Non-Hermitian Quantum Mechanics*, (Cambridge University Press)
[10] Rotter I and Bird J P 2015 *Rep. Prog. Phys.* **78** 114001
[11] Garmon S, Gianfreda M, and Hatano N 2015 *Phys. Rev.* A **92** 022125
[12] Konotop V V, Yang J and Zezyulin D A 2016 *Rev. Mod. Phys.* **88** 035002
[13] Christodoulides D and Yang J 2018 *Parity-time Symmetry and Its Applications*, (Singapore: Springer Nature)
[14] Bender C M 2019 *PT Symmetry in Quantum and Classical Systems*, (Singapore: World Scientific)
[15] Kato T 1980 *Perturbation Theory for Linear Operators*, (Berlin: Springer-Verlag)
[16] Berry M V 2004 *Czech. J. Phys.* **54** 1039
[17] Heiss W D 2012 *J. Phys. A: Math. Theor.* **45** 444016
[18] Miri M-A and Alù A 2019 *Science* **363** eaar7709
[19] Heiss W D, and Harney H L 2001 *Eur. Phys. J.* **17** 149
[20] Dembowski C, Dietz B, Gräf H D, Harney H L, Heine A, Heiss W D, and Richter A (2004) *Phys. Rev.* E **69** 056216
[21] Gilary I, Mailybaev A A, and Moiseyev N 2013, *Phys. Rev.* A **88** 010102(R)
[22] Milburn T J, Doppler J, Holmes C A, Portolan S, Rotter S, and Rabl P 2015 *Phys. Rev.* A **92** 052124
[23] Heiss D 2016, *Nat. Phys.* **12** 823
[24] Menke H, Klett, M, Cartarius H, Main J, and Wunner G 2016 *Phys. Rev.* A **93** 013401
[25] Doppler J, Mailybaev A A, Böhm J, Kuhl U, Girschik A, Libisch F, Milburn T J, Rabl P, Moiseyev N, and Rotter S 2016, *Nat.* **537** 76
[26] Xu H, Mason D, Jiang L, and Harris J G E 2016 *Nat.* **537** 80
[27] Liu W, Wu Y, Duan C-K, Rong X and Du J 2021 *Phys. Rev. Lett.* **126** 170506
[28] Wiersig J 2014 *Phys. Rev. Lett.* **112** 203901
[29] Chen W, Özdemir Ş K, Zhao G, Wiersig J, and Yang L 2017, *Nature* **548** 192
[30] Zhang M, Sweeney W, Hsu C W, Yang L, Stone A D, and Jiang L 2019 *Phys. Rev. Lett.* **123** 180501
[31] Hodaei H, Hassan A U, Wittek A, Garcia-Gracia H, El-Ganainy R, Christodoulides D N, and Khajavikhan M 2017 *Nature* **548** 187
[32] Kullig J, Yi C-H, Hentschel M, and Wiersig J 2018, *New J. Phys.* **20** 083016
[33] Nakanishi N 1958 *Prog. Theor. Phys.* **19**, 607
[34] Sudarshan E C G, Chiu C B and Gorini V 1978 *Phys. Rev.* D **18** 2914
[35] Bohm A and Gadella M 1989 *Dirac Kets, Gamow Vectors and Gelfand Triplets*, Springer Lecture Notes on Physics **348** (New York: Springer-Verlag)
[36] Petrosky T, Prigogine I and Tasaki S 1991 Phys. A **173** 175
[37] Gadella M and Pronko G P 2011 *Fortschr. Phys.* **59** 795
[38] Madrid R 2012 *J. Math. Phys.* **53** 102113
[39] Hatano N 2013 *Fortsch. Phys.* **61** 238–249
[40] Tanaka S, Garmon S, Kanki K, and Petrosky T 2016 *Phys. Rev.* A **94** 022105
[41] Kanki K, Garmon S, Tanaka S and Petrosky T 2017 *J. Math. Phys.* **58** 092101
[42] Hatano N and Ordonez G 2014 *J. Math. Phys.* **55** 122106
[43] Sasada K, Hatano N and Ordonez G 2011 *J. Phys. Soc. Jpn.* **80** 104707
[44] Garmon S and Ordonez G 2017 *J. Math. Phys.* **58** 062101

[45] Ordonez G and Hatano N 2017 *J. Phys. A: Math. Theor.* **50** 405304
[46] Ordonez G and Hatano N 2017 *Chaos* **27** 104608
[47] Dente A D, Bustos-Marún R A, and Pastawski H 2008 *Phys. Rev.* A **78** 062116
[48] Garmon S, Rotter I, Hatano N, and Segal D 2012 *Int. J. Theor. Phys.* **51** 3536
[49] Nikolopoulos G M, Bay S and Lambropoulos P 1999 *Phys. Rev.* A **60** 5079
[50] Lambropoulos P, Nikolopoulos G M, Nielson T R and Bay S 2000 *Rep. Prog. Phys.* **63** 455
[51] Dalton B J and Garraway B M 2003 *Phys. Rev.* A **68** 033809
[52] Longhi S 2006 *Phys. Rev.* A **74** 063826
[53] Garmon S, Nakamura H, Hatano N and Petrosky T 2009 *Phys. Rev.* B **80** 115318
[54] Lombardo F, Ciccarello F and Palma G M 2014 *Phys. Rev.* A **89** 053826
[55] González-Tudela A and Cirac J I 2017 *Phys. Rev. Lett.* **119** 143602
[56] González-Tudela A and Cirac J I 2017 *Phys. Rev.* A **96** 043811
[57] Sánchez-Burillo E, Zueco D, Martín-Moreno L and García-Ripoll J J 2017 *Phys. Rev.* A **96** 023831
[58] Shen H Z, Xu S, Cui H T and Yi X X 2019 *Phys. Rev.* A **99** 032101
[59] Garmon S, Ordonez G and Hatano N 2021 *Phys. Rev. Research* **3** 033029
[60] Breuer H P and Petruccione F 2002 *The Theory of Open Quantum Systems* (Oxford: Oxford University Press)
[61] Hatano N 2019 *Mol. Phys.* **117** 2121–2127
[62] Minganti F, Miranowicz A, Chhajlany R W and Nori F 2019 *Phys. Rev.* A **100** 062131
[63] Khandelwal S, Brunner N, and Haack G, Signatures of exceptional points in a quantum thermal machine, *Preprint* arXiv:2101.11553
[64] Persson E, Rotter I, Stöckmann H-J and Barth M 2000 *Phys. Rev. Lett.* **85** 2478
[65] Pastawski H M 2007 *Phys.* B **398** 278
[66] Goldberger M L and Watson K M 1964 *Phys. Rev.* **136** B1472
[67] Bell J S and Goebel C J 1965 *Phys. Rev.* **138** B1198
[68] Wiersig J, Kim S W and Hentschel M 2008 Phys. Rev. A **78** 053809
[69] Cartarius H and Moiseyev N 2011 *Phys. Rev.* A **84** 013419
[70] Heiss W D 2010 *Eur. Phys. J.* D **60** 257
[71] Fuchs J, Main J, Cartarius H and Wunner G 2014 *J. Phys. A: Math. Theor.* **47** 125304
[72] Hashimoto K, Kanki K, Hayakawa H and Petrosky T 2015 *Prog. Theor. Exp. Phys.* **2015** 023A02
[73] Hashimoto K, Kanki K, Garmon S, Tanaka S and Petrosky T 2016 *Prog. Theor. Exp. Phys.* **2016** 053A02
[74] Dietz B, Friedrich T, Metz J, Miski-Oglu M, Richter A, Schäfer F and Stafford C A 2007 *Phys. Rev.* E **75** 027201
[75] Bittner S, Dietz B, Harney H L, Miski-Oglu M, Richter A and Schäfer F 2014 *Phys. Rev.* E **89** 032909
[76] Demange G and Graefe E M 2012 *J. Phys. A: Math. Theor.* **45** 025303
[77] Heiss W D and Wunner G 2016 *J. Phys. A: Math. Theor.* **49** 495303
[78] Garmon S, Petrosky T, Simine L and Segal D 2013 *Fortschr. Phys.* **61** 261
[79] Álvarez G A, Danieli E P, Levstein P R and Pastawski H M 2006 *J. Chem. Phys.* **124** 194507
[80] Bustos-Marún R A, Coronado E A and Pastawski H M 2010 *Phys. Rev.* B **82** 035434
[81] Garmon S, Noba K, Ordonez G and Segal D 2019 *Phys. Rev.* A **99** 010102(R)
[82] Siegert A J F 1939 *Phys. Rev.* **56** 750
[83] Feshbach H 1958 *Ann. Phys.* (New York) **5** 357
[84] Feshbach H 1962 *Ann. Phys.* (New York) **19** 287
[85] Sadreev A F and Rotter I 2003 *J. Phys. A: Math. Gen.* **36** 11413
[86] F. Tisseur and K. Meerbergen, SIAM Rev. **43** 235 (2001).
[87] A. P. Seyranian and A. A. Mailybaev, *Multiparameter Stability Theory with Mechanical Applications*, World Scientific Publishing Co., Singapore (2003).
[88] Wilkinson S R, Bharucha C F, Fischer M C, Madison K W, Morrow P R, Niu Q, Sundaram B, and Raizen M G 1997 *Nature (London)* **387**, 575
[89] Crespi A, Pepe F V, Facchi P, Sciarrino F, Mataloni P, Nakazato H, Pascazio S, and Osellame R 2019 *Phys. Rev. Lett.* **122** 130401
[90] Misra B and Sudarshan E C G 1977 *J. Math. Phys.* **18** 756
[91] Chiu C B and Sudarshan E C G 1977 *Phys. Rev.* D **16** 520
[92] Kofman A G and Kurizk G 1996 *Phys. Rev.* A **54** R3750
[93] Segal D and Reichman D R 2007 *Phys. Rev.* A 012109
[94] Dreisow F, Szameit A, Heinrich M, Pertsch T, Nolte S and Tünnermann A 2008 *Phys. Rev. Lett.* **101** 143602
[95] Harrington P M, Monroe J T, and Murch K W 2017 *Phys. Rev. Lett.* **118** 240401
[96] Khalfin L A 1958 *Sov. Phys. JETP* **6** 1053
[97] Fonda L, Ghirardi G C, and Rimini A 1978 *Rep. Prog. Phys.* **41** 587

[98] Martorell J, Muga J G and Spring D W L 2009 Lect. Notes Phys. **789** 239
[99] Hack M N 1982 *Phys. Lett.* A **90** 220
[100] García-Calderón G, Mateos J L and Moshinsky M 1995 *Phys. Rev. Lett.* **74** 337
[101] Cavalcanti R M 1998 *Phys. Rev. Lett.* **80** 4353
[102] Muga J G, Delgado V and Snider R F 1995 *Phys. Rev.* B **52** 16381
[103] Hernández E, Jáuregui A and A Mondragón A 2000 *J. Phys. A: Math. Gen.* **33** 4507
[104] Heiss W D and Wunner G 2014 *Eur. Phys. J.* D **68** 284
[105] Ben-Asher A, Šimsa D, Uhlířovã T, Šindelka M and Moiseyev N 2020 *Phys. Rev. Lett.* **124** 253202
[106] Lin Z, Pick A, Lončar M, and Rodriguez A W 2016 *Phys. Rev. Lett.* **117** 107402
[107] Pick A, Zhen B, Miller O D, Hsu C W, Hernandez F, Rodriguez A W, Soljačić M and Johnson S G 2017 *Opt. Exp.* **25** 12325
[108] Dunham Y, Kanki K, Garmon S, Tanaka S, and Ordonez G 2021 *Phys. Rev.* A **103** 043513
[109] Tang W, Jiang X, Ding K, Xiao Y-X, Zhang Z-Q, Chan C T and Ma G 2020, *Science* **370** 1077
[110] Jittoh T, Matsumoto S, Sato J, Sato Y, and Takeda K 2005 *Phys. Rev.* A **71** 012109
[111] García-Calderón G, and Villavicencio J 2006 *Phys. Rev.* A **73** 062115
[112] Heiss W D and Nazmitdinov R G 2011 *Eur. Phys. J.* D **63** 369
[113] Kofman A G, Kurizki G and Sherman B 1994 *J. Mod. Opt.* **41** 353
[114] John S and Quang T 1994 *Phys. Rev.* A **50** 1764
[115] Naghiloo M, Abbasi M, Joglekar Y N, and Murch K W 2019 Nat. Phys. **15** 1232–1236
[116] Chen W, Abbasi M, Joglekar Y N, and Murch K W 2021, Quantum jumps in the non-Hermitian dynamics of a superconducting qubit, *Preprint* arXiv:2103.06274

# Classical Hamiltonian Systems with balanced loss and gain

**Pijush K Ghosh**

Department of Physics, Siksha-Bhavana, Visva-Bharati University, Santiniketan, PIN 731 235, India.

E-mail: pijushkanti.ghosh@visva-bharati.ac.in

**Abstract.** Classical Hamiltonian systems with balanced loss and gain are considered in this review. A generic Hamiltonian formulation for systems with space-dependent balanced loss and gain is discussed. It is shown that the loss-gain terms may be removed completely through appropriate co-ordinate transformations with its effect manifested in modifying the strength of the velocity-mediated coupling. The effect of the Lorentz interaction in improving the stability of classical solutions as well as allowing a possibility of defining the corresponding quantum problem consistently on the real line, instead of within Stokes wedges, is also discussed. Several exactly solvable models based on translational and rotational symmetry are discussed which include coupled cubic oscillators, Landau Hamiltonian etc. The role of $\mathcal{PT}$-symmetry on the existence of periodic solution in systems with balanced loss and gain is critically analyzed. A few non-$\mathcal{PT}$-symmetric Hamiltonian as well as non-Hamiltonian systems with balanced loss and gain, which include mechanical as well as extended system, are shown to admit periodic solutions. An example of Hamiltonian chaos within the framework of a non-$\mathcal{PT}$-symmetric system of coupled Duffing oscillator with balanced loss-gain and/or positional non-conservative forces is discussed. It is conjectured that a non-$\mathcal{PT}$-symmetric system with balanced loss-gain and without any velocity mediated interaction may admit periodic solution if the linear part of the equations of motion is necessarily $\mathcal{PT}$ symmetric —the nonlinear interaction may or may not be $\mathcal{PT}$-symmetric. Further, systems with velocity mediated interaction need not be $\mathcal{PT}$-symmetric at all in order to admit periodic solutions. Results related to nonlinear Schrödinger and Dirac equations with balanced loss and gain are mentioned briefly. A class of solvable models of oligomers with balanced loss and gain is presented for the first time along with the previously known results.

## 1. Introduction

Dissipation is an ubiquitous phenomenon in nature. There are different approaches to understand dissipation, instabilities induced by it and controlling its effect in physical systems[1, 2]. One of the earlier attempts in this direction was Hamiltonian formulation of system with dissipation. This may seem counter intuitive, since a Hamiltonian system is conservative, while dissipation induces the loss of energy to the environment. The apparent conflict can be resolved if the system plus environment is considered as a larger system. The Hamiltonian formulation of Bateman oscillator is based on this approach and describes a system comprising of a damped oscillator plus an auxiliary system of an anti-damped oscillator with the same dissipation/anti-dissipation coefficient —the dimension of the ambient space is twice that of the target space[3]. The flow in the position-velocity state-space preserves the volume, allowing a

Hamiltonian of the system, since the rate of energy-loss of the damped mode is equal to the rate of gain in energy by the anti-damped oscillator. The Bateman oscillator is an example of a system with balanced loss and gain —the flow is preserved in the position-velocity state-space although individual degrees of freedom are subjected to gain/loss.

The Bateman oscillator has been studied extensively over the last ninety years from different perspectives[4, 5, 6, 7, 8, 9, 10, 11, 12, 13, 14]. One of the undesirable features of the Hamiltonian formulation of Bateman oscillator is the introduction of the auxiliary degree of freedom. With the growing interest and relevance of $\mathcal{PT}$-symmetric theory[15], the concept of auxiliary system is abandoned in the interpretation of generalized versions of Bateman oscillator, where damped and anti-damped modes are coupled and exchange energies. It should be noted that there is no exchange of energies between the damped and anti-damped modes of the standard Bateman oscillator and a clear distinction between the two modes exists. Further, there are descriptions of quantum dissipation[16], where the energy is deposited from the system to the bath consisting of infinitely many harmonic oscillators so that the energy can not be transferred back to the system. The sole objective of adding interaction in the generalized models of Bateman oscillator is allowing a bi-directional exchange of energies between the damped and the anti-damped modes such that an equilibrium may be achieved —the energy is reverted back to the damped oscillator from the anti-damped oscillator at the same rate as it is deposited to the anti-damped oscillator from the damped oscillator. The existence of equilibrium state in a $\mathcal{PT}$-symmetric system with balanced loss and gain is worth comparing to a similar situation arising in the context of non-equilibrium thermodynamics, where the invariance of Schwinger-Keldysh action under time-reversal symmetry plus time-translation corresponds to thermodynamic equilibrium [17]. The Hamiltonian is the generator of the time-translation, while invariance under time-reversal symmetry conforms to principle of detailed balance for the quantum system. It should be noted here that $\mathcal{PT}$-symmetry may be substituted with a *non-standard time-reversal symmetry* within the realm of quantum mechanics[18]. Thus, the conditions on the Schwinger-Keldysh action for thermodynamic equilibrium and the equilibrium condition for systems with balanced loss and gain may be identified as similar.

The basic model of a generalized Bateman oscillator may be described by considering two linearly coupled identical harmonic oscillators, one of which is subjected to damping and the other to anti-damping with the same strength[19]. This describes an experimentally realized system involving coupled whispering galleries[20]. The system consists of two degrees of freedom and there is no concept of auxiliary system —the target and the ambient spaces are the same unlike in the case of the standard Bateman oscillator. The abandoning of auxiliary degrees of freedom, while considering models of generalized Bateman oscillator, is indeed a paradigm shift from the traditional treatment of the standard Bateman oscillator. The coupling allows an equilibrium state manifested as periodic solutions within some regions in the parameter-space characterizing the unbroken $\mathcal{PT}$-phase. The existence of equilibrium state is one of the novel features of a system with balanced loss and gain. The phase-transitions between broken and unbroken $\mathcal{PT}$-phases, existence of quantum bound states in an unbroken $\mathcal{PT}$-phase, role of exceptional points etc. are some of the salient aspects of a system with balanced loss and gain[15].

The central focus of this article is to review known results on classical mechanical systems with balanced loss and gain for arbitrary number of particles and physically interesting potentials. The criteria for a mechanical system to be identified as a system with balanced loss and gain is discussed in Sec. 2. The Hamiltonian formulation for a generic system with balanced loss and gain is discussed in Sec. 3 along with a representation of the matrices appearing in the Hamiltonian. The loss-gain coefficient may be taken to be constant or space-dependent. It is shown that the loss-gain terms may be removed with its effect manifested in modifying the strength of the velocity-mediated coupling through appropriate co-ordinate transformations.

This result is independent of any particular representation of the matrices. The effect of the Lorentz interaction in improving the classical stability as well as defining the quantum problem on the real line is discussed. In Sec. 4, different exactly solvable models based on translational and rotational symmetry are discussed. It is also shown by constructing $m+1$ integrals of motion for a system with $N=2m$ particles that the system is at least partially integrable for $N>2$ and completely integrable for $N=2$. The developments in the field of $\mathcal{PT}$-symmetry played a significant role in interpreting a Hamiltonian of generalized Bateman oscillator as a system without any auxiliary degrees of freedom. The role of $\mathcal{PT}$-symmetry on the existence of periodic solution is critically analyzed in Sec. 5. A non-$\mathcal{PT}$-symmetric Hamiltonian system of coupled Duffing oscillators with balanced loss-gain and positional non-conservative forces is shown to admit regular as well as chaotic dynamics. The periodic solutions along with bifurcation diagrams and Lyapunov exponents are presented. Further, a non-$\mathcal{PT}$-symmetric non-Hamiltonian system of coupled Duffing oscillators with balanced loss and gain is also shown to admit periodic solution. Other examples of non-$\mathcal{PT}$-symmetric Hamiltonian system with balanced loss and gain which admit periodic solution include Landau Hamiltonian, a dimer model and nonlinear Schrödinger equation. It is conjectured that a system with balanced loss-gain and without any velocity mediated interaction may admit periodic solution if the linear part of the equations of motion is necessarily $\mathcal{PT}$-symmetric —the nonlinear interaction may or may not be $\mathcal{PT}$-symmetric. Further, systems with velocity mediated coupling among different degrees of freedom need not be $\mathcal{PT}$ symmetric at all in order to admit periodic solutions. The subject of system with balanced loss and gain has many facets which are not included in this review and a few important omitted topics are mentioned in Sec. 6. The nonlinear Schrödinger and Dirac equations with balanced loss and gain are reviewed in this section. Further, a class of solvable models of oligomers with balanced loss and gain is presented for the first time along with the previously known results. Finally, in Sec. 7, the results are summarized along with discussions.

## 2. Overviews on Systems with balanced loss and gain

The Bateman oscillator is described by the following set of equations:

$$\begin{aligned} &\ddot{x}+2\gamma\dot{x}+\omega^2 x=0\\ &\ddot{y}-2\gamma\dot{y}+\omega^2 y=0 \end{aligned} \tag{1}$$

where $\dot{x}=\frac{dx}{dt}$ and $\ddot{x}=\frac{d^2x}{dt^2}$. The $x$ and $y$ degrees of freedom are subjected to loss and gain, respectively for $\gamma>0$. The Bateman oscillator was originally introduced as a Hamiltonian system for a dissipative simple harmonic oscillator. The basic idea is to embed the original system described by the $x$ degree of freedom in a larger system with two degrees of freedom, where the additional $y$ degree of freedom defines an auxiliary system. The ambient space is defined by taking both the $x$ and $y$ degrees of freedom together, while $x$ degree of freedom alone describes the target space. The anti-damped oscillator being the time-reversed version of the damped oscillator and the vice verse, a Hamiltonian formulation is allowed in the ambient space. The Bateman oscillator has been studied for almost ninety years from different perspectives with the central theme being a Hamiltonian description for a dissipative oscillator. A major unwanted feature of Bateman oscillator is the presence of auxiliary degree of freedom. A few other alternative approaches to study the same problem with a non-standard Hamiltonian description and without any auxiliary degree of freedom have also been proposed —(i) time-dependent Hamiltonian[3, 21], (ii) a complex Lagrangian[22], and (iii) different Hamiltonians for different parameter regimes by using modified Prelle-Singer method[23]. All these methods have their own merits and demerits with a common goal —Hamiltonian description of a dissipative system. However, abandoning the idea of auxiliary system for the case of Bateman oscillator and reinterpreting it as conservative system with balanced loss and gain for which the ambient and the target spaces are the same with two degrees of freedom has not been explored until recently.

The Bateman oscillator is one of the simplest examples of a system with balanced loss and gain. The flow in the phase space preserves its volume due to the balanced loss and gain —any imbalance leads to either growing or decaying volume. The flow in the phase-space preserves its volume even if interaction terms are added by coupling the two degrees of freedom:

$$\begin{aligned} \ddot{x} + 2\gamma\dot{x} + \omega^2 x + G_1(x,y) &= 0 \\ \ddot{y} - 2\gamma\dot{y} + \omega^2 y + G_2(x,y) &= 0, \end{aligned} \tag{2}$$

where $G_1(x,y)$ and $G_2(x,y)$ are arbitrary functions. A different viewpoint may be considered where the mechanical model described by Eq. (2) with two degrees of freedom is taken as a whole system without any auxiliary degree of freedom. In other words, the target and the ambient spaces are taken to be the same. This is a paradigm shift in the interpretation of models with balanced loss and gain and leads to a new type of interacting system with important physical consequences. The modified system defined by Eq. (2) may admit periodic solutions for specific choices of $G_{1,2}(x,y)$ which is not possible for the Bateman oscillator, i.e. $G_1(x,y) = G_2(x,y) = 0$. For example, Eq. (2) with $G_1 = \epsilon y, G_2 = \epsilon x, \epsilon \in \Re$, represents a mathematical model[19] for an experimentally realized $\mathcal{PT}$ symmetric coupled resonators and admits periodic solutions within suitable ranges of the parameters[20]. Similarly, a two particle Calogero-type model with balanced loss and gain admitting periodic solutions may be reproduced for $G_1 = \alpha(x-y)^{-3}, G_2 = -\alpha(x-y)^{-3}, \alpha \in \Re$[24]. There are many examples of systems with balanced loss and gain admitting periodic solutions in certain regions of the parameters space[25, 26, 27, 18, 28, 29, 30, 31].

The concept of a system with balanced loss and gain may be generalized for arbitrary $N$ degrees of freedom. The system is governed by the equations of motion,

$$\ddot{X} - 2\mathcal{D}\dot{X} + G(x_1, x_2, \ldots, x_N) = 0, \tag{3}$$

where the co-ordinates of the $N$ particles are denoted as $x_1, x_2, \ldots, x_N$ which are elements of the column matrix $X$. In particular, $X^T \equiv (x_1, x_2, \ldots, x_N)$, where $X^T$ denotes the transpose of $X$. The term linear in $\dot{X}$ contains information on the nature of loss-gain as well as velocity mediated coupling among the particles depending on the explicit form of the $N \times N$ matrix $\mathcal{D}$. In general, $\mathcal{D}$ may be decomposed as,

$$\mathcal{D} = D + D_{SO} + \mathcal{D}_A \tag{4}$$

where $D$ is a diagonal matrix, $D_{SO}$ is a symmetric matrix with vanishing diagonal elements and $\mathcal{D}_A$ is an anti-symmetric matrix. The symmetric matrix $\mathcal{D}_S = D + D_{SO}$ is decomposed in terms of its diagonal and off-diagonal parts, $D$ and $D_{SO}$, respectively. The Lorentz interaction is described by $\mathcal{D}_A$, while velocity mediated non-Lorentzian interaction is encoded in $D_{SO}$. The Lorentz interaction appears in the description of many physical systems[32], while velocity mediated non-Lorentzian interaction appears in the synchronization of different types of oscillators[33]. The loss and gain in the system are introduced via the matrix $D$. In general, the matrix $\mathcal{D}$ may be space-dependent and thereby, allowing space-dependent loss-gain terms in the system. The space-mediated coupling among different degrees of freedom is encoded through the $N$-component field $G(x_1, x_2, \ldots, x_N)$ with its components $G_i \equiv G_i(x_1, x_2, \ldots, x_N), i = 1, 2, \ldots, N$.

The criteria for the system described by Eq. (3) to be non-dissipative may be determined by studying the time-evolution of the flow in the $2N$ dimensional position-velocity state space.

The flow preserves the volume in the position-velocity state-space for a non-dissipative system, while the volume either grows or decays depending on whether the system is anti-damped or damped, respectively. Introducing a $2N$-component vector $\xi$ and a field $\eta$,

$$\xi \equiv \begin{pmatrix} X \\ \dot{X} \end{pmatrix}, \eta_i \equiv \xi_{N+i}, \ \eta_{N+i} = -G_i(\xi_i, \ldots, \xi_N) + \sum_{k=1}^{N} \mathcal{D}_{ik}(\xi_i, \ldots, \xi_N)\xi_{N+k} \tag{5}$$

Eq. (3) may be rewritten in terms of $2N$ first-order coupled differential equations as $\dot{\xi} = \eta$. The flow preserves the volume $V = \prod_{i=1}^{2N} d\xi_i$ in the position velocity state space provided $\eta$ is solenoidal, i. e.

$$\sum_{i=1}^{2N} \frac{\partial \eta_i}{\partial \xi_i} = 0 \ \ \Rightarrow Tr(D) = 0, \tag{6}$$

where $Tr$ denotes trace of a matrix. The two dimensional examples in Eqs. (1) and (2), when cast into the form of Eq. (3), contain $\mathcal{D} = D = \gamma \ diag(-1, 1)$ which is indeed traceless. For a system with $N > 2$ degrees of freedom, the traceless condition may be implemented in several ways depending on the actual physical scenario. The vanishing trace of $D$ is the criteria for a model governed by Eq. (3) to be identified as a system with balanced loss and gain for which the flow in the position-velocity state space preserves the volume in spite of the fact that individual degrees of freedom are subjected to loss or gain.

## 3. Hamiltonian system

The Hamiltonian formulation of a system with balanced loss and gain is a non-trivial problem. For example, the Hamiltonian of two linearly coupled identical Duffing oscillators, one of which is subjected to gain and the other with an equal amount of loss is unknown. The same system with additional nonlinear interaction may be shown to be Hamiltonian[34]. It should be mentioned here that there are well known methods to construct Lagrangian and Hamiltonian from a given set of equations of motion. The inverse variational problem was initiated by Helmholtz[34]. However, a successful implementation of the scheme becomes nontrivial for many-particle systems with nonlinear interaction. Further, a system may not admit a Lagrangian-Hamiltonian formulation at all[35]. Within this background, the Hamiltonian formulation for a class of systems with balanced loss and gain is described below.

The Hamiltonian of the Bateman oscillator has the expression,

$$\begin{aligned} H_B &= P_x P_y + \gamma \left( y P_y - x P_x \right) + \left( \omega^2 - \gamma^2 \right) xy \\ P_x &= \dot{y} - \gamma y, \ P_y = \dot{x} + \gamma x, \end{aligned} \tag{7}$$

where the canonical conjugate momenta corresponding to $x$ and $y$ are denoted as $P_x$ and $P_y$, respectively. The Hamilton's equation of motion reproduces Eq. (1). The Hamiltonian $H_B$ can be rewritten in terms of the generalized momenta $\Pi_x$ and $\Pi_y$ as,

$$H_B = \Pi_x \Pi_y + \omega^2 xy, \ \ \Pi_x = P_x + \gamma y, \Pi_y = P_y - \gamma x. \tag{8}$$

The following points may be noted:

- The expressions for $\Pi_{x,y}$ are similar to that of the generalized momenta for a particle in an uniform external magnetic field with magnitude $\gamma$ and along the direction perpendicular to the $x - y$ plane. There is no magnetic field in the system. So, $A_x := \gamma y, A_y := -\gamma x$ may be interpreted as 'fictitious gauge potentials' leading to the uniform 'fictitious magnetic field' with magnitude $\gamma$.

- The quadratic term involving the generalized momenta is not positive-definite, $\Pi_x\Pi_y = \Pi_+^2 - \Pi_-^2$, $\Pi_\pm = \frac{1}{2}(\Pi_x \pm \Pi_y)$ and the system may be interpreted as defined in the background of a pseudo-Euclidean metric with signature $(1, -1)$.

These two features are present for known Hamiltonian systems with balanced loss and gain and will be used as essential inputs for constructing Hamiltonian for a generic many-particle system with balanced loss and gain.

The Hamiltonian for a general system with balanced loss and gain is taken to be of the form,

$$H = \Pi^T \mathcal{M} \Pi + V(x_1, x_2, \ldots, x_N), \tag{9}$$

where $\mathcal{M}$ is an $N \times N$ real symmetric matrix and $\Pi = (\pi_1, \pi_2, \ldots, \pi_N)^T$ denotes $N$ component generalized momenta. The matrix $\mathcal{M}$ is non-singular so that $\mathcal{M}^{-1}$ exists and is not necessarily semi-positive definite. The matrix $\mathcal{M}$ may be interpreted as a constant background metric as in the case of Bateman oscillator. A semi-positive definite $\mathcal{M}$, if exists, may be interpreted either as a metric or a mass-matrix. The generalized momenta $\Pi$ has the expression,

$$\Pi = P + AF(X), \tag{10}$$

where $P = (p_1, p_2, \ldots, p_N)^T$ is the conjugate momentum corresponding to the coordinate$X$, $F(X) = (F_1, F_2, \ldots F_N)^T$ is $N$ dimensional column matrix whose entries are functions of coordinates and $A$ is an $N \times N$ anti-symmetric matrix. It may be noted that $a_i = \sum_{k=1}^N A_{ik} F_k$ can be interpreted as gauge potentials. In general, $a_i$ may be decomposed as $a_i = a_i^R + a_i^F$ in terms of realistic gauge potentials $a_i^R$ and fictitious gauge potentials $a_i^F$. The realistic gauge potential describes external magnetic field in the system, while the fictitious gauge potential leads to loss-gain terms as in the case of Bateman oscillator. With suitable choices of $F$, space-dependent gain-loss terms are allowed in the system. The Hamiltonian $H$ in Eq. (9) reduces to the Hamiltonian of Ref. [19] describing coupled harmonic oscillators with balanced loss-gain for the following choices of $\mathcal{M}, A, F, V$ and identification $x_1 = x, x_2 = y, p_1 = P_x, p_2 = P_y$:

$$\mathcal{M} = \frac{1}{2}\sigma_1, \ A = i\gamma\sigma_2, \ F_1 = x, \ F_2 = y, \ V(x, y) = \omega^2 xy + \frac{\beta}{2}\left(x^2 + y^2\right) \tag{11}$$

where $\sigma_i, i = 1, 2, 3$ are the Pauli matrices. The Hamiltonian $H_B$ of the Bateman oscillator is reproduced for $\beta = 0$. The Lagrangian corresponding to the Hamiltonian $H$ in Eq. (9) may be derived as follows:

$$\mathcal{L} = \frac{1}{4}\dot{X}^T \mathcal{M}^{-1} \dot{X} - \frac{1}{2}(\dot{X}^T AF + F^T A^T \dot{X}) - V(x_1, x_2, \ldots, x_N). \tag{12}$$

The kinetic energy term in the Lagrangian is not necessarily positive-definite. The presence of terms linear in velocity is essential for incorporating loss-gain in the system. These are general characteristics of Bateman oscillator and its generalized versions.

The equations of motion following from the Hamiltonian (9) or the Lagrangian (12) reads,

$$\ddot{X} - 2\mathcal{M}R\dot{X} + 2\mathcal{M}\frac{\partial V}{\partial X} = 0, \tag{13}$$

where the anti-symmetric matrix $R$, the Jacobian $J$ and $\frac{\partial V}{\partial X}$ are defined as follows:

$$R \equiv AJ - (AJ)^T, \ [J]_{ij} \equiv \frac{\partial F_i}{\partial x_j}, \ \frac{\partial V}{\partial X} \equiv \left(\frac{\partial V}{\partial x_i}, \frac{\partial V}{\partial x_2}, \cdots \frac{\partial V}{\partial x_N}\right)^T. \tag{14}$$

The matrix $R$ is anti-symmetric by definition, while $J$ has no specified symmetry. However, $J$ can be made to be symmetric or anti-symmetric with suitable choices of the $N$-component field $F$. The equations of motion (13) reduces to Eq. (3) with the following identifications:

$$\mathcal{D} = \mathcal{M}R, \quad G = 2\mathcal{M}\frac{\partial V}{\partial X}. \tag{15}$$

The criteria $Tr(\mathcal{D}) = 0$ for a balanced loss and gain system is automatically satisfied, since $\mathcal{M}$ is symmetric and $R$ is anti-symmetric. The construction of Hamiltonian for a given Eq. (3) now reduces to finding solutions for the equations in Eq. (15). It is known[36] that a matrix $\mathcal{D}$ that is similar to $-\mathcal{D}$ can always be decomposed as a product of a symmetric and an anti-symmetric matrices. However, finding expressions for $A$ and $J$ from a known $R$ is a non-trivial problem, particularly for the case of space-dependent loss-gain terms. Further, the problem of finding $V$ for a given $G$ involves solving $N$ coupled first order differential equations $\frac{\partial V}{\partial X} = \frac{1}{2}\mathcal{M}^{-1}G$ which, in general, may elude a closed form expression for $V$. Nevertheless, the Hamiltonian of a large number of physical system with balanced loss and gain may be constructed via specific representation of matrices $\mathcal{M}$, $R$ and $\mathcal{D}$ and closed form expression for $V$[18, 29, 30].

The Hamiltonian for systems with space-independent loss and gain terms for $N = 2$ has been considered earlier[3, 19]. The general formalism described above may be used to construct Hamiltonian system with space-dependent loss-gain terms. For example, a coupled Van der Pol-Duffing oscillator model with balanced loss and gain may be obtained by choosing,

$$\mathcal{M} = \sigma_1 + \alpha^2 I_2, \ A = -\frac{i\gamma}{2}\sigma_2, \ F_i = a_i x_i + b_i x_i^3, \ V(x_1, x_2) = \frac{\omega^2}{2}x_1 x_2 + \frac{\beta}{4}\left(x_1^2 + x_2^2\right) + g x_1^3 x_2, \tag{16}$$

where $I_2$ is the $2 \times 2$ identity matrix and $\omega, \alpha, \beta, \gamma, a_i, b_i, g$ are real constants. The choice of the matrix $A$ and the field $F(x_1, x_2)$ uniquely fixes the generalized momenta $\Pi$, while that of the matrix $\mathcal{M}$ and the potential $V(x_1, x_2)$ completely specifies the Hamiltonian. The expressions for the matrices $R(x_1, x_2)$ and $\mathcal{D}(x_1, x_2)$ can be computed by using Eqs. (14) and (15), respectively,

$$R = -\frac{i\gamma}{2}\sigma_2 Q(x_1, x_2), \ \mathcal{D} = \frac{\gamma}{2}Q(x_1, x_2)\left(\sigma_3 - i\alpha^2\sigma_2\right), \ Q(x_1, x_2) \equiv a_1 + a_2 + 3\left(b_1 x_1^2 + b_2 x_2^2\right) \tag{17}$$

Eq. (12) describes a coupled Van der Pol-Duffing oscillator[38],

$$\begin{aligned} &\ddot{x}_1 - \gamma Q(x_1, x_2)\left(\dot{x}_1 - \alpha^2 \dot{x}_2\right) + \omega^2 x_1 + \beta x_2 + g x_1^3 = 0 \\ &\ddot{x}_2 + \gamma Q(x_1, x_2)\left(\dot{x}_2 - \alpha^2 \dot{x}_1\right) + \omega^2 x_2 + \beta x_1 + 3g x_1^2 x_2 = 0 \end{aligned} \tag{18}$$

where $\gamma Q$ is the space-dependent loss-gain coefficient, $\gamma\alpha^2 Q$ is the external magnetic field, $\omega$ is the angular frequency of the harmonic oscillator. The strengths of the space-mediated linear and non-linear couplings are $\beta$ and $g$, respectively. The above equation may be reduced to a few known models with appropriate choices of the parameters. For example, the Hamiltonian of coupled oscillators with constant balanced loss and gain[19] is obtained for $a_1 = a_2 = 1, b_1 = b_2 = \alpha = g = 0$. The Bateman oscillator is obtained if the linear coupling is switched off additionally by taking $\beta = 0$. The Hamiltonian formulation of the above system for $\beta = \alpha = g = 0$, $a_1 = a_2 = \frac{1}{2}$ and $b_1 \to -\frac{b_1}{3}, b_2 \to -\frac{b_2}{3}$ has been considered earlier[37] which describes coupled Van der Pol oscillators. The $x_1$ degree of freedom describes a Van der Pol-Duffing oscillator for $b_2 = \beta = \alpha = 0, a_1 = a_2 = \frac{1}{2}, b_1 \to -\frac{b_1}{3}$ and the $x_2$ degree of freedom is unidirectionally coupled to it. This also provides a Hamiltonian formulation for Van der Pol-Duffing oscillator in an ambient space of two dimensions, much akin to the case of Bateman oscillator. The Hamiltonian of Ref. [31] is obtained for $a_1 = a_2 = 1, b_1 = b_2 = \alpha = 0$. The system has rich dynamical properties for the generic values of the parameters[38].

Generalizations to $N = 3$ may be achieved in several ways depending on the particular physical contexts. For example, the non-vanishing elements of the $3 \times 3$ matrices $\mathcal{M}$ and $A$ may be chosen as $\mathcal{M}_{12} = \mathcal{M}_{21} = \mathcal{M}_{33} = 1$, $\mathcal{M}_{11} = \mathcal{M}_{22} = \alpha^2$, $A_{12} = -A_{21} = -\frac{\gamma}{2}$ along with $F_3 = 0$ and $F_1, F_2$ as given in Eq. (16). The $x_3$ degree of freedom is neither subjected to gain/loss nor it is coupled to $x_1$ and $x_2$ degrees of freedom via velocity mediated coupling, since all elements of $\mathcal{D}$ are zero except for $D_{11} = -D_{22} = \frac{\gamma}{2}Q(x_1, x_2)$. The coupling among $x_1, x_2$ and $x_3$ degrees of freedom can be incorporated via the potential $V \equiv V(x_1, x_2, x_3)$. A particular choice of $V$ and the resulting equation of motions are described below,

$$\begin{aligned}
&V(x_1,x_2,x_3) = \frac{\omega^2}{4}\left(2x_1x_2 + x_3^2\right) + \frac{\beta}{4}\left(x_1^2 + x_2^2 + 2x_1x_3 + 2x_2x_3\right) + \alpha x_1^3 x_2 + \frac{\delta}{8}x_3^4\\
&\ddot{x}_1 - \gamma Q(x_1,x_2)\left(\dot{x}_1 - \alpha^2\dot{x}_2\right) + \omega^2 x_1 + \beta\left(x_2 + x_3\right) + \alpha x_1^3 = 0\\
&\ddot{x}_2 + \gamma Q(x_1,x_2)\left(\dot{x}_2 - \alpha^2\dot{x}_1\right) + \omega^2 x_2 + \beta\left(x_1 + x_3\right) + 3\alpha x_1^2 x_2 = 0\\
&\ddot{x}_3 + \omega^2 x_3 + \beta\left(x_1 + x_2\right) + \delta x_3^3 = 0
\end{aligned} \tag{19}$$

where the undamped Duffing oscillator described by $x_3$ degree of freedom linearly couples with the $x_1, x_2$ degrees of freedom. Generalizations to $N > 3$ may be continued in a similar way and is not discussed. A particular representation of matrices for arbitrary $N$ is presented in Sec. 3.2 and several examples are discussed in Refs. [18, 24, 28, 29, 30, 31].

### *3.1. Hiding the loss-gain terms*

The Bateman oscillator can be rewritten as,

$$\ddot{z}_+ + 2\gamma\dot{z}_- + \omega^2 z_+ = 0, \ddot{z}_- + 2\gamma\dot{z}_+ + \omega^2 z_- = 0, \tag{20}$$

in a rotated co-ordinate system $z_\pm = \frac{1}{\sqrt{2}}(x \pm y)$. The loss-gain terms are absent in Eq. (20), since the equation of motion for $z_+$ does not contain a term $\dot{z}_+$ and similarly the equation of motion for $z_-$ does not contain a term $\dot{z}_-$. Further, the system may be interpreted as defined in the background of a pseudo-Euclidean metric with signature $(1, -1)$ and the particle is subjected to external magnetic field proportional to $\gamma$. This is a generic feature of Hamiltonian systems with balanced loss and gain. The loss-gain terms may always be hidden in a specific coordinate system. The real symmetric matrix $\mathcal{M}$ can be diagonalized by an orthogonal matrix $\hat{O}$, i.e. $M_D = \hat{O}^T \mathcal{M} \hat{O}$. Defining a rotated co-ordinate system $\tilde{X} = \hat{O}^T X$ and $\tilde{P} = \hat{O}^T P$, eq. (3) can be rewritten as,

$$\ddot{\tilde{X}} - 2\left(M_D \tilde{R}\right)\dot{\tilde{X}} + 2M_D\frac{\partial V}{\partial \tilde{X}} = 0, \ \tilde{R} = O^T R O \tag{21}$$

The matrix $\tilde{R}$ is anti-symmetric and each diagonal element of the matrix $M_D\tilde{R}$ is zero, i.e. $[M_D\tilde{R}]_{ii} = 0 \ \forall \ i$. The loss-gain terms are hidden in the co-ordinate system $\tilde{X}$ with the effect manifested by modifying the velocity mediated coupling terms. The general formalism may be exemplified with the Van der Pol-Duffing oscillator model described by Eqs. (16,17,18) for which the matrices $\hat{O}, M_D$ and $\tilde{R}$ have the expressions,

$$\hat{O} = \frac{1}{\sqrt{2}}\left(\sigma_1 + \sigma_3\right), \ M_D = \alpha^2 I_2 + \sigma_3, \ \tilde{R} = \frac{i\gamma}{2}Q\sigma_2, \Rightarrow M_D\tilde{R} = \frac{\gamma Q}{2}\left(\sigma_1 + i\alpha^2\sigma_2\right) \tag{22}$$

It is clearly seen that the diagonal elements of the matrix $M_D\tilde{R}$ are zero, i.e. $[M_D\tilde{R}]_{11} = [M_D\tilde{R}]_{22} = 0$. Further, the off-diagonal elements are $[M_D\tilde{R}]_{12} = \frac{\gamma Q}{2}\left(1 + \alpha^2\right)$, $[M_D\tilde{R}]_{21} = \frac{\gamma Q}{2}\left(1 - \alpha^2\right)$ which can not be interpreted as describing pure Lorentz force, since $[M_D\tilde{R}]_{12} \neq -[M_D\tilde{R}]_{21}$.

The system may be defined in the background of a pseudo-Euclidean metric through a canonical scale transformation. The matrix $M_D$ and two other matrices $S$ and $\eta^a$ are defined as follows:

$$[M_D]_{ij} = \delta_{ij}\,\mathrm{sgn}(\lambda_i)|\lambda_i|, \quad [S]_{ij} = \delta_{ij}\sqrt{|\lambda_i|}, \quad [\eta^{(a)}]_{ij} = \delta_{ij}\,\mathrm{sgn}(\lambda_i), \quad a = 1, 2, \ldots, N+1 \tag{23}$$

where $\lambda_i$'s are the eigenvalues of the matrix $\mathcal{M}$ and $\mathrm{sgn}(x)$ is the signum function. The parameter space of the system may be characterized in terms of at most $N+1$ distinct regions depending on the number of negative eigenvalues of the matrix $\mathcal{M}$. The superscript $a$ in $\eta^{(a)}$ identifies the Region-$a$ in the parameter space corresponding to $a-1$ negative eigenvalues of $\mathcal{M}$. The matrix $\eta^{(a)}$ is to be interpreted as the background metric for an effective description of the system defined by the Hamiltonian $H$ and equations of motion following from it in Eqs. (9) and (13), respectively. The canonical scale transformation is defined as follows:

$$\mathcal{X} = S^{-1}\tilde{X}, \quad \mathcal{P} = S\tilde{P}, \tag{24}$$

where $\mathcal{X} \equiv (\mathcal{X}_1, \mathcal{X}_2, \ldots, \mathcal{X}_N)^T$ and $\mathcal{P} \equiv (\mathcal{P}_1, \mathcal{P}_2, \ldots, \mathcal{P}_N)^T$. The purpose of the scale transformation is to normalize the eigenvalues of $M$ to $\pm 1$ i.e. $S^{-1}\hat{O}^T\mathcal{M}\hat{O}S^{-1} = \eta^{(a)}$. The Hamiltonian and the equations of motion resulting from the transformation in different regions have the forms:

$$\begin{aligned} &H^{(a)} = \hat{\Pi}^T\eta^{(a)}\hat{\Pi} + \mathcal{V}(\mathcal{X}_1, \mathcal{X}_2, \ldots, \mathcal{X}_N), \\ &\ddot{\mathcal{X}} - 2\eta^{(a)}\mathcal{R}\dot{\mathcal{X}} + 2\eta^{(a)}\left(\frac{\partial \mathcal{V}}{\partial \mathcal{X}}\right) = 0, \quad \mathcal{V}(\mathcal{X}_1, \mathcal{X}_2, \ldots, \mathcal{X}_N) = V(x_1, x_2, \ldots, x_N), \end{aligned} \tag{25}$$

where $\mathcal{R} = S\tilde{R}S$ and the transformed generalized momenta $\Pi$ after rotation and the scale transformation is denoted as, $\hat{\Pi} \equiv S\hat{O}^T\Pi = \mathcal{P} + \frac{\mathcal{R}}{2}\mathcal{X}$. The quadratic term in momenta for Hamiltonians $H^{(1)}$ and $-H^{(N+1)}$ are semi-positive definite, while $H^{(a)}, 2 \leq a \leq N$ are not definite. The loss-gain terms of Eq. (13) are absent in Eq. (25). It may be noted that Eq. (25) contains velocity mediated Lorentzian and non-Lorentzian interaction in Region-$a, 2 \leq a \leq N$, while only Lorentzian interaction is present in Region-1 and Region-N+1.

The formalism for an effective description of the system in the background of a pseudo-Euclidean metric may be exemplified with the Van der Pol-Duffing oscillator model described by Eqs. (16,17,18). The matrix $\mathcal{M}$ has eigenvalues $\lambda_1 = \alpha^2 + 1$ and $\lambda_2 = \alpha^2 - 1$. The parameter space may be divided into two regions: (i) Region-I: $\alpha^2 > 1$ with an Euclidean metric $\eta^{(1)} = I_2$, (ii) Region-II: $\alpha^2 < 1$ with a pseudo-Euclidean metric $\eta^{(2)} = \sigma_3$. The new co-ordinates $\mathcal{X}$, the matrix $\mathcal{R}$, the space-dependent co-efficient $\mathcal{Q}(\mathcal{X}_1, \mathcal{X}_2)$ and the potential $\mathcal{V}(\mathcal{X}_1, \mathcal{X}_2)$ have the expressions:

$$\begin{aligned} &\mathcal{X}_1 = \frac{1}{\sqrt{2|\lambda_1|}}(x_1 + x_2), \quad \mathcal{X}_2 = \frac{1}{\sqrt{2|\lambda_2|}}(x_1 - x_2), \quad \mathcal{R} = \frac{i\gamma}{2}\mathcal{Q}\lambda\sigma_2 \\ &\mathcal{Q}(\mathcal{X}_1, \mathcal{X}_2) = a_1 + a_2 + \frac{3(b_1 + b_2)}{2}\left(|\lambda_1|\mathcal{X}_1^2 + |\lambda_2|\mathcal{X}_2^2\right) + 3(b_1 - b_2)\lambda\mathcal{X}_1\mathcal{X}_2 \\ &\mathcal{V}(\mathcal{X}_1, \mathcal{X}_2) = \frac{\Omega_+}{4}\mathcal{X}_1^2 + \frac{g\lambda_1^2}{4}\mathcal{X}_1^4 - \frac{\Omega_-}{4}\mathcal{X}_2^2 - \frac{g\lambda_2^2}{4}\mathcal{X}_2^4 + \frac{g\lambda}{2}\mathcal{X}_1\mathcal{X}_2\left(|\lambda_1|\mathcal{X}_1^2 - |\lambda_2|\mathcal{X}_2^2\right) \end{aligned} \tag{26}$$

where $\lambda \equiv \sqrt{|\lambda_1||\lambda_2|}$, $\Omega_+ \equiv (\omega^2 + \beta)|\lambda_1|$ and $\Omega_- \equiv (\omega^2 - \beta)|\lambda_2|$.

- Region-I: The matrix $\mathcal{M}$ is positive-definite. The effective description of the system is in the background of a two dimensional Euclidean metric. The particle is subjected to Lorentz interaction and there is no other velocity-mediated interaction.

- Region-II: The matrix $\mathcal{M}$ is indefinite. The effective description of the system is in the background of a two dimensional pseudo-Euclidean metric with the signature $(1-1)$. The particle is subjected to non-Lorentzian velocity-mediated interaction.

Another simple example with two degrees of freedom is given in Sec. 3.2.1 which explains the general idea presented above.

### *3.2. Representation of matrices*

Several representations of the matrices for a vanishing Lorentz interaction and constant loss-gain terms are presented in Ref. [18]. A few representations for space-dependent loss-gain terms are included in Refs. [29] and [30] for vanishing and non-vanishing Lorentz interaction, respectively. A particular representation for $N = 2m, m \in \mathbb{Z}^{\geq}$ with pair-wise balancing of space-dependent loss-gain terms is discussed in this article. The matrix $\mathcal{M}$ is chosen as,

$$\mathcal{M} = M + \alpha^2 I_{2m}, \alpha \in \mathbb{R}, \tag{27}$$

where $M$ is a traceless $2m \times 2m$ symmetric matrix that anti-commutes with $R$, i.e. $\{M, R\} = 0$ and $I_{2m}$ is the $2m \times 2m$ identity matrix. The substitution of $\mathcal{M}$ in eq. (27) to the expression $\mathcal{D} = \mathcal{M}R$ gives,

$$\mathcal{D} = \underbrace{MR}_{\mathcal{D}_S} + \underbrace{\alpha^2 R}_{\mathcal{D}_A} \tag{28}$$

The relations $M^T = M, R^T = -R, \{M, R\} = 0$ ensure that $\mathcal{D}_S = MR$ is a symmetric matrix. The parameter $\alpha$ controls the strength of the Lorentz interaction in the system and $\alpha = 0$ corresponds to vanishing Lorentz interaction.

The $m$ functions $Q_i \equiv Q_i(x_{2i-1}, x_{2i})$ are introduced as,

$$Q_a(x_{2a-1}, x_{2a}) = Tr(V_a^{(2)}), \quad V_a^{(2)} \equiv \begin{pmatrix} \frac{\partial F_{2a-1}}{\partial x_{2a-1}} & \frac{\partial F_{2a-1}}{\partial x_{2a}} \\ \frac{\partial F_{2a}}{\partial x_{2a-1}} & \frac{\partial F_{2a}}{\partial x_{2a}} \end{pmatrix}. \tag{29}$$

The representation of the matrices is specified as follows:

$$M = I_m \otimes \sigma_1, \quad A = \frac{-i\gamma}{2} I_m \otimes \sigma_2, \quad \mathcal{D}_S = \gamma \chi_m \otimes \sigma_3, \quad [\chi_m]_{ij} = \frac{1}{2}\delta_{ij} Q_i(x_{2i-1}, x_{2i}), \tag{30}$$

where $I_m$ is $m \times m$ identity matrix. The matrix $R$ may be determined by noting that the matrix $J$ has a block-diagonal form for the choices of $F_i \equiv F_i(x_{2i-1}, x_{2i})$:

$$R = \frac{\gamma}{2} \sum_{i=1}^{m} U_i^{(m)} \otimes \begin{pmatrix} 0 & -Q_i(x_{2i-1}, x_{2i}) \\ Q_i(x_{2i-1}, x_{2i}) & 0 \end{pmatrix}, \quad [U_a^{(m)}]_{ij} = \delta_{ia}\delta_{ja}. \tag{31}$$

This completely specifies the representation of the matrices for pair-wise balancing of loss-gain terms and vanishing non-Lorentzian velocity mediated coupling, i.e. $\mathcal{D}_S = D, D_{SO} = 0$. A representation for the case $D_{SO} \neq 0$ may be found in Ref. [30].

*3.2.1. Effect of Lorentz interaction:* The Lorentz interaction in the system vanishes for $\alpha = 0$ for which $\mathcal{M} = M$ and $\mathcal{D} = \mathcal{D}_S$. It is known[18] that the matrices $M, R$ and $\mathcal{D}_S$ anti-commute with each other and, hence,

$$\{\mathcal{M}, R\} = 0, \ \{\mathcal{M}, \mathcal{D}_S\} = 0, \{\mathcal{D}_S, R\} = 0. \tag{32}$$

An immediate consequence is that the matrix $\mathcal{M} = M$ is indefinite and corresponding to each of its $m$ positive eigenvalues $\lambda_i$, there exists an eigenvalue $-\lambda_i$. The term $\Pi^T \mathcal{M} \Pi$ in the Hamiltonian is not semi-positive definite and this has important consequences for the classical as well as quantum systems.

- The Hamiltonian is not bounded from below even for a $V$ with a well-defined lower bound. Consequently, the question of stability of the system is much more involved compared to separable Hamiltonian. In general, the Lagrange-Dirichlet theorem[39, 1] for Hamiltonian system does not give any conclusive results on the stability of the solutions.
- The quantum problem is not well-defined on the real line and a well-defined ground state does not exist. However, if the quantum Hamiltonian is defined in appropriate Stokes wedges, the system may admit well defined bound states[18, 19, 24, 28, 29, 31]. This is consistent from the viewpoint of axiomatic foundations of quantum mechanics. However, no experimentally realizable system of this type has been found so far.

The Hamiltonian of the Bateman oscillator is unbounded from below. However, for specific choices of $V$, the Hamiltonian may be bounded from below and admit periodic solutions. The Hamiltonian of Ref. [19] and examples considered in Refs. [18, 24, 28, 29, 30, 31] are bounded from below/above within some regions in the parameter-space. These examples admit stable periodic solutions at the classical level and quantum bound states.

The Lorentz interaction is switched on for $\alpha \neq 0$ and introduces external magnetic field in the system. On the other hand, the loss-gain terms may be interpreted as 'fictitious magnetic field' arising due to a 'fictitious gauge potential' containing in the generalized momenta. The eigenvalues of $\mathcal{M}$ are $\alpha^2 \pm \lambda_i, i = 1, 2, \ldots, m$ semi-positive definite for $\alpha^2 \geq max_i \lambda_i$. The stability properties of a system with balanced loss and gain is improved in this region. The famous Landau problem[32] with balanced loss and gain has been studied in Ref. [30] with various interesting results. In particular, the representation of the matrices $\mathcal{M}, R, \mathcal{D}$ may be considered as follows:

$$\mathcal{M} = \frac{1}{2}\begin{pmatrix} B+C & \gamma \\ \gamma & B-C \end{pmatrix}, R = \begin{pmatrix} 0 & 1 \\ -1 & 0 \end{pmatrix}, \mathcal{D} = \frac{1}{2}\begin{pmatrix} -\gamma & B+C \\ -(B-C) & \gamma. \end{pmatrix} \tag{33}$$

for the Hamiltonian $H$ in Eq. (9) with $N = 2$ and $V = 0$. The system is described by the equations of motion,

$$\begin{aligned} \ddot{x}_1 + \gamma \dot{x}_1 - B\dot{x}_2 - C\dot{x}_2 = 0 \\ \ddot{x}_2 - \gamma \dot{x}_2 + B\dot{x}_1 - C\dot{x}_1 = 0 \end{aligned} \tag{34}$$

where $B$ is the external uniform magnetic field, $\gamma$ is the loss-gain parameter and $C$ is the strength of non-Lorentzian velocity mediated coupling which appears in the study of synchronization of coupled oscillators[33]. The eigenvalues of the matrix $\mathcal{M}$ are,

$$\lambda_{\pm} = \frac{1}{2}(B \pm \triangle), \quad \triangle \equiv \sqrt{C^2 + \gamma^2}. \tag{35}$$

The condition for a positive-definite $\mathcal{M}$ for $C = 0$ may be interpreted as the situation in which the magnitude of the external magnetic field $B$ supersedes the magnitude of the 'fictitious magnetic field' $\gamma$, i.e. $|B| > |\gamma|$. The matrix $\mathcal{M}$ is singular for $B - \triangle$ and the parameter-space may be divided into three disjoint sectors:
(i) Region-I ( $B > \triangle$ ): The Hamiltonian is bounded from below and the system admits periodic solutions with a reduced cyclotron frequency $\omega = \sqrt{B^2 - \triangle^2}$ compared to the standard Landau system with $\omega_L = B$.
(ii) Region-II ( $-\triangle < B < \triangle$): The solutions are unbounded.
(iii) Region-III ( $B < -\triangle$): The Hamiltonian is bounded from above and in fact, it is identical with the Hamiltonian in Region-I except for an overall multiplication factor of $-1$. The system admits periodic solutions.
The system does not admit any periodic solution for a vanishing Lorentz interaction, while

periodic solutions are allowed for $B \neq 0$. The quantum Hamiltonian admits bound states in Region-I and III. Moreover, the quantum bound states are well defined on the real line without the need of defining the problem on any suitable Stokes wedges. The same feature persists for other Hamiltonian system with balanced loss and gain. The details of classical and quantum solutions of the Landau Hamiltonian with balanced loss and gain along with Hall effect and underlying supersymmetry is described in Ref. [30].

A few comments are in order before the end of the section.

- The Bateman oscillator with a modified kinetic energy term has been considered in Ref. [12] in a different context. The system, when considered on a commutative space and cast into the notation of the present article, describes the model of Ref. [19] with Lorentz interaction. It can be shown that classical periodic solutions are obtained for an extended region of parameter-space compared to that of Ref. [19]. Further, the quantum problem is well defined on the real line.
- The gyroscopic force, which has the same form as Lorentz force, has been used for long to control dissipative induce instabilities in classical systems and there are plenty of important theorems and results[1]. The relevant results as appropriate to a system with balanced loss and gain may be used successfully.

The inclusion of Lorentz interaction in order to stabilize a classical system with balanced loss and gain as well as well as to define the quantum problem on the real line may be utilized for practical applications.

## 4. Solvable Models

Several solvable models with balanced loss and gain have been considered for $N \geq 2$ degrees of freedom in Refs. [19, 24, 25, 18, 28, 29, 30]. A few solvable models are discussed below to highlight the general features of these solvable models. The case of uniform external magnetic field and constant loss-gain terms has been discussed along with its physical relevance in Sec. 3.2.1. The space-dependent loss-gain terms make the equations highly nonlinear and in general, solvable models are rare. Nevertheless, a few solvable models with space-dependent loss-gain terms are known[29] which will not be discussed in this article. A few solvable models with vanishing Lorentz interaction and constant loss-gain terms are presented below, the details of which may be found in Ref. [18]. This corresponds to $\alpha = 0$ and $F_i = x_i$ in the representation of matrices in Sec. 3.2. The matrices have the following expressions,

$$M = I_m \otimes \sigma_1, \ R = -i\gamma I_m \otimes \sigma_2, \ D_S = \gamma I_m \otimes \sigma_3 \tag{36}$$

and $\hat{O} = \frac{1}{\sqrt{2}} \left[ I_m \otimes (\sigma_1 + \sigma_3) \right]$ diagonalizes $M$ to $M_D = I_m \otimes \sigma_3$. The orthogonal transformation generates the following set of new co-ordinates:

$$z_i^- = \frac{1}{\sqrt{2}} \left( x_{2i-1} - x_{2i} \right), \ z_i^+ = \frac{1}{\sqrt{2}} \left( x_{2i-1} + x_{2i} \right), \ i = 1, 2, \ldots m \tag{37}$$

The Eqs. of motion reads,

$$\ddot{z}_i^+ - 2\gamma \dot{z}_i^- + 2\frac{\partial V}{\partial z_i^+} = 0, \ \ddot{z}_i^- - 2\gamma \dot{z}_i^+ - 2\frac{\partial V}{\partial z_i^-} = 0, \ i = 1, 2, \ldots m. \tag{38}$$

In the subsequent discussions $2V$ is replaced by $V$. The Eqs. (38) is a set of $2m$ coupled differential equations which take simple form for (i) translational and (ii) rotational invariance of the system and solvable for specific choices of potential.

*4.1. Translational invariance*

The choice of the potential $V \equiv V(z_i^-)$ or $V \equiv V(z_i^+)$ allows a decoupling of Eq. (38). The potential $V(z_i^-)$ remains invariant under the translations $x_{2i-1} \rightarrow x_{2i-1} + \eta_i, x_{2i} \rightarrow x_{2i} + \eta_i$, where $\eta_i$'s are $m$ independent parameters. The form of the potential is special in the sense that it allows $m$ independent parameters $\eta_i$ instead of a single one. The translational invariance leads to $m$ integrals of motion $\Pi_i$ which are in involution:

$$\Pi_i = 2P_{z_i^+} - \gamma z_i^-, \ \{\Pi_i, \Pi_j\}_{PB} = 0, \quad \{H, \Pi_i\}_{PB} = 0, \tag{39}$$

where $\{,\}_{PB}$ denotes the Poisson bracket and $P_{z_i^+}$ is the conjugate momenta corresponding to $z_i^+$. The Hamiltonian along with $\Pi_i$'s constitute $m+1$ integrals of motion for a system with $N = 2m$ particles, implying that the system is at least partially integrable for $N > 2$ and completely integrable for $N = 2$. A similar analysis can be performed for $V(z_i^+)$. The discussion in this article is restricted to the case $V \equiv V(z_i^-)$.

With the introduction of new co-ordinates $z_i = z_i^- + \frac{\Pi_i}{2\gamma}$, Eq. (38) can be rewritten as,

$$\begin{aligned} &\ddot{z}_i - 4\gamma^2 z_i - \frac{\partial V(z_i)}{\partial z_i} = 0, \\ &z_i^+(t) = 2\gamma \int z_i(t) dt + C_i, \quad i = 1, 2, \ldots m, \end{aligned} \tag{40}$$

where $C_i$ are $m$ integration constants. The potential depends on the gain-loss parameter via $z_i$ which may be avoided by choosing the constants of motion $\Pi_i = 0 \ \forall \ i$ leading to the initial conditions $\dot{z}_i^+(0) = 2\gamma z_i^-(0) \ \forall \ i$, where $\dot{z}_i^+(0)$ and $z_i^-(0)$ can be chosen depending on the physical requirements. The problem now lies to find $V$ for which the first equation of Eq. (40) is exactly solvable. Several examples involving coupled chain of cubic oscillators, potential solely dependent on the radial variable $r = \sum_{i=1}^m z_i^2$ in the sub-system defined by the co-ordinates $(z_1, z_2, \ldots, z_m)$, Calogero-type inverse square interaction, and Henon-Heils potential are considered in Ref. [18]. The case of cubic oscillator is discussed below.

The choice of $V$ for the simplest case of $N = 2, m = 1$ which produces a solvable system is the following:

$$\begin{aligned} &V(z_1) = -2\omega_0^2 z_1^2 - \frac{\alpha}{4} z_1^4, \ \omega, \alpha \in \Re, \\ &\ddot{z}_1 + \omega^2 z_1 + \alpha z_1^3 = 0, \ \omega^2 \equiv 4(\omega_0^2 - \gamma^2). \end{aligned} \tag{41}$$

There are three distinct regions in the parameter-space for which non-singular stable solutions can be obtained analytically in terms of Jacobi Elliptic functions —Region-I: $\omega^2 > 0, \alpha > 0$, Region-II: $\omega^2 > 0, \alpha < 0$ and Region-III: $\omega^2 < 0, \alpha > 0$. The stability requires $-\omega_0 < \gamma < \omega_0$ for region-I and Region-II, while $-\gamma < \omega_0 < \gamma$ in region-III. The nonlinear interaction allows $\gamma > \omega_0$ which has not been seen for system with linear interaction. There is an additional constraint in each region involving the amplitude and frequency of the solution, and the nonlinear coupling $\alpha$ for the existence of non-singular stable solution. For example, the solution in Region-I,

$$\begin{aligned} &z_1(t) = A \ cn(\Omega t, k), \ \ z_1^+(t) = \frac{2\gamma}{\Omega} \frac{\cos^{-1}\{dn(\Omega t, k)\} sn(\Omega t, k)}{\sqrt{1 - dn^2(\Omega t, k)}} \\ &\Omega = \sqrt{\omega^2 + \alpha A^2}, \ k^2 = \frac{\alpha A^2}{2\Omega^2}, \end{aligned} \tag{42}$$

is non-singular and stable for $-\omega_0 < \gamma < \omega_0$ and $0 < k < 1$. The solutions in other regions are given in Ref. [18]. The solvable models for higher $N$ can be constructed by using the known results of coupled cubic oscillators[40]. The case of $N = 4, m = 2$ along with its exact, non-singulars and stable solutions are also discussed in Ref. [18].

### *4.2. Rotational Invariance*

The Bateman oscillator admits a constant of motion in addition to the Hamiltonian $H_B$,

$$L_B = x\dot{y} - y\dot{x} + 2\gamma xy \tag{43}$$

thereby making it an integrable system. Their exist $m$ constants of motion similar to $L_B$ for a system with $N = 2m$ particles with the matrices $\mathcal{M}, R, D_S$ given in Eq. (36) and potential of specified form. In particular, $\hat{L} = \frac{1}{2}\left(P^T D_S X + X^T D_S P\right)$ is a constant of motion for potential satisfying the condition $X^T D_S \frac{\partial V}{\partial X} = 0$ which may be solved with the ansatz,

$$V \equiv V(\tilde{r}),\ \tilde{r}^2 \equiv X^T G X, \{G, D_s\} = 0 \tag{44}$$

where $G$ is a symmetric matrix and $D_S$ need not be restricted to any specific representation. For $N = 2$ and with the representation of matrices given by Eq. (36), $G$ is uniquely determined as $G = M = \sigma_1$. However, for $N > 2$, the choice of $G$ is not unique and there are several possibilities[18]. The variable $\tilde{r}$ takes the form

$$\tilde{r}^2 = \sum_{i=1}^{m} x_{2i-1}x_{2i} = \sum_{i=1}^{m}\left[\left(z_i^+\right)^2 - \left(z_i^-\right)^2\right], \tag{45}$$

for $G = M^{-1}$ and representation of matrices given by Eq. (36). It may be noted that $\tilde{r}$ has the interpretation of the radial variable in a pseudo-Euclidean co-ordinate with the signature of the metric as $(1, -1, 1, -1, \dots, 1, -1)$. The $m$ constants of motion in addition to the Hamiltonian have the expressions,

$$\hat{L}_i = \dot{z}_i^+ z_i^- - z_i^+ \dot{z}_i^- + \gamma\left[(z_i^+)^2 - (z_i^-)^2\right],\ i = 1, 2, \dots, m, \tag{46}$$

which are in involution,

$$\{\hat{L}_i, \hat{L}_j\}_{PB} = 0,\ \{H, \hat{L}_i\}_{PB} = 0, \tag{47}$$

implying that the system is at least partially integrable for $N > 2$ and integrable for $N = 2$ for a system of $N = 2m$ particles. The values of all the constants of motion are zero, i.e. $\hat{L}_i = 0\ \forall\ i$ and $r^2 = \sum_{i=1}^{m} q_i^2 \equiv q^2$ for the following parameterizations of the co-ordinates,

$$z_i^+(t) = q_i(t)\cosh(\gamma t),\ z_i^-(t) = q_i(t)\sinh(\gamma t) \tag{48}$$

The equations of motion can be expressed solely in terms of $q_i$,

$$\ddot{q}_i - \gamma^2 q_i + \frac{1}{q}\frac{\partial V(q)}{\partial q} q_i = 0, i = 1, 2, \dots, m, \tag{49}$$

and exactly solvable models may be constructed for suitable choices of $V(q)$. For example, $V(q) = \frac{1}{2}\omega^2 q^2 + \frac{\alpha}{4}q^4$ leads to an exactly solvable equation[40],

$$\ddot{q}_i + \Omega^2 q_i + \alpha q^2 q_i = 0,\ \Omega^2 = \omega^2 - \gamma^2. \tag{50}$$

The Bateman oscillator is reproduced for $\alpha = 0, N = 2$ for which the two modes are not coupled. The nonlinear interaction couples all the modes and the corresponding Hamiltonian is not separable like Bateman oscillator. Although the above equation (50) admits non-singular stable solutions, when expressed in terms of the original variables $x_i$, there are growing as well as decaying modes —$x_{2i}$ are always decaying in time, while that of $x_{2i-1}$ grows with time. The situation can not be saved by allowing $\Omega$ to be complex, i.e $|\gamma| > |\omega|$, since the decay and growth of the solutions are of the form $e^{\pm\gamma t}$ which is independent of $\Omega$. The volume of the flow in the phase-space is always preserved. The same behaviour has been seen for the Bateman oscillator, where the two modes are decoupled and there is no nonlinear interaction. However, inclusion of coupling among different co-ordinates via nonlinear interaction does not change the situation for $\hat{L}_i = 0$. The solutions for $\hat{L}_i \neq 0$ may give stable solutions. The addition of a term that breaks the rotational invariance may also allow stable solutions. Different quantization schemes as applied to Bateman oscillator have revealed a host of intricate issues[4, 5, 6, 7, 8, 9, 10, 11, 12], the quantization of the quartic oscillator corresponding to the $V(q)$ deserves further attention.

## 5. Role of $\mathcal{PT}$-symmetry

The discrete symmetries like parity($\mathcal{P}$), time-reversal($\mathcal{T}$) and charge-conjugation($\mathcal{C}$) play an important role in physics. The $\mathcal{PT}$-symmetric quantum mechanics is one such area where non-relativistic non-hermitian Hamiltonian with $\mathcal{PT}$ symmetry admit entirely real spectra in the unbroken $\mathcal{PT}$-regime[15]. In general, the eigen-states are not orthogonal and the unitary time-evolution is not possible. However, with a modified $\mathcal{CPT}$ norm, a consistent quantum description with unitary time-evolution is allowed[15]. There is no concept of anti-particles in non-relativistic quantum mechanics. The charge conjugation operator $\mathcal{C}$ is introduced with the expectation that at a more fundamental level it will be related to the $\mathcal{CPT}$ invariance of any Lorentz invariant local quantum field theory. It may be noted that the $\mathcal{CPT}$ theorem[42], which was originally derived for a hermitian Hamiltonian, has now been extended to the case of non-hermitian Hamiltonian[43]. There is no signature of violation of Lorentz invariance in nature so far and the $\mathcal{CPT}$ is considered to be a fundamental symmetry of nature.

The studies on $\mathcal{PT}$-symmetric systems have been diversified to many areas, including the obvious choice of $\mathcal{PT}$-symmetric classical systems. If the operator $\mathcal{C}$ of $\mathcal{CPT}$-norm of the $\mathcal{PT}$-symmetric quantum mechanics is expected to be related to the charge-conjugation operator of Lorentz invariant local quantum field theory, then the parity and the time-reversal operation for the *classical non-relativistic $\mathcal{PT}$-symmetric system* should necessarily be described by a linear transformation, since the Lorentz transformation itself is linear. Within this background, the time-reversal transformation $\mathcal{T}$ and the most general parity transformation $\mathcal{P}$ in two space dimensions may be defined as,

$$\begin{aligned} &\mathcal{T}: t \rightarrow -t,\ \tilde{P}_x \rightarrow -\tilde{P}_x,\ \tilde{P}_y \rightarrow -\tilde{P}_y \\ &\mathcal{P}: \begin{pmatrix} x \\ y \end{pmatrix} \rightarrow \begin{pmatrix} x' \\ y' \end{pmatrix} = \begin{pmatrix} x\cos\theta + y\sin\theta \\ x\sin\theta - y\cos\theta \end{pmatrix},\ \begin{pmatrix} \tilde{P}_x \\ \tilde{P}_y \end{pmatrix} \rightarrow \begin{pmatrix} \tilde{P}'_x \\ \tilde{P}'_y \end{pmatrix} = \begin{pmatrix} \tilde{P}_x\cos\theta + \tilde{P}_y\sin\theta \\ \tilde{P}_x\sin\theta - \tilde{P}_y\cos\theta \end{pmatrix} \end{aligned} \quad (51)$$

where $\theta \in (0, 2\pi)$. The Bateman oscillator in Eq. (1) is invariant under $\mathcal{PT}$ symmetry for two values of $\theta$, namely $\theta = \frac{\pi}{2}, \frac{3\pi}{2}$. However, the classical solutions are not invariant under $\mathcal{PT}$ symmetry, i.e. $\mathcal{PT}X \neq \pm X$ in any region of the parameter-space. Thus, there is no $\mathcal{PT}$-symmetric phase for the Bateman oscillator. Consider Eq. (2) with $G_1(x, y) = \epsilon y, G_2(x, y) = \epsilon x$ which describes the model considered in [19]. The system is $\mathcal{PT}$-symmetric and admits periodic solutions in certain regions in the parameter-space. There are both $\mathcal{PT}$-broken and $\mathcal{PT}$-unbroken regions for this model. The existence of periodic solutions in the $\mathcal{PT}$-unbroken phase lead further investigations on $\mathcal{PT}$-symmetric system with balanced loss and gain in various directions[24, 25, 26, 27, 18, 28, 29, 30] by including nonlinear interaction[24, 26, 27, 18, 29], many-particle system[25, 18, 28], Lorentz interaction[30], etc. The general result common to all these systems is that periodic solutions are obtained in the unbroken $\mathcal{PT}$ phase of a $\mathcal{PT}$-symmetric system.

It has been shown recently that non-$\mathcal{PT}$ symmetric classical system may also admit periodic solutions in a coupled Duffing-oscillator with balanced loss and gain[31]. This is a new result with far reaching consequences. Many non-$\mathcal{PT}$ symmetric system with balanced loss and gain may be included in the mainstream investigations which may exhibit novel features including existence of periodic solutions. It should be mentioned here that there exist many other non-$\mathcal{PT}$-symmetric systems with balanced loss and gain which admit periodic solutions in certain regions of the parameter space[38]. The examples include mechanical systems with finite degrees of freedom, models of dimer, non-relativistic field theory etc., which can be broadly classified into Hamiltonian and non-Hamiltonian systems. A few examples are described below.

### *5.1. Hamiltonian System*

*5.1.1. A coupled Duffing Oscillator model:* The equations of motion of the system are,

$$\begin{aligned} &\ddot{x} + 2\gamma\dot{x} + \omega^2 x + \beta_1 y + g x^3 = 0, \\ &\ddot{y} - 2\gamma\dot{y} + \omega^2 y + \beta_2 x + 3g x^2 y = 0 \end{aligned} \tag{52}$$

The linear coupling between the two modes are asymmetric for $\beta_1 \neq \beta_2$. The system reduces to the one considered in Ref. [19] for $g = 0, \beta_1 = \beta_2$. The equation of motion for $x$ degree of freedom decouples for $\beta_1 = 0$ and describes a damped Duffing oscillator, while the $y$ degree of freedom is unidirectionally coupled to $x$. There is no explicit forcing term. However, the term $\beta_1 y$ in the first equation of (52) acts as a driving term for the damped Duffing oscillator in a non-trivial way. The system also admits a Hamiltonian,

$$H_D = P_x P_y + \gamma\left(y P_y - x P_x\right) + \left(\omega^2 - \gamma^2\right) xy + \frac{1}{2}\left(\beta_2 x^2 + \beta_1 y^2\right) + g x^3 y, \tag{53}$$

where $P_x = \dot{y} - \gamma y$ and $P_y = \dot{x} + \gamma x$ are canonical momenta. In general, the system is non-$\mathcal{PT}$-symmetric for the $\mathcal{P}$ and $\mathcal{T}$ defined in Eq. (51). The first, second and the third terms in $H_D$ are invariant under $\mathcal{PT}$ only for $\theta = \frac{\pi}{2}, \frac{3\pi}{2}$. The fourth and the fifth terms representing linear asymmetric coupling and nonlinear interaction, respectively, are not invariant under $\mathcal{PT}$ symmetry. It can been shown that for vanishing nonlinear coupling, i.e. $g = 0$, the corresponding non-$\mathcal{PT}$-symmetric system with asymmetric linear coupling does not admit any periodic solutions. This is consistent with the general folklore that a system with balanced loss and gain admits periodic solution only for a $\mathcal{PT}$-symmetric system. However, although the system is non-$\mathcal{PT}$ symmetric for $g \neq 0, \beta_1 = \beta_2$, it admits periodic solutions[31] in some regions in the parameter-space.

The independent scales in the system may be fixed by employing the following transformations,

$$t \rightarrow \omega^{-1} t,\ x \rightarrow |\beta_2|^{-\frac{1}{2}} x,\ y \rightarrow |\beta_1|^{-\frac{1}{2}} y,\ \beta_1 \neq 0, \beta_2 \neq 0, \tag{54}$$

which allows a reduction in total number of independent parameters which is convenient for analyzing the system. The model can be described in terms of three independent parameters $\Gamma, \beta$ and $\alpha$ defined as,

$$\Gamma = \frac{\gamma}{\omega}, \beta = \frac{\sqrt{|\beta_1||\beta_2|}}{\omega^2}, \alpha = \frac{g}{|\beta_2|\omega^2}, \tag{55}$$

and the equations of motion have the following expressions:

$$\begin{aligned} &\ddot{x} + 2\Gamma\dot{x} + x + \mathrm{sgn}(\beta_1)\ \beta y + \alpha x^3 = 0, \\ &\ddot{y} - 2\Gamma\dot{y} + y + \mathrm{sgn}(\beta_2)\ \beta x + 3\alpha x^2 y = 0. \end{aligned} \tag{56}$$

The limit to the linear system $g \rightarrow 0$ now corresponds to $\alpha \rightarrow 0$. The system admits regular periodic as well as chaotic solutions for the case $\mathrm{sgn}(\beta_1)\,\mathrm{sgn}(\beta_2) = 1$ and the discussion is restricted to $\mathrm{sgn}(\beta_i) = 1, i = 1, 2$. The results for $\mathrm{sgn}(\beta_i) = -1$ can be obtained from that of results for $\mathrm{sgn}(\beta_i) = 1$ by simply allowing $\beta \rightarrow -\beta$.

The system admits five equilibrium points $P_0, P_1^{\pm}, P_2^{\pm}$ in the phase-space $(x, y, \tilde{P}_x, \tilde{P}_y)$ of the system, where $\tilde{P}_x = \dot{y} - \Gamma y$, $\tilde{P}_y = \dot{x} + \Gamma x$. The equilibrium points are,

$$P_0 = (0,0,0,0), P_1^{\pm} = (\pm\delta_+, \pm\eta_+, \mp\Gamma\eta_+, \pm\Gamma\delta_+), P_2^{\pm} = (\pm\delta_-, \pm\eta_-, \mp\Gamma\eta_-, \pm\Gamma\delta_-), \quad (57)$$

where $\delta_{\pm}$ and $\eta_{\pm}$ are defined as follows:

$$\delta_{\pm} = \frac{1}{\sqrt{3\alpha}}\left[-2 \pm \sqrt{1+3\beta^2}\right]^{\frac{1}{2}}, \quad \eta_{\pm} = -\frac{\delta_{\pm}}{3\beta}\left[1 \pm \sqrt{1+3\beta^2}\right]. \quad (58)$$

The application of Dirichlet theorem[39] is inconclusive. The linear stability analysis shows that the points $P_0$ and $P_1^{\pm}$ are stable in the following regions of the parameter space:

$$P_0 : -\frac{1}{\sqrt{2}} < \Gamma < \frac{1}{\sqrt{2}}, \quad 4\Gamma^2\left(1-\Gamma^2\right) < \beta^2 < 1$$
$$P_1^{\pm} : \beta^2 > 1, \quad \Gamma^2 \leq \frac{\sqrt{2}-1}{2}. \quad (59)$$

The points $P_2^{\pm}$ are not stable anywhere in the parameter-space. These results are confirmed by perturbative and numerical analysis[31] and the system admits periodic solutions around the equilibrium points. Regular solutions of Eq. (56) in the vicinity of the point $P_0$ and $P_1$ are plotted in Figs. 1 and 2, respectively for $\alpha = 1, \beta = .5, \Gamma = .2$.

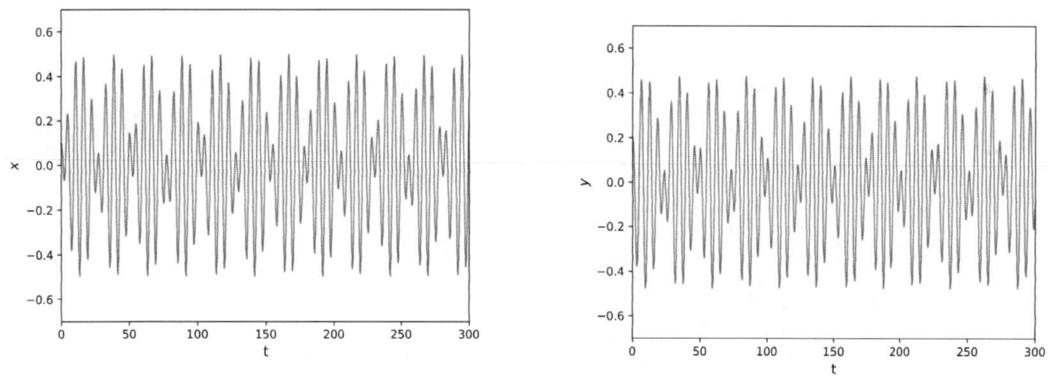


**Figure 1.** (Color online) Regular solutions of Eq. (56) in the vicinity of the point $P_0$ with the initial conditions $x(0) = .1, y(0) = 0.2, \dot{x}(0) = .03, \dot{y}(0) = .04$ and $\alpha = 1, \beta = .5, \Gamma = .2$. (*Reproduced from Ref. [31]*)

The bifurcation diagram is plotted in Fig. 3 for varying $\beta$ and $\Gamma = 0.01, \alpha = .5$. It can be seen that the chaotic regime begins for $\beta > \beta_c \equiv 1.05$. In general, the chaotic regime starts beyond $|\beta| > 1$ for a range of values of $\alpha$ and $\Gamma$. The existence of chaotic dynamics is confirmed numerically through various means —sensitivity of time-series to the initial conditions, Poincaré section, power-spectra, auto-correlation functions and computation of Lyapunov exponents. The details of the numerical investigations are given in Ref. [31], only the plots of the Lyapunov exponents for the initial condition $x(0) = .01, y(0) = .02, \dot{x}(0) = .03, \dot{y}(0) = .04$ are shown in

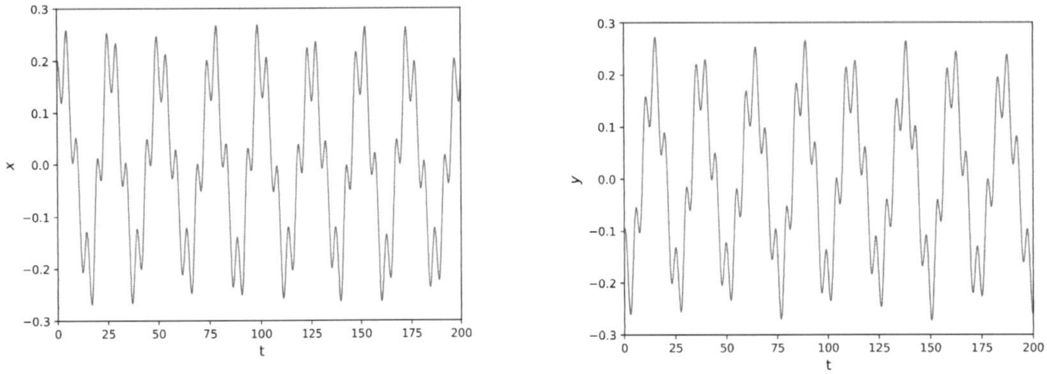


**Figure 2.** (Color online) Regular solutions of Eq. (56) in the vicinity of the point $P_1^+$ with the initial conditions $x(0) = .2, y(0) = -0.1, \dot{x}(0) = .02$, $\dot{y}(0) = .03$ and $\alpha = 1, \beta = 1.01, \Gamma = .3$.(*Reproduced from Ref. [31]*)

the left panel of the Fig. 4 which have the values $(.13248, .0015691, -.0016145, -.13244)$. The standard result that the sum of the Lyapunov exponents are zero for a Hamiltonian system may be verified within the numerical approximations by taking the values of the Lyapunov exponents up to the third decimal places with an error of the order of $10^{-4}$.

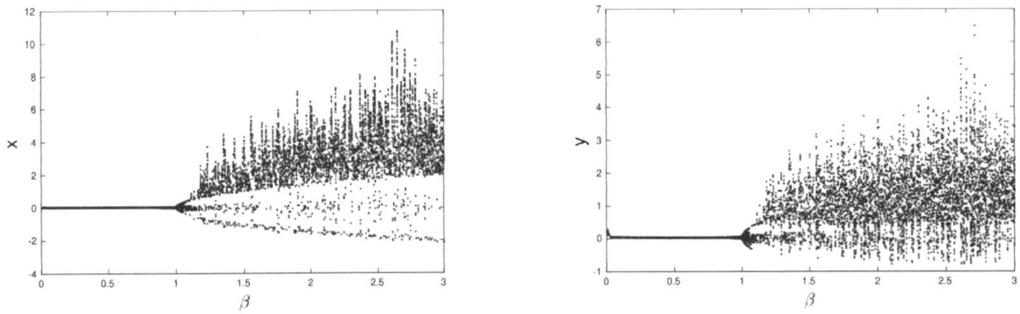


**Figure 3.** (Color online) Bifurcation diagrams for $\beta$ with $\Gamma = 0.01$ and $\alpha = .5$ with the initial conditions $x(0) = 0.01$, $y(0) = .02$, $\dot{x}(0) = .03$, $\dot{y}(0) = .04$.(*Reproduced from Ref. [31]*)

*5.1.2. Non-$\mathcal{PT}$-symmetric Positional Non-conservative force* The Hamiltonian $H_D$ shows chaotic behaviour even for $\Gamma = 0$, i.e. the system without the vanishing loss-gain terms, thereby providing an example of Hamiltonian chaos for two undamped Duffing oscillators coupled to each other in a specific way. In particular, the Hamiltonian $H_D$ can be rewritten for $\Gamma = 0$ as,

$$H_D = \left(\frac{1}{2}P_u^2 + \frac{\Omega_+}{2}u^2 + \frac{\alpha}{4}u^4\right) - \left(\frac{1}{2}P_v^2 + \frac{\Omega_-}{2}v^2 + \frac{\alpha}{4}v^4\right) + \frac{\alpha\ uv}{2}\left(u^2 - v^2\right), \Omega_\pm = 1 \pm \beta, \quad (60)$$

where the new co-ordinates and momenta are defined as,

$$u = \frac{x+y}{\sqrt{2}},\ v = \frac{x-y}{\sqrt{2}},\ P_u = \frac{\tilde{P}_x + \tilde{P}_y}{\sqrt{2}},\ P_v = \frac{\tilde{P}_x - \tilde{P}_y}{\sqrt{2}}. \quad (61)$$

This particular representation of $H_D$ describes two undamped Duffing oscillators, with different angular frequencies corresponding to the harmonic terms and identical nonlinear terms, coupled to each other through specified nonlinear interaction. The Lyapunov exponents are given in the the right panel of the Fig. 4 and have the values $(0.22685, 0.00431, -0.00431, -0.22685)$. It may be noted that the highest Lyapunov exponent for $\Gamma = 0$ is greater than the highest Lyapunov exponent for $\Gamma = .01$ with all other conditions remaining the same. The numerical results are described in detail in Ref. [31].

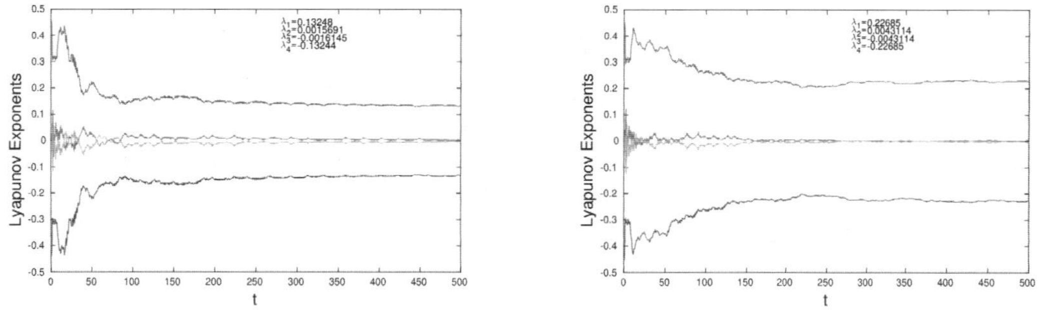


**Figure 4.** (Color online) Lyapunov exponents for $\beta = 1.5, \alpha = .5$ and (a) $\Gamma = .01$ (left panel), (b) $\Gamma = 0$ (right panel). The initial conditions for both the cases are same, $x(0) = .01, y(0) = 0.02, \dot{x}(0) = .03, \dot{y}(0) = .04$. *(Reproduced from Ref. [31])*

The standard damped Duffing oscillator with a driving term is known to show chaotic behaviour and a Hamiltonian formulation of the system is not known. However, for $H_D$, there is neither any explicit damping term nor any driving term, yet the coupling between the modes allows a chaotic regime. This may be explained as follows by analyzing the equations of motion. In particular, the nonlinear terms in Eq. (56) with $\Gamma = 0$ can not be generated as the gradient of a potential function $V$, i.e. there is no solution to the coupled partial differential equations,

$$\frac{\partial V}{\partial x} = -gx^3, \ \frac{\partial V}{\partial y} = -3gx^2y \tag{62}$$

This implies that the system contains positional non-conservative forces[1] or curl-forces[41] which are known to admit chaotic behaviour. The driving is provided by the coupling term $\beta_1 y$ in a non-standard way. This model may play a significant role in the context of investigations on curl-forces admitting a Hamiltonian formalism and quantum chaos, and further investigations in this direction are required.

*5.1.3. Landau Hamiltonian with balanced loss and gain:* The coupled Duffing oscillator model in Eq. (52) does not contain any velocity mediated coupling. The observation on a possible relation between $\mathcal{PT}$ symmetry and existence of the equilibrium state changes for system with velocity mediated coupling which may be explained by using Eq. (34) which describes Landau system with balanced loss and gain. The system is not $\mathcal{PT}$-symmetric. The second and the third terms in each equation of (34) breaks $\mathcal{PT}$ symmetry. However, the system admits periodic solutions in Region-I. The observation is that if the linear part of equations of motion of a system contains velocity mediated interaction and is non-$\mathcal{PT}$-symmetric, the system may admit periodic solutions. It may noted that the system of equations is invariant under the transformation:

$$x_1 \to x_2, x_2 \to -x_1, t \to -t, B \to -B. \tag{63}$$

The transformation on the spatial co-ordinates correspond to a rotation by $\frac{\pi}{2}$ around an axis perpendicular the '$x_1 - x_2$'-plane, which can not be identified as discrete $\mathcal{P}$ transformation. The transformation involving $t$ and $B$ correspond to time-reversal symmetry.

*5.1.4. Dimer & Non-linear Schrödinger Equation with balanced loss and gain:* The following dimer model describes time-evolution of amplitudes of $x$ and $y$ in the leading order of multiple time-scale analysis of Eq. (52) for $\Gamma \ll 1, \beta \ll 1$:[1]

$$i\frac{\partial \Psi}{\partial t} + (i\Gamma_0\sigma_3 + \beta\sigma_1)\Psi + \alpha \begin{pmatrix} |\psi_1|^2\psi_1 \\ 2|\psi_1|^2\psi_2 + \psi_1^2\psi_1^* \end{pmatrix} = 0, \quad \Psi \equiv \begin{pmatrix} \psi_1 \\ \psi_2 \end{pmatrix} \tag{64}$$

The time-reversal transformation is given by $t \to -t, i \to -i$, while $\mathcal{P} : \Psi \to \sigma_1\Psi$. The linear part of the equation, i.e. $\alpha = 0$, is invariant under $\mathcal{PT}$ transformation. However, the system is non-$\mathcal{PT}$-symmetric for $\alpha \neq 0$, yet it admits periodic solutions in some region in the parameter-space[30]. In particular, the Stokes variables,

$$Z_a = \frac{1}{2}\Psi^\dagger\sigma_a\Psi, \; R = \frac{1}{2}\Psi^\dagger\Psi = \sqrt{Z_1^2 + Z_2^2 + z_3^2}, \quad a = 1,2,3 \tag{65}$$

satisfy a solvable linear equations from which the time-periodic solutions $\Psi$ may be constructed[31]. Adding a dispersion term to Eq. (64) leads to a coupled non-linear Schrödinger equation with balanced loss and gain,

$$i\frac{\partial \Psi}{\partial t} + (i\Gamma_0\sigma_3 + \beta\sigma_1)\Psi + \frac{\partial^2\Psi}{\partial x^2} + \alpha \begin{pmatrix} |\psi_1|^2\psi_1 \\ 2|\psi_1|^2\psi_2 + \psi_1^2\psi_1^* \end{pmatrix} = 0, \tag{66}$$

the linear part of which is $\mathcal{PT}$-symmetric, while the the nonlinear term explicitly breaks $\mathcal{PT}$-symmetry. The moments $\mathcal{Z}_a = \int dx \; Z_a, \mathcal{R} = \int R \; dx$ satisfy the same time-evolution Eq. as satisfied by the Stokes variables given in Eq. (65) for well behaved fields $\Psi$ vanishing at asymptotic infinity. Thus, the moments are time-periodic for the same region for which the dimer model admits periodic solutions. The spatial degree of freedom has been integrated out and the analytic expression for $\Psi(x,t)$ can not be obtained from the time-dependence of the moments. However, the existence of periodic solutions in some region of the parameter-space is ensured.

*5.2. Non-$\mathcal{PT}$-symmetric Non-Hamiltonian System*

The central focus of this review is on Hamiltonian system. However, taking a detour from the main line of discussions, an example of a non-$\mathcal{PT}$-symmetric non-Hamiltonian System with balanced loss and gain that admits periodic solutions is presented in this section. The system is described by the equations of motion,

$$\begin{aligned} &\ddot{x} + 2\gamma\dot{x} + \omega^2 x + \beta y + \alpha_1 x^3 = 0, \\ &\ddot{y} - 2\gamma\dot{y} + \omega^2 y + \beta x + \alpha_2 y^3 = 0, \end{aligned} \tag{67}$$

which describe a damped Duffing oscillator linearly coupled to an anti-damped Duffing oscillator with different nonlinear strengths. The gain and loss terms are equally balanced and the flow in the position-velocity state space preserves its volume. However, no Hamiltonian for the system is known for $\alpha_1 \neq 0$ and/or $\alpha_2 \neq 0$. The system is Hamiltonian for $\alpha_1 = \alpha_2 = 0$ as well as $\mathcal{PT}$-symmetric[19]. The coupled Duffing oscillator model described by Eq. (67) is $\mathcal{PT}$-symmetric for

[1] Substitute $A \to \frac{1}{2}\Psi, T_2 \to t, \beta_0 \to 2\beta, \alpha \to \frac{8\alpha}{3}$ in Eq. (43) of [31] to get this particular form of the equation.

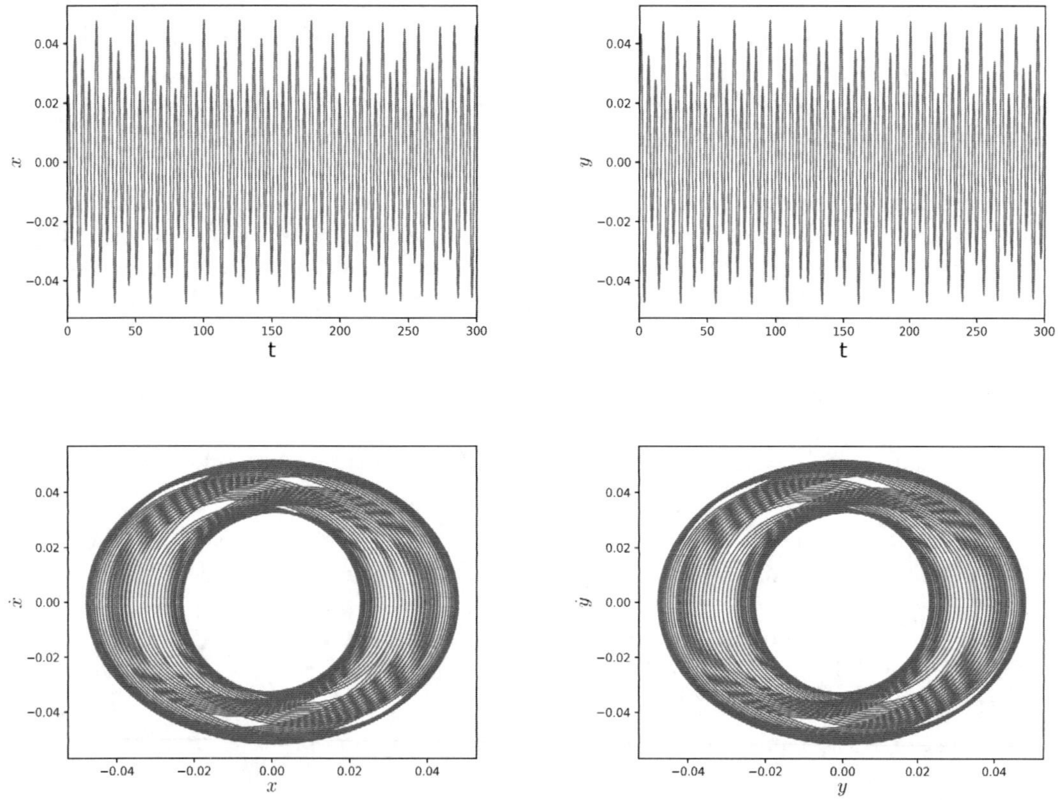


**Figure 5.** (Color online) Eq. (67) is solved with the initial conditions $x(0) = .01, y(0) = 0.02, \dot{x}(0) = .03, \dot{y}(0) = .04$ for $\alpha_1 - 0.5, \alpha_2 = 1.0, \beta = 0.5$ and $\Gamma = 0.1$. Time series of $x$ and $y$ are plotted in the first row. Plots of $\dot{x}$ vs. $x$ and $\dot{y}$ vs $y$ are given in the second row.

$\alpha_1 = \alpha_2$ and has been studied in detail in Ref. [44]. The system is non-Hamiltonian and admits periodic solutions in the unbroken $\mathcal{PT}$-regime. The nonlinear term breaks $\mathcal{PT}$-symmetry for $\alpha_1 \neq \alpha_2$. It appears that the solutions of Eq. (67) have not been investigated for $\alpha_1 \neq \alpha_2$ for which the system is non-$\mathcal{PT}$-symmetric. The linear stability analysis predicts periodic solutions around the equilibrium point $(0,0,0,0)$ in the position-velocity state space for

$$-\frac{1}{\sqrt{2}} < \Gamma < \frac{1}{\sqrt{2}}, \quad 4\Gamma^2\left(1-\Gamma^2\right) < \beta^2 < 1 \tag{68}$$

This is confirmed by numerical analysis and time-series of $x$ and $y$ for $\alpha_1 = 0.5, \alpha_2 = 1.0, \beta = 0.5, \Gamma = 0.1$ and are shown in the first row of Fig. 5. Plots of $\dot{x}$ vs $x$ and $\dot{y}$ vs. $y$ for the same values of the parameters are shown in the second row of Fig. (5). Numerical investigations show periodic solutions in a large region in the parameter space. The periodic solution of non-Hamiltonian non-$\mathcal{PT}$-symmetric coupled Duffing oscillator model is presented in this review for the first time and has not been discussed previously.

### *5.3. A Conjecture*

One important observation related to these investigations is that linear part of the equations for systems without any velocity mediated coupling is necessarily $\mathcal{PT}$-symmetric in order to have periodic solutions, while the nonlinear part may or may not be $\mathcal{PT}$-symmetric. It may be conjectured at this point that *a non-$\mathcal{PT}$ symmetric system with balanced loss-gain and without any velocity mediated coupling may admit periodic solution if the linear part of the equations of motion is necessarily $\mathcal{PT}$ symmetric —the nonlinear interaction may or may not be $\mathcal{PT}$-symmetric. Further, systems with velocity mediated coupling among different degrees of freedom need not be PT symmetric at all in order to admit periodic solutions.* The conjecture has no contradiction with the known results on $\mathcal{PT}$-symmetric systems. There is no general result in a model independent way to suggest that $\mathcal{PT}$-symmetry of a system is necessary in order to have periodic solutions. On the contrary, it is known that non-Hamiltonian dimer model without any $\mathcal{PT}$ symmetry due to imbalanced loss and gain admits stable nonlinear supermodes[45]. Similarly, within the mean field description of Bose-Einstein condensate, stationary ground state is obtained for non-$\mathcal{PT}$-symmetric confining potential[46]. The results stated above and described in detail in Refs. [31, 38] are related to mechanical systems with finite degrees of freedom, dimers and nonlinear Schrödinger equation. Thus, the conjecture is supported by known results from diverse areas of classical physics.

The conjecture is also consistent with $\mathcal{PT}$-symmetric quantum system due to the following reasons:

- Different realizations of the linear operator $\mathcal{P}$ and the anti-linear operator $\mathcal{T}$ are allowed[18, 47] in quantum mechanics as long as $|\langle\tilde{\phi}_i|\tilde{\psi}_i\rangle| = |\langle\phi|\psi\rangle|$, where $\tilde{\phi}_1 = \mathcal{P}\phi, \tilde{\psi}_1 = \mathcal{P}\psi, \tilde{\phi}_2 = \mathcal{T}\phi, \tilde{\psi}_2 = \mathcal{T}\psi$ and $\phi, \psi, \phi_i, \psi_i$ are state vectors in the relevant Hilbert space. There may exist non-trivial realizations of $\mathcal{P}$ and $\mathcal{T}$ operators in the corresponding quantum system such that it is $\mathcal{PT}$ invariant. An example of this non-standard $\mathcal{T}$-symmetry within the context of $\mathcal{PT}$-symmetric theory is discussed in Ref. [18].
- The standard $\mathcal{PT}$-symmetry may be substituted with an appropriate anti-linear symmetry for which consistent quantum description of a non-hermitian Hamiltonian admitting entirely real spectra and unitary time-evolution is allowed. A pseudo-hermitian system[48] admits anti-linear symmetry. Thus, a non-$\mathcal{PT}$-symmetric classical system with balanced loss and gain and admitting periodic solutions may, upon quantization, become a pseudo-hermitian system.

There is no analogue of anti-linear symmetry and pseudo-hermiticy in the classical physics. This leads to fixing criterion for the existence of periodic solution of a system with balanced loss and gain solely in terms of $\mathcal{PT}$ symmetry. However, any criterion based on $\mathcal{PT}$ symmetry is inadequate and incomplete. A possible resolution of the problem may be to identify appropriate $\mathcal{PT}$-symmetry/anti-linear symmetry/pseudo-hermiticity of the corresponding quantized non-hermitian Hamiltonian which may explain the entirely real spectra with unitary time-evolution. The second step is to take the classical limit of the relevant anti-linear operator, which may not necessarily reduce to the standard $\mathcal{P}$ and $\mathcal{T}$ transformation, and fix the criteria for the existence of periodic solutions for the corresponding classical Hamiltonian. It should be mentioned here that an implementation of the scheme is tricky and nontrivial, since there may be more than one quantum system for a given classical Hamiltonian based on the quantization condition. An intelligent resolution of the problem is desirable.

## 6. Omitted topics

The discussion so far is restricted to mechanical system with finite degrees of freedom. The discrete and continuum models of non-linear Schrödinger equation with balanced loss and gain or its generalized versions appear in diverse contexts and have been studied extensively in the literature with interesting results[49]. The nonlinear Dirac equation with balanced loss-gain has also been investigated[50, 51]. A brief description on these topics is presented below.

### *6.1. Non-linear Schrödinger Equation with balanced loss and gain*

The non-linear Schrödinger equation in one dimension is an integrable system with exact soliton solutions[52, 53, 54, 55]. Several generalizations of non-linear Schrödinger equation have been considered with the growing interest and relevance on $\mathcal{PT}$ symmetric theory. The $\mathcal{PT}$ symmetry motivated generalizations of non-linear Schrödinger equation may be broadly classified into three major sub-areas: (i) non-linear Schrödinger equation with $\mathcal{PT}$-symmetric confining complex potential[56], (ii) non-local non-linear Schrödinger equation[57, 58, 59] and (iii) non-linear Schrödinger Equation with balanced loss and gain[49, 60, 61, 62, 63, 64, 65, 66]. The topic non-linear Schrödinger equation with balanced loss and gain is relevant for the present review for which the literature is vast and the major results till 2016 are described in Ref. [49]. A recent result on a class of exactly solvable non-linear Schrödinger equation with balanced loss and gain is described below in brief.

The generic form of the non-linear Schrödinger equation with balanced loss and gain is given by,

$$i\left(I\frac{\partial}{\partial t}+iA\right)\Psi+\frac{\partial^2\Psi}{\partial x^2}+V(x)\Psi+G(\Psi,\Psi^*)\Psi=0,\ A:=B+iC, \tag{69}$$

where $I$ is the $N\times N$ identity matrix, $\Psi$ is an $N$-component complex scalar field and $G$ is an $N\times N$ hermitian matrix depending on the field $\Psi$. The hermitian matrix $B$ describes linear coupling among different fields, while the traceless hermitian matrix $C$ describes balanced loss and gain. The time-dependent gain-loss and linear coupling terms may be considered by taking time-dependent matrices $C$ and $B$, respectively. The matrix $G$ encodes nonlinear coupling among different fields and may be chosen depending on the physical system. The external potential $V$ is generally taken as complex and $\mathcal{PT}$-symmetric. However, the discussion in this article is restricted to real $V$. Various soliton solutions of Eq. (69) for $V=0$ and specific forms of $G$ have been obtained under certain reductions[49, 60, 61, 62, 63, 64, 65]. Recently, exactly solvable models of non-linear Schrödinger equation with balanced loss and gain have been constructed in Ref. [66] for $G=I\ \Psi^\dagger M\Psi$, where $M$ is a hermitian matrix independent of fields. The space-time modulation of the cubic nonlinearity may be introduced via the matrix $M$.

The continuity equation contains source and sink terms proportional to $\Psi^\dagger C\Psi$:

$$\frac{\partial\tilde{\rho}}{\partial t}+\frac{\partial\tilde{J}}{\partial x}=2\Psi^\dagger C\Psi,\ \ \tilde{\rho}=\Psi^\dagger\Psi,\ \ \tilde{J}=-i\left[\Psi^\dagger\frac{\partial\Psi}{\partial x}-\frac{\partial\Psi^\dagger}{\partial x}\Psi\right] \tag{70}$$

In general, the total density $\tilde{Q}=\int dx\tilde{\rho}$ is not conserved even for well-behaved $\Psi$ vanishing at asymptotic infinity, $\frac{\partial\tilde{Q}}{\partial t}=2\int dx\Psi^\dagger C\Psi$. There may be specific field configurations with additional properties, like $\Psi^\dagger C\Psi$ being an odd function of $x$, for which $\tilde{Q}$ is conserved. It is important to note that a non-standard continuity equation with the associated conserved quantity can be derived for the special case of an $\eta$-pseudo-hermitian[48] $A$, i.e. $A^\dagger=\eta A\eta^{-1}$,

$$\frac{\partial\rho}{\partial t}+\frac{\partial J}{\partial x}=0,\ \ \rho=\Psi^\dagger\eta\Psi,\ J=-i\left[\Psi^\dagger\eta\frac{\partial\Psi}{\partial x}-\frac{\partial\Psi^\dagger}{\partial x}\eta\Psi\right]. \tag{71}$$

The quantity $Q=\int dx\rho$ is a conserved quantity for the well-behaved fields $\Psi$ vanishing at infinity. The continuity equation holds irrespective of whether the matrix $\eta$ is positive-definite or indefinite. A positive-definite $\eta$ ensures a positive-definite $Q$. Further, it is known[48] that a pseudo-hermitian matrix $A$ with respect to a positive-definite $\eta$ admits entirely real eigenvalues which is required for the existence time-periodic solution of $\Psi$. It may be noted that Eq. (69) may or may not depend on $\eta$ and if it depends on $\eta$ at all, it should be through the matrix $G$.

The system described by Eq. (69) admits a Lagrangian $L = \int dx \mathcal{L}$,

$$\mathcal{L}_F = \frac{i}{2}\left[\Psi^\dagger M\left(D_0\Psi\right) - \left(D_o\Psi\right)^\dagger M\Psi\right] - \frac{\partial \Psi^\dagger}{\partial x} M \frac{\partial \Psi}{\partial x} + V(x)\Psi^\dagger M\Psi + W(\Psi, \Psi^\dagger) + \Psi^\dagger F_1 \Psi, \quad (72)$$

where the matrix $G(\Psi, \Psi^\dagger)$ and the potential $W(\Psi, \Psi^\dagger)$ are related by the equation,

$$G = M^{-1}\frac{\partial W}{\partial \Psi^\dagger} \quad (73)$$

and the hermitian matrix $M$ is independent of field $\Psi$. The operator $D_0 := I\frac{\partial}{\partial t} + iA$ has formal resemblance with the temporal component of covariant derivative with non-hermitian gauge potential $A$ and the anti-hermitian matrix $F_1 := \frac{1}{2}\left(A^\dagger M - MA\right)$. There may be alternative Lagrangian formulations of Eq. (69) for specific $G$. This particular formulation is chosen due to its conceptual similarity with $\mathcal{L}$ of Eq. (12) describing mechanical system with finite degrees of freedom —presence of fictitious gauge potential and a metric is common to both the formulations. The conjugate momenta corresponding to $\Psi$ and $\Psi^\dagger$ are $\Pi_\Psi = \frac{i}{2}\Psi^\dagger M$ and $\Pi_{\Psi^\dagger} = -\frac{i}{2}M\Psi$, respectively. The Hamiltonian density $\mathcal{H}_F$ corresponding to $\mathcal{L}_F$ has the form,

$$\mathcal{H}_F = \frac{\partial \Psi^\dagger}{\partial x} M \frac{\partial \Psi}{\partial x} - V(x)\Psi^\dagger M\Psi - W(\Psi, \Psi^\dagger) + \Psi^\dagger MA\Psi \quad (74)$$

In general, the Hamiltonian $\mathcal{H}_F$ is complex-valued and the quantized Hamiltonian is expected to be non-hermitian. However, for the case of an $M$-pseudo-hermitian $A$, $\mathcal{H}_F$ is real-valued and the corresponding quantum Hamiltonian is hermitian with suitable quantization condition. An $M$-pseudo-hermitian matrix $A$ may be used to define an $M$-pseudo-unitary[67] matrix $U := e^{-iA}$,

$$U^\dagger MU = M \Leftrightarrow A^\dagger = MAM^{-1} \quad (75)$$

The Hamiltonian $\mathcal{H}_F$ is invariant under pseudo-unitary transformation $\Psi \rightarrow U\Psi$ provided $W$ is invariant under this transformation. For example, the Hamiltonian is invariant under pseudo-unitary transformation for the choice $W \equiv W(\Psi^\dagger M\Psi)$. It should be noted that two systems connected by a unitary transformation can be considered as gauge equivalent. However, the same is not true for systems connected by pseudo-unitary transformations —the physical observable like density, square of the width of the wave-packet and its speed of growth have different expressions. The pseudo-unitary transformation can be used to construct solvable models[66].

*6.1.1. $V = 0$ and $W = \frac{\delta}{2}\left(\Psi^\dagger M\Psi\right)^2$:* The non-linear Schrödinger equation in Eq. (69) takes the form,

$$i\left(I\frac{\partial}{\partial t} + iA\right)\Psi + \frac{\partial^2\Psi}{\partial x^2} + \delta\left(\Psi^\dagger M\Psi\right)\Psi = 0, \quad (76)$$

Defining $\Psi = e^{-iA}\Phi$ and using Eq. (75), Eq. (69) is mapped[66] to the equation,

$$i\Phi_t = -\Phi_{xx} - \delta\left(\Phi^\dagger M\Phi\right)\Phi. \quad (77)$$

Eq. (77) is exactly solvable and with appropriate unitary transformation followed by a scaling[66], it can be brought to the canonical form of Manakov-Zakharov-Schulman system[52, 53]. The pseudo-unitary mapping can be used to find exact solutions of Eq. (69) with $V = 0$ and $G = \Psi^\dagger M\Psi$. Exactly solvable models with analytic expression for power-oscillation have been constructed in Ref. [66] by using this mapping. If the non-hermitian matrix $A$ is not pseudo-hermitian, then the mapping can be used to construct exact solutions for a non-autonomous system[66]. Further, the mapping can be used to construct a variety of solvable models with interesting physical properties[68] like time-dependent gain-loss terms, space-time modulated nonlinear interaction and external confining potential[68].

*6.2. Oligomer with balanced loss and gain*
The discrete non-linear Schrödinger equation with finite number of lattice points $N$ or similar equation with generalized interaction term and self-trapping phenomena is popularly known as oligomer and the $N = 2$ is termed dimer. Such equations arise in the study of $\mathcal{PT}$-symmetric lattices[49], nonlinear dynamics of molecules[69, 70], optics[71], Landau-Lifshitz equation[72] etc.. The time-evolution of amplitude of waves arising in a mechanical and/or extended system under various approximation schemes may also be modelled as oligomer. For example, a multiple time-scale analysis of Eq. (52) leads to the dimer model (64) describing the time-evolution of amplitudes in the leading order of the perturbation. The same recipe for studying the time-evolution of amplitudes of nonlinear oscillators has been used earlier in Refs. [26, 27].

The equation satisfied by an oligomer has the general form,

$$i\left(I\partial_t + iA\right)\Psi + G(\Psi, \Psi^*)\Psi = 0 \tag{78}$$

It may be noted that Eq. (69) for vanishing external potential ($V = 0$) and no dispersion term $\frac{\partial^2 \Psi}{\partial x^2}$ reduces to the oligomer model in Eq. (78). The matrices $B$, $C$ and $G(\Psi, \Psi^*)$ have the same interpretation as in the case of dimer model. The discrete non-linear Schrödinger equation with balanced loss and gain is reproduced for a tridiagonal $B$ with $[G]_{ij} = \alpha\delta_{ij}|\Psi|^2$, where the real parameter $\alpha$ denotes the strength of the nonlinear interaction. There are specific forms of $B$, $C$ and $G$ for which a Hamiltonian description is possible and the system may even be integrable[26, 27, 73, 74]. The literature on dimer with balanced loss and gain is vast and results with physical settings are nicely reviewed in Ref. [49]. A new class of exactly solvable dimer models which has not been discussed earlier is presented below.

A generic Lagrangian-Hamiltonian formulation of Eq. (78) may be presented by continuing with the general development for mechanical system with finite degrees of freedom and for the extended system. In particular, the Lagrangian and Hamiltonian are,

$$\begin{aligned} \mathcal{L}_O &= \frac{i}{2}\left[\Psi^\dagger M\left(D_0\Psi\right) - (D_o\Psi)^\dagger M\Psi\right] + W(\Psi, \Psi^\dagger) + \Psi^\dagger F_1\Psi, \\ \mathcal{H}_O &= \quad W(\Psi, \Psi^\dagger) + \Psi^\dagger M A\Psi \end{aligned}$$

where $G$ and $W$ are related via Eq. (73). The conjugate momenta corresponding to $\Psi$ and $\Psi^\dagger$ are $\Pi_\Psi = \frac{i}{2}\Psi^\dagger M$ and $\Pi_{\Psi^\dagger} = -\frac{i}{2}M\Psi$, respectively. A new class of solvable models are obtained for the choice of $W = \frac{\delta}{n+1}\left(\Psi^\dagger M\psi\right)^{n+1}$ for which Eq. (78) takes the form,

$$i\left(I\partial_t + iA\right)\Psi + \delta\left(\Psi^\dagger M\Psi\right)^n\Psi = 0 \tag{79}$$

and is exactly solvable for a $M$-pseudo-hermitian $A$. In particular, substituting $\Psi = e^{-iAt}\Phi$ in eq. (79), $\Phi$ satisfies the equation,

$$i\partial_t\Phi + \delta(\Phi^\dagger M\Phi)^n\Phi = 0. \tag{80}$$

It immediately follows that $\Phi^\dagger M\Phi$ is a conserved quantity and denoting its value at $t = 0$ as the real constant $C$, i.e. $\Phi^\dagger(0)M\Phi(0) = C$, $\Phi$ is solved as,

$$\Phi(t) = We^{i\delta C^n t}, \tag{81}$$

where $W$ is an arbitrary $N$-component constant complex vector. For the simplest case of a dimer, i.e. $N = 2$, an example of $M$-pseudo-hermitian $A$ may be presented as,

$$A = \beta^*\sigma_+ + \beta\sigma_- + i\Gamma\sigma_3, \; M = \alpha_0 I_2 + \alpha^*\sigma_+ + \alpha\sigma_- \tag{82}$$

where $I_2$ is the $2 \times 2$ identity matrix and $\sigma_\pm = \frac{1}{2}(\sigma_1 \pm \sigma_2)$. The condition for a $M$-pseudo-hermitian $A$ with positive definite $M$ has been derived in Ref. [66] in terms of the complex parameters $\beta = |\beta| e^{i\theta_\beta}, \alpha = |\alpha| e^{i\theta_\alpha}$ and the real parameters $\Gamma, \alpha_0$ as,

$$\frac{\alpha_0}{|\alpha|} = \frac{|\beta|}{\Gamma} \sin(\theta_\alpha - \theta_\beta) > 0. \tag{83}$$

Using the expression for $U := e^{-iAt}$ in Ref. [66], the solutions for $\Psi$ are obtained as,

$$\begin{aligned} \Psi_1 &= e^{i\delta C^n t} \left[ W_1 \left( \cos(\theta t) + \frac{\Gamma}{\theta} \sin(\theta t) \right) - \frac{i W_2 \beta^*}{\theta} \sin(\theta t) \right] \\ \Psi_2 &= e^{i\delta C^n t} \left[ -\frac{i W_1 \beta}{\theta} \sin(\theta t) + W_2 \left( \cos(\theta t) - \frac{\Gamma}{\theta} \sin(\theta t) \right) \right] \end{aligned} \tag{84}$$

where $\theta = \sqrt{|\beta|^2 - \Gamma^2}$. The solutions are periodic for $|\beta| > \Gamma$ and unbounded for $|\beta| \leq \Gamma$. The matrix $U$ is unitary for $\Gamma = 0$ and $\Psi^\dagger \Psi$ is independent of time. However, for $\Gamma \neq 0$, $\Psi^\dagger \Psi$ is periodic in time. Results for higher values of $N$ may be obtained in a similar way.

### *6.3. Nonlinear Dirac Equation with balanced loss and gain*

The nonlinear Dirac equation appears in diverse branches of modern science[75, 76, 77, 78, 79], although it has not been studied as much as its non-relativistic counterpart, namely non-linear Schrödinger equation. With the advent of $\mathcal{PT}$-symmetric theory, several nonlinear Dirac equations with balanced loss and gain have been considered in the literature[50, 51]. It may be recalled that the balanced loss and gain terms for the case of non-linear Schrödinger equation is introduced via a non-hermitian mass term in the Hamiltonian. The same approach is taken for the construction of nonlinear Dirac equation with balanced loss and gain. The Lagrangian density for the Dirac equation has the generic form

$$\mathcal{L}_D = \bar{\Psi} \left( i\gamma^\mu \partial_\mu - m_1 - m_2 \gamma_5 \right) \Psi - F(\bar{\Psi}, \Psi) \tag{85}$$

with the non-hermitian mass term $\bar{\Psi}(m_1 + m_2\gamma_5)\Psi$[80] and Lorentz invariant nonlinear interaction term $F$. The interaction term $F$ may be chosen depending on the physical settings. For example, $F_1 \equiv F_1(\bar{\Psi}\Psi)$ reproduces Soler model[75] for $m_2 = 0$ and in $1+1$ space-time it is also known as Gross-Neveu[76] model. The nonlinear interaction $F_2 \equiv F_2(\bar{\Psi}\gamma_5\Psi)$ depends on pseudo-scalar $\bar{\Psi}\gamma_5\Psi$. Similarly, the choice $F = J_\mu J^\mu$ with the current $J_\mu = \bar{\Psi}\gamma_\mu\Psi$ introduces vector-type self-interaction and the system describes massive Thirring Model[77] for $m_2 = 0$. The interaction term $F$ need not be Lorentz invariant for non-relativistic systems with equations having formal resemblance with the Dirac equation. The Hamiltonian is complex-valued for $m_2 \neq 0$,

$$\mathcal{H}_D = \bar{\Psi} \left( i\vec{\gamma} \cdot \vec{\nabla} + m_1 + m_2\gamma_5 \right) \Psi + F(\bar{\Psi}, \Psi), \tag{86}$$

and the quantum Hamiltonian is non-hermitian irrespective of whether $F$ is hermitian or not.

The equation of motion following from the Lagrangian reads,

$$\left( i\gamma^\mu \partial_\mu - m_1 - m_2\gamma_5 \right) \Psi - \frac{\partial F}{\partial \bar{\Psi}} = 0. \tag{87}$$

The balanced loss-gain terms can be seen explicitly with the following representation of the $\gamma$ matrices,

$$\gamma^0 = \sigma_1 \otimes I_2, \; \gamma^j = i\sigma_3 \otimes \sigma_j, \;\; \gamma_5 := i\gamma^0\gamma^1\gamma^2\gamma^3 = \sigma_2 \otimes I_2, \tag{88}$$

for which Eq. (87) reads,

$$i\left(I_4\frac{\partial}{\partial t}+i\mathcal{M}\right)\Psi=-i\sigma_2\otimes\vec{\sigma}\cdot\vec{\nabla}\Psi+\gamma^0\frac{\partial F}{\partial\bar{\Psi}},\quad \mathcal{M}:=(m_1\sigma_1+im_2\sigma_3)\otimes I_2 \tag{89}$$

The term appearing in the matrix $\mathcal{M}$ with the co-efficient $m_1$ describes linear coupling among different components of the spinor $\Psi$, while the term with the co-efficient $m_2$ describes balanced loss and gain. The loss-gain terms may be hidden by using a unitary equivalent representations of the $\gamma$ matrices. For example, the loss-gain terms are hidden in the Dirac-Pauli representation. The mass-matrix $\mathcal{M}$ is non-hermitian with doubly degenerate eigenvalues $\lambda_\pm=\pm\sqrt{m_1^2-m_2^2}$ which are real provided $|m_2|<|m_1|$. The mass-gap $2\sqrt{m_1^2-m_2^2}$ vanishes for $m_2=\pm m_1$, signalling the existence of zero mode for the linear problem, i.e. $F=0$. The ansatz $\Psi=We^{i(\vec{k}\cdot\vec{r}-\omega t)}$ with $F=0$ gives the dispersion $\omega^2=k^2+m_1^2-m_2^2$, where $W$ is a four-component constant complex vector. The frequency $\omega$ is real for real mass, i.e. $|m_2|<|m_1|$ and $\Psi$ is periodic in time.

The charge $\tilde{Q}=\int d^3x\Psi^\dagger\Psi$ is not a conserved quantity due to the non-hermitian mass term. However, a non-standard conserved charge $Q=\int d^3x\Psi^\dagger\eta\Psi$ can be introduced if a hermitian matrix $\eta$ exists satisfying the following conditions:

$$[\eta,\gamma^0\gamma^i]=0\ \forall\ i,\ \mathcal{M}^\dagger=\eta\mathcal{M}\eta^{-1},\quad W^\dagger=W,\ W\equiv\Psi^\dagger\eta\gamma^0\frac{\partial F}{\partial\bar{\Psi}} \tag{90}$$

The ansatz $\eta=\delta_1 I_4+\delta_2\sigma_2\otimes I_2$ solves the first equation for arbitrary real constants $\delta_1$ and $\delta_2$, since $\gamma^0\gamma^i=\sigma_2\otimes\sigma_i$. The second equation demands that the mass-matrix $\mathcal{M}$ is $\eta$-pseudo-hermitian, which fixes $\delta_1=1,\delta_2=\frac{m_2}{m_1}$ so that $\eta=I_4+\frac{m_2}{m_1}\gamma_5$. It may be noted that $\eta$ is positive-definite for $|m_1|>|m_2|$ and is singular for $m_2=\pm m_1$. The conserved current $J^\mu$ has the expression $J^\mu=\bar{\Psi}\gamma^\mu\eta\Psi$ leading to the conservation of $Q$ for $F=0$. This relation for the linear Dirac equation has been obtained earlier[81]. The probability density $J^0$ receives contribution only from right- or left-handed degrees of freedom for $m_2=m_1$ and $m_2=-m_1$, respectively[64]. The hermiticity of $W$ has to be ensured such that $J^\mu$ can also be taken as the conserved current for the nonlinear Dirac equation, i.e. $F\neq 0$. There are choices of $F$ for which $W$ is hermitian. For example, $W$ is hermitian for $F\equiv F(\bar{\Psi}M\Psi)$ provided the constant matrix $M$ satisfies the condition,

$$[M,\gamma^0]+\frac{m_2}{m_1}\{M,\gamma^0\gamma_5\}=0. \tag{91}$$

One important solution of the above equation is $M_1=\eta$. The nonlinear interaction $F$ contains both Lorentz scalar and pseudo-scalar interactions, since $\bar{\Psi}\eta\Psi=\bar{\Psi}\Psi+\frac{m_2}{m_1}\bar{\Psi}\gamma_5\Psi$. A more general solution can be constructed by taking $M_2=\sigma_1\otimes m$, where $m$ is an arbitrary $2\times 2$ hermitian matrix. It should be emphasized here that the existence of a conserved charge with its interpretation as the probability density is not necessary in the study of nonlinear Dirac equation, since one is dealing with relativistic field theory instead of quantum mechanics. The existence of a conserved charge corresponds to an internal symmetry of the system. However, an appropriate number operator should be defined in the corresponding quantum field theory. The investigations on nonlinear Dirac equation with balanced loss and gain are mainly restricted to its classical solutions and stability in $1+1$ dimensions. Several choices of $F$ maintaining Lorentz invariance and $\mathcal{PT}$-symmetry have been considered in Refs. [50, 51] which admit stable periodic as well as soliton solutions.

## 7. Summary & Discussions

Classical Hamiltonian systems with balanced loss and gain have been reviewed in this article. The emphasis is on mechanical system with finite degrees of freedom. The criteria for a mechanical system to be identified as a system with balanced loss and gain, irrespective of whether a Lagrangian-Hamiltonian formulation is admissible or not, has been presented in terms of the volume conservation of flow in the position-velocity state-space. The Hamiltonian formulation for systems with space-dependent balanced loss and gain has been introduced for arbitrary number of particles and generic potential. It has been shown that the loss-gain terms may be removed completely through appropriate co-ordinate transformations with its effect manifested in modifying the strength of the velocity-mediated coupling. This mapping is inherent to the generic Hamiltonian system with balanced loss and gain and does not depend on any specific form of the potential or the number of particles. In general, the quadratic term in momenta in the Hamiltonian is not positive-definite leading to instabilities. The effect of the Lorentz interaction in improving the stability of classical solutions as well as allowing a possibility of defining the corresponding quantum problem consistently on the real line, instead of within Stokes wedges, has also been discussed.

The system with $N = 2m$ particles admits at least $m + 1$ integrals of motion for a potential having specified type of translational or rotational symmetry, thereby implying that the system is at least partially integrable for $N > 2$ and completely integrable for $N = 2$. Several exactly solvable models based on translational and rotational symmetry and specific form of potentials have been discussed which include coupled cubic oscillators, Landau Hamiltonian etc. The Lorentz interaction appear naturally in the Landau Hamiltonian and there are regions in the parameter space where the quadratic term in momenta in the Hamiltonian is positive-definite. This is also the region for which stable classical solutions are obtained. The solution is same as the standard Landau problem with a reduced cyclotron frequency due to the loss-gain term.

An example of Hamiltonian chaos within the framework of a model of coupled Duffing oscillator with balanced loss and gain has been discussed. The dynamical properties of the system are rich —three out of five equilibrium points are stable and admits periodic solutions around these equilibrium points. The damped undriven Duffing oscillator is linearly coupled to another anti-damped oscillator with variable angular frequency depending on the degree of freedom of the Duffing oscillator. This coupling acts effectively as a driving term, albeit in a nontrivial way. The chaotic behaviour is seen beyond a critical value of this coupling strength. The chaotic behaviour persists even for vanishing gain-loss terms, thereby providing an example of Hamiltonian chaos for coupled Duffing oscillator without any explicit damping and driving terms. It has been argued that the system contains positional non-conservative force[1] or the curl-force[41], thereby compensating the effect of damping even if $\gamma = 0$. The coupling to the oscillator with variable frequency provides the effect of driving term. The corresponding quantum system may provide some insight into quantum chaos.

The role of $\mathcal{PT}$-symmetry on the existence of periodic solution in systems with balanced loss and gain has been critically analyzed. Examples from many-particle mechanical systems, dimer models and nonlinear Schrödinger equations without any $\mathcal{PT}$-symmetry is analyzed with the understanding that $\mathcal{P}$ corresponds to linear transformation only. This is because the Lorentz transformation is linear and no signature of its violation has been seen in nature. Moreover, $\mathcal{PT}$-symmetric quantum mechanics dwells on $\mathcal{CPT}$ norm, which at a more fundamental level is expected to correspond to the $\mathcal{CPT}$ invariance of a local hermitian Lorentz invariant theory. Based on the observations, it has been conjectured that non-$\mathcal{PT}$-symmetric system with balanced loss-gain and without any velocity mediated interaction may admit periodic solution if the linear part of the equations is necessarily $\mathcal{PT}$ symmetric —the nonlinear interaction may or may not be $\mathcal{PT}$-symmetric.Further, systems with velocity mediated coupling among different degrees of freedom need not be $\mathcal{PT}$ symmetric at all in order to admit periodic solutions. This conjecture has no contradiction with the formulation of $\mathcal{PT}$-symmetric quantum mechanics —the corresponding quantum system may be pseudo-hermitian or invariant under generalized time-reversal symmetry which has no analog in classical mechanics. The criteria for the existence of periodic solutions in terms of standard $\mathcal{PT}$-symmetry alone is not sufficient. Investigations in this direction is desirable so that a large number of non-$\mathcal{PT}$ symmetric system may be included in the mainstream of investigations with possible technological applications.

The central focus of this review is on mechanical system with finite degrees of freedom. However, there is significant advancement in the fields of oligomers, nonlinear Schrödinger and Dirac equations with balanced loss and gain. The developments in the context of oligomers and nonlinear Schrödinger equations are summarized recently in Ref. [49] and excluded for extensive discussions in this review. A very recent result on non-linear Schrödinger equation with balanced loss and gain related to the construction of exactly solvable models via non-unitary transformation has been discussed. This mapping can be used to construct a variety of solvable models with interesting physical properties[68]. The same technique is used to construct an exactly solvable dimer with balanced loss and gain which has not appeared earlier in the literature. Results related to nonlinear Dirac equations with balanced loss and gain are mentioned briefly.

**Acknowledgments**
This work is supported by a grant (**SERB Ref. No. MTR/2018/001036**) from the Science & Engineering Research Board(SERB), Department of Science & Technology, Govt. of India under the **MATRICS** scheme. The Author would like to thank Debdeep Sinha, Puspendu Roy and Supriyo Ghosh for discussions and collaboration on the topic.

## References

[1] Krechetnikov R and Marsden J E 2007 Dissipation-induced instabilities in finite dimensions *Rev. Mod. Phys.* **79** 519
[2] Morrison P J 1986 A paradigm for joined Hamiltonian and dissipative systems *Physica* **18D** 410
[3] Bateman H 1931 On Dissipative Systems and Related Variational Principles *Phys. Rev.* **38** 815
[4] Morse P M and Feshbach H 1953 *Methods of Theoretical Physics, Vol. I* (McGraw-Hill, New York)
[5] Bopp F 1973 *Sitz.-Bcr. Bayer. Akad. Wiss. Math.-naturw. KI.* **67**
[6] Feshbach H and Tikochinsky Y 1977 A Festschrift for I. I. Rabi *Trans. New York. Acad. Sci., Series 2* **38** 44
[7] Tikochinsky Y 1978 Exact propagators for quadratic Hamiltonians *J. Math. Phys.* **19** 888
[8] Dekker H 1981 Classical and quantum mechanics of the damped harmonic oscillator *Phys. Rep.* **80** 1
[9] Celeghini E, Rasetti M and Vitiello G 1992 Quantum dissipation *Ann. Phys. (N.Y)* **215** 156
[10] Banerjee R and Mukherjee P 2002 A canonical approach to the quantization of the damped harmonic oscillator *J. Phys. A: Math. Gen.* **35** 5591
[11] Chruscinski D and Jurkowski J 2006 Quantum damped oscillator I: dissipation and resonance *Ann. Phys. (N.Y.)* **321** 854
[12] Pal S K, Nandi P and Chakraborty B 2018 Connecting dissipation and noncommutativity: A Bateman system case study *Phys. Rev. A* **97** 062110
[13] Bagarello F, Gargano N and Roccati F 2020 Some remarks on few recent results on the damped quantum harmonic oscillator *Ann. Phys.* **414** 168091
Bagarello F, Gargano N and Roccati F 2019 A no-go result for the quantum damped harmonic oscillator *Phys. Lett. A***383** 2836
[14] Deguchi S and Fujiwara Y 2020 Quantization of the damped harmonic oscillator based on a modified Bateman Lagrangian *Phys. Rev A***101** 022105
Deguchi S, Fujiwara Y and Nakano K 2019 Two quantization approaches to the Bateman oscillator model *Ann. Phys.* **403** 34
[15] C M Bender 2018 *PT Symmetry: In Quantum And Classical Physics* (World-Scientific)
[16] Caldeira A O and Leggett A J 1981 Influence of Dissipation on Quantum Tunneling in Macroscopic Systems *Phys. Rev. Lett.* **46** 211
[17] Sieberer L M, Chiocchetta A, Gambassi A, Tuber U C and Diehl S 2015 Thermodynamic equilibrium as a symmetry of the Schwinger-Keldysh action *Phys. Rev. B***92** 134307
Sieberer L M, Huber S D, Altman E and Diehl S 2014 Nonequilibrium functional renormalization for driven-dissipative Bose-Einstein condensation Phys. Rev. B**89** 134310
[18] Ghosh P K and Sinha D 2018 Hamiltonian formulation of systems with balanced loss-gain and exactly solvable models *Ann. Phys. (N. Y.)* **388** 276

[19] Bender C M, Gianfreda M, Ozdemir S K, Peng B and Yang L 2013 Twofold transition in PT-symmetric coupled oscillators *Phys. Rev. A***88** 062111
[20] Peng B, Özdemir S K, Lei F, Monifi F, Gianfreda M, Long G L, Fan S, Nori F, Bender C M and Yang L 2014 Parity-time-symmetric whispering-gallery microcavities *Nat. Phys.* **10** 394
[21] Caldirola P 1941 *Nuvo Cim.* **18** 393
Kanai E 1948 On the Quantization of the Dissipative Systems *Prog. Theor. Phys.* **3** 440
[22] Dekker H 1975 On the Quantization of Dissipative Systems in the Lagrange-Hamilton Formalism *Z. Physik B***21** 295
[23] Chandrasekar V K, Senthilvelan M and Lakshmanan M 2007 On the Lagrangian and Hamiltonian description of the damped linear harmonic oscillator *J. Math. Phys.* **48** 032701
[24] Sinha D and Ghosh P K 2017 $\mathcal{PT}$-symmetric rational Calogero model with balanced loss and gain *Eur. Phys. J. Plus* **132** 460
[25] Bender C M, Gianfreda M and Klevansky S P 2014 Systems of coupled PT-symmetric oscillators *Phys. Rev. A***90** 022114
[26] Barashenkov I V and Gianfreda M 2014 An exactly solvable PT-symmetric dimer from a Hamiltonian system of nonlinear oscillators with gain and loss *J. Phys. A: Math. Theor.* **47** 282001
[27] Khare A and Saxena A 2017 Integrable oscillator type and Schrödinger type dimers *J. Phys. A: Math. Theor.* **50** 055202
[28] Sinha D and Ghosh P K 2019 On the bound states and correlation functions of a class of Calogero- type quantum many-body problems with balanced loss and gain *J. Phys. A: Math. Theor.* **52** 505203
[29] Sinha D and Ghosh P K 2019 Integrable coupled Liénard-type systems with balanced loss and gain *Ann. Phys. (N. Y.)* **400** 109
[30] Ghosh P K 2019 Taming Hamiltonian systems with balanced loss and gain via Lorentz interaction : General results and a case study with Landau Hamiltonian *J. Phys. A: Math. Theor.* **52** 415202
[31] Ghosh P K and Roy P 2020 On regular and chaotic dynamics of a non-$\mathcal{PT}$-symmetric Hamiltonian system of a coupled Duffing oscillator with balanced loss and gain *J. Phys. A: Math. Theor.* **53** 475202
[32] Landau L and Lifshitz E 1977 *Quantum Mechanics (Non-Relativistic Theory)* 3rd edn (Oxford: Pergamon)
[33] Afraimovich V S, Verichev N N and Rabinovich M I 1986 Stochastic synchronization of oscillation in dissipative system *Radiophys. Quantum Electron.* **29** 795
Ivanchenko M V, Osipov G V, Shalfeev V D and Kurths J 2004 Synchronization of two non-scalar-coupled limit-cycle oscillators *Physica D* **189** 8
Kuznetsov A P and Roman J P 2009 Properties of synchronization in the systems of non-identical coupled van der Pol and van der Pol-Duffing oscillators: broadband synchronization *Physica D* **238** 1499
[34] Talukdar B and Das U 2008 *Higher Order Systems in Classical Mechanics*(Narosa)
[35] Santilli R M 1978 *Foundations of Theoretical Mechanics* **I** (Springer-Verlag, New York)
[36] Rodman L 1997 Products of symmetric and skew-symmetric Matrices *Linear and Multilinear Algebra* **43** 19
[37] Shah T, Chattopadhyay R, Vaidya K and Chakraborty 2015 Conservative Perturbation Theory for Non-Conservative Systems *Phys. Rev. E***92** 062927
[38] Roy P and Ghosh P K (2021) *Under preparation*
[39] Dirichlet P G L 1846 *Über die Stabilität des Gleichgewichts* *Crelle* **32** 85
[40] Lakshmanan M and Sahadevan R 1993 Painlevé analysis, Lie symmetries, and integrability of coupled nonlinear oscillators of polynomial type *Phys. Rep.* **224** 1
[41] Berry M V and Shukla P 2015 Hamiltonian curl forces *Proc. R. Soc. A***471** 20150002
Berry M V and Shukla P 2016 Curl force dynamics: Symmetries, chaos and constants of motion *New Journal of Physics* **18** 063018
[42] O.W. Greenberg 2006 Why is CPT fundamental? *Found. Phys.* **36** 1535
[43] P D Manheim 2016 Extension of the CPT theorem to non-Hermitian Hamiltonians and unstable states *Phys. Lett. B* **753** 288.
[44] Cuevas J, Kevrekidis P G, Saxena A and Khare A 2013 $\mathcal{PT}$-symmetric dimer of coupled nonlinear oscillators *Phys. Rev. A***88** 032108
[45] Kominis Y, Bountis T and Flach S 2016 The Asymmetric Active Coupler: Stable Nonlinear Supermodes and Directed Transport *Sci. Rep.* **6** 33699
[46] Lunt P, Haag D, Dast D, Cartarius H and Wunner G 2017 Balanced gain and loss in Bose-Einstein condensates without PT symmetry *Phys. Rev. A***96** 023614
[47] Haake F 2010 *Quantum Signatures of Chaos, 3rd edition* (Springer)
[48] Mostafazadeh A 2010 Pseudo-Hermitian representation of quantum mechanics *Int. J. Geom. Methods in Mod. Phys.* **7** 1191
[49] Konotop V V, Yang J and Zezyulin D A 2016 Nonlinear waves in PT-symmetric systems *Rev. Mod. Phys.* **88** 035002

[50] Cuevas-Maraver J, Kevrekidis P G, Saxena A, Cooper F, Khare A, Comech A and Bender C M 2016 Solitary Waves of a $\mathcal{PT}$-Symmetric Nonlinear Dirac Equation *IEEE: J. Sel. Topics Quantum Electron.* **22** 67
[51] Alexeeva N V, Barashenkov I V and Saxena A 2019 Spinor solitons and their $\mathcal{PT}$-symmetric offspring *Ann. Phys.* **403** 198
[52] Manakov S V 1974 On the theory of two-dimensional stationary self-focusing of electromagnetic waves *Sov. Phys. JETP* **38** 248
[53] Zakharov V E and Schulman E I 1982 To the integrability of the system of two coupled nonlinear Schrödinger equation *Physica D* **4** 270
[54] Kanna T, Lakshmanan M, Dinda P T and Akhmediev N 2006 Soliton collisions with shape change by intensity redistribution in mixed coupled nonlinear Schrödinger equations *Phys. Rev. E* **73** 026604
[55] Wang D-S, Zhang D-J and Yang J 2010 Integrable properties of the general coupled nonlinear Schrödinger equations *J. Math. Phys.* **51** 023510
[56] Musslimani Z H, Makris K G, El-Ganainy R and Christodoulides D N 2008 Optical Solitons in $\mathcal{PT}$ Periodic Potentials *Phys. Rev. Lett.* **100** 030402
Shi Z, Jiang X, Zhu X and Li H 2011 Bright spatial solitons in defocusing Kerr media with $\mathcal{PT}$-symmetric potentials *Phys. Rev. A* **84** 053855
Midya B and Roychoudhury R 2013 Nonlinear localized modes in $\mathcal{PT}$-symmetric Rosen-Morse potential wells *Phys. Rev. A* **87** 045803
Khare A, Al-Marzoug S M and Bahlouli H 2012 Solitons in PT-symmetric potential with competing nonlinearity Author links open overlay panel *Phys. Lett. A* **376** 2880
[57] Ablowitz M J and Musslimani Z H 2013 Integrable Nonlocal Nonlinear Schrödinger Equation *Phys. Rev. Lett* **110** 064105
[58] Sinha D and Ghosh P K 2015 On Symmetries and Exact Solutions of a Class of Non-local Non-linear Schrödinger Equations with Self-induced $\mathcal{PT}$-symmetric Potential *Phys. Rev. E* **91** 042908
[59] Sinha D and Ghosh P K 2017 Integrable nonlocal vector nonlinear Schrödinger equation with self-induced parity-time-symmetric Potential *Phys. Lett. A* **381** 124
[60] Driben R and Malomed B A 2011 Stability of solitons in parity-time-symmetric couplers *Opt. Lett.* **36** 4323
Driben R and Malomed B A 2011 Stabilization of solitons in PT models with supersymmetry by periodic management *EPL* **96** 51001
[61] Abdullaev F. Kh., Konotop V V, Örgen M and Sørensen M P 2011 Zeno effect and switching of solitons in nonlinear couplers *Opt. Lett.* **36** 4566
[62] Barashenkov I V, Suchkov S V, Sukhorukov A A, Dmitriev S V and Kivshar Y S 2012 Breathers in $\mathcal{PT}$-symmetric optical couplers *Phys. Rev. A* **86** 053809(2012)
Barashenkov I V, Baker L and Alexeeva N V 2013 $\mathcal{PT}$-symmetry breaking in a necklace of coupled optical waveguides *Phys. Rev. A* **87** 033819
[63] Pelinovsky D E, Zezyulin D A and Konotop V V 2015 Global Existence of Solutions to Coupled $\mathcal{PT}$-Symmetric Nonlinear Schrödinger Equation *Int. J. Theor. Phys.* **54** 3920
[64] Alexeeva N V, Barashenkov I V, Sukhorukov A A and Kivshar Y S 2012 Optical solitons in $\mathcal{PT}$-symmetric nonlinear couplers with gain and loss *Phys. Rev. A***85** 063837
[65] Dias J.-P, Figueria M, Konotop V V and Zezyulin D A 2014 Supercritical blowup in coupled parity-time-symmetric nonlinear Schrödinger equations *Stud. Appl. Math.* **133** 422
[66] Ghosh P K 2021 Constructing Solvable Models of Vector Non-linear Schrödinger Equation with Balanced Loss and Gain via Non-unitary transformation *Phys. Lett. A***402** 127361
[67] Mostafazadeh A 2004 Pseudounitary operators and pseudounitary quantum dynamics *J. Math. Phys.* **45** 932
[68] Ghosh S and Ghosh P K 2021 *Under preparation*
[69] Scott A C, Lomdahl P S and Eilbeck J C 1985 Between the Local Mode and Normal Mode Limits *Chem. Phys. Letts.* **113** 29.
[70] Susanto H 2009 Few-lattice-site systems of discrete self- trapping equations *In Discrete Nonlinear Schrödinger Equations, P.G. Kevrekidis (Ed.)* pp. 249 (Springer)
[71] Morales J D H, Guerrero J, López-Aguayo S and Rodríguez-Lara B M 2016 Revisiting the Optical $\mathcal{PT}$-Symmetric Dimer *Symmetry* **8** 83
[72] Barashenkov I V and Chernyavsky A 2020 Stable solitons in a nearly $\mathcal{PT}$-symmetric ferromagnet with spin-transfer torque *Physica* D **409** 132481
[73] Barashenkov I V, Pelinovsky D E and Dubbard P 2015 Dimer with gain and loss: Integrability and PT -symmetry restoration *J. of Phys. A: Math. Theor.* **48** 325201
Barashenkov I V 2014 Hamiltonian formulation of the standard $\mathcal{PT}$-symmetric nonlinear Schrödinger dimer *Phys. Rev. A.* **90** 045802
[74] Zezyulin D A, Konotop V V 2018 Solitons in a Hamiltonian $\mathcal{PT}$-symmetric coupler *J. Phys. A: Math. Theor.* **51** 015206

[75] Soler M 1970 Classical, stable, nonlinear spinor field with positive rest energy *Phys. Rev. D* **1** 2766
[76] Gross D J and Neveu A 1974 Dynamical symmetry breaking in asymptotically free field theories *Phys. Rev.* D **10** 3235
[77] Thirring W E 1958 A soluble relativistic field theory *Ann. Phys.* **3** 91
[78] Cuevas-Maraver J, Boussaïd N, Comech A, Lan R, Kevrekidis P G and Saxena A 2018 *Solitary Waves in the Nonlinear Dirac Equation Understanding Complex Systems*(Nonlinear Systems vol 1) ed Carmona V, Cuevas-Maraver J, Fernńdez-Sánchez F and García-Medina E (Springer)
[79] Ng W and Parwani R 2009 Nonlinear dirac equations *SIGMA* **3** 023
[80] Bender C M, Jones H F and Rivers R J 2005 Dual $\mathcal{PT}$-symmetric quantum field theories *Phys. Lett.* B **625** 333
[81] Alexandre J and Bender C M 2015 Foldy–Wouthuysen transformation for non-Hermitian Hamiltonians *J. Phys. A* **48** 185403

Journal of Physics: Conference Series **2038** (2021) 012013 doi:10.1088/1742-6596/2038/1/012013

# What is the resonant state in open quantum systems?

**Naomichi Hatano**

Institute of Industrial Science, The University of Tokyo, 5-1-5 Kashiwanoha, Kashiwa, Chiba 277-8574, Japan

E-mail: hatano@iis.u-tokyo.ac.jp

**Abstract.** The article reviews the theory of open quantum system from a perspective of the non-Hermiticity that emerges from the environment with an infinite number of degrees of freedom. The non-Hermiticity produces resonant states with complex eigenvalues, resulting in peak structures in scattering amplitudes and transport coefficients. After introducing the definition of resonant states with complex eigenvalues, we answer typical questions regarding the non-Hermiticity of open quantum systems. What is the physical meaning of the complex eigenmomenta and eigenenergies? How and why do the resonant states break the time-reversal symmetry that the system observes? Can we make the probabilistic interpretation of the resonant states with diverging wave functions? What is the physical meaning of the divergence of the wave functions? We also present an alternative way of finding resonant states, namely the Feshbach formalism, in which we eliminate the infinite number of the environmental degrees of freedom. In this formalism, we attribute the non-Hermiticity to the introduction of the retarded and advanced Green's functions.

## 1. Introduction: What is the open quantum system?

Open quantum systems are called open in the sense that they exchange particle and energy fluxes with the outside world. The open system is often referred to as the system when the outside world is called the environment.

As a typical example, consider the nuclear scattering problem of radiating neutrons to a nuclide. The nuclide may absorb a neutron, becoming a radioactive isotope, and after its lifetime, it may emit a neutron, becoming back to the original nuclide. We can regard this as an open quantum system, in which the nuclide is the system and the free space with flying neutrons is the environment.

This is a typical setup of the problem of open quantum systems; the system of interest is often a complex but finite system that is embedded in the environment which is a simple and flat space with an infinite number of degrees of freedom. Another example found in condensed-matter physics is a quantum dot attached to quantum leads set in semiconductor hetero-junctions. Indeed, all quantum systems used in experiments must be open because we cannot help making a macroscopic probe interact with the system.

In conventional physics, however, theories often analyze closed systems because it is much easier to treat. As such, experimentalists who wish to reproduce the theoretical setup often must pay extra attention to keeping the system as closed and undisturbed as possible. Physics of open systems lets us take an alternative way; we try to explore novel phenomena that emerge due to the openness of the system. Experimentalists then can take advantage of the fact that they attach experimental apparatus to the system of interest.

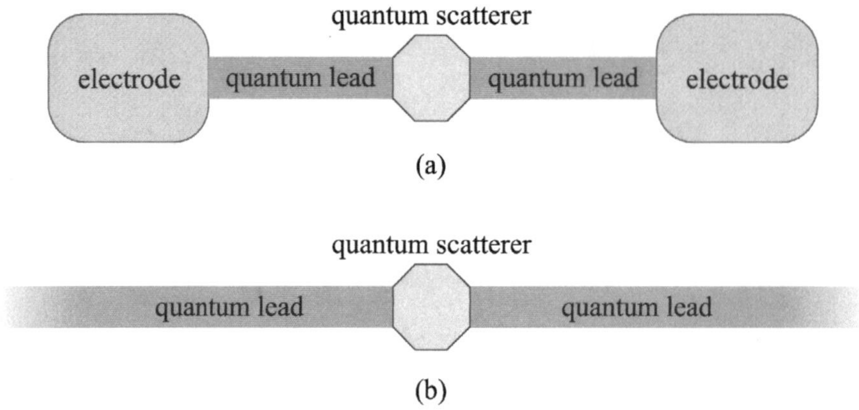


**Figure 1.** (a) Two electrodes are attached to a quantum scatterer by way of quantum leads without impurities. (b) The potential scattering problem of the quantum scatterer.

Characteristic phenomena that emerge in open quantum systems include the non-Hermiticity and resonance. The effective Hamiltonian of open systems can be non-Hermitian; indeed, the probability amplitude and the energy are not conserved due to the interaction with the environment, and hence the Hamiltonian should be non-Hermitian. The Feshbach formalism [1, 2, 3], which we will review later in Section 3, attributes the appearance of the non-Hermiticity to the elimination of the infinite degrees of freedom of the environment.

Since the Hamiltonian is non-Hermitian, it can harbor eigenstates with complex eigenvalues. In fact, such eigenstates are none other than the resonant states [4, 5, 6, 7, 8, 9, 10]. The imaginary part of the complex eigenvalue corresponds to the inverse lifetime of the resonance. In this framework of understanding open quantum systems, the radioactive isotope is indeed a resonant state of the nuclide with a complex eigenvalue whose imaginary part is equal to half the inverse lifetime of the isotope.

Another incident in which the resonant state manifests itself is the peak structure of transport coefficients. Suppose that one tries to measure the electric conductance of a quantum scatterer. One would attach two electrodes to the scatterer with quantum leads without impurities; see Fig. 1(a). Electrons that one electrode emits are transported ballistically through the quantum lead, and are scattered quantum mechanically at the scatterer. While the reflected electrons go back to the original electrode, the transmitted electrons go to the other electrode, contributing to the electric conduction. The Landauer formula [11, 12] dictates that the transmission coefficient of the quantum scattering problem gives the electric conductance $G$ in the form

$$G = \frac{2e^2}{h} T(E_{\mathrm{F}}), \tag{1}$$

where $h$ denotes the Planck constant and $T(E_{\mathrm{F}})$ is the transmission coefficient at the Fermi energy $E_{\mathrm{F}}$ of the quantum scatterer.

One might wonder why the electric conductivity of the finite section that are sandwiched by the two electrodes as in Fig. 1(a) is given by solving the scattering problem that are set in an infinite space as in Fig 1(b). It is because each electrode has by definition an infinite number of degrees of freedom, and hence electrons that are absorbed into the electrode will never come out of it coherently. This situation is the same as in the quantum scattering problem, in which the reflected and transmitted waves leave to the negative and positive infinites, and will never come back to the scattering area. For this reason, the Landauer formula (1) lets us measure the transmission coefficient in solid-state setups, which may exhibit a resonant peak whose width is proportional to the imaginary part of the complex eigenvalue of the resonant state.

We will come back to this point later in Subsection 2.6; we will attribute the fact that the resonant wave function diverges in the limit $x \to \infty$ in the setup of Fig. 1(b) to the fact that the electrodes contain a macroscopic number of electrons in the setup of Fig. 1(a).

## 2. What is the resonant state?

### *2.1. Dynamical and static pictures of resonance phenomena*

Let us first note two different ways of understanding resonance. One is dynamic and the other is static.

In the dynamic picture of resonance phenomena, one would consider the time evolution of an initial state in which a wave packet comes into a scattering potential. The wave packet would be then captured by a scattering potential, stay there during the resonance lifetime, and tunnel out of the potential. This picture is particularly effective when the system is weakly coupled to the environment. The system without coupling would have bound states. The system with weak coupling then would capture the wave packet well if the energy of the wave packet 'resonates' with a bound state. This would result in a peak in the scattering amplitude. This picture of the resonance, however, does not work well once the system-environment coupling becomes strong; the resonance peak would shift from the bound-state energy considerably.

We here instead focus on the static picture of resonance; we describe the resonance in terms of resonance eigenvalues. In order to emphasize the difference, let us give an example of coupled pendulums from classical mechanics. The equation of motion takes the form

$$\ddot{x}_1 = -\omega^2 x_1 - \alpha(x_1 - x_2), \tag{2}$$

$$\ddot{x}_2 = -\omega^2 x_2 - \alpha(x_2 - x_1), \tag{3}$$

where $x_1$ and $x_2$ denote the amplitudes of the oscillation of the respective pendulums, $\omega$ denotes their common eigenfrequency, and $\alpha$ denotes the coupling between them. In the dynamic picture, we understand the coupled pendulums in the following way. When we start the time evolution from the initial state in which only one of the pendulum, say $x_1$, oscillates, the oscillation diminishes gradually while the oscillation of the other pendulum, $x_2$, is enhanced. After the oscillation of the first pendulum stops and that of the second one reaches its maximum, the time evolution starts going back to the initial state. This is a familiar description of the double pendulum. Indeed, it is the dynamical picture of this resonance phenomenon. In fact, the initial state is *not* an eigenstate of the dynamics, and hence is unstable.

The static picture instead offers the following description. We find the eigenstates by casting Eqs. (2)-(3) into the matrix form

$$\frac{\mathrm{d}^2}{\mathrm{d}t^2}\begin{pmatrix} x_1 \\ x_2 \end{pmatrix} = -\begin{pmatrix} \omega^2+\alpha & -\alpha \\ -\alpha & \omega^2+\alpha \end{pmatrix}\begin{pmatrix} x_1 \\ x_2 \end{pmatrix}, \tag{4}$$

and by diagonalizing the matrix on the right-hand side. The eigenstates are given as follows:

$$\begin{pmatrix} 1 \\ 1 \end{pmatrix} \quad \text{for the eigenvalue } \omega^2, \tag{5}$$

$$\begin{pmatrix} 1 \\ -1 \end{pmatrix} \quad \text{for the eigenvalue } \omega^2 + 2\alpha, \tag{6}$$

The first eigenstate indicates the state in which the two pendulums oscillate with the same phase as in $x_1 = x_2$ while the second eigenstate indicates the one in which they oscillate with the phase difference $\pi$ as in $x_1 = -x_2$. Their eigenvalues tell us that the first mode oscillates at the original frequency while the second one does at a shifted frequency. *This* is the static picture of the resonance of the double pendulum. The dynamic picture is recovered by superimposing the two eigenstates accompanied by time-evolving phases.

In the following, we describe quantum resonances in terms of the latter static picture. We find resonant eigenstates and eigenvalues of the standard potential scattering problem of the time-independent Schrödingerequation in one dimension:

$$H\psi(x) = E\psi(x) \tag{7}$$

with the Hamiltonian

$$H = -\frac{\hbar^2}{2m}\frac{\mathrm{d}^2}{\mathrm{d}x^2} + V(x) \tag{8}$$

where $V(x)$ is a real function that exists only in a finite region $-\ell \le x \le \ell$.

In the following subsections, we first present in Subsect. 2.2 the definition of resonance as an eigenstate of the Schrödingerequation under a special boundary condition. The well-known eigenstates of the Schrödingerequation are the bound states and the scattering states. The bound states are normalizable, while the scattering states are box-normalizable and conserve the probability flux. In addition to these eigenstates, we here define unnormalizable eigenstates of the Schrödingerequation, called the resonant states, the anti-resonant states and the anti-bound states. The first two in particular are the solutions with broken time-reversal symmetry, which we discuss in Subsect. 2.3, and with complex eigenvalues, which we discuss in Subsect. 2.4. Finally, we consider the unnormalizability of the resonant and anti-resonant states. We will show in Subsect. 2.5 that the unnormalizability is in fact a necessary condition for the probability conservation; the wave functions of these states diverge in space but decay in time, and thereby conserves the probability. We also consider in Subsect. 2.6 the unnormalizability from the viewpoint of the Landauer formula in Eq. (1) and in Fig. 1.

### *2.2. Defining the resonance as an eigenstate of the Schrödingerequation*

A textbook definition of the resonance is the pole of the scattering amplitude. Let us set the following boundary conditions on the scattering wave function $\psi(x)$:

$$\psi(x) = \begin{cases} Ae^{ikx} + Be^{-ikx} & \text{for } x \le -\ell, \\ Ce^{ikx} & \text{for } x \ge \ell, \end{cases} \tag{9}$$

where the positive wave number $k$ is related to the incident energy $E$ as in $E = \hbar^2 k^2/(2m)$, while the amplitudes $A$, $B$ and $C$ are generally complex numbers that are fixed as functions of $E$ after solving the connection conditions of the wave functions. The reflection and transmission probabilities are then given by the energy-dependent functions $R(E) = |B/A|^2$ and $T(E) = |C/A|^2$, respectively. They must be in the range of $[0, 1]$ for $E > 0$, but can diverge at multiple points on the complex $E$ plane, which most textbooks define as the resonant states.

The poles of $R$ and $T$ (in the complex $E$ plain) are actually identical to the zeros of the incident-wave amplitude $A(E)$. This observation lets us define the boundary conditions for the resonant eigenstates by removing the incident wave from Eq. (9):

$$\psi(x) = \begin{cases} Be^{-ikx} & \text{for } x \le -\ell, \\ Ce^{ikx} & \text{for } x \ge \ell, \end{cases} \tag{10}$$

which is indeed called the Siegert boundary condition. We have outgoing waves only if $\mathrm{Re}\, k > 0$ and incoming waves only if $\mathrm{Re}\, k < 0$.

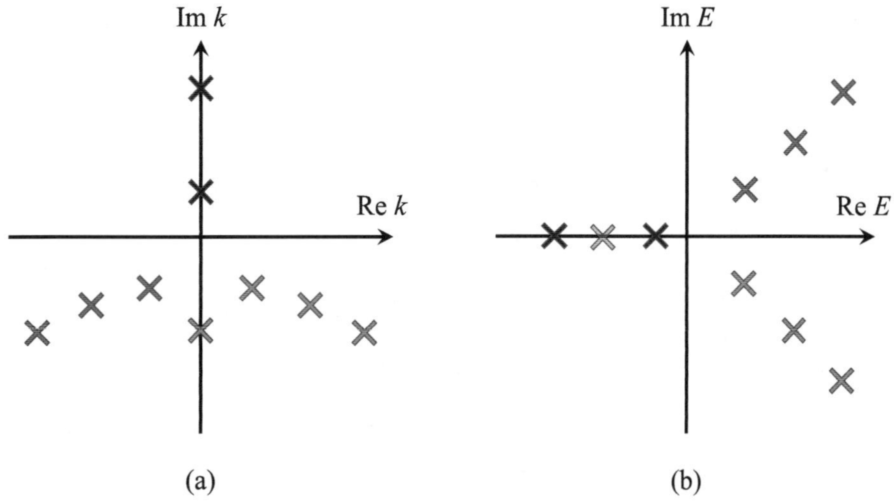


**Figure 2.** The point spectra for the scattering problem under the boundary (a) in the complex $k$ plain and (b) in the complex $E$ plain. In the complex $k$ plain (a), the bound states are located on the positive imaginary axis, the anti-bound states are on the negative imaginary axis, the resonant states are in the fourth quadrant, and the anti-resonant states are in the third quadrant. In the complex $E$ plain (b), the bound states are on the negative real axis in the first Riemann sheet, the anti-bound states are on the negative real axis in the second Riemann sheet, the resonant states are in the lower half of the second Riemann sheet, and the anti-resonant states are in the upper half of the second Riemann sheet.

In fact, the Siegert boundary condition (10) locates all point spectra, or all the discrete eigenvalues. Figure 2 shows a schematic picture of a typical distribution of discrete eigenvalues on the complex $k$ plain and the complex $E$ plain. The eigenstates located on the positive imaginary axis of the complex $k$ plain are the bound states. Indeed, inserting $k = i\kappa$ with $\kappa > 0$ into the Siegert boundary condition (10) makes it the exponentially decaying wave function of a bound state. The corresponding eigenvalue $E = -\hbar^2\kappa^2/(2m)$ is a real negative number, which is also consistent with our standard understanding of the bound state.

The eigenstates located in the fourth quadrant of the complex $k$ plain are called the resonant states, while those in the third quadrant are called the anti-resonant states, presumably for historical reasons. The energy eigenvalues of the resonant states are located in the lower half of the second Riemann sheet of the complex $E$ plain, while those of the anti-resonant states are in its upper half. We will explore the physical significance of these eigenvalues in the following subsections.

In the meantime, there are also eigenvalues on the negative imaginary axis of the complex $k$ plain. These states are often called the anti-bound states or the virtual bound states. The corresponding energy eigenvalues are located on the negative real axis of the second Riemann sheet of the complex $E$ plain. We will not discuss them much in the present article because they scarcely contribute to physical observables.

Let us mention here the fact that the wave functions of the resonant, anti-resonant and anti-bound states diverge in space. For these states, inserting $\operatorname{Im} k = i\kappa$ with $\kappa < 0$ into the Siegert boundary condition (10) reveal thethey exponential divergence of the form $|\psi(x)| \sim \exp(|\kappa||x|)$. This is presumably the reason why these states are sometime called unphysical. We hereafter try to convince the readers that they are totally physical objects.

A remark on the point spectra and the continuum spectrum is in order here. Let us assume the simplest scattering problem of a square-well potential:

$$V(x) = \begin{cases} -V_0 & \text{for } |x| \le \ell, \\ 0 & \text{for } |x| > \ell, \end{cases} \tag{11}$$

where $V_0 > 0$. We then set the wave function $Fe^{ik'x} + Ge^{-ik'x}$ with $k'^2 = k^2 + 2mV_0/\hbar^2$ for the region $[-\ell, \ell]$. In finding the scattering wave function (9), we demand two connection conditions for the continuity of the wave function and its derivative at each point of $x = \pm\ell$, which amounts to four conditions. Meanwhile we have five unknown variables, namely, the four ratios $A/B$, $C/B$, $F/B$, $G/B$ and the incident energy $E$ (which determines $k$ and $k'$). We therefore cannot fix all unknown variables. This is why we have continuous spectrum of $E$ for the scattering wave function (9). In finding the wave function (10) under the Siegert boundary condition, on the other hand, one unknown variable $A$ is missing. We demand four conditions to fix four unknown variables including $E$. This is why we have point spectra of $E$ for the wave function (10) as exemplified in Fig. 2.

Let us now analyze the resonant and anti-resonant states by adding the time-dependent factor to the wave function in Eq. (10):

$$\Psi(x,t) = e^{-iEt/\hbar}\psi(x) = e^{-iEt/\hbar} \times \begin{cases} Be^{-ikx} & \text{for } x \le -\ell, \\ Ce^{ikx} & \text{for } x \ge \ell \end{cases} \tag{12}$$

which is the solution of the time-dependent Schrödingerequation

$$i\hbar\frac{\partial}{\partial t}\Psi(x,t) = H\Psi(x,t); \tag{13}$$

note, however, that since $E$ is generally complex, the factor $e^{-iEt/\hbar}$ in Eq. (12) is generally *not* a phase.

Figure 2 tells us that the resonant states satisfy $\operatorname{Re} k > 0$ and $\operatorname{Im} E < 0$. Therefore, the wave function (12) implies the following: (i) the wave amplitude that has been captured by the potential leaves the scattering region $-\ell \le x \le \ell$ out onto the infinity $\pm\infty$; (ii) because of this, the wave amplitude decreases in time. This is the picture that the resonant state indicates as an eigenstate of the Schrödingerequation. On the other hand, the anti-resonant states satisfy $\operatorname{Re} k < 0$ and $\operatorname{Im} E > 0$, for which Eq. (12) implies that (iii) the waves gather from the infinity into the scattering region to be captured by the potential, and therefore (iv) the wave amplitude increases in time.

One might wonder in what physical situations the anti-resonant states emerge. We can answer this question by reminding ourselves of the dynamical picture of the resonance. The dynamics of the wave amplitude in the scattering region is recovered by superimposing the states of the point spectra. The resonant states dominate the dynamics in the initial-condition problem, in which we ask what will happen *after* the wave amplitude is concentrated in the scattering region. The anti-resonant states, on the other hand, dominate the dynamics in the terminal-condition problem, in which we ask what has happened *before* the wave amplitude is concentrated. Of course, it feels much more natural to us to ask the initial-condition problem, but it is only because we as human beings have the memory of the past but not the memory of the future. In other words, the quantum dynamics itself is time-reversal symmetric with respect to the exchange of the past and the future.

Indeed, K. Murch and his group, in their experiments of weak measurement, accumulated the observation data of quantum trajectories of a two-level system with leaks into the environment, and showed the following in the data analysis. Averaging the trajectories that have a common

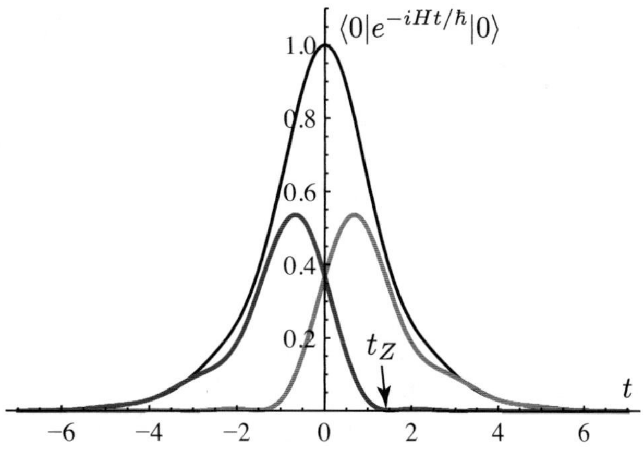


**Figure 3.** Time-dependence of the survival probability of the potential site (denoted by $|0\rangle$ here) of the tight-binding model Hamiltonian $H$. The black curve with the highest central peak is the total survival probability, while the red curve with the peak on the right is the contribution of the resonant state and the blue curve on the left is the contribution of the anti-resonant state. Taken from Ref. [13]. The contributions switch from the latter to the former within the quantum Zeno time, which is denoted by $t_Z$ here.

initial condition produces the familiar Rabi oscillation with decaying, while averaging the ones that have a common terminal condition produces the Rabi oscillation with increasing amplitude towards the terminal condition. This is a manifestation of the anti-resonant contributions in the dynamical picture.

As another way of explaining the anti-resonant states, we here mention the finding in Ref. [10] that all states with the point spectra in Fig. 2 spans the functional space if we restrict the real space $x$ inside the scattering region $-\ell \leq x \leq \ell$. We can thereby breakdown any dynamics inside the system into the contributions of the states with point spectra including resonant and anti-resonant states. For example, the survival probability [13] on the potential site at the origin of the tight-binding model of infinite size depends on time as in Fig. 3. While the decay of the survival probability for $t > 0$ is dominated by the resonant state, the survival probability grows for $t < 0$, which is dominated by the anti-resonant state. The profile is time-reversal symmetric with respect to $t = 0$, which reflects the time-reversal symmetry between the resonant and anti-resonant states as we will argue in the next subsection 2.3.

### *2.3. Broken time-reversal symmetry of the resonant state*

Although the dynamics is time-reversal symmetric, we here stress that each of the resonant and anti-resonant states does break the time-reversal symmetry. Since they appear always as a pair that is symmetric with respect to the imaginary $k$ axis and the real $E$ axis, the system of all solutions indeed keeps the time-reversal symmetry, which the original Schrödingerequation (13) accommodates.

One might ask why an eigenstate can break the time-reversal symmetry. Since the Hamiltonian $H$ in Eq. (8) is time-reversal symmetric, it commutes with the time-reversal operator, which we denote with $T$ hereafter. Namely, we have $[H, T] = 0$. One would then claim that all the eigenstates of the Hamiltonian $H$ must be the eigenstates of the time-reversal operator $T$ too, and hence all eigenstates must be time-reversal symmetric. This deduction is not correct because the time-reversal operator $T$ is *not* a linear operator.

The correct deduction is given as follows. Suppose that an eigenstate $|\psi_n\rangle$ of the Hamiltonian

$H$ has a non-degenerate eigenvalue $E_n$; that is, $H|\psi_n\rangle = E_n|\psi_n\rangle$. Applying the time-reversal operator $T$ to both sides of the equation from the left, we have

$$TE_n|\psi_n\rangle = TH|\psi_n\rangle = HT|\psi_n\rangle, \tag{14}$$

where we used the fact $[H,T] = 0$ for the second equality. If $T$ were linear, the left-hand side would be transformed to $E_nT|\psi_n\rangle$, and hence $T|\psi_n\rangle$ would be also an eigenstate of $H$ with the same eigenvalue $E_n$, so that $T|\psi_n\rangle \propto |\psi_n\rangle$. In truth, $T$ is an anti-linear operator, and hence the left-hand side is transformed to ${E_n}^*T|\psi_n\rangle$. We therefore find the following two possibilities: if $E_n$ is a real eigenvalue, indeed $T|\psi_n\rangle \propto |\psi_n\rangle$, and hence the eigenstate is time-reversal symmetric; if $E_n$ is a complex eigenvalue, Eq. (14) tells us that $T|\psi_n\rangle$ is a time-reversal eigenstate of $|\psi_n\rangle$ with a complex-conjugate eigenvalue ${E_n}^*$, which implies that we always have a pair of eigenstates that are time-reversal to each other with a complex-conjugate pair of eigenvalues. The latter possibility is none other than what we observed near the end of the previous subsection 2.2.

### *2.4. Why does a 'Hermitian' Hamiltonian have complex eigenvalues?*

One might also wonder why the seemingly Hermitian Hamiltonian $H$ in Eq. (8) can have complex eigenvalues. In fact, whether an operator is Hermitian or not depends on what functional space the operator is defined in. Indeed, the Hamiltonian $H$ is Hermitian in the functional space of bound states and scattering states, and hence these states can only have real eigenvalues, while it is non-Hermitian in the functional space of resonant and anti-resonant states, and hence these states can have complex eigenvalues.

Let us observe in Fig. 2(a) that $\operatorname{Im} k < 0$ for all resonant and anti-resonant states. Inserting this into the wave function (10), we find that it exponentially diverges in $x \to \pm\infty$. For such functions, the first term of the Hamiltonian $H$ can be non-Hermitian, which we show as follows. Consider the matrix element

$$\langle\psi|\frac{\mathrm{d}^2}{\mathrm{d}x^2}|\psi\rangle; \tag{15}$$

showing that the matrix element is real for any functions $|\psi\rangle$ would prove that the operator $\mathrm{d}^2/\mathrm{d}x^2$ is Hermitian. Note here that we must specify the functional space to find functions $|\psi\rangle$ in, which we will be the focus of the argument below.

For the sake of the argument, we first evaluate

$$\langle\psi|\frac{\mathrm{d}^2}{\mathrm{d}x^2}|\psi\rangle_L := \int_{-L}^{L}\psi(x)^*\frac{\mathrm{d}^2}{\mathrm{d}x^2}\psi(x)dx \tag{16}$$

for $L > \ell$ and then take the limit $L \to \infty$ whenever it is possible. We can transform Eq. (16) by means of partial integration:

$$\langle\psi|\frac{\mathrm{d}^2}{\mathrm{d}x^2}|\psi\rangle_L = \left[\psi(x)^*\frac{\mathrm{d}}{\mathrm{d}x}\psi(x)\right]_{x=-L}^{L} - \int_{-L}^{L}\left(\frac{\mathrm{d}}{\mathrm{d}x}\psi(x)\right)^*\left(\frac{\mathrm{d}}{\mathrm{d}x}\psi(x)\right)dx. \tag{17}$$

The second term on the right-hand side is real because the integrand $|\mathrm{d}\psi/\mathrm{d}x|^2$ is real. The first term, on the other hand, can be complex.

The imaginary part of the first term is actually the real part of

$$\psi(x)^*\hat{p}\psi(x)\Big|_{x=L} - \psi(x)^*\hat{p}\psi(x)\Big|_{x=-L}, \tag{18}$$

where $\hat{p}$ denotes the momentum operator

$$\hat{p} := \frac{\hbar}{i}\frac{\mathrm{d}}{\mathrm{d}x}. \tag{19}$$

The physical meaning of Eq. (18) becomes more evident when we evaluate it for potential scattering problems in three dimensions; Eq. (16) would then be replaced by the volume integral

$$\langle\psi|\nabla^2|\psi\rangle_\Omega := \int_\Omega \psi(\vec{x})^* \nabla^2 \psi(\vec{x}) dV \tag{20}$$

for a region $\Omega$ that contains the support of the scattering potential, while Eq. (18) would be replaced by the surface integral

$$\int_{\partial\Omega} \hat{\vec{p}} \cdot \vec{n}\, dS, \tag{21}$$

where $\vec{n}$ denotes the unit vector that is normal to the surface $\partial\Omega$ of the volume $\Omega$. Equation (21) indicates the momentum flux that is going out of the volume $\Omega$, and similarly Eq. (18) is the momentum flux that is going out of the region $[-L, L]$.

Therefore, Eq. (18) should vanish in the limit $L \to \infty$ for the bound states as well as for the scattering states (9) with the flux conservation $|A|^2 = |B|^2 + |C|^2$. Therefore, we can take the limit $L \to \infty$ in Eq. (16), and thereby prove that Eq. (15) is real for any bound states and for any scattering states. We thus conclude that the Hamiltonian $H$ is Hermitian in the functional spaces of the bound states and the scattering states, and therefore the eigenvalues of these states must be real.

On the other hand, we realize that Eq. (18) should *not* vanish for the resonant and anti-resonant states, when we remember the fact that the amplitude goes out of the scattering region for the resonant states and comes into the region for the anti-resonant states, as we discussed near the end of Subsection 2.2; namely, we have $|B|^2 + |C|^2 > |A|^2 = 0$. In fact, we cannot take the limit $L \to \infty$ in Eq. (16) because the wave functions for these states spatially diverge in $x \to \pm\infty$, and the first term on the right-hand side of Eq. (17) can have an imaginary part in general. Our conclusion in this case is that the Hamiltonian $H$ is *non*-Hermitian in the functional spaces of the resonant and anti-resonant states, and hence the eigenvalues for these states are allowed to be complex.

### *2.5. Probabilistic interpretation of the resonant state*

The answer to the question in the previous subsection 2.4 might spawn yet another question; could a spatially diverging wave function of the resonant and anti-resonant states accommodate the probabilistic interpretation of the wave function? In fact, this may be the reason why many researchers have not paid much attention to the resonant states as eigenstates. Obviously, the wave functions of the resonant and anti-resonant states are not normalizable. Yet, we here present a way of probabilistic interpretation of the diverging wave function. Indeed, the divergence is necessary for the following probabilistic interpretation, being precisely cancelled by the temporal decay.

Let us consider the probability amplitude of a resonant eigenstate $\Psi_n(x,t)$ that is contained in a region $[-L, L]$, where $L > \ell$. We now remember that for a resonant eigenstate, the probability amplitude leaks out of the scattering region onto $\pm\infty$. Therefore, in order to analyze the probability conservation for a resonant state, we must chase the leaking probability amplitude by expanding the integration range $[-L, L]$. We may evaluate the leaking speed as follows; since the real part of the momentum is $\mathrm{Re}\,\hbar k_n$, where $(\hbar k_n)^2/(2m) = E_n$, let us guess that the speed is $\mathrm{Re}\,\hbar k_n/m$. We therefore expand the integration range $[-L, L]$ as in

$$L(t) = L_0 + \frac{\mathrm{Re}\,\hbar k_n}{m} t, \tag{22}$$

where $L_0 > \ell$, so that we define

$$N_n(t) = \int_{-L(t)}^{L(t)} |\Psi_n(x,t)|^2 dx. \tag{23}$$

We are now in a position to prove $\dot{N}_n = 0$. We find

$$\dot{N}_n(t) = \int_{-L(t)}^{L(t)} \frac{\mathrm{d}}{\mathrm{d}t} |\Psi_n(x,t)|^2 dx + \dot{L}(t)\Big(|\Psi_n(L(t),t)|^2 + |\Psi_n(-L(t),t)|^2\Big) \tag{24}$$

$$= \int_{-L(t)}^{L(t)} \left(\Psi_n(x,t)^* \frac{\mathrm{d}}{\mathrm{d}t}\Psi_n(x,t) + \Psi_n(x,t)\frac{\mathrm{d}}{\mathrm{d}t}\Psi_n(x,t)^*\right) dx$$
$$+ \frac{\mathrm{Re}\,\hbar k_n}{m}\Big(|\Psi_n(L(t),t)|^2 + |\Psi_n(-L(t),t)|^2\Big). \tag{25}$$

We prove that the first term precisely cancels the second term on the right-hand side.

Utilizing the time-dependent Schrödingerequation (13), we can transform the first term as follows:

(first term on the right-hand side of Eq. (25))

$$= \frac{1}{i\hbar}\int_{-L(t)}^{L(t)} [\Psi_n(x,t)^* H\Psi_n(x,t) - (\Psi_n(x,t)^* H\Psi_n(x,t))^*] dx \tag{26}$$

$$= \frac{2}{\hbar}\,\mathrm{Im}\int_{-L(t)}^{L(t)} \Psi_n(x,t)^* H\Psi_n(x,t) dx \tag{27}$$

$$= -\frac{\hbar}{m}\,\mathrm{Im}\int_{-L(t)}^{L(t)} \Psi_n(x,t)^* \Psi_n''(x,t) dx; \tag{28}$$

in the last transformation, we used the fact that the potential $V(x)$ is a real function. After the partial integration of Eq. (28), we find

$$(\text{Eq. (28)}) = -\frac{\hbar}{m}\,\mathrm{Im}\,\big(\Psi_n(L(t),t)^*\Psi_n'(L(t),t) - \Psi_n(-L(t),t)^*\Psi_n'(-L(t),t)\big), \tag{29}$$

where the integral part vanishes because it is purely real.

We further use the fact that the resonant eigenfunction behaves as Eq. (12) in the region $|x| > \ell$, which transforms Eq. (29) further as

$$(\text{Eq. (29)}) = -\frac{\hbar}{m}\,\mathrm{Im}\left(ik_n|\Psi_n(L(t),t)|^2 + ik_n|\Psi_n(-L(t),t)|^2\right) \tag{30}$$

$$= -\frac{\hbar}{m}\,\mathrm{Re}\,k_n\Big(|\Psi_n(L(t),t)|^2 + |\Psi_n(-L(t),t)|^2\Big), \tag{31}$$

which precisely cancels the second term on the right-hand site of Eq. (25). We thereby conclude that $\dot{N}_n(t) \equiv 0$, which proves that the probability amplitude contained in the expanding range $[-L(t), L(t)]$ is conserved, offering us a probability interpretation despite that the eigenfunction itself is not normalizable.

In Eq. (24), the first term on the right-hand side describes the temporal exponential decrease of the wave amplitude, while the second term describes the spatial exponential increase. The fact that these two terms cancel each other implies that the spatial divergence is indeed necessary for the probability conservation of the resonant state; it is nothing but physical.

### *2.6. Physical meaning of the spatially divergent wave function*

In the previous subsection (2.5), we confirmed that the spatial divergence of the resonant eigenfunction is indeed necessary for the probability to be conserved. Nonetheless, one might still wonder what would be the physical meaning of the divergence.

The Landauer formula (1) gives us a hint. Of course, we cannot set up the infinite space of Fig. 1(b) for the potential scattering problem. In reality, the space for the scattering waves is terminated by electrodes and other experimental probes. The Landauer formula implies the equivalence between the two setups.

We then realize that the leaking probability fluxes in the resonant states do not really leave onto $\pm\infty$ in Fig. 1(b) but end up in the electrodes in Fig. 1(a). Indeed, when we try to measure the electric conductance of a quantum scatterer, the electrodes are macroscopic objects, and hence should contain a macroscopic number of electrons, while the quantum scatterer is a microscopic or mesoscopic object, and hence contain only a microscopic number of electrons. We can thereby understand that the spatial divergence of the resonant eigenfunction physically corresponds to the number of electrons in the electrodes of the order of the Avogadro number.

## 3. Feshbach formalism for the complex eigenvalues

### *3.1. Nonlinear eigenvalue problem and non-Markovian time evolution*

In the previous section 2, we notice that the Hamiltonian acquires non-Hermiticity because of the Siegert boundary condition (10). In the present section, we present an alternative approach to open quantum systems called the Feshbach formalism, in which we find an explicitly non-Hermitian effective Hamiltonian of the system.

Let $P$ and $Q$ denote the operators that project any states onto the subsystems of the system and the environment, respectively. If the scattering potential $V(x)$ exists only in the region $[-\ell,\ell]$, then for any wave function $\psi(x)$,

$$P\psi(x) = \begin{cases} \psi(x) & \text{for } -\ell \le x \le \ell, \\ 0 & \text{for } x < -\ell \text{ or } x > \ell, \end{cases} \tag{32}$$

$$Q\psi(x) = \begin{cases} 0 & \text{for } -\ell \le x \le \ell, \\ \psi(x) & \text{for } x < -\ell \text{ or } x > \ell. \end{cases} \tag{33}$$

We have $P^2 = P$, $Q^2 = Q$, $PQ = QP = 0$, and $P + Q = I$, where $I$ is the identity operator.

What we try to achieve is to find the Schrödingerequation for the wave function only inside the system, namely $P\left|\psi\right\rangle$. In order to do so, we operate $P$ and $Q$ to the time-independent Schrödingerequation (7):

$$PH(P+Q)\left|\psi\right\rangle = EP\left|\psi\right\rangle, \tag{34}$$

$$QH(P+Q)\left|\psi\right\rangle = EQ\left|\psi\right\rangle. \tag{35}$$

We can cast these equations into the following forms:

$$PHP(P\left|\psi\right\rangle) + PHQ(Q\left|\psi\right\rangle) = E(P\left|\psi\right\rangle), \tag{36}$$

$$QHP(P\left|\psi\right\rangle) + QHQ(Q\left|\psi\right\rangle) = E(Q\left|\psi\right\rangle). \tag{37}$$

We can represent $Q\left|\psi\right\rangle$ in terms of $P\left|\psi\right\rangle$ using the second equation (37):

$$Q\left|\psi\right\rangle = \frac{1}{E - QHQ} QHP(P\left|\psi\right\rangle). \tag{38}$$

Inserting this expression into the first equation (36), we find

$$H_{\text{eff}}(E)(P|\psi\rangle) = E(P|\psi\rangle), \tag{39}$$

where

$$H_{\text{eff}}(E) := PHP + PHQ\frac{1}{E - QHQ}QHP. \tag{40}$$

This is exactly what we tried to achieve. Note that the eigenvalue problem (39) is a nonlinear one because the Hamiltonian on its left-hand side is a nonlinear function of $E$.

We can understand the effective Hamiltonian $H_{\text{eff}}$ in Eq. (40) in the following way. The first term $PHP$ on the right-hand side of Eq. (40) is the system Hamiltonian. The second term, on the other hand, represents the effect of the environmental Hamiltonian $QHQ$. The wave amplitude in the system, namely $P|\psi\rangle$, is transferred to the environment by the coupling Hamiltonian $QHP$, propagates around the environment under the Green's function $(E - QHQ)^{-1}$ of the environmental Hamiltonian $QHQ$, and then the surviving amplitude is transferred transferred back to the system by the coupling Hamiltonian $PHQ$.

This view becomes more clear when we apply the Feshbach projection formalism to the time-dependent Schrödingerequation (13). Instead of Eqs. (36)–(37), we now have

$$i\hbar\frac{\partial}{\partial t}(P|\Psi(x,t)\rangle) = PHP(P|\Psi(x,t)\rangle) + PHQ(Q|\Psi(x,t)\rangle), \tag{41}$$

$$i\hbar\frac{\partial}{\partial t}(Q|\Psi(x,t)\rangle) = QHP(P|\Psi(x,t)\rangle) + QHQ(Q|\Psi(x,t)\rangle). \tag{42}$$

We find the formal solution of Eq. (42) in the form

$$Q|\Psi(x,t)\rangle = \int_0^t d\tau e^{iQHQ(t-\tau)/\hbar} QHP(P|\Psi(x,\tau)\rangle), \tag{43}$$

where we assumed $Q|\Psi(x,0)\rangle = 0$ for simplicity. Inserting Eq. (43) into Eq. (41), we arrive at

$$i\hbar\frac{\partial}{\partial t}(P|\Psi(x,t)\rangle) = PHP(P|\Psi(x,t)\rangle) + \int_0^t d\tau PHQ\, e^{iQHQ(t-\tau)/\hbar}\, QHP(P|\Psi(x,\tau)\rangle). \tag{44}$$

Indeed, the first term on the right-hand side gives a Markovian time evolution due to the system Hamiltonian $PHP$, while the second term gives a non-Markovian time evolution due to the environmental Hamiltonian $QHQ$; see Fig. 4 for a schematic view. The time derivative of the system state $P|\Psi(x,t)\rangle$ appears to depend on the system state in the past $P|\Psi(x,\tau)\rangle$ directly. However, the behind-the-scenes story is that the state $P|\Psi(x,\tau)\rangle$ is transferred to the environment by the coupling Hamiltonian $QHP$, evolves in time according to the environmental Hamiltonian $QHQ$ over the period of time $(t-\tau)$, and then is transferred back to the system by the coupling Hamiltonian $PHQ$ at time $t$. We emphasize that the $E$ dependence of the effective Hamiltonian $H_{\text{eff}}$ in Eq. (40) corresponds to the non-Markovian memory effect in Eq. (44).

### *3.2. Non-Hermiticity of the effective Hamiltonian $H_{eff}$*

We now argue that the effective Hamiltonian $H_{\text{eff}}$ in Eq. (40) appears to be Hermitian, but it is *not* if the environmental Hamiltonian $QHQ$ has an infinite number of degrees of freedom, and as a consequence has a continuous spectrum. The non-Hermiticity emerges when the value of $E$ falls on the continuous spectrum of $QHQ$, for which the Green's function $(E - QHQ)^{-1}$ is singular.

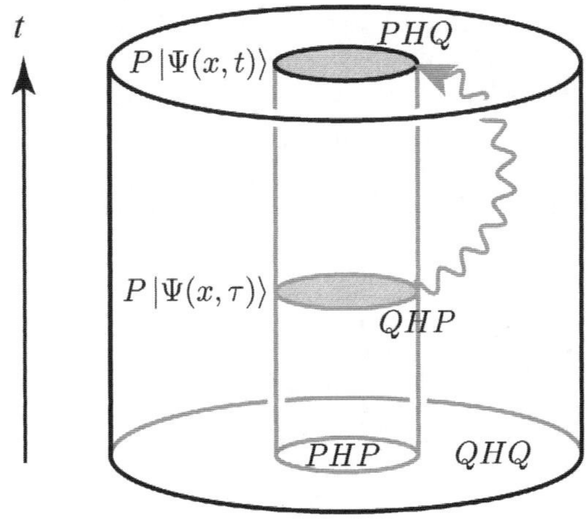


**Figure 4.** A schematic view of the non-Markovian time evolution of the system through the propagation in the environment. The system state $P\,|\Psi(x,\tau)\rangle$ is transferred to the environment by the coupling Hamiltonian $QHP$, evolves in time during the period $t-\tau$ in the environment, and is transferred back to the system by the coupling Hamiltonian $PHQ$ to be $P\,|\Psi(x,t)\rangle$, which thus depends on the state in the past indirectly.

In such cases, what we do is to introduce an infinitesimal to the denominator of the Green's function. Specifically, we have retarded and advanced Green's functions in the forms

$$G^{\mathrm{ret}}(E) := \frac{1}{E-(QHQ-i\eta)}, \tag{45}$$

$$G^{\mathrm{adv}}(E) := \frac{1}{E-(QHQ+i\eta)}, \tag{46}$$

where $\eta$ is a positive infinitesimal constant. We will now give a simple example to show [12] that the retarded Green's function describes the outgoing waves, Eq. (10) with $\mathrm{Re}\,k>0$, while the advanced Green's function describes the incoming waves, Eq. (10) with $\mathrm{Re}\,k<0$.

Let us consider the flat space for the environment, for example:

$$QHQ = -\frac{\hbar^2}{2m}\frac{\mathrm{d}^2}{\mathrm{d}x^2} \tag{47}$$

for $|x|>\ell$. For the retarded Green's function (45), the singularity now falls on the point where the eigenvalue of $QHQ$ is equal to $E+i\eta$. Assuming that the eigenvalue of the operator (47) is of the form $\hbar^2K^2/2m$ for the eigenstate $e^{iKx}$, we have

$$K^2 = k^2 + i\epsilon, \tag{48}$$

where $k^2 = 2mE/\hbar^2$ and $\epsilon = 2m\eta/\hbar^2$, which is another positive infinitesimal. Equation (48) is further transformed to

$$K = k + i\delta\,\mathrm{sgn}\,k, \tag{49}$$

where $\delta = \epsilon/|k|$ is yet another positive infinitesimal. The eigenstate of $QHQ$ is now given by the form

$$e^{iKx} = \exp(ikx - x\delta\,\mathrm{sgn}\,k). \tag{50}$$

For $x > 0$, the solution with $k > 0$ is allowed for convergence, which gives a right-going wave, while for $x < 0$ the one with $k < 0$ is allowed, which gives a left-going wave. In short, the pole of the retarded Green's function (45) is an outgoing wave. For the advanced Green's function (46), on the other hand, we flip the sign of all infinitesimals, and find that its pole is an incoming wave.

This picture is endorsed by the Fourier transform of the Green's functions. We find

$$G^{\mathrm{ret}}(t) := \frac{1}{2\pi}\int_{-\infty}^{\infty} \frac{e^{-iEt/\hbar}}{E-(QHQ-i\eta)}dE = \begin{cases} 0 & \text{for } t<0, \\ e^{-iQHQt/\hbar} & \text{for } t>0, \end{cases} \tag{51}$$

$$G^{\mathrm{adv}}(t) := \frac{1}{2\pi}\int_{-\infty}^{\infty} \frac{e^{-iEt/\hbar}}{E-(QHQ+i\eta)}dE = \begin{cases} e^{-iQHQt/\hbar} & \text{for } t<0, \\ 0 & \text{for } t>0, \end{cases} \tag{52}$$

which reveals that indeed the former describes the wave propagation out of a source for $t > 0$, while the latter describes it into a sink for $t < 0$. This also implies that the former corresponds to the case of the outgoing waves $k > 0$ in Eq. (10) due to the source $\delta(t)$, while the latter to the case of incoming waves $k < 0$ due to the sink $\delta(t)$.

We stress here that by taking the retarded and advanced Green's functions in Eq. (40), we find the effective Hamiltonians

$$H_{\mathrm{eff}}^{\mathrm{ret}}(E) := PHP + PHQ\frac{1}{E-(QHQ-i\eta)}QHP, \tag{53}$$

$$H_{\mathrm{eff}}^{\mathrm{adv}}(E) := PHP + PHQ\frac{1}{E-(QHQ+i\eta)}QHP, \tag{54}$$

explicitly non-Hermitian. Solving the nonlinear eigenvalue problem (39) with the retarded effective Hamiltonian (53) produces the resonant eigenstates, which is the solutions under the outgoing-wave boundary condition, while solving the one with the advanced effective Hamiltonian (54) produces the anti-resonant eigenstates, which is the solutions under the incoming-wave boundary conditions, both with complex eigenvalues.

In the previous section 2, the non-Hermiticity is originated in the Siegert boundary condition, Since the environmental degrees of freedom is eliminated in the Feshbach projection formalism, the non-Hermiticity hidden in the boundary condition came into view in the forms (53)–(54).

### *3.3. What is the bra vector of the resonant state?*

A critical remark is in order. We observed above that the system Hamiltonian is generally non-Hermitian in the functional spaces of the resonant and anti-resonant states. In general, a non-Hermitian operator has right- and left-eigenstates, which are generally not Hermitian conjugate to each other as in the case of the Hermitian operators. Let $J$ denote a non-Hermitian operator; in other words, $J^\dagger \neq J$. The right- and left-eigenstates are given by

$$J\left|\psi_n\right\rangle = E_n\left|\psi_n\right\rangle, \tag{55}$$

$$J^\dagger \left\langle\phi_n\right|^\dagger = E_n{}^* \left\langle\phi_n\right|^\dagger, \tag{56}$$

respectively, for which $|\psi_n\rangle \neq \langle\phi_n|^\dagger$ in general. The right- and left-eigenstates form a complete bi-orthonormal set to span the space as in

$$\langle\phi_m|\psi_n\rangle = \delta_{mn}, \qquad \sum_n |\psi_n\rangle\langle\phi_n| = I \tag{57}$$

except at the exceptional points, at which two or more eigenvectors become parallel to each other.

Then one might presume that the probability of a wave vector $|f\rangle$ would be defined as follows: (i) one would first expand the vector in terms of the right-eigenstates as in

$$|f\rangle = \sum_n f_n |\psi_n\rangle ; \tag{58}$$

(ii) one would then construct the corresponding wave vector $\langle g|$ as in

$$\langle g| = \sum_n f_n{}^* \langle \phi_n| ; \tag{59}$$

(iii) then the probability for the system to be in the state $|f\rangle$ would be given by

$$\langle g|f\rangle = \sum_n |f_n|^2. \tag{60}$$

We note that $\langle g|^\dagger \neq |f\rangle$ in general. The quantity $\left(|f\rangle^\dagger\right)|f\rangle$ on the contrary would not be a proper definition of the probability.

Nonetheless, we used in Subsection 2.5 the standard definition of the probability, that is, the square modulus of the wave amplitude, *not* the one defined in the above procedure. This is actually because the open system with the effective non-Hermitian Hamiltonian is assumed to be a part of a Hermitian closed system [14]. Since the probability in the closed system should be given by the square modulus of the wave amplitude, it should be so in a part of it too. If the system were assumed to be a non-Hermitian closed system, we would define the probability based on the biorthogonal basis [15] following the procedure given in Eqs. (58)–(60).

## 4. Summary

We have reviewed non-Hermitian quantum mechanics of open quantum systems. The non-Hermiticity emerges in two ways in such systems. It is implicit if one uses the Siegert boundary condition, but it becomes explicit when we eliminate the environmental degrees of freedom in the Feshbach projection formalism. In reviewing it, we presented answers to several typical questions that one might ask for non-Hermitian quantum mechanics. We noted that although the whole set of the eigenstates is time-reversal symmetric, each eigenstate can break the symmetry because the time-reversal operator is not a linear one. We emphasized that the seemingly Hermitian Hamiltonian can be non-Hermitian depending on the functional space on which the Hamiltonian is defined. The Hamiltonian is indeed a non-Hermitian operator in the functional space of resonant and anti-resonant states, and hence there is no contradiction in the fact that these states have complex eigenvalues. We presented a probabilistic interpretation for the resonant and anti-resonant states. This is thanks to the often misestimated fact that the wave functions of the states spatially diverge. The spatial divergence physically represents the fact that microscopic objects are attached to the ends of the free space in reality. We finally note that the use of the standard bra vector (the Hermitian conjugate of the ket vector) is legitimate for non-Hermitian Hamiltonians if they describe open systems that are parts of closed Hermitian systems.

## Acknowledgements

The study is supported by the Japan Society for Promotion of Science (JSPS) Kakenhi grant Nos. 19H00658, 19F19321 and 21H01005.

## References

[1] Feshbach H 1958 *Annual Review of Nuclear Science* **8** 49–104 URL https://doi.org/10.1146/annurev.ns.08.120158.000405
[2] Feshbach H 1958 *Annals of Physics* **5** 357–390 ISSN 0003-4916 URL https://www.sciencedirect.com/science/article/pii/0003491658900071
[3] Feshbach H 1962 *Annals of Physics* **19** 287–313 ISSN 0003-4916 URL https://www.sciencedirect.com/science/article/pii/000349166290221X
[4] Hatano N, Sasada K, Nakamura H and Petrosky T 2008 *Progress of Theoretical Physics* **119** 187–222 ISSN 0033-068X URL https://doi.org/10.1143/PTP.119.187
[5] Hatano N 2010 *Progress of Theoretical Physics Supplement* **184** 497–515 ISSN 0375-9687 URL https://doi.org/10.1143/PTPS.184.497
[6] Hatano N and Ordonez G 2011 *International Journal of Theoretical Physics* **50** 1105–1115 URL https://doi.org/10.1007/s10773-010-0576-y
[7] Sasada K, Hatano N and Ordonez G 2011 *Journal of the Physical Society of Japan* **80** 104707 URL https://doi.org/10.1143/JPSJ.80.104707
[8] Klaiman S and Hatano N 2011 *The Journal of Chemical Physics* **134** 154111 URL https://doi.org/10.1063/1.3577997
[9] Hatano N 2013 *Fortschritte der Physik* **61** 238–249 URL https://onlinelibrary.wiley.com/doi/abs/10.1002/prop.201200064
[10] Hatano N and Ordonez G 2014 *Journal of Mathematical Physics* **55** 122106 URL https://doi.org/10.1063/1.4904200
[11] Landauer R 1957 *IBM Journal of Research and Development* **1** 223–231
[12] Datta S 1995 *Electronic Transport in Mesoscopic Systems* Cambridge Studies in Semiconductor Physics and Microelectronic Engineering (Cambridge University Press)
[13] Hatano N and Ordonez G 2019 *Entropy* **21** ISSN 1099-4300 URL https://www.mdpi.com/1099-4300/21/4/380
[14] Kawabata K 2019 private communication
[15] Bender C M, Brody D C and Jones H F 2002 *Phys. Rev. Lett.* **89**(27) 270401 URL https://link.aps.org/doi/10.1103/PhysRevLett.89.270401

# Conformal bridge transformation and $\mathcal{PT}$ symmetry

**Luis Inzunza and Mikhail S. Plyushchay**

Departamento de Física, Universidad de Santiago de Chile, Casilla 307, Santiago, Chile

E-mail: `luis.inzunza@usach.cl`, `mikhail.plyushchay@usach.cl`

**Abstract.** The conformal bridge transformation (CBT) is reviewed in the light of the $\mathcal{PT}$ symmetry. Originally, the CBT was presented as a non-unitary transformation (a complex canonical transformation in the classical case) that relates two different forms of dynamics in the sense of Dirac. Namely, it maps the asymptotically free form into the harmonically confined form of dynamics associated with the $\mathfrak{so}(2,1) \cong \mathfrak{sl}(2,\mathbb{R})$ conformal symmetry. However, as the transformation relates the non-Hermitian operator $i\hat{D}$, where $\hat{D}$ is the generator of dilations, with the compact Hermitian generator $\hat{\mathcal{J}}_0$ of the $\mathfrak{sl}(2,\mathbb{R})$ algebra, the CBT generator can be associated with a $\mathcal{PT}$-symmetric metric. In this work we review the applications of this transformation for one- and two-dimensional systems, as well as for systems on a cosmic string background, and for a conformally extended charged particle in the field of Dirac monopole. We also compare and unify the CBT with the Darboux transformation. The latter is used to construct $\mathcal{PT}$-symmetric solutions of the equations of the KdV hierarchy with the properties of extreme waves. As a new result, by using a modified CBT we relate the one-dimensional $\mathcal{PT}$-regularized asymptotically free conformal mechanics model with the $\mathcal{PT}$-regularized version of the de Alfaro, Fubini and Furlan system.

## 1. Introduction

The very fact that the properties of various complex systems can be related with the properties of a free particle and obtained from it in elegant ways is just amazing. A good example of this is the connection between the free particle and the KdV hierarchy, based on the covariance of the Lax representation with respect to Darboux transformations [1]. The stationary Schrödinger equation for a one-dimensional free particle enters the game when the operators of the auxiliary spectral problem in the Lax pair representation are taken with a zero potential identified as a trivial solution of the KdV equation. Then, the iterative application of the Darboux transformation to the linear equations associated with the Lax representation combined with the Darboux dressing of the Lax operator allows ones to generate multi-soliton solutions of the equations of the KdV hierarchy. A Schrödinger system of the auxiliary spectral problem with the obtained multi-soliton potential is reflectionless being almost isospectral to the free particle, and its states are generated from the eigenstates of the free particle Hamiltonian operator by the Darboux transformation [2, 3]. At least some of reflectionless systems are converted by periodization into the finite-gap quantum systems [4, 5, 6], and their potentials can be promoted to the cnoidal type solutions by using, again, the Darboux covariance of the Lax representation [3]. One can introduce soliton defects propagating in a crystalline background with the help of the same Darboux transformations [3]. By the Miura transformation, intimately related to supersymmetry, one also can relate the free particle system with the modified KdV equation [7]. Using the same methods, one can construct $\mathcal{PT}$-regularized Calogero type quantum models with the exotic properties, whose potentials can be transformed into complex $\mathcal{PT}$-symmetric solutions of the equations of the KdV hierarchy [8, 9].

On the other hand, an important characteristic of the free particle in arbitrary case of $d$-dimensional Euclidean space is that this non-relativistic system is described by the $\mathfrak{so}(2,1)$ conformal symmetry. The non-relativistic conformal symmetry occurs naturally in a wide variety of physical phenomena, and attracted recently a lot of attention in the context of non-relativistic AdS/CFT correspondence [10, 11, 12, 13], black hole physics [14, 15, 16], cosmology [17, 18, 19, 20], AdS/CDM correspondence [21, 22, 23, 24] and QCD confinement [25, 26], to name a few. The non-relativistic conformal symmetry of the free particle and its generalizations lie in the base of the so-called conformal bridge transformation (CBT) [27, 28, 29, 30] by which the dynamics and symmetries of an asymptomatically free conformally invariant system can be mapped into those of the associated in a certain way harmonically trapped system. This corresponds to the picture described in the Dirac seminal article [31], where different forms of dynamics are studied by choosing, in the general case, a linear combination of the generators of a given symmetry as the Hamiltonian of the system. In the original work [27], it is shown that in one dimension the CBT relates the free particle and the two-body Calogero model with, respectively, the harmonic oscillator and the conformal mechanical model of de Alfaro, Fubini and Furlan (AFF) [32]. In two dimensions, the free particle system can be related with a variety of systems such as the isotropic and anisotropic harmonic oscillators, the Landau problem, and the exotic family of rotationally invariant harmonic oscillators [27, 30]. As the transformation is based on the algebraic arguments, it can be applied to systems in any conformally-invariant space-time and gauge backgrounds. In this way, it was employed to study the dynamics and hidden symmetries in backgrounds of the Dirac monopole [28] and cosmic string [29]. One of the goals of this article is to review how this transformation works and the scope of its applications.

In comparison with the Darboux transformation, which relates almost isospectral one-dimensional systems by means of finite order differential operators, the CBT is generated by a non-local operator, whose realization is not restricted by a space dimension, and it relates the systems with essentially different spectra in a non-trivial way. In fact, it is expected that the possibilities to connect the systems by the CBT expand with increasing the number of dimensions and with the conformally invariant change of the space-time metric. Additionally, some hints on a possible close relationship of the CBT with $\mathcal{PT}$-symmetric systems [33, 34, 35, 36] were indicated in the original works [27, 28, 29, 30]. They are based on the fact that at the quantum level, the transformation is realized by a non-unitary operator $\hat{\mathfrak{S}}$ that transforms a non-Hermitian operator $i\hat{D}$, where $\hat{D}$ is a generator of dilations, into the Hermitian compact generator of the $\mathfrak{sl}(2,\mathbb{R})$ symmetry which has a real discrete spectrum. In this work, we show that the connection between the one-dimensional free particle and the harmonic oscillator corresponds to a particular example of the $\mathcal{PT}$-symmetric Swanson models studied in [37, 38, 39, 40]. In this way our Hermitian generator $\hat{\mathfrak{S}}$ of the CBT can be related to a $\mathcal{PT}$-symmetric metric operator. Furthermore, since the applications of our CBT touch the systems in the spaces $\mathbb{R}^d$ with $d \geq 1$, and the models in different geometric and gauge backgrounds, new possibilities are opened for connecting $\mathcal{PT}$-symmetric systems with models that reveal interesting physical properties such as quantum anomalies, Bose-Einstein condensation, gauge symmetries, etc.

The paper is organized as follows. In Sec. 2 we present the basic properties of the CBT at the classical and quantum levels, and establish its connection with $\mathcal{PT}$-symmetric systems. In Sec. 3 we consider the explicit applications of the CBT to one-dimensional systems. As a new result we present the connection between a one-parametric family of the $\mathcal{PT}$-regularized perfectly invisible zero-gap Calogero type systems with a $\mathcal{PT}$-symmetric version of the AFF conformal mechanics. We use the relation of the former family with the free-particle by means of Darboux transformations based on a scale-invariant higher-order differential equation to build the complete set of the spectrum-generating ladder operators for the latter system. We also consider there a reinterpretation of the CBT from the point of view of the $\mathcal{PT}$-symmetric Swanson model. In Sec. 4 we consider the isotropic and anisotropic CBT in $d$-dimensions, as well as the generation of the exotic rotationally invariant harmonic oscillator in two dimensions. In Sec. 5 we study the application of the CBT for systems in two different backgrounds which correspond to the cosmic string and Dirac monopole. In Sec. 6 we show how the Darboux transformation applied appropriately to the $\mathcal{PT}$-regularized Calogero type systems allows us to produce complex $\mathcal{PT}$-symmetric solutions of the equations of the KdV hierarchy which reveal the properties typical for extreme waves. In Sec. 7 we conclude with discussion of some interesting open problems and further generalizations of the CBT in the light of the $\mathcal{PT}$ symmetry.

## 2. Conformal bridge transformation

### 2.1. *Classical case*

Consider the classical $\mathfrak{so}(2,1)$ algebra

$$\{D_0, H_0\} = H_0\,, \qquad \{D_0, K_0\} = -K_0\,, \qquad \{K_0, H_0\} = 2D_0\,, \tag{1}$$

without specifying the concrete form of the generators. Identifying $H_0$ as a Hamiltonian of a particular classical system, one sees that $D_0$ and $K_0$ cannot be true, not depending explicitly on time, integrals of motion. They, however, can easily be promoted to the dynamical, explicitly depending on time integrals of motion in the sense of the evolution equation $\dot{A} = \{A, H\} + \frac{\partial A}{\partial t} = 0$,

$$K_0 \to K = K(t) = T_{H_0}(t)(K_0)\,, \qquad D_0 \to D = D(t) = T_{H_0}(t)(D_0)\,. \tag{2}$$

Here, $T_{H_0}(t)$ indicates the Hamiltonian flux in a phase space,

$$\exp(\gamma F) \star f(q,p) := f(q,p) + \textstyle\sum_{n=1}^{\infty} \frac{\gamma^n}{n!}\{F, \{\ldots, \{F, f\underbrace{\}\ldots\}\}}_{n} := T_F(\gamma)(f)\,, \tag{3}$$

that is a canonical transformation. Obviously, $T_{H_0}(t)(H_0) = H_0$, and $H_0$, $D$ and $K$ satisfy the same $\mathfrak{so}(2,1)$ algebra, in which $D$ and $K$ are identified as generators of dilations and special conformal transformations, respectively.

The real and complex linear combinations

$$\mathcal{J}_0 = \tfrac{1}{2}(\omega^{-1}H_0 + \omega K_0)\,, \qquad \mathcal{J}_\pm = \mathcal{J}_1 \pm i\mathcal{J}_2 = -\tfrac{1}{2}\left(\omega^{-1}H_0 - \omega K_0 \pm 2iD_0\right)\,, \tag{4}$$

satisfy the classical $\mathfrak{sl}(2,\mathbb{R})$ algebra,

$$\{\mathcal{J}_0, \mathcal{J}_\pm\} = \mp i\mathcal{J}_\pm\,, \qquad \{\mathcal{J}_-, \mathcal{J}_+\} = -2i\mathcal{J}_0\,. \tag{5}$$

A constant $\omega$ of dimension of frequency is introduced to compensate the dimensions of the generators $H_0$ and $K_0$. $H_0$ has a nature of a non-compact (parabolic) generator of the $\mathfrak{so}(2,1)$ algebra [41, 42], and so, the corresponding classical system can be asymptotically free (like, e.g. a free particle, or conformal mechanics model). $\mathcal{J}_0$ is a compact (elliptic) generator of $\mathfrak{sl}(2,\mathbb{R})$, and represents a harmonically trapped (confined) version of the system $H_0$. The quantity $2\omega\mathcal{J}_0$ can be considered as a Hamiltonian of such a system, which corresponds to another form of dynamics with respect to the same conformal symmetry $\mathfrak{so}(2,1) \cong \mathfrak{sl}(2,\mathbb{R})$. Then $\mathcal{J}_\pm$ can be promoted to the dynamical integrals of motion for the harmonically trapped system by the analog of the transformation given by Eqs. (2) and (3) with $H_0$ and $t$ changed for $2\omega\mathcal{J}_0$ and $\tau$.

The quantities $2\omega\mathcal{J}_0$ and $\mathcal{J}_\pm$ generate the Newton-Hooke symmetry of the harmonically trapped system [43, 44, 45, 46, 47].

Consider now the transformation

$$\mathfrak{S}:\ (H_0, D, K) \rightarrow (-\omega\mathcal{J}_-, -i\mathcal{J}_0, \omega^{-1}\mathcal{J}_+)\,, \tag{6}$$

where we assume that $\mathcal{J}_\pm$ are the dynamical integrals of motion with respect to the evolution generated by $2\omega\mathcal{J}_0$. It is an internal automorphism of the conformal algebra $\mathfrak{so}(2,1) \cong \mathfrak{sl}(2,\mathbb{R})$ generated by the composition of the canonical transformations

$$\mathscr{T}(\tau,\beta,\gamma,\delta,t) = T_{2\omega\mathcal{J}_0}(\tau) \circ T_{\beta\gamma\delta} \circ T_H(-t)\,, \tag{7}$$

where

$$T_{\beta\gamma\delta} := T_{\omega K_0}(\beta) \circ T_{\omega^{-1}H_0}(\tfrac{1}{2}\delta) \circ T_{D_0}(\gamma) = T_{\omega K_0}(\beta) \circ T_{D_0}(\gamma) \circ T_{\omega^{-1}H_0}(\delta)\,, \tag{8}$$

$$\beta = -i\,, \qquad \gamma = -\ln 2\,, \qquad \delta = i\,. \tag{9}$$

In this composition, the first transformation $T_H(-t)$ removes dependence on $t$ in the dynamical integrals $D$ and $K$. The second transformation relates the $t = 0$ generators with the generators $\mathcal{J}_0$ and $\mathcal{J}_\pm$ of the $\mathfrak{sl}(2,\mathbb{R})$ algebra, taken at $\tau = 0$. The last transformation $T_{2\omega\mathcal{J}_0}(\tau)$ restores the $\tau$ dependence. The independent of the evolution parameters transformation $T_{\beta\gamma\delta}$ is equivalent to

$$T_{\beta\gamma\delta} = T_{G_1}(\varepsilon)\,, \qquad G_1 := \omega^{-1}H_0 - \omega K_0\,, \quad \varepsilon = \tfrac{\pi}{4}i\,. \tag{10}$$

Since the parameters $\beta$, $\delta$ and $\varepsilon$ are pure imaginary, this canonical transformation is of an unusual, complex form from the point of view of the conventional classical mechanics. It transforms, particularly, the $\mathfrak{so}(2,1)$ hyperbolic [41, 42] generator $D_0$ multiplied by $2i\omega$ into the compact real $\mathfrak{sl}(2,\mathbb{R})$ generator $\mathcal{J}_0$ multiplied by $2\omega$. This picture corresponds to the change of the form of dynamics in the sense of Dirac [31]. Both, the asymptotically free and the harmonically confined, forms of dynamics are associated to the conformal symmetry, and are related one to another by the described classical conformal bridge transformation [27, 29].

One can easily check that for the complex $\mathfrak{so}(2,1)$ automorphism (6) the following relation

$$\mathfrak{S}^2 = \mathfrak{S} \circ \mathfrak{S}\ :\ (\mathcal{J}_0, \mathcal{J}_1, \mathcal{J}_2) \rightarrow (-\mathcal{J}_0, \mathcal{J}_1, -\mathcal{J}_2) \tag{11}$$

is valid in terms of the $\mathfrak{sl}(2,\mathbb{R})$ generators, that is a rotation by $\pi$ about $\mathcal{J}_1$. Therefore, (6) is the fourth order root of the $\mathfrak{so}(2,1) \cong \mathfrak{sl}(2,\mathbb{R})$ identity automorphism, $\mathfrak{S}^4 = 1$. In terms of the $\mathfrak{so}(2,1)$ generators its eigenelement of eigenvalue 1 is $G_1 = \omega^{-1}H_0 - \omega K_0$, see Eq. (10), while linear combinations $G_{+i} = -2D_0 + \omega^{-1}H_0 + \omega K_0$ and $G_{-i} = 2D_0 + \omega^{-1}H_0 + \omega K_0$ are eigenelements of eigeinvalues $+i$ and $-i$, respectively. There is no eigenelement of the automorphism (6) of eigenvalue $-1$ to be linear in the $\mathfrak{so}(2,1)$ generators. The quadratic eigenelement of eigenvalue $-1$ is $a(G_{+i})^2 + b(G_{-i})^2$ with arbitrary coefficients $a$ and $b$. The $\mathfrak{so}(2,1)$ classical Casimir $C = H_0K_0 - D_0^2 = \frac{1}{4}G_{+i}G_{-i}$ is an eigenelement of automorphism $\mathfrak{S}$ of eigenvalue 1.

Given a particular system described by some symmetry algebra (that can be of a non-linear, $W$ type), in which conformal symmetry $\mathfrak{so}(2,1)$ appears as a subalgebra, we can always apply to this system the classical (and quantum, see below) conformal bridge transformation to relate it to its harmonically confined version. As a result, the integrals of one system can be mapped into integrals of another. Explicit examples of this are discussed in Secs. 4 and 5. Note that in this picture, the classical dynamics of the harmonically trapped system, which is generated by the compact Hamiltonian $2\omega\mathcal{J}_0$, corresponds to the Hamiltonian flow generated by the complex quantity $2i\omega D_0$ in the asymptotically free system described by the Hamiltonian $H_0$.

Journal of Physics: Conference Series **2038** (2021) 012014 doi:10.1088/1742-6596/2038/1/012014

*2.2. Quantum version*

Consider now the quantum $\mathfrak{so}(2,1)$ algebra

$$[\hat{D},\hat{H}_0]=i\hbar\hat{H}_0\,,\qquad [\hat{D},\hat{K}]=-i\hbar\hat{K}\,,\qquad [\hat{K},\hat{H}_0]=2i\hbar\hat{D}\,. \tag{12}$$

Here

$$\hat{D}=e^{-i\frac{\hat{H}_0}{\hbar}t}\hat{D}_0e^{i\frac{\hat{H}_0}{\hbar}t}\,,\qquad \hat{K}=e^{-i\frac{\hat{H}_0}{\hbar}t}\hat{K}_0e^{i\frac{\hat{H}_0}{\hbar}t}\,, \tag{13}$$

and we just note that acting on a solution of the time-dependent Schrödinger equation $\Psi(t)=e^{-i\frac{\hat{H}_0}{\hbar}t}\Psi(0)$, a generic dynamical integral operator $\hat{A}=e^{-i\frac{\hat{H}_0}{\hbar}t}\hat{A}_0e^{i\frac{\hat{H}_0}{\hbar}t}$ produces $\hat{A}(t)\Psi(t)=e^{-i\frac{\hat{H}_0}{\hbar}t}\hat{A}_0\Psi(0)$. In this work we consider only stationary eigenstates and linear combinations of them, and for this reason, at the quantum level we suppose $\hat{D}=\hat{D}_0$ and $\hat{K}=\hat{K}_0$, bearing in mind that the time dependence can be reconstructed by application of the corresponding time-evolution operator.

By introducing the rescaled by $\hbar$ quantum analogs of the linear combinations (4),

$$\hat{\mathcal{J}}_0=\tfrac{1}{2\omega\hbar}(\hat{H}_0+\omega^2\hat{K})\,,\qquad \hat{\mathcal{J}}_\pm=\hat{\mathcal{J}}_1\pm i\hat{\mathcal{J}}_2=-\tfrac{1}{2\omega\hbar}(\hat{H}_0-\omega^2\hat{K}\pm 2i\omega\hat{D})\,, \tag{14}$$

we produce the quantum $\mathfrak{sl}(2,\mathbb{R})$ algebra

$$[\hat{\mathcal{J}}_0,\hat{\mathcal{J}}_\pm]=\pm\hat{\mathcal{J}}_\pm\,,\qquad [\hat{\mathcal{J}}_-,\hat{\mathcal{J}}_+]=2\hat{\mathcal{J}}_0\,. \tag{15}$$

The quantum conformal bridge transformation (CBT) [27, 29, 30] is a similarity transformation

$$\hat{\mathfrak{S}}(\hat{H}_0)\hat{\mathfrak{S}}^{-1}=-\omega\hbar\hat{\mathcal{J}}_-\,,\qquad \hat{\mathfrak{S}}(i\hat{D})\hat{\mathfrak{S}}^{-1}=\hbar\hat{\mathcal{J}}_0\,,\qquad \hat{\mathfrak{S}}(\hat{K})\hat{\mathfrak{S}}^{-1}=\hbar\omega^{-1}\hat{\mathcal{J}}_+\,, \tag{16}$$

generated by the non-unitary, Hermitian operator

$$\hat{\mathfrak{S}}=e^{-\frac{\omega}{\hbar}\hat{K}}e^{\frac{i}{\hbar}\ln(2)\hat{D}}e^{\frac{\hat{H}_0}{\hbar\omega}}=\exp\left[\tfrac{\pi}{4\hbar}\left(\omega^{-1}\hat{H}_0-\omega\hat{K}\right)\right]\,. \tag{17}$$

Notice that all the operators $\hat{H}_0$, $\hat{K}$ and $i\hat{D}$, to which the CBT (16) is applied, as well as the CBT generator $\hat{\mathfrak{S}}$ itself commute with the $\mathcal{PT}$ operator, i.e. they are $\mathcal{PT}$-symmetric [1].

Relations (16) imply that

$$\hat{D}\,|\lambda\rangle=i\hbar\lambda\,|\lambda\rangle\quad\Rightarrow\quad \hat{\mathcal{J}}_0(\hat{\mathfrak{S}}\,|\lambda\rangle)=\lambda\hat{\mathfrak{S}}\,|\lambda\rangle\,, \tag{18}$$

$$\hat{H}_0\,|E\rangle=E\,|E\rangle\quad\Rightarrow\quad \hat{\mathcal{J}}_-(\hat{\mathfrak{S}}\,|E\rangle)=-\tfrac{E}{\hbar\omega}\hat{\mathfrak{S}}\,|E\rangle\,. \tag{19}$$

Then, to have a physical eigenstate $\hat{\mathfrak{S}}\,|\lambda\rangle$ of $\hat{\mathcal{J}}_0$, the formal state $|\lambda\rangle$ must obey the following conditions:

I. The series $\exp\left(\frac{\hat{H}_0}{2\hbar\omega}\right)|\lambda\rangle=\sum_{n=0}^{\infty}\frac{1}{n!(2\hbar\omega)^n}(\hat{H}_0)^n\,|\lambda\rangle$ has to reduce to a finite number of terms; this means that $|\lambda\rangle$ should be a Jordan state of the operator $\hat{H}_0$ corresponding to zero energy [2].

[1] Here $\mathcal{P}$ is a space reflection operator, $\mathcal{P}x=-x\mathcal{P}$, $\mathcal{P}^2=1$, and a complex conjugation operator $\mathcal{T}$ is defined by $\mathcal{T}z=-z^*\mathcal{T}$, $\mathcal{T}^2=1$, where $z\in\mathbb{C}$ is an arbitrary complex number.

[2] The wave functions of generalized Jordan states corresponding to energy $\lambda$ satisfy relations of the form $P(\hat{H})\Omega_\lambda=\psi_\lambda$, where $\hat{H}\psi_\lambda=\lambda\psi_\lambda$ and $P(\eta)$ is a polynomial [48, 49, 50]. Here we consider the Jordan states satisfying the relations $(\hat{H})^\ell\Omega_\lambda=\lambda\psi_\lambda$ with $\lambda=0$ for a certain natural number $\ell$.

II. If a wave function $\Omega_\lambda = \langle \boldsymbol{x}|\lambda\rangle$ is a rank $n$ Jordan state of $\hat{H}$ of zero energy, $\hat{H}^n\Omega_\lambda = 0$, then $\Omega_\lambda$, as well as $(\hat{H}_0)^k\Omega_\lambda$, $k = 1, \ldots, n-1$, must not have poles and have to be single-valued.

On the other hand, the eigenvectors $|E\rangle$ (physical, or non-physical, with complex eigenvalues in general case) of $\hat{H}_0$ are transformed into eigenvectors of the lowering operator $\hat{\mathcal{J}}_-$ of the $\mathfrak{sl}(2,\mathbb{R})$ algebra. Therefore, the resulting eigenstates in (19) are the coherent states of $\mathfrak{sl}(2,\mathbb{R})$ in the sense of Perelomov [51].

The formalism related to the quantum CBT admits a reinterpretation in the context of the $\mathcal{PT}$ symmetry [37, 39]. Indeed, the second relation in (16) can be written as

$$\hat{\mathfrak{S}}\hat{H}_{\mathcal{PT}} = \hat{H}_{HC}\hat{\mathfrak{S}}\,, \tag{20}$$

where $\hat{H}_{\mathcal{PT}} = 2i\omega\hbar\hat{D}$ is a $\mathcal{PT}$-invariant operator, while $\hat{H}_{HC} = 2\omega\hbar\hat{\mathcal{J}}_0$ is a Hermitian operator. From here we see that the Hermitian operator $\hat{\mathfrak{S}}$, being the generator of the complex automorphism of the conformal algebra $\mathfrak{so}(2,1) \cong \mathfrak{sl}(2,\mathbb{R})$, intertwines a non-Hermitian, but $\mathcal{PT}$-invariant Hamiltonian with a Hermitian one. By multiplying this relation from the left by $\hat{\mathfrak{S}}$ we obtain $\hat{\Theta}\hat{H}_{\mathcal{PT}} = \hat{H}^\dagger_{\mathcal{PT}}\hat{\Theta}$, where $\hat{\Theta} = \hat{\mathfrak{S}}^2$, and we have taken into account the relation $\hat{D} = -\hbar\hat{\mathcal{J}}_2$ and the quantum analog of (11). This implies that if $\hat{H}_{HC}$ represents a well defined quantum system with real eigenvalues and normalizable eigenfunctions, then $\hat{H}_{\mathcal{PT}}$ has a real spectrum with corresponding eigenstates of finite but not positive definite norm under the indefinite scalar product $\langle\lambda_1|\,\hat{\Theta}\,|\lambda_2\rangle$. With respect to this inner product, operator $\hat{H}_{\mathcal{PT}}$ is pseudo-Hermitian [35].

## 3. Applications of CBT to one-dimensional systems

In this section, based on [27], we apply the quantum CBT to one-dimensional systems. For the sake of simplicity, we use here the units $\hbar = m = 1$. In each of the examples, we consider the symmetry operators, the eigenstates, and the rank $n$ Jordan states of zero energy corresponding to an asymptotically free system. By applying the CBT, we get the symmetry operators, the eigenstates and the coherent states of the corresponding harmonically confined models.

In subsection 3.3, we give a reinterpretation of the CBT by comparing its construction with the Swanson $\mathcal{PT}$-symmetric system [37, 39] in correspondence with Eq. (20) and related comments there. This will allow us, particularly, to generalize the construction of ref. [37, 39] to the case of the AFF conformal mechanics model.

### *3.1. Example 1: The free particle - harmonic oscillator relation*

Let us start with the one-dimensional free particle symmetry generators

$$\hat{H} = -\tfrac{1}{2}\tfrac{d^2}{dx^2}\,, \qquad \hat{D} = -\tfrac{i}{2}\left[x\tfrac{d}{dx} + \tfrac{1}{2}\right]\,, \qquad \hat{K} = \tfrac{1}{2}x^2\,, \tag{21}$$

$$\hat{p} = -i\tfrac{d}{dx}\,, \qquad \hat{x} = x\,. \tag{22}$$

They produce the one-dimensional Schrödinger symmetry [52]. The eigenstates and eigenvalues of $\hat{H}$ are

$$\psi_\kappa = e^{i\kappa x}\,, \qquad E = \tfrac{1}{2}\hbar^2\kappa^2\,, \qquad \kappa \in \mathbb{R}\,. \tag{23}$$

The functions $\langle x|\lambda\rangle$ that satisfy the two conditions specified in Sec. 2.2 correspond to

$$\Omega_n(x) = \langle x|\lambda\rangle = x^n\,, \qquad n = 0, 1, 2, \ldots. \tag{24}$$

The set of states (24) as a whole is invariant under the action of the symmetry generators (21), (22),

$$\hat{H}\Omega_n = -\tfrac{1}{2}n(n-1)\Omega_{n-2}\,, \qquad \hat{K}\Omega_n = \tfrac{1}{2}\Omega_{n+2}\,, \qquad 2i\hat{D}\Omega_n = (n + \tfrac{1}{2})\Omega_n\,, \tag{25}$$

$$\hat{p}\Omega_n = -in\Omega_{n-1}\,, \qquad \hat{x}\Omega_n = \Omega_{n+1}\,. \tag{26}$$

Via the repeated application of $\hat{H}$ to $\Omega_n$, we arrive at the functions $\Omega_0 = 1$ (if $n$ is even) or $\Omega_1 = x$ (if $n$ is odd), which are the (physical and non-physical) zero energy solutions of the free particle stationary Schrödinger equation. So, $\Omega_n(x)$ are the rank $[n/2]+1$ Jordan states of zero energy of the free particle, $(\hat{H})^{[n/2]+1}\Omega_n = 0$, where $[.]$ denotes an integer part. We see that $\Omega_n$ are common formal eigenfunctions of the operators $2i\hat{D}$ and $(\hat{H})^{[n/2]+1}$.

According to Eq. (16), the application of the CBT to the free particle's $\mathfrak{so}(2,1)$ generators gives us the $\mathfrak{sl}(2,\mathbb{R})$ generators of the one-dimensional harmonic oscillator system,

$$\hat{H}_{\text{os}} = 2\omega\hat{\mathcal{J}}_0 = -\frac{1}{2}\frac{d^2}{dx^2} + \frac{\omega^2}{2}x^2\,, \qquad \hat{\mathcal{J}}_\pm = \frac{1}{4\omega}\left[\frac{d^2}{dx^2} + \omega^2x^2 \mp \omega(x\frac{d}{dx} + \frac{1}{2})\right]. \tag{27}$$

Along with them, we obtain the Heisenberg generators (ladder operators)

$$\hat{a}^\pm = \sqrt{\frac{\omega}{2}}\left(x \mp \frac{1}{\omega}\frac{d}{dx}\right), \qquad [\hat{a}^-, \hat{a}^+] = 1\,, \tag{28}$$

according to the relations $\hat{\mathfrak{S}}(\hat{p})\hat{\mathfrak{S}}^{-1} = -i\sqrt{\omega}\,\hat{a}^-$ and $\hat{\mathfrak{S}}(\hat{x})\hat{\mathfrak{S}}^{-1} = \sqrt{\frac{1}{\omega}}\,\hat{a}^+$. From here one deduces that this non-unitary (similarity) transformation can be identified as the fourth order root of the space reflection operator $\mathcal{P}$,

$$\hat{\mathfrak{S}} : (x, \hat{p}, \hat{a}^+, \hat{a}^-) \to (\hat{a}^+, -i\hat{a}^-, -i\hat{p}, x)\,, \qquad \hat{\mathfrak{S}}^2 : (x, \hat{p}, \hat{a}^+, \hat{a}^-) \to (-i\hat{p}, -ix, -\hat{a}^-, \hat{a}^+)\,,$$
$$\hat{\mathfrak{S}}^4 : (x, \hat{p}, \hat{a}^+, \hat{a}^-) \to (-x, -\hat{p}, -\hat{a}^+, -\hat{a}^-)\,. \tag{29}$$

For the sake of simplicity, we set $\omega = 1$ in (29). Notice here that the action of the CBT on generators of the Heisenberg algebra corresponds to the eighth order root of the identity automorphism, $\hat{\mathfrak{S}}^8 = 1$ [3].

The application of the operator $\hat{\mathfrak{S}}$ to the functions (24) gives us

$$\hat{\mathfrak{S}}\Omega_n = \mathcal{N}_n\psi_n\,, \qquad \psi_n = \frac{1}{\sqrt{2^n n!}}\left(\frac{\omega}{\pi}\right)^{\frac{1}{4}} H_n(\sqrt{\omega}x)e^{-\frac{\omega x^2}{2}}\,, \qquad \mathcal{N}_n = \frac{(2\pi)^{\frac{1}{4}}}{\omega^{\frac{n}{2}+\frac{1}{4}}}\sqrt{n}\,, \tag{30}$$

where we have used the Weierstrass transformation [55, 27]. Additionally, by acting from the left by the operator $\hat{\mathfrak{S}}$ on both sides of Eqs. (25) and (26) one gets the well known relations

$$\hat{H}_{\text{os}}\psi_n = \omega(n+\tfrac{1}{2})\psi_n\,, \quad \hat{\mathcal{J}}_\pm\psi_n = \sqrt{(n \pm \beta_\pm)(n + \beta_\pm \pm 1)}\psi_{n\pm 2}\,, \quad \beta_\pm = \frac{1\pm 1}{2}\,, \tag{31}$$

$$\hat{a}^-\psi_n = \sqrt{n}\psi_{n-1}\,, \qquad \hat{a}^+\psi_n = \sqrt{n+1}\psi_{n+1}\,. \tag{32}$$

Finally, the action of the operator $\hat{\mathfrak{S}}$ on the free particle eigenstates of the form (23), being simultaneously eigenfunctions of $\hat{p}$, produces

$$\phi(x,\kappa) = \hat{\mathfrak{S}}e^{i\frac{\kappa x}{\sqrt{2}}} = 2^{\frac{1}{4}}\exp\Big(-\frac{\omega^2}{2}x^2 + \frac{\kappa^2}{4\omega} + i\kappa x\Big) = \left(\frac{2\pi}{\omega}\right)^{\frac{1}{4}}\sum_{n=0}^{\infty}\left(\frac{ik}{\sqrt{2\omega}}\right)^n\frac{\psi_n}{\sqrt{n!}}\,. \tag{33}$$

These are the coherent states [56] that satisfy the relation $\hat{a}^-\phi(x,\kappa) = \frac{i\kappa}{\sqrt{2\omega}}\phi(x,\kappa)$. Under the time evolution these functions take the form

$$\phi(x,\kappa,t) = e^{-i\hat{H}_{\text{os}}t}\phi(x,\kappa) = e^{-i\frac{\omega t}{2}}\phi(x,\kappa e^{-i\omega t})\,. \tag{34}$$

To have the over-complete set of coherent states of the quantum harmonic oscillator, we allow the parameter $\kappa$ to take complex values. In the same vein one can show that the free particle

[3] See the comments related to Eq. (35) below, and refs. [53, 54] where the automorphism of the one-dimensional Heisenberg group is discussed in the context of the Stone-von Neumann theorem and the properties of the Fourier transform.

Gaussian wave packets are mapped into the squeezed states of the harmonic oscillator, see ref. [27].

Finally, we note that the described CBT formalism is close with the unitary transformation between the coordinate and the Fock-Bargmann representations. In fact, this last representation can be obtained if we formally replace the spacial variable $x$ with the complex variable $z$ in operators (21), (22) as well as in the Jordan states (24), and as an additional step, substitute the usual $L^2(\mathbb{R})$ scalar product for the inner product

$$(\psi_1, \psi_2) = \frac{1}{\pi}\int_{\mathbb{R}^2} \overline{\psi_1(z)}\psi_2(z)e^{-\bar{z}z}d^2z, \qquad d^2z = d(\operatorname{Re} z)d(\operatorname{Im} z)\,. \tag{35}$$

The kernel of the integral transformation, which is a unitary transformation from the $L^2(\mathbb{R})$ Hilbert space to the Fock-Bargmann space, can be related to the considered CBT, for further details see ref. [27].

### *3.2. Example 2: The one-dimensional Calogero model - AFF model relation*

The two-particle Calogero model admits a separation of variables in terms of the relative coordinate and the coordinate of the center of mass, which has a free dynamics. The corresponding $\mathfrak{so}(2,1)$ symmetry generators associated with the relative coordinate $x > 0$ are defined on the positive real half-line $\mathbb{R}^+$, and they correspond to

$$\hat{H}_\nu = -\frac{1}{2}\frac{d^2}{dx^2} + \frac{\nu(\nu+1)}{2x^2}\,, \qquad \hat{D} = -\frac{i}{2}\left[x\frac{d}{dx} + \frac{1}{2}\right]\,, \qquad \hat{K} = \frac{1}{2}x^2\,, \tag{36}$$

where we assume that $\nu > -\frac{1}{2}$ [50]. The eigenstates and eigenvalues of $\hat{H}_\nu$, are given by

$$\psi_{\kappa,\nu}(x) = \sqrt{x}\,J_{\nu+1/2}(\kappa x)\,, \qquad E = \frac{1}{2}\hbar^2\kappa^2\,, \qquad \kappa > 0\,, \tag{37}$$

where $J_\alpha$ are the Bessel functions of the first kind. Besides, the functions $\langle x|\lambda\rangle = \Omega_{n,\nu}$, that are well defined on $\mathbb{R}^+$ and are the zero energy Jordan states of $\hat{H}_\nu$ correspond to

$$\Omega_{n,\nu} = x^{2n+\nu+1}\,. \tag{38}$$

These functions satisfy the following set of equations,

$$\hat{H}_\nu\Omega_{n,\nu} = -n(2n+2\nu+1)\Omega_{n-1,\nu}\,, \qquad \hat{K}\Omega_{n,\nu} = \frac{1}{2}\Omega_{n+1,\nu}\,, \tag{39}$$

$$2i\hat{D}\Omega_{n,\nu} = (2n+\nu+\tfrac{3}{2})\Omega_{n,\nu}\,. \tag{40}$$

In the same vein as in the previous subsection, one can see that after the repeated application of $\hat{H}_\nu$ one gets the zero energy solution $\Omega_{0,\nu} = x^{\nu+1}$, which is a regular function on $\mathbb{R}^+$.

On the other hand, the conformal symmetry generators of the AFF model [32],

$$\hat{H}_\nu^{AFF} = 2\omega\hat{\mathcal{J}}_0 = -\frac{1}{2}\frac{d^2}{dx^2} + \frac{\nu(\nu+1)}{2x^2} + \frac{\omega^2}{2}x^2\,, \tag{41}$$

$$\hat{\mathcal{J}}_\pm = \frac{1}{4\omega}\left[\frac{d^2}{dx^2} - \frac{\nu(\nu+1)}{x^2} + \omega^2x^2 \mp \omega(x\frac{d}{dx} + \frac{1}{2})\right]\,, \tag{42}$$

are obtained by applying the CBT to generators (36). In the same way, the normalized eigenstates of $\hat{H}_\nu^{AFF}$ correspond to

$$\hat{\mathfrak{S}}\Omega_{n,\nu} = \mathcal{N}_{n,\nu}\psi_{n,\nu}\,, \qquad \psi_{n,\nu} = \sqrt{\frac{2\omega^{\nu+\frac{3}{2}}n!}{\Gamma(n+\nu+\frac{3}{2})}}x^{\nu+1}L_n^{(\nu+\frac{1}{2})}(\omega x^2)e^{-\frac{\omega x^2}{2}}\,, \tag{43}$$

where $\mathcal{N}_{n,\nu} = (-1)^n \left(\frac{2}{\omega}\right)^{\frac{\nu}{2}+n} \sqrt{\omega^{-1} n! \Gamma(n+\nu+\frac{3}{2})}$. They satisfy equations

$$\hat{H}_\nu^{AFF}\psi_{n,\nu} = E_{n,\nu}\psi_{n,\nu}\,, \qquad E_{n,\nu} = \omega(2n+\nu+\tfrac{3}{2})\,, \tag{44}$$

$$\hat{\mathcal{J}}_-\psi_{n,\nu} = -\sqrt{n(n+\nu+\tfrac{1}{2})}\psi_{n-1,\nu}\,, \quad \hat{\mathcal{J}}_+\psi_{n,\nu} = -\sqrt{(n+1)(n+\nu+\tfrac{3}{2})}\psi_{n+1,\nu}\,, \tag{45}$$

that are obtained directly from the application of $\hat{\mathfrak{S}}$ to equations (39), (40).

On the other hand, the application of the operator $\hat{\mathfrak{S}}$ to eigenstates (37) of the system $\hat{H}_\nu$ yields

$$\hat{\mathfrak{S}}\psi_{\kappa,\nu}(\tfrac{1}{\sqrt{2}}x) = 2^{\frac{1}{4}} e^{-\frac{1}{2}x^2+\frac{1}{4}\kappa^2}\sqrt{x}\, J_{\nu+1/2}(\kappa x) := \phi_\nu(x,\kappa)\,, \tag{46}$$

that are the states satisfying the relation $\hat{\mathcal{J}}_-\phi_\nu(x,\kappa) = -\frac{1}{4}\kappa^2\phi_\nu(x,\kappa)$. By changing the parameter $\kappa$ for the complex parameter $z$, one obtains coherent states that are eigenstates of operator $\hat{\mathcal{J}}_-$ with complex eigenvalue $-\frac{1}{4}z^2$. By using the evolution operator $\exp\left(-it\hat{H}_\nu^{\mathrm{AFF}}\right)$, the time-dependent coherent states are obtained,

$$\phi_\nu(x,z,t) = 2^{1/4}\sqrt{x}\, J_{\nu+1/2}(z(t)x)e^{-x^2/2+z^2(t)/4-it}\,, \qquad z(t) = ze^{-it}\,. \tag{47}$$

### *3.3. Example 3: CBT and $\mathcal{PT}$-regularized conformal systems*

The conformal symmetry operators of the $\mathcal{PT}$-regularized Calogero systems, which are defined for $x \in \mathbb{R}$, are given by [8, 9]

$$\hat{H}_{\alpha,\nu} = -\frac{1}{2}\frac{d^2}{dx^2} + \frac{\nu(\nu+1)}{2(x+i\alpha)^2}\,, \qquad \hat{D}_\alpha = -\frac{i}{2}\left[(x+i\alpha)\frac{d}{dx} + \frac{1}{2}\right]\,, \tag{48}$$

$$\hat{K}_\alpha = \tfrac{1}{2}(x+i\alpha)^2\,. \tag{49}$$

Notice that $\hat{H}_{\alpha,\nu}$ and $\hat{K}_\alpha$ are $\mathcal{PT}$-symmetric, $[\mathcal{PT}, \hat{H}_{\alpha,\nu}] = [\mathcal{PT}, \hat{K}_\alpha] = 0$, while $\hat{D}_\alpha$ is the $\mathcal{PT}$-odd, $\mathcal{PT}\hat{D}_\alpha = -\hat{D}_\alpha\mathcal{PT}$, operators. They are obtained by application of the complex translation $x \to x + i\alpha$, $\alpha \in \mathbb{R}$, $\alpha \neq 0$, generated by the Hermitian operator

$$\hat{I}_\alpha = e^{-\alpha\hat{p}} = e^{i\alpha\frac{d}{dx}}\,, \tag{50}$$

to the generators of the Hermitian Calogero system (36),

$$\hat{I}_\alpha(\hat{H}_{0,\nu})\hat{I}_{-\alpha} = \hat{H}_{\alpha,\nu}\,, \qquad \hat{I}_\alpha(\hat{D}_0)\hat{I}_{-\alpha} = \hat{D}_\alpha\,, \qquad \hat{I}_\alpha(\hat{K}_0)\hat{I}_{-\alpha} = \hat{K}_\alpha\,, \tag{51}$$

supplemented by extension of the domain for the position variable to the entire real line, that we imply in the rest of this subsection. In the same way, the eigenstates of $\hat{H}_{\alpha,\nu}$ are formally obtained by application of operator $\hat{I}_\alpha$ to the states (37), and to their corresponding linearly independent partners, which are given by the Neumann functions $Y_\nu(\kappa x)$. On the other hand, the rank $[n/2]+1$ Jordan states of zero energy are given by

$$\Omega_{n,\nu}^\alpha = \hat{I}_\alpha\Omega_{n,\nu} = (x+i\alpha)^{2n+\nu+1}\,, \qquad \Xi_{n,\nu}^\alpha = \hat{I}_\alpha\Xi_{n,\nu} = (x+i\alpha)^{2n-\nu}\,, \tag{52}$$

where $\Xi_{n,\nu} = x^{2n-\nu}$ is obtained by the transformation $\rho : \nu \to -\nu - 1$ (with respect to which generators (36) are invariant) over functions (38) [50]. Relations analogous to (39), (40) for functions $\Omega_{n,\nu}^\alpha$ are generated by applying the operator $\hat{I}_\alpha$. Additional transformation $\rho$ then yields

$$\hat{H}_{\alpha,\nu}\Xi_{n,\nu}^\alpha = -n(2n-2\nu-1)\Xi_{n-1,\nu}^\alpha\,, \qquad \hat{K}_\alpha\Xi_{n,\nu}^\alpha = \tfrac{1}{2}\Xi_{n+1,\nu}^\alpha\,, \tag{53}$$

$$2i\hat{D}_\alpha\Xi_{n,\nu}^\alpha = (2n-\nu+\tfrac{1}{2})\Xi_{n,\nu}^\alpha\,. \tag{54}$$

In the special case $\nu = m$ with $m = 1, 2, \ldots$, the system $\hat{H}_{\alpha,m}$ can be obtained from the free particle system on $\mathbb{R}$ by the Darboux transformation of the order $m$ [57, 8]. As a consequence, each such system possesses a hidden symmetry described by the Darboux-dressed generators of the translations and Galilean boosts of the free particle,

$$\hat{\mathcal{P}}_{\alpha,m} = \hat{\mathbb{A}}^{-}_{\alpha,m}\hat{p}\hat{\mathbb{A}}^{+}_{\alpha,m}\,, \qquad \hat{\mathcal{X}}_{\alpha,m} = \hat{\mathbb{A}}^{-}_{\alpha,m}(\hat{x}+i\alpha)\hat{\mathbb{A}}^{+}_{\alpha,m}\,, \tag{55}$$

$$\hat{\mathbb{A}}^{-}_{\alpha,m} = \hat{A}^{-}_{\alpha,m}\cdots\hat{A}^{-}_{\alpha,1}\,, \quad \hat{\mathbb{A}}^{+}_{\alpha,m} = \hat{A}^{+}_{\alpha,1}\cdots\hat{A}^{+}_{\alpha,m}\,, \quad \hat{A}^{\pm}_{\alpha,m} = \mp\tfrac{1}{\sqrt{2}}\left(\tfrac{d}{dx}\pm\tfrac{m}{x+i\alpha}\right)\,. \tag{56}$$

The order $(2m+1)$ differential operator $\hat{\mathcal{P}}_{\alpha,m}$ is the analog of the Lax-Novikov integral in the quantum reflectionless and finite-gap systems whose potentials are snapshots of the corresponding multi-soliton and cnoidal-type solutions to the KdV equation [7, 3]. Here the operators $\hat{\mathbb{A}}^{\pm}_{\alpha,m}$ are the higher order intertwining operators that connect the free particle Hamiltonian $\hat{H}_{\alpha,0} = \hat{H}$ with the $\mathcal{PT}$ regularized Calogero Hamiltonian $\hat{H}_{\alpha,m}$,

$$\hat{\mathbb{A}}^{-}_{\alpha,m}\hat{H}_{\alpha,0} = \hat{H}_{\alpha,m}\hat{\mathbb{A}}^{-}_{\alpha,m}\,, \qquad \hat{\mathbb{A}}^{+}_{\alpha,m}\hat{H}_{\alpha,m} = \hat{H}_{\alpha,0}\hat{\mathbb{A}}^{+}_{\alpha,m}\,. \tag{57}$$

The $\mathcal{PT}$-symmetric system $\hat{H}_{\alpha,m}$ is regular on a real line, and due to its relation to the free particle via the intertwining relations (57), it turns out to be perfectly invisible system with a unique $L^2(\mathbb{R})$ integrable state $\Xi_{0,m}$ of zero energy at the very edge of the doubly degenerate continuous part of the spectrum. The integral $\hat{\mathcal{P}}_{\alpha,m}$ separates the states of the same energy in the doubly degenerate continuous part of the spectrum as well as detects the unique bound state of the system $\hat{H}_{\alpha,m}$ by annihilating it [8, 9].

The commutation relations between the symmetry generators (55) and $\hat{H}_{\alpha,m}$, $\hat{D}_\alpha$ and $\hat{K}_\alpha$ produce an extended non-linear algebra, that includes, particularly, the Lie algebraic relations

$$[\hat{H}_{\alpha,m}\hat{\mathcal{P}}_{\alpha,m}] = 0\,, \qquad [\hat{H}_{\alpha,m}, \hat{\mathcal{X}}_{\alpha,m}] = -i\hat{\mathcal{P}}_{\alpha,m}\,, \tag{58}$$

$$[\hat{D}_\alpha, \hat{\mathcal{P}}_{\alpha,m}] = \tfrac{i}{2}(2m+1)\hat{\mathcal{P}}_{\alpha,m}\,, \qquad [\hat{D}_\alpha, \hat{\mathcal{X}}_{\alpha,m}] = \tfrac{i}{2}(2m-1)\hat{\mathcal{X}}_{\alpha,m}\,. \tag{59}$$

The action of operators (55) on Jordan states (52) with $\nu = m$ yields

$$\hat{\mathcal{P}}_{\alpha,m}\Xi^{\alpha}_{n,m} \propto \Omega^{\alpha}_{n-2m-1,m}\,, \qquad \hat{\mathcal{X}}_{\alpha,m}\Xi^{\alpha}_{n,m} \propto \Omega^{\alpha}_{n-2m,m}\,, \tag{60}$$

$$\hat{\mathcal{P}}_{\alpha,m}\Omega^{\alpha}_{n,m} \propto \Xi^{\alpha}_{n,m}\,, \qquad \hat{\mathcal{X}}_{\alpha,m}\Omega^{\alpha}_{n,m} \propto \Xi^{\alpha}_{n,m}\,, \tag{61}$$

$$\ker\hat{\mathcal{P}}_{\alpha,m} = \mathrm{span}\{\Xi^{\alpha}_{0,m},\ldots,\Xi^{\alpha}_{2m,m}\}\,, \qquad \ker\hat{\mathcal{X}}_{\alpha,m} = \mathrm{span}\{\Xi^{\alpha}_{0,m},\ldots,\Xi^{\alpha}_{2m-1,m}\}\,. \tag{62}$$

To relate a non-Hermitian $\mathcal{PT}$-symmetric asymptomatically free system like $\hat{H}_{\alpha,m}$ with its confined version, we introduce an extended CBT operator

$$\hat{\mathfrak{S}}_\alpha = \hat{I}_\alpha\hat{\mathfrak{S}}_0\hat{I}_{-\alpha}\,, \qquad \hat{\mathfrak{S}}^{-1}_\alpha = \hat{I}_\alpha\hat{\mathfrak{S}}^{-1}_0\hat{I}_{-\alpha}\,, \tag{63}$$

which in this case yields

$$\hat{\mathfrak{S}}_\alpha(\hat{H}_{\alpha,m})\hat{\mathfrak{S}}^{-1}_\alpha = -\omega\hat{\mathcal{J}}_{-,\alpha}\,, \quad \hat{\mathfrak{S}}_\alpha(i\hat{D}_\alpha)\hat{\mathfrak{S}}^{-1}_\alpha = \hat{\mathcal{J}}_{0,\alpha}\,, \quad \hat{\mathfrak{S}}_\alpha(\hat{K}_\alpha)\hat{\mathfrak{S}}^{-1}_\alpha = \tfrac{1}{\omega}\hat{\mathcal{J}}_{+,\alpha}\,, \tag{64}$$

$$\hat{\mathcal{J}}_{0,\alpha} = \tfrac{1}{2\omega}(\hat{H}_\alpha + \omega^2\hat{K}_\alpha)\,, \qquad \hat{\mathcal{J}}_{\pm,\alpha} = -\tfrac{1}{2\omega}(\hat{H}_\alpha - \omega^2\hat{K}_\alpha \pm 2i\omega\hat{D}_\alpha)\,. \tag{65}$$

Here we have $\hat{\mathcal{J}}_{0,\alpha} = \hat{I}_\alpha\hat{\mathcal{J}}_{0,0}\hat{I}_{-\alpha}$ and $\hat{\mathcal{J}}_{\pm,\alpha} = \hat{I}_\alpha\hat{\mathcal{J}}_{\pm,0}\hat{I}_{-\alpha}$, and this implies that $\hat{\mathcal{J}}_{+,\alpha}$ is not the Hermitian conjugation (with respect to the usual inner product on $\mathbb{R}$) of $\hat{\mathcal{J}}_{-,\alpha}$. Operators $(\hat{\mathcal{J}}_{0,0}, \hat{\mathcal{J}}_{\pm,0})$ correspond to the AFF model generators (41), which are singular for $x \in \mathbb{R}$.

On the other hand,

$$\hat{H}^{AFF}_{\alpha,\nu} = \hat{I}_\alpha(\hat{H}^{AFF}_{0,\nu})\hat{I}_{-\alpha} = -\frac{1}{2}\frac{d^2}{dx^2} + \frac{\nu(\nu+1)}{2(x+i\alpha)^2} + \frac{\omega^2}{2}(x+i\alpha)^2\,, \tag{66}$$

is the non-Hermitian Hamiltonian of the $\mathcal{PT}$-regularized AFF model with arbitrary value of the parameter $\nu > -1/2$. Note that for $\nu = 0$ we recover the $\mathcal{PT}$-symmetric one-dimensional harmonic oscillator with $x$ displaced for imaginary constant $i\alpha$ being the simplest case of non-Hermitian systems introduced in [33]. The corresponding eigenstates of (66) are

$$\psi^\alpha_{n,\nu} \propto \hat{\mathfrak{S}}_\alpha \Omega^\alpha_{n,\nu}\,, \qquad \phi^\alpha_{n,\nu} \propto \hat{\mathfrak{S}}_\alpha \Xi^\alpha_{n,\nu}\,, \tag{67}$$

whose explicit form is

$$\psi^\alpha_{n,\nu} = (x+i\alpha)^{\nu+1} e^{-\frac{\omega(x^2-\alpha^2)}{2}+i\omega\alpha x} L_n^{(\nu+\frac{1}{2})}(\omega(x+i\alpha)^2)\,, \qquad \phi^\alpha_{n,\nu} = \psi^\alpha_{n,-\nu-1}\,. \tag{68}$$

These functions satisfy the eigenvalue equations

$$\hat{H}^{AFF}_{\alpha,\nu}\psi^\alpha_{n,\nu} = E_{n,\nu}\psi^\alpha_{n,\nu}\,, \qquad E_{n,\nu} = \omega(2n+\nu+\tfrac{3}{2})\,, \tag{69}$$

$$\hat{H}^{AFF}_{\alpha,\nu}\phi^\alpha_{n,\nu} = \mathcal{E}_{n,\nu}\phi^\alpha_{n,\nu}\,, \qquad \mathcal{E}_{n,\nu} = \omega(2n-\nu+\tfrac{1}{2})\,, \tag{70}$$

and they are $L^2(\mathbb{R})$ normalizable, being of the form of a regular polynomial times a Gaussian term, see Figs. 1a and 1b. The notable here is that we have two different towers of states, where the distance between two consecutive energy levels in each tower is given by $\Delta E = E_{n,\nu} - E_{n-1,\nu} = \mathcal{E}_{n,\nu} - \mathcal{E}_{n-1,\nu} = 2\omega$. On the other hand, $\delta E = E_{n,\nu} - \mathcal{E}_{n,\nu} = \omega(2\nu+1)$. The last relation means that when $\nu = \ell - \frac{1}{2}$, $\delta E = \ell\Delta E$, $\ell = 1, 2, \ldots$, one could conclude that there emerges a double degeneracy in the spectrum because of the relation $E_{s-\ell,\ell-\frac{1}{2}} = \mathcal{E}_{s,\ell-\frac{1}{2}}$ with $s \geq \ell$. However, due to the Laguerre polynomial identity

$$\frac{(-\eta)^s}{s!} L_\ell^{(s-\ell)}(\eta) = \frac{(-\eta)^\ell}{\ell!} L_s^{(\ell-s)}(\eta)\,, \tag{71}$$

one can deduce that $\psi_{s-\ell,\ell-\frac{1}{2}} \propto \phi_{s,\ell-\frac{1}{2}}$, and so, such a double degeneracy does not really exist. For a similar phenomenon observed earlier in the Darboux transformations of the AFF model see ref. [50]. On the other hand, when $\nu = m$ one has $\delta E = \omega(2m+1)$. In this case the levels of the tower $E_{n,m}$ appear in the middle between the levels corresponding to $\mathcal{E}_{n,m}$, and the resulting spectrum is divided in two parts. One part corresponds to a semi-infinite equidistant part with energy levels separated by $\omega$. In another, finite part, equidistant separation between energy levels is $\Delta E = 2\omega$, see Figs. 1c-1e.

The action of the operators $\mathcal{J}_{\alpha,\pm}$ is obtained via the application of $\hat{\mathfrak{S}}_\alpha$, that yields

$$\mathcal{J}_{\alpha,\pm}\psi^\alpha_{n,\nu} \propto \psi^\alpha_{n\pm1,\nu}\,, \qquad \mathcal{J}_{\alpha,\pm}\phi^\alpha_{n,\nu} \propto \phi^\alpha_{n\pm1,\nu}\,, \qquad \ker \mathcal{J}_{\alpha,-} = \text{span}\{\psi^\alpha_{0,\nu}, \phi^\alpha_{0,\nu}\}\,. \tag{72}$$

This tells us that the states associated to each tower of energy levels can be produced by the $\mathfrak{sl}(2,\mathbb{R})$ generators starting from any fixed state, and also shows that there is no way to relate the states from the two towers when $\nu$ is not integer. In the integer case $\nu = m$ we have the operators

$$\hat{\mathcal{A}}_{\alpha,m} = \hat{\mathfrak{S}}_\alpha(\hat{\mathcal{P}}_{\alpha,m})\hat{\mathfrak{S}}_\alpha^{-1}\,, \qquad \hat{\mathcal{B}}_{\alpha,m} = \hat{\mathfrak{S}}_\alpha(\hat{\mathcal{X}}_{\alpha,m})\hat{\mathfrak{S}}_\alpha^{-1}\,, \tag{73}$$

$$[\hat{H}^{AFF}_{\alpha,m}, \hat{\mathcal{A}}_{\alpha,m}] = -\omega(2m+1)\hat{\mathcal{A}}_{\alpha,m}\,, \qquad [\hat{H}^{AFF}_{\alpha,m}, \hat{\mathcal{B}}_{\alpha,m}] = -\omega(2m-1)\hat{\mathcal{B}}_{\alpha,m}\,. \tag{74}$$

With the help of the CBT, one learns that the operator $\hat{\mathcal{A}}_{\alpha,m}$ annihilates the states $\phi_{j,m}$ with $j = 0, 1, \ldots, m$, while the operator $\hat{\mathcal{B}}_{\alpha,m}$ annihilates the states $\phi_{l,m}$ with $l = 0, 1, \ldots, m-1$.

Among these states we have all the eigenfunctions corresponding to the part separated from the infinite equidistant part of the spectrum. One finds also that they effectively relate the states of one tower with the states of another,

$$\hat{A}_{\alpha,m}\phi^{\alpha}_{n,m} \propto \psi^{\alpha}_{n-2m-1,m}\,, \qquad \hat{B}_{\alpha,m}\phi^{\alpha}_{n,m} \propto \psi^{\alpha}_{n-2m,m}\,, \tag{75}$$

$$\hat{A}_{\alpha,m}\psi^{\alpha}_{n,m} \propto \phi^{\alpha}_{n,m}\,, \qquad \hat{B}_{\alpha,m}\psi^{\alpha}_{n,m} \propto \phi^{\alpha}_{n,m}\,. \tag{76}$$

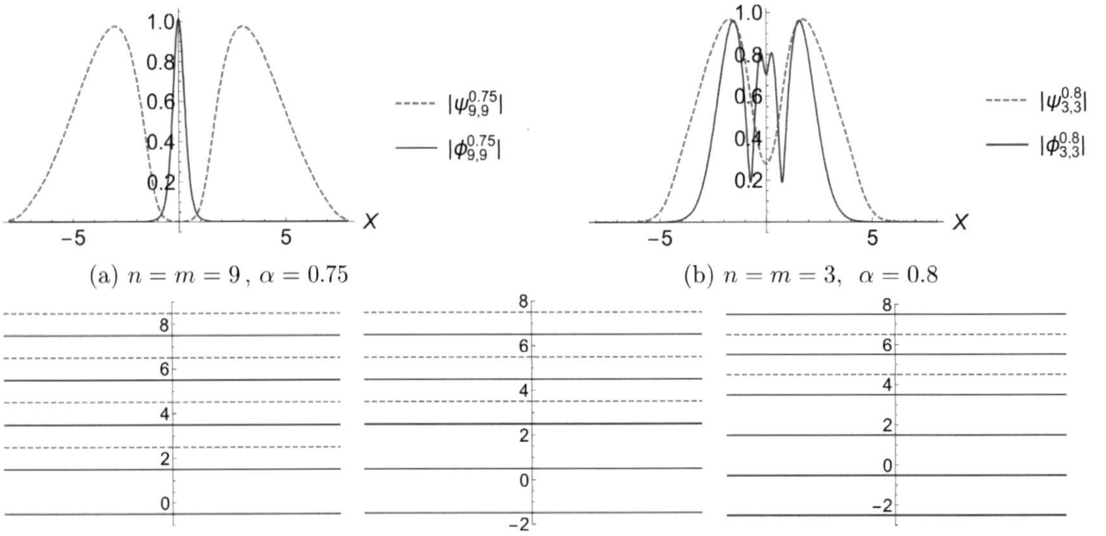


(a) $n = m = 9$ , $\alpha = 0.75$ (b) $n = m = 3$, $\alpha = 0.8$

(c) $m = 1$ (d) $m = 2$ (e) $m = 3$

Figure 1: In Figs. 1a and 1b, the modules of the states $\phi^{\alpha}_{n,m}$ and $\psi^{\alpha}_{n,m}$ (normalized by their respective maxima) with some values of $n$, $m$ and $\alpha$ are compared. On Figs. 1c, 1d and 1e, the energy levels $\mathcal{E}_{n,m}$ (continuos blue lines) and $E_{n,m}$ (dashed red lines) are shown for different values of the parameter $m$.

Finally, via application of $\hat{\mathfrak{S}}_{\alpha}$ to eigenstates of $\hat{H}_{\alpha,\nu}$ we get the overcomplete set of coherent states by allowing $\kappa$ to take complex values,

$$\hat{\mathfrak{S}}_{\alpha}\psi^{\alpha}_{\kappa,\nu}(\tfrac{1}{\sqrt{2}}x) = 2^{\frac{1}{4}}e^{-\frac{1}{2}(x+i\alpha)^2+\frac{1}{4}\kappa^2}\sqrt{x+i\alpha}\,J_{\nu+1/2}(\kappa(x+i\alpha)) := \Psi^{\alpha}_{\nu}(x,\kappa)\,, \tag{77}$$

$$\hat{\mathfrak{S}}_{\alpha}\phi^{\alpha}_{\kappa,\nu}(\tfrac{1}{\sqrt{2}}x) = 2^{\frac{1}{4}}e^{-\frac{1}{2}(x+i\alpha)^2+\frac{1}{4}\kappa^2}\sqrt{x+i\alpha}\,Y_{\nu+1/2}(\kappa(x+i\alpha)) := \Phi^{\alpha}_{\nu}(x,\kappa)\,. \tag{78}$$

Considering the issue of the inner product, one could try to take it in the form $\langle\chi_1|\,\hat{I}_{-2\alpha}\,|\chi_2\rangle$, and then we obtain $\langle\psi^{\alpha}_{n,\nu}|\,\hat{I}_{-2\alpha}\,|\psi^{\alpha}_{n,\nu}\rangle \propto \langle\psi^{0}_{n,\nu}|\psi^{0}_{n,\nu}\rangle$. However the quantity $\langle\phi^{\alpha}_{n,\nu}|\,\hat{I}_{-2\alpha}\,|\phi^{\alpha}_{n,\nu}\rangle$ diverges when $\nu \neq 0$, since functions $\phi^{0}_{n,\nu}$ are singular in the origin, see Eq. (68). Note, however, here that in the case of $\mathcal{PT}$-symmetric harmonic oscillator ($\nu = 0$) we do not have these problems since the states $\phi^{0}_{n,0}$ can be written in terms of even Hermite polynomials, which do not have singularities in the real line. So, the interesting open problem is to find a Hermitian system $\hat{\mathcal{H}}_{\nu}$, defined on the entire real line, with the same spectrum of $\hat{H}_{\alpha,\nu}$, and an operator $\hat{\mathcal{O}}_{\alpha}$ such that $\hat{\mathcal{O}}_{\alpha}\hat{\mathcal{H}}_{\nu}\hat{\mathcal{O}}^{-1}_{\alpha} = \hat{H}_{\alpha,\nu}$. One can expect that such a Hamiltonian will be a non-local operator of the nature similar to that considered in [58], see also [59].

### *3.4. A $\mathcal{PT}$-symmetric reinterpretation of CBT*

According to [37, 39], the $\mathcal{PT}$ symmetric Hamiltonian operator

$$\mathscr{H}_{\alpha,\beta,\gamma} = \alpha\hat{a}^{+}\hat{a}^{-} + \beta(\hat{a}^{-})^2 + \gamma(\hat{a}^{+})^2\,, \qquad x \in \mathbb{R}\,, \tag{79}$$

where $\hat{a}^+$ and $\hat{a}^-$ are the Hermitian conjugate raising and lowering ladder operators (28), is characterized by a purely real spectrum if the real parameters $\alpha$, $\beta$ and $\gamma$ satisfy the relation $\alpha^2 - 4\beta\gamma \geq 0$. In the particular case in which $\alpha = 0$, $\beta = \omega$ and $\gamma = -\omega$, this operator takes the form

$$\hat{\mathscr{H}}_{0,\omega,-\omega} = 2i\omega\hat{D}\,, \tag{80}$$

where $\hat{D}$ corresponds to the dilatation operator appearing in (21). The eigenstates of this Hamiltonian are the functions $\Omega_n$ presented in Eq. (24), and the eigenvalue problem corresponds to the third equation in (25). From Eq. (30) one deduces that the CBT generator $\hat{\mathfrak{S}}$ works as the operator that relates the Hamiltonian (80) with the harmonic oscillator system described by the Hermitian Hamiltonian. Indeed, from equations (16) we obtain the $\mathcal{PT}$-symmetric conjugation

$$\hat{\mathscr{H}}_{0,\omega,-\omega}\hat{\Theta} = \hat{\mathscr{H}}^{\dagger}_{0,\omega,-\omega}\hat{\Theta}\,, \qquad \hat{\Theta} = (\hat{\mathfrak{S}})^2\,, \tag{81}$$

implying that the $\mathcal{PT}$-symmetric normalization of the eigenstates $\Omega_n$ is just equivalent to normalization of eigenstates of the quantum harmonic oscillator under the usual inner product in $\mathbb{R}$. In the same vein, we note that at the classical level, the time evolution produced by the Hamiltonian $\mathscr{H}_{0,\omega,-\omega} = 2i\omega D$ in the variables $x$ and $p$ is governed by the equations $\dot{x} = \{x, \mathscr{H}_{0,\omega,-\omega}\} = i\omega x$ and $\dot{p} = \{p, \mathscr{H}_{0,\omega,-\omega}\} = -i\omega p$, which resemble the equations of motion of the classical analog $a^+$ and $a^-$ of the first order ladder operators of the harmonic oscillator system. Off course, both systems are related to each other by the classical version of the conformal bridge transformation reviewed in Section 2.1.

The model (79) can be generalised up to a concrete realization of the $\mathfrak{so}(2,1) \cong \mathfrak{sl}(2,\mathbb{R})$ generators, by changing the harmonic oscillator operators by

$$\hat{\mathcal{H}}_{\alpha,\beta,\gamma} = \alpha\hat{\mathcal{J}}_0 + \beta\hat{\mathcal{J}}_- + \gamma\hat{\mathcal{J}}_+ = \tfrac{\alpha-\beta-\gamma}{2\omega}\hat{H} + \tfrac{\omega(\alpha+\beta+\gamma)}{2}\hat{K} + i(\beta-\gamma)\hat{D}\,. \tag{82}$$

Relation (81) holds since it is true by the conformal algebraic arguments. This also means that the $\mathcal{PT}$-symmetric normalization of the considered physical states of the system, which are also the rank $[n/2]+1$ Jordan states of zero energy of $\hat{H}$, correspond to the normalization of the eigenstates of the system given by the Hamiltonian $\hat{\mathcal{J}}_0$. In this way, if we select the realization (36), where the generators are defined on $\mathbb{R}^+$, the physical eigenstates of (82) with $\alpha = 0$, $\beta = -\gamma$ correspond to (38), and the corresponding eigenvalue equation is given by Eq. (40).

## 4. CBT for higher-dimensional Euclidean systems

Let us start with realization of the operators $\hat{H}$, $\hat{D}$ and $\hat{K}$ in higher-dimensional systems. In the simplest case of a free particle in $\mathbb{R}^d$, the generators of its $\mathfrak{so}(2,1)$ conformal symmetry are given by (in this section we restore the dimensional constants):

$$\hat{H} = \textstyle\sum_{i=1}^{d}\hat{H}_i\,, \qquad \hat{D} = \sum_{i=1}^{d}\hat{D}_i\,, \qquad \hat{K} = \sum_{i=1}^{d}\hat{K}_i\,, \tag{83}$$

$$\hat{H}_i = \tfrac{-\hbar^2}{2m}\tfrac{\partial^2}{\partial x_i^2}\,, \qquad \hat{D}_i = -i\tfrac{\hbar}{2}\left(x_i\tfrac{\partial}{\partial x_i} + \tfrac{1}{2}\right), \qquad \hat{K}_i = \tfrac{m}{2}x_i^2\,. \tag{84}$$

These symmetry generators are complemented by the Heisenberg algebra generators

$$\hat{p}_j = -i\hbar\tfrac{\partial}{\partial x_j}\,, \qquad \hat{\xi}_j = mx_j\,, \qquad [\hat{\xi}_j,\hat{\xi}_k] = [\hat{p}_j,\hat{p}_k] = 0\,, \qquad [\hat{\xi}_j,\hat{p}_k] = i\hbar m\delta_{jk}\,, \tag{85}$$

and the angular momentum tensor $\hat{M}_{ij} = \frac{1}{m}(\hat{\xi}_i\hat{p}_j - \hat{\xi}_j\hat{p}_i)$. Together, all these generators produce a $d$-dimensional Schrödinger symmetry of a free particle [52].

The commutation relations in (85) imply that different conformal bridge transformations can be applied for each spatial direction, and each of them works in the same way as in the one-dimensional case considered in Sec. 3.1.

First, let us consider the isotropic CBT produced by the operator

$$\hat{\mathfrak{S}} = \Pi_{i=1}^{d}\hat{\mathfrak{S}}_i\,, \qquad \hat{\mathfrak{S}}_i = e^{-\frac{\omega}{\hbar}\hat{K}_i} e^{\frac{\hat{H}_i}{2\hbar\omega}} e^{\frac{i}{\hbar}\ln(2)\hat{D}_i}\,, \qquad [\hat{\mathfrak{S}}_i, \hat{\mathfrak{S}}_j] = 0\,. \tag{86}$$

This operator generates a composed CBT with equal frequencies in each direction. For this reason, it commutes with the angular momentum tensor, and so, is rotationally invariant. Applying the similarity transformation given by (86) to the free particle, we obtain, in accordance with Eqs. (16) and (14), the $d$-dimensional isotropic harmonic oscillator, with the $\mathfrak{sl}(2,\mathbb{R})$ generators $\hat{\mathcal{J}}_\pm$ to be quadratic radial ladder operators. Also one gets

$$\hat{\mathfrak{S}}(\hat{p}_j)\hat{\mathfrak{S}}^{-1} = -i\sqrt{m\hbar\omega}\,\hat{a}_j^-\,, \qquad \hat{\mathfrak{S}}(\hat{\xi}_j)\hat{\mathfrak{S}}^{-1} = \sqrt{\tfrac{m\hbar}{\omega}}\,\hat{a}_j^+\,, \tag{87}$$

where $\hat{a}_i^+$ are the first order ladder operators for each direction,

$$\hat{a}_i^\pm = \sqrt{\tfrac{m\omega}{2\hbar}}\left(x_i \mp \tfrac{\hbar}{m\omega}\tfrac{\partial}{\partial x_i}\right)\,, \qquad [\hat{a}_i^\pm, \hat{a}_j^\pm] = 0\,, \qquad [\hat{a}_i^-, \hat{a}_j^+] = \delta_{ij}\,. \tag{88}$$

The second option is the anisotropic CBT composed from generators with different values of frequencies $\omega_i > 0$,

$$\hat{\mathfrak{S}}_{\omega_1,\ldots\omega_d} = \Pi_{i=1}^{d}\hat{\mathfrak{S}}_{\omega_i}\,, \qquad \hat{\mathfrak{S}}_{\omega_i} = e^{-\frac{\omega_i}{\hbar}\hat{K}_i} e^{\frac{\hat{H}_i}{2\hbar\omega_i}} e^{\frac{i}{\hbar}\ln(2)\hat{D}_i}\,, \qquad [\hat{\mathfrak{S}}_{\omega_i}, \hat{\mathfrak{S}}_{\omega_j}] = 0\,. \tag{89}$$

Via the similarity transformation, this operator and its inverse transform the linear combination $2i\hat{\mathcal{D}}^{\epsilon_1,\ldots,\epsilon_d}_{\omega_1,\ldots,\omega_d} = 2i\sum_{i=1}^{d}\omega_i\epsilon_i\hat{D}_i$ into the $d$-dimensional anisotropic oscillator Hamiltonian $\hat{H}^{\epsilon_1,\ldots,\epsilon_d}_{\omega_1,\ldots,\omega_d} = \sum_i^d \epsilon_i \hat{H}^{\omega_i}_{\text{os}}$ with $\hat{H}^{\omega_i}_{\text{os}} = \hbar\omega_i(\hat{a}^+_{\omega_i}\hat{a}^-_{\omega_i} + \frac{1}{2})$, and $\hat{a}^\pm_{\omega_i} = \sqrt{\frac{m\omega_i}{2\hbar}}\left(x_i \mp \frac{\hbar}{m\omega_i}\frac{\partial}{\partial x_i}\right)$, where each $\epsilon_i$ can be chosen as 1 or $-1$. On the other hand, up to multiplicative constants, one also gets $\hat{\mathfrak{S}}_{\omega_1,\ldots\omega_d} : (\hat{\xi}_i, \hat{p}_i) \to (\hat{a}^+_{\omega_i}, \hat{a}^-_{\omega_i})$.

Systems in $d$ Euclidean dimensions, such as the free particle and isotropic or anisotropic harmonic oscillators, can have hidden symmetries generated by higher order integrals of motion [60]. Since these integrals are always written in terms of $\hat{x}$ and $\hat{p}$ (or in terms of (88)), the CBT maps the symmetry generators of one system into those of another. In practical terms, application of the CBT to the operators that commute with $2i\hat{D}$ ($2i\hat{\mathcal{D}}$) produces the Hamiltonian symmetries of the isotropic (anisotropic) case. This scheme allows a reinterpretation from the perspective of $\mathcal{PT}$ symmetry, in the spirit of Sec. 3.4, since the operators $2i\hat{D}$ and $2i\hat{\mathcal{D}}$ are a generalization of the Hamiltonian (80) to $d$-dimensions, in the isotropic and anisotropic case, respectively.

To see some concrete applications of the higher dimensional CBT in detail, we will consider only the case $d = 2$. In particular, we focus our attention on the exotic rotationally invariant harmonic oscillator (ERIHO) system [30], which corresponds to the planar isotropic harmonic oscillator extended by a Zeeman type term. This model is generated by the isotropic CBT by generalizing the already considered constructions. It represents a one parametric family of systems revealing different phases, two of which correspond to the Landau problem. Additionally, in [30] it also was shown that in spite of the explicit rotationally invariant nature, the model is unitary equivalent to the planar anisotropic harmonic oscillator (AHO) via the application of a certain $\mathfrak{su}(2)$ rotation accompanied by an anisotropic $\mathfrak{so}(1,1)$ Bogolyubov transformation [61].

### *4.1. The ERIHO system: classical case*

Starting from the two-dimensional free particles system, let us consider the following complex combination of its symmetry generators

$$2iD_0 + gp_\varphi = x_j\Delta_{jk}p_k \qquad g \in \mathbb{R}\,, \tag{90}$$

$$\Delta_{jk} = i\delta_{jk} + g\epsilon_{jk}\,, \qquad \Delta_{jk}\Delta_{jl} = (g^2-1)\delta_{jl}\,, \qquad \det\Delta = g^2 - 1\,. \tag{91}$$

The generator $p_\varphi = M_{12} = \epsilon_{ij} x_i p_j$ of $\mathfrak{so}(2)$ rotations is invariant under the classical isotropic CBT. As a result, (90), multiplied by $\omega$, is transformed into the classical Hamiltonian of the ERIHO system,

$$H_g = H_{\rm osc} + g\omega p_\varphi\,, \qquad H_{\rm osc} = \tfrac{1}{2m} p_i p_i + \tfrac{1}{2} m\omega^2 x_i x_i\,. \tag{92}$$

System (92) admits the following three different physical interpretations [30].

First, $H_g$ corresponds to the Hamiltonian of a planar particle in a non-inertial frame rotating with angular velocity $\Omega = g\omega$ and subjected to the action of the isotropic harmonic trap $U = \frac{1}{2} k x_i x_i$. The cases $k > m\Omega^2$, $k = m\Omega^2$ and $0 < k < m\Omega^2$ correspond, respectively, to the phases $0 < g^2 < 1$, $g^2 = 1$ and $g^2 > 1$ of the system (92), while the inertial case $\Omega = 0$, $k = m\omega^2$ corresponds to the phase of the isotropic oscillator of (92) with $g = 0$.

Second, in the cases $g = +1$ and $g = -1$, (92) takes the form of the Hamiltonian of Landau problem in symmetric gauge with different orientation (sign) of the magnetic field $B$ and $\omega = g\omega_B$, $\omega_B = qB/2mc$, where $q$ is the charge of a particle. Then the phases with $0 \leq g^2 < 1$ and $g^2 > 1$ of (92) correspond to the extended Landau problem in the presence of the additional harmonic potential term $\frac{1}{2} m\Lambda x_i x_i$ with $\Lambda > 0$ and $0 > \Lambda > -\omega_B^2$, respectively, where $\omega = \sqrt{\Lambda + \omega_B^2}$, and $g = \omega_B/\omega$. The repulsive critical, $\Lambda = -m\omega_B^2$, and supercritical, $\Lambda < -m\omega^2$, cases of the extended Landau problem have no analogs in the system (92).

Finally, in terms of the classical analogues of the circular ladder operators,

$$b_1^- = \tfrac{1}{\sqrt{2}}(a_1^- - i a_2^-)\,, \quad b_1^+ = (b_1^-)^*\,, \quad b_2^- = \tfrac{1}{\sqrt{2}}(a_1^- + i a_2^-)\,, \quad b_2^+ = (b_2^-)^*\,, \tag{93}$$

$$a_i^\pm = \sqrt{\tfrac{m\omega}{2}}\left(x_i \mp \tfrac{i}{m\omega} p_i\right)\,, \tag{94}$$

Hamiltonian (92) takes the form

$$H_g = \omega\left(\ell_1 b_1^+ b_1^- + \ell_2 b_2^+ b_2^-\right)\,, \qquad \ell_1 = 1 + g\,, \qquad \ell_2 = 1 - g\,. \tag{95}$$

It looks like the anisotropic harmonic oscillator Hamiltonian, but system (92) is manifestly rotational invariant.

In correspondence with relations (91) and the comments on different interpretations, it is expected that the system (92) should have essentially different physical properties and symmetries in the cases $g^2 < 1$ and $g^2 > 1$, as well as when $g = \pm 1$. Indeed, the system corresponds to the planar isotropic harmonic oscillator when $g = 0$, meanwhile, as it was already mentioned, the model at $g = \pm 1$ represents the Landau problem in the symmetric gauge. In the case $|g| < 1$, the Hamiltonian (92) formally looks like the Euclidean AHO with different frequencies $\omega_1 \neq \omega_2$, $\omega_i = \ell_i \omega$, contrary to the case of $|g| > 1$, when (92) has instead the form of a Hamiltonian of the Minkowskian AHO with frequencies of two different signs. This last family of systems resembles the Pais-Uhlenbeck oscillator, which recently attracted a considerable attention in relation to the $\mathcal{PT}$-symmetry, see Refs. [62, 63, 64]. Finally, in the limit $g \to \infty$, one has

$$g^{-1} H_g \to \omega p_\varphi = \omega(b_1^+ b_1^- - b_2^+ b_2^-) = b_i^+ \eta_{ij} b_j^-\,, \qquad \eta = \mathrm{diag}(1, -1), \tag{96}$$

which can be interpreted as the isotropic Minkowskian oscillator.

From the point of view of the $\mathcal{PT}$ symmetry, the generator (90) is a generalization of the classical analogue of the $\mathcal{PT}$ invariant Hamiltonian of the form (80), extended now by the angular momentum taken with arbitrary coupling constant. Here, the isotropic CBT provides us the transformation that connects this system with its real (Hermitian in the quantum case) counterpart $H_g$.

By solving the equations of motion for $b_j^\pm$, $\dot{b}_j^\pm = \pm i\omega\ell_j b_j^\pm$, $j = 1, 2$, and using the relation $\sqrt{m\omega}(x_1 + ix_2) = b_1^+ + b_2^-$, we get the trajectories of the system,

$$z(t) = x_1(t) + ix_2(t) = R_1 e^{i\gamma_1} e^{i\omega\ell_1 t} + R_2 e^{-i\gamma_2} e^{-i\omega\ell_2 t}\,, \tag{97}$$

where $R_i \geq 0$ and $\gamma_i \in \mathbb{R}$ are the integration constants. The energy and angular momentum of the system are given by $E_g = m\omega^2(\ell_1 R_1^2 + \ell_2 R_2^2)$, $p_\varphi = m\omega(R_1^2 - R_2^2)$. Notice that for $g^2 < 1$ the exponents in (97) evolve in opposite directions, while in the case of $g^2 > 1$ they change in the same direction that depends on the sign of $g$. On the other hand, at $g = +1$ $(g = -1)$, one gets $\omega_2 = 0$ $(\omega_1 = 0)$, and the orbit is a circumference of radius $R_1$ $(R_2)$ centered at $(X_1, X_2)$ with $Z = X_1 + iX_2 = R_2 e^{-i\gamma_2}$ $(Z = R_1 e^{i\gamma_1})$.

In general case, the trajectory is closed for arbitrary choice of the integration constants iff the condition $\ell_1/\ell_2 = q_2/q_1$ with $q_1, q_2 \in \mathbb{Z}$ is fulfilled. This implies rational values for the parameter $g = (q_2 - q_1)/(q_1 + q_2)$. Some trajectories for rational values of $g$ are shown in Figs. 2 and 3. In the case of Minkowskian isotropic oscillator (96), the trajectories are obtained by applying the transformation $\omega \to \omega/|g|$, and taking the limit $|g| \to \infty$ in (97). As a result one gets a circle centered in the origin of the coordinate system, $z(t) = e^{i\epsilon\omega t}(R_1 e^{i\gamma_1} + R_2 e^{i\gamma_2})$, where $\epsilon = \pm 1$ for $g \to \pm\infty$ [30].

The closed character of the trajectories for rational values of the parameter $g$ indicates that some additional true integrals of motion have to appear in the corresponding systems. To obtain them, let us employ the classical CBT.

In the case in which we select $g$ as the irreducible rational number

$$g_<^{s_1,s_2} = (s_2 - s_1)/(s_1 + s_2)\,, \qquad s_1, s_2 = 1, 2, \ldots\,, \qquad |g_<^{s_1,s_2}| < 1\,, \tag{98}$$

it is easy to see that the phase space functions of the classical free particle

$$S_{s_1,s_2}^+ = (\xi_+)^{s_1}(p_+)^{s_2}\,, \qquad \hat{S}_{s_1,s_2}^- = (p_-)^{s_1}(\xi_-)^{s_2}\,, \tag{99}$$

where $p_\pm = p_1 \pm ip_2$ and $\xi_\pm = \xi_1 \pm i\xi_2$, Poisson commute with (90). After the application of the classical CBT we get (up to certain constant multiplicative factors)

$$\mathcal{L}_{s_1,s_2}^+ = (b_1^+)^{s_1}(b_2^-)^{s_2}\,, \qquad \mathcal{L}_{s_1,s_2}^- = (\mathcal{L}_{s_1,s_2}^+)^*\,. \tag{100}$$

These new generators are the true integrals of motion for the system (92). Together with $p_\varphi$ and $H_g$, they generate a non-linear deformation of the $\mathfrak{u}(2) \cong \mathfrak{su}(2) \oplus \mathfrak{u}(1)$ algebra [30], which in the case of $s_1 = s_2 = 1$, $g = 0$, reduces to the $\mathfrak{u}(2) \cong \mathfrak{su}(2) \oplus \mathfrak{u}(1)$ Lie algebraic symmetry of the isotropic oscillator. As $\{\mathcal{L}_{s_1,s_2}^+, \mathcal{L}_{s_1,s_2}^-\}$ is a polynomial of $H_g$ and $p_\varphi$ (of order $s_1 + s_2$), effectively (100) provides us with only one new integral independent from $H_g$ and $p_\varphi$.

If instead of (98) we chose now the irreducible fraction

$$g_>^{s_1,s_2} = (s_2 + s_1)/(s_2 - s_1)\,, \qquad |g_>^{s_1,s_2}| > 1\,, \tag{101}$$

one can show that the polynomials of the free particle Heisenberg generators

$$\Xi_{s_1,s_2}^+ = (\xi_+)^{s_1}(\xi_-)^{s_2}\,, \qquad \Xi_{s_1,s_2}^- = (p_+)^{s_1}(p_-)^{s_2} \tag{102}$$

Poisson commute with (90), and after the application of the conformal bridge transformation, we obtain (up to certain multiplicative constants) the true integrals of motion for our system,

$$\mathcal{J}_{s_1,s_2}^+ = (b_1^+)^{s_1}(b_2^+)^{s_2}\,, \qquad \mathcal{J}_{s_1,s_2}^- = (\mathcal{J}_{s_1,s_2}^+)^*\,. \tag{103}$$

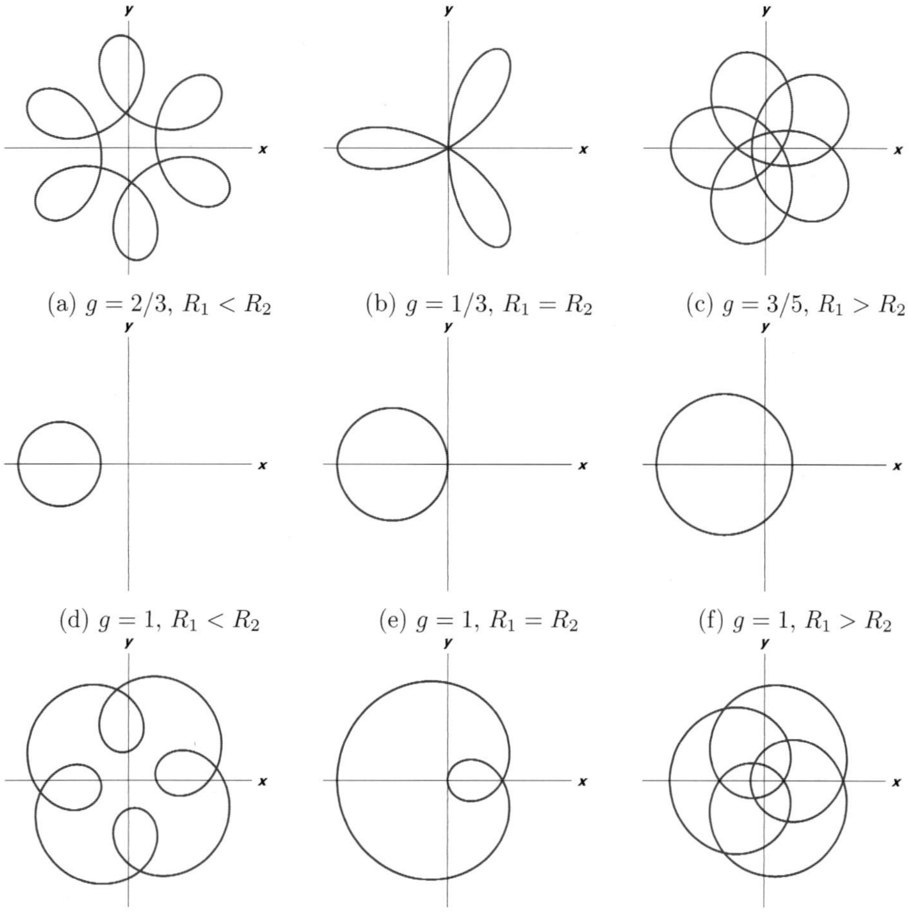


(a) $g = 2/3$, $R_1 < R_2$ (b) $g = 1/3$, $R_1 = R_2$ (c) $g = 3/5$, $R_1 > R_2$

(d) $g = 1$, $R_1 < R_2$ (e) $g = 1$, $R_1 = R_2$ (f) $g = 1$, $R_1 > R_2$

(g) $g = 3/2$, $R_1 < R_2$ (h) $g = 3$, $R_1 = R_2$ (i) $g = 5/3$, $R_1 > R_2$

Figure 2: Trajectories for some rational values of $g$. In cases b), e) and h), $p_\varphi = 0$ and trajectories pass through the origin. For $R_1 \neq R_2$, $\text{sign}\,(p_\varphi) = \text{sign}\,(R_1 - R_2)$.

As in the previous case, these integrals generate a non-linear algebra, which corresponds here to a deformation of the $\mathfrak{gl}(2,\mathbb{R}) = \mathfrak{sl}(2,\mathbb{R}) \oplus \mathfrak{u}(1)$ algebra [30].

The symmetries of the phases $g = 1$ and $g = -1$ of the Landau problem as well as of the isotropic Minkowskian oscillator, $|g| = \infty$, can also be reproduced by the described CBT, see ref. [30].

We do not discuss here the complete algebraic structure generated by computing the Poisson brackets between the true integrals and the rest of dynamical symmetries of the system. Nevertheless we notice, that in the case $g = g_<^{s_1,s_2}$ ($g = g_>^{s_1,s_2}$) the integrals $\mathcal{J}^\pm_{s_1,s_2}$ ($\mathcal{L}^\pm_{s_1,s_2}$) are dynamical, and can be generated via the Poisson brackets between the integrals $\mathcal{L}^\pm_{s_1,s_2}$ ($\mathcal{J}^\pm_{s_1,s_2}$) and the quantities $\mathcal{L}^\pm_{1,1} := \mathcal{L}_\pm$ ($\mathcal{J}^\pm_{1,1} := \mathcal{J}_\pm$), which also are dynamical integrals of the system. In this process we generate a large but still finite number (depending on the values of the integer parameters $s_1$ and $s_2$) of dynamical integrals, which together with four true integrals generate a finite non-linear algebra of the $W$ type. From this point of view we also have a kind of transmutation of symmetries for the "dual" pairs of the systems with $g = g_<^{s_1,s_2}$ and $g = g_>^{s_1,s_2} = 1/g_<^{s_1,s_2}$, where the non-linearly deformed $\mathfrak{u}(2)$ and $\mathfrak{gl}(2,\mathbb{R})$ subalgebras generated

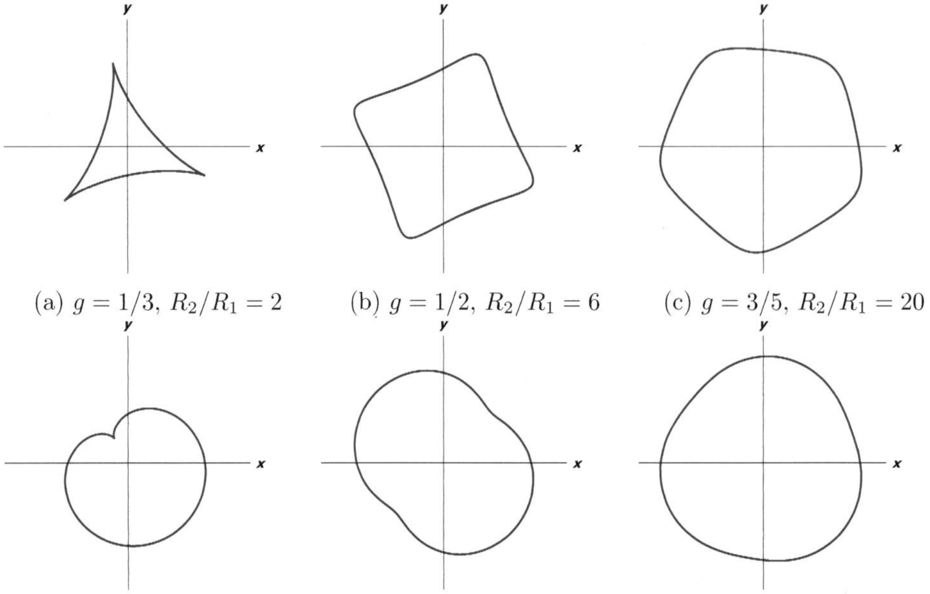

(a) $g = 1/3$, $R_2/R_1 = 2$ (b) $g = 1/2$, $R_2/R_1 = 6$ (c) $g = 3/5$, $R_2/R_1 = 20$

(d) $g = 3$, $R_2/R_1 = 2$ (e) $g = 2$, $R_2/R_1 = 6$ (f) $g = 5/3$, $R_2/R_1 = 20$

Figure 3: Trajectories for some rational values of $g$ and $R_1/R_2$. The "dual" figures (a) and (d), see below, correspond to a general case $R_1|\ell_1| = R_2|\ell_2|$ of the trajectories with cusps, in which velocity turns into zero.

by the sets $(H_g, \mathcal{L}_2, \mathcal{L}^{\pm}_{s_1,s_2})$ and $(H_g, \mathcal{J}_0, \mathcal{J}^{\pm}_{s_1,s_2})$ change their role in the sense of the true and dynamical sub-symmetries. For more details see ref. [30].

*4.2. Quantum case of the ERIHO system*
At the quantum level we have

$$\hat{H}_g = \hat{H}_{\rm osc} + \omega\hat{p}_\varphi = \hat{\mathfrak{S}}\omega(2i\hat{D} + g\hat{p}_\varphi)\hat{\mathfrak{S}}^{-1} = \hbar\omega(\ell_1\hat{b}_1^+\hat{b}_1^- + \ell_2\hat{b}_2^+\hat{b}_2^- + 1)\,, \tag{104}$$

$$\ell_1 = 1 + g\,, \qquad \ell_2 = 1 - g\,, \tag{105}$$

$$\hat{b}_1^- = \tfrac{1}{\sqrt{2}}(\hat{a}_1^- - i\hat{a}_2^-)\,, \quad \hat{b}_1^+ = (\hat{b}_1^-)^\dagger\,, \quad \hat{b}_2^- = \tfrac{1}{\sqrt{2}}(\hat{a}_1^- + i\hat{a}_2^-)\,, \quad \hat{b}_2^+ = (\hat{b}_2^-)^\dagger\,. \tag{106}$$

Notice here that the operator $2i\hat{D} + g\hat{p}_\varphi$, to which we apply the CBT, is $\mathcal{PT}$ symmetric if $\mathcal{P}$ is identified as a spatial reflection operator in two dimensions, $\mathcal{P}x_1 = -x_1\mathcal{P}$, $\mathcal{P}x_2 = x_2\mathcal{P}$. To obtain the eigenstates and the spectrum of this system, analogously to the procedure described in Sec. 3, we first have to solve the eigenvalue equation

$$(2i\hat{D} + g\hat{p}_\varphi)\phi_\lambda = \hbar\left((1+g)z\tfrac{\partial}{\partial z} + (1-g)z^*\tfrac{\partial}{\partial z^*}\right)\phi_\lambda = \lambda\phi_\lambda\,, \qquad z = x_1 + ix_2\,. \tag{107}$$

The well defined in $\mathbb{R}^2$ solutions of this equation correspond to $\phi_{n_1,n_2} = z^{n_1}(z^*)^{n_2}$, where $n_1$ and $n_2$ are non-negative integers. These are the Jordan states of the two-dimensional free particle that satisfy the equations $\hat{H}\phi_{n_1,n_2} = -\frac{2\hbar}{m}n_1n_2\phi_{n_1-1,n_2-1}$, $2i\hat{D}\phi_{n_1,n_2} = \hbar(n_1+n_2+1)\phi_{n_1,n_2}$, $\hat{p}_\varphi\phi_{n_1,n_2} = \hbar(n_1-n_2)\phi_{n_1,n_2}$, which imply that $(2i\hat{D} + g\hat{p}_\varphi)\phi_{n_1,n_2} = \hbar(n_1\ell_1 + n_2\ell_2 + 1)\phi_{n_1,n_2}$. The isotropic two-dimensional CBT produces (up to multiplicative constants) a map

$$\hat{\mathfrak{S}} : (\hat{\xi}_+, \hat{\xi}_-, \hat{p}_+, \hat{p}_-) \quad \rightarrow \quad (\hat{b}_1^+, \hat{b}_2^+, \hat{b}_2^-, \hat{b}_1^-)\,, \tag{108}$$

where $\hat{p}_\pm = \hat{p}_1 \pm i\hat{p}_2$ and $\hat{\xi}_\pm = \hat{\xi}_1 \pm i\hat{\xi}_2$, as well the map (16) with $\hat{H}$, $\hat{D}$ and $\hat{K}$ given by (83) with $d = 2$, while $\hat{p}_\varphi$ is left invariant. By computing the action of the generators $(\hat{H}, \hat{D}, \hat{K}, \hat{p}_\varphi, \hat{\xi}_+, \hat{\xi}_-, \hat{p}_+, \hat{p}_-)$ on the states $\phi_{n_1,n_2}$, and with the subsequent application of the CBT generator $\hat{\mathfrak{S}}$ from the left, one obtains the equations

$$\hat{b}_1^\pm \Psi_{n_1,n_2} = \sqrt{n_1 + \beta_\pm}\Psi_{n_1\pm1,n_2}\,, \qquad \hat{b}_2^\pm \Psi_{n_1,n_2} = \sqrt{n_2 + \beta_\pm}\Psi_{n_1,n_2\pm1}\,, \tag{109}$$

$$\hat{\mathcal{J}}_\pm \Psi_{n_1,n_2}\sqrt{(n_1 + \beta_\pm)(n_2 + \beta_\pm)}\Psi_{n_1\pm1,n_2\pm1}\,, \tag{110}$$

$$\hat{H}_g \Psi_{n_1,n_2} = E_{n_1,n_2}\Psi_{n_1,n_2}\,, \qquad \hat{p}_\varphi \Psi_{n_1,n_2} = \hbar(n_1 - n_2)\Psi_{n_1,n_2}\,, \tag{111}$$

$$E_{n_1,n_2} = \hbar\omega(\ell_1 n_1 + \ell_2 n_2 + 1)\,, \qquad \beta_\pm = \tfrac{1\pm1}{2}\,. \tag{112}$$

Here the physical eigenstates $\Psi_{n_1,n_2}(x_1, x_2)$ are given by

$$\hat{\mathfrak{S}}\phi_{n_1,n_2} = \mathcal{N}_{n_1,n_2}\Psi_{n_1,n_2}\,, \qquad \mathcal{N}_{n_1,n_2} = \left(\tfrac{2\hbar}{m\omega}\right)^{\frac{n_1+n_2}{2}}\sqrt{n_1!n_2!\pi}\,, \tag{113}$$

$$\Psi_{n_1,n_2} = \sqrt{\tfrac{m\omega}{\hbar\pi n_1!n_2!}}\, H_{n_1,n_2}\left(\sqrt{\tfrac{m\omega}{\hbar}}x_1, \sqrt{\tfrac{m\omega}{\hbar}}x_2\right) e^{-\frac{m\omega}{2\hbar}(x_1^2+x_2^2)}\,, \tag{114}$$

where the functions are the generalized Hermite polynomials of two indexes [65].

From equations (109) one deduces that the operators $\hat{b}_i^\pm$ are the spectrum generating ladder operators of the system for arbitrary values of $g$. Eqs. (111) and (112) yield the energy spectrum of the system and the angular momentum value of each stationary state. In dependence on the value of $g$, the spectrum has the following properties. It is degenerate iff $g$ is a rational number, that we assume from now on. The spectrum is positive, has a finite degeneracy, and the ground state is not degenerate when $|g| < 1$. In the case $|g| > 1$, it is not bounded from below, and has infinite degeneracy in each energy level. Finally, we have the spectrum of the Landau problem when $|g| = 1$, see [27].

In the case in which $g$ is equal to (98) one gets that the integrals

$$\hat{\mathcal{L}}^+_{s_1,s_2} = (\hat{b}_1^+)^{s_1}(\hat{b}_2^-)^{s_2}\,, \qquad \hat{\mathcal{L}}^-_{s_1,s_2} = (\hat{\mathcal{L}}^-_{s_1,s_2})^\dagger\,, \tag{115}$$

which are the direct quantum analogs of $\mathcal{L}^\pm_{s_1,s_2}$, act as follows,

$$\hat{\mathcal{L}}^\pm_{s_1,s_2}\Psi_{n_1,n_2} = \sqrt{\tfrac{\Gamma(n_1+\beta_\pm s_1+1)\Gamma(n_2+\beta_\mp s_2+1)}{\Gamma(n_1-\beta_\mp s_1+1)\Gamma(n_2-\beta_\pm s_2+1)}}\Psi_{n_1\pm s_1,n_2\mp s_2}\,. \tag{116}$$

Besides, when $g$ corresponds to the case (101), the action of the quantum analogs of the integrals $\hat{\mathcal{J}}^\pm_{s_1,s_2}$,

$$\hat{\mathcal{J}}^+_{s_1,s_2} = (\hat{b}_1^+)^{s_1}(\hat{b}_2^+)^{s_2}\,, \qquad \hat{\mathcal{J}}^-_{s_1,s_2} = (\hat{\mathcal{J}}^-_{s_1,s_2})^\dagger\,, \tag{117}$$

yields

$$\hat{\mathcal{J}}^\pm_{s_1,s_2}\Psi_{n_1,n_2} = \sqrt{\tfrac{\Gamma(n_1+\beta_\pm s_1+1)\Gamma(n_2+\beta_\pm s_2+1)}{\Gamma(n_1-\beta_\mp s_1+1)\Gamma(n_2-\beta_\mp s_2+1)}}\Psi_{n_1\pm s_1,n_2\pm s_2}\,. \tag{118}$$

All the normalizable eigenfunctions with the same energy can be obtained by repeated application of these operators to some fixed state $\Psi_{n_1,n_2}$. When considering the case (98), the action of the integrals $\hat{\mathcal{L}}^\pm_{s_1,s_2}$ with both upper signs produces a finite list of states. This happens due to obligatorily appearance of the poles in the Gamma function in a denominator of some coefficients. In contrast, when $g$ is given by Eq. (101), equations (118) imply that the repeated action of $\hat{\mathcal{J}}^-_{s_1,s_2}$ at some step annihilates a state, but the repeated application of $\hat{\mathcal{J}}^+_{s_1,s_2}$ will never produces zero. The described properties of the quantum integrals reflect the properties of the spectrum in dependence on the corresponding rational value of $g$.

As in the previous section, we can construct the coherent states of the system. The way to obtain them is to apply the CBT operator to the eigenstates of the free particle Hamiltonian. For this, we consider the plane wave $e^{\frac{1}{\sqrt{2}}(\alpha_1 z+\alpha_2 z^*)}$, which, in dependence on the values of the parameters $\alpha_1, \alpha_2 \in \mathbb{C}$ can be a physical or non-physical, formal eigenstate of $\hat{H}$. The resulting $L^2(\mathbb{R}^2)$ integrable functions are eigenstates of operators $\hat{b}_i^-$ with eigenvalues $\sqrt{\frac{\hbar}{m\omega}}\alpha_i$, $i=1,2$, and they hold their shape under the time translations and rotations [30].

## 5. CBT in cosmic strings and Dirac monopole backgrounds

Here we discuss applications of CBT with non-trivial realizations of conformal generators in more than one dimension. In the first subsection we consider the relationship between the free particle and the harmonic oscillator on a cosmic string background [29]. In the second subsection we comment on the three-dimensional example in the Dirac monopole background [28]. This second example corresponds to a direct generalization of the relationship between the one-dimensional Calogero type model and the AFF conformal mechanics studied in Sec. 3.2 to the case of three-dimensional spaces.

### *5.1. CBT in a cosmic string background*

The metric corresponding to the $(2+1)$ cosmic string space-time is given by [66, 67]

$$dS^2 = -c^2dt^2 + ds^2\,, \quad ds^2 = \left(1-\tfrac{8\mu G}{c^2}\ln\left(\tfrac{r}{r_0}\right)\right)(dr^2+r^2d\varphi^2)\,, \tag{119}$$

where $G$ is Newton constant, $c$ is the speed of light, $\mu$ is the linear mass density of the cosmic string and $r_0$ corresponds to the cosmic string radius. By introducing the new coordinate

$$r'^2 = \left(1-\tfrac{8\mu G}{c^2}\ln\left(\tfrac{r}{r_0}\right)\right)r^2\,, \quad \alpha^2dr'^2 = \left(1-\tfrac{8\mu G}{c^2}\ln\left(\tfrac{r}{r_0}\right)\right)dr^2\,, \quad \alpha=\frac{1}{1-\frac{4\mu G}{c^2}}>0\,, \tag{120}$$

one gets (renaming $r' \to r$)

$$ds^2 = \alpha^2dr^2 + r^2d\varphi^2\,. \tag{121}$$

When $\alpha>1$, which implies $\mu>0$, metric (121) is obtained from the three-dimensional Euclidean metric reduced to the conic surface $z=\lambda_E r$. On the other hand, when $0\leq\alpha<1$, that means $\mu<0$, metric (121) is obtained by reducing a $(2+1)$ dimensional Minkowski space metric $ds^2=-c^2d\tau^2+dr^2+r^2d\varphi^2$ to the surface $c\tau=\lambda r$, $0<\lambda<1$. Such metric also appears in condensed matter systems [68, 69, 70, 71, 72].

The non-relativistic action of a free particle in this space is $I=\int Ldt$, $L=\frac{m}{2}g_{ij}\frac{dx_i}{dt}\frac{dx_j}{dt}=\frac{m}{2}\left(\alpha^2\dot{r}^2+r^2\dot{\varphi}^2\right)$, and the classical Hamiltonian corresponds to

$$H^{(\alpha)} = \frac{1}{2m}\left(\frac{p_r^2}{\alpha^2}+\frac{p_\varphi^2}{r^2}\right)\,. \tag{122}$$

The formal analogs of the momenta integrals and the Galilean boosts generators are given by

$$\Pi_\pm = \Pi_1 \pm i\Pi_2 = \left(\tfrac{p_r}{\alpha}\pm i\tfrac{p_\varphi}{r}\right)e^{\pm i\frac{\varphi}{\alpha}}\,, \tag{123}$$

$$\Xi_\pm = \Xi_1 \pm i\Xi_2 = \left[\alpha m r - t\left(\tfrac{p_r}{\alpha}\pm i\tfrac{p_\varphi}{r}\right)\right]e^{\pm i\frac{\varphi}{\alpha}}\,. \tag{124}$$

These are well defined phase space functions only when $\alpha^{-1}$ is an integer, while in the general case they are multi-valued. Despite this obstacle, we can use these formal conserved quantities to construct the well defined integrals for the system. In the general case of $\alpha$, we have the

$\mathfrak{sl}(2,\mathbb{R}) \oplus \mathfrak{u}(1)$ generators which are the Hamiltonian $H^{(\alpha)}$, the dilatations generator $D$, the generator of the special conformal transformations $K$, and the generator of rotations $J_0$,

$$H^{(\alpha)} = \tfrac{1}{2m}\Pi_+\Pi_-\,, \qquad D = \tfrac{1}{4m}(\Xi_+\Pi_- + \Pi_+\Xi_+)\,, \tag{125}$$

$$K = \tfrac{1}{2m}\Xi_+\Xi_-\,, \qquad J_0 = \tfrac{i}{4m}(\Xi_+\Pi_- - \Pi_+\Xi_+) = \tfrac{\alpha}{2}p_\varphi\,. \tag{126}$$

For the case of rational values of $\alpha = q/k$, with $q, k = 1, 2, \ldots,$ one can construct

$$\mathcal{O}^\pm_{\mu,\nu} = (\Xi_\pm)^\mu(\Pi_\pm)^\nu\,, \qquad \mu = 0, 1, \ldots, q, \qquad \nu = q - \mu\,, \tag{127}$$

$$\mathcal{S}^\pm_{\mu',\nu'} = (\Xi_\pm)^{\mu'}(\Pi_\pm)^{\nu'}\,, \qquad \mu' = 0, 1, \ldots, 2q, \qquad \nu' = 2q - \mu'\,. \tag{128}$$

Here, the generators $\mathcal{O}^\pm_{\mu,\nu}$ ($\mathcal{S}^\pm_{\mu',\nu'}$) have the angular dependence $e^{\pm ik\varphi}$ ($e^{\pm i2k\varphi}$), and therefore, they are well defined phase space functions. The finite sets of generators (127) and (128) are obtained by taking repeated Poisson brackets between $K$ (or $H^{(\alpha)}$) with $\mathcal{O}^\pm_{0,q}$ ($\mathcal{O}^\pm_{q,0}$) and $\mathcal{S}^\pm_{0,2q}$ ($\mathcal{S}^\pm_{2q,0}$) respectively. On the other hand, the brackets $\{\mathcal{O}^+_{\mu,\nu}, \mathcal{O}^-_{\lambda,\sigma}\}$ and $\{\mathcal{S}^+_{\mu',\nu'}, \mathcal{S}^-_{\lambda',\sigma'}\}$ are polynomial functions of $m$, $D$, $J_0$, and $H^{(\alpha)}$ only. These properties imply that the sets $\mathcal{U}_1 = \{H^{(\alpha)}, K, D, J_0, \mathcal{O}^\pm_{\mu,\nu}\}$ and $\mathcal{U}_2 = \{H^{(\alpha)}, K, D, J_0, \mathcal{S}^\pm_{\mu',\nu'}\}$ generate independent non-linear subalgebras. The complete symmetry algebra of the system corresponds to $\mathcal{U}_1 \cup \mathcal{U}_2$ and also one can show that $\mathcal{U}_1$ is an ideal subalgebra [29]. For subsequent application of the conformal bridge transformation, it is useful to write down explicitly the brackets

$$\{D, \mathcal{O}^\pm_{\mu,\nu}\} = \tfrac{\nu-\mu}{2}\mathcal{O}^\pm_{\mu,\nu}\,, \qquad \{D, \mathcal{S}^\pm_{\mu',\nu'}\} = \tfrac{\nu'-\mu'}{2}\mathcal{S}^\pm_{\mu',\nu'}\,. \tag{129}$$

From them one sees that in the case $q = 2n$, the integrals that Poisson commute with $D$ correspond to ($\mathcal{O}^\pm_{n,n}, \mathcal{S}^\pm_{2n,2n} = (\mathcal{O}^\pm_{n,n})^2$), while in the case $q = 2n+1$, only the integral $\mathcal{S}^\pm_{2n+1,2n+1}$ are dilatation invariant.

These properties associated with the parameter $\alpha$ can be predicted by analyzing the classical trajectories

$$r(\varphi) = \frac{r_*}{\cos((\varphi-\varphi_*)/\alpha)}\,, \qquad r_* = \frac{p_\varphi}{\sqrt{2mH^{(\alpha)}}}\,, \qquad -\tfrac{\pi}{2}\alpha \le \varphi - \varphi_* \le \tfrac{\pi}{2}\alpha\,, \tag{130}$$

from where we learn that the scattering angle is $\varphi_{\text{scat}} = \alpha\pi$. Some examples of the trajectories are shown on Fig 4.

Though in a free case special values of the parameter $\alpha$ associated with existence of additional non-trivial integrals of motion reveal themselves in dynamics only in rational values of the scattering angle in comparison with a flat case where $\varphi_{\text{scat}} = \pi$, they will explicitly be detected in the dynamics after applying the conformal bridge transformation.

After quantization, the Hamiltonian operator, its eigenstates and its spectrum are given by

$$\hat{H}^{(\alpha)} = -\frac{\hbar^2}{2m}\frac{1}{\sqrt{g}}\frac{\partial}{\partial x^i}\sqrt{g}g^{ij}\frac{\partial}{\partial x^j} = -\frac{\hbar^2}{2m}\left(\frac{1}{\alpha^2 r}\frac{\partial}{\partial r}\left(r\frac{\partial}{\partial r}\right) + \frac{1}{r^2}\frac{\partial^2}{\partial\varphi}\right)\,, \tag{131}$$

$$\psi^\pm_{\kappa,l}(r,\varphi) = \sqrt{\frac{\kappa}{2\pi\alpha}}J_{\alpha l}(\kappa r)e^{\pm il\varphi}\,, \qquad E = \frac{\hbar^2\kappa^2}{2m\alpha^2}\,, \quad \kappa \ge 0\,, \quad l = 0, 1, \ldots\,. \tag{132}$$

The eigenfunctions satisfy $\left\langle \psi^\pm_{\kappa,l} \middle| \psi^\mp_{\kappa',l'} \right\rangle = \delta_{ll'}\delta(\kappa - \kappa')$, where $\langle\Psi_1|\Psi_2\rangle = \int_V \Psi_1^*\Psi_2\sqrt{g}dV = \int_0^\infty \alpha r dr \int_0^{2\pi} d\varphi \Psi_1^*\Psi_2$. The quantum versions of the formal integrals $\Pi_\pm$ are given by

$$\hat{\Pi}_\pm = e^{\pm\frac{i\varphi}{2\alpha}}\left(\tfrac{1}{\alpha}\hat{p}_r \pm \tfrac{i}{r}\hat{p}_\varphi\right)e^{\pm\frac{i\varphi}{2\alpha}} = -i\hbar\tfrac{1}{\alpha}e^{\pm i\frac{\varphi}{\alpha}}\left(\frac{\partial}{\partial r} \pm i\frac{\alpha}{r}\frac{\partial}{\partial\varphi}\right)\,, \tag{133}$$

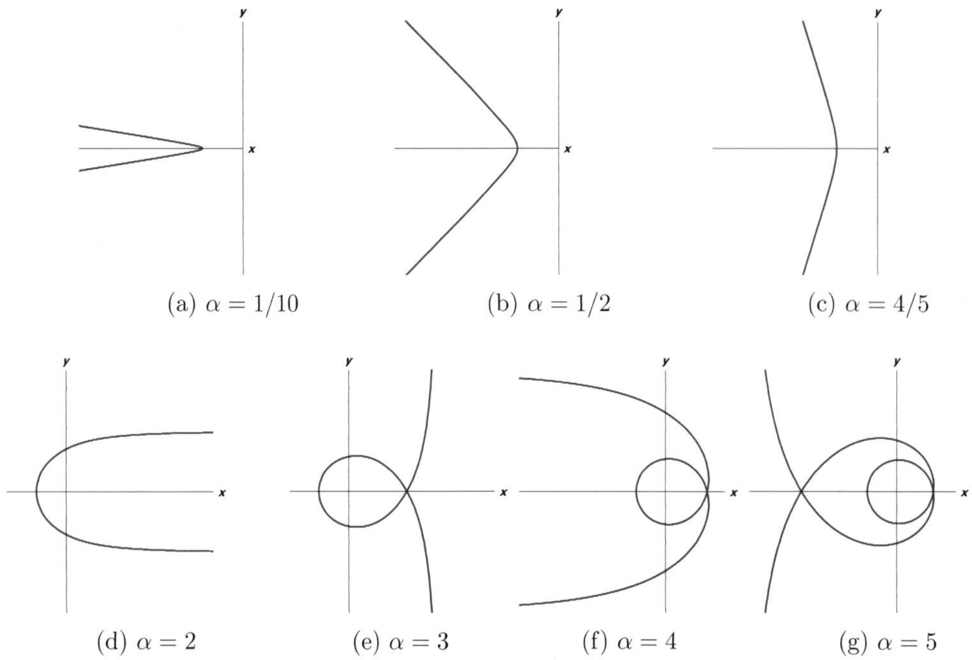


Figure 4: Some examples of the geodesic motion in the conical geometry in coordinates $x = r\cos\varphi$, $y = r\sin\varphi$. From the first three figures one sees that for $0 < \alpha < 1$, the dynamics resembles that of the repulsive Kepler-Coulomb problem. When $\alpha > 1$ and is even, $\alpha = 2n$, the particle experiences a backscattering. When $\alpha$ is odd, $\alpha = 2n + 1$, the particle approaches the initial direction asymptotically after $n$ times circling the vertex of the cone.

and from the exponential factors one deduces that the action of these operators on eigenstates produce non-physical solutions in the general case. Explicitly we have

$$\hat{\Pi}_{\pm}\psi^{\pm}_{\kappa,l}(r,\varphi) = i\frac{\hbar\kappa}{\alpha}\sqrt{\frac{\kappa}{2\pi\alpha}}J_{\alpha l+1}(\kappa r)e^{\pm i(l+\frac{1}{\alpha})\varphi}\,, \tag{134}$$

$$\hat{\Pi}_{\pm}\psi^{\mp}_{\kappa,l}(r,\varphi) = -i\frac{\hbar\kappa}{\alpha}\sqrt{\frac{\kappa}{2\pi\alpha}}J_{\alpha l-1}(\kappa r)e^{\pm i(l-\frac{1}{\alpha})\varphi}\,. \tag{135}$$

Quantum analogs of the generators (125) and (126) can be constructed straightforward for arbitrary values of $\alpha$. But this is not the case for the integrals corresponding to rational values of this geometrical parameter. In fact, with the help of expressions (134) and (135) one can show that the well defined symmetry operators that are the quantum analogs of the integrals (127), (128) can only be constructed for the special case of integer values of $\alpha = q$. This reveals a kind of the quantum anomaly in the system, since this is the only case in which the action of the corresponding operators do not produce functions outside the Hilbert space constructed from the eigenstates (132), see [29] for more details.

By applying the classical conformal bridge transformation to generators (125) and (126) we obtain the $\mathfrak{sl}(2,\mathbb{R})\oplus\mathfrak{u}(1)$ generators of the harmonic oscillator in the geometry defined by (121),

$$\mathcal{J}_0 = \frac{1}{2}\mathfrak{b}^+_a\mathfrak{b}^-_a = \frac{1}{2\omega}H^{(\alpha)}_{\text{os}}\,, \quad \mathcal{J}_{\pm} = \mathfrak{b}^{\pm}_1\mathfrak{b}^{\pm}_2\,, \quad \mathcal{L}_2 = \frac{1}{2}(\mathfrak{b}^+_1\mathfrak{b}^-_1 - \mathfrak{b}^+_2\mathfrak{b}^-_2) = \frac{1}{2}\alpha p_\varphi\,, \tag{136}$$

$$\mathfrak{b}^-_1 = \frac{1}{2}e^{i(\omega t-\frac{\varphi}{\alpha})}\left(\alpha\sqrt{m\omega}r + \frac{p_\varphi}{\sqrt{m\omega}r} + \frac{ip_r}{\alpha\sqrt{m\omega}}\right)\,, \qquad \mathfrak{b}^+_1 = (\mathfrak{b}^-_1)^*\,, \tag{137}$$

$$\mathfrak{b}^-_2 = \frac{1}{2}e^{i(\omega t+\frac{\varphi}{\alpha})}\left(\alpha\sqrt{m\omega}r - \frac{p_\varphi}{\sqrt{m\omega}r} + \frac{ip_r}{\alpha\sqrt{m\omega}}\right)\,, \qquad \mathfrak{b}^+_2 = (\mathfrak{b}^-_2)^*\,, \tag{138}$$

where

$$H_{\rm os}^{(\alpha)} = \frac{1}{2m}\left(\frac{p_r^2}{\alpha^2} + \frac{p_\varphi^2}{r^2}\right) + \frac{m\omega^2\alpha^2}{2}r^2\,, \tag{139}$$

is the Hamiltonian of the isotropic harmonic oscillator in a cosmic string background, and the formal dynamical integrals $\mathfrak{b}_i^\pm$ correspond to the mapping

$$\hat{\mathfrak{S}} : (\hat{\Xi}_+, \hat{\Xi}_-, \hat{\Pi}_+, \hat{\Pi}_-) \;\to\; \left(\sqrt{\tfrac{2m\hbar}{\omega}}\hat{\mathfrak{b}}_1^+, \sqrt{\tfrac{2m\hbar}{\omega}}\hat{\mathfrak{b}}_2^+, -i\sqrt{2m\omega\hbar}\hat{\mathfrak{b}}_2^-, -i\sqrt{2m\omega\hbar}\hat{\mathfrak{b}}_1^-\right). \tag{140}$$

By solving the associated equations of motion of the system one gets

$$r^2(\varphi) = \frac{p_\varphi^2}{mH_{\rm os}^{(\alpha)}}\left(1 + \delta\cos\left(\tfrac{2}{\alpha}(\varphi - \varphi_*)\right)\right)^{-1}\,, \quad \delta = \sqrt{1 - \left(\frac{\omega\alpha p_\varphi}{H_{\rm os}^{(\alpha)}}\right)^2}\,, \tag{141}$$

from where one finds that the closed trajectories are possible only in the rational case $\alpha = q/k$, see Fig. 5.

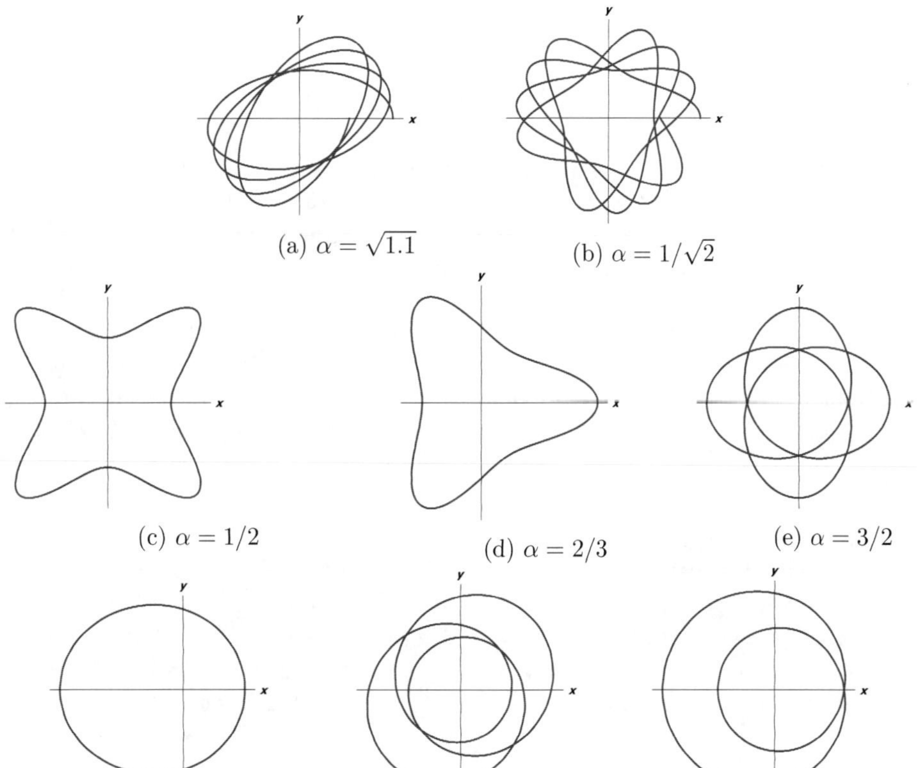

(a) $\alpha = \sqrt{1.1}$ (b) $\alpha = 1/\sqrt{2}$

(c) $\alpha = 1/2$ (d) $\alpha = 2/3$ (e) $\alpha = 3/2$

(f) $\alpha = 2$ (g) $\alpha = 3$ (h) $\alpha = 4$

Figure 5: Images of the trajectory for some irrational and rational values of $\alpha$. One can show in particular that $r(\varphi) = r(\varphi + \alpha l\pi)$, $l = 1, 2, \ldots$. From here one deduces that in the case $\alpha = q/k$ the number of maxima/minima of $r$ on the orbit is $\mathscr{N}_{\rm max/min} = k(q \bmod 2 + 1)$.

In correspondence with (141), there are globally well defined in the phase space integrals of motion that control the periodic behaviour of the trajectory iff $\alpha$ is rational. To find these integrals we use the relations (140) to transform the quantities (127) and (128) (up to inessential multiplicative constant factors) into

$$\mathcal{G}^{+}_{\mu,\nu} = (\mathfrak{b}^{+}_{1})^{\mu}(\mathfrak{b}^{-}_{2})^{\nu}\,, \quad (\mathcal{G}^{+}_{\mu,\nu})^{*} = \mathcal{G}^{-}_{\nu,\mu}\,, \quad \mathcal{F}^{+}_{\mu',\nu'} = (\mathfrak{b}^{+}_{1})^{\mu'}(\mathfrak{b}^{-}_{2})^{\nu'}\,, \quad (\mathcal{F}^{+}_{\mu',\nu'})^{*} = \mathcal{F}^{-}_{\nu',\mu'}\,. \tag{142}$$

Since the classical CBT is a canonical transformation, the algebraic properties of the free particle algebra are inherit by the harmonic oscillator algebra (with $2i\omega D$ as a pre-image of the harmonic oscillator Hamiltonian). This implies that in the case $q = 2n$ with $n = 1, 2, \ldots$, the true integrals of the harmonic oscillator system in the cosmic string background are $(\mathcal{G}^{\pm}_{n,n}, \mathcal{F}^{\pm}_{2n,2n} = (\mathcal{G}^{\pm}_{n,n})^2)$ and in the case $q = 2n+1$, the true integrals are $\mathcal{F}^{\pm}_{2n+1,2n+1}$. This is due to the dilatation invariance of their corresponding pre-images, see (129) and comments below.

At the quantum level, the corresponding Hamiltonian operator, the eigenstates and the spectrum are given by

$$\hat{H}^{(\alpha)}_{\text{os}} = -\frac{\hbar^2}{2m}\left(\frac{1}{\alpha^2 r}\frac{\partial}{\partial r}\left(r\frac{\partial}{\partial r}\right) + \frac{1}{r^2}\frac{\partial^2}{\partial \varphi}\right) + \frac{\alpha^2 m\omega^2}{2}r^2\,, \tag{143}$$

$$\psi^{\pm}_{n_r,l}(r,\varphi) = \left(\frac{m\omega\alpha^2}{\hbar}\right)^{\frac{1}{2}}\sqrt{\frac{n_r!}{2\pi\alpha\Gamma(n_r+\alpha l+1)}}\,\zeta^{\alpha l}L^{(\alpha l)}_{n_r}(\zeta^2)e^{-\frac{\zeta^2}{2}\pm il\varphi}\,, \quad \zeta = \sqrt{\frac{m\alpha^2\omega}{\hbar}}\,r\,, \tag{144}$$

$$E_{n,l} = \hbar\omega(2n_r + \alpha l + 1)\,, \quad n_r\,, l = 0, 1, \ldots\,. \tag{145}$$

Eigenstates (144) and the spectrum can be obtained by applying the corresponding realization of the operator $\hat{\mathfrak{S}}$ on the zero energy Jordan states of the free particle Hamiltonian (131), which are simultaneously eigenstates of $2i\omega\hat{D}$. In this case these Jordan states are given by $\Omega^{\pm}_{n_r,l}(r,\varphi) = r^{2n_r+\alpha l}e^{\pm l\varphi}$. In the same vein, the application of $\hat{\mathfrak{S}}$ to functions (132) gives us the coherent states of the system. Due to this connection one deduces that the quantum anomaly mentioned above for the free system is also present in the harmonically confined one. We refer for the details to [29].

In correspondence with the spectral properties of the confined system, one notes that it acquires a special degeneracy when $\alpha = q/k$, however, due to the presence of the quantum anomaly, only in the case $\alpha = q$ one can construct well defined operators that correctly reflect the degeneracy in the spectrum [29]. When $\alpha = 2n$ ($\alpha = 2n+1$), these operators are $\hat{\mathcal{G}}^{\pm}_{n,n}$ ($\hat{\mathcal{F}}^{\pm}_{2n+1,2n+1}$).

### 5.2. CBT in a Dirac monopole background

The non-trivial three dimensional example we present here corresponds to the dynamics of a particle with electric charge $e$, which is coupled to a Dirac magnetic monopole with magnetic charge $g$ and is subjected to the central potential $V(r) = \frac{m\omega^2 r^2}{2} + \frac{\alpha}{2mr^2}$. Parameter $\alpha$ is a real numerical constant and $\omega > 0$ is a frequency associated to the harmonic trap. The model and its supersymmetric extensions were extensively studied in [28] and here we just consider the system in relation to CBT.

The Hamiltonian of the system is

$$H_\omega = \frac{\boldsymbol{\pi}^2}{2m} + \frac{m\omega^2 r^2}{2} + \frac{\alpha}{2mr^2}\,, \quad \boldsymbol{\pi} = \boldsymbol{p} - e\boldsymbol{A}\,, \quad \nabla\times\boldsymbol{A} = \boldsymbol{B} = g\frac{\boldsymbol{r}}{r^3}\,. \tag{146}$$

By considering the Poincaré integral of the system

$$\boldsymbol{J} = \boldsymbol{r}\times\boldsymbol{\pi} - \nu\boldsymbol{n}\,, \qquad J^2 = \boldsymbol{J}^2\,, \qquad \boldsymbol{n}\cdot\boldsymbol{J} = -\nu\,, \qquad eg = \nu\,, \tag{147}$$

we note that the Hamiltonian (146) in spherical coordinates admits an 'AFF model representation',

$$H_\omega = \frac{\pi_r^2}{2m} + \frac{\mathscr{L}^2}{2mr^2} + \frac{m\omega^2 r^2}{2}\,, \qquad \mathscr{L}^2 := \boldsymbol{J}^2 - \nu^2 + \alpha\,, \tag{148}$$

and by using the fact that $\{r, \pi_r\} = 1$, it is deduced that the generators

$$\mathcal{J}_0 = \tfrac{1}{2\omega^2} H_\omega \,, \qquad \mathcal{J}_\pm = -\tfrac{1}{2\omega}(H_0 - \omega^2 K_0 \pm i2\omega D_0)\,, \tag{149}$$

produce the $\mathfrak{sl}(2,\mathbb{R})$ algebra. In these last relations we have introduced the $\mathfrak{so}(2,1)$ symmetry generators of the system without the harmonic trap

$$H_0 = \frac{\pi_r^2}{2m} + \frac{\mathscr{L}^2}{2mr^2}\,, \qquad D_0 = \frac{1}{2} r\pi_r - H_0 t\,, \qquad K_0 = \frac{mr^2}{2} - Dt - H_0 t^2\,. \tag{150}$$

Both forms of dynamics are connected by the classical conformal bridge transformation (6). The model $\hat{H}_0$ and its supersymmetric extensions were studied in details in [73]. It is worth to mention that the Poincaré integral $\boldsymbol{J}$ Poisson commutes with all generators (149) and (150), and plays the role of angular momentum of the system.

After solving the trajectory equation for the system (146), one gets $\boldsymbol{r} = r(\varphi)\boldsymbol{n}$, where

$$r^2(\varphi) = \tfrac{\mathscr{L}^2}{mH_\omega}[1 - \rho\cos(\tfrac{2\mathscr{L}}{J}\varphi)]^{-1}\,, \qquad \rho = \sqrt{1 - \tfrac{\omega^2\mathscr{L}^2}{H_\omega^2}}\,, \tag{151}$$

$$\boldsymbol{n} = \tfrac{-\nu \boldsymbol{J}}{J} + \boldsymbol{n}_\perp(\varphi)\,, \qquad \boldsymbol{n}_\perp(\varphi) = \boldsymbol{n}_\perp(0)\cos(\varphi) + J^{-1}\boldsymbol{J} \times \boldsymbol{n}_\perp \sin(\varphi)\,, \tag{152}$$

and $\boldsymbol{n}_\perp \cdot \boldsymbol{J} = 0$. After that, the angle $\varphi = \varphi(t)$ is obtained by substitution of $r^2(\varphi)$ into the $\mathfrak{sl}(2,\mathbb{R})$ generators (149). From (152) one concludes that the dynamics occurs on the surface of a dynamical cone defined by the equation $\boldsymbol{r}\cdot\boldsymbol{J} = -\nu r$. Also from these solutions, with taking into account the definition (148) of $\mathscr{L}$, we find that the trajectories are closed for arbitrary values of angular momentum $J$ only when $\alpha = \nu^2$. On the other hand, for $\alpha \neq \nu^2$, the trajectories are closed only for special values of $J$ given by the equation

$$\alpha = \nu^2 + \left(\tfrac{1}{4}\tfrac{l_a^2}{l_r^2} - 1\right) J^2\,, \qquad l_a, l_r = 1, 2, \ldots\,. \tag{153}$$

The special properties at $\alpha = \nu^2$ are expected to be reflected in the presence of the hidden symmetry associated with additional non-trivial integrals of motion in the system [28]. In Fig. 6 examples of the trajectories are shown for some irrational and rational values of the parameter $\alpha$.

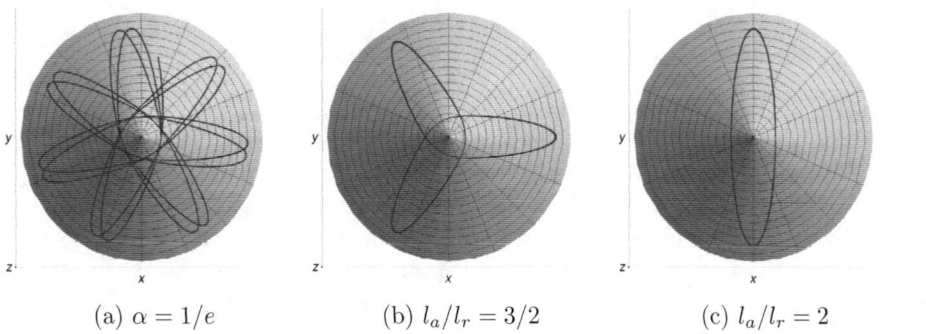

(a) $\alpha = 1/e$ (b) $l_a/l_r = 3/2$ (c) $l_a/l_r = 2$

Figure 6: Examples of some non-closed and closed trajectories are shown. The last relation $l_a/l_r = 2$ corresponds to the special case $\alpha = \nu^2$.

To find the already anticipated hidden integrals of the system we employ the classical CBT. For this aim, let use introduce the analogs of the Laplace-Runge-Lentz vector and generator of the Galilean boost transformations for the asymptomatically free system governed by the conformal generators (150), that are only available when $\alpha = \nu^2$ [73],

$$\boldsymbol{V} = \boldsymbol{\pi} \times \boldsymbol{J}\,, \qquad \boldsymbol{G} = (m\boldsymbol{r} - \boldsymbol{\pi} t) \times \boldsymbol{J}\,. \tag{154}$$

The components of these vector quantities satisfy the relations

$$\{H_0, G_i\} = -V_i\,, \qquad \{K_0, V_i\} = G_i\,, \qquad \{H_0, V_i\} = \{K_0, G_i\} = 0 \tag{155}$$

$$\{D_0, V_i\} = \tfrac{1}{2}V_i\,, \qquad \{D_0, G_i\} = -\tfrac{1}{2}G_i\,, \qquad \{D_0, V_iG_j\} = 0\,, \tag{156}$$

and the classical conformal bridge transformation corresponds to the mapping

$$\mathfrak{S} : (\boldsymbol{V}, \boldsymbol{G}) \;\rightarrow\; (-i\sqrt{m\omega}\boldsymbol{a}, \sqrt{\tfrac{m}{\omega}}\boldsymbol{a}^*)\,, \quad \boldsymbol{a} = \sqrt{\tfrac{m\omega}{2}}(\boldsymbol{r} + \tfrac{i}{m\omega}\boldsymbol{\pi}) \times \boldsymbol{J}\, e^{i\omega t}\,, \tag{157}$$

$$\mathfrak{S} : (-i\omega(G_iV_j + G_jV_i), \omega(G_iV_j - G_jV_i)) \;\rightarrow\; (T_{(ij)},, T_{[ij]})\,, \tag{158}$$

$$T_{(ij)} = m\omega(a_i^* a_j + a_j^* a_i)\,, \quad T_{[ij]} = -im\omega(a_i^* a_j - a_j^* a_i)\,. \tag{159}$$

Here, $T_{(ij)}$ is the symmetric tensor integral of the system (146), being the analog of the Fradkin tensor integral of the three-dimensional isotropic harmonic oscillator [74], while $T_{[ij]}$ is the anti-symmetric tensor proportional to the Poincaré integral. In terms of $\boldsymbol{r}$ and $\boldsymbol{\pi}$, the explicit form of the components of these tensors are

$$2T_{(ij)} = (\boldsymbol{\pi} \times \boldsymbol{J})^i(\boldsymbol{\pi} \times \boldsymbol{J})^j + m^2\omega^2(\boldsymbol{r} \times \boldsymbol{J})^i(\boldsymbol{r} \times \boldsymbol{J})^j\,, \qquad 2T_{[ij]} = \epsilon_{ijk} m\omega(J^2 - \nu^2)J_k\,, \tag{160}$$

from where we explicitly see that these are the higher order integrals of motion of the hidden symmetries. It turns out that the complete geometric information on the trajectory that appears in Fig 6c is encoded in the symmetric tensor, see [28].

As in the previous examples, here we can also obtain all the information for the quantum version of the model by applying the quantum CBT to the asymptotically free version without harmonic potential. This time it is necessary to take into account the quantization prescription for the Dirac monopole, where the parameter $\nu$ has to be an integer or half integer number [75, 76, 77], and the eigenstates are given in terms of the monopole harmonics [78, 79]. Due to the presence of the integral (160), the spectrum, which is discrete and bounded from below, has a special degeneracy depending on the choice of the quantized parameter $\nu$. Furthermore, the corresponding spectrum generation operators can be constructed from the quantum version of the complex vectors $\boldsymbol{a}$ and $\boldsymbol{a}^*$ introduced in (159). For more details, see ref. [28].

## 6. $\mathcal{PT}$-symmetric systems and extreme waves

The conformal bridge transformation presented in all the previous sections shows how to derive the properties of harmonically confined systems from the associated model whose dynamics is asymptotically free. In this section, with the help of the generalized Darboux transformations [1, 80] we construct reflectionless $\mathcal{PT}$-symmetric systems with rational potentials of the type we already considered in section 3.3. Then we promote the obtained stationary potentials to the complex $\mathcal{PT}$-symmetric solutions to the KdV equation and higher equations of its hierarchy, whose peculiar evolution reveals the properties typical for extreme waves. We follow here refs. [8, 9], and work in the units $\hbar = 1$, $m = 1/2$.

First, we remind that the usual Darboux transformation and its generalizations allow ones to generate from a given one-dimensional quantum system $\hat{H}_0 = -\frac{d^2}{dx^2} + V(x)$ another system described by the Hamiltonian

$$\hat{H}_{[n]} = -\frac{d^2}{dx^2} + V(x) - 2\frac{d^2}{dx^2}\ln(W(\phi_1, \ldots, \phi_n))\,. \tag{161}$$

Here functions $(\phi_1, \ldots, \phi_n)$ are the so-called seed states, which are physical or formal, non-physical eigenfunctions of $\hat{H}_0$ with different eigenvalues $\lambda_j$, and $W(\ldots)$ is the Wronskian. If the seed states are chosen so that $W(\phi_1, \ldots, \phi_n) \neq 0$ in the domain where the potential $V(x)$ of the initial system $\hat{H}_0$ is regular, then the potential of the generated system $\hat{H}_{[n]}$ will be nonsingular in the same domain. The Darboux transformation ensures that any, physical or non-physical, eigenfunction $\psi$ of $\hat{H}_0$ of eigenvalue $E$ not included in the set of the seed states can be mapped into the corresponding eigenfunction $\Psi_{[n]}$ of $\hat{H}_{[n]}$,

$$\Psi_{[n]} = \frac{W(\phi_1 \ldots, \phi_n, \psi)}{W(\phi_1, \ldots \phi_n)}, \qquad \hat{H}_{[n]}\Psi_{[n]} = E\Psi_{[n]} . \tag{162}$$

These relations can be verified by employing the intertwining relations $\hat{\mathbb{A}}_n^- \hat{H}_0 = \hat{H}_{[n]}\hat{\mathbb{A}}_n^-$ and $\hat{\mathbb{A}}_n^\dagger \hat{H}_{[n]} = \hat{H}_0 \hat{\mathbb{A}}_n^\dagger$ , where the operator

$$\hat{\mathbb{A}}_n^- = \hat{A}_n^- \ldots \hat{A}_1^- , \qquad \hat{A}_k^- = \hat{\mathbb{A}}_{k-1}\phi_i \frac{d}{dx}\left(\frac{1}{\hat{\mathbb{A}}_{k-1}^- \phi_k}\right), \qquad \hat{\mathbb{A}}_0^- = 1 , \tag{163}$$

is constructed iteratively. With the help of this operator, eigenstate (162) can be presented in the form $\Psi_{[n]} = \hat{\mathbb{A}}_n^- \psi$. In particular case in which $n = 1$, the function $W(x) = -(\ln(\phi_1(x))'$ is called super-potential, and the systems $\hat{H}_0$ and $\hat{H}_{[n]}$ can also be presented in the equivalent, up to an additive common shift, form

$$\hat{H}_\pm = -\frac{d^2}{dx^2} + V_\pm , \qquad V_\pm = W^2 \pm W' . \tag{164}$$

The confluent Darboux transformation follows the same rules but now Jordan states of $\hat{H}_0$ can appear in the set of the seed states, see [8, 9, 48, 49, 50]. In Sec. 3.3 we showed how the $\mathcal{PT}$-regularized Calogero system with integer coupling constant $\nu = m$ can be related with the free particle by means of an appropriate Darboux transformation.

It worth to note here a similarity of the non-unitary CBT we described in the previous sections with the Darboux transformations. As in the CBT construction, the system $\hat{H}_{[n]}$ produced from the initial system $\hat{H}_0$ in general case is not completely isospectral to it. However, the generated system $\hat{H}_{[n]}$ inherits some important properties of $\hat{H}_0$. This happens, for instance, in the case when $\hat{H}_0$ corresponds to the free particle, from which reflectionless quantum systems are produced. They represent snapshots of the soliton solutions to the KdV equation. This can be compared with the CBT that transforms an asymptotically free system possessing conformal symmetry into the harmonically trapped system with the same conformal symmetry but realized in another form. The essential difference between the two transformations is that the Darboux transformation, being generated by an operator of finite differential order, is local. The generator of the CBT $\mathfrak{S}$ is, however, essentially non-local since it includes in its structure the exponent of the second order differential Hamiltonian operator of the initial conformally invariant system. In the next, concluding section, we will return to some aspect of similarity between the CBT, based on the conformal symmetry, and supersymmetry generated by the Darboux transformations of the second order with $n = 2$.

Let us start with the $\mathcal{PT}$-regularized Calogero type Hamiltonian

$$2\hat{H}_{\alpha,1} = -\frac{d^2}{dx^2} + \frac{2}{\xi^2} , \qquad \xi = x + i\alpha . \tag{165}$$

Putting $\nu = 1$ in Eqs. (56) and (57), one deduces that the system (165) can be obtained from the free particle via the first order Darboux transformation by selecting its formal zero energy eigenfunction $\Omega^\alpha_{0,0} = \xi = x + i\alpha x$ as the seed state. Eq. (55) yields us then the Lax-Novikov integral of the system (165),

$$2\hat{\mathcal{P}}_{\alpha,1} = -\left(\tfrac{d}{dx} - \tfrac{1}{\xi}\right)\hat{p}\left(\tfrac{d}{dx} + \tfrac{1}{\xi}\right) = -i\tfrac{1}{4}\hat{M}\,, \quad \hat{M} = -4\tfrac{d^3}{dx^3} + 6u\tfrac{d}{dx} + 3u\,, \tag{166}$$

where $u(\xi) = \frac{2}{\xi^2}$. The condition of commutativity of the third order operator (166) with Hamiltonian (165) means that the potential $u(\xi) = \frac{2}{\xi^2}$ satisfies the stationary KdV equation $-6uu_x + u_{xxx} = 0$. Using the Galilean invariance of the KdV equation, one finds then that the function

$$U(x,\tau) = -\tfrac{1}{6}c + \tfrac{2}{(x+i\alpha-c\tau)^2}\,, \quad c \in \mathbb{R}\,, \tag{167}$$

will satisfy the dynamical KdV equation, which can be presented equivalently in the Lax form

$$\partial_\tau \hat{L} = [\hat{L}, \hat{M}] \quad \Leftrightarrow \quad U_\tau - 6UU_x + U_{xxx} = 0\,, \tag{168}$$

where $L = -\frac{d^2}{dx^2} + U(x,\tau)$ and $\hat{M}$ is given by Eq. (166) with $u$ there changed for $U(x,\tau)$. Notice that if we extend the definition of the time reflection operator $\mathcal{T}$ by requiring additionally $\mathcal{T}\tau = -\tau\mathcal{T}$, the time-dependent KdV equation will be invariant under the $\mathcal{PT}$ transformation if $U(x,\tau)$ is $\mathcal{PT}$-symmetric: $[U(x,\tau), \mathcal{PT}] = 0$. The real and imaginary part of such a field $U(x,\tau) = v(x,\tau) + iw(x,\tau)$, like this happens in the particular simplest case (167), will satisfy the system of coupled non-linear dynamical equations

$$v_\tau - 3(v^2 - w^2)_x + v_{xxx} = 0\,, \qquad w_\tau - 6(vw)_x + w_{xxx} = 0\,. \tag{169}$$

To construct a more interesting $\mathcal{PT}$-symmetric solution to the KdV equation, let us use system (165) as a starting point for a new Darboux transformation. To this aim we select as the seed state the function $\psi^1_{\alpha,\gamma} = \gamma\Xi^\alpha_{0,1} + \Omega^\alpha_{0,1} = \gamma\xi^{-1} + \xi^2$. This is a zero energy eigenfunction of (165). With pure imaginary parameter $\gamma$, function $\psi^1_{\alpha,\gamma}$ is $\mathcal{PT}$-invariant. A further restriction $\gamma = i\varrho\alpha^3$ with $\varrho \in \mathbb{R}$, $\varrho \neq -8, 1$, guarantees that $\mathcal{PT}$-odd superpotential $W = -\frac{d}{dx}\ln(\psi^1_{\alpha,\gamma}) = \frac{1}{\xi} - \frac{3\xi^2}{\xi^3+\gamma}$ does nor take zero value. The generated potentials of the corresponding supersymmetric partner systems (164),

$$V_+ = \tfrac{6}{\xi^2} - \tfrac{6\gamma(4\xi^3+\gamma)}{\xi^2(\xi^3+\gamma)^2} := V^{(1)}_+(x;\alpha,\gamma)\,, \qquad V_- = \tfrac{2}{\xi^2}\,, \tag{170}$$

are $\mathcal{PT}$-symmetric non-singular functions. Potential $V_+$ is a stationary solution to the higher order equation of the KdV hierarchy

$$U_\tau + 30U^2U_x - 20U_xU_{xx} - 10UU_{xxx} + U_{xxxxx} = 0\,. \tag{171}$$

The substitution $\gamma \to \gamma(\tau) = 12\tau + i\varrho\alpha^2$ with $\varrho > 1$ transforms function $V_+$ into the dynamical field $V^{(1)}_+(x,\tau;\alpha,\varrho) = V^{(1)}_+(x;\alpha,\gamma(\tau))$ which is a $\mathcal{PT}$-symmetric solution of the KdV equation to be non-singular for $\tau \in (-\infty,\infty)$ [8]. Some plots of the real and imaginary parts of the inverted field $-V^{(1)}_+(x,\tau;\alpha,\varrho)$ are shown on Fig. 7.

In the same way we provide a second example based on the $\mathcal{PT}$-symmetric system

$$2\hat{H}_{\alpha,2} = -\tfrac{d^2}{dx^2} + \tfrac{6}{\xi^2}\,. \tag{172}$$

This system can be obtained via a second order confluent Darboux transformation from the free particle system, where the corresponding seed states are $\Omega_{0,0} = \xi$ and $\Omega_{1,0} = \xi^3$ [8]. The Lax-Novikov integral of this model is given by the five order differential operator

$$4\hat{\mathcal{P}}_{\alpha,2} = \left(\tfrac{d}{dx} - \tfrac{2}{\xi}\right)\left(\tfrac{d}{dx} - \tfrac{1}{\xi}\right)\hat{p}\left(\tfrac{d}{dx} + \tfrac{1}{\xi}\right)\left(\tfrac{d}{dx} + \tfrac{2}{\xi}\right)\,. \tag{173}$$

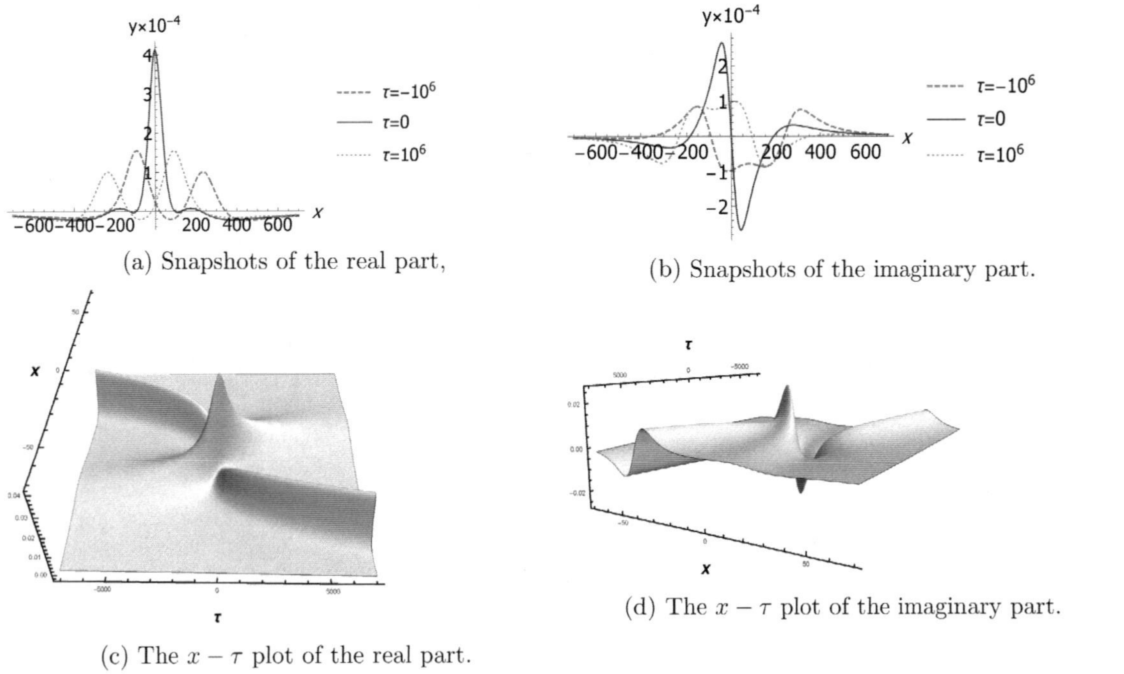


(a) Snapshots of the real part,

(b) Snapshots of the imaginary part.

(c) The $x-\tau$ plot of the real part.

(d) The $x-\tau$ plot of the imaginary part.

Figure 7: The plots of the real and imaginary parts of $-V_+^{(1)}(x,\tau;\alpha,\varrho)$ with $\alpha=10$ and $\varrho=5$.

As in the previous example, one can show that this operator is written in terms of derivatives and the potential $u=\frac{6}{x^2}$ only, which is a stationary solution to the higher order equation of the KdV hierarchy (171).

As a seed state we choose the following nodeless zero energy eigenfunction of (172), $\psi^2_{\alpha,\gamma}=\gamma\Xi^{\alpha}_{0,2}+\Omega^{\alpha}_{0,2}=\gamma\xi^{-2}+\xi^3$, with $\gamma=i\varrho\alpha^5$, ans now $\varrho$ is a real numerical parameter different from $-1$ and $-4$. With this choice one gets the $\mathcal{PT}$-odd superpotential $W=-\frac{d}{dx}\ln(\psi^2_{\alpha,\gamma})=\frac{2}{\xi}-\frac{5\xi^4}{\xi^5+\gamma}$, and the $\mathcal{PT}$-symmetric potentials of the super-partner systems are given by

$$V_+=\frac{12}{\xi^2}-\frac{10\gamma(6\xi^5+\gamma)}{\xi^2(\xi^5+\gamma)^2}:=V_+^{(2)}(x;\alpha,\gamma)\,,\qquad V_-=\frac{6}{\xi^2}\,. \tag{174}$$

Similarly to the previous example, after the substitution $\gamma\to\gamma(\tau)=-720\tau+i\varrho\alpha^5$ with $\varrho>24$, potential $V_+^{(2)}(x;\alpha,\gamma(\tau))$ satisfies the higher order non-linear field equation (171). The real and imaginary parts of the inverted function $-V_+^{(2)}(x;\alpha,\gamma(\tau))$ are shown in Fig 8.

Finally, we note that near a critical value of the parameter $\varrho$, which in the first example is $\varrho=1$ and in the second case it is $\varrho=24$, the real part of the potentials, defined respectively in (170) and (174), have a $\delta$-function like behaviour, while the corresponding imaginary parts have a form similar to $\delta'$-function [8, 9]. This behaviour is typical for the extreme (or, the so-called rogue) waves, that corresponds to soliton type waves with extreme values of the amplitude emerging in the process of their evolution.

## 7. Discussion and outlook

In conclusion, we indicate some open questions and problems related to the considered topics that deserve a further attention.

**1.** The two systems presented in Sec. 3.3 have interesting spectral properties. The first one is asymptotically free, and corresponds to a $\mathcal{PT}$-regularized Calogero type model. It is a perfectly

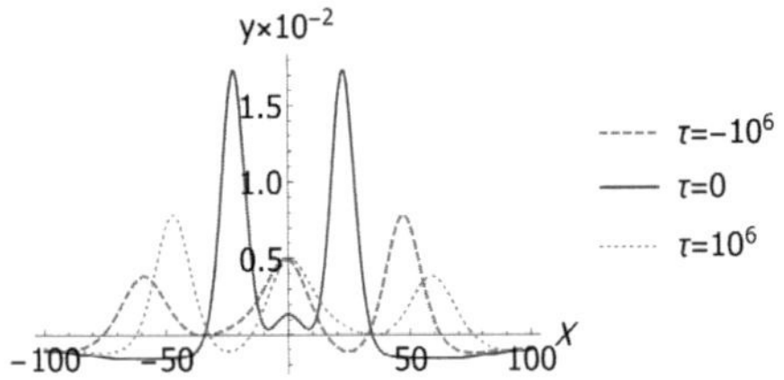


(a) Snapshot of the real part.

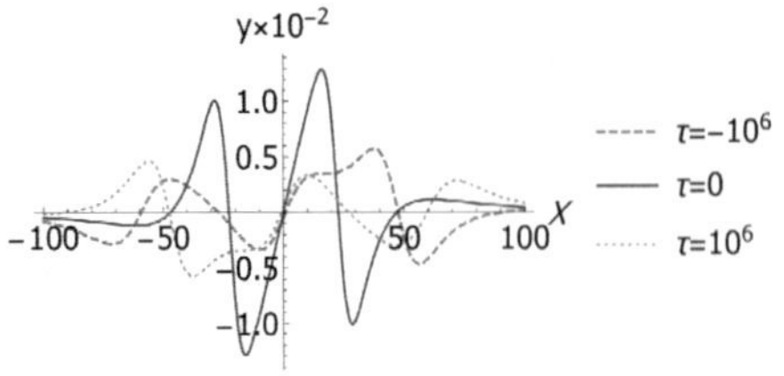


(b) Snapshot of the imaginary part.

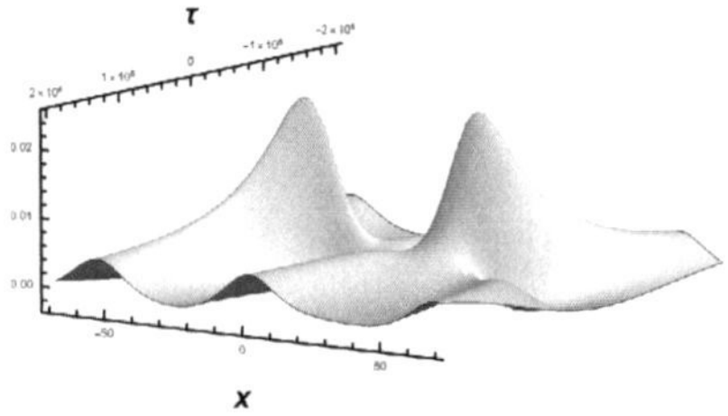


(c) The $x - \tau$ plot of the real part.

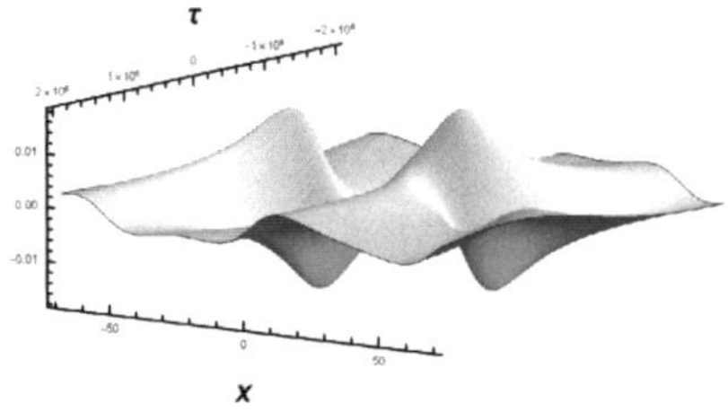


(d) The $x - \tau$ plot of the imaginary part.

Figure 8: The plots of the real and imaginary parts of $-V_+^{(2)}(x;\alpha,\gamma(\tau))$ with $\alpha = 20$ and $\varrho = 25$.

invisible zero-gap system with the unique $L^2(\mathbb{R})$ integrable eigenfunction of zero energy when the parameter $\nu$ in the potential term $\nu(\nu+1)(x+i\alpha)^{-2}$ is an integer number, $\nu = m$, and the system can be related to a free particle by the Darboux transformation. These spectral properties are coherently reflected by the presence (available only at $\nu = m$) of a well defined on all the real line Lax-Novikov integral [8, 9]. On the other hand, the harmonically confined $\mathcal{PT}$-regularized AFF model has two spectral towers that do not touch each other for any value of the parameter $\nu$, and for which the complete set of the spectrum generating ladder operators can only be constructed when, again, the parameter $\nu$ is integer. As it was indicated at the end of that section, the well defined indefinite scalar product for those $\mathcal{PT}$-symmetric systems is unknown for us. We also do not know the equivalent Hermitian systems into which they can be transformed. However, due to the similarity of the spectrum of the $\mathcal{PT}$-regularized AFF model to the spectrum of a non-local model presented in ref. [58], one can expect that those two families of the $\mathcal{PT}$-regularized models can be related somehow to the non-local models with Hamiltonians that include in the structure the spatial reflection operator $\mathcal{P}$.

**2.** The higher order confluent Darboux transformations, which appeared in the $\mathcal{PT}$-regularized Calogero type systems with $\nu = m$, are directly related to the construction of higher order quantum supersymmetry [2, 6, 50, 80, 81] since the higher order intertwining operators can be promoted to higher order supercharges. In general this kind of systems, including the non-Hermitian ones, are described by non-linear superalgebras [2, 6, 8, 9, 81]. The generators of the higher order Darboux transformations can be factorized into generators of Darboux transformations of the corresponding lower orders, and such factorization is non-unique [82, 83]. In dependence on the choice of the factorization, the initial and final Darboux-related Hermitian systems can be related via "virtual" systems which can be non-Hermitian. This happens, in particular, in the case of the second order supersymmetry with Darboux generators to be differential operators of the second order [83]. In some examples considered there, the $\mathcal{PT}$-symmetric systems like (165) do appear in the form of non-Hermitian virtual systems.. An interesting question is whether other non-Hermitian systems obtained in this way can fit into the scheme of the $\mathcal{PT}$-symmetric models and CBT.

**3.** Consider the following similarity transformation of the operator $i\hat{D}$,

$$\hat{\mathfrak{S}}_{a,b}(i\hat{D})\hat{\mathfrak{S}}_{a,b}^{-1} = i(1+2ab)\hat{D} + b\hat{H} - a(1+ab)\hat{K} := i\hat{D}_{a,b}\,, \qquad \hat{\mathfrak{S}}_{a,b} = e^{\frac{a}{\hbar}\hat{K}}e^{\frac{b}{\hbar}\hat{H}}\,, \tag{175}$$

where $a = \alpha\omega$, $b = \beta\omega^{-1}$, and $\alpha$ and $\beta$ can be complex in general case dimensionless parameters. Our CBT generator (17) corresponds here to the particular choice $\alpha = -1$, $\beta = 1/2$. The operator $\hat{\mathfrak{S}}_{a,b}$ is a generator of the internal automorphism of $\mathfrak{so}(2,1) \cong \mathfrak{sl}(2,\mathbb{R})$ of the most general form since the inclusion of the operator $e^{\frac{c}{\hbar}\hat{D}}$ in its structure reduces just to the change of the parameters $\alpha$ and $\beta$. The operator $i\hat{D}_{a,b}$ has the structure of the operator (82) with corresponding identification of the parameters. From (175) we obtain the relation

$$\hat{\mathfrak{S}}\hat{\mathfrak{S}}_{a,b}^{-1}(i\hat{D}_{a,b})\hat{\mathfrak{S}}_{a,b}\hat{\mathfrak{S}}^{-1} = \hbar\hat{\mathcal{J}}_0\,, \tag{176}$$

where $\hat{\mathfrak{S}}$ is the $\mathcal{PT}$-symmetric generator (17) of our CBT. In our CBT scheme, the eigenvectors $|\lambda\rangle$ that are transformed into the physical eigenstates of $\hbar\hat{\mathcal{J}}_0$ correspond to the zero energy Jordan states of $\hat{H}$, which, in turn, are also eigenstates of $2i\hat{D}$ with real eigenvalues. Then, in this extended scheme, the eigenstates of the operator $i\hat{D}_{a,b}$, that have to be transformed into the physical eigenstates of $\hbar\hat{\mathcal{J}}_0$, are $e^{\frac{a}{\hbar}\hat{K}}e^{\frac{b}{\hbar}\hat{H}}\,|\lambda\rangle$. However, in the general case we cannot say too much about the behaviour of the resulting functions in the coordinate representation for particular choice of the initial asymptotically free system described by the Hamiltonian $\hat{H}$. One can expect that the detailed analysis of this aspect should restrict the choice of the parameters $\alpha$ and $\beta$. We just note that in our CBT with $\mathcal{PT}$-symmetric generator $\hat{\mathfrak{S}}$, the Jordan states of $\hat{H}$, which are transformed into eigenstates of $\hat{\mathcal{J}}_0$, satisfy the equation $\hat{H}^n\,|\lambda\rangle = 0$ with some integer $n$. This is a scale-invariant equation for asymptotically free examples of the systems considered by us here. It is related with a unique peculiarity of our CBT: its generator $\hat{\mathfrak{S}}$, having the property $\hat{\mathfrak{S}}^4 = \mathcal{P}$, and so, being internal automorphism of the $\mathfrak{so}(2,1) \cong \mathfrak{sl}(2,\mathbb{R})$ algebra, maps the first order scale-invariant differential operator $2i\hat{D}$ into the second order differential operator $\hbar\hat{\mathcal{J}}_0$.

**4.** For our $\mathcal{PT}$-symmetric CBT it does not matter if the number of degrees of freedom is greater than two, or if we are working in some exotic geometry. The only decisive factor is to have the generators of the $\mathfrak{so}(2,1)$ conformal symmetry of the initial asymptotically free system to be able to construct the CBT operator. In this way, the operator $i\hat{D}$, to which we apply the CBT to get the Hamiltonian operator of the associated harmonically trapped system, can be extended by a $\mathcal{PT}$-symmetric 'Zeeman type' term $\hat{\mathcal{Z}}$ that commutes with the $\mathfrak{so}(2,1)$ generators. In this case the operator $(2i\hat{D}+g\hat{\mathcal{Z}})$ will be mapped into $2\hbar\hat{\mathcal{J}}_0+g\hat{\mathcal{Z}}$. For example if we consider a free particle in the cosmic string background, and take the combination $2\omega(i\hat{D}+g\hat{J}_0)$, with $\hat{D}$ and $\hat{J}_0$ to be the corresponding quantum analogues of the quantities (125), (126), the application of the corresponding CBT yields us

$$\hat{H}_g^{(\alpha)} = -\frac{\hbar^2}{2m}\left(\frac{1}{\alpha^2 r}\frac{\partial}{\partial r}\left(r\frac{\partial}{\partial r}\right) + \frac{1}{r^2}\frac{\partial^2}{\partial\varphi}\right) + \frac{\alpha^2 m\omega^2}{2}r^2 - i\hbar\omega g\alpha\frac{\partial}{\partial\varphi}\,. \tag{177}$$

This is a direct analog of the ERIHO quantum Hamiltonian (see Sec. 4) in the cosmic string metric (conical background from the viewpoint of condensed matter physics [69, 71]). Results related to this particular system will be presented by us soon.

In the case of two dimensions we also can add fermionic degrees of freedom by taking a term of the form $\mathcal{Z} = \omega g\sigma_3 p_\varphi$. In particular, the application of the isotropic CBT to the generator $\omega(2i\hat{D} + \sigma_3\hat{p}_\varphi)$ produces Hamiltonian of the supersymmetric Landau problem [80]. This is an important indication how the CBT can be generalized for supersymmetric case. In the same vein, the scheme of the Swanson model can be extended. The indicated generalizations can be interesting, particularly, from the point of view of physics of Bose-Einstein condensates [84, 85, 86, 87] and physics of anyons [88, 89, 90].

## Acknowledgments
The work was partially supported by the FONDECYT Project 1190842 and the DICYT Project 042131P_POSTDOC.

## References
[1] Matveev V B and Salle M A 1991 *Darboux transformations and solitons* (Springer, Berlin)
[2] Arancibia A, Guilarte J M and Plyushchay M S 2013 Effect of scalings and translations on the supersymmetric quantum mechanical structure of soliton systems *Phys. Rev.* D **87** 045009 (*Preprint* arXiv: 1210.3666 [math-ph])
[3] Arancibia A and Plyushchay M S 2015 Chiral asymmetry in propagation of soliton defects in crystalline backgrounds *Phys. Rev. D* **92** 105009 (*Preprint* arXiv: 1507.07060 [hep-th])
[4] Dunne G V and Rao K 2000 Lame instantons *JHEP* **01** 019 (*Preprint* arXiv: hep-th/9906113)
[5] Correa F and Plyushchay M S 2007 Peculiarities of the hidden nonlinear supersymmetry of the Poschl-Teller system in the light of the Lame equation *J. Phys.* A **40** 14403 (*Preprint* arXiv: 0706.1114 [hep-th])
[6] Correa F, Jakubský V, Nieto L M and Plyushchay M S 2008 Self-isospectrality, special supersymmetry, and their effect on the band structure *Phys. Rev. Lett.* **101** 030403 (*Preprint* arXiv: 0801.1671 [hep-th])
[7] Arancibia A, Guilarte J M and Plyushchay M S 2013 Fermion in a multi-kink-antikink soliton background, and exotic supersymmetry *Phys. Rev.* D **88** 085034 (*Preprint* arXiv: 1309.1816 [hep-th])
[8] Guilarte J M and Plyushchay M S 2017 Perfectly invisible $\mathcal{PT}$-symmetric zero-gap systems, conformal field theoretical kinks, and exotic nonlinear supersymmetry *JHEP* **12** 061 (*Preprint* arXiv: 1710.00356 [hep-th])
[9] Guilarte J M and Plyushchay M S 2019 Nonlinear symmetries of perfectly invisible $\mathcal{PT}$-regularized conformal and superconformal mechanics systems *JHEP* **01** 194 (*Preprint* arXiv: 1806.08740 [hep-th])
[10] Leiva C and Plyushchay M S 2003 Conformal symmetry of relativistic and nonrelativistic systems and AdS/CFT correspondence *Annals Phys.* **307** 372 (*Preprint* arXiv: hep-th/0301244)
[11] Barbon J L and Fuertes C A 2008 On the spectrum of nonrelativistic AdS/CFT *JHEP* **0809** 030 (*Preprint* arXiv: 0806.3244 [hep-th])
[12] Duval C, Hassaine M and Horvathy P A 2009 The geometry of Schrödinger symmetry in gravity background/non-relativistic CFT *Annals Phys.* **324** 1158 (*Preprint* arXiv: 0809.3128 [hep-th])
[13] Chamon C, Jackiw R, Pi S Y and Santos L 2011 Conformal quantum mechanics as the $CFT_1$ dual to $AdS_2$ *Phys. Lett.* B **701** 503 (*Preprint* arXiv: 1106.0726 [hep-th])
[14] Claus P, Derix M, Kallosh R, Kumar J, Townsend P K and Van Proeyen A 1998 Black holes and superconformal mechanics *Phys. Rev. Lett.* **81** 4553 (*Preprint* arXiv: hep-th/9804177)
[15] De Azcárraga J A, Izquierdo J M, Bueno J P and Townsend P K 1999 Superconformal mechanics, black holes, and nonlinear realizations *Phys. Rev.* D **59** 084015 (*Preprint* arXiv: hep-th/9810230)
[16] Gibbons G W and Townsend P K 1999 Black holes and Calogero models *Phys. Lett.* B **454** 187 (*Preprint* arXiv: hep-th/9812034)
[17] Duval C, Gibbons G W and Horvathy P A 1991 Celestial mechanics, conformal structures and gravitational waves *Phys. Rev.* D **43** 3907 (*Preprint* arXiv: hep-th/0512188)
[18] Britto-Pacumio R, Michelson J, Strominger A and Volovich A 2000 Lectures on superconformal quantum mechaics and multi-black hole moduli spaces *NATO Sci. Ser.* C **556** 255 (*Preprint* arXiv: hep-th/9911066)
[19] Pioline B and Waldron A 2003 Quantum cosmology and conformal invariance *Phys. Rev. Lett.* **90** 031302 (*Preprint* arXiv: hep-th/0209044)
[20] Cariglia M, Galajinsky A, Gibbons G W and Horvathy P A 2018 Cosmological aspects of the Eisenhart-Duval lift *Eur. Phys. J. C* **78** 314 (*Preprint* arXiv: 1802.03370 [gr-qc])
[21] Son D T 2008 Toward an AdS/could atoms correspondence: a geodesic realization of Schrödinger symmetry *Phys. Rev.* D **78** 046003 (*Preprint* arXiv: 0804.3972 [hep-th])
[22] Balasubramanian K and McGreevy J 2008 Gravity duals for nonrelativistic CFTs *Phys. Rev. Lett.* **101** 061601 (*Preprint* arXiv: 0804.4053 [hep-th])
[23] Prain A, Fagnocchi S and Liberati S 2010 Analogue cosmological particle creation: quantum correlations in expanding Bose-Einstein condensates *Phys. Rev.* D **82** 105018 (*Preprint* arXiv: 1009.0647 [gr-qc])
[24] Ohashi K, Fujimori T and Nitta M 2017 Conformal symmetry of trapped Bose-Einstein condensates and massive Nambu-Goldstone modes *Phys. Rev.* A **96** 051601 (*Preprint* arXiv: 1705.09118 [cond-mat.quant-gas])
[25] Brodsky S J, de Téramond G F, Dosch H G and Erlich J 2015 Light-front holographic QCD and emerging confinement *Phys. Rept.* **584** 1 (*Preprint* arXiv: 1407.8131 [hep-ph])
[26] Deur A, Brodsky S J and de Téramond G F 2015 Connecting the hadron mass scale to the fundamental mass scale of quantum chromodynamics *Phys. Lett.* B **750** 528 (*Preprint* arXiv: 1409.5488 [hep-ph])

[27] Inzunza L, Plyushchay M S and Wipf A 2020 Conformal bridge between asymptotic freedom and confinement *Phys. Rev.* D **101** 105019 (*Preprint* arXiv: 1912.11752 [hep-th])
[28] Inzunza L, Plyushchay M S and Wipf A 2020 Hidden symmetry and (super)conformal mechanics in a monopole background *JHEP* **04** 028 (*Preprint* arXiv: 2002.04341 [hep-th])
[29] Inzunza L and Plyushchay M S 2020 Conformal bridge in a cosmic string background *JHEP* **21** 165 (*Preprint* arXiv: 2012.04613 [hep-th])
[30] Inzunza L and Plyushchay M S 2021 Conformal generation of an exotic rotationally invariant harmonic oscillator *Phys. Rev.* D **103** 106004 (*Preprint* arXiv: 2103.07752 [quant-ph])
[31] Dirac P A M 1949 Forms of relativistic dynamics *Rev. Mod. Phys.* **21** 392
[32] de Alfaro V, Fubini S and Furlan G 1976 Conformal invariance in quantum mechanics *Nuovo Cim.* A **34** 569
[33] Bender C M and Boettcher S 1998 Real spectra in non-Hermitian Hamiltonians having PT symmetry *Phys. Rev. Lett.* **80** 5243 (*Preprint* arXiv: physics/9712001)
[34] Bender C M, Brody D C and Jones H F 2004 Complex extension of quantum mechanics *Phys. Rev. Lett.* **89** 270401 (*Preprint* arXiv: quant-ph/0208076)
[35] Mostafazadeh A 2002 Pseudo-Hermiticity versus PT symmetry. The necessary condition for the reality of the spectrum *J. Math. Phys.* **43** 205 (*Preprint* arXiv: math-ph/0107001)
[36] Bender C M 2007 Making sense of non-Hermitian Hamiltonians *Rept. Prog. Phys.* **70** 947 (*Preprint* arXiv: hep-th/0703096 [hep-th])
[37] Swanson M 2004 Transition elements for a non-Hermitian quadratic Hamiltonian *J. Math. Phys.* **45** 585
[38] de Morisson Faria C F and Fring A 2006 Time evolution of non-Hermitian Hamiltonian systems *J. Phys.* A **39** 9269 (*Preprint* arXiv: quant-ph/0604014)
[39] Musumbu D P, Geyer H B and Heiss W D 2006 Choice of a metric for the non-Hermitian oscillator *J. Phys. A 40 F75* (*Preprint* arXiv: quant-ph/0611150)
[40] Assis P E and Fring A 2008 Non-Hermitian Hamiltonians of Lie algebraic type *J. Phys. A* **42** 015203 (*Preprint* arXiv: 0804.4677 [quant-ph])
[41] Bargmann V 1947 Irreducible unitary representations of the Lorentz group *Ann. Math.* **48** 568
[42] Plyushchay M S 1993 Quantization of the classical $SL(2,R)$ system and representations of $\overline{SL(2,R)}$ group *J. Math. Phys.* **34** 3954
[43] Niederer U 1973 The maximal kinematical invariance group of the harmonic oscillator *Helv. Phys. Acta* **46** 191
[44] Alvarez P D, Gomis J, Kamimura K and Plyushchay M S 2007 (2+1)D Exotic Newton-Hooke symmetry, duality and projective phase *Annals Phys.* **322** 1556 (*Preprint* arXiv: hep-th/0702014)
[45] Galajinsky A 2010 Conformal mechanics in Newton-Hooke spacetime *Nucl. Phys.* B **832** 586 (*Preprint* arXiv: 1002.2290 [hep-th])
[46] Papadopoulos G 2013 New potentials for conformal mechanics *Class. Quant. Grav.* **30** 075018 (*Preprint* arXiv: 1210.1719 [hep-th])
[47] Andrzejewski K 2014 Conformal Newton-Hooke algebras, Niederer's transformation and Pais-Uhlenbeck oscillator *Phys. Lett.* B **738** 405 (*Preprint* arXiv: 1409.3926 [hep-th])
[48] Correa F, Jakubský V and Plyushchay M S 2015 PT-symmetric invisible defects and confluent Darboux-Crum transformations *Phys. Rev.* A **92** 023839 (*Preprint* arXiv: 1506.00991 [hep-th])
[49] Cariñena J F and Plyushchay M S 2016 Ground-state isolation and discrete flows in a rationally extended quantum harmonic oscillator. *Phys. Rev.* D **94** 105022 (*Preprint* arXiv: 1611.08051 [hep-th])
[50] Inzunza L and Plyushchay M S 2019 Klein four-group and Darboux duality in conformal mechanics *Phys. Rev.* D **99** 125016 (*Preprint* arXiv: 1902.00538 [hep-th])
[51] Perelomov A 1986 *Generelized coherent states and their applications* (Springer-Verlag, Berlin)
[52] Niederer U 1972 Maximal kinematical invariance group of the free Schrödinger equation *Helv. Phys. Acta* **45** 802
[53] Howe R 1980 On the role of the Heisenberg group in harmonic analysis *Bull. Amer. Math. Soc.* **3** 821
[54] Pinsky M A 2001 *Introduction to Fourier analysis and wavelets* (Brooks/Cole)
[55] Bilodeau G G 1962 The Weierstrass transform and Hermite polynomials *Duke Math. J.* **29** 293
[56] Gazeau J P 2009 *Coherent states in quantum physics* (WileyVCH, New York)
[57] Correa F, Del Olmo M A and Plyushchay M S 2005 On hidden broken nonlinear superconformal symmetry of conformal mechanics and nature of double nonlinear superconformal symmetry *Phys. Lett.* B **628** 157 (*Preprint* arXiv: hep-th/0508223)
[58] Plyushchay M S 2000 Hidden nonlinear supersymmetries in pure parabosonic systems *Int. J. Mod. Phys.* A **15** 3679 (*Preprint* arXiv: hep-th/9903130)
[59] Assis P E and Fring A 2008 Metrics and isospectral partners for the most generic cubic $\mathcal{PT}$-symmetric non-Hermitian Hamiltonian *J. Phys.* A **41** 244001 (*Preprint* arXiv: 0708.2403 [quant-ph])
[60] Cariglia M 2014 Hidden symmetries of dynamics in classical and quantum physics *Rev. Mod. Phys.* **86** 1283

(*Preprint* arXiv: 1411.1262 [math-th])
[61] Bogolubov N 1947 On the theory of superfluidity *J. Phys. (Moscow)* **11** 23
[62] Bender C M and Mannheim Ph D 2008 No-ghost theorem for the fourth-order derivative Pais-Uhlenbeck oscillator model *Phys. Rev. Lett.* **100** 110402 (*Preprint* arXiv: 0706.0207 [hep-th])
[63] Smilga A V 2009 Comments on the dynamics of the Pais-Uhlenbeck oscillator *SIGMA* **5** 017 (*Preprint* arXiv: 0808.0139 [quant-ph])
[64] Mostafazadeh A 2010 A Hamiltonian formulation of the Pais-Uhlenbeck oscillator that yields a stable and unitary quantum system *Phys. Lett.* A **375** 93 (*Preprint* arXiv: 1008.4678 [hep-th])
[65] Ghanmi A 2013 Operational formulae for the complex Hermite polynomials $H_{p,q}(z,\bar{z})$ *Integral Transforms Spec. Funct.* **24** 884 (*Preprint* arXiv: 1211.5746 [math.CA])
[66] Sokolov D D and Starobinsky A A 1977 On the structure of curvature tensor on conical singularities *Dokl. Akad. Nauk* **234** 1043 [*Sov. Phys. - Dokl.* **22** 312]
[67] Vilenkin A 1985 Cosmic strings and domain walls *Phys. Rept.* **121** 263
[68] Visser M 1989 Traversable wormholes: some simple examples *Phys. Rev.* D **39** 3182
[69] Katanaev M O and Volovich I V 1992 Theory of defects in solids and three-dimensional gravity *Annals Phys.* **216** 1
[70] Cramer J G, Forward R L, Morris M S, Visser M, Benford G and Landis G A 1995 Natural wormholes as gravitational lenses *Phys. Rev.* D **51** 3117
[71] Volovik G E 2003 *The universe in a helium droplet* (*Preprint* Oxford Science Publications)
[72] Manton N S 2017 Five vortex equations *J. Phys.* A **50** 125403 (*Preprint* arXiv: 1612.06710 [hep-th])
[73] Plyushchay M S and Wipf A 2014 Particle in a self-dual dyon background: hidden free nature, and exotic superconformal symmetry *Phys. Rev.* D **89** 045017 (*Preprint* arXiv: 1311.2195 [hep-th])
[74] Fradkin D M 1965 Three-dimensional isotropic harmonic oscillator and $SU_3$ *Am. J. Phys.* **33** 207
[75] Goddard P and Olive D I 1978 Magnetic monopoles in gauge field theories *Rept. Prog. Phys.* **41** 1357
[76] Plyushchay M S 2000 Monopole Chern-Simons term: Charge monopole system as a particle with spin *Nucl. Phys.* B **589** 413 (*Preprint* hep-th/0004032)
[77] Plyushchay M S 2001 Free conical dynamics: Charge-monopole as a particle with spin, anyon and nonlinear fermion-monopole supersymmetry *Nucl. Phys. Proc. Suppl.* **102** 248 (*Preprint* hep-th/0103040)
[78] Wu T T and Yang C N 1976 Dirac monopole without strings: monopole harmonics *Nucl. Phys.* B **107** 365
[79] Wu T T and Yang C N 1977 Some properties of monopole harmonics *Phys. Rev.* D **16** 1018
[80] Cooper F, Khare A and Sukhatme U 1995 Supersymmetry and quantum mechanics *Phys. Rept.* **251** 267 (*Preprint* arXiv: hep-th/9405029)
[81] Plyushchay M S 2019 Nonlinear supersymmetry as a hidden symmetry *Integrability, Supersymmetry and Coherent States (Spain)* CRM Series in Mathematical Physics ed S. Kuru, J. Negro *et al* (*Preprint* Cham: Springer) p 163 (*Preprint* arXiv: 1811.11942 [hep-th])
[82] Klishevich S M and Plyushchay M S 2001 Nonlinear supersymmetry, quantum anomaly and quasiexactly solvable systems. *Nucl. Phys.* B **606** 583 (*Preprint* arXiv: hep-th/0012023 [hep-th])
[83] Plyushchay M S 2017 Schwarzian derivative treatment of the quantum second-order supersymmetry anomaly, and coupling-constant metamorphosis *Annals Phys.* **377** 164 (*Preprint* arXiv: 1602.02179 [hep-th])
[84] Cooper N R 2008 Rapidly rotating atomic gases *Adv. Phys.* **57** 539 (*Preprint* arXiv: 0810.4398 [cond-mat.mes-hall])
[85] Fetter A L 2009 Rotating trapped Bose-Einstein condensates *Rev. Mod. Phys.* **81** 647 (*Preprint* arXiv: 0801.2952 [cond-mat.stat-mech])
[86] Goldman N, Juzeliūnas G, Öhberg P and Spielman I B 2014 Light-induced gauge fields for ultracold atoms *Rept. Prog. Phys.* **77** 126401 (*Preprint* arXiv:1308.6533 [cond-mat.quant-gas])
[87] Haugset T, Haugerud H and Andersen J O 1997 Bose-Einstein condensation in anisotropic harmonic traps *Phys. Rev.* A **55** 2922
[88] Horvathy P A and Plyushchay M S 2002 Non-relativistic anyons, exotic Galilean symmetry and noncommutative plane *JHEP* **06** 033 (*Preprint* arXiv: hep-th/0201228)
[89] Horvathy P A and Plyushchay M S 2004 Anyon wave equations and the noncommutative plane *Phys. Lett.* B **595** 547 (*Preprint* arXiv: hep-th/0404137)
[90] Alvarez P D, Gomis J, Kamimura K and Plyushchay M S 2008 Anisotropic harmonic oscillator, non-commutative Landau problem and exotic Newton-Hooke symmetry *Phys. Lett.* B **659** 906 (*Preprint* arXiv: 0711.2644 [hep-th])

# Solitons in a *PT*-symmetric grating-assisted co-directional coupler

**Hugh Jones**[1] **and Mykola Kulishov**[2]
[1]Physics Department, Imperial College, London SW7 2BZ, UK
[2]InPhenix Inc. 250 North Mines Rd., Livermore, CA 94551, USA
E-mail: [1]h.f.jones@imperial.ac.uk, [2]mkulishov@inphenix.com

**Abstract.** We explore a co-directional coupling arrangement between two waveguides mediated by a *PT*-symmetric sinusoidal grating characterized by an index-modulation parameter $\kappa$ and a gain/loss parameter $g$. We show that the device supports soliton-like solutions for both the *PT*-conserving regime $g < \kappa$ and the *PT*-broken regime $g > \kappa$. In the first case the coupler exhibits a gap in wave-number $k$, and the solitons can be regarded as an extension of a previous solution found for pure index modulation. In the second case the coupler exhibits a gap in frequency $\omega$ and the solutions are entirely new.

## 1. Introduction

Since soliton propagation in an optical fiber was first demonstrated 40 years ago, we have seen an explosion of new ideas, materials and designs for new all-optical devices and systems based on the robustness of soliton signals [1, 2]. Soliton properties rely strongly on the interplay between structural and/or material dispersions on the one hand and refractive index nonlinearity on the other. Under the proper propagation conditions, it takes a certain distance to achieve this dispersion-nonlinearity balance, a distance that varies from a few hundred meters to several kilometers in the case of an unperturbed communication fiber. With sufficiently high signal power, gap solitons in reflective fiber Bragg gratings can achieve this balance within a few centimeters. The unique dispersion characteristics of fiber or waveguide Bragg gratings allow one to control soliton velocities from zero to the speed of light in the bare fiber or waveguide [3, 4].

While Bragg gratings with sub-micron periods provide coupling between waves propagating in opposite directions, gratings with longer pitch (from tens of microns to hundreds of microns) can be resonant also with co-propagating modes, leading to forward coupling. Such periodical structures are called long-period gratings (LPGs) [5, 6]. Although originally proposed for such purposes as band-rejection filters, due to their broad-band operation (compared with short-period Bragg gratings), LPGs are perfect devices for ultra-short optical signal processing, like, for example, temporal differentiation [7] and integration of sub-picosecond optical waveforms [8].

Inside reflective Bragg gratings strong interaction between forward and backward waves produces mode dispersion, which can be depicted in a dispersion diagram showing the relation between the frequency $\omega$ and the wave propagation constant, $k(\omega)$, when nonlinearity is neglected. As was shown by Kogelnik and Shank [9] in the case of reflective gratings with index modulation, their mode interaction generates a band of frequencies (stop-band), centered at the Bragg frequency $\omega_B$, in which propagation is forbidden, i.e. a forbidden frequency gap ($\omega$-gap) (Fig. 1(a)). Outside the stop-band the dispersion curves are hyperbolas. The vicinity of the stop-band is a spectral area where different types of Bragg soliton may exist [10] when nonlinearity is taken into account.

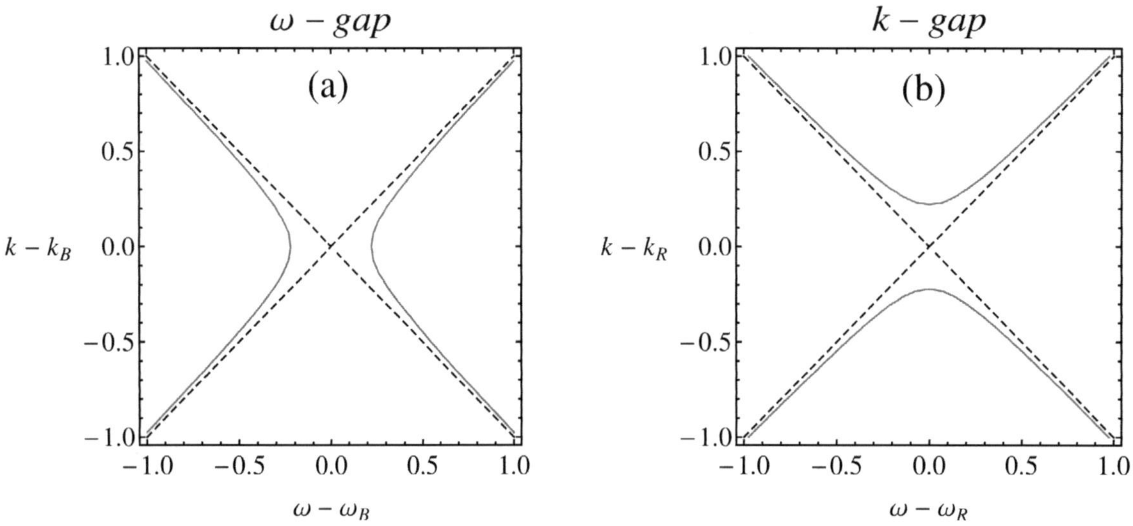


**Figure 1.** Band-gap diagrams (arbitrary scale) for reflective Bragg grating: (a) index grating ($\omega$-gap) and (b) gain/loss grating ($k$-gap).

On the other hand, in the case of a reflective grating produced by gain/loss modulation the dispersion diagram has the same hyperbolic dispersion curves with the same asymptotes as before, except that the curves are rotated by $90^o$ (Fig. 1 (b)). Instead of a stop-band in frequency, we now have a forbidden band of propagation constants, i.e. a gap in wave number ($k$-gap), centered at the resonance frequency $\omega_R$.

When it comes to band-gap diagrams for long-period gratings providing co-propagation coupling, the situation is now reversed: LPGs that couple through index modulation belong to $k$-gap devices [11, 12, 13], while those coupling through gain/loss modulation are $\omega$-gap structures.

The duality between contra-propagating ($k$-gap) and co-propagating ($\omega$-gap)mode coupling is well established for index Bragg grating and index LPGs [12, 14, 15]. Indeed the long-period co-propagation coupling was considered among the first examples of a periodic structure yielding trapping through quadratic nonlinearity [16]. The coupled-wave equations governing the long-period co-directional coupling are very similar to those of contra-directional coupling but with space and time variables interchanged. Although solitons of these models are related to the existence of a spectral gap (though in wave-number space or $k$-gap), they are more naturally named resonance solitons [12]. The duality between the description of gap solitons in reflective nonlinear Bragg index gratings and resonance solitons implies an important physical difference: solitons of the co-propagating modes cannot have zero velocity in the lab frame [13] if they are coupled through index modulation.

Recently nonlinear optics has been applied to new artificial optical materials that exhibit very unusual dispersion properties. Plasmonic meta-materials, materials with zero and negative refractive index, and $PT$-symmetric materials have all been studied from the point of view of nonlinear solitary wave propagation [17, 18, 19]

The situation becomes especially interesting and unusual in the case of $PT$-symmetric gratings [20] that combine refractive index modulation with periodic loss and gain modulation of the same pitch shifted by a half period, $\Lambda \equiv 2\pi/K$:

$$\Delta n(z) = \Delta n_0 \cos(Kz) + i(\alpha_0/k_0)\sin(Kz), \tag{1}$$

where $\Delta n_0$ and $\alpha_0$ are the amplitudes of the index and gain/loss modulations and $k_0 \equiv 2\pi/\lambda_0$ the wave number in vacuum. $PT$-symmetric co-directional couplers were first proposed by Greenberg and Orenstein [21]. They demonstrated that such a combination of two gratings makes the coupler asymmetrical, with complete asymmetry occurring when the grating amplitudes are equal: $\Delta n_0 = \alpha_0/k_0$, so that $\Delta n(z) = \Delta n_0 \exp(\pm iKz)$. The coupler in such a balanced mode exhibits very unusual behavior [22]. A signal launched into an input port of WG#1, will be coupled into WG#2 with amplification provided at a certain value of the coupling coefficient ($\kappa$) and the grating length ($L$) product, $\kappa L > 1$ (Fig. 2(a)). At the same time, if this signal launched into an input port of WG#2, it will not be coupled into WG#1 at all. It will propagate through WG#2 as if the grating were invisible. This unique property of $PT$-symmetric co-directional coupling can be used to trap light if we connect the output of the receptor waveguide WG#2 with its input to form a ring, as shown in Fig. 2(b). It also can be used to design lasers with very low threshold [23] and many other applications.

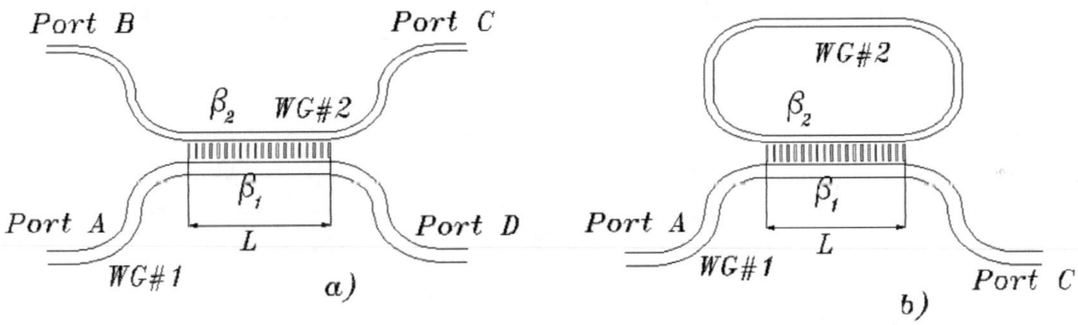


**Figure 2.** (a) A grating-assisted co-directional coupler, (b) Use of such a coupler to trap light in a ring.

$PT$-symmetry provides a means of controlling the dispersion characteristics of the grating-assisted coupler. With only index modulation ($\alpha_0 = 0$) it is a $k$-gap structure, as shown in Fig. 3(a). However, as the amplitude of the gain/loss modulation in Eq. (1) gradually increases, the wave-number gap shrinks and finally disappears in the case of a balanced $PT$-symmetric grating ($\alpha_0 = k_0\Delta n_0$, Fig. 3(b)). Finally in the broken $PT$-symmetry situation, when the gain/loss modulation amplitude exceeds the index amplitude ($\alpha_0 > k_0\Delta n_0$), Fig. 3(c)), the coupler becomes an $\omega$-gap structure, similar to a reflective index grating. The larger the gain/loss amplitude, the larger the forbidden gap.

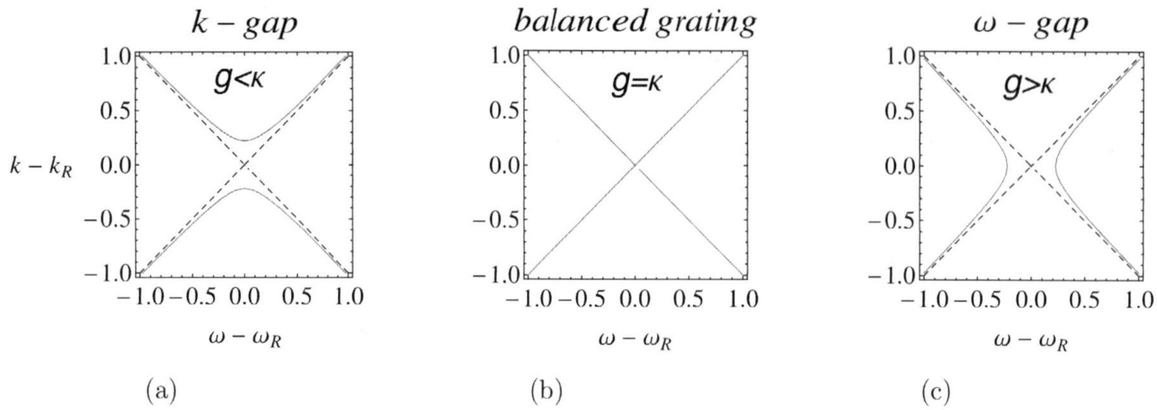


**Figure 3.** Band-gap diagrams (arbitrary scale) in the vicinity of the resonance frequency $\omega_R$ for a grating-assisted co-directional coupler with (a) unbroken $PT$ symmetry ($g < \kappa$), (b) balanced $PT$ symmetry ($g = \kappa$), (c) broken $PT$ symmetry ($g > \kappa$). The parameters $g$ and $\kappa$ are proportional to $\alpha_0$ and $\Delta n_0$ respectively, with $g/\kappa = \alpha_0/(k_0 \Delta n_0)$.

$PT$-symmetric grating-assisted co-directional couplers with a large grating period (tens or hundreds of microns) are much easier to manufacture using standard optical photolithography than similar contra-directional couplers with sub-micron periodicity, which involve complex *e*-beam lithography. Due to grating-assisted coupling there is considerable freedom of design, whereby WG#1 and WG#2 could be quite different, with different nonlinearity and dispersion. Also, as already mentioned, co-directional couplers can be built with rather broad spectral transmission capable of processing very short optical waveforms.

## 2. Coupled-mode equations for the coupler

Let us consider a nonlinear grating-assisted co-directional coupler with two dissimilar parallel waveguides WG#1 and WG#2 with different propagation constants $\beta_1(\omega)$ and $\beta_2(\omega)$, as shown in Fig. 2(a). We neglect any evanescent coupling, so optical signals propagating along the two waveguides in the same direction ($z$) can interact with one another only through a $PT$-symmetric grating with combined modulation of refractive index and loss/gain shifted by a half period, $\Lambda$, as in Eq. (1).

In linear mode the grating enables coupling between the two co-propagating modes in the vicinity of the resonance frequency, $\omega_R$, where $\beta_1(\omega_R) - \beta_2(\omega_R) = K$. We therefore write the longitudinal part of the electromagnetic field in the respective waveguides as $E_1 = A_1(z,t)\mathrm{e}^{i(\beta_1 z - \omega t)}$ and $E_2 = A_2(z,t)\mathrm{e}^{i(\beta_2 z - \omega t)}$, where $A_1(z,t)$ and $A_2(z,t)$ are slowly-varying functions of $z$ and $t$.

For $\omega$ near to $\omega_R$ we use the Taylor expansion for $\beta_1$ and $\beta_2$:

$$\beta_r(\omega) = \beta_r(\omega_R) + \beta_r'(\omega_R)(\omega - \omega_R) + \frac{1}{2}\beta_r''(\omega_R)(\omega - \omega_R)^2 + \ldots, \quad (2)$$

($r$ =1, 2), in which the coefficients are related to the group velocity and dispersion by $V_{Gr} = 1/\beta_r'(\omega_R)$ and $\beta_{2r} = \beta_r''(\omega_R)$.

The coupled-wave equations for $A_1$ and $A_2$, including nonlinearity, then take the form

$$i\left(\frac{\partial}{\partial z} + \frac{1}{V_{G1}}\frac{\partial}{\partial t}\right)A_1 + \frac{1}{2}\beta_{21}\frac{\partial^2 A_1}{\partial t^2} + (\kappa - g)A_2 e^{-2i\delta z} + \gamma_1|A_1|^2 A_1 \;=\; 0 \quad (3a)$$

$$i\left(\frac{\partial}{\partial z}+\frac{1}{V_{G2}}\frac{\partial}{\partial t}\right)A_2+\frac{1}{2}\beta_{22}\frac{\partial^2 A_2}{\partial t^2}+(\kappa+g)A_1e^{+2i\delta z}+\gamma_2|A_2|^2A_2 \;=\; 0 \qquad (3b)$$

where $\kappa=\frac{1}{2}k_0\Delta n_0$, $g=\frac{1}{2}\alpha_0$, and $\delta$ is the detuning parameter $\delta(\omega)\equiv\frac{1}{2}(\beta_1(\omega)-\beta_2(\omega)-K)$, with $\delta(\omega_R)=0$.

In the following we will neglect dispersion in the waveguides themselves, setting $\beta_{12}=\beta_{22}=0$: dispersion is provided by the grating itself.

The phase factors $e^{\pm 2i\delta z}$ can be removed by defining $F_1(z,t)=A_1(z,t)e^{i\delta z}$ and $F_2(z,t)=A_2(z,t)e^{-i\delta z}$, giving the following equations for $F_1$ and $F_2$:

$$i\left(\frac{\partial}{\partial z}+\frac{1}{\Delta V_G}\frac{\partial}{\partial T}\right)F_1+\delta F_1+(\kappa-g)F_2+\gamma_1|F_1|^2F_1 \;=\; 0 \qquad (4a)$$

$$i\left(\frac{\partial}{\partial z}-\frac{1}{\Delta V_G}\frac{\partial}{\partial T}\right)F_2-\delta F_2+(\kappa+g)F_1+\gamma_2|F_2|^2F_2 \;=\; 0, \qquad (4b)$$

where $\Delta V_G$ is defined by

$$\frac{1}{\Delta V_G}=\frac{1}{2}\left(\frac{1}{V_{G1}}-\frac{1}{V_{G2}}\right)$$

and $F_1$ and $F_2$ are now regarded as functions of $z$ and $T$, where

$$T=t-\frac{1}{2}\left(\frac{1}{V_{G1}}+\frac{1}{V_{G2}}\right)z.$$

As a final simplification we can scale $T$ according to $T=\tau/\Delta V_G$, so that the equations become

$$i\left(\frac{\partial}{\partial z}+\frac{\partial}{\partial \tau}\right)F_1+\delta F_1+(\kappa-g)F_2+\gamma_1|F_1|^2F_1 \;=\; 0 \qquad (5a)$$

$$i\left(\frac{\partial}{\partial z}-\frac{\partial}{\partial \tau}\right)F_2-\delta F_2+(\kappa+g)F_1+\gamma_2|F_2|^2F_2 \;=\; 0, \qquad (5b)$$

It is these equations which when linearized give the band-structure diagrams of Fig. 3. In order to solve them in the nonlinear case it is convenient to write the complex $F_r$ in polar form: $F_r=a_re^{i\varphi_r}$, with $a_r$ real and positive. We then have four equations for the unknown functions $a_r$ and $\varphi_r$:

$$\left(\frac{\partial}{\partial z}+\frac{\partial}{\partial \tau}\right)a_1 \;=\; (\kappa-g)a_2\sin\Phi \qquad (6a)$$

$$\left(\frac{\partial}{\partial z}-\frac{\partial}{\partial \tau}\right)a_2 \;=\; -(\kappa+g)a_1\sin\Phi \qquad (6b)$$

$$\left(\frac{\partial}{\partial z}+\frac{\partial}{\partial \tau}\right)\varphi_1 \;=\; \delta+(a_2/a_1)(\kappa-g)\cos\Phi+\gamma_1a_1^2 \qquad (6c)$$

$$\left(\frac{\partial}{\partial z}-\frac{\partial}{\partial \tau}\right)\varphi_2 \;=\; -\delta+(a_1/a_2)(\kappa+g)\cos\Phi+\gamma_2a_2^2 \qquad (6d)$$

where $\Phi=\varphi_1-\varphi_2$.

## 3. Soliton solutions

Solitary waves correspond to solutions of Eqs. (6) that depend on a single variable. The choice of variable will determine whether the solutions belong to the $PT$-symmetric regime $g<\kappa$ or the broken-symmetry regime $g>\kappa$.

### *3.1. PT-symmetric regime $g<\kappa$, gap in wave-number*

A natural choice (cf. Ref. [12]) is

$$\zeta=\frac{\tau-qz}{\sqrt{1-q^2}}, \tag{7}$$

where the free parameter $q$ is the "velocity" of the soliton in the $(\tau,z)$ plane.

Defining the scaled amplitudes $u_1$ and $u_2$ by

$$a_1(\zeta)=\frac{u_1(\zeta)}{\sqrt{1-q}} \quad \text{and} \quad a_2(\zeta)=\frac{u_2(\zeta)}{\sqrt{1+q}}, \tag{8}$$

Eqs. (6) then become

$$\frac{du_1}{d\zeta} = (\kappa-g)u_2\sin\Phi \tag{9a}$$

$$\frac{du_2}{d\zeta} = (\kappa+g)u_1\sin\Phi \tag{9b}$$

$$\frac{d\varphi_1}{d\zeta} = \sqrt{\frac{1+q}{1-q}}\,\delta+\frac{u_2}{u_1}(\kappa-g)\cos\Phi+\gamma_1\frac{u_1^2}{1-q}\sqrt{\frac{1+q}{1-q}} \tag{9c}$$

$$\frac{d\varphi_2}{d\zeta} = \sqrt{\frac{1-q}{1+q}}\,\delta-\frac{u_1}{u_2}(\kappa+g)\cos\Phi-\gamma_2\frac{u_2^2}{1+q}\sqrt{\frac{1-q}{1+q}} \tag{9d}$$

From Eqs. (9a) and (9b) we see that

$$\frac{u_1^2}{\kappa-g}=\frac{u_2^2}{\kappa+g}+C_1 \tag{10}$$

We will take the constant $C_1$ to be zero, which will certainly be the case for solutions that go to zero at infinity. Then, since $u_1$ and $u_2$ are both real and positive, this implies that we must have $g<\kappa$, as advertised, and

$$u_2=u_1\sqrt{\frac{\kappa+g}{\kappa-g}}, \tag{11}$$

It is convenient to define $\eta=u_1^2$, in which case Eq. (9a) becomes

$$\frac{d\eta}{d\zeta}=2\alpha\eta\sin\Phi, \tag{12}$$

where $\alpha=\sqrt{\kappa^2-g^2}$.

Combining Eqs. (9c) and (9d) we obtain the following equation for $\Phi$:

$$\frac{d\Phi}{d\zeta}=2q\Delta+2\alpha\cos\Phi-\theta\eta \tag{13}$$

where $\Delta=\delta/\sqrt{1-q^2}$ and

$$\theta=-\frac{\kappa+g}{\kappa-g}\frac{\gamma_2}{1+q}\sqrt{\frac{1-q}{1+q}}-\frac{\gamma_1}{1-q}\sqrt{\frac{1+q}{1-q}}.$$

Equations (12) and (13) are very similar to those considered by Conti and Trillo [10] when studying the existence of gap solitons in a completely different physical set-up. We give here a brief account of their equations and their solutions, which can readily be adapted to the present situation. Similar equations for were also derived by Kazantseva et al. [17] for yet another physical set-up, but for the restricted case $\Delta = 0$.

The two equations of Ref. [10], in our notation, are

$$\frac{d\eta}{d\zeta} = 2\eta \sin\Phi, \tag{14}$$

$$\frac{d\Phi}{d\zeta} = 2\Delta + 2\cos\Phi - \eta. \tag{15}$$

These equations can be derived from the conserved Hamiltonian

$$H = (2\cos\Phi + 2\Delta - \eta/2)\eta \tag{16}$$

in the sense that $d\eta/d\zeta = -\partial H/\partial\Phi$ and $d\Phi/d\zeta = \partial H/\partial\eta$. As a result, $H$ is a constant of the motion.

There are two basic types of soliton supported by these equations, those where $\eta$ goes to zero at infinity, and those where it tends to a finite constant value. These values are given by the stationary points $(\eta_s, \Phi_s)$ of the equations, where $d\eta/d\zeta = d\Phi/d\zeta = 0$.

It is worth noting that $H_s \equiv H(\eta_s, \Phi_s) = \eta_s^2/2$. Then the equation $H = H_s$ can be solved for $\cos\Phi$, giving

$$\cos\Phi = -\Delta + \frac{\eta^2 + \eta_s^2}{4\eta}, \tag{17}$$

from which $\sin\Phi$ in Eq. (14) can be derived, giving an equation in terms of $\eta$ only.

As can readily be seen from Eqs. (14) and (15), one stationary point is $\eta_s = 0$, $\cos\Phi_s = -\Delta$, which requires $|\Delta| \leq 1$. These solutions, which were termed bright (B) solitons in Ref. [10], are given by a suitable choice of the arbitrary additive constant in $\zeta$ as

$$\eta_B = \frac{4(1-\Delta^2)}{\cosh(2\sqrt{1-\Delta^2}\,\zeta) - \Delta} \tag{18}$$

$$\Phi_B = \arccos\left\{\frac{1 - \Delta\cosh(2\sqrt{1-\Delta^2}\,\zeta)}{\cosh(2\sqrt{1-\Delta^2}\,\zeta) - \Delta}\right\} \tag{19}$$

The other stationary points are given by $\eta_s = 2(\Delta - 1)$ and $\Phi_s = \pi$, which, since $\eta$ must be positive, are only valid for $\Delta > 1$, i.e. for $\delta > \sqrt{1-\beta^2}$, outside the so-called dynamical gap [10]. They give rise to two solutions, termed dark (DK) and antidark (AK), which have the same relative phase $\Phi$, but different amplitudes. In fact, as can be seen from Eq. (17), $\eta = y\eta_s$ and $\eta = \eta_s/y$ give the same value of $\cos\Phi$. The AK solutions are

$$\eta_{AK} = 2(\Delta - 1)\frac{\sqrt{\Delta}\cosh(2\sqrt{\Delta-1}\,\zeta) + 1}{\sqrt{\Delta}\cosh(2\sqrt{\Delta-1}\,\zeta) - 1}) \tag{20}$$

$$\Phi_{AK} = \arccos\left\{-\Delta + (\Delta - 1)\frac{\Delta\cosh^2(2\sqrt{\Delta-1}\,\zeta) + 1}{\Delta\cosh^2(2\sqrt{\Delta-1}\,\zeta) - 1}\right\} \tag{21}$$

They represent a localized peak above the constant background $\eta_s = 2(\Delta - 1)$.

As has been mentioned, the dark soliton solutions, which were given incorrectly in Ref. [10], have $\Phi_{DK} = \Phi_{AK}$ and

$$\eta_{DK} = 2(\Delta - 1)\frac{\sqrt{\Delta}\cosh(2\sqrt{\Delta-1}\,\zeta) - 1}{\sqrt{\Delta}\cosh(2\sqrt{\Delta-1}\,\zeta) + 1} \tag{22}$$

They represent a localized dip below the same constant background.

We are now in a position to adapt these results to our own situation. The equations (14) and (15) can be brought to the form of (12) and (13) by the substitutions

$$\zeta \to \alpha\zeta, \quad \Delta \to q\Delta/\alpha, \quad \eta \to \eta\theta/\alpha. \tag{23}$$

Thus the bright soliton solutions to Eqs. (12) and (13) are

$$\eta_B = \frac{4}{\theta}\frac{(\alpha^2 - \tilde{\Delta}^2)}{\alpha\cosh(2\sqrt{\alpha^2 - \tilde{\Delta}^2}\,\zeta) - \tilde{\Delta}} \tag{24}$$

$$\Phi_B = \arccos\left\{\frac{\alpha - \tilde{\Delta}\cosh(2\sqrt{\alpha^2 - \tilde{\Delta}^2}\,\zeta)}{\alpha\cosh(2\sqrt{\alpha^2 - \tilde{\Delta}^2}\,\zeta) - \tilde{\Delta}}\right\}, \tag{25}$$

where $\tilde{\Delta} = q\Delta = q\delta/\sqrt{1-q^2}$, which are valid as long as $|\tilde{\Delta}| < \alpha$. Moreover the overall factor $\theta$ in Eq. (24) must be positive, since $\eta = u_1^2$ is positive.

The anti-dark solutions, with $\tilde{\Delta} > \alpha$, are now

$$\eta_{AK} = \frac{2}{\theta}(\tilde{\Delta} - \alpha)\frac{\sqrt{\tilde{\Delta}}\cosh(2\sqrt{\alpha(\tilde{\Delta} - \alpha)}\,\zeta) + \sqrt{\alpha}}{\sqrt{\tilde{\Delta}}\cosh(2\sqrt{\alpha(\tilde{\Delta} - \alpha)}\,\zeta) - \sqrt{\alpha}} \tag{26}$$

$$\Phi_{AK} = \arccos\left\{-\frac{\tilde{\Delta}}{\alpha} + \left(\frac{\tilde{\Delta}}{\alpha} - 1\right)\frac{\tilde{\Delta}\cosh^2(2\sqrt{\alpha(\tilde{\Delta} - \alpha)}\,\zeta) + \alpha}{\tilde{\Delta}\cosh^2(2\sqrt{\alpha(\tilde{\Delta} - \alpha)}\,\zeta) - \alpha}\right\}, \tag{27}$$

representing a localized peak above the background value $\eta_s = 2(\tilde{\Delta} - \alpha)/\theta$. The dark solutions are trivially derived from these.

In the limit as $\tilde{\Delta} \to \alpha$ the $\eta_B$ and $\eta_{AK}$ solutions merge, to give a Lorentzian shape, namely

$$\eta_B \to \frac{\alpha}{\theta}\left(\frac{8}{1 + 4\alpha^2\zeta^2}\right) \tag{28}$$

$$\Phi_B \to \arccos\left(\frac{1 - 4\alpha^2\zeta^2}{1 + 4\alpha^2\zeta^2}\right), \tag{29}$$

while $\eta_{DK} \to 0$.

### 3.2. *Broken PT-symmetry regime $g > \kappa$, gap in frequency*

As we have seen, the choice of variables (7) meant that soliton solutions could only be found for the $PT$-symmetric regime $g < \kappa$. However, such solutions can be found in the broken-symmetry regime if we essentially interchange the roles of $z$ and $\tau$, defining $\zeta$ instead as

$$\zeta = \frac{z - v\tau}{\sqrt{1 - v^2}}, \tag{30}$$

where the free parameter $v$, with $|v| < 1$, does now have the meaning of a velocity.

Defining the scaled amplitudes $u_1$ and $u_2$ by

$$a_1(\zeta) = \frac{u_1(\zeta)}{\sqrt{1 - v}} \quad \text{and} \quad a_2(\zeta) = \frac{u_1(\zeta)}{\sqrt{1 + v}}, \tag{31}$$

we obtain the following equations:

$$\frac{du_1}{d\zeta} = (\kappa - g)u_2 \sin\Phi \tag{32a}$$

$$\frac{du_2}{d\zeta} = -(\kappa + g)u_2 \sin\Phi \tag{32b}$$

$$\frac{d\varphi_1}{d\zeta} = \sqrt{\frac{1+v}{1-v}}\,\delta + \frac{u_2}{u_1}(\kappa - g)\cos\Phi + \gamma_1\frac{u_1^2}{1-v}\sqrt{\frac{1+v}{1-v}} \tag{32c}$$

$$\frac{d\varphi_2}{d\zeta} = -\sqrt{\frac{1-v}{1+v}}\,\delta + \frac{u_1}{u_2}(\kappa + g)\cos\Phi + \gamma_2\frac{u_2^2}{1+v}\sqrt{\frac{1-v}{1+v}} \tag{32d}$$

From Eqs. (32a) and (32b) we see that

$$\frac{u_1^2}{g-\kappa} = \frac{u_2^2}{g+\kappa}, \tag{33}$$

taking the additive constant to be zero, as before. Then, since $u_1$ and $u_2$ are both real and positive, this implies that we must now have $g > \kappa$, and

$$u_2 = u_1\sqrt{\frac{g+\kappa}{g-\kappa}}, \tag{34}$$

Again defining $\eta$ as $\eta = u_1^2$, Eq. (32a) becomes

$$\frac{d\eta}{d\zeta} = -2\alpha\eta\sin\Phi, \tag{35}$$

where $\alpha = \sqrt{g^2 - \kappa^2}$.

Combining Eqs. (32c) and (32d) we obtain the following equation for $\Phi$:

$$\frac{d\Phi}{d\zeta} = 2\Delta - 2\alpha\cos\Phi - \theta\eta, \tag{36}$$

where in this case $\Delta = \delta/\sqrt{1-v^2}$ and $\theta$ is the same as previously defined, with $q \to v$.

The substitutions required to bring Eqs. (14) and (15) to the form of (35) and (36) are

$$\zeta \to \alpha\zeta, \quad \Delta \to \Delta/\alpha, \quad \eta \to \eta\theta/\alpha, \quad \Phi \to \Phi + \pi. \tag{37}$$

Thus the soliton solutions in this case take essentially the same form as those of the previous section, Eqs. (24) - (29), except that $\Phi$ is shifted by $\pi$ and $\tilde{\Delta}$ is replaced by $\Delta$. However, it is very important to recall that $\alpha$ is now defined as $\alpha = \sqrt{g^2 - \kappa^2}$, and the variable $\zeta$ is defined as in Eq. (30) rather than (7). The parameter $\theta$ must still be positive.

## 4. Discussion

The *PT*-symmetric case, $g < \kappa$, seems a straightforward extension of the pure index coupling situation investigated by Wabnitz [12], except that in that paper the nonlinear terms were different, allowing for a true soliton solution. It is worth noting that the lab. frame velocity of the soliton in this case, dictated by the choice of variable $\zeta \propto \tau - qz$ is

$$V_s = \frac{2V_{G1}V_{G2}}{V_{G1}(1-q) + V_{G2}(1+q)}, \tag{38}$$

which lies between $V_{G1}$ and $V_{G2}$, as might be expected.

The broken $PT$ case, $g > \kappa$, has some unusual features. First, there there is no sign of the exponential growth one might expect when the $PT$ symmetry is broken, and second, the choice of variable $\zeta \propto z - v\tau$, with $\tau = \Delta V_G T$ and $T = t - \frac{1}{2}(1/V_{G1} + 1/V_{G2})z$, gives the lab. frame velocity of the soliton as

$$V_s = \frac{2vV_{G1}V_{G2}}{V_{G2}(1+v) - V_{G1}(1-v)}. \tag{39}$$

This has the nice property that the soliton can be stationary, when $v = 0$, the mechanism for which may be related to the stopping of photons at the exceptional point in a $PT$-symmetric medium[24]. At first sight it appears that $V_s$ has the less attractive property that it is unlimited if $v$ is allowed to vary freely, even becoming infinite when $V_{G2} = ((1-v)/(1+v))V_{G1}$. However, it is worth noting that the way that $V_s$ enters the expression for $\zeta$ according to

$$\zeta = \frac{-v\Delta V_G}{\sqrt{1-v^2}}\left(t - \frac{z}{V_s}\right), \tag{40}$$

so that as $V_s$ increases, the soliton peak becomes wider and less well defined as a function of $z$, becoming a function of $t$ only when $V_s \to \infty$.

Recall that the choice of variable for $g > \kappa$ was dictated by the fact that we took the constant of integration $C_1$ to be zero in Eq. (33), as we did in Eq. (10), in order that solutions could be found that went to zero at infinity. However, if $C_1 \neq 0$ other forms of solution can be found for both $g < \kappa$ and $g > \kappa$, along the lines of Refs. [25] and [26].

We should also mention that a soliton solution can be found [27] at the exceptional point $g = \kappa$ for specially constrained values of the parameters of the waveguides and their couplings. The soliton velocity in this case has the same general features as in the case $g > \kappa$.

## References

[1] Maimistov A I 2010 *Quantum Electron.* **40** 756
[2] Aceves A B 2000 *Chaos* **10** 584
[3] Christodoulides D N and Joseph R I 1989 *Phys. Rev. Lett.* **62** 1746
[4] Aceves A B and Wabnitz S 1989 *Phys. Lett.* **A** 141
[5] Hill K O, Malo B, Vineberg K A, F. Bilodeau, Johnson D C and I. Skinner 1990 *Electron. Lett.* **26** 1270
[6] Vengsarkar A N, Lemair P J, Judkins J B, Bhatia V, Erdogan T and Sipe J E 1996 *J. Lightwave Technol.* **14** 58
[7] Kulishov M and Azaña J 2005 *Opt. Lett.* **30** 2700
[8] Park Y, Kulishov M, Slavik R and Azaña J 2006 *Opt. Express* **14** 12670
[9] Kogelnik H and Shank C V 1972 *J. Appl. Phys.* **43** 2327
[10] Conti C and S. Trillo S (2001) *Phys. Rev. E* **64** 036617
[11] Van Simaeys G, Coen S, Haelterman M and Trillo S 2004 *Phys. Rev. Lett.* **92** 223902
[12] Wabnitz S 1989 *Opt. Lett.* **14** 1071
[13] Buryak A V, Di Trapani P, Skryabin D V, and Trillo S 2002 *Phys. Rep.* **370** 63
[14] Yariv A 1973 *IEEE J. Quantum Electron.* **9** 919
[15] de Sterke C M and Sipe J E 1994 *Prog. Opt.* **33** Ch 3 (Amsterdam: Elsevier)
[16] Trillo S 1996 *Opt. Lett.* **21** 1732
[17] Kazantseva E V, Maimistov A I and Ozhenko S S 2009 *Phys. Rev. A* **80** 043833
[18] Miri M A, Aceves A B, Kottos T, Kovanis T V and Christodoulides D N 2012 *Phys. Rev. A* **86** 033801
[19] Raja S V, Govindarajan A, Mahalingam A and Lakshmanan M 2019 *Phys. Rev. A* **100**, 033838
[20] Kulishov M, Laniel J M, Bélanger N, Azaña J and Plant D V 2005 *Opt. Express* **13** 3068
[21] Greenberg M and Orenstein M 2004 *Opt. Express* **12**, 4013
[22] Kulishov M, Laniel J M, Bélanger N and Plant D V 2005 *Opt. Express* **13**, 3567
[23] Kulishov M and Kress B 2013 *Opt. Express* **21** 22327
[24] Goldzak T, Mailybaev A A and Moiseyev N 2018 *Phys. Rev. Lett.* **120** 013901
[25] Gupta S K and Sarma A K 2014 *Europhys. Lett.* **105** 44001
[26] Kudryashov N A, Maimistov A I and Sinelshchikov D I 2012 *Phys. Lett. A* **376** 3658
[27] Fring A, private communication.

# Emergence of parity time symmetric quantum critical phenomena

**Ranjith R Kumar[1,2], S Rahul[1,2], Y R Kartik[1,2] and Sujit Sarkar[1]**
[1]Theoretical Sciences Division, Poornaprajna Institute of Scientific Research, Bidalur, Bengaluru, 562164, India.
[2]Graduate Studies, Manipal Academy of Higher Education, Madhava Nagar, Manipal-576104, India.

E-mail: sujit.tifr@gmail.com

**Abstract.** Emergence of different interesting and insightful phenomena at different length scale is the heart of quantum many-body system. We show that the physics of parity-time (PT) symmetry is one new addition to them. We show explicitly that the emergence of different topological excitation at different length scale for the PT symmetry system through the analysis of renormalization group (RG) flow lines. We observe that the higher order RG process favour the emergence of asymptotic freedom like behaviour and also show the effect of strong correlation on the emergent phases. Interestingly, the asymptotic freedom like behaviour is favoured by PT symmetry phase of the system. Moreover, we also derive the scaling relation for the couplings in RG equations. These findings can be tested experimentally in ultracold atoms.

## 1. Introduction

In quantum many-body physics, emergent phenomena are an essential aspect. In this view, fundamentally new kinds of phenomena emerge within the complex assemblies of particles which can not be anticipated from a priori knowledge of the microscopic laws of nature. One can raise the question at the fundamental level: what emergent principles and laws develop as we proceed from the microscopic scale to the macro scale? P. W. Anderson first introduced the concept of "emergent phenomena" into physics [1, 2]. The behavior of large and complex aggregation of elementary particles cannot be understood in terms of simple extrapolation properties of a few particles. Instead, at each level of complexity entirely new properties appear and the understanding of the new behavior triggers a new front of the research area.

Laughlin and Pines [3] have explained the emergent phenomena very nicely in their article on "The Theory of Everything". The emergent physical phenomena regulated by higher organizing principles have a property, namely their insensitivity to microscopic scale that is directly relevant to the broad question of what is knowable in the deepest sense of the term. One can also understand the emergent phenomena from the low energy excitations of conventional superconductors. It is completely generic and is characterized by a handful of parameters that may be obtained experimentally but not from the first principle.

It has been known since the early 1970s that renormalizability is an emergent property of ordinary matter either in stable quantum phases, such as the superconducting state, or at particular zero-temperature phase transitions between such states called quantum critical points. In either case the low-energy excitation spectrum becomes more and more generic and less and

less sensitive to microscopic details as the energy scale of the measurement is lowered, until in the extreme limit of low energy, all evidence of the microscopic equations vanishes away.
In this research paper, we study the emergent physics of parity-time (PT) symmetric quantum mechanical system in more advanced level. In the non-hermitian quantum system, PT symmetric quantum mechanics is an extension of conventional quantum mechanics into complex domain [4, 5, 6, 7, 8]. Initially, PT symmetry quantum mechanics has started as an interesting mathematical discovery and a good theoretical exercise in theoretical physics [4, 5, 6, 7, 8]. But in current literature, it has been expanded experimentally in different field of science like open quantum systems [8, 9], physics of gain and loss (as found in photonics [10, 11, 12]) or systems where the non-Hermiticity models the finite lifetime [13, 14], localization–delocalization of correlated many-body system [15, 16, 17], biological systems [18, 19, 20], Weyl semi-metals [21, 22, 23, 24, 25], topology and dissipation [26, 27] and PT symmetric circuit QED [28]. One of the most important feature of non-hermitian quantum mechanics is the exceptional point (EP) [3, 4, 5, 6, 29, 30]. This EP has a dramatic effect on the system, leading to the nontrivial physics with interesting counterintuitive features, which we will observe in this study.
We use the renormalization group (RG) method to study this PT symmetry quantum criticality problem. The mathematical structure and results of the RG theory are a significant conceptual advancement in the quantum field theory of both high-energy and condensed matter physics [31, 32] in the last several decades. The need for RG is really transparent in condensed matter physics. RG theory is a formalism that relates the physics at different length scales in condensed matter physics and the physics at different energy scales in high-energy physics [33, 34]. One of the successes of RG theory is in the study of classical and quantum Berezinskii–Kosterltz–Thouless (QBKT) transition [35] in XY model and topological insulator respectively. The classical two dimensional XY model was found to have a power-law correlation function at low-temperature regime and exponential at the high-temperature regime. The phase transition associated with it was not due to the well-known spontaneous symmetry breaking mechanism. Berezinskii [31], in the year 1971 and Kosterlitz and Thouless [32], in the year 1973 have explained this new kind of phase transition in terms of topological non-trivial vortex using the RG method. The high-temperature regime favours the thermal generation of vortices that unbounded, while in the low-temperature regime vortices are bounded and vortex and anti-vortex are always found in pairs. This binding and unbinding of vortex pairs facilitate the phase transition which is now called BKT transition. QBKT at zero temperature was also found to appear in the interacting helical liquid system at the one-dimensional edge of a two-dimensional topological insulator, coupled to an external magnetic field and s-wave superconductor. Through the RG flow analysis of this model, QBKT was identified between Luttinger liquid phase and Ising-ferromagnetic and gaped superconducting phase. The direction of the RG flow distinguishes between distinct quantum phases.
**Motivation:** We present emergent physics of quantum criticality at different length scales. We will see that the interplay between many-body correlations and PT symmetry leads to the emergence of quantum critical phenomena beyond the Hermitian paradigm of quantum many-body physics [36, 37]. We also raise the question of how the topology related to the PT-symmetry physics at different length-scales of the system emerges.
A successful part of the RG theory is the observation of asymptotic freedom for high-energy physics and the other is the Berezinskii–Kosterltz–Thouless transition physics for condensed matter physics, i.e, the two extreme ends of theoretical physics. In this study we are successful in unifying these two observations of RG theory in a single framework of physics PT symmetry for interacting quantum many body system.

## 2. The Model Hamiltonian and Renormalization Group Equation

We consider a class of one-dimensional quantum systems described by the sine-Gordon field theory

$$H = H_0 + V(\phi), \tag{1}$$

where $H_0$ is

$$H_0 = \left(\frac{hv}{2\pi}\right) \int dx[K(\partial_x \theta(x))^2 + \frac{1}{K}(\partial_x \phi(x))^2]. \tag{2}$$

The Hamiltonian, $H$ gives a universal framework for describing one dimensional interacting bosons and fermionic system and $V(\phi(x))$ is the sine-Gordon potential. Here $\phi(x)$ is field with the parity operation, $P\phi P^{-1} = -\phi$. The $\theta(x)$ is the dual field of $\phi(x)$ and satisfies the commutation relation, $[\phi(x), \partial_x \theta(x')] = -i\pi\delta(x - x')$. $H_0$ is the Tomonoga Luttinger liquid (TLL) Hamiltonian. We will see that the value of $K$ will play an important role to determine length-scale dependent quantum criticality for this model Hamiltonian system. The relevance of this term gives the gapped phase in the system, this transition from the gapless Luttinger liquid phase to the Mott insulating phase in the system.

A generalization to the PT symmetric case by adding an imaginary contribution [36, 37] to the potential term is as follows

$$V(\phi) = \frac{\alpha_r}{\pi} cos(2\phi) - \frac{i\alpha_i}{\pi} sin(2\phi), \tag{3}$$

where $\alpha_r$ and $\alpha_i$ are the real and imaginary part of the potential. The imaginary part of the potential, which introduces physics of spectral singularity occurs when the real and imaginary part of the potential are same [28].

We will observe that when $\alpha_r$ becomes relevant, a stable gapped phase, i.e, the fluctuation of $\phi$ gets suppressed. But when $\alpha_i$ becomes relevant, it facilities the fluctuations of $\phi$. Then the imaginary of the potential profile are not the same as a consequence of it and the imaginary potential enhances the correlation of conjugate field $\theta$. To get the correct physical picture of quantum criticality for this model Hamiltonian, RG study is essential [33]. The analytical expressions for the second and third order RG equations are the following:

The analytical expressions for the second order RG equations are [36, 37],

$$\begin{aligned} \frac{dg_r}{dl} &= (2 - K)\, g_r \\ \frac{dg_i}{dl} &= (2 - K)\, g_i \\ \frac{dK}{dl} &= ({g_i}^2 - {g_r}^2)K^2. \end{aligned} \tag{4}$$

The analytical expressions for the third order RG equations are,

$$\begin{aligned} \frac{dg_r}{dl} &= (2 - K)\, g_r + 5{g_r}^3 - 5{g_i}^2 g_r \\ \frac{dg_i}{dl} &= (2 - K)\, g_i - 5{g_i}^3 + 5{g_r}^2 g_i \\ \frac{dK}{dl} &= ({g_i}^2 - {g_r}^2)K^2. \end{aligned} \tag{5}$$

Here $l$ is the logarithmic RG scale and $g_{r,i} = \frac{\alpha_{r,i} a^2}{hv}$ are the dimensionless coupling constants with $a$ being a short distance cut-off and $v$ is the velocity of the collective excitations.

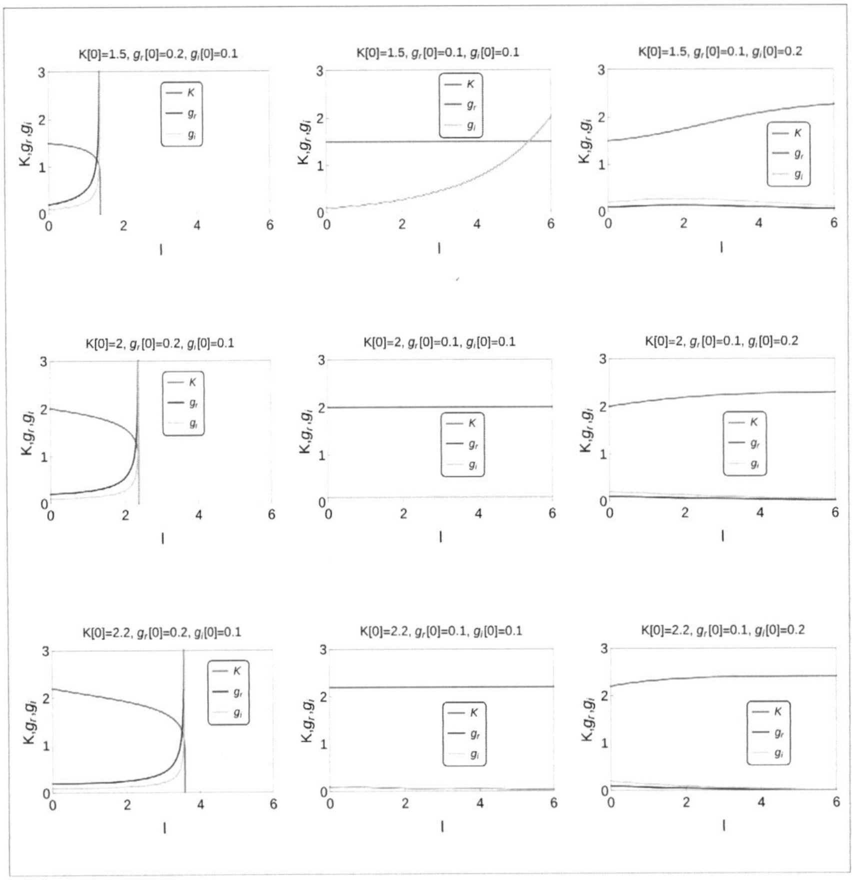


**Figure 1.** (Color online.) RG flow diagram for the couplings $g_r$, $g_i$ and $K$ with length scale ($l$) for third order RG equations (Eq. 5). Three different panels are for the different values of $K[0]$ (1.5, 2 and 2.2 are for the upper, middle and lower panels respectively) and three different figures in each panel for different values of $g_r[0]$ and $g_i[0]$. The initial values of $g_r[0], g_i[0]$ and $K[0]$ are depicted in the figure captions. In this plot we use the magnitude of the imaginary part of the coupling.

## 3. Derivation of scaling relation for the PT symmetry RG equations

It is well known that the critical theory is invariant under the rescaling. Then the singular part of the free energy density satisfies the following scaling relations,

$$f_s[g_r, g_i] = e^{-2l} f_s[e^{(2-K)l} g_r, e^{(2-K)l} g_i]. \tag{6}$$

The scale $l$ can be fixed from the analytical relation, $e^{(2-K)l^*} g_r = 1$. Finally, after few steps of calculation, we arrive at

$$f_s[g_r, g_i] = {g_r}^{2/(2-K)} f_s[1, {g_r}^{-(2-K)/(2-K)} g_i]. \tag{7}$$

The scaling relation between the real and imaginary part of the sine-Gordon potential is

$${g_r}^{-1} g_i \sim 1. \tag{8}$$

It reveals from this analytical relation that the $g_r$ and $g_i$ are proportional to each other and are independent of $K$.

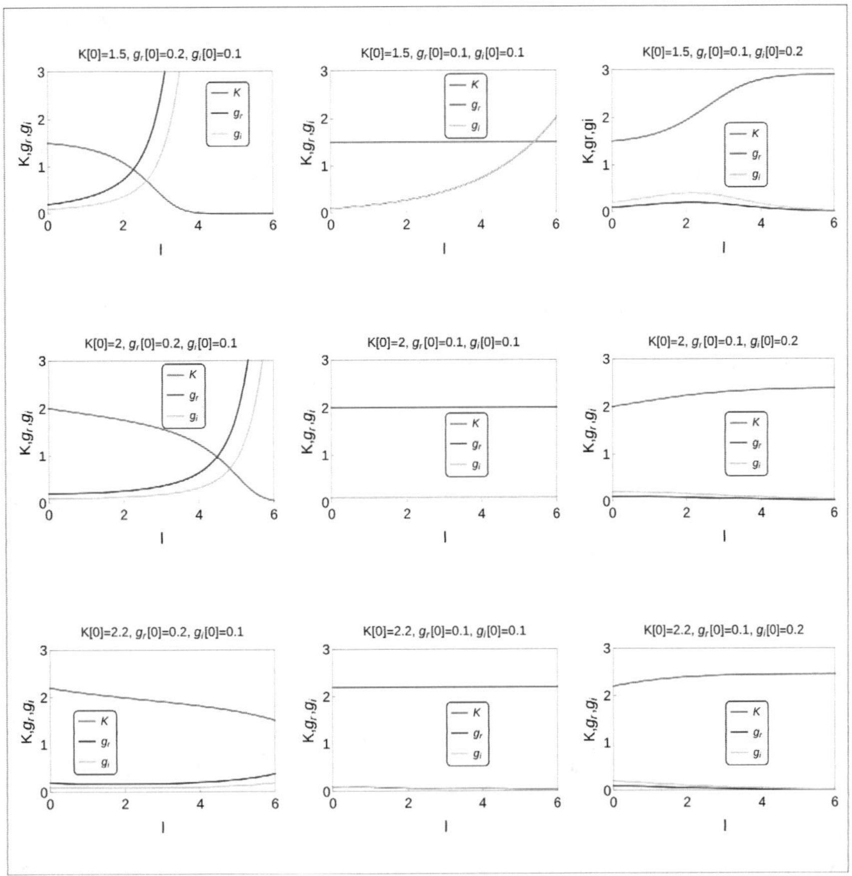


**Figure 2.** (Color online.) RG flow diagram for the couplings $g_r$, $g_i$ and $K$ with length scale $(l)$ for second order RG equations (eq. 4). Three different panels are for the different values of $K[0]$ (1.5, 2 and 2.2 are for the upper, middle and lower panel respectively ) and three different figures in each panel for different values of $g_r[0]$ and $g_i[0]$. The initial values of $g_r[0], g_i[0]$ and $K[0]$ are depicted in the figure captions. In this plot we use the magnitude of the imaginary part of the coupling.

## 4. Results along with physical interpretations

### *4.1. Emergence of quantum criticality: length scale dependent PT symmetry study from the whole set of RG equations*

The author of Ref.[34] have solved the double frequencies dual field sine-Gordon Hamiltonian from the perspective of interacting Helical liquid and also for the topological states of interacting quantum matter. The most interesting features of the PT symmetry quantum criticality is that there are two couplings ($g_r$ and $g_i$) but there is only single field $\phi(x)$. When the PT symmetry is broken, the fluctuation of the $\phi(x)$ facilitate the correlation of conjugate field. Therefore, PT symmetry broken phase generate the physics of sine-Gordon dual field theory for the system. This is the uniqueness of PT symmetry quantum criticality over the dual field double frequencies sine-Gordon field theory.

The authors of Ref.[38] and Ref.[39] have shown that in QCD, the gauge theory of quarks and gluons are asymptotically free, i.e., the coupling vanishes at very short distance (large momentum) and grows at large distance (small momentum). This allowed us to understand

Journal of Physics: Conference Series **2038** (2021) 012016 doi:10.1088/1742-6596/2038/1/012016

why quarks seemed to be free inside the nucleons in deep inelastic scattering and are confined at large distance. However, the present problem is not QCD. Here we solve a sine-Gordon model Hamiltonian with a generalization of PT symmetry. At the same time we are not interpreting our results in terms of quark and gluon physics, rather in terms of the real ($g_r$) and imaginary ($g_i$) part of the potential. Therefore, we interpret our results from the length-scale dependent asymptotic freedom like behaviour for this model Hamiltonian system.

Length-scale dependence study brings out the concept of asymptotic freedom like behaviour in the RG flow sense. This asymptotic freedom is a feature of all RG flows with a marginally relevant perturbation. We would like to explain it explicitly through the $\beta$ function explanation. This can be written as $\beta_\lambda = \frac{d\lambda}{dlnL} = C\lambda^2$ (here $\beta_\lambda$ is the $\beta$ function of the RG theory from where one can predict the nature of the RG flow lines of coupling constant $\lambda$; in quantum field theory flow lines are defined in energy scale but here we define RG flow lines in length scale. Here $\lambda$ is the coupling constant and $C > 0$ is a constant. As a result, the effective coupling is weak at short distances and strong at long distances. We use this behaviour of coupling constant of QCD in our present study.

Recently, asymptotic freedom like behaviour has found in different context of quantum matter system. The author of Ref.[39], has found the asymptotic freedom like behaviour for the topological state of interacting quantum matter. However, for the present study, we are interested to find the asymptotic freedom like behaviour for the PT symmetry system.

Now we present the emergent quantum criticality of PT symmetry physics at different length-scales of the system and also obtain the signature of asymptotic freedom like behaviour.

In Fig.1, we present the results for 3rd order RG equations from the whole set of RG equations. This figure consists of three panels for different initial values of $K[0]$. The upper, middle and lower panels are respectively for $K[0] = 1.5$, 2 and 2.2. Each panel consists of three figures for different values of the coupling $g_r[0]$ and $g_i[0]$. For each panel: left, middle and right figures are respectively for different initial values satisfying the conditions $g_r[0] > g_i[0]$, $g_r[0] = g_i[0]$ (spectral singularity condition) and $g_r[0] < g_i[0]$. It is found from this study that in the left figure for each panel, both the couplings increases with the length scale very sharply while the $K$ decreases. This is the emergence of anomalous phase with the length scale, i.e., both of the couplings increase with in the length scale for the smaller values of $K$. We observe that as the initial values of $K[0]$ increases (as shown in the middle and lower panel), the variation of the couplings ($g_r$ and $g_i$) with length scale are not sharp.

The middle figure for each panel is for $g_r[0] = g_i[0]$, i.e., at the spectral singularity point. We observe that $K$ remains constant. For $K[0] = 1.5$ (the first panel), both the couplings increase with the length scale for the lower values of $K$, i.e., the system is in the anomalous phase region. As the value of $K$ increases for the other panels both the couplings ($g_r$ and $g_i$) decreases with length scale,i.e., there is no anomalous phase region for this phase. For this regime of parameter space system is in the Luttinger liquid phase.

In the right figures of each panel, $K$ increases with the length scale for the smaller initial values of $K$ otherwise it is saturate. But the RG flow lines for the couplings ($g_r$ and $g_i$) are flowing off to zero. In this region, system is in the critical phase.

In Fig.2, we present the results of second order RG study for the same regime of parameter space as we did for third order (Fig.1). It reveals from this study , variation of the couplings constants with the length scale is not so sharp as we observed for third order RG. For the higher initial values of $K[0]$, almost there is no variation of $K$ with length scale. The asymptotic freedom like behaviour disappears for the higher values of $K$. We observe that there is no evidence of asymptotic freedom for the higher values of $K[0] (> 2.2)$ for the PT symmetry phase ( $g_r[0] > g_i[0]$). Thus it is clear from this study that lower value $K[0]$ and also the higher initial value of $g_r[0]$ from the asymptotic like behaviour. The higher order RG process favours asymptotic freedom like behaviour.

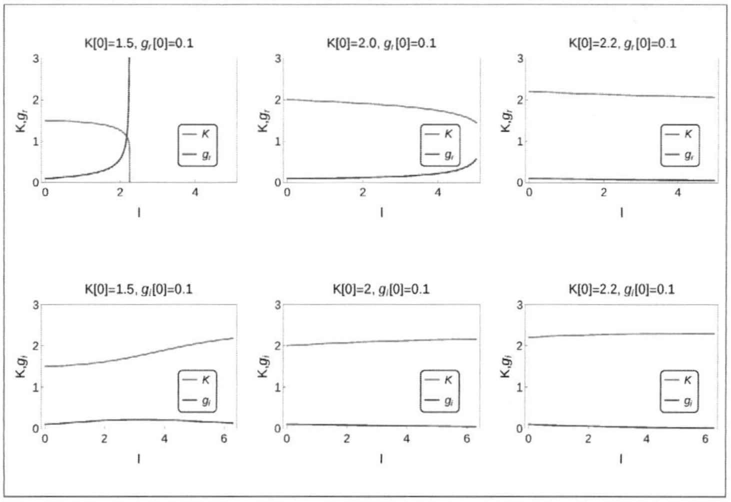


**Figure 3.** (Color online.) RG flow diagram for the couplings $g_r$, $g_i$ and $K$ with length scale ($l$) for the 3rd order RG equations (eq.8 and eq.9 ). The upper and lower panels are respectively for $g_r$,$K$ and $g_i$,$K$ with length scale ($l$). The initial values of $g_r[0], g_i[0]$ and $K[0]$ are depicted in the figures. In this plot we use the magnitude of the imaginary part of the coupling.

We have presented the scaling relation between the couplings $g_r$ and $g_i$ in Eq.6. We notice that this scaling relation is independent of $K$. Both of the couplings are proportional to each other, therefore the dependence of $g_r$ and $g_i$ for different initial values of $K[0]$ is not possible from this scaling relation, what we have found from the numerical solution of RG equations. One similarity that we have noticed from this study is that for a particular value $K[0]$ both of the couplings are showing the same behaviour with the length scale.

### *4.2. Emergence of quantum criticality: length scale dependent PT symmetry study from the conventional quantum Berezinskii-Kosterlitz-Thoulesss RG equations.*

Here we present the four sets of conventional QBKT equations from the whole sets of RG equations upto second and third order terms. Second order QBKT RG equations are,

$$\frac{dg_r}{dl} = (2-K)\,g_r$$
$$\frac{dK}{dl} = -{g_r}^2 K^2. \tag{9}$$

$$\frac{dg_i}{dl} = (2-K)\,g_i$$
$$\frac{dK}{dl} = {g_i}^2 K^2. \tag{10}$$

Third order QBKT RG equations are,

$$\frac{dg_r}{dl} = (2-K)\,g_r + 5{g_r}^3$$
$$\frac{dK}{dl} = -{g_r}^2 K^2. \tag{11}$$

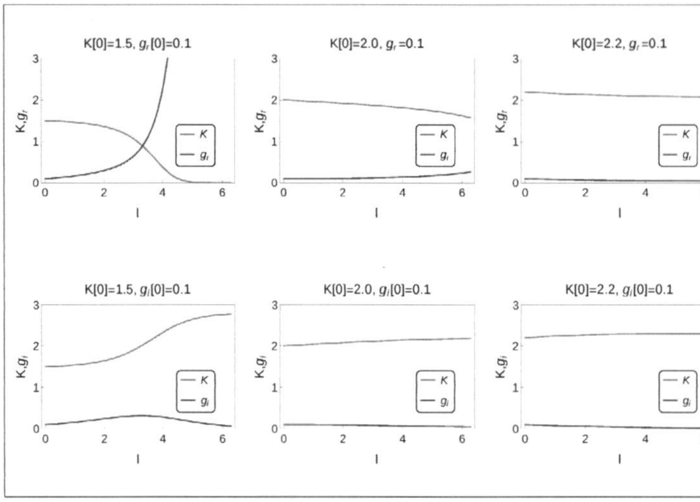


**Figure 4.** (Color online.) RG flow diagram for the couplings $g_r$, $g_i$ and $K$ with length scale ($l$) for the 2nd order RG equations. The upper and lower panels are respectively for $g_r$,$K$ and $g_i$,$K$ with length scale ($l$). The initial values of $g_r[0], g_i[0]$ and $K[0]$ are depicted in the figures. In this plot we use the magnitude of the imaginary part of the coupling.

$$\frac{dg_i}{dl} = (2 - K)\, g_i - 5{g_i}^3$$
$$\frac{dK}{dl} = {g_i}^2 K^2. \tag{12}$$

In Fig.3, we present the length scale dependent variation of couplings ($g_r$ and $g_i$ ) for the third order quantum BKT equations. It consists of two panels, the upper and lower panels are respectively for $g_r$ and $g_i$. Each panel consists of three figures for three different initial values of $K[0]$. The left, middle and the right figures are respectively for $K = 1.5, 2$ and 2.2. It reveals from this study that the $K$ decreases rapidly with the length scale for the lower values of $K[0]$ ($< 2$) and it almost saturates for the higher initial values of $K[0]$. The coupling constant $g_r$ increases as along as $K[0] < 2$ and decreases for $K[0] > 2$. In the lower panel we observe $K$ increases with the length scale but the RG flow lines for the coupling constant, $g_i$ goes to zero. In Fig.4, we present the length scale dependent variation of couplings in the quantum BKT equation with the second order terms. It consists of two panels where the upper and lower panels are respectively for $g_r$ and $g_i$. Each panel consists of three figures for three different values of $K$. The left, middle and the right figures are respectively for $K = 1.5, 2$ and 2.2. Thus it is clear from the studies of Fig. 3 and Fig. 4 that for only $K = 1.5$, $g_r$ with $l$ shows the asymptotic freedom like behaviour. This findings can be tested experimentally in ultracold atoms.

## 5. Discussion

We have observed that the presence of this quantum criticality for both the whole set renormalization group equations and also from the conventional quantum BKT equation. We have also presented the scaling equation for these coupling constants. We have observed the asymptotic freedom like behaviour favour for the PT symmetric phase only. We have also presented scaling relation for the coupling constants.

**Acknowledgments**
The author would like to acknowledge Prof. Y. Ashida for useful discussions during the progress of this work and also Prof. Chandan Dasgupta critically reviewing the manuscript. The author would like to acknowledge Admar Mutt Education Foundation, DST (EMR /2017 /000898) for the support and RRI library for books and journals. The authors also acknowledge different academic activities of ICTS. Finally author acknowledge virtual PHHQP seminar series.

## References

[1] Anderson P W 1972 *Science* **177** 393-396
[2] Anderson P W 1984 *Basic Notion of Condensed Matter Physics* (Benjamin Cummings)
[3] Laughlin R B and Pines D 2000 *Proceedings of the national academy of sciences of the United States of America* **97**, 28
[4] Bender C M 2019 *PT Symmetry: In Quantum and Classical Physics* (World Scientific, Singapore)
[5] Heiss D 2012 *J. Phys. A: Math. Theor.* **45** 444016
[6] Rotter I 2009 *J. Phys. A: Math. Theor.* **42** 153001
[7] Schindler S T and Bender C M 2018 *J. Phys. A: Math. Theor.* **51** 055201
[8] Moiseyev N 2011 *Non-Hermitian Quantum Mechanics* (Cambridge University Press, New York).
[9] Bender C M and Boettcher S 1998 *Phys. Rev. Lett.* **80** 5243
[10] Rüter C E, Makris K G, El-Ganainy R, Christodoulides D N, Segev M, and Kip D 2010 *Nat. Phys.* **6**, 192-5.
[11] Longhi S 2017 *Europhys. Lett.* **120** 64001
[12] Ozawa T *et al* 2019 *Rev. Mod. Phys.* **91** 015006
[13] Kozii V and Fu L 2017 Non-Hermitian Topological Theory of Finite-Lifetime Quasiparticles: Prediction of Bulk Fermi Arc Due to Exceptional Point *arXiv:1708.05841*
[14] Yoshida T, Peters R, Kawakami N 2018 *Phys. Rev.* B **98** 035141
[15] Hatano N and Nelson D R 1996 *Phys. Rev. Lett.* **77** 570
[16] Hatano N and Nelson D R 1997 *Phys. Rev.* B **56** 8651
[17] Hatano N and Nelson D R 1998 *Phys. Rev.* B **58** 8384
[18] Nelson D R and Shnerb N M 1998 *Phys. Rev.* E **58** 1383
[19] Amir N, Hatano N and Nelson D R 2016 *Phys. Rev.* E **93** 042310
[20] Murugan A and Vaikuntanathan S 2017 *Nat. Commun.* **8** 13881
[21] Xu Y, Wang S T and Duan L M 2017 *Phys. Rev. Lett.* **118** 045701
[22] Zyuzin A A and Zyuzin A Y 2018 *Phys. Rev.* B **97** 041203
[23] Cerjan A, Xiao M, Yuan L and Fan S 2018 *Phys. Rev.* B **97** 075128
[24] González J and Molina R A, *Phys. Rev.* B **96** 045437
[25] Molina R A and González J 2018 *Phys. Rev. Lett.* **120** 146601
[26] Bardyn C E, Baranov M A, Kraus C V, Rico E, İmamoğlu A, Zoller P and Diehl A 2013 *New J. Phys.* **15** 085001
[27] Budich J C and Diehl S 2015 *Phys. Rev.* B **91** 165140
[28] Quijandría F, Naether U, Özdemir S K, Nori F and Zueco D 2018 *Phys. Rev.* A **97** 053846
[29] Heiss D 2016 *Nat. Phys.* **12** 823-824
[30] Kato T 1966 *Perturbation Theory for Linear Operators* (Springer)
[31] Berezinskii V L 1971 *Sov. Phys. JETP* **32** 493
[32] Kosterlitz J M and Thouless D J 1973 *Journal of Physics C: Solid State Physics* **6** 1181
[33] Zee A 2013 *Quantum Field Theory in a NutShell* (Universities Press, Hyderabad)
[34] Shankar R 1994 *Rev. Mod. Phys.* **66** 129
[35] Kumar R R, Rahul S, Sahoo S N and Sarkar S 2020 *Phase Transitions* **93** 606-629
[36] Ashida Y, Furukawa S and Ueda M 2017 *Nat. Commun.* **8** 15791
[37] Sarkar S 2021 *Sci. Rep.* **11** 5510
[38] Sarkar S 2020 *Sci. Rep.* **10** 2299
[39] Gross D J and Wilczek F 1973 *Phys. Rev. Lett.* **30** 1343

# Examples of PT Phase Transition : QM to QFT

**Bhabani Prasad Mandal**

Department of Physics, Institute of Science, Banaras Hindu University, Varanasi -221005, India

E-mail: bhabani.mandal@gmail.com

**Abstract.**
Parity Time Reversal (PT) phase transition is a typical characteristic of most of the PT symmetric non-Hermitian (NH) systems. Depending on the theory, a particular system and spacetime dimensionality PT phase transition has various interesting features. In this article we review some of our works on PT phase transitions in quantum mechanics (QM) as well as in Quantum Field theory (QFT). We demonstrate typical characteristics of PT phase transition with the help of several analytically solved examples. In one dimensional QM, we consider examples with exactly as well as quasi exactly solvable (QES) models to capture essential features of PT phase transition. The discrete symmetries have rich structures in higher dimensions which are used to explore the PT phase transition in higher dimensional systems. We consider anisotropic SHOs in two and three dimensions to realize some connection between the symmetry of original hermitian Hamiltonian and the unbroken phase of the NH system. We consider the 2+1 dimensional massless Dirac particle in the external magnetic field with PT symmetric non-Hermitian spin-orbit interaction in the background of the Dirac oscillator potential to show the PT phase transition in a relativistic system. A small mass gap, consistent with the other approaches and experimental observations is generated only in the unbroken phase of the system. Finally we develop the NH formulation in an SU(N) gauge field theoretic model by using the natural but unconventional Hermiticity properties of the ghost fields. Deconfinement to confinement phase transition has been realized as PT phase transition in such a non-hermitian model.

## 1. Introduction

The use of NH theories to address various problems in physics has a very long and rich history [1]. However no attempts were made to develop a consistent quantum theory with NH hamiltonians until towards the end of last century. It was shown that the NH systems which respect PT - symmetry can yield a complete real spectrum [2]. Consequently it has been shown that a fully consistent quantum theory with real spectrum, a complete set of orthonormal eigenfunctions with positive definite norms and unitary time evolution can be developed for certain class of NH systems in a modified Hilbert space [3, 4, 5, 6]. These exciting theoretical works towards the consistency of NH quantum mechanics (NHQM) strongly paved the way forward for NHQM to be the topic of frontier research in different areas in and interdisciplinary branches of physics during the last two decades [7, 8, 9, 10, 11, 12, 13, 14, 15, 16, 17, 18, 19, 20, 21, 22, 23, 24, 25, 26, 27, 28, 29, 30, 31, 32, 33, 34, 35, 36]. Due to the analogy of the Schrodinger equation with certain wave equations in optics, the phenomena of NHQM can also be mapped to the analogous phenomena in optics. This leads to the possibility of experimental observation of the theoretical predictions of NHQM. This has been indeed the case and some of the predictions of NHQM have been observed in optics [37, 38, 39, 40]. The realizations of NHQM phenomena have ignited huge interest to study the subject both theoretically and experimentally.

PT symmetric NH systems generally exhibit a phase transition[1] that separates two parametric regions (i) region of the unbroken PT symmetry in which the entire spectrum is real and eigenstates of the system respect PT symmetry and (ii) a region of broken PT symmetry in which the whole spectrum (or a part of it) appears as complex conjugate pairs and eigenstates of the Hamiltonian do not respect PT symmetry. Very rich physics is associated with this phase transition point. Even though most of the PT symmetric NH systems pass through PT phase transition, there are systems which do not show any such transition and remain either in broken or in unbroken phase for the entire parameter region. Many of the well known phase transitions have also been realized as PT phase transition in the NH version of those systems. This particular topic has been developed enormously over the past two decades [41, 42, 43, 44, 45, 46, 47, 48, 49, 50, 51, 52, 53, 54, 55, 56, 57, 58, 59, 60, 61, 62, 63, 64, 65]. It is impossible to attempt a review of all those in this work. In this article we intend to review some typical features of PT phase transition in non-relativistic as well as in relativistic quantum mechanics and in NH quantum field theoretical models. We mainly will focus on our own works [61, 62, 63, 64, 65] in these topics, which are solved analytically. In the first model in one dimensional quantum mechanics, we consider a system of a spin half particle in external magnetic field coupled to an oscillator through PT symmetric NH interaction.We found that, except the ground state, all other states occur in doublet due to the interaction of oscillator with two levels system. Such doublets form invariant subspace of the system.The energy eigenvalues and associated eigenfunctions corresponding to each doublet are solved analytically. Under certain conditions on the parameters ( frequency of the oscillator, strength of the magnetic field and strength of the NH coupling) we found that we have real energy eigenvalues and the corresponding eigenstates respect the PT symmetry in each invariant subspace. Thus we observe PT phase transition in each invariant subspace of the system in this exactly solvable model [61]. However PT phase transition is also observed in quasi exactly solvable (QES) models[62, 66, 67]. To demonstrate this we consider a couple of real QES systems and analyse them after making them NH in a PT symmetric way. We explicitly show the occurrence of PT phase transition even in QES systems through analytic calculations [62].

Even though numerous NH systems have been studied in 1-d QM, the study of NH system in higher dimensions is restricted to few models [68, 69, 70, 71, 72, 73, 63]. Therefore next we consider a couple of oscillator models in higher dimensional QM [63]. In the first model, we consider an anisotropic SHO with a simple PT symmetric NH interaction which can be solved analytically. Very interestingly we observe the system to undergo PT phase transition as long as the oscillator is anisotropic in $x$ and $y$ direction. The isotropic oscillator with the same NH interaction remains always in the broken phase and it is not possible to find any parameter space for the unbroken phase. Exactly similar conclusion is reached with the three dimensional model. In this case we start with an isotropic SHO in the external magnetic field. External magnetic field induces anisotropy in the system and again we have PT phase transition in the 3-d model. These two examples provide enough indications about some hidden connection between

[1] It is more appropriate to call this as "PT symmetry breaking transition" since "phase transition" is typically reserved for non-analyticity of free energy in classical statistical mechanics and applies to thermodynamic-limit systems.

the symmetry of the hermitian part of the hamiltonian and the PT phase transition.

(2+1) dimensional relativistic QM has become extremely important due to discovery of graphene , a single layer of carbon atoms arranged in a honeycomb lattice[74, 75, 76, 77]. To investigate PT phase transition in a relativistic model, we consider a massless Dirac particle in (2+1) dimension in the external transverse magnetic field in the background of Dirac oscillator (DO)[78, 79, 80, 81, 82, 83]. Further we make it NH by introducing a Rasbha type spin-orbit coupling with imaginary coupling constant. By using the various structure of parity and time reversal transformation in relativistic QM, we show that such model is PT symmetric and can be solved exactly with analytic calculations. The time reversal of this system represents the other valley of the massless particle. We show the existence of PT phase transition in both the valleys depending on the strength of the magnetic field and value of the spin orbit coupling. More interestingly a small mass gap which is consistent with other approaches is generated only in the unbroken phase of the system. Mass gap vanishes at the phase transition point. Details of this work is provided in Ref [64]

Finally we discuss NH theories in QFT. In particular we discuss how PT phase transition can occur in a model of free charged scalar with NH mass matrix. However it is more interesting to consider NH models in gauge theories. Using the natural but unconventional Hermiticity properties of ghost fields, SU(N) QCD is cast as NH model of QCD. A new quadratic gauge was recently introduced in Ref [84] to study Abelian projection. In this gauge SU(N) gauge theoretic model has two distinct phases, namely the normal or deconfinement phase and ghost condensed phase or confinement phase. The PT properties of this model have been investigated in both phases to realize that deconfiment to confinement phase transition is actually a PT phase transition [65]. Appropriate C operator for this theory has been constructed.

Now we indicate the plan of this present manuscript. In the next section, we discuss various forms parity and time reversal symmetry in one, two and three dimensional quantum theory, in (2+1) dimensional relativistic quantum mechanics and in quantum field theory. PT phase transition in 1-d exactly solvable as well as in QES systems are reviewed with explicit examples in Sec III. In section IV. we demonstrate the connection between symmetry and broken phases with the help of anisotropic oscillators in two and three dimensional quantum mechanics. PT phase transition in (2+1) dimensional relativistic system is considered Sec V. Section VI. is devoted to discuss phase transition in SU(N) QCD. Summary and conclusions are presented in Sec VII.

## 2. Parity and Time Reversal symmetries

In this section we will discuss the various form of parity and time reversal symmetry applicable in the different contexts.

### *2.1. PT transformations in 1-d QM*

Parity and time reversal transformations are defined in 1-d QM as

$$P: \quad x \to -x, \quad p \to -p; \quad T: \quad x \to x, \quad p \to -p \quad i \to -i \tag{1}$$

Here parity is simply space refection about the origin and time reversal is anti-linear operator. In general if we consider reflection about $x = a$, then parity transformation will be $x \to 2a - x, \;\; p \to -p.$. This general parity transformation will be used in Sec. III.B

### 2.2. PT transformations in higher dimensional QM

The parity transformation in 2-d, has typical characteristics. Since, parity transformation is an improper Lorentz transformation, in two dimensions it can not be written as $x' = -x, \ \ y' = -y$. The determinant of the transformation matrix for this is $+1$. The above transformation represents rotation about z-axis with rotation angle $\theta = \pi$ [85]. The discrete parity transformation in 2-d QM therefore can be defined in few alternative ways, as for example,

$$P_1 \ : \ x' = -x, \ \ y' = y; \ \ p_x' = -p_x, \ \ p_y' = p_y \tag{2}$$
$$P_2 \ : \ x' = x, \ \ y' = -y; \ \ p_x' = p_x, \ \ p_y' = -p_y. \tag{3}$$

Both of these forms $P_1$ and $P_2$ are equivalent and one can use either of these while checking $PT$ symmetry of a NH system in two dimensions. In 2-d parity transformation has also been realized as [86]

$$P_3 : x' = y, \ y' = x; \ \ p_x' = p_y, \ \ p_y' = p_x \tag{4}$$

In fact in the study of NH theories, one only requires $P$ to be an involution such that overall $PT$ becomes an anti-linear operator.

The structure of parity transformation in 3-d QM is much more rich. Various different versions of the anti-linear $PT$ transformation due to the rich structure of parity transformation in 3-d have been addressed in details in Ref. [87]. However in three or in any odd dimension commonly used parity transformation is given as space inversion.

$$x' = -x, \ \ y' = -y, \ \ z' = -z; \ \ p_x' = -p_x, \ \ p_y' = -p_y, \ \ p_z' = -p_z \tag{5}$$

In the even dimension, one has to be careful in defining parity transformation as the determinant of transformation matrix should be -1. We will using these transformation properties in Sec IV. Time reversal has same structure in higher dimension as in 1-d.

### 2.3. PT transformations in relativistic QM

Now we discuss the symmetry properties in relativistic QM. In (3+1) dimension the P and T which leave Dirac equation invariant, for the Dirac wave function are given by

$$P\psi(x,y,z,t) = \gamma_0\psi(-x,-y,-z,t); \quad T\psi(x,y,z,t) = -i\gamma_1\gamma_3\psi^*(x,y,z,-t) \tag{6}$$

where $\gamma_i \ \ i = 0,1,2,3$ are well known Dirac gamma matrices . Since the P transformation in 2-d is bit different ( as discussed in Sec II.B) the Dirac wavefunctions in (2+1) dimension transform as

$$P_1\psi(x,y,t) = \sigma_y\psi(-x,y,t); \quad P_2\psi(x,y,t) = \sigma_x\psi(x,-y,t). \tag{7}$$

Free Dirac equation remains invariant under such $P_1$ and $P_2$ transformations. The T transformation ($t \longrightarrow -t, \ \ i \longrightarrow -i, \ \ p_x \longrightarrow p_x, \ \ p_y \longrightarrow -p_y$) in (2+1) dimension Dirac theory is defined as $T = i\sigma_y\tilde{K}$, where $\tilde{K}$ is complex conjugation operation such that,

$$T\psi(x,y,t) = i\sigma_y\tilde{K}\psi(x,y,-t) = i\sigma_y\psi^\star(x,y,-t). \tag{8}$$

We will be using these in the later section to study PT phase transition in (2+1) relativistic QM.

### 2.4. PT transformations in QFT

Detail discussion in discrete symmetries for various quantum field theoretic systems can be found in standard text books [88]. In this section we list down only those equations which will be used in Section VI for the investigation of PT phase transition in QFT. For charged scalar field $\Phi = \begin{bmatrix}\phi_1\\ \phi_2\end{bmatrix}$, P and T respectively are defined on the doublet as follows

$$\Phi \xrightarrow{\text{P}} P_{2\times 2}\Phi; \quad \Phi \xrightarrow{\text{T}} T_{2\times 2}\Phi^* \tag{9}$$

From the analogy of the P transformation in 2-d, where $x \to x$ and $y \to -y$ we can say that the field $\phi_1$ transforms as a scalar and the other, $\phi_2$ transforms as a pseudo scalar. Thus the $P_{2\times 2}$ is

$$P_{2\times 2} = \begin{bmatrix}1 & 0\\ 0 & -1.\end{bmatrix} = \sigma_z \tag{10}$$

This implies the choice for the time reversal, $T_{2\times 2}$ such that the Lagrangian for free complex scalar field is PT-invariant. We must choose $T_{2\times 2} = \mathbf{1}_2$ [89, 91]. One can in principle swap the roles of $P_{2\times 2}$ and $T_{2\times 2}$ as long as symmetry is concern, however in order to interpret this PT-symmetric theory in terms of a coupled system with gain and loss, one should take T $= \mathbf{1}_2$ [89, 91].

Now we discuss these transformation properties of fields in SU(N) field theory as we will be focusing that for our NH study. Parity and time reversal properties of the gluons are well defined and is given by

$$A_i^a(\text{x}, t) \xrightarrow{\text{P}} -A_i^a(-\text{x}, t); \quad A_0^a(\text{x}, t) \xrightarrow{\text{P}} A_0^a(-\text{x}, t). \tag{11}$$

The rule for parity is same for all gluons as it is a linear operator. The SU(N) QCD Lagrangian is invariant under P if we choose ghosts to be scalars/pseudo scalars. However, we consider ghost fields transform as pseudo-scalar under P is more appropriate as ghosts are anti-commuting scalar

$$c^a(\text{x}, t) \xrightarrow{\text{P}} -c^a(-\text{x}, t); \quad \overline{c^a}(\text{x}, t) \xrightarrow{\text{P}} -\overline{c^a}(-\text{x}, t). \tag{12}$$

Since some of the generators of $SU(N)$ are purely imaginary, the time reversal (which is an anti-linear transformation) property is not same for all gluons. Suppose $a = \{q\}$ corresponds to purely imaginary generators and $a = \{p\}$ corresponds to real generators. Then for SU(3) $q = 2, 5, 7$ and $p = 1, 3, 4, 6, 8$. And T for SU(N) gluons is given by

$$\begin{aligned} A_i^p(\text{x}, t) &\xrightarrow{\text{T}} -A_i^p(\text{x}, -t); \quad A_0^p(\text{x}, t) \xrightarrow{\text{T}} A_0^p(\text{x}, -t), \\ A_i^q(\text{x}, t) &\xrightarrow{\text{T}} A_i^q(\text{x}, -t); \quad A_0^q(\text{x}, t) \xrightarrow{\text{T}} -A_0^q(\text{x}, -t), \end{aligned} \tag{13}$$

Thus the field strength with any spacetime and group indices can utmost change up to overall negative sign i.e., $F^a_{\mu\nu} \xrightarrow{\text{T}} \pm F^a_{\mu\nu}$. The time reversal property for ghosts is then defined in the following manner,

$$\begin{aligned} c^p(\text{x}, t) &\xrightarrow{\text{T}} ic^p(\text{x}, -t); \quad \overline{c^p}(\text{x}, t) \xrightarrow{\text{T}} i\overline{c^p}(\text{x}, -t) \\ c^q(\text{x}, t) &\xrightarrow{\text{T}} c^q(\text{x}, -t); \quad \overline{c^q}(\text{x}, t) \xrightarrow{\text{T}} \overline{c^q}(\text{x}, -t) \end{aligned} \tag{14}$$

such that $SU(N)$ QCD Lagrangian remains invariant under PT.

## 3. PT Phase transition in 1-d QM

In this section we will demonstrate PT phase transition in 1-d QM with the help of two simple models with exact analytical calculations. We chose one exactly solvable and another one QES models.

### *3.1. Non-Hermitian interaction of Spin-half particle with SHO*

A spin $\frac{1}{2}$ particle in the external magnetic field, $\vec{B}$ coupled to an oscillator through some NH interaction is described by the Hamiltonian

$$H = \mu\vec{\sigma}\cdot\vec{B} + \hbar\omega a^\dagger a + \rho(\sigma_+ a - \sigma_- a^\dagger). \tag{15}$$

. Where $\vec{\sigma}$'s are Pauli matrices, $\rho$ is some arbitrary real parameter and $\sigma_\pm \equiv \frac{1}{2}[\sigma_x \pm i\sigma_y]$ are spin projection operators. Usual creation and annihilation operators for the oscillator states $a, a^\dagger$ are defined as $a = \frac{p-im\omega x}{\sqrt{2m\omega\hbar}}, \quad a^\dagger = \frac{p+im\omega x}{\sqrt{2m\omega\hbar}}$ with $a|n>= \sqrt{n}|n-1>$, and $a^\dagger|n>= \sqrt{n+1}|n+1>$ where the notation $|n>$ for number eigenvectors for the oscillator has been adopted. We choose the external magnetic field in z-direction, $\vec{B} = B_0\hat{z}$ and the Hamiltonian for the system as given in Eq.(15) is reduced to ,

$$H = \frac{\epsilon}{2}\sigma_z + \hbar\omega a^\dagger a + \rho(\sigma_+ a - \sigma_- a^\dagger), \tag{16}$$

where $\epsilon = 2\mu B_0$. Note that this Hamiltinian is not hermitian as,

$$H^\dagger = \frac{\epsilon}{2}\sigma_z + \hbar\omega a^\dagger a - \rho(\sigma_+ a - \sigma_- a^\dagger), \quad \neq H \tag{17}$$

as $\sigma_\pm^\dagger = \sigma_\mp$. However this hamiltonian is invariant under PT transformation. To see the PT phase transition in this system we need to find the energy eigenvalues and corresponding eigenvectors of the system. We adopt the notation for the state as, $|n, \frac{1}{2}m_s>$ where $n$ is eigenvalue for the number operator $a^\dagger a$ i.e. $a^\dagger a|n>= n|n>$ and $m_s = \pm 1$ are the eigenvalues of the operator $\sigma_z$ i.e. $\sigma_z|\frac{1}{2}m_s>= m_s|\frac{1}{2}m_s>$. It is readily seen that $|0, -\frac{1}{2}>$ is a ground state of the Hamiltonian with eigenvalue $-\frac{\epsilon}{2}$ and it is non-degenerate.

$$H|0, -\frac{1}{2}>= -\frac{\epsilon}{2}|0, -\frac{1}{2}> . \tag{18}$$

We observe that the next possible states $|0, \frac{1}{2}>$ is not a eigenstate of the Hamiltonian,

$$H|0, \frac{1}{2}>= \frac{\epsilon}{2}|0, \frac{1}{2}> -\rho|1 - \frac{1}{2}> . \tag{19}$$

The spin projection operators $\sigma_\pm$ have the usual properties $\sigma_\pm|n, \pm\frac{1}{2}>= 0; \quad \sigma_\pm|n, \mp\frac{1}{2}>= |n, \pm\frac{1}{2}>$. However this state along with the state $|1, -\frac{1}{2}>$ close under the action of the Hamiltonian and form a invariant subspace in the space of states as,

$$H|1, -\frac{1}{2}>= (\hbar\omega - \frac{\epsilon}{2})|1, -\frac{1}{2}> +\rho|0, \frac{1}{2}> . \tag{20}$$

First two excited states belong to this sector spanned by these two states, $|0, \frac{1}{2}>$ and $|1, -\frac{1}{2}>$ wherein the Hamiltonian matrix is given by

$$H_1 = \begin{bmatrix} \frac{\epsilon}{2} & \rho \\ -\rho & -\frac{\epsilon}{2} + \hbar\omega \end{bmatrix}.$$

The eigenvalues of this Hamiltonian matrix are given by $\lambda_1^{I,II} = \frac{1}{2}\left[\hbar\omega \pm \sqrt{(\hbar\omega-\epsilon)^2 - 4\rho^2}\right]$. Note these eigenvalues are real provided $|\hbar\omega - \epsilon| \geq 2\rho$. Putting $2\rho = (\hbar\omega - \epsilon)\sin\theta_1$ we find the eigenvectors corresponding to this doublet are

$$
\begin{aligned}
|\Psi_1^I> &= \cos\frac{\theta_1}{2}|0,\frac{1}{2}> + \sin\frac{\theta_1}{2}|1,-\frac{1}{2}>, \quad \text{for } \lambda_1^I = \frac{\hbar\omega}{2}(1+\cos\theta_1) - \frac{\epsilon}{2}\cos\theta_1, \\
|\Psi_1^{II}> &= \sin\frac{\theta_1}{2}|0,\frac{1}{2}> + \cos\frac{\theta_1}{2}|1,-\frac{1}{2}>, \quad \text{for } \lambda_2^{II} = \frac{\hbar\omega}{2}(1-\cos\theta_1) + \frac{\epsilon}{2}\cos\theta_1.
\end{aligned}
$$

It may be observed that these two states are not orthogonal to each other nor do they have to be as $H \neq H^\dagger$. The system can be solved by considering the subsequent subspaces and the full spectrum is obtained. The general states of the sector spanned by $|n, \frac{1}{2}>$ and $|n+1, -\frac{1}{2}>$ wherein the Hamiltonian matrix is given by,

$$
H_{n+1} = \begin{bmatrix} \frac{\epsilon}{2} + n\hbar\omega & \rho\sqrt{n+1} \\ -\rho\sqrt{n+1} & -\frac{\epsilon}{2} + (n+1)\hbar\omega \end{bmatrix}.
$$

Now we have the eigenvalues of this Hamiltonian matrix are given by

$$
\lambda_{n+1}^{I,II} = \frac{1}{2}\left[(2n+1)\hbar\omega \pm \sqrt{(\hbar\omega-\epsilon)^2 - 4\rho^2(n+1)}\right]. \tag{21}
$$

These eigenvalues are real provided $|\hbar\omega - \epsilon| \geq 2\rho\sqrt{n+1}$. Now putting $2\rho\sqrt{n+1} = (\hbar\omega - \epsilon)\sin\theta_{n+1}$, we find the eigenvectors corresponding to this doublet are

$$
\begin{aligned}
|\Psi_{n+1}^I> &= \cos\frac{\theta_{n+1}}{2}|n,\frac{1}{2}> + \sin\frac{\theta_{n+1}}{2}|n+1,-\frac{1}{2}>, \\
&\quad \text{for } \lambda_{n+1}^I = \frac{\hbar\omega}{2}(2n+1+\cos\theta_{n+1}) - \frac{\epsilon}{2}\cos\theta_{n+1}, \\
|\Psi_{n+1}^{II}> &= \sin\frac{\theta_{n+1}}{2}|n,\frac{1}{2}> + \cos\frac{\theta_{n+1}}{2}|n+1,-\frac{1}{2}>, \\
&\quad \text{for } \lambda_{n+1}^{II} = \frac{\hbar\omega}{2}(2n+1-\cos\theta_{n+1}) + \frac{\epsilon}{2}\cos\theta_{n+1}.
\end{aligned} \tag{22}
$$

We observe that these eigenstates respect PT symmetry $PT|\Psi_{n+1}^{I,II}>= (-1)^n|\Psi_{n+1}^{I,II}>$ as $P|n, \pm\frac{1}{2}>= (-1)^n|n, \pm\frac{1}{2}>$ and $\sigma_z|n, \pm\frac{1}{2}>= \pm|n, \pm\frac{1}{2}>$. Thus we have real eigenvalues when the symmetry is not broken. In the regime $|\hbar\omega - \epsilon| < 2\rho\sqrt{n+1}$ the eigenvalues for a particular doublet become complex conjugate. For $n+1$ th doublet, the complex eigenvalues are $\frac{1}{2}\left[\hbar\omega(2n+1) + \frac{i}{2}\sqrt{4\rho^2(n+1) - (\hbar\omega-\epsilon)^2}\right]$.

### 3.2. *PT Phase transition in 1-d QES system*

There are only limited number of systems in QM which can be solved exactly. However there are Hamiltonians which preserve the finite dimensional Hilbert space and the spectrums for those systems are calculated exactly for a finite part in a closed form. Such systems are known as QES[95, 96, 97] . We consider the QES system described by the NH but PT-invariant Hamiltonian

$$
H = p^2 - (\zeta\cosh 2x - iM)^2, \tag{23}
$$

where the parameter $\zeta$ is real, parameter $M$ has only integer values and $\hbar = 2m = 1$. This NH system is PT symmetric. Here we consider parity as $x \to i\pi/2 - x,\ p \to -p$. Like corresponding Hermitian model ( removing $i$ from the Hamiltonian) , this NH system is also shown to be QES system. The QES levels for this systems are calculated using Bender and Dunne [92] method for different integer values of $M$. For positive integer values of $M(= J$, say), first $J$ levels for this systems can be calculated exactly in the closed form. We observed very different scenarios for the even and odd values of $J$. Hence the cases for even and odd M are presented separately.

**Case-I** $M = 2$ **:** The QES eigenvalues and corresponding eigenvalues are given as

$$\begin{aligned} E_\pm &= 3 - \zeta^2 \pm 2i\zeta\,. \\ \psi_+ = e^{i\frac{\zeta}{2}\cosh 2x}\cosh x; &\qquad \psi_-(x) = e^{i\frac{\zeta}{2}\cosh 2x}\sinh x \end{aligned} \tag{24}$$

QES eigenvalues are in complex conjugate pair and can never be real for non-zero $\zeta$ and PT symmetry is broken spontaneously as the QES eigenfunctions are not eigenstate of PT. Because under the PT transformation as defined above, $\cosh x \to -i\sinh x, \sinh x \to i\cosh x$ while $i\cosh 2x$ remains invariant so that the two wave functions are not invariant under PT.

**Case-II:** $M = 4$**:** The four QES eigenvalues are

$$E^1_\pm = 11 - \zeta^2 - 2i\zeta \pm 4\sqrt{1 - i\zeta - \zeta^2};\;\; E^2_\pm = 11 - \zeta^2 + 2i\zeta \pm 4\sqrt{1 + i\zeta - \zeta^2}\,. \tag{25}$$

and the corresponding eigenfunctions are

$$\begin{aligned} \psi^1_\pm &= e^{i\frac{\zeta}{2}\cosh 2x}\phi(x)(A\sinh 3x + B\sinh x)\,,\; \frac{B}{A} = \frac{E - 7 + \zeta^2}{2i\zeta}; \\ \psi^2_\pm &= e^{i\frac{\zeta}{2}\cosh 2x}\phi(x)(A\cosh 3x + B\cosh x)\,,\; \frac{B}{A} = \frac{E - 7 + \zeta^2}{2i\zeta}\,, \end{aligned} \tag{26}$$

It is easy to check that PT symmetry is broken spontaneously. This is always true for any value of even $M$. However on the other hand, if $M$ is an odd integer, then the eigenvalues are real provided $\zeta^2 \le \zeta_c^2$ and PT symmetry is respected by the eigenstates .

**Case-III M=1:** The QES level is at $E = 1 - \zeta^2$ with the eigenfunction $\psi(x) = e^{i\frac{\zeta}{2}\cosh 2x}$. Here $PT\psi = \psi$ and hence PT symmetry remains unbroken for any value of $\zeta$.This is expected as there is no possibility to have a complex conjugate pair of eigenvalues for $M = 1$.

**Case IV:** $M = 3$**:** The three QES eigenvalues and eigenfunctions are

$$\begin{aligned} E = 5 - \zeta^2 &\qquad E_\pm = 7 - \zeta^2 \pm 2\sqrt{1 - 4\zeta^2} \\ \psi = e^{i\frac{\zeta}{2}\cosh 2x}\sinh 2x &\qquad \psi_\pm = e^{i\frac{\zeta}{2}\cosh 2x}(A\cosh 2x + iB) \end{aligned} \tag{27}$$

where $\frac{B}{A} = \frac{4\zeta}{E - 9 + \zeta^2}$. These QES levels are real if $\zeta^2 \le \zeta_c^2 = 1/4$ and at $\zeta = \zeta_c$, the two highest QES levels are degenerate. It is easily checked that the corresponding wavefunction $\psi$ is indeed an eigenfunctions of PT with eigenvalue 1 while the other two wave functions ($\psi_\pm$) are eigenfunctions with eigenvalue -1. Thus the PT symmetry remains unbroken. It has been checked for higher odd vales of $M$ numerically and same result is true. Thus we have PT phase transition in this QES system for odd integers of $M$. For even $M$ the QES system always in broken phase.

Further, using the anti-isospectral transformation QES eigenvalues and eigenfunctions of the PT-invariant potential, $V(\theta) = (\zeta\cos 2\theta - iM)^2$ are related to those of the $(\zeta\cosh 2x - iM)$

by $\bar{E}_k = -E_{M-1-k}$, $\bar{\psi}_k(\theta) = \psi_{M-1-k}(ix)$. In particular, the QES eigenvalues of this periodic potential are real if M is an odd integer and $\zeta \leq \zeta_c$ and further in this case QES eigenstates are also eigenstate of PT. Therefore like the earlier case we have PT phase transition in tis NH QES case. On the other hand, when M is an even integer, then the eigenvalues are complex conjugate pairs and PT symmetry is spontaneously broken.The system remains in broken phase always.

There are many examples of PT phase transition in 1-d quantum mechanics but now we will focus on some models in two, three dimension to get some new features of PT phase transition.

## 4. PT Phase transition in higher dimension

In this section we consider two examples, one in two dimensions and other one in three dimensions to demonstrate the PT phase transition in higher dimension. Few other works in PT phase transition in higher dimension can also be found in[68, 69, 70, 71, 72, 73]

### *4.1. Role of anisotropy in PT Phase transition in 2-d*

We consider an PT symmetric NH anisotropic oscillator in 2-d described by the Hamiltonian,

$$H = \frac{p_x^2}{2m} + \frac{p_y^2}{2m} + \frac{1}{2}m\omega_x^2 x^2 + \frac{1}{2}m\omega_y^2 y^2 + i\lambda xy, \; = H_0 + H_{nh} \tag{28}$$

$\lambda$ is real and $\omega_x \neq \omega_y$. The NH interaction term is invariant under $P_1T$, $P_2T$ and $P_3T$. However we will not use $P_3T$ as $H_0$ is not invariant under it. $P_1T(i\lambda xy) = P_2T(i\lambda xy) = i\lambda xy$

The energy eigenvalues and eigenfunctions for this system are written as

$$\begin{aligned}
E_{n_1,n_2} &= (n_1+\frac{1}{2})\hbar C_1 + (n_2+\frac{1}{2})\hbar C_2, \;\; C_1^2 = \frac{1}{2}\left[\omega_+^2 - \frac{\omega_-^2}{k}\right]; \;\; C_2^2 = \frac{1}{2}\left[\omega_+^2 + \frac{\omega_-^2}{k}\right]. \\
\psi_{n_1,n_2}(x,y) &= N\exp\left[-\frac{m}{2\hbar}[(C_1+C_2)(x^2+y^2) + (C_2-C_1)2i\lambda kxy\right] \\
& H_{n_1}[\alpha_1(\sqrt{\frac{k+1}{2}}x - i\sqrt{\frac{k-1}{2}}y)]H_{n_2}[\alpha_2(\sqrt{\frac{k-1}{2}}x + i\sqrt{\frac{k+1}{2}}y)]
\end{aligned} \tag{29}$$

$\alpha_1^2 = \frac{mC_1}{\hbar}$ and $\alpha_2^2 = \frac{mC_2}{\hbar}$. And where, $\omega_\pm^2 \equiv \omega_x^2 \pm \omega_y^2$; $\quad k^{-1} = \sqrt{1 - \frac{4\lambda^2}{m^2\omega_-^4}}$ ;

Now we consider the case when $k$ is real i.e. $|\lambda| \leq |\frac{m\omega_-^2}{2}|$. In this case,

$$C_1^2 = \frac{1}{2}[\omega_x^2(1+\frac{1}{k}) + \omega_y^2(1-\frac{1}{k})] > 0, \qquad C_2^2 = \frac{1}{2}[\omega_x^2(1-\frac{1}{k}) + \omega_y^2(1+\frac{1}{k})] > 0, \tag{30}$$

as $k \geq 1$. This further leads to entire real spectrum. Further the eigenfunctions respect both $P_1T$ and $P_2T$ i.e. $P_iT\psi_{n_1,n_2}(x,y) = (-1)^{s_i}\psi_{n_1,n_2}(x,y)$; $\;\; i = 1,2. \;\; s_1 = n_1+n_2, \;\; s_2 = 1.$ Therefore $PT$ symmetry is unbroken as long as $|\lambda| \leq |\frac{m\omega_-^2}{2}|$. On the other hand for $|\lambda| > |\frac{m\omega_-^2}{2}|$, $k$ is imaginary and hence $P_iT\psi_{n_1,n_2}(x,y) \neq \pm\psi_{n_1,n_2}(x,y)$ for $i = 1,2$ and the spectrum is no longer real. Some of the eigenvalues occur in complex conjugate pairs, the spectrum in this situation is written as

$$E_{n_1,n_2} = (n_1+\frac{1}{2})\hbar(A - iB) + (n_2+\frac{1}{2})\hbar(A+iB) \tag{31}$$

Where A and B are real. It is clear from the above equation that $E_{n_1,n_2}$ and $E_{n_2,n_1}$ are complex conjugate to each other for $n_1 \neq n_2$ and energy eigenvalue are real when $n_1 = n_2$. The critical value of the coupling $\lambda_c(=\frac{m\omega^2}{2})$ depends on the anisotropic of the system. If the system is more anisotropic the span of the $PT$ unbroken phase is longer. When the system becomes isotropic, i.e. $\omega_x = \omega_y$, i.e. $\lambda_c = 0$, the system will always lie in the broken $PT$ phase and it will not be possible to have entire spectrum real for any condition on the parameters. This result leads to a very important realization for this particular systems. Rotational symmetry which leads to isotropy of original Hermitian system $H_0$ is complementary to unbroken $PT$ symmetry of the $PT$ symmetric NH system $H$. As long as the original system $H_0$ has rotational invariance, $H$ can not have unbroken $PT$ symmetry. The moment rotational symmetry of $H_0$ breaks the NH system becomes capable of going through a $PT$ phase transition.The role of degeneracies in the Hermitian part of the PT Hamiltonian in determining a nonzero threshold has been extensively studied in Refs [70, 71].

### *4.2. Three-dimensional isotropic harmonic oscillator in an external imaginary magnetic field*

We consider an isotropic simple harmonic oscillator in external imaginary magnetic field ($iB$) with $i\mu_l.B$ coupling in 3-d as

$$H = \frac{1}{2m}(\vec{p} - \frac{iq\vec{A}}{c})^2 + \frac{1}{2}mw^2(x^2+y^2+z^2) + i\vec{\mu}_l.\vec{B} \tag{32}$$

where $q$ is the charge of the oscillator. The vector potential in symmetric gauge for a uniform magnetic field in $z$ direction is written as $\vec{A} = \{-\frac{By}{2}, \frac{Bx}{2}, 0\}$ . Then the above Hamiltonian is reduced to.

$$H = \frac{1}{2m}(p_x + \frac{iqyB}{2c})^2 + \frac{1}{2m}(p_y - \frac{iqxB}{2c})^2 + \frac{1}{2m}(p_z)^2 + \frac{1}{2}m\omega^2(x^2+y^2+z^2) + i\mu_{lz}B \tag{33}$$

This $PT$ invariant Hamiltonian further simplified to

$$H = \frac{p_x^2}{2m} + \frac{p_y^2}{2m} + \frac{p_z^2}{2m} + \frac{1}{2}m\omega_1^2(x^2+y^2) + \frac{1}{2}m\omega^2 z^2 \tag{34}$$

with $\omega_1^2 = \omega^2 - \frac{q^2B^2}{4m^2c^2} = \omega^2 - \frac{\omega_c^2}{4}$ $\omega_c = \frac{qB}{mc}$ is usual cyclotron frequency. Anisotropy is induced in the system due to external magnetic field in some specific direction. The energy eigenvalue and eigenfunction for this systems are calculated as,

$$E_{n_xn_yn_z} = (n_x+n_y+1)\hbar\omega_1 + (n_z+\frac{1}{2})\hbar\omega \tag{35}$$

$$\psi_{n_xn_yn_z}(x,y,z) = \exp\left[-\frac{\alpha^2}{2}(x^2+y^2) + \frac{\alpha_1^2 z^2}{2}\right] H_{n_x}(\alpha x)H_{n_y}(\alpha y)H_{n_z}(\alpha_1 z) \tag{36}$$

where, $\alpha^2 = \frac{m\omega_1}{\hbar}$, $\alpha_1^2 = \frac{m\omega}{\hbar}$. Now we will show that this system undergoes a PT phase transition, if $B \leq \frac{2m\omega c}{q}$ for fixed $\omega$ or for a fixed magnetic field $\omega \geq \frac{\omega_c}{2}$, then $\omega_1$ is real and hence the entire spectrum is real and $PT\psi_{n_xn_yn_z}(x,y,z) - (-1)^{n_x+n_y+n_z}\psi_{n_xn_yn_z}(x,y,z)$, indicating the system is in unbroken $PT$ phase. Alternatively if $B > B_c = \frac{2m\omega c}{q}$ or $\omega \leq \frac{\omega_c}{2}$, then $\omega_1 = \pm\sqrt{\omega^2 - \frac{q^2B^2}{4m^2c^2}}$, becomes complex, we have pairs of complex conjugate eigenvalues given as $E_{n_xn_yn_z} = \pm i(n_x+n_y+1)\hbar\tilde{\omega} + (n_z+\frac{1}{2})\hbar\omega$ and system is in broken phase of $PT$. This means if we gradually change the strength of the external magnetic field for fixed oscillator frequency or if we change the oscillator frequency,$\omega$ for fixed magnetic field, the system undergoes a PT phase transition. The details on this can be found in Ref. [63].

## 5. PT Phase transition in Relativistic QM

In the previous sections we have considered several examples to show the PT phase transitions in non-relativistic quantum mechanics in one and higher dimensions. Now we are going to consider PT symmetric NH systems in relativistic quantum mechanics to highlight some features of PT phase transition.

### 5.1. *The Model*

Let us consider a massless Dirac particle in $(2+1)d$ with imaginary spin-orbit interaction in the transverse uniform magnetic field in the background of DO potential as described by the Hamiltonian

$$H = v_f[\vec{\sigma}.(\vec{p} + \frac{e\vec{A}}{c} - iK_1\vec{r}\beta) + i\lambda(\vec{\sigma}\times\vec{\Pi}).\hat{z}, \tag{37}$$

where $K_1$, $\lambda$ are real constants and $v_f$ is Fermi velocity. The term linear in $\vec{r}$ is DO potential [79] as in the non-relativistic limit, it reduces to SHO potential with strong spin-orbit coupling. Discussions on DO have been found in [80, 81, 82, 83].We assume the magnetic field along z-direction, $B = B_0\hat{k}$ for simplicity and choose the vector potential in the symmetric gauge as $\vec{A} = (-\frac{B_0 y}{2}, \frac{B_0 x}{2}, 0)$. The Hamiltonian in component form is written as

$$H = v_f(\sigma_x\Pi_x + \sigma_y\Pi_y) - iK_1v_f(\sigma_x x + \sigma_y y)\beta + i\lambda(\sigma_x\Pi_y - \sigma_y\Pi_x) \tag{38}$$

where, $\Pi_x = p_x - \frac{eB_0 y}{2c}$ and $\Pi_y = p_y + \frac{eB_0 x}{2c}$. It is straight forward to check that

$$H^\dagger = v_f(\sigma_x\Pi_x + \sigma_y\Pi_y) - iK_1v_f(\sigma_x x + \sigma_y y)\beta - i\lambda(\sigma_x\Pi_y - \sigma_y\Pi_x) \neq H$$

This system, however is invariant under both the $P_iT$ as $P_iTH(P_iT)^{-1} = H, \quad i = 1, 2$. The graphene has two distinct valleys in its electronic structure in which the electrons have the same energy. Similar to that, the time reversal of this Hamiltonian for massless particle defines the system to other valley and is given by,

$$\tilde{H} = THT^{-1} = v_f(\sigma_x\tilde{\Pi}_x + \sigma_y\tilde{\Pi}_y) - iK_1v_f(\sigma_x x + \sigma_y y)\beta - i\lambda(\sigma_x\tilde{\Pi}_y - \sigma_y\tilde{\Pi}_x) \tag{39}$$

with $\tilde{\Pi}_x = p_x + \frac{eB_0 y}{2c}$ and $\tilde{\Pi}_y = p_y - \frac{eB_0 x}{2c}$. Further note $\tilde{H} \neq \tilde{H}^\dagger$ and $[\tilde{H}, P_1T] = 0 = [\tilde{H}, P_2T]$. Now we proceed to show the PT phase transition and mass gap generation as a consequence of PT phase transition in both valleys of the system..

### 5.2. *PT phase transition and generation of mass gap*

In this section we analytically solve the Dirac equations corresponding to H and $\tilde{H}$ for the massless particle explicitly to obtain energy levels and eigenfunctions. This will help us to demonstrate PT phase transition in (2+1) dimensional in this relativistic system. We analytically showed that mass gap which is consistent with experimental finding is generated due to imaginary Rashba interaction in the background of DO potential as long as the system is in the unbroken phase. We further show how the different solutions of Dirac equation for both the valleys in this particular model are interrelated. For this purpose we write the Hamiltonians in Eqs.(38) and (39) in a compact form on a complex plane $z = (x + iy)$ as,

$$H = \begin{pmatrix} 0 & A\Pi_z + iC_1\bar{z} \\ B\Pi_{\bar{z}} + iC_2 z & 0 \end{pmatrix}, \quad \tilde{H} = \begin{pmatrix} 0 & B\Pi_z + iC_2\bar{z} \\ A\Pi_{\bar{z}} + iC_1 z & 0 \end{pmatrix} \tag{40}$$

with, $A = 2(v_f - \lambda)$, $B = 2(v_f + \lambda)$, $C_1 = K_1 v_f - (v_f - \lambda)\frac{B_0 e}{2c}$ and $C_2 = -K_1 v_f + (v_f + \lambda)\frac{B_0 e}{2c}$ are constants and canonical conjugate momenta in complex plane $\Pi_z$ and $\Pi_{\bar{z}}$ are defined as $\Pi_z = -i\hbar\frac{d}{dz} = \frac{1}{2}(p_x - ip_y)$, $\Pi_{\bar{z}} = -i\hbar\frac{d}{d\bar{z}} = \frac{1}{2}(p_x + ip_y)$ . It can be verified in a straight forward manner as

$$[\bar{z}, \Pi_{\bar{z}}] = i\hbar; \quad [z, \Pi_{\bar{z}}] = 0; \quad [\bar{z}, \Pi_z] = 0; \quad [\Pi_{\bar{z}}, \Pi_z] = 0; \quad [z, \Pi_z] = i\hbar \tag{41}$$

### *5.3. Solution for the system $H$ and $\tilde{H}$*

We present the solutions for DE corresponding to the Hamiltonian $H$ only and solution for $\tilde{H}$ can be written using symmetry properties. For this purpose we assume a two component trial solution as

$$\psi = \begin{pmatrix} \xi(z,\bar{z})e^{d_1 z\bar{z}} \\ i\eta(z,\bar{z})e^{d_1 z\bar{z}} \end{pmatrix} \tag{42}$$

The solutions are given as,

$$E_n^2 = (n+1)(AC_2 - BC_1) \tag{43}$$

$$\psi_n^I(z,\bar{z},t) = a_n^I \begin{pmatrix} z^n \\ iz^{n+1} \end{pmatrix} e^{\frac{C_1}{A\hbar}z\bar{z}} e^{-\frac{i}{\hbar}E_n t} \text{ for } d_1 = \frac{C_1}{A\hbar} \tag{44}$$

$$\text{and } E'^2{}_n = (n+1)(BC_1 - AC_2) = -E_n^2 \tag{45}$$

$$\psi_n^{II}(z,\bar{z},t) = a_n^{II} \begin{pmatrix} \bar{z}^{n+1} \\ i\bar{z}^n \end{pmatrix} e^{\frac{C_2}{B\hbar}z\bar{z}} e^{-\frac{i}{\hbar}E'_n t} \text{ for } d_1 = \frac{C_2}{B\hbar} \tag{46}$$

We observe the following parameter symmetry to obtain $\tilde{H}$ from $H$

$$H \xrightarrow[C_1 \leftrightarrow C_2]{A \leftrightarrow B} \tilde{H}$$

Interchanging $A \leftrightarrow B, C_1 \leftrightarrow C_2$ is equivalent to $\lambda \to -\lambda, K_1 \to -K_1$ and $B_0 \to -B_0$. Using these changes of parameters in the solutions for Hamiltonian H. the solutions for the time reversal system, $\tilde{H}$ are obtained

$$E_n^2 = (n+1)(AC_2 - BC_1) \tag{47}$$

$$\tilde{\psi}_n^I(z,\bar{z},t) = \tilde{a}_n^I \begin{pmatrix} \bar{z}^n \\ i\bar{z}^{n+1} \end{pmatrix} e^{\frac{C_1}{A\hbar}z\bar{z}} e^{-\frac{i}{\hbar}E_n t} \text{ for } \tilde{d}_1 = \frac{C_1}{A\hbar} \tag{48}$$

$$\text{and } E'^2{}_n = (n+1)(BC_1 - AC_2) = -E_n^2 \tag{49}$$

$$\tilde{\psi}_n^{II}(z,\bar{z},t) = a_n^{II} \begin{pmatrix} z^{n+1} \\ iz^n \end{pmatrix} e^{\frac{C_2}{B\hbar}z\bar{z}} e^{-\frac{i}{\hbar}E'_n t} \text{ for } \tilde{d}_1 = \frac{C_2}{B\hbar} \tag{50}$$

Now will proceed to discuss PT phase transition in the relativistic system with these solutions.

### *5.4. PT phase transition*

Putting the values of the constant we see that the energy eigenvalues for the solutions corresponding to $d_1 = \tilde{d}_1 = \frac{C_1}{A\hbar}$ in both the valleys are,

$$E_n = \pm\sqrt{(n+1)\left[2(v_f^2 - \lambda^2)\frac{B_0 e\hbar}{c} - 4K_1 v_f^2\hbar\right]} \tag{51}$$

The mass gap between positive and negative energy solutions in the system is given by

$$\Delta_0 = \sqrt{2(v_f^2 - \lambda^2)\frac{B_0 e\hbar}{c} - 4K_1 v_f^2\hbar} \tag{52}$$

The condition for real spectrum

$$\text{(i) } \lambda^2 \le v_f^2\left(1 - \frac{2K_1 c}{B_0 e}\right) \equiv \lambda_c^2 \quad \text{OR}$$
$$\text{(ii) } B_0 > 2K_1\frac{v_f^2 c}{(v_f^2 - \lambda^2)e} \equiv B_0^c \tag{53}$$

And under this condition, $\psi_n^I$ and $\tilde{\psi}_n^I$ respect af both $P_1T$ and $P_2T$ symmetry as $P_jT\psi_n^I(z,\bar{z},t) = -i(-1)^n\psi_n^I(z,\bar{z},t)$; and $P_jT\tilde{\psi}_n^I(z,\bar{z},t) = -i(-1)^n\tilde{\psi}_n^I(z,\bar{z},t)$; $j = 1,2$.

Now we have few interesting comments regarding the phase transition of this system.

- Under the condition in ( 53 ) $d_1 = \tilde{d}_1 = \frac{C_1}{A\hbar}$ system corresponding to both the valleys are in unbroken PT phase and solutions for $d_1 = \tilde{d}_1 = \frac{C_2}{B\hbar}$) correspond to the broken PT phase as $P_iT\psi_n^{II} \neq a_i\psi_n^{II}$, $P_iT\tilde{\psi}_n^{II} \neq b_i\tilde{\psi}_n^{II}$ for $i = 1,2$
- Reverse things happen, when the condition in Eq. 53 is not satisfied i.e. $\lambda > \lambda_c$, for fixed $B_0$ or $B_0 < B_0^c$ for fixed $\lambda$. That is the solutions for $d_1 = \tilde{d}_1 = \frac{C_2}{B\hbar}$ correspond to the unbroken PT phase and the $d_1 = \tilde{d}_1 = \frac{C_1}{A\hbar}$ will be in broken PT phase.
- Thus the system in both the valleys undergo PT phase transition for either of the values $d_1 = \tilde{d}_1 = \frac{C_1}{A\hbar}$ or $d_1 = \tilde{d}_1 = \frac{C_2}{B\hbar}$
- For $d_1 = \tilde{d}_1 = \frac{C_1}{A\hbar}$, when $\lambda < \lambda_c$ or $B_0 > B_0^c$ we see from Eq.(52) mass gap is always positive and becomes imaginary when $\lambda > \lambda_c$ or $B_0 < B_0^c$. Thus a mass gap is generated as long as system is unbroken PT phase and given by the expression in Eq. (52). Further in unbroken phase $\Delta_0$ varies as $\sqrt{B_0}$ for large enough magnetic field. This result is consistent with other approaches.
- For $d_1 = \tilde{d}_1 = \frac{C_2}{B\hbar}$ when $\lambda > \lambda_c$ or $B_0 < B_0^c$ the system is in unbroken phase and again real mass gap is generated which is consistent with other approaches. Thus mass gap is always generated in this theory which is consist with other approaches.

**Special cases** :

(i) I ($K_1 = 0$), i.e. No DO potential, the system passes to broken phase when strength of Rashba interaction exceeds Fermi velocity $\lambda^2 > v_f^2$. The mass gap in the unbroken PT phase is given by $\Delta_0 = \sqrt{2(v_f^2 - \lambda^2)\frac{B_0 e\hbar}{c}}$ which is real as $\lambda^2 < v_f^2$ for the unbroken phase. Interestingly the mass gap vanishes at the transition point $\lambda^2 = v_f^2$.

(ii) In the absence of spin-orbit coupling ($\lambda = 0$), the system is Hermitian, we can have real energy eigenvalues as long as the magnetic field $B_0 > \frac{2K_1 c}{e}$, the mass gap in this case is $\Delta_0 = \sqrt{2v_f^2\hbar(\frac{B_0 e}{c} - 2K_1)}$ which is always real for $B_0 > \frac{2K_1 c}{e}$. In both the cases for sufficiently large magnetic field $\Delta_0 \propto \sqrt{B_0}$, which is consistent with other approaches. In this case also mass gap vanishes at the transition point, $B_0 = \frac{2K_1 c}{e}$. This critical value of magnetic field is same as obtained in [82]. The factor of 2 is due to the choice of vector potential in the symmetric gauge.

## 6. PT Phase transition in gauge theories

Before we go into the discussion of NH SU(N) gauge field theoretic model, we want to point out some technique, which is equally applicable for the non-Hermitian scalar field theoretic models.

### 6.1. *Non-Hermitian complex scalars*

We consider the following simple complex scalar NH model discussed in Refs. [89, 91].

$$L = \partial_\mu\phi_1^*\partial_\mu\phi_1 + \partial_\mu\phi_2^*\partial_\mu\phi_2 + [\phi_1^* \ \ \phi_2^*]\begin{bmatrix} m_1^2 & \mu^2 \\ -\mu^2 & m_2^2. \end{bmatrix}\begin{bmatrix} \phi_1 \\ \phi_2 \end{bmatrix} \tag{54}$$

Here $m_1^2, m_2^2, \mu^2 \geq 0$ and the mass term in the above Lagrangian density is not Hermitian. However considering the PT transformations for field theoretic models as discussed in Sec. II it is straightforward to check that the Lagrangian in Eq. 54 is PT symmetric. The eigenvalues of the mass matrix are given as,

$$\Lambda_\pm^2 = \frac{1}{2}(m_1^2 + m_2^2) \pm \sqrt{(m_1^2 - m_2^2)^2 - 4\mu^4} \tag{55}$$

For $|m_1^2 - m_2^2| \geq 2\mu^2$, we are in the phase of unbroken PT symmetry and when $|m_1^2 - m_2^2| < 2\mu^2$ happens, we step into the region of broken PT-symmetry as eigenvalues turn complex and $\mathrm{PT}\psi_\pm = \pm\psi_\pm$ is no longer valid, where $\psi_\pm$ are eigenfunctions of the mass matrix corresponding to eigenvalues $\Lambda_\pm^2$. We shall encounter similar NH mass matrix for gluons in our model with SU(N) gauge theories.. The hidden C-symmetry can be associated with the fact, that eigenvalues in Eq. (55) do not change under $\mu^2 \to -\mu^2$. The charge conjugation is defined as follows $\Phi \xrightarrow{\mathrm{C}} C\Phi^*$ and in this model C=P [89, 91]. The theory in the region $|m_1^2 - m_2^2| < 2\mu^2$ violates CP also but preserves CT symmetry. Such charge conjugation symmetry exist in the non Abelian model also.

### 6.2. *SU(N) QCD in the quadratic gauge*

The resulting effective Lagrangian density with a new quadratic gauge fixing $F^a[A^\mu(x)] = A_\mu^a(x)A^{\mu a}(x)$ , [84] for each $a$ is given as,

$$\mathcal{L}_Q = -\frac{1}{4}F_{\mu\nu}^a F^{\mu\nu a} - \frac{1}{2\zeta}(A_\mu^a A^{\mu a})^2 - 2\overline{c^a}A^{\mu a}(D_\mu c)^a \tag{56}$$

. $\zeta$ is an arbitrary gauge fixing parameter, the field strength $F_{\mu\nu}^a = \partial_\mu A_\nu^a(x) - \partial_\nu A_\mu^a(x) - gf^{abc}A_\mu^b(x)A_\nu^c(x)$ and $(D_\mu c)^a = \partial_\mu c^a - gf^{abc}A_\mu^b c^c$. The summation over an index $a$ is understood when it appears repeatedly, including when occurred thrice in the ghost term. The resulting Lagrangian is BRST invariant which is essential for the ghost independence of the green functions and unitarity of the $S$-matrix. This quadratic gauge is very useful to discuss the abelian projection of SU(N) gauge theory.

### *6.3. Two phases of the theory*

This theory has two different phases (i) the normal or deconfined phase described by the Lagrangian in Eq. (56) and (ii) the ghost condensed phase showing the confinement. In this ghost fields are auxiliary fields as they do not have kinetic terms. However play an important role in the IR regime. To demonstrate the significance of ghosts we expand the ghost Lagrangian, $-\overline{c^a}A^{\mu a}(D_\mu c)^a = -\overline{c^a}A^{\mu a}\partial_\mu c^a + gf^{abc}\overline{c^a}c^c A^{\mu a}A^b_\mu$ where the summation over indices $a$, $b$ and $c$ each runs independently over 1 to $N^2-1$. Now if the ghosts freeze they amount to a non-zero mass matrix for the gluons as follows

$$(M^2)^{ab}_{\text{dyn}} = 2g\sum_{c=1}^{N^2-1} f^{abc}\langle\overline{c^a}c^c\rangle. \tag{57}$$

We would get masses of gluons by diagonalizing the matrix and finding its eigenvalues. The ghost condensation as a concept was introduced in Ref. [93]. In an $SU(N)$ symmetric state, where all ghost-anti-ghost condensates are identical

$$\langle\overline{c^1}c^1\rangle = ... = \langle\overline{c^1}c^{N^2-1}\rangle = ... = \langle\overline{c^{N^2-1}}c^1\rangle = ... = \langle\overline{c^{N^2-1}}c^{N^2-1}\rangle \equiv K. \tag{58}$$

The mass matrix becomes

$$(M^2)^{ab}_{\text{dyn}} = 2gK\sum_{c=1}^{N^2-1} f^{abc} \tag{59}$$

which is an antisymmetric matrix i.e., non-Hermitian, $(M^2)^\dagger \neq M^2$ due to the antisymmetry of the structure constants.

The mass matrix has $N(N-1)$ non-zero eigenvalues only and thus has nullity $N-1$ which implies that $N(N-1)$ off-diagonal gluons obtain masses and the $N-1$ diagonal gluons remain massless. Because of the antisymmetry, eigenvalues of $M^2$ are purely imaginary and in conjugate pairs. The massive off-diagonal gluons are inferred as evidence of Abelian dominance, which is one of signatures of quark confinement. Further, mass squared of the off-diagonal gluon is purely imaginary, hence the pole of the off-diagonal gluon propagator is on imaginary $p^2$ axis which is another important signature of color confinement. The mass for gluons generated through a given dynamical mechanism breaks the gauge symmetry as usual. We note that there exist other mechanisms where the mass can consistently be given to gluons in a gauge invariant manner, a thorough overview of such mechanisms is found in Refs. [94] . Thus, we see that the ghost condensation acts as the QCD vacuum. Therefore, in the ghost condensed phase the Lagrangian can effectively be given as follows

$$\mathcal{L}_{GC} = -\frac{1}{4}F^a_{\mu\nu}F^{\mu\nu a} - \frac{1}{2\zeta}(A^a_\mu A^{\mu a})^2 + M^2_a A^a_\mu A^{\mu a} \tag{60}$$

Here for the diagonal gluons $M^2_a = 0$, and off diagonal elements occur as $\pm im^2$. For $SU(3)$, $M^2_3 = M^2_8 = 0$ and the off-diagonal gluons, $M^2_1 = +im^2_1, M^2_2 = -im^2_1$, $M^2_4 = +im^2_2, M^2_5 =$

$-im_2^2$, $M_6^2 = +im_3^2, M_7^2 = -im_3^2$ ($m_1^2, m_2^2, m_3^2$ are positive real). So the gluons 1 and 2 can be considered as conjugate of each other. The same is true for other pairs. Hence for $SU(3)$, the last term of the effective Lagrangian in Eq. 60 would be

$$M_a^2 A_\mu^a A^{\mu a} = +im_1^2(A_\mu^1 A^{\mu 1} - A_\mu^2 A^{\mu 2}) + im_2^2(A_\mu^4 A^{\mu 4} - A_\mu^5 A^{\mu 5}) + im_3^2(A_\mu^6 A^{\mu 6} - A_\mu^7 A^{\mu 7}) \quad (61)$$

Now to show the PT phase transition we need to discuss hermiticity properties of the system as PT symmetry becomes meaningful only in non-Hermitian systems. The hermiticity property of fields $A_\mu^a$ is well defined since they describe real degrees of freedom. Fields must be Hermitian in order to define the real degrees of freedom i.e., $A_\mu^{a\dagger} = A_\mu^a$. However, this is not the case for ghosts as the hermiticity properties remain unclear. As the operation of conjugation in principle transforms particle to its anti-particle, the following is the natural choice of hermiticity property for ghosts [90]

$$c^{a\dagger} = \overline{c^a}; \quad \overline{c^a}^\dagger = c^a \quad (62)$$

With these hermiticity properties the Lagrangian in the normal phase is not-hermitian since ghost term is not Hermitian as $(\overline{c^a}\partial_\mu c^a)^\dagger = (\partial_\mu \overline{c^a})c^a \neq \overline{c^a}\partial_\mu c^a$. The effective Lagrangian in the ghost condensed (confinement) phase (60) is also not Hermitian as the mass term for gluons is purely imaginary as explained. Important point here is to note that non hermiticity of the Lagrangian in this ghost condensed phase is free of the hermiticity convention for ghosts as they do not appear in this phase and thus the non hermiticity of the ghost condensed phase is profound. Anti-linearity makes two sets of ghosts transform in a completely different manner. Thus, the theory in normal phase is individually both parity and time reversal invariant. This PT symmetry breaks down spontaneously in the confined phase as we explain now.

It is easy to check that parity is still a symmetry of the theory in the confined phase as given in Eq. (60). However, the time reversal is broken due to pure complex nature of the mass term, $M_a^2 A_\mu^a A^{\mu a} = \xrightarrow{\mathrm{T}} -M_a^2 A_\mu^a A^{\mu a}$ and also PT$\psi \neq \pm\psi$, where $\psi$s are eigenfunctions of the mass matrix (59). The first two terms of $\mathcal{L}_{GC}$ remain unaffected by the time-reversal. Thus, PT symmetry is broken in this phase. We can see that the anti symmetric nature of structure constant appearing in the mass matrix has led to this breaking. The PT symmetry breaking in the confined phase is profound as it is independent of the convention for ghosts. Therefore, the transition from the normal phase to the confinement phase with $SU(N)$ symmetric ghost condensates is identified as PT phase transition from unbroken to broken PT phase. The association between colour confinement and spontaneous PT breaking is model and mechanism independent even though in this model the link is through ghost condensation since one prime signature of quark confinement, the pole of the propagator on purely imaginary $p^2$ axis, inevitably breaks PT symmetry. Usefulness of a consistent model such as one in this paper lies in that it gives valuable insight into a process through which the link can take place.

There is a crucial difference between the non Abelian model and the toy model of complex scalars. Complex scalar theory has the parameter $\eta \equiv \frac{2\mu^2}{|m_1^2 - m_2^2|}$ whose value separates two phases of the PT symmetry in the theory. There is no such single order parameter in the non Abelian theory which governs the phase transition. Different ghost bilinears $\overline{c^a}c^c$ ($a$ and $c$ runs over 1 to $N^2 - 1$ independently) gradually condensing to the stated $SU(N)$ symmetric vacuum give rise to the PT phase transition in this non Abelian model. Thus, we have provided a gauge theory in which PT phase transition is explicitly shown for the first time[65].

Finally we would like to comment on the hidden C symmetry, which satisfy $[H, C]\psi = 0$, $[PT, C]\psi = 0$, $C^2 = \mathbf{1}$. So far, no explicit representation of the C-symmetry is known in the framework of gauge theories. We realize that the inner automorphism exchanges group indices i.e. for SU(3), $1 \leftrightarrow 2$, $4 \leftrightarrow 5$, $6 \leftrightarrow 7$, $3 \leftrightarrow 8$ plays the role of C symmetry [84] It is clear that the theory in both the phases is invariant under CPT. In the broken PT phase, the theory also violates CP symmetry but preserves the CT, in complete analogy with the scalar model described in sec. II.

## 7. Conclusion

We review some of our works on PT phase transition through analytically solvable examples in various space time dimension in quantum mechanics and in quantum field theory. Our examples capture typical features of this phase transitions. In one dimension we realized PT phase transition very early through a QES system. QES levels have real energy eigenvalues and corresponding eigenvectors respect PT symmetry in the unbroken phase. However interestingly for even values of the QES parameter, this particular system always remain in broken phase. PT symmetric NH interaction between two level system and SHO was considered. In this model, except ground state all levels are doubly degenerate at the transition point. PT phase transition is shown to occur in all invariant subspaces of the system. Interesting connection between rotational symmetry of the original Hamiltonian and unbroken phase is observed in 2 and 3 dimensional quantum mechanical systems. Graphene acquires small mass gap due to spin orbit coupling. By considering PT symmetric NH spin orbit coupling for a massless Dirac particle in the background of DO potential, we show that mass gap is only generated in the unbroken PT phase of the system. SU(N) gauge field theoretic models is cast as NH model by adopting the natural but unconventional hermiticity properties of the ghost fields. By using a recently introduced quadratic gauge the phase transition in SU(N) QCD has been studied. We realize deconfinment to confinement phase transition as a PT phase transition in such a model.

## References


[1] Dirac P M 1942 *Proc. Roy. Soc.* A **180** 1; Pauli W 1943 *Revs. Mod. Phys.* **15** 175; Feynman R 1949 *Phys. Rev.* **76** 769; Wu T T 1959 *Phys. Rev. Lett.* **15** 1390; Sudarshan E C G 1961 *Phys. Rev.* **123** 2183; Lee T D and Wick G C 1969 *Nucl. Phy.* B **9** 209; Fisher M F 1978, *Phys. Rev. Lett.* **40** 1610; Cardy J L 1985 *Phys. Rev. Lett.* **54** 1354; Berry M V 1998 *J. Phys.* A **31** 3493

[2] Bender C M and Boettcher S 1998 *Phys. Rev. Lett.* **80** 5243

[3] Bender C M 2007 *Rep. Prog. Phys.* **70** 947 and references therein

[4] Mostafazadeh A 2010 *Int. J. Geom. Meth. Mod. Phys.* **7**1191 and references therein

[5] Moiseyev N 2011 *Non-Hermitian quantum mechanics* (Cambridge University Press)

[6] Bender C M 2019 *PT Symmetry: In Quantum and Classical Physics* (World Scientific, Singapore)

[7] Bender C M and Boettcher S 2002 *Phys. Rev. Lett.* **89** 270401-1

[8] Mostafazadeh A 2009 *Phys. Rev. Lett.* **102** 220402.

[9] Basu-Mallick B and Mandal B P 2001 *Phys. Lett.* A **284** 231

[10] Levai G 2008 *J. Phys.* A **41** 244015

[11] Rotter I and Bird J P 2015 *Rep. Prog. Phys.* **78** 114001 and references therein

[12] Wan W, Chong Y, Ge L, Noh H, Stone A D and Cao H 2011*Science* **331** 889

[13] Liu N, Mesch M, Weiss T, Hentschel M and H Giessen 2010 *Nano Lett.* **10**2342

[14] Noh H, Chong Y, Stone A D and Cao H 2011 *Phys. Rev. Lett.* **108** 6805

[15] Gmachl C F 2010 *Nature* **467** 37

[16] Fring A and T Taira 2020 *Phys. Rev.* D **101** 045014; *Nucl. Phys.* B **950** 114834 *Phys. Lett.* B **807** 135583

[17] Fring A and Frith T 2019 *Phys. Rev.* A **100** 010102

[18] Bender C M, Hassanpour N, Klevansky S P and Sarkar S 2018 *Phys. Rev.* D **98**, 125003

[19] Beygi A, Klevansky S P and Bender C M 2019 *Phys. Rev.* A **99** 062117

[20] Bender C M and Boettcher S 1998 *J. Phys.* A **31**, L273 ; Bender C M, Boettcher S and Meisinger P N 1999 *J. Math. Phys.* **40** 2210; Bender C M and Dunne G V 1999 *J. Math. Phys.* **40** 4616; Bender C M, Dunne

G V and Meisinger P N 1999 *Phys. Lett.* A **252** 272 ; Bender C M, Milton K A and Meisinger P N1999, *J. Math. Phys.* **40** 2201
[21] Ghatak A, RayMandal R D and Mandal B P 2013 *Annl. of Phys.* **336** 540; Ghatak A, Nathan J A, Mandal B P and Ahmed Z 2012 *J. Phys. A: Math. Theor.* **45** 465305
[22] Hasan M, Ghatak A and Mandal B P 2014 *Annl. of Phys.* **344** 17
[23] Ghatak A and Mandal B P 2012 *J. Phys. A: Math. Theor.* **45** 355301
[24] Mandal B P and Ghatak A 2012 *J. Phys. A: Math. Theor.* **45** 444022
[25] Ghatak A, Hasan M and Mandal B P 2015 *Phys. Lett.* A **379** 1326
[26] Yadav R K, Khare A, Bagchi B, Kumari N and Mandal B P 2016, *J. of Math. Phys.* **57** 062106
[27] Dwivedi A and Mandal B P 2021 *Annl. of Phys.* **425** 168382
[28] Hasan M, Singh V N and Mandal B P 2021 *Eur. Phys. J. Plus* **135** 640
[29] Hasan M and Mandal B P 2020 *J. of Math. Phys.* **61** 032104
[30] Mandal B P, Maurya B K and Singh A K 2020 *Eur. Phys. J. Plus.* **135** 327
[31] Hasan M and Mandal B P 2020 *Eur. Phys. J. Plus.* **135** 248
[32] Hasan M and Mandal B P 2020 *Eur. Phys. J. Plus.* **135** 84
[33] Hassan M and Mandal B P 2018 *Annl. of Phys.* **396** 371
[34] Ghatak A, Hasan M and Mandal B P 2015 *Phys. Lett.* A **379** 1326
[35] Hasan M, Ghatak A and Mandal B P 2014, *Annl. of Phys.* **344** 17
[36] Kumari N, Yadav R K, Khare A, Bagchi B and Mandal B P 2016 *Annl. of Phys.* **373** 163
[37] Musslimani Z H, Makris K G, El-Ganainy R and Christodoulides D N 2008 *Phys. Rev. Lett.* **100** 030402
[38] Ruter C E, Makris K G, El-Ganainy R, Christodulides D N, Segev M and Kip D 2010 *Nature Phys.* **6** 192
[39] Guo A and Salamo G J 2009 *Phys. Rev. Lett.***103** 093902
[40] West C T, Kottos T and Prosen T 2010 *Phys. Rev. Lett.* **104** 054102
[41] Liu T, Cheng S, Guo H and Xianlong G 2021 *Phys. Rev.* B **103** 104203
[42] Schiffer S, Liu X, Hu H , and Wang J 2021 *Phys. Rev.* A **103** 011302
[43] Znojil M 2020 *Scientific Reports* **10(1)**18523
[44] Li J, Harter A K, Liu J, Melo L, Joglekar Y N and Luo L 2019 *Nature Communications* **10** 855
[45] Joglekar Y N and Barnett J L 2011 *Phys. Rev.* A **84** 024103
[46] Iorsh I, Poshakinskiy A and Poddubny A 2020 *Phys. Rev. Lett.* **125** 183601
[47] Zhang X H H and Baranger H U 2021 *Phys. Rev.* A **103** 033711
[48] Longhi S 2020 *Opt. Lett.* **45** 1591
[49] Suchkov S V, Churkin D V and Sukhorukov A A 2020 *Opt. Exp.* **28** 30348
[50] Wang K, Dai J, Shao L B Yang S A and Zhao Y X 2020 *Phys. Rev. Lett* **125** 126403
[51] Tuniz A, Wieduwilt T and Schmidt M A 2019 *Phys. Rev. Lett.* **123** 213903
[52] Zhang F, Feng Y, Ge L and Wan W 2020 *Phys. Rev. Lett.* **124** 053901
[53] Longhi S 2019 *Phys. Rev. Lett.* **122** 237601
[54] Chitsazi M, Li H, Ellis F M and Kottos T 2017 *Phys. Rev. Lett.* **119** 093901
[55] Modak R and Mandal B P 2021 *Phy. Rev.* A **103** 062416
[56] Liu T, Cheng S, Guo H and Xianlong G 2021*Phys Rev.* B **103** 104203
[57] Schiffer S, Liu X, Hu H and Wang J 2021 *Phys Rev.* A **103** 011302
[58] Gopalakrishnan S and Gullans M 2021 *Phys. Rev. Lett.* **126** 170503
[59] Cai X 2021 Anderson localization and topological phase transitions in non-Hermitian Aubry-André-Harper models with p-wave pairing *Preprint* arXiv:2103.04107
[60] Nakanishi Y and Sasamoto T 2021 PT phase transition in open quantum systems with Lindblad dynamics *Preprint* arXiv:2104.07349
[61] Mandal B P 2005 *Mod. Phys. Lett.* A **20** 655
[62] Khare A and Mandal B P 2000 *Phys. Lett.* A **272** 53
[63] Mandal B P, Mourya B K and Yadav R K 2013 *Physics Letters* A **377** 1043
[64] Mandal B P, Mourya B K, Ali K and Ghatak A 2015 *Annl. of Phys.* **363** 185
[65] Rawal H and Mandal B P 2019 *Nucl. Phys.* B **946** 114699
[66] Khare A and Mandal B P 1998 *J. Math. Phys.* **39** 3476
[67] Khare A, Mandal B P 2009 *Pramana* **73** 387
[68] Bender C M, Dunne G V, Meisinger P N and Simsek M 2001 *Phys. Lett.* A **281** 311
[69] Bender C M, Berry M and Mandilara A 2002 *J. Phys.* **35** L467
[70] Ge L and Stone A 2014 *Phy. Rev.* X **4**, 031011
[71] Agarwal K S, Pathak R K and Joglekar Y N 2015 *Europhysics Lett.* **112** 31003

[72] M Znojil 2003 *J. of Phys.* A **36** 7825
[73] Levai G 2008 *J. of Phys. A* **41** 244015
[74] Novoselov K S, Geim A K, Morozov S V, Jiang D, Katsnelson M I , Grigorieva I V, Dubonos S V and Forsov A A 2005 *Nature* **438** 197
[75] Novoselov K S et al 2004 *Science* **306** 666
[76] Castro Neto A H et al 2009 *Rev. Mod. Phys.* **81** 109
[77] Katsnelson M I , Novoselov K S and A K Geim A K 2006 *Nature Phys.* **2** 620
[78] M Moshinsky M and Szczepaniak A1989 J. of Phys. A **22** 1817
[79] Ito D, Mori K and E Carrieri E 1967 *Nuovo Cimento* A **51** 1119
[80] Mandal B P and Verma S 2010 *Phys. Lett.* A **374** 1021
[81] Mandal B P and Rai S K 2012 *Phys. Lett.* A **376** 2467
[82] Panella O and Roy P 2014 *Phy. Rev.* A **90** 042111
[83] Tyagi V, Rai S K and Mandal B P 2019 *Europhysics Lett.* **128** 30004
[84] Raval H and Yajnik U A 2015 *Phys. Rev.* D **91** 085028
[85] Bazeia D, Das A and Losano L 2009 Phys. Lett. B **673** 283
[86] Wigner E 1960 *J. Math. Phys.* **1** 409
[87] Dey S, Fring A and Gouba L 2012 *J. of Phys.* A **45** 385302
[88] See for example Das A 2008 *Quantum Field Theory* (Singapore : World Scientific)
[89] Alexandre J, Millington P and Seynaeve D 2017 *Phys. Rev.* D **96** 065027
[90] Kugo T and Ojima I 1979 *Prog. Theor. Phys. Suppl.* **66** 1
[91] Alexandre J, Millington P and Seynaeve D 2017 Consistent description of field theories with non-Hermitian mass terms *Preprint* arXiv:1710.01076v1
[92] Bender C M and Dunne G V 1996 *J. Math. Phys.* **37** 6
[93] Kondo K I and Shinohara T 2000 *Phys. Lett.* B **491** 263
[94] Aguilar A C and Papavassiliou J 2006 *JHEP* **12** 012 and references therein
[95] A G Ushveridze A G 1995 *Quasi exactly solvable models in Quantum mechanics* (Bristol: IOP Publishing)
[96] Turbiner A V 1988 *Comm. Math. Phys.* **118** 467
[97] Khare A and Mandal B P 1998 *Physics Letters* A **239** 197

# Antilinear symmetry and the ghost problem in quantum field theory

**Philip D. Mannheim**

Department of Physics, University of Connecticut, Storrs, CT 06269, USA

E-mail: `philip.mannheim@uconn.edu`

**Abstract.**
The recognition that the eigenvalues of a non-Hermitian Hamiltonian could all be real if the Hamiltonian had an antilinear symmetry such as $PT$ stimulated new insight into the underlying structure of quantum mechanics. Specifically, it led to the realization that Hilbert space could be richer than the established Dirac approach of constructing inner products out of ket vectors and their Hermitian conjugate bra vectors. With antilinear symmetry one must instead build inner products out of ket vectors and their antilinear conjugates, and it is these inner products that would be time independent in the non-Hermitian but antilinearly symmetric case even as the standard Dirac inner products would not be. Moreover, and in a sense quite remarkably, antilinear symmetry could address not only the temporal behavior of the inner product but also the issue of its overall sign, with antilinear symmetry being capable of yielding a positive inner product in situations such as fourth-order derivative quantum field theories where the standard Dirac inner product is found to have ghostlike negative signature. Antilinear symmetry thus solves the ghost problem in such theories by showing that they are being formulated in the wrong Hilbert space, with antilinear symmetry providing a Hilbert space that is ghost free. Antilinear symmetry does not actually get rid of the ghost states. Rather, it shows that the reasoning that led one to think that ghosts were present in the first place is faulty. Implications of our results for constructing unitary quantum theories of gravity are presented.

## 1. Introduction

Interest in non-Hermitian Hamiltonians was sparked by the work of Bender and collaborators [1, 2] on the non-Hermitian Hamiltonian $H = p^2 + ix^3$, as it turned out to have a completely real energy eigenspectrum. Ever since the work of Dirac the requirement of Hermiticity of the Hamiltonian had been foundational for quantum mechanics. However quantum mechanics did not actually need to rely on Hermiticity because it is only a sufficient condition for the reality of the eigenvalues of a Hamiltonian but not a necessary one. Thus while Hermiticity implies reality there is no converse theorem that says that if a Hamiltonian is not Hermitian then its eigenvalues cannot all be real. However, by and large the identification of a necessary condition had not been pursued prior to the work of [1, 2], and there was even a reason for this, since within linear algebra no such condition had actually been found. This is not surprising since when the necessary condition was eventually identified it turned out not to be a linear condition at all. Rather, the necessary condition is that the Hamiltonian must possess an antilinear symmetry, viz. a symmetry that does not just act on operators but one which conjugates complex numbers as well.

To see how antilinearity works it is instructive to look at the eigenvector equation

$$i\frac{\partial}{\partial t}|\psi(t)\rangle = H|\psi(t)\rangle = E|\psi(t)\rangle. \tag{1.1}$$

On replacing $t$ by $-t$ and then multiplying by a general antilinear operator $A$ we obtain

$$i\frac{\partial}{\partial t}A|\psi(-t)\rangle = AHA^{-1}A|\psi(-t)\rangle = E^*A|\psi(-t)\rangle. \tag{1.2}$$

From (1.2) we see that if $H$ has an antilinear symmetry so that $AHA^{-1} = H$, then, as first noted by Wigner in his study of time reversal invariance, energies can either be real and have eigenfunctions that obey $A|\psi(-t)\rangle = |\psi(t)\rangle$, or can appear in complex conjugate pairs that have conjugate eigenfunctions ($|\psi(t)\rangle \sim \exp(-iEt)$ and $A|\psi(-t)\rangle \sim \exp(-iE^*t)$). To establish a converse, suppose we are given that the energy eigenvalues are real or appear in complex conjugate pairs. In such a case not only would $E$ be an eigenvalue but $E^*$ would be too. Hence, we can set $HA|\psi(-t)\rangle = E^*A|\psi(-t)\rangle$ in (1.2), and obtain

$$(AHA^{-1} - H)A|\psi(-t)\rangle = 0. \tag{1.3}$$

Then if the eigenstates of $H$ are complete, (1.3) must hold for every eigenstate, to yield $AHA^{-1} = H$ as an operator identity, with $H$ thus having an antilinear symmetry.

The above analysis thus demonstrates that antilinearity is necessary to ensure that all energies are real or that some or all of them appear in complex conjugate pairs, and without it it is guaranteed that not all of the eigenvalues could be real. Once given antilinearity, to enforce reality alone we note that if $|\psi(-t)\rangle$ is also an eigenstate of $A$ so that up to an overall phase $A|\psi(-t)\rangle = |\psi(t)\rangle$, then from (1.1) and (1.2) it follows that $E = E^*$, while if $A|\psi(-t)\rangle \neq |\psi(t)\rangle$ it follows that $E \neq E^*$. As discussed in [3], the necessary and sufficient condition for the reality of the energy eigenvalues then is that $H$ have an antilinear symmetry and that all of its eigenstates be eigenstates of the antilinear operator.

Returning now to $H = p^2 + ix^3$ itself, we note that it is left invariant under a particular antilinear symmetry, namely $PT$ symmetry, where $P$ is the parity operator (a linear operator) while $T$ is the antilinear time-reversal operator. (Under $PT$ $p \to p$, $x \to -x$ and $i \to -i$.) It was the realization [1, 2] that $PT$ symmetry underlay the $H = p^2 + ix^3$ theory that opened up the study of $PT$ symmetric systems, with the field having then grown rapidly (see e.g. [4, 5, 6, 7, 8, 9] and references therein.)

Once one considers non-Hermiticity one is no longer able to use the Dirac inner product, viz. the overlap of a ket vector with a bra vector that is constructed as the Hermitian conjugate of the ket vector, as the Dirac inner product is no longer time independent (viz. $\langle\psi(t)|\psi(t)\rangle = \langle\psi(0)|e^{iH^\dagger t}e^{-iHt}|\psi(0)\rangle$ is not equal to $\langle\psi(0)|\psi(0)\rangle$ if $H^\dagger \neq H$). Rather, one should use the overlap of a ket vector with a bra vector that is constructed as the conjugate of the ket vector with respect to the antilinear symmetry, as this then does give a time-independent inner product if $H$ has an antilinear symmetry. As discussed in [10, 11, 3, 12] this particular inner product is also equal to the overlap $\langle\psi(t)|V|\psi(t)\rangle$ where the $V$ operator effects the pseudo-Hermitian condition

$$VHV^{-1} = H^\dagger. \tag{1.4}$$

That this condition is equivalent to antilinear symmetry can be seen from the fact that under it $H$ and $H^\dagger$ possess a common set of eigenvalues, and thus all the eigenvalues of $H$ are either real or in complex conjugate pairs, i.e., precisely the same outcome as when $H$ has an antilinear symmetry. With $|\psi_i(t)\rangle$ and $\langle\psi_i(t)|$ obeying $i\partial_t|\psi_i(t)\rangle = H|\psi_i(t)\rangle$, $-i\partial_t\langle\psi_i(t)| = \langle\psi_i(t)|H^\dagger$ we obtain

$$i\frac{\partial}{\partial t}\langle\psi_i(t)|V|\psi_j(t)\rangle = \langle\psi_i(t)|(VH - H^\dagger V)|\psi_j(t)\rangle. \tag{1.5}$$

Thus as long as $V$ is invertible, the $\langle\psi_i(t)|V|\psi_j(t)\rangle$ inner product is time independent when (1.4) is imposed, just as required. Moreover, for the converse we note that if all $\langle\psi_i(t)|V|\psi_j(t)\rangle$ matrix elements are time independent, then completeness of the $|\psi_i(t)\rangle$ and $\langle\psi_i(t)|$ states entails that $VH - H^\dagger V$. Thus through their joint connection to the $VHV^{-1} = H^\dagger$ condition, antilinearity implies the time independence of the $V$-based inner products, while the time independence of these inner products implies antilinearity [3].

With the recognition that with antilinearity a Hamiltonian can have real eigenvalues even if it is not Hermitian, and that with antilinearity one can have a time-independent inner product even if the Hamiltonian is not Hermitian, we turn now to the third key aspect of antilinearity, namely the role of boundary conditions. Once we have done this we can address the issue of the ghost problem associated with the sign of the inner product.

## 2. What is meant by Hermiticity

To understand the role of boundary conditions, we note that for a second-order differential operator $D$ in the self-adjoint form $D = -p(x)d^2/dx^2 - p'(x)d/dx + q(x)$ that acts on wave functions $\phi(x)$, $\psi(x)$, one can show (Green's theorem) that

$$\int_a^b dx[\phi^* D\psi - (\psi^* D\phi)^*] = \int_a^b dx[\phi^* D\psi - [D\phi]^*\psi] = \left[p\psi\phi^{*\prime} - p\phi^*\psi'\right]\Big|_a^b. \tag{2.1}$$

Hermiticity in the form $\langle\phi|D|\psi\rangle = \langle\psi|D|\phi\rangle^*$ is then only secured by the vanishing of the surface term at the spatial boundary. While quantum mechanics is based on commutators of operators, they are not defined until a basis on which the operators act is specified. For such a basis to be viable it needs to be well-behaved at the spatial boundary Thus for the $[\hat{x}, \hat{p}] = i$ commutator, we can only set $\hat{p} = -id/dx$ when it acts on a good, i.e., normalizable, test function according to

$$\left[\hat{x}, -i\frac{d}{dx}\right]\psi(x) = i\psi(x). \tag{2.2}$$

Thus for a harmonic oscillator $\hat{H} = \hat{p}^2 + \hat{x}^2$ for instance we have the following two typical solutions:

$$\left[-\frac{d^2}{dx^2} + x^2\right]e^{-x^2/2} = e^{-x^2/2}, \qquad \left[-\frac{d^2}{dx^2} + x^2\right]e^{+x^2/2} = -e^{+x^2/2}, \tag{2.3}$$

with the eigenvalue of $e^{-x^2/2}$ being positive and the eigenvalue of $e^{+x^2/2}$ being negative. Of these two solutions only the $e^{-x^2/2}$ wave function is normalizable (cf. vanishing of the surface term), with $\int dx\psi^*(x)\psi(x)$ being finite. And when acting on it we can indeed represent $\hat{p}$ as $\hat{p} = -id/dx$. Here $x$ is real and we are working in the coordinate basis in which $\hat{x}$ is Hermitian, has real eigenvalues $x$, and is diagonal in this basis.

But what of the $e^{+x^2/2}$ solution? It is not normalizable and we cannot represent $\hat{p}$ as $-id/dx$ when acting on it since we cannot throw away the surface term in an integration by parts. However, that does not mean that we can ignore this solution, it means only that when acting on it one cannot represent $\hat{p}$ as $-id/dx$. So maybe we could represent $\hat{p}$ by some alternate form.

Thus suppose we make $x$ pure imaginary. Then $e^{+x^2/2}$ is normalizable on the imaginary axis. Thus we can take both $\hat{x}$ and $\hat{p}$ to be anti-Hermitian and represent $[\hat{x},\hat{p}] = i$ as $[-i\hat{x}, i\hat{p}] = i$. This is equivalent to implementing the similarity transformation $\hat{S} = \exp(\pi\hat{p}\hat{x}/2)$ that effects

$$\hat{S}\hat{x}\hat{S}^{-1} = -i\hat{x} = \hat{y}, \quad \hat{S}\hat{p}\hat{S}^{-1} = i\hat{p} = \hat{q}, \tag{2.4}$$

with such a similarity transformation preserving both the commutation relation $[\hat{x},\hat{p}] = [\hat{y},\hat{q}] = i$ and the eigenvalues of any Hamiltonian $\hat{H}(\hat{x},\hat{p})$ that is built out of $\hat{x}$ and $\hat{p}$. We thus have

$$[\hat{y},\hat{q}]\,\psi(y) = \left[\hat{y}, -i\frac{d}{dy}\right]\psi(y) = i\psi(y), \tag{2.5}$$

and now $e^{+x^2/2} = e^{-y^2/2}$ is a good test function. Thus $e^{-x^2/2}$ is a good test function when $x$ is real, while $e^{+x^2/2}$ is a good test function when $x$ is pure imaginary. When $x$ is pure imaginary we can set

$$[\hat{p}^2 + \hat{x}^2]e^{+x^2/2} = -[\hat{q}^2 + \hat{y}^2]e^{-y^2/2} = \left[\frac{d^2}{dy^2} - y^2\right]e^{-y^2/2} = -e^{-y^2/2}. \tag{2.6}$$

Thus while the eigenvalues of $\hat{p}^2 + \hat{x}^2$ would be positive if $\hat{p}$ and $\hat{x}$ are both Hermitian, the eigenvalues of $\hat{p}^2 + \hat{x}^2$ would be negative if $\hat{p}$ and $\hat{x}$ are both anti-Hermitian.

Now taking $\hat{x}$ and $\hat{p}$ to be anti-Hermitian might not be conventional, but to take these two operators to be Hermitian in the first place has a hidden assumption, namely that they are acting on their own eigenstates. However, we are not interested in how they act on their own eigenstates, since in order to determine the eigenstates and eigenvalues of a Hamiltonian $\hat{H}$ that is built out of them what matters is how $\hat{x}$ and $\hat{p}$ act on the eigenstates of $\hat{H}$. Thus while both $\hat{x}$ and $\hat{p}$ might be Hermitian when acting on their own eigenstates that does not necessarily make them Hermitian when acting on the eigenstates of $\hat{H}$. This then is the secret of *PT* symmetry, with there potentially being a mismatch between the eigenstates of a Hamiltonian and the eigenstates of the operators out of which it is built.

To be general, we note that quantum mechanics allows us to make a similarity transformation through any angle of the form $\hat{S} = \exp(-\theta\hat{p}\hat{x})$, to thus effect

$$\hat{S}\hat{x}\hat{S}^{-1} = \hat{x}\exp(i\theta), \quad \hat{S}\hat{p}\hat{S}^{-1} = \hat{p}\exp(-i\theta) \tag{2.7}$$

while preserving the $[\hat{x},\hat{p}]$ commutator. Ordinarily this is not of any significance since we work with Hermitian operators that have normalizable wave functions on the real axis, and we have no need to go into the complex plane. But if the wave functions are not normalizable on the real axis, we may be able to continue into so-called "Stokes wedges" in the complex plane where they then are normalizable, and cross over "Stokes lines" that divide the different regions ($\theta = \pi/4$ and $\theta = -\pi/4$ in the harmonic oscillator case). The two sides of the Stokes line represent different, inequivalent, realizations of the quantum theory. And not only that, they represent formulations in different Hilbert spaces. Specifically, for the harmonic oscillator with Hamiltonian and commutator

$$\hat{H} = a^\dagger a + \frac{1}{2}, \quad [a, a^\dagger] = 1, \tag{2.8}$$

the positive energy sector is associated with a Hilbert space that is built on a vacuum that obeys $a|\Omega_1\rangle = 0$, while the negative energy sector is associated with a Hilbert space that is built on a vacuum that obeys $a^\dagger|\Omega_2\rangle = 0$.

For the $PT$ symmetry prototype $\hat{H} = \hat{p}^2 + i\hat{x}^3$, we note that its energy eigenvalues would indeed be complex if the eigenstates of $\hat{x}$ were in the same Hilbert space as the eigenstates of $\hat{H}$. However, it turns out [1, 2, 4, 8] that the eigenstates of $\hat{H}$ are not normalizable on the real $x$ axis. Nonetheless, one can find appropriate Stokes wedges in the complex $x$ plane in which these eigenfunctions are normalizable. And in these wedges the eigenvalues of $\hat{H}$ are all real. It is in this way then that $PT$ symmetry makes $\hat{H} = \hat{p}^2 + i\hat{x}^3$ viable. At the same time it serves a warning: one cannot determine an energy eigenspectrum by inspection. One has to first identify appropriate boundary conditions.

In fact $PT$ symmetry serves a second warning as well. One cannot even tell if a Hamiltonian is Hermitian by inspection. Specifically, consider as an example the relativistic action

$$I_{\rm S} = \int d^4x \left[\frac{1}{2}\partial_\mu\phi\partial^\mu\phi - \frac{1}{2}m^2\phi^2\right] \tag{2.9}$$

with Hamiltonian

$$H = \int d^3x \frac{1}{2}[\dot{\phi}^2 + \vec{\nabla}\phi\cdot\vec{\nabla}\phi + m^2\phi^2]. \tag{2.10}$$

Solutions to the wave equation obey

$$\phi(\vec{x},t) = \sum[a(\vec{k})\exp(-i\omega_{\vec{k}}t + i\vec{k}\cdot\vec{x}) + a^\dagger(\vec{k})\exp(+i\omega_{\vec{k}}t - i\vec{k}\cdot\vec{x})], \qquad \omega_{\vec{k}}^2 = \vec{k}^2 + m^2, \tag{2.11}$$

and the Hamiltonian is given by

$$H = \sum\frac{1}{2}[\vec{k}^2 + m^2]^{1/2}[a^\dagger(\vec{k})a(\vec{k}) + a(\vec{k})a^\dagger(\vec{k})]. \tag{2.12}$$

If $m^2 > 0$ all energies are real, and both $H$ and $\phi(\vec{x},t)$ are Hermitian. However, if $m^2 = -n^2 < 0$, then

$$\omega_{\vec{k}}^2 = \vec{k}^2 - n^2. \tag{2.13}$$

Thus now the $k < n$ energies come in complex conjugate pairs, and neither $H$ nor $\phi(\vec{x},t)$ is Hermitian. And yet inspection of (2.9) would have suggested that even for negative $m^2$ both $H$ and $\phi(\vec{x},t)$ are Hermitian.

As a second example consider the non-relativistic Hamiltonian

$$H_{\rm PU}(\alpha,\beta) = \frac{p_x^2}{2} + p_z x + (\alpha^2-\beta^2)x^2 - \frac{1}{2}(\alpha^2+\beta^2)^2 z^2 \tag{2.14}$$

with real $\alpha$ and $\beta$. This Hamiltonian is associated with the Pais-Uhlenbeck ($PU$) fourth-order derivative oscillator theory [13] that we discuss below. This Hamiltonian was constructed in [14] and explored in detail in [15, 16]. It also appears to be Hermitian. However, as we show below, the energy eigenvalues of its two first excited states are the complex $2\alpha + i\beta$ and $2\alpha - i\beta$. Thus to determine if a Hamiltonian is indeed Hermitian one needs to solve the theory and get the eigenvalues first.

As a third example consider the non-relativistic Hamiltonian

$$H_{\rm PU}(\omega) = \frac{p_x^2}{2} + p_z x + \omega^2 x^2 - \frac{1}{2}\omega^4 z^2, \tag{2.15}$$

with real $\omega$. It also appears to be Hermitian. However, as we show below, the Hamiltonian only has a single one-particle state, with energy equal to the $2\omega$. And while this energy eigenvalue is even real, the Hamiltonian actually turns out to be a non-diagonalizable Jordan-block Hamiltonian as there is only one first excited state, with there not being any second eigenvector, to thus leave the Hamiltonian with an incomplete set of eigenvectors. With the Hamiltonian not being diagonalizable it cannot be Hermitian, and cannot be even though its eigenvalues are real. Thus to determine if a Hamiltonian is indeed Hermitian one needs to solve the theory and show that the set of eigenvectors is complete.

As a fourth example consider the non-relativistic Hamiltonian

$$H_{\rm PU}(\omega_1,\omega_2) = \frac{p_x^2}{2} + p_z x + \frac{1}{2}\left(\omega_1^2+\omega_2^2\right)x^2 - \frac{1}{2}\omega_1^2\omega_2^2 z^2 \tag{2.16}$$

with real $\omega_1 > \omega_2 > 0$. Now for this example, while the energy eigenvalues of its two first excited states are the real $\omega_1$ and $\omega_2$, and while this time the set of energy eigenvectors is complete, its eigenfunctions are not normalizable on the real axis. And so yet again it is not Hermitian. However, as we discuss below, there does exist a Stokes wedge in which its eigenvectors are normalizable [15], and this wedge does not contain the real axis. Thus to determine if a Hamiltonian is indeed Hermitian one needs to solve the theory and show that there is a domain in which its eigenvectors are normalizable.

While it is not Hermiticity that these four examples have in common, there is some other attribute that they do share. In the first two of the cases one has a complex conjugate pair of energy eigenvalues. Thus, as noted in Sec. 1, both Hamiltonians must have an antilinear symmetry. In [15, 16] it was shown that the third and fourth examples have an antilinear symmetry too. Thus all four of the Hamiltonians have an antilinear symmetry, with it thus being antilinearity rather than Hermiticity that they all have in common. Now in [3] it was shown on general grounds that under only two requirements, namely invariance under the complex Lorentz group and conservation of probability, the Hamiltonian must be CPT invariant, where $C$ is the charge conjugation operator. And it must be so regardless of whether or not it might be Hermitian. (The standard proof of the $CPT$ theorem required Hermiticity.) Now while one cannot a priori tell whether a Hamiltonian might be Hermitian, one can tell a priori if it has an antilinear symmetry as that requires no need to solve the theory, and is fixed solely by the behavior of the operators and coefficients that appear in the Hamiltonian. Now for our four examples $C$ just happens to be separately conserved, and thus for all of them $CPT$ reduces to $PT$. In fact this is a general requirement for any non-relativistic theory that descends from a covariant relativistic theory (which any theory must do if it is to be physically relevant). Thus, if one is below the threshold for pair creation $C$ is separately conserved and $CPT$ defaults to $PT$. This thus puts $PT$ symmetry on a quite secure theoretical footing.

Both (2.14) and (2.15) can be derived by setting $\omega_1 = \alpha + i\beta$, $\omega_2 = \alpha - i\beta$ and $\omega_1 = \omega_2 = \omega$ in (2.16). Now in Sec. 1 we had noted that with antilinearity one could have real eigenvalues or eigenvalues that are in a complex conjugate pair. However, we also noted that in the complex pair realization the eigenvectors are complex conjugates of each other. Thus if we switch off the complex part by setting $\beta$ to zero in (2.14) the eigenvalues not only become equal, so do the eigenvectors. Thus in this limit we must lose an eigenvector. Thus the transition from (2.16) to (2.14) must go through a singular point (known in the $PT$ literature as an exceptional point) in which the Hamiltonian becomes of non-diagonalizable Jordan-block form. Thus for antilinear symmetry there are actually three realizations: eigenvalues all real and eigenvectors complete, eigenvalues all real and but eigenvectors incomplete, eigenvalues in complex conjugate pairs and eigenvectors complete. As discussed in [15, 16, 3], in all cases one can construct a time-independent probability.

Turning back now to the harmonic oscillator, we note that while we can find a Stokes wedge in which its $e^{+x^2/2}$ type eigenfunctions are normalizable, no amount of similarity transforming can convert the associated negative energy eigenvalues into positive ones. And it is for this reason that we have to abandon this sector of the oscillator theory. For our purposes here we note that there is a second disease in this sector, namely with $a^\dagger$ annihilating the vacuum according to $a^\dagger|\Omega_2\rangle = 0$, it is the operator $a$ that serves as the creator, with the one-particle matrix element $\langle\Omega_2|a^\dagger a|\Omega_2\rangle = -\langle\Omega_2|\Omega_2\rangle$ then being of ghostlike negative signature. As we now show, a similar ghost problem occurs for the $PU$ oscillator as well, but using the techniques of $PT$ symmetry that we have described above, this ghost problem proves to be solvable.

## 3. The ghost problem of fourth-order derivative theories

In quantum field theory radiative corrections generate ultraviolet divergences. In renormalizable field theories such as quantum electrodynamics one needs to regularize these infinities, while in non-renormalizable theories such as Einstein gravity one tries to augment the theory so as to lower the degree of divergence. The prototype for regularizing divergences in quantum field theories while maintaining gauge invariance is the Pauli-Villars regularization scheme proposed in [17] in which one replaces the standard one-particle $1/(k^2 - M^2 + i\epsilon)$ propagator by the two-particle

$$D_{\rm PV}(x) = \int \frac{d^4k}{(2\pi)^4} e^{-ik\cdot x} D_{\rm PV}(k), \quad D_{\rm PV}(k) = \frac{1}{k^2 - M_1^2 + i\epsilon} - \frac{1}{k^2 - M_2^2 + i\epsilon}, \tag{3.1}$$

so that a $1/k^2$ asymptotic ultraviolet behavior at large $k^2$ would be replaced by the more convergent $1/k^4$. As conceived in [17], the two particles would be associated with a second-order derivative action of the form

$$I_{S_1} + I_{S_2} = \int d^4x \left[\tfrac{1}{2}\partial_\mu\phi_1\partial^\mu\phi_1 - \tfrac{1}{2}M_1^2\phi_1^2\right] + \int d^4x \left[\tfrac{1}{2}\partial_\mu\phi_2\partial^\mu\phi_2 - \tfrac{1}{2}M_2^2\phi_2^2\right], \tag{3.2}$$

with the associated wave operators requiring the components of the $D_{\rm PV}$ propagator to obey

$$\begin{aligned}(\partial_t^2 - \vec{\nabla}^2 + M_1^2)(-i)\langle\Omega_1|T[\phi_1(x)\phi_1(0)]|\Omega_1\rangle &= -\delta^4(x),\\ (\partial_t^2 - \vec{\nabla}^2 + M_2^2)(-i)\langle\Omega_2|T[\phi_2(x)\phi_2(0)]|\Omega_2\rangle &= +\delta^4(x).\end{aligned} \tag{3.3}$$

If we apply the wave operator $(\partial_t^2 - \nabla^2 + M_2^2)$ to $-i\langle\Omega_2|T[\phi_2(x)\phi_2(0)]|\Omega_2\rangle$ we generate a term of the form $-i\langle\Omega_2|\delta(t)[\dot{\phi}_2(x),\phi_2(0)]|\Omega_2\rangle$. With the conjugate of $\phi_2$ being $\delta I_{S_2}/\delta\dot{\phi}_2 = +\dot{\phi}_2$, to recover (3.3) we would need to set $[\dot{\phi}_2(x,t=0),\phi_2(0)] = +i\delta^3(x)$ rather than the standard $[\dot{\phi}_1(x,t=0),\phi_1(0)] = -i\delta^3(x)$ used for $\phi_1$.

To confirm this choice of quantization, we note that following the insertion of

$$\sum |n_1\rangle\langle n_1| - \sum |n_2\rangle\langle n_2| = I \tag{3.4}$$

with its ghostlike relative minus sign into

$$D_{\rm PV}(x) = -i\langle\Omega_1|T[\phi_1(x)\phi_1(0)]|\Omega_1\rangle - i\langle\Omega_2|T[\phi_2(x)\phi_2(0)]|\Omega_2\rangle \tag{3.5}$$

we also recover (3.1). Thus ghost states could reduce asymptotic divergences, but at the price of loss of probability and loss of unitarity.

In order to gain further insight into the Pauli-Villars regulator scheme Pais and Uhlenbeck [13] replaced the two-field action by a one-field fourth-order derivative action

$$I_S = \tfrac{1}{2}\int d^4x \Big[\partial_\mu\partial_\nu\phi\partial^\mu\partial^\nu\phi - (M_1^2 + M_2^2)\partial_\mu\phi\partial^\mu\phi + M_1^2M_2^2\phi^2\Big], \tag{3.6}$$

with a fourth-order derivative equation of motion given by

$$(\partial_t^2 - \vec{\nabla}^2 + M_1^2)(\partial_t^2 - \vec{\nabla}^2 + M_2^2)\phi(x) = 0, \tag{3.7}$$

and an associated propagator of the same form as the Pauli-Villars propagator:

$$D_{\rm PU}(k) = \frac{1}{(k^2 - M_1^2 + i\epsilon)(k^2 - M_2^2 + i\epsilon)} = \frac{1}{(M_1^2 - M_2^2)}\left(\frac{1}{k^2 - M_1^2 + i\epsilon} - \frac{1}{k^2 - M_2^2 + i\epsilon}\right). \tag{3.8}$$

If we identify the $D_{\rm PU}(k)$ propagator as

$$D_{\rm PU}(k) = -i\int d^4x e^{ik\cdot x}\langle\Omega|T[\phi(x)\phi(0)]|\Omega\rangle, \tag{3.9}$$

then the insertion of (3.4) into $-i\langle\Omega|T[\phi(x)\phi(0)]|\Omega\rangle$ would generate $D_{\rm PU}(k)$. So with the relative minus sign in (3.4) again there appear to be ghost states and loss of unitarity.

Although the study of [13] was motivated by the Pauli-Villars regularization scheme, in trying to construct renormalizable quantum theories of gravity one also encounters fourth-order derivative theories. Specifically, up to metric indices the action given in (3.6) describes the coupling of the second-order derivative Einstein theory to a fourth-order derivative theory, which unlike the Einstein theory itself, is based not on the Ricci scalar alone but also on its square. To make such theories viable one thus needs to find a way to eliminate the ghost states that they are thought to possess [18], and to do so while retaining their beneficial role for renormalizability. This we shall do below. However, before doing so we should point out that the identification of the c-number propagator given in (3.8) with a matrix element such as that given in (3.9) cannot actually be made prior to solving the underlying quantum theory. While one can go from a quantum-mechanical Hilbert space to c-number propagators one cannot infer the structure of the underlying q-number theory from a knowledge of the c-number propagator alone. And when we now do construct the underlying Hilbert space we will find that the propagator given in (3.8) should be identified with a completely different matrix element, one that involves no Hilbert space with any ghost states at all.

## 4. Quantizing the Pais-Uhlenbeck oscillator theory

Since only time derivatives are relevant for quantization, Pais and Uhlenbeck simplified the discussion by restricting to one momentum state $k$, and with real and positive $\omega_1 = (\bar{k}^2 + M_1^2)^{1/2}$ and real and positive $\omega_2 = (\bar{k}^2 + M_2^2)^{1/2}$ the $I_S$ action given in (3.6) reduces to the acceleration-dependent Pais-Uhlenbeck action

$$I_{\rm PU} = \tfrac{1}{2}\int dt\left[\ddot{z}^2 - \left(\omega_1^2 + \omega_2^2\right)\dot{z}^2 + \omega_1^2\omega_2^2 z^2\right], \tag{4.1}$$

with equation of motion and propagator

$$\frac{d^4z}{dt^4} + (\omega_1^2 + \omega_2^2)\frac{d^2z}{dt^2} + \omega_1^2\omega_2^2 z = 0,$$
$$G_{\rm PU}(E) = \frac{1}{(E^2 - \omega_1^2 + i\epsilon)(E^2 - \omega_2^2 + i\epsilon)} = \frac{1}{(\omega_1^2 - \omega_2^2)}\left(\frac{1}{E^2 - \omega_1^2 + i\epsilon} - \frac{1}{E^2 - \omega_2^2 + i\epsilon}\right). \tag{4.2}$$

This action is a constrained action since with only $z$, $\dot{z}$ and $\ddot{z}$, there are too many canonical variables for one oscillator but not enough for two. So on setting set $x = \dot{z}$ and using the method of Dirac Constraints we obtain [14] the two-oscillator Hamiltonian

$$H_{\rm PU}(\omega_1, \omega_2) = \tfrac{1}{2}p_x^2 + p_z x + \tfrac{1}{2}\left(\omega_1^2 + \omega_2^2\right)x^2 - \tfrac{1}{2}\omega_1^2\omega_2^2 z^2, \quad [z, p_z] = i, \quad [x, p_x] = i, \tag{4.3}$$

with canonical variables $z$, $p_z$, and $x$, $p_x$. And before even quantizing the theory (i.e., before implementing the $[z,p_z]=i$, $[x,p_x]=i$ commutation relations) we immediately encounter a separate problem in addition to the ghost problem. Specifically, because of the overall minus sign in the $-\frac{1}{2}\omega_1^2\omega_2^2z^2$ term, the energy spectrum is unbounded from below. The existence of this kind of classical instability was first identified by Ostrogradski [19] as being characteristic of higher-derivative theories.

To now quantize the theory we make the standard substitutions

$$\begin{aligned} z &= a_1+a_1^\dagger+a_2+a_2^\dagger, \quad p_z=i\omega_1\omega_2^2(a_1-a_1^\dagger)+i\omega_1^2\omega_2(a_2-a_2^\dagger), \\ x &= -i\omega_1(a_1-a_1^\dagger)-i\omega_2(a_2-a_2^\dagger), \quad p_x=-\omega_1^2(a_1+a_1^\dagger)-\omega_2^2(a_2+a_2^\dagger), \end{aligned} \tag{4.4}$$

and obtain a Hamiltonian and commutator algebra [14]

$$\begin{aligned} H_{\rm PU}(\omega_1,\omega_2)&=2(\omega_1^2-\omega_2^2)(\omega_1^2a_1^\dagger a_1-\omega_2^2a_2^\dagger a_2)+\tfrac{1}{2}(\omega_1+\omega_2), \\ [a_1,a_1^\dagger]&=\frac{1}{2\omega_1(\omega_1^2-\omega_2^2)}, \quad [a_2,a_2^\dagger]=-\frac{1}{2\omega_2(\omega_1^2-\omega_2^2)}. \end{aligned} \tag{4.5}$$

We note that on taking $\omega_1>\omega_2>0$ for definitiveness, the sign of the $[a_2,a_2^\dagger]$ commutator is negative.

If we define the vacuum according to

$$a_1|\Omega_1\rangle=0, \quad a_2|\Omega_1\rangle=0, \tag{4.6}$$

we find that the energy spectrum is bounded from below with $|\Omega_1\rangle$ being the ground state with energy $(\omega_1+\omega_2)/2$. But the excited state $a_2^\dagger|\Omega_1\rangle$, which lies at energy $\omega_2$ above the ground state, has a Dirac norm $\langle\Omega_1|a_2a_2^\dagger|\Omega_1\rangle=-1/[2\omega_2(\omega_1^2-\omega_2^2)]$ that is negative.

On the other hand if we define the vacuum according to

$$a_1|\Omega_2\rangle=0, \quad a_2^\dagger|\Omega_2\rangle=0, \tag{4.7}$$

the theory is then free of negative-norm states, but the energy spectrum is unbounded from below, with negative energy states propagating forward in time.

As we see, the discussion parallels the discussion we gave in Sec. 2 for the single harmonic oscillator as we again have to deal with negative energies or negative norms. However, unlike that case we note that now the two problems occur in different Hilbert spaces and not in the same one, one built on a vacuum that obeys $a_2|\Omega_1\rangle=0$ and the other built on a vacuum that obeys $a_2^\dagger|\Omega_2\rangle=0$. While having either negative norms or negative energies would be problematic, in no single Hilbert space does one have to deal with both problems. There is nothing one could do with a theory whose energy spectrum is unbounded from below. Thus the only chance for the theory is if we can make sense of the $a_2|\Omega_1\rangle=0$ sector with its seeming states of negative norm. As we now show, using the techniques of $PT$ theory we actually can do that, with the solution involving the one thing we have not yet discussed for the $PU$ oscillator theory, namely boundary conditions. And that $PT$ symmetry might even be relevant, we note that all the eigenvalues of $H_{\rm PU}$ are real (standard two-oscillator spectrum, with all the poles of (3.8) and (4.2) being on the real frequency axis). As our discussion in Sec. 1 indicates, the Hamiltonian is either Hermitian or has an antilinear symmetry. Boundary conditions will then tell us which one it is.

As to whether $PT$ might actually be able to solve this ghost problem at all, we recall that it has already had success in solving another ghost problem, namely the one associated with the Lee model [20]. The Lee model was itself introduced in order to address renormalization issues as in it one could determine the relation between the bare and renormalized coupling constants in a closed form. This theory also turned out to have a negative Dirac norm ghost problem, and given its relevance to Pauli-Villars this aspect of it was much studied at the time [21, 22]. However, it was only after the advent of $PT$ theory that this problem was finally solved [23]. In the Lee model the ghost state is only found to occur for a specific range of values of the bare coupling constant, and in this range the bare coupling constant is complex. This meant that the theory was not Hermitian in this range, and the authors of [23] showed that it instead was $PT$ symmetric. One then has to use the $PT$ inner product rather than the Dirac one, and the $PT$ norm was found to be positive. The resolution of this ghost problem should be regarded as a considerable triumph for $PT$ theory, especially since it had not at all been developed for that purpose. So we now apply the same techniques to the Pais-Uhlenbeck oscillator theory.

## 5. Solving the ghost problem

To identify the appropriate boundary conditions for the Pais-Uhlenbeck theory we set $p_z = -i\partial_z$, $p_x = -i\partial_x$ in (4.3) and solve the associated Schrödinger equation. For the state $|\Omega_1\rangle$ with energy $(\omega_1+\omega_2)/2$ we obtain the wave function [14]

$$\psi_0(z,x) = \exp\left[\tfrac{1}{2}(\omega_1+\omega_2)\omega_1\omega_2 z^2 + i\omega_1\omega_2 zx - \tfrac{1}{2}(\omega_1+\omega_2)x^2\right], \tag{5.1}$$

with analogous wave functions being found for the excited states that are built out of $|\Omega_1\rangle$. The wave function $\psi_0(z,x)$ is a somewhat unusual wave function as it is convergent in one of its variables ($x$) while being divergent in its other one ($z$). Since it is divergent in $z$ the wave function is not normalizable on the real $z$ axis. Hence the Dirac norm of $|\Omega_1\rangle$, viz. $\langle\Omega_1|\Omega_1\rangle = \int dzdx\langle\Omega_1|z,x\rangle\langle z,x|\Omega_1\rangle = \int dzdx\psi_0^*(z,x)\psi_0(z,x)$, is not finite. Thus even before we ask what to do about this divergence, we note immediately that the closure relation given in (3.4), viz. $\sum|n_1\rangle\langle n_1| - \sum|n_2\rangle\langle n_2| = I$, could not hold as its validity assumes that all the states that appear in it are normalizable. Hence any conclusions that might be drawn from it such as the existence of ghost states are not valid. This of course does not mean that there are no ghost states, it means only that this particular line of reasoning is invalid.

To address whether or not there might still be ghost states anyway, we need to deal with the divergence of $\psi_0(z,x)$ at large $z$. Noting that it would become normalizable if $z$, but not $x$, were to be taken to be pure imaginary, we thus identify two complex $z$ plane Stokes wedges in which $\psi_0(z,x)$ is normalizable, namely the upper and lower quadrants of a letter $X$ shape drawn in the complex $z$ plane. The quantum theory does not live in the left and right quadrants of this letter $X$, and while the $-\frac{1}{2}\omega_1^2\omega_2^2 z^2$ term might give an Ostrogradski instability with real $z$, it does not do so with imaginary $z$. To implement making $z$ pure imaginary we follow (2.7) and implement the similarity transformation [15]

$$y = e^{\pi p_z z/2} z e^{-\pi p_z z/2} = -iz, \quad q = e^{\pi p_z z/2} p_z e^{-\pi p_z z/2} = ip_z, \quad [y,q]=i, \qquad [x,p]=i, \tag{5.2}$$

with $p_x = p$, and obtain

$$e^{\pi p_z z/2} H_{\rm PU}(\omega_1,\omega_2) e^{-\pi p_z z/2} = \bar{H} = \frac{p^2}{2} - iqx + \frac{1}{2}\left(\omega_1^2+\omega_2^2\right)x^2 + \frac{1}{2}\omega_1^2\omega_2^2 y^2. \tag{5.3}$$

With $\psi_0(z=iy,x)$ taking the form

$$\begin{aligned}\psi_0(y,x) &= \exp\left[-\tfrac{1}{2}(\omega_1+\omega_2)\omega_1\omega_2 y^2 - \omega_1\omega_2 yx - \tfrac{1}{2}(\omega_1+\omega_2)x^2\right], \\ &= \exp\left\{-\frac{1}{2(\omega_1+\omega_2)}\left[\omega_1\omega_2\left[(\omega_1+\omega_2)y+x\right]^2 + x^2(\omega_1^2+\omega_2^2+\omega_1\omega_2)\right]\right\},\end{aligned} \tag{5.4}$$

and with $\omega_1$ and $\omega_2$ both positive, the wave function is now well-behaved for all $x$ and $y$. With $p$, $q$, $x$ and $y$ now acting as Hermitian operators on $\psi_0(y,x)$, through its $i$ factor the Hamiltonian $\bar{H}$ is manifestly not Hermitian, but all of its eigenvalues are still real since one cannot change eigenvalues by a similarity transformation, and they already were all real before we made the transformation. Given the similarity of its $i$ dependence to that of $H = p^2 + ix^3$, we find [15] that $\bar{H}$ thus falls into the class of non-Hermitian theories that have a PT symmetry ($x$ has $P$, $T$ and $PT$ phases $(-,+,-)$, $p$ is $(-.-.+)$, $q$ is $(+,+,+)$, $y$ is $(+,-,-)$) as realized in the PT symmetry realization in which all energy eigenvalues are real.

With $\bar{H}$ not being Hermitian we must distinguish between its left-eigenvectors and right-eigenvectors (as defined in (5.11) below), and for the ground state they take the form

$$\begin{aligned}\psi_0^R(y,x,t) &= \exp\left[-\tfrac{1}{2}(\omega_1+\omega_2)\omega_1\omega_2 y^2 - \omega_1\omega_2 yx - \tfrac{1}{2}(\omega_1+\omega_2)x^2 - \tfrac{i}{2}(\omega_1+\omega_2)t\right],\\ \psi_0^L(y,x,t) &= \exp\left[-\tfrac{1}{2}(\omega_1+\omega_2)\omega_1\omega_2 y^2 + \omega_1\omega_2 yx - \tfrac{1}{2}(\omega_1+\omega_2)x^2 + \tfrac{i}{2}(\omega_1+\omega_2)t\right],\end{aligned} \tag{5.5}$$

with a normalization

$$\int dx dy \psi_0^L(y,x,t)\psi_0^R(y,x,t) = \frac{\pi}{(\omega_1+\omega_2)(\omega_1\omega_2)^{1/2}} \tag{5.6}$$

that is finite, time independent and real.

To remove the $-iqx$ cross term from $\hat{H}$ and decouple the oscillators we introduce a Hermitian operator $Q$ according to

$$Q = \alpha(pq + \omega_1^2\omega_2^2 xy) = Q^\dagger, \qquad \alpha = \frac{1}{\omega_1\omega_2}\log\left(\frac{\omega_1+\omega_2}{\omega_1-\omega_2}\right), \tag{5.7}$$

with an additional similarity transformation bringing $\hat{H}$ to the form [15]

$$e^{-Q/2}\bar{H}e^{Q/2} = \bar{H}' = \frac{p^2}{2} + \frac{q^2}{2\omega_1^2} + \frac{1}{2}\omega_1^2 x^2 + \frac{1}{2}\omega_1^2\omega_2^2 y^2. \tag{5.8}$$

We recognize $\bar{H}'$ as being a fully acceptable standard, positive norm two-dimensional oscillator system.

In addition, we note that with its phase being $-Q/2$ rather than $-iQ/2$, the $e^{-Q/2}$ operator is not unitary. The transformation from $\bar{H}$ to $\bar{H}'$ is thus not a unitary transformation, but is a transformation from a skew basis with eigenvectors $|n\rangle$ to an orthogonal basis with eigenvectors

$$|n'\rangle = e^{-Q/2}|n\rangle, \quad \langle n'| = \langle n|e^{-Q/2}. \tag{5.9}$$

Then since $\langle n'|m'\rangle = \delta_{mn}$, the eigenstates of $\bar{H}$ obey

$$\begin{aligned}&\langle n|e^{-Q}|m\rangle = \delta_{mn}, \quad \sum_n |n\rangle\langle n|e^{-Q} = I, \quad \bar{H} = \sum_n |n\rangle E_n \langle n|e^{-Q},\\ &\bar{H}|n\rangle = E_n|n\rangle, \quad \langle n|e^{-Q}\bar{H} = \langle n|e^{-Q}E_n.\end{aligned} \tag{5.10}$$

We thus recognize the inner product as being not $\langle n|m\rangle$ but $\langle n|e^{-Q}|m\rangle$, with the conjugate of $|n\rangle$ being $\langle n|e^{-Q}$, where according to (5.10) $\langle n|e^{-Q}$ is a left eigenstate $\langle L|$ of $\bar{H}$.

Moreover, the $e^{-Q}$ transformation effects $e^{-Q}\bar{H}e^{Q} = \bar{H}^\dagger$. The inner product $\langle n|e^{-Q}|m\rangle$ thus emerges as the time-independent $V$ norm $\langle n|V|m\rangle$ that we introduced in Sec. 1, with $e^{-Q}$ serving as $V$ in the $PU$ case, and with $V$ effecting $V\bar{H}V^{-1} = \bar{H}^\dagger$. We can thus write the norm as $\langle L|R\rangle$ where $|R\rangle$ is a right-eigenvector of $\bar{H}$, and $\langle L| = \langle R|V$ is a left-eigenvector since

$$i\partial_t|R\rangle = \bar{H}|R\rangle, \quad -i\partial_t\langle L| = -i\partial_t\langle R|V = \langle R|\bar{H}^\dagger V = \langle R|V\bar{H} = \langle L|\bar{H}. \tag{5.11}$$

In addition, the state $\langle n|e^{-Q}$ is also the $PT$ conjugate of $|n\rangle$, so that the inner product is the overlap of a state with its $PT$ conjugate rather than of that with its Hermitian conjugate, just as we had noted earlier. And as such this inner product is positive definite since $\langle n'|m'\rangle = \delta_{mn}$ is. The $PU$ oscillator theory is thus a fully viable unitary theory with no states of negative norm in its Hilbert space.

However, now that we have obtained a very conventional looking $\bar{H}'$ we have to ask what happened to the propagator and whether we may have lost the relative minus sign in $D_{\rm PU}(k)$ and its non-relativistic limit as given in (3.8) and (4.2). Since all we have done is make a similarity transformation we could not change the signs in the propagator. Given (5.11) and the discussion given above, we thus have to conclude that the propagator had been misidentified in setting it equal to the standard matrix element given in (3.9). Instead, we must associate it with

$$D_{\rm PU}(k) = -i\int d^4x e^{ik\cdot x}\langle\Omega|VT[\phi(x)\phi(0)]|\Omega\rangle, \tag{5.12}$$

with use of the relativistic generalization of $\sum_n |n\rangle\langle n|e^{-Q} = I$ for the insertion of strictly positive norm intermediate states then, as explicitly shown in [16] (and in [24] for the $PU$ case itself), enabling us to recover the required minus sign in a theory possessing no states of negative norm.

In Sec. 3 we identified two different theories, one based on the two-field action $I_{S_1} + I_{S_2}$ given in (3.2), and the other based on the one-field $I_S$ given in (3.6). Both theories lead to the propagator given in (3.8). And since the theory based on $I_{S_1} + I_{S_2}$ explicitly contains the ghost states that appear in the closure relation given in (3.4) it is generally thought that the theory based on $I_S$ must have ghost states too. However, one cannot draw such a conclusion since the theories based on $I_{S_1} + I_{S_2}$ and $I_S$ are different theories. Thus ghosts in one of the theories does not necessitate ghosts in the other one as well. And as we have seen, if we associate the propagator given in (3.8) with a fourth-order derivative theory rather than a second-order one, then one not only has a theory that is unitary, in addition none of its good behavior in the ultraviolet is impaired.

To provide some further insight into our resolution of the ghost problem, we note that when we introduced creation and annihilation operators $a_1, a_2, a_1^\dagger, a_2^\dagger$ for the original variables $z, p_z, x, p_x$ in (4.4), we were led to a matrix element $\langle\Omega_1|a_2a_2^\dagger|\Omega_1\rangle$ that is negative. Since such a matrix element is in the form of the modulus squared of $a_2^\dagger|\Omega_1\rangle$ it should have been positive, and so it was proposed to resolve this conflict by quantizing the theory with an indefinite Hilbert space metric. As formulated, this analysis presupposed that the original variables $z, p_z, x, p_x$ are all Hermitian, so that $a_2^\dagger$ is then the Hermitian conjugate of $a_2$. However, by a study of boundary conditions we have found that this is not in fact the case. Consequently, $a_2^\dagger$ as defined in (4.4) is not in fact the Hermitian conjugate of $a_2$, and there is then no contradiction in obtaining a negative value for $\langle\Omega_1|a_2a_2^\dagger|\Omega_1\rangle$, and thus no need to invoke a negative metric at all. On constructing a creation and annihilation algebra for the transformed $y, q, x, p$ variables the commutation algebra given in (4.5) is replaced by one in which all commutators have positive signature [16].

Also we note that while $\bar{H}'$ is a straightforward free theory two-oscillator Hamiltonian, interactions are not as straightforward. If we add a $\lambda z^4$ term onto $H_{\rm PU}(\omega_1, \omega_2)$ given in (4.3) this is equivalent to adding a term $\lambda y^4$ onto $\bar{H}$ given in (5.3). Thus in $\bar{H} + \lambda y^4$ the $\bar{H}$ term is not Hermitian while the $\lambda y^4$ term is. On the other hand if we now make the transformation to the Hermitian $\bar{H}'$ given in (5.8) we have to transform the $\lambda y^4$ term as well. On transforming $y$ we obtain [16]

$$y' = e^{-Q/2}ye^{Q/2} = y\cosh\theta + \frac{i}{\omega_1\omega_2}p\sinh\theta, \quad \tanh\theta = \frac{\omega_2}{\omega_1}. \tag{5.13}$$

Thus now it is $\lambda y'^4$ that is not Hermitian. Thus in neither case is the total Hamiltonian Hermitian. As we now show, this has consequences for interacting processes.

## 6. Interactions

**Figure 1.** First radiative correction in scalar field theory.

Once we have shown that there are no negative norms associated with the free scalar-field theory action $I_S$ given in (3.6), it must be the case that when we add on an interaction such as $-\lambda\phi^4$ to $I_S$ no negative norm states can be generated by perturbative radiative corrections since one cannot change the signature of a Hilbert space in perturbation theory. Despite this, if we insert the propagator given in (3.8) into the one loop Feynman diagram shown in Fig. (1) and take the vertices to be of the $L(\text{interacting}) = -\lambda\phi^4$ form, then because of the relative minus sign in (3.8) some of the discontinuities in the graph will be negative, and thus would on their own violate unitarity. Explicit calculation shows that the total discontinuity is given by [24]

$$\begin{aligned}\Pi(p_0) = &- \frac{i\lambda^2}{8\pi(M_1^2-M_2^2)^2}\biggl(\theta(p_0-2M_2)\frac{(p_0^2-4M_2^2)^{1/2}}{2p_0}+\theta(p_0-2M_1)\frac{(p_0^2-4M_1^2)^{1/2}}{2p_0}\\ &- \theta(p_0-M_1-M_2)\frac{[p_0^2-(M_1+M_2)^2]^{1/2}[p_0^2-(M_1-M_2)^2]^{1/2}}{p_0^2}\biggr),\end{aligned} \tag{6.1}$$

where the total incoming four-vector is $(p_0,0,0,0)$. As we can see from (6.1), the contributions with positive discontinuity cannot cancel those with negative discontinuity since the various discontinuities start at different thresholds and have different mass dependences.

However, this is not the complete situation, though it would have been had (the relativistic generalization of) either $\bar{H}+\lambda y^4$ or $\bar{H}'+\lambda y'^4$ been Hermitian. But neither is. Nonetheless, since both have a $PT$ structure, they can be brought to a Hermitian form by the application of yet another similarity transformation of the form $S'=\exp(-Q'/2)$, so that the operator that is now needed for the $V$-norm is $V'=\exp(-Q')$. On doing this we change the conjugate of a right eigenvector $|R\rangle$ to $\langle L|=\langle R|\exp(-Q')$, where the left eigenvector was introduced in Sec. 5. In lowest order this $Q'$ transformation will contain a term of order $\lambda$. Now the negative discontinuity given in (6.1) is of order $\lambda^2$. Thus to cancel it we need some other term of the same order. As noted in [24], this term is provided by the tree approximation point vertex diagram (shaped like the letter $X$), since while it is nominally of order $\lambda$ it is evaluated between states that have been adjusted by $V'$. Since the modification is of order $\lambda$ the net contribution of the tree graph is of order $\lambda^2$. As shown in [24], when this modified tree diagram is taken in conjunction with $\Pi(p_0)$ there is no net negative discontinuity in the scattering amplitude, and the net discontinuity is unitary.

To see why the $V$ modification is needed for unitarity, we note that the S-matrix is given not by $\sum|R_i\rangle\langle R_i|$ but by

$$S=\sum_i|R_i\rangle\langle L_i|=\sum|R_i\rangle\langle R_i|V. \tag{6.2}$$

With the unitary time evolution operator being of the form $U=e^{-iHt}$, then with $H$ obeying $VHV^{-1}=H^\dagger$ as per (1.4), we find that

$$U^\dagger=e^{iH^\dagger t}=Ve^{iHt}V^{-1}=VU^{-1}V^{-1}, \tag{6.3}$$

so that instead of the standard (i.e., Hermitian-based) $UU^\dagger = I$ we have

$$UV^{-1}U^\dagger V = I. \tag{6.4}$$

As we see, the operator $V$ is central to unitarity considerations in scattering processes, and must always be taken into account in non-Hermitian theories that have an antilinear symmetry.

## 7. Jordan-block realization of the Pais-Uhlenbeck theory

Things change quite radically when we take the two frequencies to be equal. On setting $\omega_1 = \omega_2 = \omega$, the $\omega$-dependent Hamiltonian given in (2.16) becomes the Hamiltonian given in (2.15). Applying the frequency-independent (5.2) to (2.15) then brings it to the form

$$e^{\pi p_z z/2} H_{\rm PU}(\omega) e^{-\pi p_z z/2} = \frac{p^2}{2} - iqx + \omega^2 x^2 + \frac{1}{2}\omega^4 y^2. \tag{7.1}$$

As previously the right-eigenvectors and left-eigenvectors are distinct, and for the ground state with energy $\omega$ the eigenfunctions given in (5.5) take the form [16]

$$\begin{aligned}
\psi_0^R(y,x,t) &= \exp\left[-\omega^3 y^2 - \omega^2 yx - \omega x^2 - i\omega t\right], \\
\psi_0^L(y,x,t) &= \exp\left[-\omega^3 y^2 + \omega^2 yx - \omega x^2 + i\omega t\right],
\end{aligned} \tag{7.2}$$

with finite, positive, time-independent normalization

$$\int dydx \psi_0^L(y,x,t)\psi_0^R(y,x,t) = \frac{\pi}{2\omega^2}. \tag{7.3}$$

As we see, the $\omega_1 \neq \omega_2$ ground state wave function given in (5.5), being non-degenerate, continues directly into the $\omega_1 \neq \omega_2$ ground state given in (7.3).

However, this is not the case for the two one-particle states. Before we set $\omega_1 = \omega_2$ the two one-particle excited states with eigenvalues $3\omega_1/2 + \omega_2/2$ and $\omega_1/2 + 3\omega_2/2$ have wave functions and normalizations that take the form [16]

$$\begin{aligned}
&\psi_1^R(y,x,t) = (x+\omega_2 y)\exp\left[-\tfrac{1}{2}(\omega_1+\omega_2)\omega_1\omega_2 y^2 - \omega_1\omega_2 yx - \tfrac{1}{2}(\omega_1+\omega_2)x^2 - \tfrac{i}{2}(3\omega_1+\omega_2)t\right], \\
&\psi_1^L(y,x,t) = \eta_1(x-\omega_2 y)\exp\left[-\tfrac{1}{2}(\omega_1+\omega_2)\omega_1\omega_2 y^2 + \omega_1\omega_2 yx - \tfrac{1}{2}(\omega_1+\omega_2)x^2 + \tfrac{i}{2}(3\omega_1+\omega_2)t\right], \\
&\psi_2^R(y,x,t) = (x+\omega_1 y)\exp\left[-\tfrac{1}{2}(\omega_1+\omega_2)\omega_1\omega_2 y^2 - \omega_1\omega_2 yx - \tfrac{1}{2}(\omega_1+\omega_2)x^2 - \tfrac{i}{2}(\omega_1+3\omega_2)t\right], \\
&\psi_2^L(y,x,t) = \eta_2(x-\omega_1 y)\exp\left[-\tfrac{1}{2}(\omega_1+\omega_2)\omega_1\omega_2 y^2 + \omega_1\omega_2 yx - \tfrac{1}{2}(\omega_1+\omega_2)x^2 + \tfrac{i}{2}(\omega_1+3\omega_2)t\right], \\
&\int dydx \psi_1^L(y,x,t)\psi_1^R(y,x,t) = \frac{\eta_1\pi(\omega_1-\omega_2)\omega_2}{2(\omega_1+\omega_2)^2(\omega_1\omega_2)^{3/2}}, \quad \int dydx \psi_1^L(y,x,t)\psi_2^R(y,x,t) = 0, \\
&\int dydx \psi_2^L(y,x,t)\psi_2^R(y,x,t) = -\frac{\eta_2\pi(\omega_1-\omega_2)\omega_1}{2(\omega_1+\omega_2)^2(\omega_1\omega_2)^{3/2}}, \int dydx \psi_2^L(y,x,t)\psi_1^R(y,x,t) = 0,
\end{aligned} \tag{7.4}$$

where the $\eta_1 = 1$, $\eta_2 = -1$ phases of $\psi_1^L(y,x,t)$ and $\psi_2^L(y,x,t)$ are fixed in accordance with (5.10). (These phases are also the $PT$ phases of $\psi_1^R(y,x,t)$ and $\psi_2^R(y,x,t)$ [12].) In the equal-frequency limit both energies become $2\omega$, and both wave functions collapse onto the common [16]

$$\begin{aligned}
\psi_1^R(y,x,t) &= (x+\omega y)\exp\left[-\omega^3 y^2 - \omega^2 yx - \omega x^2 - 2i\omega t\right], \\
\psi_1^L(y,x,t) &= \eta_1(x-\omega y)\exp\left[-\omega^3 y^2 + \omega^2 yx - \omega x^2 + 2i\omega t\right].
\end{aligned} \tag{7.5}$$

Thus only one right-eigenvector and one left-eigenvector survive, and the Hamiltonian becomes of non-diagonalizable Jordan-block form. As is characteristic of Jordan-block Hamiltonians, with $\int dydx\psi_1^L(y,x,t)\psi_2^R(y,x,t)=0$ as per (7.4), in the $\psi_2^R(y,x,t)\to\psi_1^R(y,x,t)$, $\psi_2^L(y,x,t)\to\psi_1^L(y,x,t)$ limit the norm becomes zero, with

$$\int dydx\psi_1^L(y,x,t)\psi_1^R(y,x,t)=0. \tag{7.6}$$

To see what happened to the other wave functions we set $\omega_1=\omega+\epsilon$, $\omega_2=\omega-\epsilon$, and on taking the $\epsilon\to 0$ limit with a singular $1/2\epsilon$ prefactor we obtain

$$\begin{aligned}\psi_{1a}^R(y,x,t)&=\lim_{\epsilon\to 0}\frac{1}{2\epsilon}(\psi_2^R(y,x,t)-\psi_1^R(y,x,t))\\&=[(x+\omega y)it+y]\exp\left[-\omega^3y^2-\omega^2yx-\omega x^2-2i\omega t\right],\\\psi_{1a}^L(y,x,t)&=\lim_{\epsilon\to 0}\frac{1}{2\epsilon}(\eta_2\psi_2^L(y,x,t)-\eta_1\psi_1^L(y,x,t))\\&=-[(x-\omega y)it+y]\exp\left[-\omega^3y^2+\omega^2yx-\omega x^2+2i\omega t\right],\end{aligned}\tag{7.7}$$

with the resulting linear in $t$ dependence indicating that these two wave functions are not stationary. Thus between (7.5) and (7.7) we still have the same number of wave functions that we had in (7.4). These four wave functions still form a complete basis for the one-particle sector of the equal-frequency Hilbert space and its dual space. It is just the eigenbasis of the equal-frequency Hamiltonian that is not complete.

Even though the wave functions $\psi_{1a}^R(y,x,t)$ and $\psi_{1a}^L(y,x,t)$ are not eigenstates of the equal-frequency Hamiltonian, they are solutions to the time-dependent Schrödinger equation. Then, since they have the same good asymptotic spatial behavior as the eigenstates themselves one can still construct time-independent probabilities for them. These are of the form [16]

$$\begin{aligned}\int dydx\psi_1^L(y,x,t)\psi_{1a}^R(y,x,t)=\int dydx\psi_{1a}^L(y,x,t)\psi_1^R(y,x,t)&=-\frac{\pi}{8\omega^4},\\\int dydx\psi_{1a}^L(y,x,t)\psi_{1a}^R(y,x,t)&=-\frac{\pi}{8\omega^5}.\end{aligned}\tag{7.8}$$

Now initially it does not appear to be possible that these probabilities could be time independent since both $\psi_{1a}^R(y,x,t)$ and $\psi_{1a}^L(y,x,t)$ are growing in time. And yet they have to be since they both obey the Schrödinger equation, and the time derivatives of these probabilities can be written as asymptotic spatial surface terms that vanish. Mathematically, the reason why these probabilities are time independent is because each term that grows with $t$ is multiplied by a coefficient that is zero since the states in (7.6) have zero norm. With this same pattern repeating for the higher excited states as well, and with the stationary and non-stationary states together forming a complete basis, any wave packet that is built out of them will preserve its probability in time. Despite it being radically different from the way unitarity is implemented for Hamiltonians that have a complete basis of eigenstates, this is how unitarity is achieved for non-diagonalizable, Jordan-block Hamiltonians; with it being the basis for the Hilbert space on which the Hamiltonian acts rather than that of the Hamiltonian itself that needs to be complete. And as such, Jordan-block Hamiltonians are completely viable quantum-mechanically, showing that quantum mechanics is far richer than conventional Hermitian quantum mechanics with Dirac inner products.

To gain more insight into this analysis we consider the wave equations given in (3.7) and (4.2). As long as $\omega_1\neq\omega_2$ the solutions to (4.2) are of the form $\exp(-i\omega_1 t)$, $\exp(-i\omega_2 t)$ and their conjugates. However, when we set $\omega_1 = \omega_2 = \omega$ (or analogously $M_1 = M_2 = M$), the solutions take a different form, viz. $\exp(-i\omega_1 t)$, $t\exp(-i\omega_1 t))$ and their conjugates, since

$$(\partial_t^2 + \omega^2)(\partial_t^2 + \omega^2)(ae^{-i\omega t} + bte^{-i\omega t}) = (\partial_t^2 + \omega^2)[-2ib\omega e^{-i\omega t}] = 0. \tag{7.9}$$

Thus the second class of solutions possesses a term that is linear in $t$. Since such a term is not stationary, already in the classical theory wave equation prior to quantization we can see that the quantized version of the theory will be Jordan block. However, we should also note that in the classical theory, while the appropriate fourth-order theory Hamiltonian may have an Ostrogradski instability, it is still a time-independent constant of motion in these solutions (since that is what Ostrogradski set out to construct), so there is no runaway in energy. And even in the classical theory, if formulated with pure imaginary $z$ there is no instability. That we can continue coordinates into the complex plane even in the classical theory is allowed since symplectic transformations with complex angles preserve the Poisson bracket algebra. In fact, the complex similarity transformations that we made in the quantum theory parallel complex symplectic transformations in the classical theory. And if they both transform through the same angles, there is even a correspondence principle at each of those angles, with Poisson brackets mapping into commutators [3].

We conclude this section by noting three additional features of the equal-frequency limit. First we note that in the limit both $D_{\rm PU}(k)$ given in (3.8) and $G_{\rm PU}(E)$ given in (4.2) become singular. Consequently, the partial fraction decomposition breaks down, and any insight that it might provide into the structure of the eigenspectrum does not carry over to the equal-frequency case. Secondly, the canonical commutator structure given in (4.5) also becomes singular in the limit. And thirdly, the similarity transformation required to diagonalize $H_{\rm PU}$ as given in (5.7) and (5.8) becomes singular in the limit as well, so that no diagonalization could be effected. The equal-frequency theory thus represents a whole branch of quantum mechanics that is simply inaccessible with Hermitian Hamiltonians. As we will see in Sec. 9 this could play a central role in constructing a viable quantum theory of gravity.

## 8. Complex conjugate energy pair realization of the Pais-Uhlenbeck theory

For the complex conjugate energy pair realization with $\omega_1 = \alpha + i\beta$, $\omega_2 = \alpha - i\beta$ with real and positive $\alpha$ and $\beta$ we apply the similarity transform given in (5.2) to (2.14) and obtain

$$e^{\pi p_z z/2} H_{\rm PU}(\alpha,\beta) e^{-\pi p_z z/2} = \frac{p}{2} - iqx + (\alpha^2 - \beta^2)x^2 + \frac{1}{2}(\alpha^2+\beta^2)^2 y^2. \tag{8.1}$$

The ground state energy $(\omega_1 + \omega_2)/2 = \alpha$ is still real, and the associated eigenfunctions are of the form

$$\begin{aligned}\psi_0^R(y,x,t) &= \exp\left[-\alpha(\alpha^2+\beta^2)y^2 - (\alpha^2+\beta^2)yx - \alpha x^2 - i\alpha t\right],\\ \psi_0^L(y,x,t) &= \exp\left[-\alpha(\alpha^2+\beta^2)y^2 + (\alpha^2+\beta^2)yx - \alpha x^2 + i\alpha t\right],\end{aligned} \tag{8.2}$$

with a normalization

$$\int dxdy \psi_0^L(y,x,t)\psi_0^R(y,x,t) = \frac{\pi}{2\alpha(\alpha^2+\beta^2)^{1/2}} \tag{8.3}$$

that is finite, time independent and real.

The one-particle state wave functions with respective energies $2\alpha + i\beta$ and $2\alpha - i\beta$ are

$$\psi_1^R(y,x,t) = (x + (\alpha - i\beta)y)\exp\left[-\alpha(\alpha^2+\beta^2)y^2 - (\alpha^2+\beta^2)yx - \alpha x^2 - i(2\alpha + i\beta)t\right],$$

$$\psi_1^L(y,x,t) = (x-(\alpha-i\beta)y)\exp\left[-\alpha(\alpha^2+\beta^2)y^2+(\alpha^2+\beta^2)yx-\alpha x^2+i(2\alpha+i\beta)t\right],$$
$$\psi_2^R(y,x,t) = (x+(\alpha+i\beta)y)\exp\left[-\alpha(\alpha^2+\beta^2)y^2-(\alpha^2+\beta^2)yx-\alpha x^2-i(2\alpha-i\beta)t\right],$$
$$\psi_2^L(y,x,t) = (x-(\alpha+i\beta)y)\exp\left[-\alpha(\alpha^2+\beta^2)y^2+(\alpha^2+\beta^2)yx-\alpha x^2+i(2\alpha-i\beta)t\right]. \quad (8.4)$$

As can be shown, and as follows from (1.5), the overlaps $\int dydx\psi_1^L(y,x,t)\psi_1^R(y,x,t) = i\pi\beta(\alpha-i\beta)/(4\alpha^2(\alpha^2+\beta^2)^{3/2})$ and $\int dydx\psi_2^L(y,x,t)\psi_2^R(y,x,t) = -i\pi\beta(\alpha+i\beta)/(4\alpha^2(\alpha^2+\beta^2)^{3/2})$ are both finite and time independent. And the overlaps $\int dydx\psi_2^L(y,x,t)\psi_1^R(y,x,t)$ and $\int dydx\psi_1^L(y,x,t)\psi_2^R(y,x,t)$ where the time dependences would not cancel are both zero on account of the spatial integrations.

It should be emphasized that the $\int dydx\psi_1^L(y,x,t)\psi_1^R(y,x,t)$ and $\int dydx\psi_2^L(y,x,t)\psi_2^R(y,x,t)$ overlaps are time independent. Thus despite the existence of states that both grow and decay exponentially in time one can still construct transition matrix elements that are constant in time. And in fact this is enforced by (1.5), since in the complex energy pair realization of the theory (1.4) still holds. The pseudo-Hermitian condition $VHV^{-1} = H^\dagger$ thus takes care of the time dependence of matrix elements. This should be contrasted with the standard discussion of decaying states. There one only allows for exponentials that decay, and one does so by actually taking the Hamiltonian to have a non-Hermitian component, one that necessarily (as per Sec. 1) could not possess an antilinear symmetry. One does not allow for exponentials that grow. In the antilinear case exponential behavior is natural, and not only are both growing and decaying exponentials present, they regulate each other to give time-independent transitions. That these transitions are time independent is due to the fact that the decaying system and the system that it decays into are both involved, with a decrease in population of the decaying system necessitating an increase in population of the system into which it decays. The treatment of decays in the antilinear case is thus more comprehensive than the standard treatment.

One additional interesting feature of $H_{\rm PU}(\alpha,\beta)$ is that it contains no ad hoc, non-antilinearly symmetric, dissipative term of the type one uses for the decay of a single harmonic oscillator, and yet still describes decays. This is because not only is the decaying system taken into consideration, the decay products are included in the analysis too, i.e., a two-oscillator system.

## 9. Implications for quantum gravity

As a quantum theory the standard second-order derivative Einstein gravitational theory is not renormalizable since its $1/k^2$ propagator generates uncontrollable infinities in the ultraviolet. To attempt to ameliorate this difficulty one can add on terms that are quadratic in the Ricci tensor $R_{\mu\nu}$ and the Ricci scalar $R^\alpha{}_\alpha$, to give an action of the form

$$I_{\rm GRAV} = \int d^4x(-g)^{1/2}\left[-AR^\alpha{}_\alpha + B(R^\alpha{}_\alpha)^2 + CR^{\mu\nu}R_{\mu\nu}\right], \quad (9.1)$$

where $A$, $B$ and $C$ are constants, with $A$ having the dimension of inverse length squared, and $B$ and $C$ being dimensionless. One has no need to include any $R_{\lambda\mu\nu\kappa}R^{\lambda\mu\nu\kappa}$ term as it can be traded for the $(R^\alpha{}_\alpha)^2$ and $R^{\mu\nu}R_{\mu\nu}$ terms because $(-g)^{1/2}\left[R_{\lambda\mu\nu\kappa}R^{\lambda\mu\nu\kappa} - 4R_{\mu\kappa}R^{\mu\kappa} + (R^\alpha{}_\alpha)^2\right]$ is a total divergence. On adding on a matter source with energy-momentum tensor $T_{\mu\nu}$, variation of the $I_{\rm GRAV}$ action with respect to the metric generates a gravitational equation of motion of the form

$$AG_{\mu\nu} + BW^{(1)}_{\mu\nu} + CW^{(2)}_{\mu\nu} = -\frac{1}{2}T_{\mu\nu}. \quad (9.2)$$

Here $G_{\mu\nu}$ is the Einstein tensor and $W^{(1)}_{\mu\nu}$ and $W^{(2)}_{\mu\nu}$ may for instance be found in [25], with these various tensors being of the form

$$G^{\mu\nu} = R^{\mu\nu} - \frac{1}{2}g^{\mu\nu}g^{\alpha\beta}R_{\alpha\beta},$$

$$W^{\mu\nu}_{(1)} = 2g^{\mu\nu}\nabla_\beta\nabla^\beta R^\alpha{}_\alpha - 2\nabla^\nu\nabla^\mu R^\alpha{}_\alpha - 2R^\alpha{}_\alpha R^{\mu\nu} + \frac{1}{2}g^{\mu\nu}(R^\alpha{}_\alpha)^2,$$
$$W^{\mu\nu}_{(2)} = \frac{1}{2}g^{\mu\nu}\nabla_\beta\nabla^\beta R^\alpha{}_\alpha + \nabla_\beta\nabla^\beta R^{\mu\nu} - \nabla_\beta\nabla^\nu R^{\mu\beta} - \nabla_\beta\nabla^\mu R^{\nu\beta} - 2R^{\mu\beta}R^\nu{}_\beta + \frac{1}{2}g^{\mu\nu}R_{\alpha\beta}R^{\alpha\beta}. \tag{9.3}$$

If we now linearize about flat spacetime with background metric $\eta_{\mu\nu}$ and fluctuation metric $g_{\mu\nu} = \eta_{\mu\nu} + h_{\mu\nu}$, to first perturbative order we obtain

$$\begin{aligned}
\delta G^{\mu\nu} &= \frac{1}{2}\left(\partial_\alpha\partial^\alpha h_{\mu\nu} - \partial_\mu\partial^\alpha h_{\alpha\nu} - \partial_\nu\partial^\alpha h_{\alpha\mu} + \partial_\mu\partial_\nu h\right) - \frac{1}{2}\eta_{\mu\nu}\left(\partial_\alpha\partial^\alpha h - \partial^\alpha\partial^\beta h_{\alpha\beta}\right),\\
\delta W^{(1)}_{\mu\nu} &= [2\eta_{\mu\nu}\partial_\alpha\partial^\alpha - 2\partial_\mu\partial_\nu][\partial_\beta\partial^\beta h - \partial_\lambda\partial_\kappa h^{\lambda\kappa}],\\
\delta W^{(2)}_{\mu\nu} &= \frac{1}{2}\partial_\beta\partial^\beta[\partial_\mu\partial_\nu h - \partial_\mu\partial_\lambda h^\lambda{}_\nu - \partial_\nu\partial_\lambda h^\lambda{}_\mu + \partial_\lambda\partial^\lambda h_{\mu\nu}]\\
&+ \frac{1}{2}[g_{\mu\nu}\partial_\beta\partial^\beta - 2\partial_\mu\partial_\nu][\partial_\lambda\partial^\lambda h - \partial_\kappa\partial_\lambda h^{\kappa\lambda}].
\end{aligned} \tag{9.4}$$

Here $h = \eta^{\mu\nu}h_{\mu\nu}$, and $\delta W^{\mu\nu}_{(1)}$ and $\delta W^{\mu\nu}_{(2)}$ are given in [26]. On taking the trace of the fluctuation around a background (9.2) we obtain

$$[-3A + (6B + 2C)\partial_\beta\partial^\beta]\left(\partial_\lambda\partial^\lambda h - \partial_\kappa\partial_\lambda h^{\kappa\lambda}\right) = -\frac{1}{2}\eta^{\mu\nu}\delta T_{\mu\nu}. \tag{9.5}$$

In the convenient transverse gauge where $\partial_\mu h^{\mu\nu} = 0$, the propagator for $h$ is given by

$$D(h,k^2) = \frac{1}{(6B+2C)(k^2+i\epsilon)(k^2-M^2+i\epsilon)} = -\frac{1}{M^2(6B+2C)}\left(\frac{1}{k^2+i\epsilon} - \frac{1}{k^2-M^2+i\epsilon}\right),$$
$$M^2 = \frac{-3A}{6B+2C}. \tag{9.6}$$

As we see, in this case the graviton propagator for $h$ that would be associated with the Einstein tensor $\delta G_{\mu\nu}$ alone is replaced by the Pauli-Villars type $D_{\rm PV}(k) = 1/(k^2+i\epsilon) - 1/(k^2-M^2+i\epsilon)$ propagator given in (3.1). And now the leading behavior at large momenta is $1/k^4$. In consequence, and with the same behavior occurring for the other components of $h_{\mu\nu}$, the theory is now renormalizable [18]. However, as such the theory suffers some shortcomings, the possibility of ghosts due to the relative minus sign in $D_{\rm PV}(k)$ or the possibility of negative energies, and the existence of additional particles due to the presence of poles in the $1/(k^2 - M^2 + i\epsilon)$ component of the propagator [18]. That there would be additional particles in the theory is actually an indirect consequence of a theorem due to Weinberg [27], namely that if one has a single massless spin two particle in a standard positive definite Hilbert space it must couple via the Einstein tensor. Thus any departure from pure Einstein would necessitate new particles (string theory also has this concern). As we had noted above, one can avoid negative energies as they lie in a different Hilbert space than the ghost states. And then one can even avoid negative Dirac norm ghost states since one instead has to use the $PT$ symmetry theory inner product. The work of [15] thus makes the theory based on (9.1) unitary. However, one cannot get rid of the extra states, and making them heavy enough to have evaded detection may only be postponing the inevitable.

There is however an alternate fourth-order derivative theory that is renormalizable, has no negative energy particles, is unitary, and yet does not have additional particles, namely the conformal gravity theory that has been advocated and explored in [25, 28, 29] and references therein. In the theory the requirement of local conformal invariance under $g_{\mu\nu}(x) \to e^{2\alpha(x)}g_{\mu\nu}(x)$, where $\alpha(x)$ is a local function of the coordinates, requires a polynomial action to have the unique form

$$I_{\rm W} = -\alpha_g \int d^4x\,(-g)^{1/2} C_{\lambda\mu\nu\kappa}C^{\lambda\mu\nu\kappa} \equiv -2\alpha_g \int d^4x\,(-g)^{1/2}\left[R_{\mu\kappa}R^{\mu\kappa} - \frac{1}{3}(R^{\alpha}{}_{\alpha})^2\right]. \qquad (9.7)$$

Here $\alpha_g$ is a dimensionless gravitational coupling constant, and $C_{\lambda\mu\nu\kappa}$ is the conformal Weyl tensor. Functional variation with respect to the metric $g_{\mu\nu}(x)$ generates fourth-order derivative gravitational equations of motion of the form (see e.g. [25])

$$\begin{aligned}\frac{1}{(-g)^{1/2}}\frac{\delta I_{\rm W}}{\delta g_{\mu\nu}} &= -2\alpha_g W^{\mu\nu} = -2\alpha_g\left[2\nabla_{\kappa}\nabla_{\lambda}C^{\mu\lambda\nu\kappa} - R_{\kappa\lambda}C^{\mu\lambda\nu\kappa}\right]\\ &= -2\alpha_g\left[W^{\mu\nu}_{(2)} - \frac{1}{3}W^{\mu\nu}_{(1)}\right] = -\frac{1}{2}T^{\mu\nu},\end{aligned} \qquad (9.8)$$

where the functions $W^{\mu\nu}_{(1)}$ and $W^{\mu\nu}_{(2)}$ are given above.

As such, the theory admits of a vacuum solution in which the Ricci tensor vanishes, and is thus able to meet the same solar system tests as the Ricci flat solution to Einstein gravity. However, the conformal theory has other solutions beyond the Newtonian potential, such as linear and quadratic potentials, and is then able to fit galactic rotation curves and the accelerating universe data without dark matter and with a cosmological constant that is under control because of the underlying conformal symmetry of the theory (see [29] and references therein).

For our purposes here we note that when $W_{\mu\nu}$ is linearized about a flat background, to first perturbative order one obtains [30]

$$\begin{aligned}\delta W_{\mu\nu} &= \frac{1}{2}(\eta^{\rho}{}_{\mu}\partial^{\alpha}\partial_{\alpha} - \partial^{\rho}\partial_{\mu})(\eta^{\sigma}{}_{\nu}\partial^{\beta}\partial_{\beta} - \partial^{\sigma}\partial_{\nu})K_{\rho\sigma}\\ &- \frac{1}{6}(\eta_{\mu\nu}\partial^{\gamma}\partial_{\gamma} - \partial_{\mu}\partial_{\nu})(\eta^{\rho\sigma}\partial^{\delta}\partial_{\delta} - \partial^{\rho}\partial^{\sigma})K_{\rho\sigma},\end{aligned} \qquad (9.9)$$

where $K_{\mu\nu} = h_{\mu\nu} - \frac{1}{4}\eta_{\mu\nu}\eta^{\alpha\beta}h_{\alpha\beta}$ with $\eta^{\mu\nu}K_{\mu\nu} = 0$. Now we had shown above that both $\delta W^{\mu\nu}_{(1)}$ and $\delta W^{\mu\nu}_{(2)}$ depend on the trace $h$. However, in the particular linear combination $\delta W^{\mu\nu}_{(2)} - (1/3)\delta W^{\mu\nu}_{(1)}$ the trace drops out and $\delta W^{\mu\nu}$ only depends on the traceless $K_{\mu\nu}$. The distinction here is that in (9.1) the $B(R^{\alpha}{}_{\alpha})^2$ and $CR^{\mu\nu}R_{\mu\nu}$ terms with dimensionless $B$ and $C$ are separately globally scale invariant, i.e., invariant under $g_{\mu\nu}(x) \to e^{2\alpha}g_{\mu\nu}(x)$ where $\alpha$ is a constant. However, the action based on the particular combination $R^{\mu\nu}R_{\mu\nu} - (1/3)(R^{\alpha}{}_{\alpha})^2$ has a local symmetry for all local $\alpha(x)$. With $W_{\mu\nu}$ being traceless, as is appropriate to a conformal theory, it follows that $\delta[\eta^{\mu\nu}W_{\mu\nu}] = \eta^{\mu\nu}\delta W_{\mu\nu} - h^{\mu\nu}W_{\mu\nu} = 0$. But for a flat background $W_{\mu\nu} = 0$, and thus for fluctuations around flat it follows that, as can readily be checked, $\eta^{\mu\nu}\delta W_{\mu\nu} = 0$. Thus even though there are ten $h_{\mu\nu}$ fluctuations, because $\delta W_{\mu\nu}$ is traceless it can only depend on the nine degrees contained in the traceless $K_{\mu\nu}$.

From the structure of (9.9), we find that in the convenient transverse gauge $\partial_{\mu}K^{\mu\nu} = 0$ the perturbative equation of motion (9.9) simplifies to [30]

$$4\alpha_g\delta W_{\mu\nu} = 2\alpha_g(\partial_t^2 - \vec{\nabla}^2)^2K_{\mu\nu} = \delta T_{\mu\nu}. \qquad (9.10)$$

With the perturbative equation of motion being diagonal in the $(\mu,\nu)$ indices, we can treat each component of $K_{\mu\nu}$ independently. Noting the similarity with (7.9), we see that the solutions to a source-free (9.10) are of the form $\exp(-ik\cdot x)$ and $t\exp(-ik\cdot x)$ together with their conjugates. As our discussion in Sec. 7 shows, with there being such non-stationary solutions, the quantum Hamiltonian of the theory must be of Jordan-block form, and in [30] this was confirmed explicitly.

In momentum space the propagator associated with (9.10) behaves as $1/k^4$, to thus have power counting renormalizable behavior in the ultraviolet. This $1/k^4$ propagator is often thought of as being a limit of a massive particle Pauli-Villars propagator, but as we have shown above this cannot be the case since the limit of equal masses or of setting $A = 0$, i.e., $M^2 = 0$, in (9.6) is singular. Moreover, we note that the $D(h, k^2)$ propagator given in (9.6) becomes undefined if $6B + 2C = 0$. This is precisely the conformal gravity case where $B = -C/3$, and this has to be so since the trace $h$ decouples from the conformal gravity fluctuation equation given in (9.9). Thus as we see, not only is the limit from second order plus fourth order to pure fourth order singular, the limit from a globally scale invariant fourth-order theory to a locally conformal invariant fourth-order theory is singular too.

To the extent that one might want to relate the $1/k^4$ propagator to a limit of a propagator with mass we could write it as

$$\frac{1}{(k^2+i\epsilon)^2} = \lim_{M^2\to 0}\frac{d}{dM^2}\left(\frac{1}{k^2-M^2+i\epsilon}\right). \tag{9.11}$$

This shows that the number of poles is the same as the number of poles that occur in a second-order theory rather than in a massive fourth-order theory, while also showing that there are no relative minus signs in the propagator. The particle spectrum of pure fourth-order theories is totally different from that of a second-order plus fourth-order theory. Thus even while conformal gravity is based on fourth-order derivative equations it does not have a proliferation of particle states. First, it loses particle states since some of the states are not eigenstates, and second the eigenstates that it does have are zero norm states that would leave no imprint in a detector. Even though the theory only contains one graviton eigenstate despite not being based on the Einstein equations, it evades Weinberg's theorem by not being formulated in a positive definite Hilbert space but in one with zero norm states. Since (9.10) is a wave equation, conformal gravity does have gravitational waves even as the quantized form of these waves have zero norm. Quantum conformal gravity is renormalizable, is unitary in the way described in Sec. 7 for Jordan-block systems, and has no extraneous particles. All of this is achievable because conformal gravity is a $PT$ theory rather than an Hermitian one, and could not have been achieved otherwise.

Beyond this there is one further aspect to the theory, which initially might appear to be of concern. Since we need to make the transformation in (5.2) we would be led to a metric of the form $ig_{\mu\nu}$. However, since $g^{\mu\nu}g_{\nu\tau} = \delta^{\mu}_{\tau}$, we would have to make a change $-ig^{\mu\nu}$ in the contravaraint component of the metric. Then, since the Levi-Civita connection $\Gamma^{\lambda}_{\mu\nu} = \frac{1}{2}g^{\lambda\alpha}(\partial_{\mu}g_{\nu\alpha} + \partial_{\nu}g_{\mu\alpha} - \partial_{\alpha}g_{\nu\mu})$ contains both contravariant and covariant metric components, it would be left unchanged under this transformation. Since the Riemann tensor $R^{\lambda}{}_{\mu\nu\kappa}$ is built out of the connection and its derivatives the Riemann tensor is also not affected by the change in phase of the metric, though it could be affected by raising or lowering indices. Then, with (9.7) being quadratic in the Riemann tensor the $I_{\rm W}$ action would still be real.

Now a reader might think that higher-derivative quantum gravity is only of academic interest. To see that this is not in fact the case consider the Dirac action for a massless fermion coupled to a background geometry of the form

$$I_{\rm D} = \int d^4x(-g)^{1/2}i\bar{\psi}\gamma^c V^{\mu}_c(\partial_{\mu} + \Gamma_{\mu})\psi, \tag{9.12}$$

where the $V^{\mu}_a$ are vierbeins and $\Gamma_{\mu} = -(1/8)[\gamma_a, \gamma_b](V^b_{\nu}\partial_{\mu}V^{a\nu} + V^b_{\lambda}\Gamma^{\lambda}_{\nu\mu}V^{a\nu})$ is the spin connection. In (9.12) the spin connection has expressly been introduced in order to make $I_{\rm D}$ be locally Lorentz invariant. Now while the spin connection $\Gamma_{\mu}$ was not at all introduced for any other purpose, nonetheless, in its presence $I_{\rm D}$ has an additional symmetry, namely that it is locally conformal invariant under

$$V^{\mu}_a \to e^{-\alpha(x)}V^{\mu}_a(x), \quad \psi(x) \to e^{-3\alpha(x)/2}\psi(x), \quad g_{\mu\nu}(x) \to e^{2\alpha(x)}g_{\mu\nu}(x). \tag{9.13}$$

We thus get local conformal invariance for free. In fact, other than the double-well Higgs potential, the entire $SU(3) \times SU(2) \times U(1)$ standard model is locally conformal invariant. Thus if fermion masses are generated dynamically [29], then the entire standard model would be locally conformal invariant. 't Hooft [31] has also argued that there should be an underlying local conformal symmetry in nature.

We now introduce the path integral $\int D[\psi]D[\bar{\psi}] \exp iI_{\rm D} = \exp(iI_{\rm EFF})$, and on performing the path integration on $\psi$ and $\bar{\psi}$ obtain an effective action with leading term [32]

$$I_{\rm EFF} = \int d^4x(-g)^{1/2}\frac{C}{20}\left[R_{\mu\nu}R^{\mu\nu} - \tfrac{1}{3}(R^\alpha{}_\alpha)^2\right], \tag{9.14}$$

where $C$ is a logarithmically divergent constant. Thus when the standard model is coupled to gravity we generate the fourth-order derivative conformal gravity action. But the standard model is ghost free. Since this particular fermion path integration is equivalent to a one loop Feynman diagram, and since one cannot change the signature of a Hilbert space in perturbation theory, conformal gravity must be ghost free too. And if it were not, then the standard model would not be unitary either. When coupled to gravity, radiative corrections to $SU(3) \times SU(2) \times U(1)$ generate fourth-order conformal gravity whether we like it or not. We thus have to deal with conformal gravity one way or another, and we do deal with it by using the $PT$ symmetry inner product. Now the steps leading from the flat space Dirac action to the curved space (9.12) and then to (9.14) are all completely standard. Thus the steps leading to the conformal gravity action given in (9.14) are all completely standard and beyond reproach. But conformal gravity is a $PT$ theory. We are thus led, essentially inexorably, to gravity being a $PT$ theory, and to $PT$ theory thus being relevant to fundamental physics.

## 10. Final comments

In this paper we have explored Hamiltonians that are not Hermitian but instead have an antilinear symmetry. As such, there is an intrinsic difference between Hermiticity and antilinearity. Specifically, how an operator might behave under an antilinear transformation is an intrinsic property of the operator alone. However, whether or not an operator might be Hermitian is not defined a priori but depends on the nature of the basis on which it acts. To enable theories with antilinear symmetry to give time-independent probabilities we have had to replace the standard Dirac inner product by the antilinear symmetry inner product in which the bra vector is no longer the Hermitian conjugate of the ket vector but is instead its antilinear symmetry conjugate. That we can do this at all is because the Schrödinger equation only involves the ket vectors. There is thus some freedom in picking the appropriate bra vectors for the dual space. This freedom has been present in quantum mechanics since its inception but had been overlooked for many years until the emergence of the $PT$ symmetry program. This modification of the inner product is not in any way a change in quantum mechanics, it only uses what quantum mechanics had always allowed, and shows how rich it can actually be. Antilinear symmetry has three distinct realizations: eigenvalues all real and eigenvectors complete, eigenvalues all real but eigenvectors incomplete, eigenvalues in complex conjugate pairs and eigenvectors complete. Of these, only the first one could be realized by Hermitian operators, with the latter two providing options for quantum mechanics that go beyond the Hermitian case. While it had not at all been developed for the purpose, the change in the inner product allows for a solution to the negative ghostlike Dirac inner product that had plagued higher-derivative quantum field theories. Should it turn out that quantum gravity is associated with a fourth-order derivative theory, that theory would have to be a $PT$ theory rather than a Hermitian one. If one of the four fundamental forces in nature turns out to be a $PT$ theory that would constitute a quite considerable achievement for the $PT$ symmetry program.

## References

[1] Bender C M and Boettcher S 1998 *Phys. Rev. Lett.* **80** 5243
[2] Bender C M, Boettcher S and Meisinger P N 1999 *J. Math. Phys.* **40** 2201
[3] Mannheim P D 2018 *J. Phys. A: Math. Theor.* **51** 315302
[4] Bender C M 2007 *Rep. Prog. Phys.* **70** 947
[5] *Special issue on quantum physics with non-Hermitian operators*, ed C. Bender, A. Fring, U. Günther and H. Jones 2012 *J. Phys. A: Math. Theor.* **45** 444001 - 444036
[6] *Theme issue on PT quantum mechanics*, ed C M Bender, M DeKieviet and S P Klevansky 2013 *Phil. Trans. R. Soc.* A **371** issue 1989
[7] El-Ganainy R, Makris K G, Khajavikhan M, Musslimani Z H, Rotter S and Christodoulides D N 2018 *Nature Phys.* **14** 11
[8] Bender C M 2018 *PT Symmetry in Quantum and Classical Physics* (Singapore: World Scientific Press)
[9] *Virtual seminar series on Pseudo-Hermitian Hamiltonians in Quantum Physics*, ed C M Bender, F Correa and A Fring 2021 *J. Phys.: Conf. Series*
[10] Mostafazadeh A 2002 *J. Math. Phys.* **43** 205; **43** 2814; **43** 3944
[11] Solombrino L 2002 *J. Math. Phys.* **43** 5439
[12] Mannheim P D 2018 *Phys. Rev. D* **97** 045001
[13] Pais A and Uhlenbeck G E 1950 *Phys. Rev.* **79** 145
[14] Mannheim P D and Davidson A 2000 *Preprint* hep-th/0001115; 2005 *Phys. Rev. A* **71** 042110
[15] Bender C M and Mannheim P D 2008 *Phys. Rev. Lett.* **100** 110402
[16] Bender C M and Mannheim P D 2008 *Phys. Rev. D* **78** 025022
[17] Pauli W and Villars F 1949 *Rev. Mod. Phys.* **21** 434
[18] Stelle K S 1977 *Phys. Rev. D* **16** 953; 1978 *Gen. Relativ. Gravit.* **9** 353
[19] Ostrogradski M *Memoires sur les equations differentielles relatives au probleme des isoperimetres* 1850 *Mem. Ac. St. Petersbourg, VI* **4** 385
[20] Lee T D 1954 *Phys. Rev.* **95** 1329
[21] Källén G and Pauli W 1955 *Mat. Fys. Medd. Dan. Vid. Selsk.* **30** 7
[22] Heisenberg W 1957 *Nucl. Phys.* **4** 532
[23] Bender C M, Brandt S F, Chen J-H and Wang Q 2005 *Phys. Rev.* D **71** 025014
[24] Mannheim P D 2018 *Phys. Rev.* D **98** 045014
[25] Mannheim P D 2006 *Prog. Part. Nucl. Phys.* **56** 340
[26] Mannheim P D 2012 *Phys. Rev.* D **85** 124008
[27] Weinberg S 1965 *Phys. Rev.* **138** B988
[28] Mannheim P D 2012 *Found. Phys.* **42** 388
[29] Mannheim P D 2017 *Prog. Part. Nucl. Phys.* **94** 125
[30] Mannheim P D 2011 *Gen. Relativ. Gravit.* **43** 703
[31] 't Hooft G 2015 Singularities, horizons, firewalls, and local conformal symmetry *Preprint* arXiv:1511.04427 [gr-qc]
[32] 't Hooft G 2010 Probing the small distance structure of canonical quantum gravity using the conformal group *Preprint* arXiv:1009.0669 [gr-qc]

Journal of Physics: Conference Series **2038** (2021) 012019 doi:10.1088/1742-6596/2038/1/012019

# Non-Hermitian Yukawa interactions of fermions with axions: potential microscopic origin and dynamical mass generation

**Nick E. Mavromatos**

King's College London, Department of Physics, Theoretical Particle Physics and Cosmology Group, Strand, London WC2R 2LS, UK

E-mail: nikolaos.mavromatos@kcl.ac.uk

**Abstract.** In this mini review, we discuss some recent developments regarding properties of (quantum) field-theory models containing anti-Hermitian Yukawa interactions between pseudoscalar fields (axions) and Dirac (or Majorana) fermions. Specifically, we first motivate physically such interactions, in the context of string-inspired low-energy effective field theories, involving right-handed neutrinos and axion fields. Then we proceed to discuss their formal consistency within the so-called Parity-Time-reversal(PT)-symmetry framework. Subsequently, we review dynamical mass generation, induced by the Yukawa interactions, for both fermions and axions. The Yukawa couplings are assumed weak, given that they are conjectured to have been generated by non-perturbative effects in the underlying microscopic string theory. The models under discussion contain, in addition to the Yukawa terms, also anti-Hermitian anomalous derivative couplings of the pseudoscalar fields to axial fermion currents, as well as interactions of the fermions with non-Hermitian axial backgrounds. We discuss the role of such additional couplings on the Yukawa-induced dynamically generated masses. For the case where the fermions are right-handed neutrinos, we compare such masses with the radiative ones induced by both, the anti-Hermitian anomalous terms and the anti-Hermitian Yukawa interactions in phenomenologically relevant models.

## 1. Introduction

The Parity-Time-reversal(PT)-symmetry framework [1, 2, 3, 4] is an innovative approach to quantum theory, with a plethora of theoretical and experimental applications. These span various branches of physics, and are rapidly expanding to embrace new phenomena, even as this review is being written (for a partial but indicative list of such applications, the reader can consult the mini review [5].) PT symmetry guarantees the self consistency of quantum mechanical models with non-Hermitian Hamiltonians, characterised by real energy eigenvalues. The reality of the energy eigenvalues, despite the lack of Hermiticity of the Hamiltonian, can be understood [6, 7, 8, 9, 10] by means of the *antilinear* nature of PT symmetry. In fact, as argued in [10], PT symmetry constitutes only a special case of non-Hermitian Hamiltonians with real eigenvalues. If a quantum system is characterised by an *antilinear symmetry*, this is the most general condition that one can impose on a quantum theory for which one can have a time-independent inner product and a self-adjoint Hamiltonian with real energy eigenvalues. For each of the above properties Hermiticity is only a sufficient condition but not a necessary one.

Hermiticity is then a special case, in which the Hamiltonian of the system has both antilinearity and Hermiticity.[1]

Extension of the methods of PT-symmetric non-Hermitian quantum-mechanics towards the formulation of non-Hermitian quantum field theories within the PT-symmetry framework is at present in its initial stages, but with very encouraging results [12, 13, 14, 15, 16, 17, 18, 19, 20, 21, 22]. There are studies in this framework associated with Dirac fermion field theories [23], neutrino models [24, 25], spontaneous breaking of global and local (gauge) symmetries in non-Hermitian field theories [26, 27, 28, 29, 30, 31, 32], discrete symmetries [33] and supersymmetry [34] in such models, chiral magnetic effect in non-Hermitian fermionic systems [35], (1+1)-dimensional time-like Liouville conformal field theories [36], a discussion on the potential rôle of non-Hermitian Hamiltonians for the stability of the Higgs vacuum and other field theories of interest to particle physics [37], as well as studies of non-Abelian magnetic monopole solutions [38], and more generally complex Bogomol'nyi-Prasad-Sommerfield (BPS) solitons with real energies [39].

All the above field-theoretic systems are relativistic, for which the reality of the energy eigenvalues can be understood by the extension of the quantum mechanical ideas of the existence of an underlying antilinear symmetry [8, 9, 10] to quantum field theories with non-Hermitian Hamiltonians. The antilinear symmetry is uniquely identified with $\mathcal{C}$PT [10], where $\mathcal{C}$ is an appropriate definition of the charge conjugation, which may differ from the standard definition of the Dirac conjugation operator [21]. Indeed, as shown in [10], requiring the existence of time-independent inner products and invariance under *complex* Lorentz transformations, forces the antilinear symmetry to be *uniquely* $\mathcal{C}$PT [10]. In this way, the standard CPT theorem (with C denoting the standard Dirac charge-conjugation operator), which is based on locality, unitarity and Lorentz invariance of the corresponding field-theoretic Lagrangian densities, can be extended to appropriate field-theoretic systems with non-Hermitian Hamiltonians. In this latter $\mathcal{C}$PT-invariant framework, PT-symmetric systems are charactrerised by a separate invariance under charge conjugation $\mathcal{C}$ [10].

In the above examples, the non-Hermiticity of the Hamiltonians has been assumed, without an attempt to provide microscopic explanations. In [40, 41, 42] we have discussed possible microscopic explanations for a particular kind of non-Hermitian interactions, that of *anti-Hermitian Yukawa* interactions between pseudoscalar (axion-like) and fermion fields, within the framework of certain string-inspired models [43]. The underlying $\mathcal{C}$PT symmetry that characterises the model is responsible for the reality of the energy eigenvalues, according to the general arguments of [6, 7, 8, 9, 10], mentioned above. Moreover, we have studied dynamical mass generation for the pseudoscalar and fermion fields in such systems, within a non-perturbative Schwinger-Dyson (SD) treatment. It is the purpose of this article, to briefly review these studies.

The structure of the article is as follows: in the next section, 2, we review the string-inspired model [43] and explain how the non-Hermitian Yukawa interactions emerge, along with non-Hermitian anomalous couplings of the pseudoscalar (axion-like) fields. We note at this point that the axion-like particles in this model are associated with stringy excitations and are in general different from the QCD axion. In the literature such fields are sometimes called axion-like particles (ALPs), a terminology used in [41, 42] where this review is partly based on. For notational brevity, though, in what follows, we shall refer to them simply as axions. We motivate the use of non-Hermitian Yukawa models (embedded in appropriately generalised-PT ($\mathcal{C}$PT) frameworks) as providers of alternative ways of dynamical-mass generation for *both* axions and fermions. In the string-inspired model in which the fermions are right-handed neutrinos, both these fields and the axions in their massive phase could provide candidates for dark matter.

[1] We also note that *spontaneous breaking* of PT symmetry, as for instance is the case in the (1+1)-dimensional quantum mechanical lattice system studied in [11], leads to the existence of an energy spectrum with complex branches.

In section 3, we discuss the SD mass generation for axions and fermions, in the absence of anomalous axion couplings. In section 4, we include such anomalous interactions of axions, and study their effects on the anti-Hermitian-Yukawa-interaction-induced mass generation. We also include external non-Hermitian axial backgrounds, which are motivated within the context of the string-inspired model of section 2. Conclusions and outlook are presented in section 5.

## 2. String-inspired Models and a potential origin of non-Hermitian Interactions

In this section we shall motivate the origin of non-Hermitian interactions of axions in low-energy effective actions inspired from string theory. To this end, we shall first review some basic features of the spectrum of string theories. We shall be dealing with perturbative effective actions of string theory [44, 45, 46, 47], restricting ourselves to at-most-quadratic order in space-time derivatives. This suffices when we discuss physics at energy scales much below the string mass scale, which serves as an ultraviolet (UV) momentum-cut-off scale of the point-like effective field theories stemming from strings [44, 45].

### 2.1. The Bosonic massless gravitational mutliplet of strings and anomalies

In superstring theory [44, 45, 46, 47], after compactification to four space-time dimensions, the bosonic ground state of the closed-string sector consists of *massless* fields in the so-called *gravitational multiplet*, which contains a spin-0 (scalar) dilaton $\Phi(x)$, a spin-2 traceless symmetric tensor field, $g_{\mu\nu}(x)$, which is uniquely identified as the (3+1)-dimensional graviton, and a spin-1 antisymmetric tensor gauge field $B_{\mu\nu}(x) = -B_{\nu\mu}(x)$, known as the Kalb-Ramond (KR) field. In what follows, for brevity and concreteness, we shall set the four-dimensional dilaton field to a constant, $\Phi(x) = \Phi_0$. This will fix the string coupling

$$g_s = \exp(\Phi) = \exp(\Phi_0) \ . \tag{1}$$

There are always consistent solutions of the four-dimensional string theory with such a configuration, and this suffices for our purposes in this review.

There is a U(1) gauge symmetry of the closed-string (3+1)-dimensional target-space-time effective-field-theory action, associated with the KR $B$-field transformations

$$B_{\mu\nu}(x) \to B_{\mu\nu}(x) + \partial_{[\mu}\theta_{\nu]}(x), \quad \mu,\nu = 0,\dots 3, \quad \theta_\mu(x) \in \mathbb{R}, \tag{2}$$

where Greek indices from now on denote space-time indices, taking on the values $0,\dots 3$, and the symbol $[\dots]$ denotes antisymmetrization of the respective indices.

The U(1) gauge symmetry of the closed-string sector implies that the corresponding effective action will be expressed only in terms of the field strength of the $B$-field:

$$\mathcal{H}_{\mu\nu\rho}(x) = \partial_{[\mu}B_{\nu\rho]}(x). \tag{3}$$

This is subject to the following Bianchi identity

$$\mathcal{H}_{[\nu\rho\sigma\,;\,\mu]} = \partial_{[\mu}\mathcal{H}_{\nu\rho\sigma]} = 0 \ , \tag{4}$$

where, from now on, we omit the space-time-coordinate arguments of the fields $(x)$ for brevity. The semicolon denotes covariant derivative with respect to the standard Christoffel connection $\Gamma^\alpha_{\ \mu\nu} = \Gamma^\alpha_{\ \nu\mu}$ of the metric $g_{\mu\nu}$. It should be stressed that, due to the total antisymmetry of $\mathcal{H}$, the terms involving the gravitational Christoffel connection drop out from (4).

In superstring theory, anomaly cancellation requirements imply a modification of the KR field strength (3) by appropriate gauge (Yang-Mills ($\Omega_{3Y}$)) and Lorentz ($\Omega_{3L}$) (gravitational) Chern-Simons terms (Green-Schwarz (GS) mechanism) [45]

$$\mathcal{H} = \mathbf{dB} + \frac{\alpha'}{8\,\kappa}\Big(\Omega_{3\rm L} - \Omega_{3\rm Y}\Big),$$
$$\Omega_{3\rm L} = \omega^a_{\ c} \wedge \mathbf{d}\omega^c_{\ a} + \frac{2}{3}\omega^a_{\ c} \wedge \omega^c_{\ d} \wedge \omega^d_{\ a}, \quad \Omega_{3\rm Y} = \mathbf{A} \wedge \mathbf{dA} + \mathbf{A} \wedge \mathbf{A} \wedge \mathbf{A}, \tag{5}$$

where $\alpha' = M_s^{-2}$, with $M_s$ the string mass scale, which is in general different from the four-dimensional Planck mass scale $M_P = 1.22 \times 10^{19}$ GeV $\equiv \sqrt{8\,\pi}\,\kappa^{-1}$. For notational brevity, in (5) we used differential-form language [48], with $\mathbf{d}$ denoting the exterior-derivative one-form, $\mathbf{d} = dx^\mu\,\partial_\mu$, and $\wedge$ the exterior ("wedge") product among differential forms, such that $\mathbf{f}^{(k)} \wedge \mathbf{g}^{(\ell)} = (-1)^{k\,\ell}\,\mathbf{g}^{(\ell)} \wedge \mathbf{f}^{(k)}$, where $\mathbf{f}^{(k)}$, and $\mathbf{g}^{(\ell)}$ are $k-$ and $\ell-$ forms, respectively. Above, $\mathbf{A}$ is the Yang-Mills gauge field one-form, and $\omega^a_{\ b}$ the spin-connection one-form (the Latin indices $a, b, c, d$ are (3+1)-dimensional tangent-space (*i.e.* Lorentz-group-SO(1,3)) indices, referring to the Minkowski manifold which is tangent to the space-time manifold at a coordinate point $x$).

The addition of the Chern-Simons terms in (5) leads to a modification of the Bianchi identity (4), which can now be written as [45]

$$\varepsilon^{\mu\nu\rho\sigma}\,\mathcal{H}_{[\nu\rho\sigma\,;\,\mu]} = \varepsilon_{abc}{}^{\mu}\,\mathcal{H}^{abc}{}_{;\mu} = \frac{\alpha'}{32\,\kappa}\,\sqrt{-g}\,\Big(R_{\mu\nu\rho\sigma}\,\widetilde{R}^{\mu\nu\rho\sigma} - \mathbf{F}_{\mu\nu}\,\widetilde{\mathbf{F}}^{\mu\nu}\Big)$$
$$\equiv \sqrt{-g}\,\mathcal{G}(\omega, \mathbf{A}) = \partial_\mu\,\mathcal{K}^\mu(\omega, \mathbf{A}), \tag{6}$$

where the right-hand side denotes the mixed anomaly, due to chiral fermions in the theory circulating in the anomalous loop [49, 50], $g$ denotes the determinant of the metric tensor, $\mathbf{F} = \mathbf{dA} + \mathbf{A} \wedge \mathbf{A}$ is the two-form corresponding to the Yang-Mills field strength (we use form notation for brevity here), $R_{\mu\nu\rho\sigma}$ is the Riemann space-time curvature tensor[2] and

$$\varepsilon_{\mu\nu\rho\sigma} = \sqrt{-g}\,\epsilon_{\mu\nu\rho\sigma}, \quad \varepsilon^{\mu\nu\rho\sigma} = \frac{{\rm sgn}(g)}{\sqrt{-g}}\,\epsilon^{\mu\nu\rho\sigma}, \tag{7}$$

with $\epsilon^{0123} = +1$, etc., are the gravitationally covariant Levi-Civita tensor densities, totally antisymmetric in their indices. The symbol $\widetilde{(\dots)}$ over the curvature- or gauge-field-strength tensors denotes the corresponding dual, defined as

$$\widetilde{R}_{\mu\nu\rho\sigma} = \frac{1}{2}\varepsilon_{\mu\nu\lambda\pi}R^{\lambda\pi}{}_{\rho\sigma}, \quad \widetilde{\mathbf{F}}_{\mu\nu} = \frac{1}{2}\varepsilon_{\mu\nu\rho\sigma}\,\mathbf{F}^{\rho\sigma}. \tag{8}$$

The non-zero quantity on the right hand side of (6) is the "mixed (gauge and gravitational) quantum anomaly" [50, 49], which is known to be a total divergence of a function $\mathcal{K}^\mu$ (containing the Chern-Simons forms in (5)).

To lowest order in the string Regge slope $\alpha'$, the (3+1)-dimensional effective action of the closed-string bosonic sector is then given by [44, 45, 46, 47]:

$$S_B = -\int d^4x\sqrt{-g}\Big(\frac{1}{2\kappa^2}\,R + \frac{1}{6}\,\mathcal{H}_{\lambda\mu\nu}\,\mathcal{H}^{\lambda\mu\nu} + \dots\Big). \tag{9}$$

[2] Our conventions and definitions used throughout this work are: signature of metric $(+,-,-,-)$, Riemann Curvature tensor $R^\lambda{}_{\mu\nu\sigma} = \partial_\nu\,\Gamma^\lambda{}_{\mu\sigma} + \Gamma^\rho{}_{\mu\sigma}\,\Gamma^\lambda{}_{\rho\nu} - (\nu \leftrightarrow \sigma)$, Ricci tensor $R_{\mu\nu} = R^\lambda{}_{\mu\lambda\nu}$, and Ricci scalar $R = R_{\mu\nu}g^{\mu\nu}$.

The KR field strength terms $\mathcal{H}^2$ in (9) can be absorbed [45] (up to an irrelevant total divergence) into a contorted generalised curvature $\overline{R}(\overline{\Gamma})$, with a "torsional connection" [51] $\overline{\Gamma}$, corresponding to a contorsion tensor proportional to the totally antisymmetric $\mathcal{H}^{\rho}_{\mu\nu}$ field strength,

$$\overline{\Gamma}^{\rho}_{\mu\nu} = \Gamma^{\rho}_{\mu\nu} + \frac{\kappa}{\sqrt{3}}\, \mathcal{H}^{\rho}_{\mu\nu} \neq \overline{\Gamma}^{\rho}_{\nu\mu} \,, \tag{10}$$

where $\Gamma^{\rho}_{\mu\nu} = \Gamma^{\rho}_{\nu\mu}$ is the torsion-free (Riemannian) Christoffel symbol.

The reader should notice that the modifications (5) and the right-hand-side of (6) contain the *torsion-free* spin connection. In fact, it can be shown [52, 53] that any potential contributions from the (totally-antisymmetric) torsion $\mathbf{H}$ three-form in the anomaly equation can be removed by adding to the string effective action appropriate counterterms order by order in perturbation theory.

Since the anomaly $\mathcal{G}(\omega, \mathbf{A})$ is an exact one loop result, one may implement the Bianchi identity (6) as a $\delta$-functional constraint in the quantum path integral of the action (9) over the fields $\mathcal{H}$, $\mathbf{A}$, and $g_{\mu\nu}$, and express the latter in terms of a Lagrange multiplier (pseudoscalar) field [54, 55, 56, 57, 58] $b(x)/\sqrt{3}$ (where the normalisation factor $\sqrt{3}$ is inserted so that the field $b(x)$ will acquire a canonical kinetic term, as we shall see below) :

$$\begin{aligned}
&\Pi_x\, \delta\Big(\varepsilon^{\mu\nu\rho\sigma}\, \mathcal{H}_{\nu\rho\sigma}(x)_{;\mu} - \mathcal{G}(\omega, \mathbf{A})\Big) \Rightarrow \\
&\int \mathcal{D}b\, \exp\Big[i \int d^4x\sqrt{-g}\, \frac{1}{\sqrt{3}}\, b(x)\Big(\varepsilon^{\mu\nu\rho\sigma}\, \mathcal{H}_{\nu\rho\sigma}(x)_{;\mu} - \mathcal{G}(\omega, \mathbf{A})\Big)\Big] \\
&= \int \mathcal{D}b\, \exp\Big[-i \int d^4x\sqrt{-g}\, \Big(\partial^{\mu}b(x)\, \frac{1}{\sqrt{3}}\, \epsilon_{\mu\nu\rho\sigma}\, \mathcal{H}^{\nu\rho\sigma} + \frac{b(x)}{\sqrt{3}}\, \mathcal{G}(\omega, \mathbf{A})\Big)\Big]
\end{aligned} \tag{11}$$

where the second equality has been obtained by partial integration, upon assuming that the KR field strength dies out at spatial infinity. Inserting (11) into the (*Euclidean*) path integral with respect to the action (9), and integrating over the $\mathcal{H}$ field, one obtains [54, 55] an effective action in terms of the anomaly and a dynamical, *massless*, KR-axion field $b(x)$, with canonically-normalised kinetic terms.

### *2.2. Ambiguities in the KR-axion effective action and non-Hermitian interactions*

There is a known ambiguity [59] in analytically continuing the resulting effective action back to Minkowski space time, which stems from the following fact: the $\mathcal{H}$-field-strength Euclidean path integration results in the presence of a $b(x)$-dependent, quadratic in space-time derivatives, term in the bosonic (B) Euclidean (E) string effective action of the form:

$$S^{(\mathrm{E})}_{\mathrm{eff\,B}} \ni \int d^4x\, \sqrt{g^{(\mathrm{E})}}\, \frac{1}{12}\varepsilon^{(\mathrm{E})}_{\mu\nu\rho\lambda}\, \varepsilon^{\mu\nu\rho\sigma\,(\mathrm{E})}\, \partial^{\lambda}b\, \partial_{\sigma}b. \tag{12}$$

where the covariant Levi-Civita tensor density $\varepsilon^{(\mathrm{E})}_{\mu\nu\rho\lambda}$ has been defined in (7), but here the notation (E) indicates that this quantity is evaluated in a Euclidean-signature metric (with the convention (+,+,+,+)). The ambiguity concerns the stage at which we analytically continue the quantity (12) back to Minkowski space-time.

If we *first* use the following property of the Levi-Civita tensor density in four space-time dimensions with Euclidean metric ("**Scheme I**"):

$$\varepsilon^{(\mathrm{E})}_{\mu\nu\rho\lambda}\, \varepsilon^{\mu\nu\rho\sigma\,(\mathrm{E})} = +6\, \delta^{\sigma}_{\lambda} \,, \tag{13}$$

where $\delta_\lambda^\sigma$ denotes the Kronecker delta, and *then* analytically continue to Minkowski space time, we obtain a real effective action for the dynamical field $b(x)$ (which plays the rôle of the KR gravitational axion): [54]

$$S_{\rm B}^{\rm eff(I)} = \int d^4x\sqrt{-g}\Big[-\frac{1}{2\kappa^2}\,R + \frac{1}{2}\,\partial_\mu b\,\partial^\mu b + \sqrt{\frac{2}{3}}\,\frac{\alpha'}{96\,\kappa}\,b(x)\,\Big(R_{\mu\nu\rho\sigma}\,\widetilde{R}^{\mu\nu\rho\sigma} - \mathbf{F}_{\mu\nu}\,\widetilde{\mathbf{F}}^{\mu\nu}\Big) + \dots\Big], \tag{14}$$

where the ellipsis denote gauge, as well as higher derivative, terms appearing in the string effective action, that we ignore for our discussion here. In this construction the KR axion appears as a *standard pseudoscalar field*, with a canonically-normalised kinetic term with the *correct sign* relative to the space-time curvature (Eisntein-Hilbert) terms in (14). We also observe that, in view of the anomaly, the KR axion field couples to the gravitational and gauge fields. This latter interaction is P and T violating, and hence in view of the overall CPT invariance of the relativistic, local and unitary (quantum) field theory (14), also CP violating (we remind the reader that C denotes the standard Dirac charge-conjugation operator).

On the other hand, if one *first* analytically continues (12) to a Minkowski-signature space time, and *then* uses the Minkowski version of (13) ("**Scheme II**"):

$$\varepsilon_{\mu\nu\rho\lambda}\,\varepsilon^{\mu\nu\rho\sigma} = -6\,\delta_\lambda^\sigma\;, \tag{15}$$

where the minus sign on the right-hand side is due to the dependence of the contravariant Levi-Civita tensor density (7) on the (Minkowski) signature of the metric tensor, then, one obtains an effective action in which the kinetic terms of the $b$ field have the *wrong* sign relative to the space-time-curvature terms in the effective action, and thus the KR axion would behave like a *ghost* field:

$$S_{\rm B}^{\rm eff(II)} = \int d^4x\sqrt{-g}\Big[-\frac{1}{2\kappa^2}\,R - \frac{1}{2}\,\partial_\mu b\,\partial^\mu b + \sqrt{\frac{2}{3}}\,\frac{\alpha'}{96\,\kappa}\,b(x)\,\Big(R_{\mu\nu\rho\sigma}\,\widetilde{R}^{\mu\nu\rho\sigma} - \mathbf{F}_{\mu\nu}\,\widetilde{\mathbf{F}}^{\mu\nu}\Big) + \dots\Big], \tag{16}$$

Naively, one would ignore this second construction, because of the common perception that a ghost axion field does not carry any physical significance. However, in view of the PT-symmetric framework [1, 2, 3, 4, 8] and its field-theory extensions, one should reconsider this point of view. Indeed, by viewing the $b$ field in (16) as purely imaginary,

$$b(x) \to i\,b(x), \quad b(x) \in \mathbb{R}, \tag{17}$$

one may write the corresponding effective action as

$$S_{\rm B}^{\rm eff\,(II)} = \int d^4x\sqrt{-g}\Big[-\frac{1}{2\kappa^2}\,R + \frac{1}{2}\,\partial_\mu b\,\partial^\mu b + i\,\sqrt{\frac{2}{3}}\,\frac{\alpha'}{96\,\kappa}\,b(x)\,\Big(R_{\mu\nu\rho\sigma}\,\widetilde{R}^{\mu\nu\rho\sigma} - \mathbf{F}_{\mu\nu}\,\widetilde{\mathbf{F}}^{\mu\nu}\Big) + \dots\Big], \tag{18}$$

which has a canonical kinetic term for the (redefined) axion field $b$, which no longer behaves as a ghost, but it is now characterised by non-Hermitian anomalous interactions. The latter make sense in a generalised PT framework, as has been discussed in [42] and will be reviewed below. Such non-Hermitian effective actions are characterised by a generalised antilinear $\mathcal{C}$PT symmetry, with the charge conjugation operation $\mathcal{C}$ defined appropriately [10, 21], to be discussed below, which guarantees the reality of the energy eigenvalues of the system.

Although the parameters $\alpha'$ and $\kappa^2$ are independent in generic string models [46, 47], especially in view of the possibility of large-extra-dimension compactifications, nonetheless for concretreness in what follows, we shall set [40, 42]

$$\sqrt{\alpha'} = M_s^{-1} \sim \kappa = M_{\rm Pl}^{-1} = \sqrt{8\,\pi}\, M_P^{-1} \sim (2.4 \times 10^{18})^{-1}\, {\rm GeV}^{-1}. \tag{19}$$

where the reader should recall that $M_s$ is the string mass scale, and $M_{\rm Pl}$ is the reduced Planck mass in (3+1)-dimensions. It goes without saying that this restriction on string parameters does not affect the qualitative conclusions of our analysis on mass generation, and one can straightforwardly extend it to include more general cases in which $\sqrt{\alpha'} \neq \kappa$.

### *2.3. Inclusion of fermions*

Upon the inclusion of (Dirac) fermions, the torsion interpretation (10) of the KR three-form $\mathcal{H}$, implies that the curved-space Dirac terms read [54, 60, 56, 57]:

$$\begin{aligned} S_{Dirac} &= \int d^4x\sqrt{-g}\,\Big[\frac{i}{2}\Big(\overline{\psi}_j\gamma^\mu\overline{\mathcal{D}}(\overline{\omega})_\mu\,\psi_j - (\overline{\mathcal{D}}(\overline{\omega})_\mu\,\overline{\psi}_j)\,\gamma^\mu\,\psi_j\Big) - m^{(j)}\,\overline{\psi}_j\,\psi_j\Big], \\ &= \int d^4x\sqrt{-g}\bar{\psi}_j\Big(\frac{i}{2}\Gamma^a\overleftrightarrow{\partial_a} - m^{(j)}\Big)\psi_j - \int d^4x\sqrt{-g}\,(\mathcal{F}_a + B_a)\,\bar{\psi}_j\gamma^5\Gamma^a\psi_j \\ &\equiv S_{Dirac}^{Free} + \int d^4x\sqrt{-g}\,(B_a + \mathcal{F}_a)\,J^{5\,a}\,, \end{aligned} \tag{20}$$

where $\overline{\omega}_{ab\mu} = \omega_{ab\mu} + K_{ab\mu}$, $K_{abc} = \frac{1}{2}(\mathcal{H}_{cab} - \mathcal{H}_{abc} - \mathcal{H}_{bca}) = -\frac{1}{2}\mathcal{H}_{abc}$, is the generalised spin-connection with (totally-antisymmetric) torsion $\mathcal{H}$. As before, Latin indices $a, b, c, \dots$ denote tangent-space indices, raised and lowered with the help of the Minkowski metric $\eta^{ab}$ of the tangent space (at a point with coordinates $x^\mu$) of a space-time with metric $g_{\mu\nu}(x) = e^a_\mu(x)\,\eta_{ab}\,e^b_\nu(x)$, with $e^a_\mu(x)$ the vielbeins and $e^\mu_a(x)$ their inverse. $\Gamma^a$ is a tangent-space Dirac matrix, such that the space-time Dirac matrices $\gamma^\mu(x)$ are given by $\gamma^\mu(x) = e^\mu_a(x)\,\Gamma^a$, and we used the standard notation for $\overline{\chi}\,\overleftrightarrow{\partial_a}\,\psi = \overline{\chi}\partial_a\psi - \overline{\partial_a\chi}\,\psi$. The (gravitational) covariant derivative is given by $\overline{\mathcal{D}}_a = \partial_a - \frac{i}{4}\,\overline{\omega}_{bca}\,\sigma^{bc}$, $\sigma^{ab} = \frac{i}{2}[\Gamma^a, \Gamma^b]$.

In (20) we have defined the quantities $\mathcal{F}^d = \varepsilon^{abcd}\,e_{b\lambda}\,\partial_a\,e^\lambda_c$, $B^d = -\dfrac{1}{4}\,\varepsilon_{abc}{}^{d}\,\mathcal{H}^{abc}$. In flat or Friedmann-Robertson-Walker space time backgrounds, of interest to us in this review, $\mathcal{F}^a = 0$, and thus it will not play any rôle in our analysis.

The axial fermionic current is given by : $J^{5\,\mu} = \bar{\psi}_j\,\gamma^\mu\,\gamma^5\psi_j$, and correspondingly $J^{5\,a} = \bar{\psi}_j\,\Gamma^a\,\gamma^5\psi_j$, with the repeated index $j = 1, \dots, N_f$ summed over the fermionic matter species $\psi_j$ of the model. The matrix $\gamma^5 = i\,\gamma^0\,\gamma^1\dots\gamma^3$ is the standard chirality matrix. The term involving the interactions of the $b$-field with the axial current in (20) can be partially integrated to give

$$\int d^4x\,\sqrt{-g}\,\frac{\kappa}{2}\sqrt{\frac{3}{2}}\,\partial_\mu b\,J^{5\mu} = \int d^4x\,\sqrt{-g}\,\frac{\kappa}{2}\sqrt{\frac{3}{2}}\,b(x)\,J^{5\mu}{}_{;\mu} \tag{21}$$

where the covariant divergence of the axial current is non-zero in a theory of massless chiral fermions with chiral mixed anomalies, as in (20):

$$J^{5\mu}{}_{;\mu} \propto \mathcal{G}(\omega, \mathbf{A}) \neq 0 \tag{22}$$

where $\mathcal{G}(\omega, \mathbf{A})$ is defined in (6). The proportionality factors involve the number of chiral fermionic degrees of freedom circulating in the anomalous loop [49], which is model dependent. In some models, e.g. the string-inspired cosmology models of [57, 58], one may encounter a cancellation of the gravitational anomalies, during the radiation and matter epochs of the Universe, but chiral and QCD- anomalies (where the gauge field **A** represents the gluon) survive.

It should be understood in what follows that one may extend the above considerations to include chiral as well as *Majorana* chiral fermions [41, 42], as is the case of the model of [43], which will be our main motivation for the non-Hermitian models discussed in this review.

Adding the fermionic action (20) to the bosonic one (9), implementing the constraint (11), performing the $\mathcal{H}$-path integration (in Euclidean formalism), and analytically continuing back to Minkowski space time, we obtain the following effective action in the "**Scheme I**" (13):

$$S^{\rm eff\,(I)} = \int d^4x\sqrt{-g}\Big[-\frac{1}{2\kappa^2}\,R + \frac{1}{2}\,\partial_\mu b\,\partial^\mu b - \sqrt{\frac{2}{3}}\frac{\kappa}{96}\,\partial_\mu b(x)\,\mathcal{K}^\mu\Big] \\ + S_{Dirac}^{Free} + \int d^4x\sqrt{-g}\,\Big(\mathcal{F}_\mu + \frac{\kappa}{2}\sqrt{\frac{3}{2}}\,\partial_\mu b\Big)\,J^{5\mu} - \frac{3\kappa^2}{16}\,\int d^4x\sqrt{-g}\,J^5_\mu J^{5\mu} + \dots\Big] + \dots, \tag{23}$$

where $\mathcal{K}^\mu$ has been defined in (6), and is related to the mixed anomalies. The ellipsis in (23) indicate gauge field kinetic terms, as well as terms of higher order in derivatives, of no direct relevance to us here. The four-fermion axial-current-current *repulsive* term in (23), is characteristic of Einstein-Cartan fermionic theories on non-Riemannian spaces with torsion [51, 61], which in our case is provided by the field strength of the KR field, $\mathcal{H}_{\mu\nu\rho}$ (10). The action (23) is *Hermitian*.

On the other hand, in the "**Scheme II**" (15), upon using (17), one arrives at the following *non-Hermitian (complex)* effective action:

$$S^{\rm eff\,(II)} = \int d^4x\sqrt{-g}\Big[-\frac{1}{2\kappa^2}\,R + \frac{1}{2}\,\partial_\mu b\,\partial^\mu b - i\sqrt{\frac{2}{3}}\frac{\kappa}{96}\,\partial_\mu b(x)\,\mathcal{K}^\mu\Big] \\ + S_{Dirac}^{Free} + \int d^4x\sqrt{-g}\,\Big(\mathcal{F}_\mu + i\frac{\kappa}{2}\sqrt{\frac{3}{2}}\,\partial_\mu b\Big)\,J^{5\mu} - \frac{3\kappa^2}{16}\,\int d^4x\sqrt{-g}\,J^5_\mu J^{5\mu} + \dots\Big] + \dots, \tag{24}$$

which is embeddable in a generalised PT-framework [42], and we shall discuss it below.

The reader should have noticed the invariance of the actions (23) and (24) under the shift symmetry of the KR axion field

$$b(x) \to b(x) + \text{constant}, \tag{25}$$

which is characteristic of axion physics [62, 63]. In our string model this is associated [45, 54, 55] with the U(1) gauge invariance (2) of the closed-string actions (9) and (20).

### *2.4. Non-perturbative effects and breaking of axionic shift symmetry*

In the presence of non-Abelian gauge fields in the anomalous couplings of the axions, e.g. QCD-type gluon fields, non-perturbative instantons are responsible for the generation of *shift-symmetry-breaking* axion potentials [62, 63]. In particular, for a generic axion field $a(x)$ with anomalous coupling to gluons (or in general non-Abelian gauge fields) of the (generic) form:

$$S_{\rm axion-QCD} \ni \int d^4x\,\sqrt{-g}\,\frac{1}{f_a}a(x)\mathbf{F}_{\mu\nu}\,\widetilde{\mathbf{F}}^{\mu\nu}\,, \tag{26}$$

where $f_a$ denotes the pertinent axion coupling (with mass dimension +1), and $\mathbf{F}_{\mu\nu}$ is the non-Abelian gauge-field strength, non-perturbative instanton effects are responsible for generating an axion potential of the form

$$\mathcal{U}(a) = \Lambda_{\rm inst}^4 \left[1 - \cos\left(\frac{a(x)}{f_a}\right)\right], \tag{27}$$

where $\Lambda_{\rm inst}$ is the appropriate scale, e.g. the QCD scale ($\Lambda_{\rm QCD} \sim 200$ MeV), if the gauge field pertains to gluons. The potential (27), therefore, breaks the shift symmetry (25), which is now restricted to [62, 63]

$$a(x) \rightarrow a(x) + 2\pi\, f_a \ . \tag{28}$$

In our string-inspired model, it is the KR field field that may couple to such instanton effects, through the anomalous terms in (23). In such a case, the corresponding axion coupling $f_a$ is given by

$$f_b \equiv \sqrt{\frac{8}{3}}\frac{\kappa}{\alpha'} = \sqrt{\frac{8}{3}}\left(\frac{M_s}{M_{\rm Pl}}\right)^2 M_{\rm Pl}\ , \tag{29}$$

where $M_{\rm Pl}$ is defined in (19), and we have used the notation $f_b$ to denote the axion coupling pertaining to the KR axion $b(x)$. Here we keep the $M_s$ and $M_{\rm Pl}$ different in magnitude, following the phenomenological study in [57, 58]. We shall come back to applying (19) later. We stress that the generation of a potential for the KR field implies that a torsion interpretation of this field is no longer possible in the massive phase.

From (29), we thus observe that the range of the axion coupling $f_b$ depends on the allowed range of the string mass scale $M_s$. For the model of, e.g., ref. [57, 58], the allowed range of the latter scale is

$$M_{\rm Pl} \gtrsim M_s \gtrsim 10^{-3}\, M_{\rm Pl}\ , \tag{30}$$

where the upper bound of $M_s$ corresponds to (19). This implies

$$3.9 \times 10^{12}\ {\rm GeV} \lesssim f_b \lesssim 3.9 \times 10^{18}\ {\rm GeV}. \tag{31}$$

On the other hand, the generic axion coupling constant $f_a$ is constrained to lie in the range [63]

$$10^9\ {\rm GeV} < f_a < 10^{12}\ {\rm GeV}\ . \tag{32}$$

We note, though, that astrophysical constraints [64, 65, 66, 67, 68] may extend the upper bound up to $10^{17}$ GeV. We thus observe that, even if the astrophysical constraints are ignored, there is still a marginal overlap (in order of magnitude) between the minimally allowed region of $f_b$, (31), and the maximally allowed phenomenological region of the QCD axion coupling constant, (32).

From (27), for the case of the KR axion $b(x)$, we infer that the instanton-induced KR-axion mass is [58]

$$m_b = \sqrt{\left.\frac{\partial^2 V_b^{\rm QCD}}{\partial b^2}\right|_{b=0}} = \frac{\Lambda_{\rm QCD}^2}{f_b} = \sqrt{\frac{3}{8}}\left(\frac{\Lambda_{\rm QCD}}{M_s}\right)^2 M_{\rm Pl} = \sqrt{\frac{3}{8}}\left(\frac{\Lambda_{\rm QCD}}{M_{\rm Pl}}\right)^2\left(\frac{M_{\rm Pl}}{M_s}\right)^2 M_{\rm Pl}\ , \tag{33}$$

which, in view of (30), (31), lies in the range

$$1.17 \times 10^{-11}\ {\rm eV} \lesssim m_b \lesssim 1.17 \times 10^{-5}\ {\rm eV}, \tag{34}$$

well within the range calculated in lattice QCD approaches (see, e.g. ref. [69] and references therein): $m_a \sim 5.7\,(\frac{10^{12}\ \text{GeV}}{f_a}) \times 10^{-6}$ eV.

The above considerations pertain to the Hermitian effective action (23) in "**Scheme I**". In this review we shall discuss the phenomenology of models with non-Hermitian anomalous couplings, (24), in "**Scheme II**". Such models correspond to purely imaginary axion coupling constants for which the aforementioned mass generation through instantons is not clear (for, instance, for purely imaginary axion couplings in the potential (27), $f_a = i f_b$, $f_b \in \mathbb{R}$ (with $f_b$ defined in (29)), there are no mass terms generated, as the corresponding quadratic terms in the axion field $a(x)$ have the wrong sign). It is the point of this review to discuss alternative ways for mass generation for axions (and fermions) dynamically, which is done in sections 3 and 4. We use specific non-Hermitian effective actions which, as we shall discuss, complement (24) by appropriate *anti-Hermitian Yukawa interactions* of axions with the fermions [43]. The latter constitute the main source of dynamical-mass generation.

Before doing this, though, it is necessary to motivate the pertinent non-Hermitian models carefully within the microscopic string theory framework. This necessitates some discussion first on the various types of axions existing in string models, which we now proceed to review.

### 2.5. Stringy Model-independent (KR) and Model-dependent axions: kinetic mixing and Yukawa interactions

In ref. [43] it was postulated that non-perturbative physics might also be responsible for a further breaking of the shift symmetry by means of Yukawa interactions between axions and fermions, specifically right-handed neutrinos, in the context of the above-described Hermitian string-inspired model (23) in the "**Scheme I**". To this end, one exploits the fact that in string theory [55, 70] there are many axion fields [55, 70], associated with the Kaluza-Klein (KK) zero modes of appropriate $p$-forms in the spectrum of strings compactified to four space-time dimensions, *i.e.* formulated on a target-space-time manifold of the form $M_{1,3} \times \mathcal{X}_6$. Here, $M_{1,3}$ denotes the uncompactified (3+1)-dimensional space time, and $\mathcal{X}_6$ the extra-dimensional space, assumed to be a smooth compact manifold. For instance, in heterotic string theory [45, 47], one has the (Neveu-Schwarz(NS)-type) two-form $\mathcal{B}$ of the Kalb-Ramond field in ten dimensions, which, upon compactification on an appropriate (say Calabi-Yau [45]) six-dimensional compact space, $\mathcal{X}_6$, can be written as:

$$\mathcal{B} = B_{\mu\nu}(x)\, dx^\mu\, dx^\nu + \frac{1}{2\pi}\, b^I(x)\, \omega^I_{ij}(z,\bar{z})\, dz^i\, d\bar{z}^j\ , \quad \mu,\nu = 0,\dots 3, \quad i,j = 1,2,3 \qquad (35)$$

where $z^i$, $i = 1,2,3$ are complex coordinates parametrising the compact manifold. The $B_{\mu\nu}(x)$ field yields, upon the dualisation procedure associated with the implementation of the corresponding Bianchi identity for its field strength (6) via a Lagrange multiplier, the KR axion $b(x)$, as discussed above. In the language of string theory this is the so-called *model independent* axion, as it is present in all string theories. The quantities $\omega^I_{ij}(z,\bar{z})$, $I = 1,\dots h^{1,1}$ in (35), represent harmonic (1,1) forms that depend only on the coordinates of the complex manifold, and are linked to the aforementioned KK zero modes. One uses the normalisation [55]

$$\int_{\mathcal{C}^J} \omega^I = \delta^{IJ} \qquad (36)$$

where $\mathcal{C}^I$ is a 2-cycle in the compact manifold. In other words, the harmonic forms $\omega^I$ span the integer (1,1) cohomology group of the target space [48].

The quantities $b^I(x)$, $I = 1,\dots h^{1,1}$ represent dimensionless pseudoscalar fields on the uncompactified space-time, and the factor $\frac{1}{2\pi}$ has been inserted so that the fields $b^I(x)$ have a period $2\pi$, as is conventional for axions[3]. These four-dimensional fields correspond to the so-called *model-dependent axions* [55]. The kinetic terms of the two-form yield the four-space-time-dimensional kinetic terms of the $b^I(x)$ fields. Indeed, to this end, and taking into account the structure and the space-time dependences of the field components in (35), we need only to consider the following components of the field strength $\mathcal{H}_{MNP} = \partial_{[M}B_{NP]}$ (with capital Greek letters denoting indices referring to the ten-dimensional space-time of the superstring/brane theory, $M, N, P = 0, \ldots 9$):

$$\mathcal{H}_{\mu ij} = \partial_\mu B_{ij} = (\partial_\mu b^I(x))\, \omega^I_{ij}(z, \bar{z}) \ , \quad \mu = 0, \ldots 3, \quad i, j = 1, 2, 3 \ .$$

These components are the only ones associated with the $b^J$-model-dependent-axion terms on the right-hand side of (35), which are non-vanishing (the terms associated with the first, $B_{\mu\nu}$-dependent terms yield, of course, the purely four-dimensional KR antisymmetric field strength (3)).

The pertinent kinetic terms for the axions $b^I(x)$ in (3+1)-dimensions stem from $\mathcal{H}_{MNP}\,\mathcal{H}^{MNP}$-structures in the ten-dimensional effective action, which, upon compactification down to (3+1)-dimensions, yield terms of the form

$$\begin{aligned} S_{10-\rm dim} \ni \int \sqrt{-g}\, d^4x \int_{\mathcal{X}_6} \partial_\mu B_{ij}\, \partial^\mu B^{ij} &= \int \sqrt{-g}\, d^4x\, \partial_\mu b^I(x) \partial^\mu b^J(x) \int_{\mathcal{X}_6} \omega^I_{ij}(z, \bar{z})\, \omega^{J\, ij}(z, \bar{z}) \ , \\ &\equiv \int \sqrt{-g}\, d^4x\, \partial_\mu b^I(x) \partial^\mu b^J(x)\, \gamma^{IJ} \quad \mu = 0, \ldots 3, \quad I, J = 1, \ldots h^{1,1}, \end{aligned} \tag{37}$$

where, for brevity, we only indicated the structures, omitting numerical coefficients. The reader should observe the non-trivial kinetic mixing $\gamma^{IJ} \neq \delta^{IJ}$ of the model-dependent stringy axions $b(x)^I$.[4]

It is important to stress that the kinetic terms of the model-depenent stringy axions do *not* suffer any ambiguities and are the *same* in both "**Schemes I and II**", corresponding to standard axion kinetic terms, since they do not contain the Levi-Citiva tensor density $\varepsilon_{\mu\nu\rho\sigma}$ (7) (with $\mu, \nu, \rho, \sigma = 0, \ldots 3$ denoting indices in the (3+1)-dimensional space-time manifold $M_{1,3}$).

The axion coupling for the fields $b^I(x)$ can be determined by looking [55] at the (one loop) counterterms required for the GS anomaly-cancellation mechanism in string theory [45]. As a concrete example, we may consider the $E_8 \times E_8$ heterotic string, formulated on $M_{1,3} \times \mathcal{X}_6$, with the Standard Model gauge group $SU(3)_c \times SU(2) \times U_Y(1)$ embedded, say, in the first $E_8$ group factor. For brevity and concreteness, we take the uncompactified space-time manifold $M_{1,3}$ to be Minkowski flat. As shown in [55], then, one obtains the following effective-action anomaly terms for the axion-$b^I(x)$ fields:

$$S_{\rm anom\ string\ axion} = -\Big(\frac{1}{16\,\pi^2} \int_{\mathcal{X}_6} \omega^I(z, \bar{z}) \wedge \Big[{\rm Tr}_1 \mathbf{F} \wedge \mathbf{F} - \frac{1}{2} \mathbf{R} \wedge \mathbf{R}\Big]\Big) \int d^4x\, b^I(x) \frac{1}{16\pi^2} {\rm Tr}_1 \mathbf{F} \wedge \mathbf{F} \tag{38}$$

where, in order to arrive at (38) starting from the original form of the GS counterterms, the Bianchi identity (6) has been used. The term inside the parentheses on the right-hand side of the above relation expresses mixed anomalies in the compact manifold $\mathcal{X}_6$, with $\mathbf{F}$ ($\mathbf{R}$) the appropriate gauge-field (compact-space-$\mathcal{X}_6$ curvature) two-form over the compact space, and $\wedge$ the appropriate exterior product among differential forms [48], as mentioned previously; the trace ${\rm Tr}_1$ pertains to the first $E_8$ gauge group.

[3] In this formalism, the corresponding axion fields with mass dimension +1 and canonically-normalised kinetic terms in the effective action are given by $f_{b^J}\, b^J$, where $f_{b^J}$ are the corresponding axion couplings, of mass dimension +1, appearing in the (gauge and gravitational) anomalous, CP-violating, interactions of the axion moduli fields [55].

[4] Such mixing has been exploited in [71, 72] to discuss a widening of the allowed window of stringy-axion coupling constants, and also study large-field axion-induced inflation.

We also remark at this point that in generic string or D-brane models, compactified or projected to four space-time dimensions, *model-dependent* axion fields $a^I(x)$ are also obtained as KK zero models of other appropriate $p$-form fields, $C_p$, in the string spectrum on $M_{1,3} \times \mathcal{X}_6$, for instance, the Ramond-Ramond(RR)-type $p = 0, 2, 4$-forms of type IIB string theory, or the $p = 1, 3$-forms of type IA [55]. Similarly to the NS 2-form $B$ (35), the corresponding *model-dependent*, dimensionless, (3+1)-dimensional axion fields are then given by

$$a^I(x) = \frac{1}{2\pi}\int_{\mathcal{C}_I^{(p)}} C_p\ , \quad I = 1, \dots M, \tag{39}$$

where $\mathcal{C}_I^{(p)} \subset \mathcal{X}_6$ are appropriate homologically-non-equivalent $p$-cycles in the compact manifold, and we normalised again the axion so as to have period $2\pi$. We stress that qualitatively similar anomaly terms to (38) occur for the axions $a^I(x)$.

In the model of [43], we may assume that there is an anomaly cancellation between gauge and gravitational anomaly terms in the compact manifold $\mathcal{X}_6$ in (38), which means that such terms *vanish*. This is a special case, sufficient for our purposes here. In this case, the stringy model-dependent axions are characterised by their kinetic terms only and have *no potential*.

However, even in such a case, there is an induced coupling between the axions $b^I(x)$ (or $a^I(x)$) and anomaly terms in the (3+1)-dimensional effective action, if there is *kinetic mixing* of such model-dependent axions with the model-independent KR $b$-axion field:

$$S_{\text{kin. mix.}} = \gamma_J \int d^4x\sqrt{-g}\, \partial_\mu b\, \partial^\mu a_J, \quad J = 1, \dots, M, \tag{40}$$

where the repeated index $J = 1, \dots M$, is summed over the stringy axion species. One could also include standard QCD axions in such a summation that may co-exist with stringy axions. Such a mixing has been assumed in the effective field theory of [43].

The term (40) could be added to (23), along with the following Hermitian shift-symmetry-breaking Yukawa interactions [43]

$$S_{\text{Yukawa}} = i\sum_{j=1}^{N_f}\sum_{J=1}^{M} \lambda_{Jj} \int d^4x\sqrt{-g}\, a_J\, \overline{\psi}_j\, \gamma_5\, \psi_j\ . \tag{41}$$

where $j$ runs over fermion species $N_f$ in the model.

Before proceeding in examining the consequences of (41) for mass generation, which is the main topic of our work here, we should make some important remarks, distinguishing the *genuinely quantum* Yukawa structures (41) from *classical* non-derivative Yukawa structures associated with derivative interactions of axions after the use of fermion equations of motion. The latter characterise any axion model, and are used, for instance, in experimental nuclear-physics searches of axions, as they describe interactions of axions to nucleons [63]. As such, they are also present in the *classical* effective action of the (string)model-independent KR axion $b(x)$, due to its torsion nature [54].

Let us explain now how such classical Yukawa interactions arise from derivative, shift-symmetry-respecting interactions. To this end, we first recall that a derivative coupling of a generic axion field $a(x)$ to matter fermions has the form (*cf.* (20)):

$$S^{\text{eff}}_{a-F} \ni \int d^4x\sqrt{-g}\, \psi_j\gamma^\mu\gamma^5\psi_j\, \partial_\mu a(x) = -\int d^4x\sqrt{-g}\, a(x)\, \nabla_\mu\, (\psi_j\gamma^\mu\gamma^5\psi_j), \tag{42}$$

where in the right-hand side we performed partial integration, assuming as usual that the fields and their derivatives vanish in the boundaries of space-time. If the fermions are massless and chiral, then there might be quantum anomalies that could spoil the conservation of the chiral current, $\overline{\psi}_i\,\gamma^\mu\,\gamma^5\,\psi_j$, so this term yields non trivial anomalous contributions to the effective action. If, however, the fermions $\psi_j$ are massive, of mass $m^{(j)}$, then the divergence of the chiral current is non-zero already classically, as follows from the equations of motion

$$\nabla_\mu\left(\psi_j\gamma^\mu\,\gamma^5\psi_j\right) = 2im^{(j)}\overline{\psi}_j\,\gamma^5\,\psi_j\;, \quad j = \dots N_f. \tag{43}$$

In an anomalous quantum theory, of course, the right-hand side of (43) also receives anomalous contributions of the form $\mathcal{G}(\omega, \mathbf{A})$ (defined in (6)), see (22).

Substituting (43) onto the effective action term (42), one then obtains non-derivative Yukawa coupling terms in the effective action of the form (41). But this constitutes a classical argument, since the fermion equations of motion have been used. In contrast, the structures (41) represent *fully quantum* interactions, for which the fermion equations of motion should *not* be used. This is why we cannot apply (41) to the model-independent KR axion, because the latter, as we have seen, arises from a dualisation procedure of the gauge-invariant (under the symmetry (2)) KR field strength $\mathcal{H}_{\mu\nu\rho}$, and, as such, it should appear in the local low-energy effective Lagrangian of string theory through its derivatives.

At this point we remark that the reader might be tempted, following [54, 55], to redefine the field strength $\mathcal{H}_{\mu\nu\rho}$ by appropriate counterterms involving fermions, so as to ensure the validity of the modified Bianchi identity

$$\varepsilon^{\mu\nu\rho\sigma}\,\mathcal{H}_{[\nu\rho\sigma\,;\,\mu]} - \sqrt{-g}\,\mathcal{G}(\omega, \mathbf{A}) - i\,\lambda\,\sqrt{-g}\,\overline{\psi}\,\gamma^5\,\psi = 0\;, \quad \lambda \in \mathbb{R}\;, \tag{44}$$

order-by-order in perturbation theory of the (Hermitian) string-inspired effective field theory described by the actions (9) and (20). As we have mentioned previously, classically (43) one may obtain the structures $\overline{\psi}\,\gamma^5\,\psi$ from the divergence of the axial fermion current, related to the anomalies $\mathcal{G}(\omega, \mathbf{A})$, upon use of the fermion equations of motion for massive fermions. By postulating the constraint (44) one ensures the appearance of such structures at a quantum level as independent operators in the effective action. This would formally ensure the presence of Yukawa interactions $\lambda\,b(x)\,\overline{\psi}(x)\,\gamma^5\,\psi(x)$ in the string inspired field theory action obtained after the dualisation procedure, by means of which one implements the constraint (44) in the path integral of the effective low-energy field theory via the Lagrange-multiplier field $b(x)$. In such a case, under the "**Scheme II**", one would obtain effective actions with anti-Hermitian anomaly and Yukawa terms for the model-independent KR axion $b(x)$, embeddable directly in a generalised-PT ($\mathcal{C}$PT) framework [17, 10, 40, 41, 42]. The problem with this approach, however, is that the constraint (44) would imply a redefinition of the field strength $\mathcal{H}_{\mu\nu\rho}$, which, apart from the standard Chern-Simons local counterterms (5), required for the GS anomaly-cancellation mechanism [45], would involve additional *non-local* fermionic counterterms of the form (up to numerical coefficients, not relevant for our discussion): $\propto\; i\,\lambda\,\varepsilon_{\mu\nu\rho\sigma}\,\nabla^\sigma\,\frac{1}{\Box}\,(\overline{\psi}\,\gamma^5\,\psi)$, where $\Box$ denotes the gravitationally-covariant D'Alembertian, $\Box \equiv \nabla_\mu\,\nabla^\mu$, with $\nabla_\mu$ the gravitational covariant derivative. Such non-local terms are not acceptable in the framework of local effective field theories obtained from strings, given that the perturbative-scattering-matrix approach that defines the perturbative-string-theory limit [44] would not be valid. Hence we shall not consider such a procedure here, but we felt like mentioning it, in case one is prepared to speculate on the existence of such non-local counterterms in a more general quantum-gravity framework.

### *2.6. Non-perturbative effects and various types of Majorana neutrino masses*

In [43] we therefore considered the addition of the non-derivative interactions (41) in the sector of model-dependent axions, which are not related to the aforementioned dualisation procedure of $\mathcal{H}_{\mu\nu\rho}$. In this respect, we have offered an alternative interpretation of the origin of such terms.

It has been conjectured in [43] that instanton or other non-perturbative effects in string/brane theories can generate such shift-breaking Yukawa couplings, which are therefore characterised by Yukawa couplings $\lambda_{Jj}$ very small in magnitude (*cf.* see (47) below).

It is pointed out at this point that non-perturbatively generated masses for right-handed Majorana neutrinos in string/brane theories have been considered in the literature in the past [73, 74, 75, 76], which are associated with (effective (3+1)-dimensional) field operators of the form

$$M_s\, e^{-U}\, \overline{N}(x)\, N(x)\ , \tag{45}$$

where $M_s$ is the string mass scale, $N$ is the Majorana field of the neutrino, and $U$ denotes linear combinations of complex string moduli, whose imaginary parts correspond to axion-like particles (which thus are characterised by the standard restricted shift symmetry (28), corresponding to periodic shifts by $2\pi$ for the dimensionless fields; the Majorana-neutrino-mass operator (45) is also invariant under the standard-model gauge group and a $\mathrm{U(1)_{B-L}}$ gauge symmetry, where B(L) denotes Baryon (Lepton) quantum numbers). The string/D-brane instanton zero modes imply right-handed neutrino Majorana masses suppressed by $\propto \sum_r \mathcal{D}_r\, e^{-S_r}$, where $\mathcal{D}_r$ are coefficients depending on flavour couplings of the neutrinos, and $S_r$ is the (real part of the) instanton action, for the type-$r$ stringy instanton (we note that there are many types of instantons in string /brane theories [73, 74, 75, 76]).

The interaction (41) we are conjecturing in [43], and reviewing here, is rather different from (45), as it depends *linearly* on the model-dependent axion, and thus is not characterised by a periodic-shift symmetry, but breaks the shift symmetry completely. It remains to be seen whether microscopic mechanisms in concrete string/brane theories exist that yield (41). This falls beyond the scope of the current work, where the model with this extra interaction is considered phenomenologically [43] as a model beyond the standard model where dynamical right-handed neutrino masses can be generated in non-conventional ways. Moreover, the kinetic mixing (40) itself, could be the result of the same non-perturbative physics [40, 41], *e.g.* through one loop-effects involving heavy fermions, existing only in internal lines of graphs, coupled to axions $b$ and $a^I$ via the (non-perturbatively generated) Yukawa couplings (41).

In the model of [43], the non-perturbative Yukawa coupling couples to right-handed Majorana neutrinos $\psi_R$ only, hence one considers adding the following Hermitian terms in the Hermitian effective string action (23), within the "**Scheme I**" framework:

$$\mathcal{S}_{a-b} = \int d^4x\sqrt{-g}\,\Big[\frac{1}{2}(\partial_\mu a(x))^2 + \gamma\,\partial_\mu b\,\partial^\mu a + i\,\lambda\, a(x)\,\Big(\overline{\psi}_R^{\ C}\,\psi_R - \overline{\psi}_R\psi_R^C\Big) + \dots\Big]\ , \tag{46}$$

where $\psi_R^C = (\psi_R)^C$ is the (Dirac) charge-conjugate of the right-handed fermion $\psi_R$, which in [43] is considered to be a sterile neutrino $\nu_R$ (the reader should notice that, due to the chiral nature of the fermions $\psi_R$ , the $\lambda$-dependent terms in (46) correspond to the axial form in (41)). In (46), for brevity, we did not write explicitly the kinetic terms of $\psi_R$. These are understood to be included in the .... Due to the non-perturbative nature of $\lambda$ one may assume

$$\lambda \sim \exp(-S_{\rm instanton}) \ll 1, \tag{47}$$

where $S_{\rm instanton} \gg 1$ is the (real part of an appropriate) large and positive instanton action.[5]

Redefining the $b$-field as $b \to b' = b + \gamma\, a$, one diagonalises the kinetic axion terms in (46), and can readily perform the $b'$ field integration in the path integral corresponding to the effective action obtained from the sum of (23) and (46). In this way one arrives at an effective

[5] In view of the above discussion on the existence of many types of instantons in string theory, one might consider $\lambda$ as proportional to the appropriate sum involving many exponential suppression factors of the (real parts of the) actions of individual instantons.

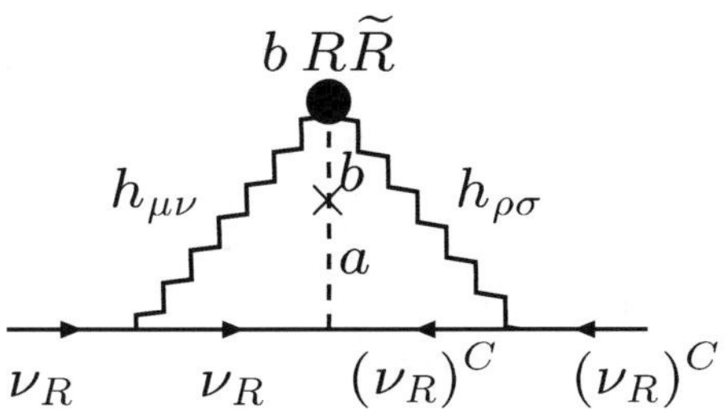


**Figure 1.** *Feynman graph pertaining to the anomalously generated Majorana mass for the right-handed fermions $\psi_R$, identified with sterile neutrinos $\nu_R$ in [43]. The dark blob denotes the operator $b(x)\,R_{\mu\nu\lambda\rho}\widetilde{R}^{\mu\nu\lambda\rho}$ (up to numerical coefficients), which is the only part of the anomaly relevant for sterile neutrinos, which do not couple to gauge fields; $b(x)$ denotes the KR (gravitational) axion and $a(x)$ an axion-like particle. The wavy lines are gravitons $h_{\mu\nu}$, and dashed lines denote axions. The cross "$\times$" indicates the $b(x)$-$a(x)$ kinetic mixing (40). Figure taken from [43].*

(3+1)-dimensional action for the dynamics of the stringy axion field $a(x)$ [43]. Upon rescaling appropriately the $a(x)$ field, so as to have canonical kinetic terms after the above procedure, one arrives at the following effective action involving the $a$-field:

$$\mathcal{S}_a^{\rm eff}\Big|_{\rm Scheme\ I} = \int d^4x\sqrt{-g}\,\Big[\frac{1}{2}(\partial_\mu a(x))^2 - \frac{\gamma\,a(x)}{f_b\,\sqrt{1-\gamma^2}}\,\mathcal{G}(\omega,\mathbf{A}) + \frac{i\,\lambda}{\sqrt{1-\gamma^2}}\,a(x)\,\Big(\overline{\psi}_R^{\ C}\psi_R - \overline{\psi}_R\psi_R^{\ C}\Big) + \dots\Big]\,, \quad (48)$$

where the KR-axion coupling $f_b$ is defined in (29). The ellipsis includes terms independent of $a(x)$, including the kinetic terms of the $\psi_R$ fermions, as well as anomaly-dependent terms stemming from the $b'(x)$-field path integration, of the form $\mathcal{K}_\mu\mathcal{K}^\mu$, with $\mathcal{K}^\mu$ defined in (6). The reader should notice the induced anomalous interactions of the (model-dependent stringy) axion field $a(x)$ field in (48), proportional to the kinetic-mixing parameter $\gamma/\sqrt{1-\gamma^2}$. It should be stressed that, to avoid the axion behaving as a ghost, one needs to impose the condition [43]

$$0 < \gamma^2 < 1\ . \quad (49)$$

From the effective action (48), one may deduce novel ways of generating *radiatively* Majorana masses for the right-hended fermions (sterile neutrinos in the model of [43]), which are based on diagrams depicted in figure 1. The radiatively-generated Majorana sterile-neutrino mass is $M_R$ found to be [43]:

$$M_R \sim \frac{\sqrt{3}\,\lambda\,\gamma\,\kappa^5\Lambda^6}{49152\sqrt{8}\,\pi^4(1-\gamma^2)}\,, \quad (50)$$

where $\Lambda$ is the Ultra-Violet (UV) momentum cutoff. The computation involved the use of graviton fluctuations $h_{\mu\nu}(x)$ about flat Minkowski space-time backgrounds. These yield non-trivial contributions to the gravitational-anomaly term $R_{\mu\nu\rho\sigma}\,\widetilde{R}^{\mu\nu\rho\sigma}$, which are computed using perturbative (quantum) gravity methods developed in [77] (note that the gravitational-anomaly term vanishes for flat or cosmological Friedmann-Lemaitre-Robertson-Walker background space-times). In the context of string theory, $\Lambda$ and $\kappa^{-1}$ are related [43] via the string mass scale and compactification radii of the extra-dimensional spatial manifold. For a generic quantum gravity model, independent of string theory, one may use simply $\Lambda \sim \kappa^{-1}$. The sterile-fermion mass (50) is *independent* of the axion-$a$ potential [43], and thus its mass. As the reader can notice, the mass $M_R$ is generated for arbitrarily small $\gamma$ and $\lambda$ (real) couplings. This covers the case where both $\gamma$ and $\lambda$ are generated non-perturbatively, say by instanton effects, as discussed previously, and thus are very small, of order (47).

We now remark that, in the "**Scheme II**", in which there are purely-imaginary anomalous couplings of the KR model-independent axions $b(x)$ in the effective action (24), a redefinition of the KR axion field $b \to b' = b + i\,\gamma\, a$, decouples the model-independent axion from the model-dependent one, $a(x)$, and, as the reader can easily verify, upon integrating out formally in the path integral $b'(x)$, one obtains the following effective action for the model-dependent axion field $a(x)$:

$$\mathcal{S}_a^{\rm eff}\Big|_{\rm Scheme\ II} = \int d^4x\sqrt{-g}\,\Big[\frac{1}{2}(\partial_\mu a(x))^2 + \frac{\gamma\, a(x)}{f_b\,\sqrt{1+\gamma^2}}\,\mathcal{G}(\omega, \mathbf{A}) + \frac{i\,\lambda}{\sqrt{1+\gamma^2}}\,a(x)\left(\overline{\psi}_R^{\ C}\psi_R - \overline{\psi}_R\psi_R^{\ C}\right) + \dots\Big]\,, \tag{51}$$

where the ellipsis includes anomaly-dependent terms stemming from the $b'$ integration of the form $-\mathcal{K}^\mu\mathcal{K}^\mu$ (note the opposite sign as compared with the corresponding terms in the Hermitian case (48)). The reader should notice that there is no restriction in the range of $\gamma$ in this case. It should also be noticed that the action (51) is still Hermitian, despite the non-Hermiticity of the KR-$b$-axion Lagrangian (24), in "**Scheme II**". Through the anomalous (radiative) graphs of fig. 1, one generates a right-handed-neutrino Majorana mass in this case of the form

$$M_R \sim \frac{\sqrt{3}\,\lambda\,\gamma\,\kappa^5\Lambda^6}{49152\sqrt{8}\,\pi^4(1+\gamma^2)}, \tag{52}$$

for the entire range of $\gamma \in \mathbb{R}$ (in computing the graph of fig. 1 care should be taken so that the product $\frac{\lambda\gamma}{\sqrt{1+\gamma^2}}$ is not too big, otherwise our perturbative analysis [43] breaks down; this is guaranteed for sufficiently small $|\lambda\,\gamma| < 1$).

### 2.7. *Motivation for embedding the model* (51) *in a generalised-PT ($\mathcal{C}PT$) framework*

We next remark that, although from a string point of view, the resulting effective actions involving model-dependent axions and non-derivative Yukawa interaction with fermions are real (*cf.* (48), (51)), nonetheless we *may embed* them in a non-Hermitian PT-symmetry framework by *analytically continuing both* coefficients $\gamma$ and $\lambda$ to purely imaginary values [40, 41]

$$\gamma \to i\,\gamma, \quad \lambda \to i\,\lambda, \quad \gamma, \lambda \in \mathbb{R}. \tag{53}$$

Formally, such a procedure will leave the radiatively generated masses (50), (52) *real*, but now the corresponding effective actions will have non-Hermitian interactions. This result, however, is only formal, given that the presence of *non-Hermitian anomaly terms* are yet to be understood. Hence, the methods developed for the Hermitian case to deal with graviton fluctuations of the gravitational anomaly terms [77] which lead [43] to the radiative mass (50) might not apply.

We therefore seek for alternative mass generation in such non-Hermitian cases embeddable in a generalised PT-symmetric framework [17], which is actually [40, 41, 42] $\mathcal{C}PT$ symmetric, under an appropriate definition of the charge-conjugation $\mathcal{C}$ operator [21], as we shall discuss in the next section. As already mentioned, such an embedding guarantees the reality of the pertinent energy spectra [10] and, thus, potentially of the masses that can be generated dynamically.

The motivation for the analytic continuation (53), lies on the fact that the Hermitian Yukawa interactions cannot lead to dynamical generation of axion and fermion masses, unless there is a *bare* axion mass present [41, 42]. In constrast, in the non-Hermitian case (53) one can obtain *truly dynamical* masses for *both* axions and fermions, in the absence of bare masses for such fields, but in the presence of attractive four-fermion interactions, as we shall discuss below. The physical interest in having massive axions is linked with the potential rôle of such fields as dark matter.

It is the purpose of this article to review briefly the dynamical generation of masses for axions and fermions in this latter (non-Hermitian) set up. In the next section 3, we examine the issue of dynamical mass generation in theories with non-Hermitian (actually, anti-Hermitian) Yukawa couplings in the absence of anomalous terms, i.e. $\gamma = 0$, $\lambda \neq 0$. Motivated by our arguments about the potential generation of such Yukawa interactions through non-perturbative effects, we restrict ourselves to very small Yukawa coupling $|\lambda| \ll 1$. We incorporate a non zero $\gamma \neq 0$, $|\gamma < 1|$, as well as non-Hermitian axial backgrounds, in section 4. Despite the smallness of the couplings $\lambda$ and $\gamma$, we use non-perturbative Schwinger-Dyson (SD) analysis.

In what follows we shall use only *prototype models*, involving one Dirac fermion and one axion. This suffices to demonstrate the main conclusions. The extension to Majorana fermions is straightforward and has been done in [41, 42], where we refer the interested reader for details. These references also contain an analysis of SD dynamical mass generation for the Hermitian effective actions (48) (or (51)).

We remark at this point that, in addition to the $\lambda$- and $\gamma$-dependent terms in (48) (or (51)), which are analytically continued, as per (53), to yield the corresponding anti-Hermitian, $\mathcal{C}$PT-invariant theories of interest to us here [40, 41, 42], we shall also include *attractive Hermitian* four-fermions interactions. Such interactions are present in any low-energy effective string theory with fermions, as a consequence of the exchange of heavy string states among fermions [78]. Integrating out such heavy states results in effective low-energy actions with contact four-fermion interactions of various types. It suffices for our purposes to consider the addition to our prototype models of four-fermion attractive interactions of the form

$$-\frac{1}{2\,f_4^2}\Big(\overline{\psi}\,\gamma^5\,\psi\Big)^2\,, \tag{54}$$

where $f_4$ is an appropriate coupling with dimensions of [mass]. As we shall discuss below, the rôle of such four-fermion interactions is crucial for fermion dynamical mass generation. Therefore, in view of the quantum-torsion-induced four-fermion *repulsive* interactions involving the covariant square of the axial fermion current $J^5_\mu$ in (23) (or (24)), which upon use of Fierz identities, yield - among other terms - repulsive structures of the form (54), one must have *sufficiently strong* attractive interactions (54) to *overcome* the corresponding torsion-induced *repulsions*. This will be assumed in what follows, and from now on we ignore the repulsive four-fermion interactions due to torsion, as we assume that the dimensionful coupling $f_4^{-2}$ of the interaction (54) describes an effective interaction where the effects of repulsion have been subtracted appropriately.

As we shall see in the following sections, the dynamically-generated masses will be much smaller than the ultraviolet (UV) momentum cut-off scale $\Lambda$ in the effective, string-inspired, low-energy four-dimensional theory [43, 40, 41]. It is natural to consider $\Lambda \lesssim M_s$, where $M_s$ is the string scale, above which stringy effects appear, which would jeopardize the validity of any point-like string-inspired effective action. As we have demonstrated in [41, 42], and shall review below, the four-fermion dimensionful coupling $f_4$ and UV cut-off scale $\Lambda$, will be determined self-consistently by the SD analysis in the massive phase of the models [41].

## 3. Dynamical Mass generation induced by non-Hermitian Yukawa interactions

We consider the following model consisting of a single pseudoscalar (axion) field $\phi(x)$ and a single Dirac fermion, $\psi(x)$, interacting with each other via a non-Hermitian Yukawa (specifically, anti-Hermitian) interaction with a real coupling $\lambda \in \mathbb{R}$. Both axion and fermion fields have zero bare masses. We restrict ourselves to Minkowski space time for concreteness. The action reads

$$\mathcal{S}_{\rm Yuk}^{\rm non-Herm} = \int d^4x \Big( \frac{1}{2}\, \partial_\mu \phi\, \partial^\mu \phi + \overline{\psi}\, i\, \partial\!\!\!/\, \psi + \lambda\, \phi\, \overline{\psi}\, \gamma^5\, \psi - \frac{g^2}{2\, f_4^2}\, (\overline{\psi}\gamma^5\, \psi)^2 \Big), \quad \lambda \in \mathbb{R}, \tag{55}$$

where we use the standard notation $\partial\!\!\!/ \equiv \gamma^\mu\, \partial_\mu$, with $\gamma^\mu$, $\mu = 0, \dots 3$, the four-dimensional Dirac matrices. Following the discussion after (39) in section 2.5, we include *no axion-$\phi$ potential* in the model. The model is $\mathcal{C}$PT invariant, provided one uses the following definitions for the discrete transformations for parity P, time-reversal T and charge conjugation $\mathcal{C}$, acting on fermions [40]:

$$\begin{aligned} \text{parity} \quad P \quad &: \quad \psi(t,\vec{r}) \to \gamma^0 \psi(t, -\vec{r}) \\ \text{time-reversal} \quad T \quad &: \quad \psi(t,\vec{r}) \to \gamma^1\gamma^3 \psi(-t, \vec{r}) \\ \text{charge conjugation} \quad \mathcal{C} \quad &: \quad \psi(t,\vec{r}) \to -i\gamma^2 \psi^*(t, \vec{r}) \; , \end{aligned} \tag{56}$$

such that the anti-Hermitian interaction $\phi\, \overline{\psi}\, \gamma^5\, \psi$ is $\mathcal{C}PT$-even under the above transformations if $\phi$ is a pseudo-scalar.[6] It is the existence of this anti-linear $\mathcal{C}PT$ symmetry that guarantees the existence of real energy eigenvalues, and thus real dynamical masses [40, 41], for this anti-Hermitian system, according to the general arguments of [6, 7, 8, 9, 10].

In (55) the parameter $g$ is set to take on the values $g = 0, 1$, depending on whether we consider the absence or the presence of attractive four-fermion interactions, respectively, and $f_4$ is an effective dimensionful coupling, with mass dimensions of [mass], which is determined self-consistently in a SD analysis [41], as we shall review below. As already mentioned, in the context of microscopic string-inspired models, the attractive four-fermion interactions in (55) express the net result, after appropriate subtraction of the corresponding repulsive interactions of the same form, obtained after applying the Fierz identities to the repulsive axial-current-current four-fermion interaction $\frac{3\kappa^2}{16}\, J^5_\mu\, J^{5\,\mu}$ in (24). Indeed, for a single fermion we are considering here, one has[7]

$$\frac{3\kappa^2}{16}\, J^5_\mu\, J^{5\,\mu} = \frac{3\kappa^2}{16}\, (\overline{\psi}\, \gamma^5\, \gamma^\mu\, \psi)\, (\overline{\psi}\, \gamma^5\, \gamma_\mu\, \psi) = \frac{3\kappa^2}{16}\, (\overline{\psi}\, \gamma^\mu\, \psi)\, (\overline{\psi}\, \gamma_\mu\, \psi) + \frac{3\kappa^2}{8}\, (\overline{\psi}\, \psi)^2 - \frac{3\kappa^2}{8}\, (\overline{\psi}\, \gamma^5\, \psi)^2. \tag{57}$$

We shall come back to this point later on in the article, after we estimate the strength of $f_4$ necessary for a consistent SD dynamical mass generation [41].

At the moment we remark that, motivated by the stringy axion models discussed above, we consider very small Yukawa couplings $|\lambda| \ll 1$ in (55). We are interested in dynamical mass generation, due to the coupling $\lambda$, as a way of mass generation alternative to the radiative anomalous generation (52), and independent of the anomalous coupling $\gamma$, since in the model (55)

[6] We note for completion that, in case $\phi$ is a scalar, one can also guarantee the $\mathcal{C}$PT invariance of the pertinent action, by defining appropriately the corresponding $\mathcal{C}$ operator.

[7] The reader should have noticed that, upon bringing the vector-current-current interaction onto the left-hand-side of (57), one would obtain a combination of four-fermion interactions which, if attractive, would coincide with the chiral-symmetry-invariant Nambu-Jona-Lasinio four-fermion interactions [79]. However, as we have seen above, the quantum torsion generates only the axial-current-current part of this and is repulsive. In string effective theories [78], as already mentioned, one has various vector-current-current induced terms with coefficients inversely proportional to the square of the mass of the exchanged massive string states. In our work here we are not interested in a detailed study of such four-fermion interactions. In this review, we shall only use (57) to make some generic phenomenological observations on the magnitude of $f_4$, consistent with SD mass generation, which would allow us to ignore the repulsive torsion-induced interactions.

$\gamma = 0$. We shall follow a Schwinger-Dyson (SD) approach to study dynamical mass generation for both axions and fermion fields [40].

Before doing this, we would like to remark that in the absence of attractive four-fermion interactions, i.e. when $g = 0$, there is *no* dynamical mass generation for *fermions*, but there could be for axions (pseudoscalar) $\phi(x)$ fields, as follows from *energetics arguments* [40] that we review in the next subsection.

### *3.1. Energetics arguments against dynamical generation of fermion mass in non-Hermitian Yukawa models without four-fermion self interactions*

To this end [40], let us first assume a non-zero *bare* mass for the scalar $M_0 \neq 0$. This will serve as a regulator for the definition of inverse propagators, as discussed in [40]. Our arguments on mass generation will not be affected by its presence, and thus $M_0$ can eventually be removed, as we shall see. To discuss dynamical mass generation for the fermions in non-Hermitian models a Euclidean path-integral quantisation is necessary (with a Euclidean-metric signature convention (+,+,+,+), which is used throughout this work), in which the anti-Hermitian Yukawa interaction appears as a *phase*:

$$Z_\lambda[j,\bar\eta,\eta] = \int \mathcal{D}[\phi,\psi,\bar\psi] \exp\left(-S_{Herm} - S_{anti-Herm} - S_{sources}\right) , \tag{58}$$

where

$$\begin{aligned} S_{Herm} &= \int d^4x \left(\frac{1}{2}\partial_\mu\phi\partial^\mu\phi + \frac{M_0^2}{2}\phi^2 + \bar\psi i\partial\!\!\!/\psi\right), \quad S_{sources} = \int d^4x(J\phi + \bar\eta\psi + \bar\psi\eta), \\ S_{anti-Herm} &= -\lambda\int d^4x\ \phi\bar\psi\gamma^5\psi \ = \ i\lambda\int d^4x\ \phi\,\Phi\ , \quad \text{with}\ \ \Phi \equiv i\bar\psi\gamma^5\psi\ \ \text{a phase.} \end{aligned}$$

The source-$J,\eta,\bar\eta$ terms in (59) have been inserted for the correct definition of the path integral.

From a basic property of complex calculus we obtain the inequality [40]

$$\begin{aligned} \exp(-S^{ferm}_{eff}) &\leq \int \mathcal{D}[\phi] \left|\exp\left(-\int_x \bar\psi i\partial\!\!\!/\psi + \bar\eta\psi + \bar\psi\eta + \frac{1}{2}\phi G^{-1}\phi + i\lambda\phi\Phi\right)\right| \\ &= \int \mathcal{D}[\phi]\ \exp\left(-\int_x \bar\psi i\partial\!\!\!/\psi + \bar\eta\psi + \bar\psi\eta + \frac{1}{2}\phi G^{-1}\phi\right)\ , \end{aligned} \tag{59}$$

We note that $S^{ferm}_{eff}$ plays the role of the (Euclidean) *vacuum energy functional.* hence, as a consequence of (59), it appears that this quantity is *larger* in the non-Hermitian Yukawa theory than in the free theory. This precludes energetically fermion dynamical mass generation [80], in contrast to the usual Hermitian case, where such a dynamical mass lowers the energy of the system [41]. This can also be seen explicitly by integrating out the massive scalar field $\phi$. In such a case, one obtains an effective fermionic action $S^{ferm}_{eff}$:

$$\begin{aligned} \exp(-S^{ferm}_{eff}) &\equiv \exp\left(-\int_x \bar\psi i\partial\!\!\!/\psi + m\bar\psi\psi + \bar\eta\psi + \bar\psi\eta\right)\int\mathcal{D}[\phi]\exp\left(\int_x \frac{1}{2}\phi G^{-1}\phi + i\lambda\phi\ \Phi\right) \\ &= \exp\left(-\int_x \bar\psi i\partial\!\!\!/\psi + m\bar\psi\psi + \bar\eta\psi + \bar\psi\eta + \frac{\lambda^2}{2}\Phi G\Phi\right)\ , \end{aligned} \tag{60}$$

where $G^{-1} = -\Box + M_0^2$ and $\Phi$ is defined in (59) (here the reader recognises the importance of assuming a non-zero bare mass for the scalar, $M_0 \neq 0$, so as to have a well-defined operator $G^{-1}$). Ignoring higher-order derivatives, which are irrelevant for our analysis, we obtain

$$S^{ferm}_{eff} \simeq \int_x \bar\psi i\partial\!\!\!/\psi + m\bar\psi\psi + \frac{\lambda^2}{2M_0^2}(\bar\psi\gamma^5\psi)^2\ . \tag{61}$$

The reader should then notice that this effective action includes a *repulsive* four-fermion interaction, for all values of the regulator $M_0$, which increases the energy of the system, in agreement with the generic argument (59), and thus dynamical mass generation for the fermions is impossible. Moreover, the form of the induced repulsive interaction implies the absence also of the chiral mass $\mu$, as the latter would be associated with a condensate of the form $\langle i\bar{\psi}\gamma^5\psi\rangle$, which is also not formed due to the above energetics reasons.[8]

In constrast, one may have axion mass generation, which can be established upon considering the effective action $S^{scal}_{eff}$ for the pseudoscalar field $\phi$, obtained after integrating out massive fermions (of bare mass $m_0 \neq 0$)

$$\exp(-S^{scal}_{eff}) \equiv \exp\Big(-\int_x \frac{1}{2}\partial_\mu\phi\partial^\mu\phi + \frac{M_0^2}{2}\phi^2 + J\phi\Big)\int \mathcal{D}[\psi,\bar{\psi}]\exp\Big(-\int_x \bar{\psi}(i\partial\!\!\!/ + m_0 - \lambda\phi\gamma^5)\psi\Big) .$$

For a constant scalar field configuration $\phi_0$, which suffices for demonstrating our main argument, the effective potential reads [40]

$$U_{eff}(\phi_0) = \frac{M_0^2}{2}\phi_0^2 - \mathrm{Tr}\left\{\ln(p\!\!\!/ + m_0 - \lambda\phi_0\gamma^5)\right\} , \tag{62}$$

which implies

$$\frac{dU_{eff}}{d\phi_0} = M_0^2\,\phi_0 + \frac{\lambda^2\phi_0}{4\pi^2}\left(\Lambda^2 - (m_0^2 - \lambda^2\phi_0^2)\ln\left(\frac{\Lambda^2 + m_0^2 - \lambda^2\phi_0^2}{m_0^2 - \lambda^2\phi_0^2}\right)\right) , \tag{63}$$

where $\Lambda$ is the UV cut off. The energies are therefore *real*, and the non-Hermitian theory is self consistent, for $m^2 \geq \lambda^2\phi_0^2$, as expected from the study in [17]. The possibility of real energies is attributed, of course, to the underlying $\mathcal{C}$PT antilinear symmetry [10, 40]. In the limit $\lambda^2\phi_0^2 \to m^2$, the effective potential becomes a mass term

$$U_{eff} \to \frac{1}{2}(M^{(1)})^2\phi_0^2 \quad \text{with} \quad (M^{(1)})^2 = M_0^2 + \frac{\lambda^2}{4\pi^2}\Lambda^2 . \tag{64}$$

From eq.(64) it is evident that dynamical axion-mass generation is possible in the anti-Hermitian-Yukawa-interaction model (55) (with $g = 0$), since on setting the bare mass to zero, $M_0 = 0$, one obtains a non-zero result from (64)

$$(M^{(1)})_0^2 = \frac{\lambda^2}{4\pi^2}\Lambda^2 > 0, \tag{65}$$

which is the same as the result of a one-loop calculation.

In [41] the same conclusions as above are reached by a detailed SD analysis, which avoids the use of regulator bare masses. We refer the interested reader to that work for details. Below we shall only describe the main results of such an analysis.

[8] The terminology "chiral mass" $\mu$ (or, equivalently, non-Hermitian mass) is used here for the coefficient accompanying the (non-Hermitian) fermion bilinear term $\bar{\psi}\gamma^5\psi$ in the Lagrangian, extending appropriately the corresponding definition in the Hermitian case. The four-fermion interactions in (61) can be written in a Hartree-Fock (mean field) approximation, up to an irrelevant constant, as: $(\bar{\psi}\gamma^5\psi)^2 = -\langle i\bar{\psi}\gamma^5\psi\rangle\, i\bar{\psi}\gamma^5\psi +$ four − fermion − interactions of pure quantum fluctuations. This implies the identification of $\mu$ with the chiral condensate $\langle i\bar{\psi}\gamma^5\psi\rangle$. In view of the repulsive nature of the four-fermion interactions in (61), such a condensate cannot form, and hence dynamical generation of $\mu \neq 0$ is not possible in the anti-Hermitian model.

### *3.2. Schwinger-Dyson Analysis of non-Hermitian Yukawa interactions, in the absence of four-fermion interactions*

To formulate properly our quantum non-Hermitian prototype models, one necessarily uses, as already mentioned, a Euclidean ("$E$") path-integral formalism. For the model (55) with $g = 0$, of interest in this subsection, the appropriate Euclidean generating functional is given by

$$Z^E[J,\eta,\overline{\eta}] = \int \mathcal{D}[\phi\psi\overline{\psi}] \exp\left\{-\int d^4x \left[\frac{1}{2}\,\partial_\mu\phi\partial^\mu\phi + \overline{\psi}i\partial\!\!\!/\psi - \lambda\phi\,\overline{\psi}\gamma^5\psi\right] - \int d^4x[J\phi + \overline{\psi}\eta + \overline{\eta}\psi]\right\}. \tag{66}$$

Taking into account that we are dealing with very small Yukawa couplings, $|\lambda| \ll 1$, possibly generated by non-perturbative effects in the underlying microscopic string/brane theories, and following standard SD methods, one arrives at the following SD equations for the pseudoscalar ($s$) and fermion ($f$) propagators in Fourier space [41]:

$$\begin{aligned} G_s^{-1}(k) - S_s^{-1}(k) &= \lambda\,\mathrm{tr}\left[\gamma^5 \int \frac{d^4p}{(2\pi)^4} G_f(p)\Gamma^{(3)}(p,k)G_f(p-k)\right] \\ G_f^{-1}(k) - S_f^{-1}(k) &= -\lambda\gamma^5 \left(\int \frac{d^4p}{(2\pi)^4} G_f(p)\Gamma^{(3)}(p,k)G_s(p-k)\right), \end{aligned} \tag{67}$$

where the dressed inverse propagators are given by: $G_f^{-1}(p) = p\!\!\!/ + m + \mu\gamma^5$, with $\mu$ the non-Hermitian (or "chiral") fermion mass, defined previously in footnote 8, $G_s^{-1}(p) = p^2 + M^2$, and we used the bare inverse propagators $S_f^{-1}(p) = p\!\!\!/$, $S_s^{-1}(p) = p^2$. The vertex function is in the rainbow approximation [41], appropriate for weak Yukawa interactions, $|\lambda| \ll 1$, and reads:

$$\Gamma^{(3)}(p,k) \simeq \lambda\gamma^5\ . \tag{68}$$

In [41] and here, we restrict ourselves to the case of real $\mu \in \mathbb{R}$. This stems from the fact that we are interested in the (physically relevant) case where the *energies* of the system in the massive phase are *real*, for which one must have the following condition among the (dynamically generated) mass parameters [17, 40]

$$|\mu| \leq |m|. \tag{69}$$

The reality of the energy eigenspectrum in this case may be traced back [40, 41] to the underlying $\mathcal{C}$PT (antilinear)symmetry of the field theory (55), with $\mathcal{C}$ appropriately defined ((56)) [21], according to the general arguments of [6, 7, 8, 9, 10]. As discussed in [20, 21], the condition (69) guarantees unitarity, in the sense of the existence of a well-defined, *conserved*, probability density for the non-Hermitian fermionic system, which is less than unity.

Upon assuming vanishing external momenta $k = 0$ in the SD equations (67), and performing the Euclidean momentum integrations with an UV cutoff $\Lambda$, we arrive at [40]

$$M^2 = \frac{\lambda^2}{4\pi^2}\left[\frac{(\Lambda^2 + m^2 - 3\mu^2)\Lambda^2}{\Lambda^2 + m^2 - \mu^2} - (m^2 - 3\mu^2)\ln\left(1 + \frac{\Lambda^2}{m^2 - \mu^2}\right)\right] \tag{70}$$

$$m + \mu\gamma^5 = -\frac{\lambda^2}{16\pi^2}\frac{(m - \mu\gamma^5)}{M^2 - m^2 + \mu^2}\left[M^2 \ln\left(1 + \frac{\Lambda^2}{M^2}\right) - (m^2 - \mu^2)\ln\left(1 + \frac{\Lambda^2}{m^2 - \mu^2}\right)\right] \tag{71}$$

Following our previous discussion in section 3.1 on the energetics of the action (61), implying non-formation of a dynamical chiral mass $\mu \neq 0$, we set from now on $\mu = 0$ [41].

On setting $m = \mu = 0$, which is consistent with (69), we observe from (70) that there is now a consistent solution for the dynamically generated axion mass $M$, since (70) leads to

$$M^2 = \frac{\lambda^2}{4\pi^2}\Lambda^2 \ll \Lambda^2, \quad \lambda^2 \ll 1, \tag{72}$$

which is the same as the expression (65). On account of (69), the case where $m = 0$ but $\mu \neq 0$ is not allowed, as it would lead to unphysical situations with complex energies.

For $\mu = 0$, and $m\,M \neq 0$, we have the following system of SD equations

$$M^2 = \frac{\lambda^2}{4\pi^2}\left[\Lambda^2 - m^2 \ln\left(1 + \frac{\Lambda^2}{m^2}\right)\right] \tag{73}$$

$$1 = -\frac{\lambda^2}{16\pi^2}\frac{1}{M^2 - m^2}\left[M^2 \ln\left(1 + \frac{\Lambda^2}{M^2}\right) - m^2 \ln\left(1 + \frac{\Lambda^2}{m^2}\right)\right] \tag{74}$$

We will now consider partial solutions $0 \neq m \simeq M \ll \Lambda$. From (74):

$$-\frac{16\pi^2}{\lambda^2} = \ln\left(1 + \frac{\Lambda^2}{M^2}\right), \tag{75}$$

which is inconsistent. It is also readily seen that the case $\mu = M = 0$ also does not lead to fermion mass generation.

Hence dynamical fermion mass is not possible for pure anti-Hermitian Yukawa interactions, only scalar mass can be generated dynamically, in agreement with the generic energetics arguments [40] provided in the previous subsection 3.1.

### *3.3. Non-Hermitian Yukawa Interactions in the presence of attractive four-fermion interactions*

We next proceed to discuss explicitly the rôle on dynamical-mass generation played by additional *Hermitian attractive* four-fermion interactions in the non-Hermitian Yukawa model [41], *i.e.* we consider the case $g = 1$ in (55). The pertinent Euclidean generating functional reads:

$$\begin{aligned} Z^E[J,\eta,\overline{\eta}] &= \int \mathcal{D}[\phi\psi\overline{\psi}] \exp\left\{-\int d^4x \left[\frac{1}{2}\partial_\mu\phi\partial^\mu\phi + \overline{\psi}i\partial\!\!\!/\psi - \lambda\phi\,\overline{\psi}\gamma^5\psi + \frac{1}{2f_4^2}(\overline{\psi}\,\gamma^5\psi)^2\right] \right. \\ &\quad \left. - \int d^4x[J\phi + \overline{\psi}\eta + \overline{\eta}\psi]\right\} \\ &= \int \mathcal{D}[\sigma\phi\psi\overline{\psi}] \exp\left\{-\int d^4x \left[\frac{1}{2}\partial_\mu\phi\partial^\mu\phi + \overline{\psi}i\partial\!\!\!/\psi - \lambda\phi\overline{\psi}\gamma^5\psi + \frac{f_4^2}{2}\sigma^2 + i\,\sigma\overline{\psi}\gamma^5\psi\right] \right. \\ &\quad \left. - \int d^4x[J\phi + \overline{\psi}\eta + \overline{\eta}\psi + K\sigma]\right\}, \end{aligned} \tag{76}$$

where in the second equality we have linearised the four-fermion interactions using an auxiliary pseudoscalar field $\sigma$. The additional ingredient with respect to the analysis in the previous subsection is the $\sigma$ propagator, which in momentum space is given by $G_\sigma(p) = 1/f_4^2$.

As a result of the extra field $\sigma(x)$, the SD equations with vanishing external momenta read now [41]:

$$G_f^{-1}(0) - S_f^{-1}(0) = -\lambda^2\gamma^5\left(\int \frac{d^4p}{(2\pi)^4} G_f(p)\gamma^5 G_s(p)\right) - \gamma^5\left(\int \frac{d^4p}{(2\pi)^4} G_f(p)\gamma^5 G_\sigma(p)\right) \tag{77}$$

$$G_s^{-1}(0) - S_s^{-1}(0) = \lambda^2 \,\mathrm{tr}\left[\gamma^5 \int \frac{d^4p}{(2\pi)^4} G_f(p)\gamma^5 G_f(p)\right] \tag{78}$$

Using the dressed inverse propagators, as in the previous subsection, and performing the integrals using an UV cutoff $\Lambda$, we arrive at the following system of algebraic equations:

$$\begin{aligned} m + \mu\gamma^5 = &-\frac{\lambda^2}{16\pi^2}\frac{(m-\mu\gamma^5)}{M^2 - m^2 + \mu^2}\left[M^2 \ln\left(1+\frac{\Lambda^2}{M^2}\right) - (m^2-\mu^2)\ln\left(1+\frac{\Lambda^2}{m^2-\mu^2}\right)\right] \\ &+\frac{(m-\mu\gamma^5)}{16\pi^2 f_4^2}\left(\Lambda^2 - (m^2-\mu^2)\ln\left(1+\frac{\Lambda^2}{m^2-\mu^2}\right)\right) \end{aligned} \tag{79}$$

$$M^2 = \frac{\lambda^2}{4\pi^2}\left[\frac{(\Lambda^2+m^2-3\mu^2)\Lambda^2}{\Lambda^2+m^2-\mu^2} - (m^2-3\mu^2)\ln\left(1+\frac{\Lambda^2}{m^2-\mu^2}\right)\right] \tag{80}$$

As the SD pseudoscalar mass equation is independent of $f_4$, it is straightforward to see from (80) that, for $m = \mu = 0$, one obtains the dynamically generated (pseudo)scalar mass (72).

If the case $\mu = 0$ but $m\,M \neq 0$, the system of SD equations reads

$$\begin{aligned} 1 &= -\frac{\lambda^2}{16\pi^2}\frac{1}{M^2-m^2}\left[M^2\ln\left(1+\frac{\Lambda^2}{M^2}\right) - m^2\ln\left(1+\frac{\Lambda^2}{m^2}\right)\right] + \frac{1}{16\pi^2 f_4^2}\left(\Lambda^2 - m^2\ln\left(1+\frac{\Lambda^2}{m^2}\right)\right) \\ M^2 &= \frac{\lambda^2}{4\pi^2}\left[\Lambda^2 - m^2\ln\left(1+\frac{\Lambda^2}{m^2}\right)\right], \end{aligned} \tag{81}$$

which for the specific case

$$\Lambda \gg m \simeq M \neq 0, \tag{82}$$

leads to

$$\begin{aligned} 1 &= -\frac{\lambda^2}{16\pi^2}\ln\left(1+\frac{\Lambda^2}{M^2}\right) + \frac{m^2}{4\,f_4^2\,\lambda^2} \Rightarrow m^2 \simeq M^2 \simeq 4\,\lambda^2\,f_4^2 + \left|\mathcal{O}(\lambda^4\,\ln\lambda^2)\right|, \\ M^2 &\simeq m^2 \simeq \frac{\lambda^2}{4\pi^2}\Lambda^2\,. \end{aligned} \tag{83}$$

For consistency we then obtain

$$f_4 \simeq \frac{\Lambda}{4\pi} - \left|\mathcal{O}(\lambda^2\,\ln\lambda^2)\right|, \quad \lambda^2 \ll 1. \tag{84}$$

Thus, in non-Hermitian Yukawa interactions, upon the inclusion of sufficiently strong four-fermion attractive interactions, one can obtain dynamical mass for both fermions and axions.

Some important remarks are in order at this point. The reader should have noticed from (83) that there is SD dynamical mass for *arbitrarily small values* of the Yukawa coupling $\lambda$, which thus includes the case where such couplings are generated by stringy instanton effects, as discussed in section 2.5 (*cf.* (47)). We stress that this is a consequence of the fact that the prototype model does not contain self-interactions of the axion-$\phi$ field, which is motivated by the stringy considerations of section 2.5 (*cf.* discussion following (39)). We note that, in the presence of such self-interactions, *e.g.* a $\phi^4$-term, thIn constraste situation concerning fermion mass generation can be very different. For instance, in the specific (Hermitian) model of [81], where there is a common coupling between the Yukawa and $\phi^4$ self-interactions, there is a critical value of $g$ above which dynamical fermion mass is possible. We do not consider axion self-interactions neither in this review nor in [40, 41, 42].

We also note that for our purposes it suffices that we considered the special case (82) for the dynamical masses of axions and fermions of approximately equal magnitude. More general solutions of course may exist, but they are not relevant for our purposes, which are to demonstrate explicitly a non-trivial concrete solution of the SD system entailing dynamical masses for both axions and fermions.

We remark at this point that, although above we view the interaction (54) as an effective attractive interaction which incorporates the subtraction of repulsive terms of similar form stemming from the quantum-torsion-induced repulsive interactions (24) upon applying the Fierz identities (57), nonetheless one may also consider the case in which this interaction genuinely dominates the torsion terms, for an appropriate value of the cut-off scale $\Lambda$, which defines the momentum scale above which the validity of the low-energy effective field theory breaks down. In such a case one would justify ignoring the repulsive interactions altogether. To determine in this case the scale $\Lambda$, we should compare the coefficient $1/(2f_4^2)$ of the attractive terms (54), with $f_4$ determined by (84), with the corresponding coupling $\sqrt{3\kappa^2/8}$ of the repulsive quantum-torsion-induced terms (57). The reader observes that dominance of the attractive term (54) by, say, at least one order of magnitude, over the corresponding term in (57), would imply a cut-off momentum scale

$$\frac{1}{2\,f_4^2} \gtrsim \mathcal{O}(10) \times \frac{3\,\kappa^2}{8} \quad \overset{\text{cf.}\,(84)}{\Rightarrow} \quad \Lambda \lesssim \mathcal{O}\Big(\sqrt{\frac{4\,\pi}{15}}\Big)\, M_P \sim 0.9\, M_P \ , \tag{85}$$

which is phenomenologically reasonable for an effective theory embedded in an UV complete quantum gravity theory, such as strings. In fact, the range for $\Lambda$ in (85) is compatible with (19), which is satisfied for

$$\Lambda \sim M_s \simeq 0.2\, M_P \ , \tag{86}$$

implying that $M_s$ acts as an effective cut-off energy scale above which the validity of the point-like string-effective theory breaks down. Under (86), the dynamically generated mass (83) becomes

$$M \simeq m \simeq \frac{|\lambda|}{2\pi}\,\Lambda \simeq 0.03\,|\lambda|\, M_P \simeq 0.04\,|\lambda| \times 10^{19}\ \text{GeV}. \tag{87}$$

If $\lambda$ is generated by string instantons (47), then, in order to have axion masses larger than $10^{-24}$ eV, which is a lower-bound for the mass of an axion in case the latter plays the rôle of a dominant dark-matter species [67], one needs the (real parts of the) stringy-instanton actions to be of order: $0 < S_{\text{instanton}} < 116.5$ (in natural units of $\hbar c = 1$). Smaller axion masses $> 10^{-32}$ eV are allowed if the axion constitutes only a small (*e.g.* $< 5\%$) part of dark matter in the Universe, in which case one obtains for the instanton actions $0 < S_{\text{instanton}} < 134.9$. Such considerations allow for a rather wide variety of axion masses in our scenarios, which can thus play a rôle as dark matter candidates, including ultralight ones, with sensitivity to be falsified at immediate-future experimental facilities.[9]

We also note that the presence of Yukawa interactions, coupling the axion to right-handed neutrinos in the model of [43], under the assumption (82), (83) for the dynamically generated masses, can be reconciled with phenomenologically realistic situations of particle physics by

[9] Such instanton actions are smaller in magnitude than the ones required for axion quintessence in Hermitian stringy axion cosmologies, in which the contributions of axions to dark energy are non-negligible; in such cases one needs $S_{\text{instanton}} \sim 200 - 300$ [67]. We remark at this point, however, that, strictly speaking, one cannot directly compare the phenomenological constraints discussed in the context of our non-Hermitian models with the phenomenological constraints for generic Hermitian stringy axion models based on instanton-induced axion potentials [55], as a consequence of the anomalous couplings of the (stringy) axions. The dynamical axion masses (87) pertain to a model (55) without anomalous terms ($\gamma = 0$) and no potential for the pseudoscalar fields. Although non-Hermitian anomaly terms ($\gamma \neq 0$) are discussed in the next section 4, nonetheless for such non-Hermitian models the concept of instanton-induced potentials is not completely understood, as already mentioned. Hence, the considerations in [55] might not apply even to that case.

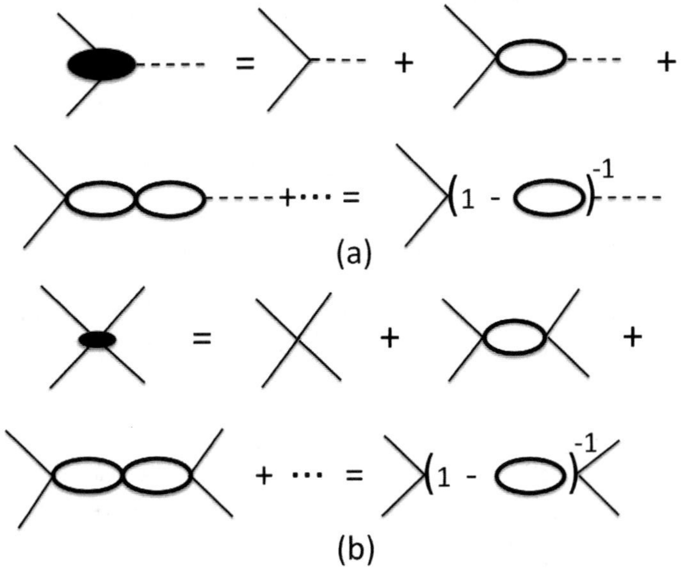


**Figure 2.** *Resummation (dark elliptical blobs) of leading (quadratic) UV divergences due to four-fermion 'bubble' diagrams (ellipse without filling in the figures). Upper figure* (a)*: effects on Yukawa interaction vertex. Lower figure* (b)*: effects on four-fermion vertex. Continuous lines refer to fermions, dashed lines pertain to pseudoscalar fields. Figures taken from [41].*

considering such ultralight right-handed neutrinos as being almost decoupled from the standard-model matter sector. In other words, the dominant coupling of such right-handed neutrinos with masses given by (83) to matter is the Yukawa one, while their mixing through Higgs-portal-type interactions with standard-model matter is negligible. Of course, such restrictions are lifted if one views the prototype models discussed here and in [40, 41, 42] as purely phenomenological, not necessarily connected to microscopic string models, in which case $\lambda$ is a free parameter, which need not be generated by string instanton effects, and hence need not be that small.

### *3.4. Resummation effects of strong four-fermion interactions*

Before closing this section we would like to comment briefly on how resummation of the four-fermion interactions affects qualitatively the above conclusions. Indeed, this is an important aspect, given the strong coupling nature of the dimensionless four-fermion coupling (84)

$$\frac{\Lambda}{\sqrt{2}\,f_4} \sim 4\pi > 1, \tag{88}$$

which implies the necessity for a resummations of the loop "bubble" diagrams associated with quantum corrections of the four-fermion vertex, as indicated in fig. 2. This issue was discussed in detail in [41], where we refer the interested reader for details. For our purposes here, we only mention that the result of such a resummation, as far as the leading quadratic UV divergences are concerned, of interest to us here, is that the relevant quantum corrections can be absorbed in "renormalised" Yukawa and four-fermion couplings:

$$f_{4R} \sim \frac{1}{\sqrt{2}} f_4 \,, \quad \lambda_R \sim 2\,\lambda \,, \tag{89}$$

which are thus of the same order as the bare couplings. Therefore, the validity of the above SD analysis and the pertinent conclusions are guaranteed upon replacing the bare couplings $\lambda$, $f_4$ by the 'renormalised' ones (89) [41]. This is to be understood in all the previous formulae as well as the ones that follow. For notational convenience, we omit from now on the suffix $R$.

## 4. Inclusion of non-Hermitian anomaly terms and non-Hermitian axial backgrounds

A final topic we would like to discuss in this mini-review is the rôle of *non-hermitian anomaly and axial background* terms on the dynamical mass generation induced by the non-Hermitian Yukawa interactions in the model. The presence of the anomalous terms have been responsible, as we have discussed in section 2, for the radiative generation of Majorana masses for the right-handed neutrinos in the Hermitian model of [43]. Formally, when both Yukawa and anomalous interactions are characterised by purely imaginary couplings (53), *i.e.* when such interactions are anti-Hermitian, the expressions for the radiative fermion masses (52) retain their reality. However, the concept of anomalies, and the relevant studies, are not yet developed for non-Hermitian theories, except making formal analogies, and therefore it would be interesting to examine alternative ways for fermion (and axion) mass generation, through the anti-Hermitian Yukawa interaction, and scrutinise the rôle of the presence of anomalous terms on such a phenomenon. This has been done in detail in [42]. Below we shall only review the basic conclusions of that work.

In addition, we shall also examine in this section the rôle of non-Hermitian interactions of fermions with a constant axial background $\mathcal{B}_\mu$

$$\mathcal{S}^{antiherm}_{\rm axial-backgr} = \int d^4x\sqrt{-g}\, i\, \mathcal{B}_\mu \overline{\psi}\, \gamma^\mu\, \gamma^5\, \psi\ , \quad \mathcal{B}_\mu \in \mathbb{R}\ . \tag{90}$$

In the context of microscopic string theories, discussed in section 2, the microscopic origin of the non-Hermitian background can be traced back to the non-Hermitian anomalous interactions of the KR gravitational axion of the massless bosonic string multiplet in some formulations of the theory ( see "**Scheme II**", (24)). To understand this, we mention that, in Hermitian theories, i.e. when $i\mathcal{B}_\mu \in \mathbb{R}$, such constant backgrounds (with temporal components only in a given Lorentz frame, specifically the cosmological Robertson-Walker (RW) frame) have been argued to characterise cosmological configurations of the KR gravitational axion field, $b(x)$, exhibiting a linear dependence in the cosmic time, $t$, in the RW frame [60, 57],

$$b(x) = (\text{constant}) \times t\ . \tag{91}$$

Non-hermitian constant axial backgrounds (90), then, can arise in the "**Scheme II**" (24) of the effective string theory actions, in the way discussed in section 2. Moreover, in Hermitian models, constant axial backgrounds provide alternative ways for tree-level Leptogenesis [60, 56] (*i.e.* the generation of Lepton-number asymmetry between particles and anti-particles), in theories involving massive right-handed neutrinos, as a result of the asymmetric decays of the latter to standard-model leptons and antileptons in the presence of such backgrounds. It is yet to be seen whether non-Hermitian axial backgrounds (90) can lead to similar Leptogenesis phenomena in non-Hermitian theories.

Their contribution to mass generation, though, is one such aspect that can be settled already. In the work of [42], which will be reviewed briefly here, we shall study the rôle of such interactions on the masses of axions and fermions, generated dynamically by non-Hermitian Yukawa couplings. We also note that in [82] such non-Hermitian constant backgrounds have been examined from the point of view of their rôle in fermion mass generation in the context of

Hermitian Nambu-Jona-Lasinio four-fermion models [79], which however did not contain Yukawa interactions, in contrast to our case here, where the latter constitute the dominant source for mass generation.

Therefore, in this section we shall study the following prototype action:

$$\begin{aligned} \mathcal{S}^{anti-Herm}_{axial-back+anom} &= \int d^4x\sqrt{-g}\,\Big(\frac{1}{2}\partial_\mu\phi\partial^\mu\phi + \bar{\psi}i\partial\!\!\!/\psi - \frac{i\,\gamma}{f_b}(\partial_\mu\phi)\bar{\psi}\gamma^\mu\gamma^5\psi \\ &+ i\,\mathcal{B}_\mu\bar{\psi}\gamma^\mu\gamma^5\psi + \lambda\,\phi\,\bar{\psi}\gamma^5\psi - \frac{g^2}{2f_b^2}(\bar{\psi}\gamma^5\psi)^2\Big). \end{aligned} \tag{92}$$

We remark at this stage that, in the context of the model [43], $f_b$ in (92) is given by (29), but here we consider it as an arbitrary mass scale to be determined self consistently in terms of the UV cutoff $\Lambda$ in the phase where dynamical mass generation occurs [42].

In (92) we have included non-trivial four-fermion attractive interactions, since, from arguments given previously (section 3.1), we know that their absence in anti-Hermitian Yukawa models disfavours dynamical mass generation for fermions. The coupling $g$ is inserted to demonstrate the difference (in general) of the strength of the four-fermion attractive interactions from that of the anomalous terms. The anomalies are encoded in the four-divergence of the axial current (*cf.* (22)), and appear after partial integration of the relevant (third) term on the right-hand-side of (92), with coefficient $-i\,\gamma/f_b$. In what follows, consistently with the discussion so far, we shall work in a model formulated in a Minkowski spacetime.

The action (92) is $\mathcal{CPT}$ invariant, under the transformations (56) for the fermions, which guarantees the reality of the energy spectrum, and hence the dynamical masses [42], according to the generic analysis of [6, 7, 8, 9, 10]. We note that a constant anti-Hermitian axial background would break *spontaneously* the (complex) Lorentz symmetry, and thus the generalised $\mathcal{C}$PT symmetry, as in the Hermitian case [60, 56]. Nonetheless, the reality of the dynamically-generated masses is guaranteed [42], as the condition (69) is maintained.

### *4.1. Models with non-Hermitian Yukawa and anomalous interactions - no axial background*

We commence our study with the case of anti-Hermitian Yukawa interactions in the presence of anti-Hermitian anomalous couplings to fermions, in the absence of axial background, i.e. setting $\mathcal{B}_\mu = 0$ in (92). In this case, the Euclidean partition function, under the linearization of the four-fermion interactions by means of the auxiliary pseudoscalar $\sigma$, reads (upon the standard inclusion of appropriate sources $\eta, \bar{\eta}, K, J$ for the fields $\psi$, $\overline{\psi}$, $\sigma$ and $\phi$, repsectively):

$$\begin{aligned} Z[K,J,\eta,\bar{\eta}] &= \int \mathcal{D}[\sigma\phi\psi\bar{\psi}]\exp\Big(-\int d^4x\Big[\frac{1}{2}\partial_\mu\phi\partial^\mu\phi + \bar{\psi}i\partial\!\!\!/\psi \\ &+ i\frac{\gamma}{f_b}(\partial_\mu\phi)\bar{\psi}\gamma^\mu\gamma^5\psi - \lambda\,\phi\,\bar{\psi}\gamma^5\psi + \frac{f_b^2}{2g^2}\sigma^2 - i\sigma\bar{\psi}\gamma^5\psi\Big] \\ &- \int d^4x[J\phi + \bar{\psi}\eta + \bar{\eta}\psi + K\sigma]\Big) \end{aligned} \tag{93}$$

In the way written, the anomalous interactions with coupling $\gamma$ in (92) imply that the Feynman rule for the (bare) axion-fermion vertex $\phi\,\bar{\psi}\psi$ is given by $\left(\lambda - \frac{\gamma}{f_b}p\!\!\!/\right)\gamma^5$, in the convention where all the momenta are incoming to the vertex.

The details of the SD analysis are given in [42] and will not be repeated here, where we only state the final result. We seek a partial solution of the SD equations for dynamical mass generation stemming from (93), for which there is an approximate equality of fermion ($m$) and axion ($M$) dynamical masses, $m \simeq M$, with the chiral mass $\mu = 0$, which suffices for our purposes, for reasons explained above. Using the rainbow approximation for the vertices, appropriate for small Yukawa and anomalous couplings $|\gamma| \sim |\lambda| \ll 1$, assumed here, and taking into account the effects of resummation of the four-fermion interactions, as described in the previous section 3.4, we arrive at the following system of SD equations describing the dynamical generation of fermion and pseudoscalar masses [42]

$$\begin{aligned} 1 &= -\frac{\lambda^2}{16\pi^2}\frac{1}{M^2-m^2}\left[M^2\ln\left(1+\frac{\Lambda^2}{M^2}\right)-m^2\ln\left(1+\frac{\Lambda^2}{m^2}\right)\right] \\ &+\frac{\gamma^2}{16\pi^2 f_b^2}\frac{1}{M^2-m^2}\left[-\Lambda^2(M^2-m^2)+M^4\ln\left(1+\frac{\Lambda^2}{M^2}\right)-m^4\ln\left(1+\frac{\Lambda^2}{m^2}\right)\right] \\ &+\frac{g^2}{16\pi^2 f_b^2}\left(\Lambda^2-m^2\ln\left(1+\frac{\Lambda^2}{m^2}\right)\right) \end{aligned} \tag{94}$$

and

$$M^2 = \frac{\lambda^2}{4\pi^2}\left[\Lambda^2 - m^2\ln\left(1+\frac{\Lambda^2}{m^2}\right)\right], \tag{95}$$

where $\Lambda$ is the UV cut-off, as in previous sections. Seeking solutions with $m \simeq M \neq 0$ (*cf.* ((82)) we obtain

$$1 = -\frac{1}{16\pi^2}\left(\lambda^2 - \frac{\gamma^2 M^2}{f_b^2}\right)\ln\left(1+\frac{\Lambda^2}{M^2}\right) + \frac{g^2-\gamma^2}{16\pi^2 f_b^2}\left[\Lambda^2 - m^2\ln\left(1+\frac{\Lambda^2}{M^2}\right)\right] \tag{96}$$

and

$$\frac{4\pi^2}{\lambda^2}M^2 = \left[\Lambda^2 - m^2\ln\left(1+\frac{\Lambda^2}{m^2}\right)\right] \tag{97}$$

Considering $\Lambda^2 \gg M^2$, $|\lambda|, |\gamma| \ll 1$, with $g \neq \gamma$, and substituting (97) in (96), we arrive at [42]:

$$M^2 \simeq m^2 \simeq \frac{4 f_b^2 \lambda^2}{g^2-\gamma^2} + |\mathcal{O}\Big((\lambda^4, \lambda^2\gamma^2)\,\ln(\lambda^2)\Big)| \tag{98}$$

Hence, consistency between (98) and (97) requires

$$\frac{f_b}{\sqrt{g^2-\gamma^2}} \simeq \frac{\Lambda}{4\pi} - |\mathcal{O}(\lambda^2)| \quad \Rightarrow \quad M^2 \simeq m^2 \simeq \frac{\lambda^2}{4\pi^2}\Lambda^2 + |\mathcal{O}\Big((\lambda^4, \lambda^2\gamma^2)\,\ln(\lambda^2)\Big)|\,, \tag{99}$$

for small $\gamma, \lambda$ couplings. In order to have dynamical mass generation one must have[10]

$$g^2 > \gamma^2, \tag{100}$$

The reader should have noticed that bare axion and fermion masses are not necessary, which implies that dynamical mass generation is possible under the above restriction (100). Thus in this case, contrary to the Hermitian one (which is studied in [42]), it is a sufficiently strong attractive four-fermion interaction that catalyses the Yukawa-induced dynamical mass generation, since the non-Hermitian anomaly term *resists* mass generation.

In [40, 42] the case of Majorana fermions has also been studied, which are of relevance to the model of [43]. The case is qualitatively similar to the Dirasc case studied above, apart from some

[10] For a discussion on the limiting case $g^2 = \gamma^2 + \epsilon$, $\epsilon \to 0^+$, we refer the reader to [42].

numerical factors in the expression for the dynamical mass and the one providing the connection of the four-fermion couplings $f_b$ to the cut-off $\Lambda$. In such a case, the analogue of (98) reads:

$$M^2_{\rm Dyn\ Maj} \simeq \frac{2f_b^2\,\lambda^2}{g^2-\gamma^2} + \dots, \quad \text{with} \quad \frac{f_b}{\sqrt{g^2-\gamma^2}} \simeq \frac{\Lambda}{2\sqrt{2}\pi}, \tag{101}$$

implying:

$$M^{anti-herm}_{\rm Dyn\ Maj} \sim \frac{\lambda}{2\pi}\,\Lambda, \quad |\lambda| \ll 1. \tag{102}$$

Upon comparison of (102) with (52) (with $|\gamma|$, $|\lambda| \ll 1$, as required for consistency of our SD treatment, which also characterises the physically motivated case in which these interactions are generated by non-perturbative stringy effects, as discussed in section 2), one obtains:

$$M_R^{\rm anti-herm} \sim \frac{\sqrt{3/2}\,\gamma\,(\kappa\,\Lambda)^5}{49152\,\pi^3}\, M^{\rm anti-herm}_{\rm Dyn\ Maj} \simeq 8\cdot 10^{-7}\,\gamma\,(\kappa\,\Lambda)^5\, M^{\rm anti-herm}_{\rm Dyn\ Maj} \ll M^{\rm anti-herm}_{\rm Dyn\ Maj}\,, \tag{103}$$

for any value of a sub-planckian cutoff $\kappa\,\Lambda \lesssim 1$. Hence in the non-Hermitian case, the anomaly-and-Yukawa-induced dynamical mass dominates the radiative anomalously-induced mass, provided the existence of the latter is rigorously proven of course in this case, something which is currently pending.

### *4.2. Non-Hermitian axial background and Yukawa interactions - no anomalies*

As a final topic for discussion in this review, we consider the case where an anti-Hermitian axial background (90) is present for an anomaly-free fermion model, *i.e.* upon setting $\gamma = 0$ in (92). The incorporation of both non-Hermitian axial background and anomalous interactions is straightforward and presents no conceptual difficulties, only algebraically more elaborate analysis. Hence it will not be discussed here.

For details of the SD study of models with both anti-Hermitian background and Yukawa interactions we refer the reader to [42]. Below we shall only give the final results. The pertinent solutions for SD mass generation for axions and fermions (with approximately equal masses $m \simeq M(cf.(82))$ and chiral mass $\mu = 0$) read:

$$\begin{aligned} 1 \;=\; & -\frac{\lambda^2}{16\pi^2\bar{M}^2\bar{\mathcal{B}}^2}\Bigg(1+2(\bar{M}^2+\bar{\mathcal{B}}^2)-\sqrt{(1+\bar{M}^2-\bar{\mathcal{B}}^2)^2+4\bar{M}^2\bar{\mathcal{B}}^2} \\ & -\left(\sqrt{(1+\bar{M}^2-\bar{\mathcal{B}}^2)^2+4\bar{M}^2\bar{\mathcal{B}}^2}-(\bar{M}^2+\bar{\mathcal{B}}^2)\right)(\bar{M}^2+7\bar{\mathcal{B}}^2) \\ & +4\bar{\mathcal{B}}^2(\bar{M}^2-2\bar{\mathcal{B}}^2)\log\left[\frac{1+\bar{M}^2-\bar{\mathcal{B}}^2+\sqrt{(1+\bar{M}^2-\bar{\mathcal{B}}^2)^2+4\bar{M}^2\bar{\mathcal{B}}^2}}{2\bar{M}^2}\right]\Bigg). \end{aligned} \tag{104}$$

and

$$
\begin{aligned}
1 \;=\; & \frac{\lambda^2}{32\pi^2\bar{\mathcal{B}}^2}\left(1+\bar{\mathcal{B}}^2\left(1-\frac{\bar{\mathcal{B}}^2}{\sqrt{\bar{\mathcal{B}}^4+4\bar{M}^2\bar{\mathcal{B}}^2}}\right)\log\left(1+\frac{1}{\bar{M}^2}\right)-\left(\sqrt{(1+\bar{M}^2-\bar{\mathcal{B}}^2)^2+4\bar{M}^2\bar{\mathcal{B}}^2}-(\bar{M}^2+\bar{\mathcal{B}}^2)\right)\right. \\
& -3\bar{\mathcal{B}}^2\log\left[\frac{1+(\bar{M}^2-\bar{\mathcal{B}}^2)+\sqrt{(1+\bar{M}^2-\bar{\mathcal{B}}^2)^2+4\bar{M}^2\bar{\mathcal{B}}^2}}{2\bar{M}^2}\right] \\
& \left.+\frac{\bar{\mathcal{B}}^4}{\sqrt{\bar{\mathcal{B}}^4+4\bar{M}^2\bar{\mathcal{B}}^2}}\log\left[\frac{\left(\sqrt{\bar{\mathcal{B}}^4+4\bar{M}^2\bar{\mathcal{B}}^2}+\sqrt{(1+\bar{M}^2-\bar{\mathcal{B}}^2)^2+4\bar{M}^2\bar{\mathcal{B}}^2}\right)^2-(1+\bar{M}^2)^2}{\left((\bar{M}^2+\bar{\mathcal{B}}^2)+\sqrt{\bar{\mathcal{B}}^4+4\bar{M}^2\bar{\mathcal{B}}^2}\right)^2-\bar{M}^4}\right]\right) \\
& -\frac{\bar{g}^2}{64\pi^2\bar{\mathcal{B}}^2}\left(1+2(\bar{M}^2+\bar{\mathcal{B}}^2)-\sqrt{(1+\bar{M}^2-\bar{\mathcal{B}}^2)^2+4\bar{M}^2\bar{\mathcal{B}}^2}\right. \\
& -\left(\sqrt{(1+\bar{M}^2-\bar{\mathcal{B}}^2)^2+4\bar{M}^2\bar{\mathcal{B}}^2}-(\bar{M}^2+\bar{\mathcal{B}}^2)\right)(\bar{M}^2+7\bar{\mathcal{B}}^2) \\
& \left.+4\bar{\mathcal{B}}^2(\bar{M}^2-2\bar{\mathcal{B}}^2)\log\left[\frac{1+(\bar{M}^2-\bar{\mathcal{B}}^2)+\sqrt{(1+\bar{M}^2-\bar{\mathcal{B}}^2)^2+4\bar{M}^2\bar{\mathcal{B}}^2}}{2\bar{M}^2}\right]\right), \qquad (105)
\end{aligned}
$$

where $\bar{M} = M/\Lambda$ and $\bar{\mathcal{B}} = \mathcal{B}/\Lambda$ and $\bar{g} = (g\Lambda)/f_b$.

Figure 3 plots the right hand side of (104) as a function of the axion mass $\bar{M}$ for different values of $\mathcal{B}$. All the curves intersect the solid line, corresponding to the fixed value 1, which represents the left hand side of (104). The reader should observe that, for a non-Hermitian axial background, the larger the background, the bigger the dynamical mass. This is to be contrasted with the case of a Hermitian axial background, for which, as we have discussed in [42], one faces the opposite situation, that is, the larger the background, the smaller the dynamical mass.

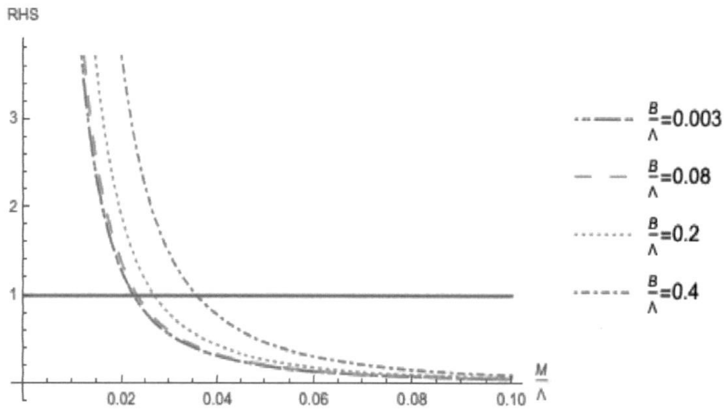


**Figure 3.** *Plot of the right-hand side of* (104) *versus the dynamical axion mass* $M$ *in units of the cut-off* $\Lambda$*, for the fixed value* $\lambda^2 = 0.02$ *for definiteness. All curves intersect the solid line, which represents the left hand side of* (104)*, implying dynamical mass generation for the pseudoscalar field, without the need to introduce a bare mass for it. Figure taken from [42].*

From (104) one obtains a value for $M$, which, upon being inserted in (105), determines the value of the four-fermion coupling $g/f_b$ for which there is a consistent dynamical mass for the fermion, of approximately equal magnitude to the pseudoscalar mass, (82). For concrete examples, we refer the reader to figure 4. We observe from the figure that the dashed line, corresponding to the right hand side of (104), intersects the constant dotted line at 1, representing the left-hand side of the equation. This implies the existence of a non-trivial solution for $M/\Lambda = 0.159$. Using this value in (105), we then obtain a consistent solution for $\bar{g} = 12.58$, in the sense that for this value of $\bar{g}$ the three curves have a common intersection at $M/\Lambda = 0.159$. This demonstrates that there is dynamical mass generation with $m \simeq M < \Lambda$ in this case for strong (dimensionless) four-fermion couplings $\bar{g} > 1$.

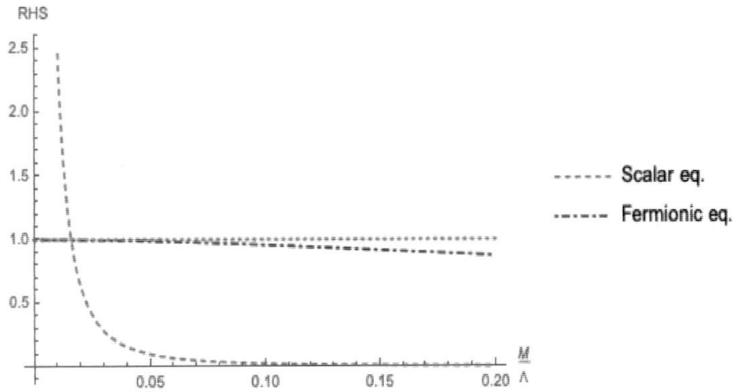


**Figure 4.** *Plot of the right-had-side of* (104) *(dashed line), for* $\lambda^2 = 0.01$ *and* $\bar{\mathcal{B}} = 0.0004$. *The dotted-dashed line represents the right hand side of* (105) *for the above values of* $\lambda^2$ *and* $\bar{\mathcal{B}}$, *and* $\bar{g} = 12.58$. *The constant dotted line at 1 represents the left hand side of the equations* (104) *and* (105). *The existence of a common intersection point for all three curves demonstrates the existence of dynamical mass generation, of approximately equal magnitude, for both fermions and axions. Figure taken from [42].*

Although in our analysis above, and in [40, 42] we restricted ourselves to the partial solution $m \simeq M$, we remark that solutions beyond this restriction also exist. For instance, if we consider the case in which the Yukawa interactions are absent $\lambda^2 \to 0$, and only the (attractive) four-fermion interaction is present, $g \neq 0$, in (92), then we recover the result of [82] for dynamical mass generation only for fermions, for which the fermionic SD equation reads [42]

$$
\begin{aligned}
1 \;=\; & -\frac{g^2}{64\pi^2 f_b^2 \mathcal{B}^2}\Big(\Lambda^4 + \Lambda^2\Big(2(m^2+\mathcal{B}^2) - \sqrt{(\Lambda^2+m^2-\mathcal{B}^2)^2+4m^2\mathcal{B}^2}\Big) \\
& -(m^2+7\mathcal{B}^2)\Big(\sqrt{(\Lambda^2+m^2-\mathcal{B}^2)^2+4m^2\mathcal{B}^2} - (m^2+\mathcal{B}^2)\Big) \\
& +4\mathcal{B}^2(m^2-2\mathcal{B}^2)\log\left[\frac{(m^2-\mathcal{B}^2)+\Lambda^2+\sqrt{(\Lambda^2+m^2-\mathcal{B}^2)^2+4m^2\mathcal{B}^2}}{2m^2}\right]\Big). \qquad (106)
\end{aligned}
$$

This is essentially the same as equation (34) in ref. [82], where the Nambu Jona Lasinio four-fermion model was studied in the presence of a constant non-Hermitian axial background. The study of (106) indicates [82] that the inclusion of the background increases the dynamical fermion mass due to the strong four-fermion interactions alone.

## 5. Conclusions and Outlook

In this mini-review we discussed some aspects of non-Hermitian (specifically, anti-Hermitian) Yukawa models of interacting fermion and axion fields associated with their consistency as quantum field theoretic models, stemming, e.g., from microscopic models embedded in string theory [43]. Apart from the anti-Hermitian Yukawa interactions, the models included anti-Hermitian anomaly terms, as well as interactions of the fermions with constant anti-Hermitian axial backgrounds. The underlying $CPT$ symmetry (56) of these models guarantee [10, 40, 41, 42] the reality of the energy spectra, and thus the dynamical masses.

We have addressed the issue of dynamical mass generation for both fermions and axions, with the conclusion that in the absence of attractive four-fermion (Hermitian) interactions, only an axion mass can be generated dynamically, due to energetics. In contrast, in the presence of sufficiently strong four-fermion attractive interactions, Yukawa-interaction-induced dynamical mass generation for *both* fermions and axions is possible. The role of the (sufficiently strong) four-fermion interactions, therefore, is that of a catalyst for the dynamical mass generation in this case. The anomaly terms, on the other hand, resist dynamical generation of fermion masses. The (constant) axial background assists the Yukawa-coupling-induced mass generation, in the presence of strong four-fermion interactions, in the sense that the larger the value of the background field, the larger the magnitude of the anti-Hermitian-Yukawa-interaction-induced dynamical mass.

In the context of microscopic string theories, the origin of the non-Hermitian background can be traced back to the non-Hermitian anomalous interactions of the KR gravitational axion of the massless bosonic string multiplet in some formulations of the theory ( see "**Scheme II**", (24), in section 2). It would be interesting to investigate further the consequences of such anti-Hermitian axial backgrounds for Leptogenesis, as per the scenario of [60, 56]. In general, non-Hermitian Yukawa interactions, of the type discussed here, may play a rôle in neutrino physics, including neutrino oscillations, as well as non-Hermitian extensions of the standard model.

We conclude, by mentioning the recent work of [83], according to which anti-Hermitian Yukawa interactions, of the type discussed in [40, 41] and reviewed here, may also arise as a result of linearising Hermitian four-fermion interactions, of Nambu-Jona-Lasinio type [79], using appropriate complex auxiliary (pseudo)scalar fields. According to such an analysis, one has *spontaneous* appearance of *non-Hermiticity* in these systems. In addition, four-fermion interactions of Nambu-Jona-Lasinio type with *complex couplings*, whose imaginary part represents dissipation, have been discussed in [84], with the conclusion that the imaginary coupling tends to enhance chiral-symmetry breaking up to a given threshold, above which the symmetry is restored. Such results could be testable in systems of ultracold atomic gases, or other materials that are characterised by Dirac-(Majorana-, or Weyl-)fermion-like excitations [85, 86, 87].

As an outlook, we mention that, from a pure field-theoretic view point, the non-Hermitian models we have discussed here should be studied in a more rigorous way as far as their path-integral quantization is concerned. To this end, we should combine our SD treatment for mass generation with an appropriate extension to fermionic models of the methods developed in [22] for defining properly the path-integral measure of pseudoscalar fields in PT-symmetric pseudoscalar field theories, leading to their renormalization. Such a combined study might lead to a renormalization-group improved SD analysis for our models.

Moreover, an understanding of the concept of anomalies in such non-Hermitian quantum field theories is pressing. When, and if, this is accomplished, one could obtain a geometric understanding of the anti-Hermitian anomalous interactions that we focus our attention on in this review, and in [42]. A fully microscopic understanding of such terms in the context of UV-complete theories of quantum gravity, such as string/brane theory, is also desirable, but this might be a rather long shot.

Another potentially interesting topic is the application of the non-Hermitian anomalous Yukawa models in the presence of (approximately constant) non-Hermitian axial backgrounds to the generation of a Lepton-number asymmetry in the early Universe (Leptogenesis), by extending appropriately the corresponding scenarios of the Hermitian case [60, 56], provided, of course, that such an extension is possible. The dynamical fermion-mass generation in such non-Hermitian models, studied in [40, 42] and reviewed above, when applied to Majorana neutrinos, might lead to Lepton-number asymmetries due to the asymmetric decays of such massive fields to standard-model leptons and antileptons in the presence of the non-Hermitian axial backgrounds. We hope to be able to come back to a study of some of these important issues in the near future.

We close this mini review with an optimistic note. We believe that, although several aspects of non-Hermitian quantum field theories embeddable in a generalised-PT($\mathcal{C}$PT) framework are yet to be understood, nonetheless the rapidly-growing interest in such theories is a strong indicator that soon they might found themselves at comparable levels of rigour and physical applicability with their Hermitian counterparts. In this respect, it would be interesting to see whether there are fundamental aspects of particle physics models, perhaps in the dark sector, that could be explained with such non-Hermitian interactions, which otherwise would remain a mystery. Such an example constituted the topic of this review, namely the novel mechanisms involving anti-Hermitian Yukawa interactions for dynamical generation of masses of right-handed neutrinos and axions, both of which can play the rôle of dark matter candidates in models beyond the standard model of particle physics.

## Acknowledgments

The author wishes to thank C.M. Bender, F. Correa and A. Fring for the invitation to write this mini-review for the Proceedings of the virtual series on Pseudo-Hermitian Hamiltonians in Quantum Physics (XIX<vPHHQP<XX), which the author contributed to as a speaker of a seminar on the topics presented ihere. He also thanks Jean Alexandre, Sarben Sarkar and Alex Soto for discussions and collaboration. This work is supported in part by the UK Science and Technology Facilities research Council (STFC) under the research grant ST/T000759/1. The author also acknowledges participation in the COST Association Action CA18108 *Quantum Gravity Phenomenology in the Multimessenger Approach (QG-MM).* He also thanks the Institute of Particle Physics (IFIC-CSIC) of the University of Valencia (Valencia, Spain ) for a scientific associateship (*Doctor Vinculado*).

## References

[1] Bender C M and Boettcher S 1998 *Phys. Rev. Lett.* **80** 5243–5246 (*Preprint* physics/9712001)
[2] Bender C M, Boettcher S and Meisinger P 1999 *J. Math. Phys.* **40** 2201–2229 (*Preprint* quant-ph/9809072)
[3] Bender C M 2005 *Contemp. Phys.* **46** 277–292 (*Preprint* quant-ph/0501052)
[4] Bender C M 2007 *Rept. Prog. Phys.* **70** 947 (*Preprint* hep-th/0703096)
[5] Bender C M 2015 *J. Phys. Conf. Ser.* **631** 012002
[6] Mostafazadeh A 2002 *J. Math. Phys.* **43** 205–214 (*Preprint* math-ph/0107001)
[7] Mostafazadeh A 2002 *J. Math. Phys.* **43** 2814–2816 (*Preprint* math-ph/0110016)
[8] Mostafazadeh A 2010 *Int. J. Geom. Meth. Mod. Phys.* **7** 1191–1306 (*Preprint* 0810.5643)
[9] Mannheim P D 2018 *Phys. Rev. D* **97** 045001 (*Preprint* 1708.01247)
[10] Mannheim P D 2018 *J. Phys. A* **51** 315302 (*Preprint* 1512.04915)
[11] Chernodub M 2017 *J. Phys. A* **50** 385001 (*Preprint* 1701.07426)
[12] Bender C M, Milton K A and Savage V M 2000 *Phys. Rev. D* **62** 085001 (*Preprint* hep-th/9907045)
[13] Bender C M and Jones H 2000 *Found. Phys.* **30** 393–411
[14] Bender C M, Brody D C and Jones H F 2004 *Phys. Rev. D* **70** 025001 [Erratum: Phys.Rev.D 71, 049901 (2005)] (*Preprint* hep-th/0402183)
[15] Bender C M, Brandt S F, Chen J H and Wang Q h 2005 *Phys. Rev. D* **71** 065010 (*Preprint* hep-th/0412316)
[16] Bender C M and Klevansky S 2010 *Phys. Rev. Lett.* **105** 031601 (*Preprint* 1002.3253)
[17] Bender C M, Jones H and Rivers R 2005 *Phys. Lett. B* **625** 333–340 (*Preprint* hep-th/0508105)
[18] Bender C M, Branchina V and Messina E 2012 *Phys. Rev. D* **85** 085001 (*Preprint* 1201.1244)
[19] Bender C M, Branchina V and Messina E 2013 *Phys. Rev. D* **87** 085029 (*Preprint* 1301.6207)
[20] Alexandre J and Bender C M 2015 *J. Phys. A* **48** 185403 (*Preprint* 1501.01232)
[21] Alexandre J, Millington P and Seynaeve D 2017 *Phys. Rev. D* **96** 065027 (*Preprint* 1707.01057)
[22] Bender C M, Hassanpour N, Klevansky S and Sarkar S 2018 *Phys. Rev. D* **98** 125003 (*Preprint* 1810.12479)
[23] Beygi A, Klevansky S and Bender C M 2019 *Phys. Rev. A* **99** 062117 (*Preprint* 1904.00878)
[24] Alexandre J, Bender C M and Millington P 2015 *JHEP* **11** 111 (*Preprint* 1509.01203)

[25] Alexandre J, Bender C M and Millington P 2017 *J. Phys. Conf. Ser.* **873** 012047 (*Preprint* 1703.05251)
[26] Alexandre J, Ellis J, Millington P and Seynaeve D 2018 *Phys. Rev. D* **98** 045001 (*Preprint* 1805.06380)
[27] Alexandre J, Ellis J, Millington P and Seynaeve D 2019 *Phys. Rev. D* **99** 075024 (*Preprint* 1808.00944)
[28] Alexandre J, Ellis J, Millington P and Seynaeve D 2020 *Phys. Rev. D* **101** 035008 (*Preprint* 1910.03985)
[29] Mannheim P D 2019 *Phys. Rev. D* **99** 045006 (*Preprint* 1808.00437)
[30] Fring A and Taira T 2020 *Nucl. Phys. B* **950** 114834 (*Preprint* 1906.05738)
[31] Fring A and Taira T 2020 *Phys. Rev. D* **101** 045014 (*Preprint* 1911.01405)
[32] Fring A and Taira T 2020 (*Preprint* 2004.00723)
[33] Alexandre J, Ellis J and Millington P 2020 (*Preprint* 2006.06656)
[34] Alexandre J, Ellis J and Millington P 2020 *Phys. Rev. D* **101** 085015 (*Preprint* 2001.11996)
[35] Chernodub M N and Cortijo A 2020 *Symmetry* **12** 761 (*Preprint* 1901.06167)
[36] Bender C M, Hook D W, Mavromatos N E and Sarkar S 2014 *Phys. Rev. Lett.* **113** 231605 (*Preprint* 1408.2432)
[37] Bender C M, Hook D W, Mavromatos N E and Sarkar S 2016 *J. Phys. A* **49** 45LT01 (*Preprint* 1506.01970)
[38] Fring A and Taira T 2020 *Phys. Lett. B* **807** 135583 (*Preprint* 2006.02718)
[39] Fring A and Taira T 2020 *J. Phys. A* **53** 455701 (*Preprint* 2007.15425)
[40] Alexandre J and Mavromatos N E 2020 *Phys. Lett. B* **807** 135562 (*Preprint* 2004.03699)
[41] Alexandre J, Mavromatos N E and Soto A 2020 *Nucl. Phys. B* **961** 115212 (*Preprint* 2004.04611)
[42] Mavromatos N E and Soto A 2020 *Nucl. Phys. B, in press* (*Preprint* 2006.13616)
[43] Mavromatos N E and Pilaftsis A 2012 *Phys. Rev. D* **86** 124038 (*Preprint* 1209.6387)
[44] Green M B, Schwarz J H and Witten E 2012 *Superstring Theory Vol. 1: 25th Anniversary Edition* Cambridge Monographs on Mathematical Physics (Cambridge University Press) ISBN 978-1-139-53477-2, 978-1-107-02911-8
[45] Green M B, Schwarz J H and Witten E 2012 *Superstring Theory Vol. 2: 25th Anniversary Edition* Cambridge Monographs on Mathematical Physics (Cambridge University Press) ISBN 978-1-139-53478-9, 978-1-107-02913-2
[46] Polchinski J 2007 *String theory. Vol. 1: An introduction to the bosonic string* Cambridge Monographs on Mathematical Physics (Cambridge University Press) ISBN 978-0-511-25227-3, 978-0-521-67227-6, 978-0-521-63303-1
[47] Polchinski J 2007 *String theory. Vol. 2: Superstring theory and beyond* Cambridge Monographs on Mathematical Physics (Cambridge University Press) ISBN 978-0-511-25228-0, 978-0-521-63304-8, 978-0-521-67228-3
[48] Eguchi T, Gilkey P B and Hanson A J 1980 *Phys. Rept.* **66** 213
[49] Alvarez-Gaume L and Witten E 1984 *Nucl. Phys. B* **234** 269
[50] Weinberg S 2013 *The quantum theory of fields. Vol. 2: Modern applications* (Cambridge University Press) ISBN 978-1-139-63247-8, 978-0-521-67054-8, 978-0-521-55002-4
[51] Hehl F, Von Der Heyde P, Kerlick G and Nester J 1976 *Rev. Mod. Phys.* **48** 393–416
[52] Hull C 1986 *Phys. Lett. B* **167** 51–55
[53] Mavromatos N 1988 *J. Phys. A* **21** 2279
[54] Duncan M J, Kaloper N and Olive K A 1992 *Nucl. Phys. B* **387** 215–235
[55] Svrcek P and Witten E 2006 *JHEP* **06** 051 (*Preprint* hep-th/0605206)
[56] Bossingham T, Mavromatos N E and Sarkar S 2018 *Eur. Phys. J. C* **78** 113 (*Preprint* 1712.03312)
[57] Basilakos S, Mavromatos N E and Solà Peracaula J 2020 *Phys. Rev. D* **101** 045001 (*Preprint* 1907.04890)
[58] Basilakos S, Mavromatos N E and Solà Peracaula J 2020 *Phys. Lett. B* **803** 135342 (*Preprint* 2001.03465)
[59] Giddings S B and Strominger A 1988 *Nucl. Phys. B* **306** 890–907
[60] de Cesare M, Mavromatos N E and Sarkar S 2015 *Eur. Phys. J. C* **75** 514 (*Preprint* 1412.7077)
[61] de Berredo-Peixoto G, Freidel L, Shapiro I and de Souza C 2012 *JCAP* **06** 017 (*Preprint* 1201.5423)
[62] Kim J E 1987 *Phys. Rept.* **150** 1–177
[63] Kim J E and Carosi G 2010 *Rev. Mod. Phys.* **82** 557–602 [Erratum: Rev.Mod.Phys. 91, 049902 (2019)] (*Preprint* 0807.3125)
[64] Raffelt G G 2008 *Lect. Notes Phys.* **741** 51–71 (*Preprint* hep-ph/0611350)
[65] Arvanitaki A and Dubovsky S 2011 *Phys. Rev. D* **83** 044026 (*Preprint* 1004.3558)
[66] Arvanitaki A, Baryakhtar M and Huang X 2015 *Phys. Rev. D* **91** 084011 (*Preprint* 1411.2263)
[67] Marsh D J E 2016 *Phys. Rept.* **643** 1–79 (*Preprint* 1510.07633)
[68] Arvanitaki A, Baryakhtar M, Dimopoulos S, Dubovsky S and Lasenby R 2017 *Phys. Rev. D* **95** 043001 (*Preprint* 1604.03958)
[69] Grilli di Cortona G, Hardy E, Pardo Vega J and Villadoro G 2016 *JHEP* **01** 034 (*Preprint* 1511.02867)
[70] Arvanitaki A, Dimopoulos S, Dubovsky S, Kaloper N and March-Russell J 2010 *Phys. Rev. D* **81** 123530 (*Preprint* 0905.4720)

[71] Shiu G, Staessens W and Ye F 2015 *Phys. Rev. Lett.* **115** 181601 (*Preprint* 1503.01015)
[72] Shiu G, Staessens W and Ye F 2015 *JHEP* **06** 026 (*Preprint* 1503.02965)
[73] Ibanez L and Uranga A 2007 *JHEP* **03** 052 (*Preprint* hep-th/0609213)
[74] Ibanez L, Schellekens A and Uranga A 2007 *JHEP* **06** 011 (*Preprint* 0704.1079)
[75] Blumenhagen R, Cvetic M and Weigand T 2007 *Nucl. Phys. B* **771** 113–142 (*Preprint* hep-th/0609191)
[76] Cvetic M, Richter R and Weigand T 2007 *Phys. Rev. D* **76** 086002 (*Preprint* hep-th/0703028)
[77] Donoghue J F 1994 *Phys. Rev. D* **50** 3874–3888 (*Preprint* gr-qc/9405057)
[78] Antoniadis I, Benakli K and Laugier A 2001 *JHEP* **05** 044 (*Preprint* hep-th/0011281)
[79] Nambu Y and Jona-Lasinio G 1961 *Phys. Rev.* **122** 345–358
[80] Vafa C and Witten E 1984 *Nucl. Phys. B* **234** 173–188
[81] Bashir A and Diaz-Cruz J 1999 *J. Phys. G* **25** 1797–1805 (*Preprint* hep-ph/9906360)
[82] Felski A, Beygi A and Klevansky S 2020 *Phys. Rev. D* **101** 116001 (*Preprint* 2004.04011)
[83] Chernodub M, Cortijo A and Ruggieri M 2020 (*Preprint* 2008.11629)
[84] Kanazawa T 2020 (*Preprint* 2009.13363)
[85] Vafek O and Vishwanath A 2014 *Ann. Rev. Condensed Matter Phys.* **5** 83–112 (*Preprint* 1306.2272)
[86] Wilczek F 2014 (*Preprint* 1404.0637)
[87] Armitage N, Mele E and Vishwanath A 2018 *Rev. Mod. Phys.* **90** 015001 (*Preprint* 1705.01111)

# The Role of Symmetry in Non-Hermitian Scattering[1]

**Andreas Ruschhaupt[1], Miguel A Simón[2], Anthony Kiely[3] and J Gonzalo Muga[2]**
[1]Department of Physics, University College Cork, Cork T12 YN60, Ireland
[2]Departamento de Química Física, UPV/EHU, Apdo. 644, 48080 Bilbao, Spain
[3]School of Physics, University College Dublin, Dublin 4, Ireland

**Abstract.** We review recent work on asymmetric scattering by Non-Hermitian (NH) Hamiltonians. Quantum devices with an asymmetric scattering response to particles incident from right or left in effective 1D waveguides will be important to develop quantum technologies. They act as microscopic equivalents of familiar macroscopic devices such as diodes, rectifiers, or valves. The symmetry of the underlying NH Hamiltonian leads to selection rules which restrict or allow asymmetric response. NH-symmetry operations may be organized into group structures that determine equivalences among operations once a symmetry is satisfied. The NH Hamiltonian posseses a particular symmetry if it is invariant with respect to the corresponding symmetry operation, which can be conveniently expressed by a unitary or antiunitary superoperator. A simple group is formed by eight symmetry operations, which include the ones for Parity-Time symmetry and Hermiticity as specific cases. The symmetries also determine the structure of poles and zeros of the $S$ matrix. The ground-state potentials for two-level atoms crossing properly designed laser beams realize different NH symmetries to achieve transmission or reflection asymmetries.

## 1. Introduction

It is hardly necessary to emphasize the importance of devices with asymmetric response, such as valves, diodes, or rectifiers, which appear in biology and various technologies to manage and control macroscopic currents and flows. We expect that these asymmetries will similarly play a key role in the microscopic realm to develop quantum technologies.

Among the microscopic asymmetric devices, the Maxwell demon is possibly the best known, [1, 2, 3]. The demon was first discussed theoretically by prominent physicists such as Maxwell, Smoluchowski, Szilard, Feynman, or Landauer [2]. A frequent attitude has been to *exorcise* the (malignant) demon, either to save the second law, if the demon does not work; or to show that, if it works, the 2nd law is not violated. A modern perspective that we embrace here is more pragmatic, and instead aims at *exercising* the demon, namely, building up demonic devices, i.e., one-way barriers, for practical applications [3], such as particle trapping or cooling. Skordos and Zurek [4] made an important observation: Two gas chambers of equal volume were connected via an asymmetric trapdoor that could slide on rails on one of the chambers only. When the door motion was cooled a differential of pressure between the two parts occurred, but not otherwise. In general automated demons like this function thanks to some dissipation. In a 1D barrier, if an incident particle dissipates its energy only on one side, right and left transmission probabilities

[1] This paper is largely based on the talks by J. G. Muga and A. Ruschhaupt within the PHHQP seminar series.

may differ. Otherwise, in a conventional process with a Hermitian potential the probabilities are equal.

One-way barriers ("atom diodes") have been designed using lasers interacting with two- or three-level atoms [5, 6, 7, 3, 8]. These ideas have been used in conjunction with the group of Mark Raizen, which has implemented them experientially [9, 10, 11, 12]. This work evolved into an isotope separation technique for medical use [13].

In an ideal one-way barrier, particles are fully transmitted if they are incident from one side (say the left), and fully reflected from the other side. We could encode this asymmetric response as $\mathcal{T}/\mathcal{R}$, where $\mathcal{T}$ stands for full transmission and $\mathcal{R}$ stands for full reflection. In fact there are several other asymmetric devices. For example, an ideal one-way detector may function as $\mathcal{R}/\mathcal{A}$, absorbing -i.e., detecting- all particles impinging from one side only [14] ($\mathcal{A}$ stands for "absorption"). Taking as a reference maximal all-or-nothing responses (1 or 0 scattering coefficients), six basic asymmetric responses were identified [15], namely $\mathcal{TR}/\mathcal{A}, \mathcal{T}/\mathcal{R}, \mathcal{T}/\mathcal{A}, \mathcal{TR}/\mathcal{R}, \mathcal{R}/\mathcal{A}, \mathcal{TR}/\mathcal{T}$. All these responses would macroscopically correspond to different rectifier types and may have useful applications. Can we construct them? It turns out that the different asymmetric responses may or may not be realized depending on scattering selection rules that follow from Hamiltonian symmetries. In particular, Hermiticity, which may be understood as one of these symmetries, forbids any of the asymmetric responses, whereas PT-symmetry forbids responses that are transmission-asymmetric. However, other Hamiltonian symmetries applicable to Non-Hermitian Hamiltonians are able to produce the different scattering asymmetries. In order to fully understand and exploit the connection between Hamiltonian symmetry and scattering properties, we first review NH scattering in 1D. Furthermore, we also outline NH symmetries, how they are most conveniently formulated, and their consequences.

## 2. A pinch of 1D scattering theory for Non-Hermitian potentials

We consider scattering Hamiltonians that can be written as the sum of a kinetic energy operator $H_0 = p^2/(2m)$ for a particle of mass $m$ and a potential energy operator $V$,

$$H = H_0 + V. \tag{1}$$

The potential function in position representation $V(q, q') = \langle q|V|q'\rangle$ is assumed to decay fast enough to 0 when $q$ or $q'$ go to infinity so that the usual operators of scattering theory are well defined and the Hilbert space is biorthogonally decomposed into a continuum part with real eigenvalues and a discrete part, see [16]. Notably, the potential is generally considered to be non-local, in the sense that elements $\langle q|V|q'\rangle$ may be nonzero for $q \neq q'$. These non-local interactions can arise as effective interactions for some subspace even if underlying interactions for the larger space are local [17, 18]. A simple example that we shall use later on is a two-level atom interacting with a semiclassical laser field [19, 20].

We use for scattering transmission and reflection amplitudes from left and right incidence, the symbols $T^l, R^l$, and $T^r, R^r$.[2] As well, $\widehat{T}^l$, $\widehat{R}^l$, and $\widehat{T}^r$, $\widehat{R}^r$ denote corresponding amplitudes for $H^\dagger$ [16]. The S matrix, related to the scattering operator $S$ that connects asymptotic states,[3] is nothing but the scattering amplitudes for $H$ in matrix form

$$\mathsf{S} = \begin{pmatrix} T^l(p) & R^r(p) \\ R^l(p) & T^r(p) \end{pmatrix}, \tag{2}$$

[2] We adopt the convention of calling the squared moduli of the amplitudes "coefficients". In NH Physics these are not necessarily "probabilities", since they may not add to one.

[3] $\mathsf{S}$ is the on-shell part of the matrix for $S$ in momentum representation factoring out an energy Dirac delta and a $p/m$ factor, see [16] for further details and precise definitions.

which, for NH Hamiltonians, is generically not unitary. However, generalized unitarity relations exist that relate $S$ and the corresponding operator for $H^\dagger$, $\hat{S}$, as $\hat{S}^\dagger S = S\hat{S}^\dagger$ [16]. Generalized unitarity relates hatted and unhatted amplitudes and helps to set the "scattering selection rules" [15]. These are recipes which use the Hamiltonian symmetries to determine whether scattering asymmetries in transmission and reflection are attainable or unattainable. Our next task will be to define and study these symmetries.

## 3. Non-Hermitian symmetries

We adopt here the perspective that defines symmetries by the simple prescription "change without changing" [21], namely, if an operation is performed on an object -in our case the Hamiltonian- and the object looks the same before and after the operation, that object possesses the symmetry associated with that symmetry operation (a technical refinement, following Wigner, will be discussed below). For example, rotating a square by $\pi/2$ about a perpendicular axis across its center gives the same square, so it is symmetrical with respect to $\pi/2$-rotations. Note that the operation (rotating by $\pi/2$) makes sense and can be carried out irrespective of whether or not the object possesses the corresponding symmetry. Some of the results we shall discuss, specifically group structures, refer to the symmetry operations, rather than to the symmetries.

Coming back to Hamiltonians, consider the following operations on $H$: $A^{-1}HA$, $A^{-1}H^\dagger A$, where $A$ is unitary or antiunitary. The first operation could be phrased as: "multiply $H$ by $A^{-1}$ from the left and by $A$ from the right". The second operation would start with the instruction "take the adjoint of $H$", and then perform the multiplications. If one of these operations leaves the Hamiltonian invariant, $H$ possesses that particular symmetry, i.e.,

$$\text{Operation on } H: \; A^{-1}HA. \;\; \text{Symmetry}: \; A^{-1}HA = H \text{ or } AH = HA, \tag{3}$$

$$\text{Operation on } H: \; A^{-1}H^\dagger A. \;\; \text{Symmetry}: \; A^{-1}H^\dagger A = H \text{ or } AH = H^\dagger A. \tag{4}$$

The first type of symmetry is the "standard" one. The second one is termed $A$-pseudohermiticity [22, 15]. For a physical discussion of the two possibilities and their relation to dynamical invariants see [23]. In practice we shall limit $A$ to the following set of operators that form Klein's four-group

$$A \in K_4 = \{1, \Pi, \Theta, \Pi\Theta\}, \tag{5}$$

where $\Pi$ is parity, and $\Theta$ time reversal. The action on a position basis state multiplied by a complex constant $c$ is $\Pi c|q\rangle = c|-q\rangle$, $\Theta c|q\rangle = c^*|q\rangle$. Klein's group is an Abelian group with two generators. It has many applications, among others it is the basis for dodecaphonic music [24]. Fascinating as this is, the reason for us to restrict the possible $A$ to this group is that, when combined with the two operations in (3) and (4), an Abelian group[4] $\mathbb{Z}_2 \times \mathbb{Z}_2 \times \mathbb{Z}_2$ of eight symmetry operations is formed that may be considered a complete set of operations combining transposition, inversion, and complex conjugation. Table 1 gives the corresponding eight symmetry operations, which are assigned a roman numeral in column 1,[5] and a group-theory symbol in terms of three generators $(x, y, z)$ in column 2. Column 4 gives the explicit transformation on $H$ which may be expressed as a superoperator (column 3) acting on $H$, generically $\mathcal{S}$, see the next section, which also discusses the relation to Wigner's characterization of quantum symmetries. Column 5 gives $\langle q|(\mathcal{S}H)|q'\rangle$. $H$ will posses the symmetry associated with a particular $\mathcal{S}$ if $\langle q|H|q'\rangle = \langle q|(\mathcal{S}H)|q'\rangle$ for all $q$, $q'$. In practice this condition must be

[4] This "elementary Abelian group" is sometimes denoted as $E8$, not to be confused with the exceptional Lie group.

[5] The roman numeral is used to design both the symmetry operation or the corresponding symmetry, the context should clarify its exact meaning.

**Table 1.** Group elements (transformations) in different notations. 1: Roman number code; 2: Group theory notation of group elements in terms of generators $x$, $y$, $z$; 3: Superoperators $\mathcal{S}$; 4: Explicit action of the superoperators on $H$; 5: Matrix element $\langle q|(\mathcal{S}H)|q'\rangle$.

| 1 | 2 | 3 | 4 | 5 |
|---|---|---|---|---|
| I | $e$ | $\mathcal{L}_1$ | $H$ | $\langle q\|H\|q'\rangle$ |
| II | $x$ | $\mathcal{D}$ | $H^\dagger$ | $\langle q'\|H\|q\rangle^*$ |
| III | $z$ | $\mathcal{L}_\Pi$ | $\Pi H\Pi$ | $\langle -q\|H\|-q'\rangle$ |
| IV | $xz$ | $\mathcal{L}_\Pi\mathcal{D}$ | $\Pi H^\dagger\Pi$ | $\langle -q'\|H\|-q\rangle^*$ |
| V | $y$ | $\mathcal{L}_\theta$ | $\theta H\theta$ | $\langle q\|H\|q'\rangle^*$ |
| VI | $xy$ | $\mathcal{L}_\theta\mathcal{D}$ | $\theta H^\dagger\theta$ | $\langle q'\|H\|q\rangle$ |
| VII | $zy$ | $\mathcal{L}_{\Pi\theta}$ | $\Pi\theta H\Pi\theta$ | $\langle -q\|H\|-q'\rangle^*$ |
| VIII | $zxy$ | $\mathcal{L}_{\Pi\theta}\mathcal{D}$ | $\Pi\theta H^\dagger\Pi\theta$ | $\langle -q'\|H\|-q\rangle$ |

satisfied by the potential since the kinetic energy is symmetrical with respect to all operations. The multiplication table of the group of operations and its "presentation" (set of relations among the generators) is given in Table 2.

## 4. Symmetry transformations as superoperators

Let us define first superoperators $\mathcal{L}_A$ acting on $H$ as

$$\mathcal{L}_A H = A^{-1}HA, \tag{6}$$

and a "dagger" superoperator $\mathcal{D}$ as [23], $\mathcal{D}H = H^\dagger$. Hermiticity is the invariance upon this superoperator. $\mathcal{L}_A$ and $\mathcal{D}$ commute when $A^{-1} = A^\dagger$, as it happens for our basic operators 1, $\Pi$, $\theta$ and $\theta\Pi$ in Klein's group. Note that $\mathcal{L}_A$ alone, with $A \in K_4 = \{1, \Pi, \Theta, \Pi\Theta\}$, constitute a Klein group with two generators, whereas including $\mathcal{D}$ provides the third generator needed for $\mathbb{Z}_2 \times \mathbb{Z}_2 \times \mathbb{Z}_2$.

Superoperators, just like ordinary operators, are linear if they leave complex constants invariant and antilinear if they transform them to their complex conjugates. In particular [23],

$$\begin{aligned}
\mathcal{L}_A(cH) &= cA^\dagger HA, \quad A\,\text{unitary},\\
\mathcal{L}_A(cH) &= c^*A^\dagger HA, \quad A\,\text{antiunitary},\\
\mathcal{D}(cH) &= c^*H^\dagger,\\
\mathcal{L}_A\mathcal{D}(cH) &= \mathcal{D}\mathcal{L}_A(cH) = c^*A^\dagger H^\dagger A, \quad A\,\text{unitary},\\
\mathcal{L}_A\mathcal{D}(cH) &= \mathcal{D}\mathcal{L}_A(cH) = cA^\dagger H^\dagger A, A\,\text{antiunitary}.
\end{aligned} \tag{7}$$

The superoperators in Table 1 form a group $\mathbb{Z}_2 \times \mathbb{Z}_2 \times \mathbb{Z}_2$. As noted before, we denote a superoperator in this group by a generic notation $\mathcal{S}$. In formal manipulations we shall later on use distinguishing subscripts, e.g. $\mathcal{S}_j$, where $j = 1, 2, ..., 8$ mapping $I \to 1$, $II \to 2$, etc...

### *4.1. Valid symmetry operations.*

The transformations that in quantum physics lead to possible symmetries, i.e. symmetry transformations (operations), are not arbitrary. Wigner stated that the transformations that act on wave functions should leave the modulus of the scalar product of two states, their "transition probability", invariant, which restricts the corresponding operators to be unitary or antiunitary

[25]. Our superoperators imply a mild generalization of Wigner's definition, as they leave the scalar product of two density operators, the most general way of expressing a state, invariant [23].

We will denote a scalar product of two given (linear) operators $F$ and $G$ as the standard Hilbert-space inner product $\langle\langle F, G\rangle\rangle = Tr[F^\dagger G]$ [23]. Expectation values for an observable $F$ and a density operator $\rho$, both Hermitian, take the form $\langle F\rangle = Tr[F\rho] = Tr[F^\dagger\rho] = \langle\langle F, \rho\rangle\rangle$.

We define the adjoint of a given superoperator $\mathcal{S}$ as the superoperator $\mathcal{S}^\dagger$ which fulfills [23]

$$\langle\langle G, \mathcal{S}F\rangle\rangle = \langle\langle F, \mathcal{S}^\dagger G\rangle\rangle^* \text{ for } \mathcal{S} \text{ linear}, \tag{8}$$

$$\langle\langle G, \mathcal{S}F\rangle\rangle = \langle\langle F, \mathcal{S}^\dagger G\rangle\rangle \text{ for } \mathcal{S} \text{ antilinear}. \tag{9}$$

For unitary or antiunitary operators $A$, i.e., $A^{-1} = A^\dagger$, we have [23]

$$\begin{aligned} \mathcal{L}_A^\dagger(\cdot) &= \mathcal{L}_{A^\dagger}(\cdot) \equiv A(\cdot)A^\dagger, \\ \mathcal{D}^\dagger(\cdot) &= \mathcal{D}(\cdot), \\ (\mathcal{L}_A\mathcal{D})^\dagger(\cdot) &= \mathcal{L}_{A^\dagger}\mathcal{D}(\cdot). \end{aligned} \tag{10}$$

All eight superoperators satisfy $\mathcal{S}^\dagger = \mathcal{S}^{-1}$, so these superoperators are unitary or antiunitary. Thus they keep invariant the scalar product of two density operators,

$$\langle\langle \rho_1, \rho_2\rangle\rangle = \langle\langle \mathcal{S}\rho_1, \mathcal{S}\rho_2\rangle\rangle. \tag{11}$$

Our definition of symmetry operation is indeed broader than Wigner's, which, for example, does not include relations of the form (4).

**Table 2.** Multiplication table of $\mathbb{Z}_2 \times \mathbb{Z}_2 \times \mathbb{Z}_2$ which may be summarized in the "presentation" $\langle x, y, z \mid x^2 = y^2 = z^2 = e, xy = yx, zx = xz, zy = yz\rangle$.

| | $e$ | $x$ | $z$ | $zx$ | $y$ | $xy$ | $zy$ | $zxy$ |
|---|---|---|---|---|---|---|---|---|
| $e$ | $e$ | $x$ | $z$ | $zx$ | $y$ | $xy$ | $zy$ | $zxy$ |
| $x$ | $x$ | $e$ | $zx$ | $z$ | $xy$ | $y$ | $zxy$ | $zy$ |
| $z$ | $z$ | $zx$ | $e$ | $x$ | $zy$ | $zxy$ | $y$ | $xy$ |
| $zx$ | $zx$ | $z$ | $x$ | $e$ | $zxy$ | $zy$ | $xy$ | $y$ |
| $y$ | $y$ | $xy$ | $zy$ | $zxy$ | $e$ | $x$ | $z$ | $zx$ |
| $xy$ | $xy$ | $y$ | $zxy$ | $zy$ | $x$ | $e$ | $zx$ | $z$ |
| $zy$ | $zy$ | $zxy$ | $y$ | $xy$ | $z$ | $zx$ | $e$ | $x$ |
| $zxy$ | $zxy$ | $zy$ | $xy$ | $y$ | $zx$ | $z$ | $x$ | $e$ |

## 5. Equivalences among symmetry operations

From the group multiplication table, it follows that once one symmetry is satisfied, automatically the symmetry operations are equivalent by pairs. More precisely, we will say that, conditioned to an existing symmetry, $\mathcal{S}_i H = H$, two symmetry operations represented by $\mathcal{S}_k$ and $\mathcal{S}_l$ are equivalent, $\mathcal{S}_k \sim \mathcal{S}_l$, if $\mathcal{S}_k H = \mathcal{S}_l H$.[6] As $\mathcal{S}_i$ represents a symmetry by hypothesis, the operations represented by the pair $\mathcal{S}_l$ and $\mathcal{S}_k = \mathcal{S}_l\mathcal{S}_i$ (found from the multiplication table) are equivalent. An example of equivalence that may be familiar to some is that, conditioned on $xyH = H$ (which

[6] In [15] the stronger assumption $\mathcal{S}_k H = \mathcal{S}_l H = H$ was made, namely, that the equivalent pairs were also symmetries. Note that equivalence can be formulated more generally as done here or in [26] without that assumption.

**Table 3.** Equivalences among symmetry operations for the potential elements. Given the symmetry of the upper row, the table provides the equivalent symmetry operations. The trivial identity (I) is excluded as this symmetry is satisfied by all potentials.

| II | III | IV | V | VI | VII | VIII |
|---|---|---|---|---|---|---|
| III∼IV | II∼IV | II∼III | II∼VI | II∼V | II∼VIII | II∼VII |
| V∼VI | V∼VII | V∼VIII | III∼VII | III∼VIII | III∼V | III∼VI |
| VII∼VIII | VI∼VIII | VI∼VII | IV∼VIII | IV∼VII | IV∼VI | IV∼V |

is satisfied in particular by all local potentials in coordinate representation; more generally this symmetry implies that the matrix is complex-symmetric or, equivalently, self-transpose), then $zyH = zxH$. In the alternative operator language this means that, conditioned on $\theta H^\dagger \theta = H$, we have that $\Pi\theta H \Pi\theta = \Pi H^\dagger \Pi$. In words, with the proper conditioning ($\theta$-pseudohermiticity, i.e., the symmetry of complex-symmetric matrices), the symmetry transformations for PT-symmetry and for parity-pseudohermiticity give the same result when acting on the Hamiltonian. If it happens that $H$ is indeed PT-symmetrical (i.e., $\Pi\theta H \Pi\theta = H$), then it will also be parity-pseudo-Hermitian ($\Pi H^\dagger \Pi = H$) and vice versa [22, 27]. These pairs were worked out systematically in [15] conditioned on a given (primary) symmetry, see Table 3, and follow directly from the group multiplication table.

## 6. Selection rules

Putting together the effect of the symmetries on $S$-matrix elements and generalized unitarity relations, selection rules may be found, see Table 4. It turns out that symmetries II (Hermiticity) and III (parity) are not capable of any, reflection or transmission, scattering asymmetry; symmetries VI (time reversal pseudohermiticity) and VII (PT symmetry) allow for reflection asymmetry but not for transmission asymmetry; symmetries V (time-reversal symmetry) and VIII (PT pseudohermiticity) allow for transmission asymmetry but not for reflection asymmetry, whereas I (trivial symmetry) and IV (parity pseudohermiticity) allow for both scattering asymmetries. Note also the importance on non-locality for asymmetric transmission: All local potentials do satisfy automatically symmetry VI, and are therefore necessarily transmission reciprocal. All these results are for linear (Schrödinger) dynamics. Nonlinearity could break down these selection rules [28, 29, 30].

Table 5 connects these results with the six asymmetry types mentioned in the Introduction.

## 7. Pole symmetries

The symmetries which imply real or complex-conjugate pairs of energy eigenvalues for bound eigenstates are II, IV,V and VII. These complex-conjugate pairs have been previously discussed in [31, 32, 33, 34] for a general class of diagonalizable Hamiltonians that posses discrete spectrum. A novel aspect uncovered in [35] is that whenever one of the above-mentioned four symmetries are satisfied, not only the complex eigenvalues representing the bound states come in conjugate-complex pairs, but all the complex poles of the $S$-matrix eigenvalues have this property. This applies in particular to antibound or virtual states, resonances, and antiresonances. The four Hamiltonian symmetries could also be formulated as the commutation of the Hamiltonian with specific antilinear operators, which are evident for V (where the antilinear operator is $\Theta$) and VII ($\Theta\Pi$), and can be constructed for II and IV [35]. In fact, the results for discrete Hamiltonians in [31, 32, 33] generalize to scattering Hamiltonians also in the sense that the pseudohermitian symmetries in $\mathbb{Z}_2 \times \mathbb{Z}_2 \times \mathbb{Z}_2$ can be expressed as ordinary commutations with operators $B$ (linear for $A$ antilinear and antilinear for $A$ linear) that may be explicitly constructed [35].

**Table 4.** Hamiltonian symmetries, relation between matrix elements of $S$ and/or $\widehat{S}$ matrices ($S$ is for $H$ and $\widehat{S}$ for $H^\dagger$). From them the next four columns relate the scattering amplitudes. Together with generalized unitarity relations they also imply relations for the moduli and phases (not shown). The last two columns summarize if perfect asymmetric transmission or reflection are possible: "✓" means possible (but not necessary), "✗" means impossible. In some cases "✓" is accompanied by a condition that must be satisfied.

| Code | Symmetry | $\langle p\|S\|p'\rangle =$ | $T^l =$ | $T^r =$ | $R^l =$ | $R^r =$ | Moduli relations | $\|T^l\|=1$ $\|T^r\|=0$ | $\|R^l\|=1$ $\|R^r\|=0$ |
|---|---|---|---|---|---|---|---|---|---|
| I | $1H = H1$ | $\langle p\|S\|p'\rangle$ | $T^l$ | $T^r$ | $R^l$ | $R^r$ | | ✓ | ✓ |
| II | $1H = H^\dagger 1$ | $\langle p\|\widehat{S}\|p'\rangle$ | $\widehat{T}^l$ | $\widehat{T}^r$ | $\widehat{R}^l$ | $\widehat{R}^r$ | $\|T^l\|=\|T^r\|,\ \|R^l\|=\|R^r\|$ | ✗ | ✗ |
| III | $\Pi H = H\Pi$ | $\langle -p\|S\|-p'\rangle$ | $T^r$ | $T^l$ | $R^r$ | $R^l$ | $\|T^l\|=\|T^r\|, \|R^l\|=\|R^r\|$ | ✗ | ✗ |
| IV | $\Pi H = H^\dagger\Pi$ | $\langle -p\|\widehat{S}\|-p'\rangle$ | $\widehat{T}^r$ | $\widehat{T}^l$ | $\widehat{R}^r$ | $\widehat{R}^l$ | | ✓ $R^r R^{l*} = 1$ | ✓ $T^r T^{l*} = 1$ |
| V | $\Theta H = H\Theta$ | $\langle -p'\|\widehat{S}\|-p\rangle$ | $\widehat{T}^r$ | $\widehat{T}^l$ | $\widehat{R}^l$ | $\widehat{R}^r$ | $\|R^l\| = \|R^r\|$ | ✓ $\|R^{r,l}\| = 1$ | ✗ |
| VI | $\Theta H = H^\dagger\Theta$ | $\langle -p'\|S\|-p\rangle$ | $T^r$ | $T^l$ | $R^l$ | $R^r$ | $\|T^l\| = \|T^r\|$ | ✗ | ✓ |
| VII | $\Theta\Pi H = H\Theta\Pi$ | $\langle p'\|\widehat{S}\|p\rangle$ | $\widehat{T}^l$ | $\widehat{T}^r$ | $\widehat{R}^r$ | $\widehat{R}^l$ | $\|T^l\| = \|T^r\|$ | ✗ | ✓ $\|T^{r,l}\| = 1$ |
| VIII | $\Theta\Pi H = H^\dagger\Theta\Pi$ | $\langle p'\|S\|p\rangle$ | $T^l$ | $T^r$ | $R^r$ | $R^l$ | $\|R^l\| = \|R^r\|$ | ✓ | ✗ |

**Table 5.** Device types for transmission and/or reflection asymmetry in the first row (see main text for nomenclature, binary values (0 or 1) for the transmission and reflection coefficients are considered here as an ideal case). The second row gives the corresponding symmetries that allow each device.

| $\mathcal{TR}/\mathcal{A}$ | $\mathcal{T}/\mathcal{R}$ | $\mathcal{T}/\mathcal{A}$ | $\mathcal{TR}/\mathcal{R}$ | $\mathcal{R}/\mathcal{A}$ | $\mathcal{TR}/\mathcal{T}$ |
|---|---|---|---|---|---|
| I | I | I,VIII | I,VIII | I,VI | I,IV,VI,VII |

For Hermitian Hamiltonians with scattering (and possibly bound) eigenstates, their $S$-matrix poles are symmetric in the complex momentum plane with respect to the imaginary axis. In the upper half-plane, which corresponds to the first energy Riemann sheet, the poles are on the imaginary axis and represent bound states in the point spectrum of the Hamiltonian. In the lower half-plane they come in symmetrical resonance and antiresonance pairs, and may also lie on the imaginary axis as "virtual states". A further symmetry is the occurrence of a zero at the complex-conjugate momentum of a given pole. These properties are well known for partial wave scattering by spherical potentials but also hold for the $S$-matrix eigenvalues in one dimensional scattering [16]. For NH-Hamiltonians the above pole- and pole/zero-symmetries do not hold in general. The point spectrum of $H$ may include, apart from "ordinary bound states" on the imaginary axis, also eigenvalues ($S$-matrix poles) in the first and second quadrant. However the symmetries of zeros and poles, which are characteristic of Hermitian potentials, also hold for the symmetries IV, V, and VII. These relations had only been shown for PT-symmetry (symmetry VII) in [16].

The configuration with poles located on the imaginary axis or as symmetrical pairs implies important consequences. In particular, it provides stability of the real energy eigenvalues with respect to parameter variations of the potential. While a simple pole on the imaginary axis can move along that axis when a parameter is changed, it cannot move off this axis (since this would violate the pole-pair symmetry) or bifurcate. The formation of pole pairs occurs near special parameter values for which two poles on the imaginary axis collide.

## 8. Physical realization

Whereas Hermiticity does not allow for any asymmetry in transmission and reflection probabilities, PT symmetry or the symmetry of local potentials, technically "pseudohermiticy with respect to time reversal" [15], do not allow for asymmetric transmission [16, 36], see symmetries II (Hermiticity), VII (PT symmetry), and VI (time-reversal pseudohermiticity) in Tables 4 and 5. Thus non-local, non-PT, and non-Hermitian potentials are needed to implement a rich set of scattering asymmetries, and in particular asymmetric transmission.

In [15] potentials with proper symmetries were inverse engineered formally from the wanted scattering responses using polynomial ansatzes. Similarly in [35] we designed separable potentials with different symmetries to study and illustrate pole motions and structure. While the mathematical modelling was clearly possible, the question of a possible physical realization was open. Finally [20] put forward a physical realization of effective NH, non-local Hamiltonians which do not posses PT symmetry.

The system considered in [20] is a two-level atom with ground level $|1\rangle$ and excited state $|2\rangle$ impinging onto a laser illuminated region [19, 14]. The motion is assumed to be one dimensional, and the wavefunction accounts for atoms before the first spontaneous emission, i.e. it is assumed that no resetting to the ground state occurs. The (two-component) state $\mathbf{\Phi}_k = \left(\phi_k^{(1)}\ \phi_k^{(2)}\right)^T$ for the atom impinging with wavenumber $k$ obeys, in an interaction picture, an effective stationary Schrödinger equation with a time-independent Hamiltonian [19, 14] $\mathcal{H}\mathbf{\Phi}_k(x) = E\mathbf{\Phi}_k(x)$, where

$$\mathcal{H} = H_0\mathbf{1} + \mathcal{V} = \frac{1}{2m}\begin{pmatrix} p^2 & 0 \\ 0 & p^2 \end{pmatrix} + \mathcal{V}(x), \tag{12}$$

$$\mathcal{V}(x) = \frac{\hbar}{2}\begin{pmatrix} 0 & \Omega(x) \\ \Omega(x)^* & -(2\Delta + i\gamma) \end{pmatrix}. \tag{13}$$

$E = \hbar^2k^2/2m$ is the energy, and $\Omega(x)$ is the position-dependent, on-resonance Rabi frequency, where real and imaginary parts may be controlled independently using two laser field quadratures; $\gamma$ is the inverse of the life time of the excited state; $\Delta = \omega_L - \omega_{12}$ is the detuning (laser angular frequency minus the atomic transition angular frequency $\omega_{12}$); $p = -i\hbar\partial/\partial x$ is the momentum operator; and $\mathbf{1} = |1\rangle\langle 1| + |2\rangle\langle 2|$ is the unit operator for the internal-state space. Complementary projectors $P = |1\rangle\langle 1|$ and $Q = |2\rangle\langle 2|$ are defined to select ground and excited state components. Using the partitioning technique [17, 18], it is found that the ground state amplitude $\phi_k^{(1)}$ satisfies the equation

$$E\phi_k^{(1)}(x) = H_0\phi_k^{(1)}(x) + \int dy\, \langle x, 1|\mathcal{W}(E)|y, 1\rangle \phi_k^{(1)}(y), \tag{14}$$

where $\mathcal{W}(E) = P\mathcal{V}P + P\mathcal{V}Q(E + i0 - Q\mathcal{H}Q)^{-1}Q\mathcal{V}P$, is non local and energy dependent. This provides a a physical realization of an effective (in general) non-local, non-Hermitian potentials

$$V(x, y) = \langle x, 1|\mathcal{W}(E)|y, 1\rangle = \frac{m}{4}\frac{e^{i|x-y|q}}{iq}\Omega(x)\Omega(y)^*, \tag{15}$$

where $q = \sqrt{2mE}(1 + \mu)^{1/2}/\hbar$, $\operatorname{Im} q \geq 0$, and $\mu = \hbar(2\Delta + i\gamma)/(2E)$. The reflection and transmission amplitudes $R^{r,l}$ and $T^{r,l}$ may be calculated directly using the potential (15) or as corresponding amplitudes for transitions from ground state to ground state in the full two-level theory.

Imposing symmetries on $\Omega(x)$ leads to different Hamiltonian symmetries. We are mostly interested in the ones leading to asymmetric scattering which excludes II and III. The structure in (15) implies that symmetries IV, V and VII also imply Hermiticity. We are thus left with symmetries I, VI and VIII as the interesting ones in this physical setting. Rabi frequencies for transmission asymmetric devices of one-way filter of types $\mathcal{T}/\mathcal{A}$ and $\mathcal{R}/\mathcal{A}$ are worked out in [20] making use of simply and realistic Gaussian shapes. A $\mathcal{TR}/\mathcal{A}$ device with equal 1/2 transmission and reflection coefficients is also designed. An optical analog would be a darkish one-way mirror, where an observer on the left $O_L$ sees himself/herself reflected, and an observer on the right $O_R$ would see the person standing on the left $O_L$ through the mirror but not himself/herself reflected in the mirror, and $O_R$ cannot be seen by the observer on the left $O_L$.

## 9. Discussion

This review focuses on our recent results on how Non-Hermitian symmetries determine the possible asymmetric scattering of single particles by 1D potentials. While PT-symmetry [37] has got most of the attention in NH Physics, in particular because its applications in optics [38, 39, 40], Non-Hermitian symmetries beyond PT-symmetry have been considered recently in different contexts [41, 42, 43, 15, 23, 35, 26, 44, 45], offering a richer set of properties and applications. In particular the group structure described here for scattering systems can be generalized for finite matrices, where a larger non-Abelian group of symmetry operations naturally emerges [26]. Symmetry has always played a prominent role in physics and we expect it to be very relevant in Non-Hermitian Physics as well, for which some interesting symmetry-operation groups, relations and consequences are being found.

## Acknowledgments

We are grateful to A. Alaña, S. Martínez Garaot, A. Buendía, S. Longhi, A. Mostafazadeh, M. Znojil, and V. V. Konotop for useful comments. This work was supported by the Basque Country Government (Grant No. IT986-16) and PGC2018-101355-B-I00 (MCIU/AEI/FEDER,UE). M.A.S. acknowledges financial support by the Basque Government predoctoral program (Grant No. PRE-2018-2-0177).

## References

[1] Maxwell J C, Theory of Heat, 4th ed. Longmans, Green and Co., London, 1875 , pp. 328-329.
[2] Maxwell's Demon: Entropy, Information, Computing, edited by H. S. Leff and A. Rex Princeton University Press, Princeton, 1990.
[3] Ruschhaupt A, Muga J G and Raizen M G 2006 *J. Phys. B* **39** 3833
[4] Skordos P and Zurek W 1992 *Am. J. Phys.* **60** 876
[5] Ruschhaupt A and Muga J G 2004 *Phys. Rev. A* **70** 061604(R)
[6] Ruschhaupt A and Muga J G 2006 *Phys. Rev. A* **73** 013608
[7] Ruschhaupt A, Muga J G and M. G. Raizen 2006 *J. Phys. B* **39** L133
[8] Ruschhaupt A and Muga J G 2007 Phys. Rev. A **76** 013619
[9] Raizen M G, Dudarev A M, Qian Niu and Fisch N J 2005 *Phys. Rev. Lett.* **94** 053003
[10] Dudarev A M, Marder M, Qian Niu, Fisch N J and Raizen M G 2005 *EPL* **70** 761
[11] Raizen M G 2009 Science **324** 1403
[12] Jerkins M, Chavez I, Even U and Raizen M G 2010 *Phys. Rev. A* **82** 033414
[13] Raizen M G and Klappauf B 2012 *New J. Phys.* **14** 023059
[14] Ruschhaupt A, Muga J G and Hegerfeldt G C 2009 *Lect. Notes Phys.* **789** 65
[15] Ruschhaupt A, Dowdall T, Simón M A and Muga J G 2017 EPL **120** 20001
[16] Muga J G, Palao J P, Navarro B and Egusquiza I L 2004 *Phys. Rep.* **395** 357. Eq. (113) should read $\langle x|V|x'\rangle = \langle -x|V|-x'\rangle^*$.
[17] Feshbach H 1958 *Ann. Phys. (N.Y.)* **5** 357
[18] Feshbach H 1962 *Ann. Phys. (N.Y.)* **19** 287
[19] Ruschhaupt A, Damborenea J A, Navarro B, Muga J G and Hegerfeldt G C 2004 *EPL* **67** 1
[20] Ruschhaupt A, Kiely A, Simón M A and Muga J G 2020 *Phys. Rev. A* **102** 053705 (2020)
[21] Wilczek, F 2015 *A beautiful question* (New York: Penguin Press)
[22] Mostafazadeh A 2010 *Int. J. Geom. Methods Mod. Phys.*, **07** 1191
[23] Simón M A, Buendía A and Muga J G 2018 *Mathematics* **6** 111

[24] Andreatta M 2004 in *Perspectives in mathematical and computational music theory*, edited by G Mazzola, Thomas Noll, Emilio Lluis-Puebla (Electronic Publishing Osnabrück)
[25] Wigner E. P. 1959, *Group Theory and Its Application to the Quantum Mechanics of Atomic Spectra* (New York, NY, USA: Academic Press) pp 233-236
[26] Alaña A, Martínez-Garaot S, Simón M A and Muga J G 2020 J. Phys. A **53** 135304
[27] Znojil M 2015, in *Non-Selfadjoint Operators in Quantum Physics*, ed F Bagarello, J P Gazeau, F H Szafraniec and M Znojil (Hoboken, NJ: Wiley, ) Chapt. 1, p. 7
[28] Lin Z, Ramezani H, Eichelkraut T, Kottos T, Cao H and Christodoulides D N 2011 *Phys. Rev. Lett.* **106** 213901
[29] Peng B et al. 2014 *Nat. Phys.* **10** 394
[30] Xu Y L, Feng L, Lu M H, Chen Y F 2014 *IEEE Photonics J.* **6**, 0600507
[31] Mostafazadeh A 2002 *J. Math. Phys.* **43** 205
[32] Mostafazadeh A 2002 *J. Math. Phys.* **43** 2814
[33] Mostafazadeh A 2002 *J. Math. Phys.* **43** 3944
[34] Bender C M and Mannheim P D 2010 *Phys. Lett. A* **374**, 1616
[35] Simón M A, Buendía A, Kiely A, Mostafazadeh A and Muga J G 2019 *Phys. Rev. A* **99** 052110
[36] Mostafazadeh A, Scattering theory and PT-symmetry, in *Parity-time Symmetry and Its Applications*, Christodoulides D., Yang J. (eds). Springer Tracts in Modern Physics, vol 280. (Springer, Singapore, 2018).
[37] Bender C M and Boettcher S 1998 *Phys. Rev. Lett.* **80** 5243
[38] Ruschhaupt A, Delgado F and Muga J G 2005 *J. Phys. A* **38** L171
[39] Longhi S 2017 *EPL* **120** 64001
[40] Konotop V V, Yang J and Zezyulin D A 2016 *Rev. Mod. Phys.* **88** 035002
[41] Nixon S and Yang J 2016 *Phys. Rev. A* **93** 031802(R)
[42] Nixon S and Yang J 2016 *Opt. Lett.* **41** 2747
[43] Chen P and Chong Y D 2017 *Phys. Rev. A* **95** 062113
[44] Bernard D and LeClair A 2002, A Classification of Non- Hermitian Random Matrices, in *Statistical Field Theories*, NATO Science Series vol 73, ed A. Cappelli and G. Mussardo (Dordrecht: Springer) pp 207-214
[45] Kawabata K, Shiozaki K, Ueda M and Sato M 2019 *Phys. Rev. X* **9** 041015

Journal of Physics: Conference Series **2038** (2021) 012021 doi:10.1088/1742-6596/2038/1/012021

# Conserved quantities, exceptional points, and antilinear symmetries in non-Hermitian systems

**Frantisek Ruzicka[1†], Kaustubh S. Agarwal[2] and Yogesh N. Joglekar[2‡]**

[1]Institute of Nuclear Physics, Czech Academy of Sciences, Rez 250 68, Czech Republic

[2]Department of Physics, Indiana University Purdue University Indianapolis (IUPUI), Indianapolis, Indiana 46202, United States

E-mail: [†]`fruzicka@gmail.com`, [‡]`yojoglek@iupui.edu`

**Abstract.** Over the past two decades, open systems that are described by a non-Hermitian Hamiltonian have become a subject of intense research. These systems encompass classical wave systems with balanced gain and loss, semiclassical models with mode selective losses, and minimal quantum systems, and the meteoric research on them has mainly focused on the wide range of novel functionalities they demonstrate. Here, we address the following questions: Does anything remain constant in the dynamics of such open systems? What are the consequences of such conserved quantities? Through spectral-decomposition method and explicit, recursive procedure, we obtain all conserved observables for general $\mathcal{PT}$-symmetric systems. We then generalize the analysis to Hamiltonians with other antilinear symmetries, and discuss the consequences of conservation laws for open systems. We illustrate our findings with several physically motivated examples.

## 1. Introduction

Discrete or continuous symmetries and conservation laws that result from them have been instrumental to the theoretical and experimental developments in physics over the past few centuries [1, 2]. Identifying the quantities that remain constant during temporal evolution of a system allows us to severely constraint their global dynamics. Starting from Kepler's law (equal-area-swept in equal-times for planar orbits) [3], these time-invariants have underpinned numerous scientific advances; they include the prediction of (electron) neutrinos based on the requirement of energy, momentum, and angular momentum conservation in beta-decay processes [4]. On the theoretical front, conservation laws have played critical role in the development of approximate methods that are indispensable for studying interacting, many-body systems. These self-consistent or conserving approximations characterize a special set of Feynman diagrams, used to calculate response functions or susceptibilities, such that the approximate answers still obey relevant conservation laws [5, 6].

The dynamics of isolated systems are governed by static Hamiltonians. For classical systems, this Hamiltonian (energy) function, defined over the phase space, is real. In this case, all conserved quantities are those whose Poisson bracket with the Hamiltonian vanishes, i.e. they commute with the Hamiltonian [1]. For quantum systems, the same criterion is applicable although now the Hamiltonian $H_0$ is a Dirac-Hermitian operator, $H_0 = H_0^\dagger$, and so are "observables" in the quantum theory [7]. A full set of conserved quantities for such a system are therefore obtained by identifying all independent observables that obey the commutation property. Some of the conserved observables are immediately apparent. An obvious one is the identity operator, $\mathbb{1}$, that gives rise to the conservation of the Dirac inner-product between two states, and in particular, the norm of a given state. A second observable is the Hamiltonian

Journal of Physics: Conference Series **2038** (2021) 012021 doi:10.1088/1742-6596/2038/1/012021

itself, and it expresses the conservation of energy for a given, isolated system. In both classical and quantum cases, the dynamics generated by a Hamiltonian are reversible and thus do not evolve into a steady state.

In contrast, the dynamics of a classical system coupled to an environment become irreversible in the thermodynamic limit, and the system evolves towards the ground state or a non-equilibrium steady state. In such cases, although conservation laws apply to the system+environment, the existence or implications of time-invariant quantities for a classical system alone are unclear. When a small quantum system is coupled to an environment, it loses its coherence. This process is traditionally described by the Gorini-Kossakowski-Sudarshan Lindblad equation [8, 9, 10] for the (reduced) density matrix of the system, and results in a completely-positive trace preserving (CPTP) map on the space of physical density matrices [11]. In such cases, the conserved quantities are determined by observables that simultaneously commute with the system Hamiltonian and all relevant Lindblad dissipators [12, 13]. Absent special symmetries, it implies that the *state-norm or the trace of the system density-matrix is the only conserved quantity* for such quantum channels.

In this article, we will review symmetries and conservation laws for a new class of systems that are governed by non-Hermitian Hamiltonians [14]. Their coherent, non-unitary, non-norm-preserving dynamics are qualitatively different from the coherent, unitary dynamics of isolated systems as well as the incoherent, non-unitary, trace-preserving dynamics of systems coupled to an environment. Since Bender and coworkers' discovery [15, 16], over the past two decades, non-Hermitian Hamiltonians that are invariant under combined operations of parity and time-reversal ($\mathcal{PT}$) have emerged as new research frontier [17]. Its meteoric growth was driven first by theoretical interest in the generalization of a quantum theory to non-Hermitian, self-adjoint Hamiltonians [18, 19], and then by experiments on open, classical wave systems [20] with balanced gain and loss that are faithfully described by $\mathcal{PT}$-symmetric Hamiltonians [21]. A prototypical $\mathcal{PT}$-symmetric Hamiltonian $H(\gamma) \neq H^{\dagger}(\gamma)$ has a purely real spectrum when the gain-loss strength $\gamma$ is smaller than a threshold value $\gamma_{\rm PT}$. At $\gamma = \gamma_{\rm PT}$, two (or more) eigenvalues of $H(\gamma)$ coincide and the corresponding eigenvectors coalesce, giving rise to an exceptional point (EP) degeneracy [22, 23]. In contrast to the diabolic point (DP) degeneracies of self-adjoint operators, the eigenvectors of $H(\gamma_{\rm PT})$ do not span the space. When $\gamma$ exceeds the threshold, the degenerate eigenvalues of $H(\gamma)$ turn into complex conjugate pairs and the non-orthogonal eigenvectors of $H(\gamma)$ span the space again. This transition from real to complex-conjugate eigenvalues is called $\mathcal{PT}$-symmetry breaking transition. We remind the reader that the time-evolution operator $G(t) \equiv \exp[-iH(\gamma)t)]$ ($\hbar = 1$) is not unitary due to the non-orthogonal nature of the eigenstates of $H(\gamma)$ irrespective of whether the spectrum of $H(\gamma)$ is real or complex-conjugate pairs. In physical terms, the system changes from having purely oscillatory eigenmodes and a bounded, oscillatory state-norm to amplifying and decaying eigenmodes that, in the absence of nonlinearities, lead to an exponentially growing state-norm. Given this qualitatively different behaviors across the $\mathcal{PT}$-symmetry breaking transition at the EP, it makes sense to ask, "Is there anything that remains time-invariant in these open systems with balanced gain and loss"? As we will show below, this question can be completely answered with a recursive, analytical construction.

After the extensive realizations of $\mathcal{PT}$-symmetric Hamiltonians in classical platforms that include coupled mechanical [24] and electrical [25, 26, 27, 28, 29] oscillators, coupled optical waveguides [30, 31], fiber loops [32, 33], microring resonators [34], acoustics [35], optomechanical systems [36], and so on, past two years have seen tremendous efforts to extend these ideas into the semiclassical or purely quantum domains. These efforts have led to the realizations of passive $\mathcal{PT}$-symmetric Hamiltonians with mode-selective losses in ultracold atoms in a two-level system [37] and momentum-space lattice [38], a single NV center [39], single [40] and correlated photons [41], and a superconducting transmon circuit [42]. As we will show below, although these quantum systems are governed by lossy Hamiltonians, conservation laws apply to them as well and lead to observable consequences that arise from the difference between decay rate for the slow mode that emerges after the passive $\mathcal{PT}$-symmetry breaking transition [43] and the average decay rate for the lossy system.

The plan for the paper is as follows. In Sec. 2 we present an algebraic approach to obtain conserved observables and generalize the results from those for a $\mathcal{PT}$-symmetric Hamiltonian to Hamiltonians with a broader class of symmetries. We show that the nature of the conservation laws is purely algebraic, not restricted to the Hermiticity of the system in any way. Instead, it depends heavily on the locations of the eigenvalues, itself is a reflection of some underlying symmetry of the system. We prove existence conditions for the conserved quantities. In Sec. 3, we review a recursive procedure that allows explicit calculation of the conserved, intertwining operators without using the spectral decomposition approach of Sec. 2, and illustrate our construction with a few examples. We generalize the ideas of conserved quantities to non-$\mathcal{PT}$-symmetric Hamiltonians in Sec. 4. Section 5 concludes the paper with remarks about consequences for lossy $\mathcal{PT}$-symmetric systems, generalization to time-periodic, non-Hermitian systems and systems undergoing non-unitary, discrete-time quantum walks, and a summary.

As an aside, we note that in this work, we *only consider Dirac-Hermiticity as the defining criterion* for all observables; this is due to fact that nature seems only consistent with a Born-rule in the quantum theory that uses Dirac Hermitian-conjugate to obtain the probability density. Therefore, we will not consider formally consistent mathematical models with non-Hermitian Hamiltonians or "observables" that are self-adjoint [19, 44, 45]. We will also restrict ourselves to finite-dimensional systems, thereby circumventing the issues of domains, boundedness, and invertibility [46].

## 2. Conserved quantities in non-Hermitian systems

In classical mechanics, observables that are conserved during time evolution are most easily defined in the Hamiltonian formalism, where they are determined by a vanishing Poisson bracket [1]. Therefore, it should come at no surprise that their quantum counterparts become by far most apparent in the Heisenberg picture. Let the system under consideration be governed by a static Hamiltonian $H$ that may not be Hermitian. It satisfies equations of motion $i\partial_t|\psi(t)\rangle = H|\psi(t)\rangle$ and $-i\partial_t\langle\psi(t)| = \langle\psi(t)|H^\dagger$. By definition, a linear operator $\eta$ is a constant of motion if and only if $\langle\psi(t)|\eta|\psi(t)\rangle = \mathrm{Tr}[\eta\rho_\psi(t)]$ remains constant for any arbitrary state $|\psi\rangle$ (or a density matrix $\rho_\psi$). In the absence of intrinsic time-dependence, this constraint translates into

$$i\frac{d}{dt}\langle\psi(t)|\eta|\psi(t)\rangle = \langle\psi(t)|\eta H - H^\dagger\eta|\psi(t)\rangle = 0. \quad (1)$$

Due to the linearity of the constraint in Eq.(1), without loss of generality, we can choose $\eta$ to be a Hermitian matrix. When $H = H^\dagger$ (an isolated system), the observable conservation is therefore equivalent to commutation, just as expected. A Hermitian system trivially leaves two important operators conserved, namely the identity (state-norm) and the Hamiltonian itself (energy). We emphasize that many more independent, conserved operators can be constructed very easily; an example is the set of Hermitian projectors onto the eigenspace for each real eigenvalue. These are often disregarded in many applications since they are not connected to fundamental symmetries of the system. However, they fulfill all requirements for a genuine conserved observable and will play an important role in the following analysis. When $H$ is not Hermitian, Eq.(1) leads to the following intertwining constraint,

$$\eta H = H^\dagger\eta. \quad (2)$$

This characterization of conserved observables as intertwining operators, Eq.(2), has appeared in the literature in the context of pseudo-Hermitian operators ($\eta$ is invertible) [18, 19, 44] and the $\mathcal{CPT}$-inner product ($\eta$ is positive definite) [16]. Here, we only focus on them as conserved observables for an open system with gain and loss.

### *2.1. Conserved quantities via spectral decomposition*

To show the existence of conserved operators, we proceed by direct construction using spectral decomposition techniques [18]. It is a simple observation that we can restrict our attention to $n$-dimensional Hermitian operators in the following analysis. Since they form the basis of the underlying operator Hilbert space of dimension $n^2$, they can serve as generators of all the other, possibly non-Hermitian, operators by complex linear combinations.

For a given $n$-dimensional Hamiltonian $H$ and an unknown $\eta$, solving the intertwining relation Eq.(2) is equivalent to solving a set of $n^2$ linear equations with $n^2$ unknowns. The question posed here—identification of all conserved quantities—concerns the dimension of its solution space. For this section, we will assume that the Hamiltonian $H$ with non-degenerate eigenvalues $\epsilon_k$ has a complete set of right eigenvectors $\{|R_k\rangle\}$ defined by $H|R_k\rangle = \epsilon_k|R_k\rangle$ and let $\{\langle L_k|\}$ be the left-eigenvectors that satisfy $\langle L_k|H = \epsilon_k\langle L_k|$. In the spirit of choosing our conventions close to the experiments, we will normalize these eigenvectors according to the Dirac norm, $\langle R_k|R_k\rangle = 1 = \langle L_m|L_m\rangle$. We also remind the reader that right-eigenvectors are not orthogonal to each other, $\langle R_k|R_m\rangle \neq 0$, and neither are the left eigenvectors. However, they form a biorthogonal basis, $\langle L_k|R_m\rangle \propto \delta_{km}$. We note here that the general case where the eigenvalue $\epsilon_m$ has a degeneracy $d_m$ has been discussed, albeit with different normalization, in Ref. [18]. The Hamiltonian $H \neq H^\dagger$ admits a spectral decomposition

$$H = \sum_{k=1}^{N} \epsilon_k \frac{|R_k\rangle\langle L_k|}{\langle L_k|R_k\rangle} \quad , \quad H^\dagger = \sum_{k=1}^{N} \epsilon_k^* \frac{|L_k\rangle\langle R_k|}{\langle R_k|L_k\rangle} \tag{3}$$

in such a basis where the denominator $\langle L_k|R_k\rangle \neq 1$ due to our choice of Dirac normalization. Equation (1) implies that the eigenvalues of $H$ are either purely real or occur in complex conjugate pairs for a nonzero intertwining operator to exist. This means the characteristic equation for eigenvalues of $H$ has real coefficients or, equivalently, there is an antilinear operator $\mathcal{A}$ that commutes with $H$ [18]. Without loss of generality, we will call this antilinear operator $\mathcal{A} = \mathcal{PT}$. From the definitions, it also follows that $\eta|R_k\rangle \propto |L_k\rangle$ provided $\epsilon_k$ is real, and $\eta|R_\alpha\rangle \propto |L_{\alpha^*}\rangle$ provided $\epsilon_{\alpha^*} \neq \epsilon_\alpha$ is a complex energy. By sandwiching the intertwining relation between $\langle R_k|$ and $|R_j\rangle$, we also get $(\epsilon_j - \epsilon_k^*)\langle R_k|\eta|R_j\rangle = 0$ [18]. It is now easy to explicitly show that the following Hermitian operators

$$\eta_k = |L_k\rangle\langle L_k|, \tag{4}$$
$$\eta_{s\alpha} = (|L_\alpha\rangle\langle L_{\alpha^*}| + |L_{\alpha^*}\rangle\langle L_\alpha|)/2, \tag{5}$$
$$\eta_{a\alpha} = i(|L_\alpha\rangle\langle L_{\alpha^*}| - |L_{\alpha^*}\rangle\langle L_\alpha|)/2, \tag{6}$$

are intertwining operators. Here the index $k$ $(\alpha, \alpha^*)$ spans over real (complex conjugate) eigenvalues. These are analogues of conserved eigenvector projections in traditional Hermitian quantum mechanics, but their physical meaning is certainly much more subtle. Further analysis of this equation will depend on the degeneracy of the spectrum of the Hamiltonian under investigation. We are going to separately address the cases of non-degenerate spectrum, pure eigenvalue degeneracy and the exceptional points.

### *2.2. Non-degenerate spectrum*

If the Hamiltonian under consideration has a non-degenerate spectrum, there are no more remaining solutions to be found. We can thus conclude that the number of conserved operators is equal to the number of eigenvalues of $H$, i.e. the count of independent conserved quantities reaches $n$. The completely general formula for $\eta$ in such a case is given by

$$\eta = \sum_{k \in \text{real}} A_k \eta_k + \sum_{\alpha \in \text{complex}} (A_{s\alpha}\eta_{s\alpha} + A_{a\alpha}\eta_{a\alpha}). \tag{7}$$

As a concrete example, let us consider $H_3 = JS_x + i\gamma S_z$ where $S_x$ and $S_z$ are the 3-dimensional representation of SU(2) [47]. This non-Hermitian Hamiltonian commutes with $\mathcal{P}_3\mathcal{T}$ operator where $\mathcal{P}_3$ = antidiag(1, 1, 1) and $\mathcal{T} = *$, the complex-conjugation operator. Classically, the Hamiltonian $H_3$ represents a $\mathcal{PT}$-symmetric trimer and it has a third-order EP at $\gamma_{\text{EP}} = J$ [48]. Since $H_3$ is transpose-symmetric, its left-eigenvectors are just transpose of the right ones, i.e. $\langle L_k| = |R_k\rangle^T$. When $\gamma/J = \sin\theta \leq 1$, the purely real eigenvalues of $H_3$ are given by $\epsilon_k = \{\pm\cos\theta, 0\}$ and the corresponding Dirac-normalized, right eigenvectors are given by

$$|R_\pm(\theta)\rangle = \frac{1}{2}\begin{bmatrix} e^{\pm i\theta} \\ \pm\sqrt{2} \\ e^{\mp i\theta} \end{bmatrix}, \quad |R_0(\theta)\rangle = \frac{1}{\sqrt{2(1+\sin^2\theta)}}\begin{bmatrix} -1 \\ \sqrt{2}i\sin\theta \\ 1 \end{bmatrix}. \tag{8}$$

We leave it to the reader to check that (i) the right eigenvectors are not orthogonal; (ii) the left and right eigenvectors satisfy $\langle L_k|R_j\rangle = 0$ when $k \neq j$; (iii) the left-right inner products $\langle L_k|R_k\rangle$ are not unity, and (iv) the resolution of identity is given by

$$\frac{|R_+\rangle\langle L_+|}{\langle L_+|R_+\rangle} + \frac{|R_-\rangle\langle L_-|}{\langle L_-|R_-\rangle} + \frac{|R_0\rangle\langle L_0|}{\langle L_0|R_0\rangle} = \mathbb{1}_3. \tag{9}$$

When $\gamma/J = \cosh\beta \geq 1$, spectrum of $H_3$ is given by $\{\pm i\sinh\beta, 0\}$ and the corresponding Dirac-normalized, right eigenvectors are given by

$$|R_\pm(\beta)\rangle = \frac{1}{\sqrt{2(1+\cosh 2\beta)}}\begin{bmatrix} -e^{\pm\beta} \\ \sqrt{2}i \\ e^{\mp\beta} \end{bmatrix}, \quad |R_0(\beta)\rangle = \frac{1}{\sqrt{2(1+\cosh^2\beta)}}\begin{bmatrix} -1 \\ \sqrt{2}i\cosh\beta \\ 1 \end{bmatrix} \tag{10}$$

These results for eigenvectors allow us to explicitly construct the intertwining operators, Eqs.(4)-(6) when the spectrum is purely real or has a complex-conjugate pair. For both cases, we leave as an exercise to the reader to check that the three $\eta$ matrices obtained in this manner satisfy the intertwining relation, Eq.(2).

We see that the conserved eigenvector projections generalize naturally from real to complex-conjugate eigenvalues but no further. This provides a fixed limit of existence for those quantities, and shows that their existence is not so ubiquitous as it may have looked on the first sight. In our formalism, the building blocks of a general $\eta$ are projections, and therefore all $n$ of them would be needed to result into an invertible operator, just as expected. However, our results hold more generally even for Hamiltonians having one or more dissipation channels. Such Hamiltonians may still admit some conserved operators, though not as many as a Hamiltonian with full rank.

### *2.3. Spectrum with diabolic-point degeneracies*

When the spectrum of $H \neq H^\dagger$ becomes degenerate, the sufficiency part in the above proof ceases to hold, and there might exist more conserved quantities than those found explicitly so far. Their exact count depends on the nature of the degeneracy. For eigenvalue-only or diabolic-point degeneracy, we now show that it's possible to explicitly construct additional independent conserved quantities in a very simple fashion.

Let the $k_D$-dimensional subspace associated with an eigenvalue $\epsilon_k$ be spanned by $k_D$ orthonormal, right-eigenvectors $\{|R_k, a\rangle | a = 1, \cdots, k_D\}$, with the corresponding set of the left-eigenvectors denoted by $\{\langle L_k, a|\}$. It is easy to check that the operators

$$\eta_k^{ab} = |L_k, a\rangle\langle L_k, b| \tag{11}$$

are also conserved. We note that the $\eta_k^{ab}$ defined in Eq.(11) are not Hermitian; however, they can be used to create symmetric and antisymmetric Hermitian combinations. Therefore, the total number of independent, conserved, Hermitian operators becomes $\sum_k k_D^2$. As is expected, this expression turns out to be equal to $n$ when the spectrum is non-degenerate. Also, it is equal to $n^2$ when it is maximally degenerate, i.e. the Hamiltonian is an identity matrix and trivially conserves any operator whatsoever. We stress that this result is purely algebraic, and does not rely on the existence of special symmetries for the underlying model.

### *2.4. Spectrum with exceptional-point degeneracies*

The remaining scenario concerns a degeneracy where both eigenvalues and their underlying eigenvectors coalesce, i.e. EP degeneracies. As a result of the eigenvector coalescence, the spectral resolution approach used above ceases to work. We can however bring those ideas back in a slightly different form using the notion of generalized right eigenvectors. These are vectors $|v_{Rm}\rangle$ that satisfy the equation $(H-\lambda)^m|v_{Rm}\rangle = 0$ for $m \geq 2$. Note that at $m=1$, this definition gives the regular right eigenvector at the EP. The set of generalized eigenvectors always form a basis of the underlying Hilbert space. Moreover, the spectral resolution results can be extended to non-diagonalizable Hamiltonians at an EP by using additional terms of the form $|v_{Rm}\rangle\langle v_{Lm}|$ for each generalized eigenvector of the Hamiltonian. Therefore, for an $n$-dimensional Hamiltonian, an exceptional point of order $N \leq n$ gives rise to $N$ conserved operators, in addition to the $n-N$ operators that are generated by eigenvectors corresponding to the non-degenerate eigenvalues.

As an example, let us consider $H_3$ at the EP. At this point, the sole right-eigenvector satisfies the equation $H_3|v_{R1}\rangle = 0$ and is given by $|v_{R1}\rangle = (-1, \sqrt{2}i, 1)^T/2$. We also note that

$$|v_{R1}\rangle = \lim_{\theta\to\pi/2} |R_{\pm,0}(\theta)\rangle = \lim_{\beta\to 0} |R_{\pm,0}(\beta)\rangle. \tag{12}$$

The second (generalized) eigenvector $|v_{R2}\rangle$ is defined by the equation $H_3^2|v_{R2}\rangle = 0$. A convenient method to define $|v_{R2}\rangle$ is by equation $H_3|v_{R2}\rangle = |v_{R1}\rangle$, and leads to $|v_{R2}\rangle = i(1,0,1)^T/2$. The third vector is defined by $H_3^3|v_{R3}\rangle = 0$ and can be chosen as any vector that is linearly independent of $|v_{R1}\rangle$ and $|v_{R2}\rangle$. In each case, the left-eigenvectors are given by the transpose of the right ones.

Thus, an $n$-dimensional Hamiltonian whose spectrum is either non-degenerate or has only EP degeneracies always admits $n$ conserved, intertwining operators, no matter how many and what order the EPs are. The fact that there are exactly $n$ linearly independent intertwining operators is quite surprising and shows that the number of conserved, independent operators is a stable quantity, governed by simple algebraic laws. This is because EP degeneracies are far more common than diabolic-point (level crossing) degeneracies. We also note that the conserved operators themselves remain smooth across an exceptional point.

## 3. Recursive procedure for generating conserved operators

The spectral-decomposition approach provides explicit, analytical formulae for constructing all conserved operators of a given Hamiltonian in terms of its eigenvectors. Unfortunately, that is not of much practical help because obtaining analytical expressions for eigenvectors is a highly nontrivial task even for small dimensions $n \sim 6$. However, we will show that there is a simple but powerful workaround to this issue: the conserved operators may be constructed iteratively by using the prescription

$$\eta_{k+1} = \eta_k H \tag{13}$$

starting from an initial $\eta_1$ that is determined from the outset. Note that if $\eta_k = \eta_k^\dagger$ is Hermitian, the intertwining relation Eq.(2) implies that recursively obtained $\eta_{k+1}$ is also Hermitian. It also follows that the different $\eta$ operators do not commute with each other, and the commutator is proportional to the anti-Hermitian part of $H$. Finally, since $H$ obeys a characteristic polynomial equation of order $n$, it follows that $\eta_{N+1} = \eta_1 H^N$ can be written as a linear combination of lower-order operators $\eta_{k\leq N}$. Thus, for an $n$-dimensional system, this recursive procedure gives rise to $n$ non-commuting, linearly independent conserved observables [14].

For a broad class of transpose-symmetric Hamiltonians $\eta_1$ is found as follows. For conserved quantities to exist, the Hamiltonian must have purely real or complex-conjugate eigenvalues, i.e. an antilinear operator $\mathcal{A}$ commutes with it. Let us write $\mathcal{A} = \mathcal{PT} = \mathcal{L}*$ where $\mathcal{L}$ denotes *the entire linear part of* $\mathcal{A}$ and $*$ is purely the complex-conjugation operation. (In general, the operator $\mathcal{T}$ has a nontrivial unitary part; that has been absorbed into the operator $\mathcal{L}$.) It is easy to see that if $H = H^T$, then $\mathcal{L}$ is an intertwining operator. Once $\eta_1 = \mathcal{L}$ is identified, the rest are determined by the recursive procedure.

To illustrate the simplicity and power of this method, let us consider the Hamiltonian $H_3$. It follows that the three, linearly independent intertwining operators are given by

$$\eta_1 = \mathcal{P}_3, \tag{14}$$
$$\eta_2 = \mathcal{P}_3 H_3, \tag{15}$$
$$\eta_3 = \mathcal{P}_3 H_3^2. \tag{16}$$

The three-dimensional model and the subsequent analysis can be generalized $H_D = JS_x + i\gamma S_z$ where $S_x, S_z$ are $D$-dimensional representations of SU(2) [47, 49]. Although the spectral decomposition process becomes more and more cumbersome as $D$ increases, the recursive procedure maintains its simplicity. We emphasize that since Eq.(2) is a linear relation, the choice of intertwining operators is not unique. Any linear combination of $\{\eta_1, \ldots, \eta_D\}$s is also a conserved quantity. In general, the eigenvalues of $\eta_k$ are not positive definite. However, only in the $\mathcal{PT}$-symmetric region, one can create positive-definite $\eta$s that, in the literature, have been used to create a complex extension of quantum mechanics in the $\mathcal{PT}$-symmetric region [16, 18, 50].

We note that the transpose-symmetric-Hamiltonian constraint is satisfied by many models of balanced gain and loss systems. However, there are two major exceptions. The first is generalized Hatano-Nelson models that span classical, semiclassical, or quantum systems. The second is purely classical systems such as coupled mechanical or electrical oscillators, where the dynamics of (square-root of) the energy density is governed by a Hamiltonian with purely imaginary entries. Both cases are discussed in the following paragraphs.

### *3.1. Asymmetric Hatano Nelson model*

Generalized Hatano-Nelson models are characterized by an asymmetric, nearest-neighbor tunneling profile [51, 52]. They show non-Hermitian skin effect [53], generalized bulk-boundary correspondence [54], and lead to novel applications such as the topological funneling of light [33]. Here, we construct the first intertwining operator $\eta_1$, which then permits the recursive construction of remaining conserved observables.

The prototypical Hatano-Nelson Hamiltonian with dimension $D$ is given by

$$H_{\rm HN} = JS_x + i\gamma S_y \neq H_{\rm HN}^T. \tag{17}$$

Note that Eq.(17) can either represent a small system with few sites, or alternatively it can represent the block-diagonal, momentum-space Hamiltonian for an infinite lattice with a unit cell of size $D$. The Hatano-Nelson model has asymmetric tunneling characterized by two energy scales $J \pm \gamma$ respectively. We remind the reader that for $D \geq 4$, the tunneling elements in $H_{\rm HN}$ depend on the site index. The Hamiltonian $H_{\rm HN}$, Eq.(17), is equivalent to the traditional $\mathcal{PT}$-symmetric Hamiltonian $H_D$,

$$H_{\rm HN} = R_x(-\pi/2) H_D R_x(\pi/2), \tag{18}$$

due to the unitary transformation $R_x(-\pi/2)$ where $R_x(\theta) = \exp(-iS_x\theta) = R_x(-\theta)^\dagger$ is a rotation about the $x$-axis through angle $\theta$. Since the first intertwining operator for $H_D$ is known, $\eta_1 = \mathcal{P}_D = \text{antidiag}(1, \ldots, 1)$, we can construct the one for the Hatano Nelson Hamiltonian as

$$\eta_1 = R_x(-\pi/2) \mathcal{P}_D R_x(\pi/2), \tag{19}$$

and the remaining $D-1$ operators then follow trivially. It is easy to check that when $D = 2$, since the parity operator $\mathcal{P}_2 = 2\sigma_x$ commutes with the rotation operator, the so-called "$\mathcal{PT}$-inner product" [55, 56] is a constant of motion for both $H_D$ and $H_{\rm HN}$. It is also important to note that for higher dimensions $D \geq 3$, the first intertwining operator $\eta_1$, Eq.(19), does not commute with the corresponding operator $\mathcal{P}_D$ for $H_D$.

*3.2. Conserved quantities for $\mathcal{PT}$ electrical circuit models*

An inductor-capacitor ($LC$) circuit is the simplest electrical analog of a mass+spring pendulum where the energy oscillates between the kinetic form (inductor) and the potential form (capacitor). In contrast to the loss or gain in a mechanical system, it is easy to implement controlled loss and gain in electrical circuits by using resistors and operational amplifiers. Two such $LC$ oscillators, one with a resistor $R$ (loss) and second with negative resistor $-R$ (gain), form the loss and gain "sites" of a $\mathcal{PT}$-symmetric dimer. These "sites" can be coupled through mutual inductance [25, 26, 29], a coupling capacitor [27], or a coupling inductor [28].

Let us consider two $LC$ circuits coupled through mutual inductance [29]. Kirchhoff laws govern the dynamics of voltages across capacitors $V_{1,2}(t)$ and currents across the inductors $I_{1,2}(t)$, and the resulting set of linear, coupled equations can be mapped onto Schrodinger-like equation for a state vector $|\psi(t)\rangle$ such that its norm is the instantaneous circuit energy, i.e. $\mathcal{E}(t) = \langle\psi(t)|\psi(t)\rangle$. The non-Hermitian Hamiltonian describing the dynamics of the real state vector $|\psi(t)\rangle$ is given by [29]

$$\frac{H(\gamma)}{(\omega_0/2)} = i\begin{bmatrix} -2\gamma & 0 & \gamma_0 & -\gamma_{\rm PT} \\ 0 & 2\gamma & -\gamma_{\rm PT} & \gamma_0 \\ -\gamma_0 & \gamma_{\rm PT} & 0 & 0 \\ \gamma_{\rm PT} & -\gamma_0 & 0 & 0 \end{bmatrix} = -\sigma_y \otimes (\gamma_0 \mathbb{1}_2 - \gamma_{\rm PT}\sigma_x) - i\gamma(\mathbb{1}_2 + \sigma_z) \otimes \sigma_z. \quad (20)$$

Here $\omega_0 = 1/\sqrt{LC}$ is the fundamental frequency of each oscillator, $\gamma = \omega_0(L/R)$ is the dimensionless measure of gain or loss, $\gamma_0 = (1/\sqrt{1-\mu} + 1/\sqrt{1+\mu})$ and $\gamma_{\rm PT} = (1/\sqrt{1-\mu} - 1/\sqrt{1+\mu})$ denote the two values of $\gamma$ at which the Hamiltonian $H(\gamma)$ has an EP, and $0 \leq \mu = M/L \leq 1$ denotes the dimensionless mutual inductance coupling between the gain and loss circuits. To obtain conserved quantities for this $\mathcal{PT}$ electric dimer, we need to obtain the first intertwining operator for $H(\gamma) \neq H^T(\gamma)$. Note that the Hamiltonian $H(\gamma)$ has the antilinear symmetry $\mathcal{A} = \mathcal{PT}$ with $\mathcal{P} = \mathbb{1}_2 \otimes \sigma_x$ and $\mathcal{T} = (\sigma_z \otimes \mathbb{1}_2)*$. However, since $H(\gamma)$ is not transpose-symmetric, the linear part $\mathcal{L}$ of $\mathcal{A} = (\sigma_z \otimes \sigma_x)*$ is not an intertwining operator.

Looking at the tensor product structure of Eq.(20), it is clear that the source of the transpose asymmetry is the $\sigma_y$ term, and the unitary transformation $U = \exp(+i\pi\sigma_z/4) \otimes \mathbb{1}_2$ will convert the Hamiltonian (20) into a transpose-symmetric matrix $\mathbb{H} = UHU^\dagger = \mathbb{H}^T$. Under the same transformation, the antilinear operator $\mathcal{A}$ maps into $\mathcal{A}' = U\mathcal{A}U^\dagger = [\exp(+i\pi\sigma_z/2)\sigma_z \otimes \sigma_x]*$. Here, we have used the complex conjugation property $U^* = U^\dagger$. Comparing the result with $\mathcal{A}' = \mathcal{L}'*$ gives $\mathbb{1}_2 \otimes \sigma_x$ as the first intertwining operator for the matrix $\mathbb{H}$, and thereby the first intertwining operator for $H(\gamma)$, Eq.(20), is given by inverse $U$ transformation, i.e.

$$\eta_1 = \mathbb{1}_2 \otimes \sigma_x. \quad (21)$$

The remaining operators are then trivially obtained through the recursive construction. We leave the analysis of two $LC$ circuits connected by a coupling inductor $L_c$ [28] as an exercise to the reader.

## 4. Hamiltonians with $\mathcal{PT}$, anti-$\mathcal{PT}$, and other symmetries

The robust algebraic nature of the results about conserved observables for Hamiltonians with real or complex-conjugate spectra suggests several possibilities for further generalization [57, 58]. The antilinear symmetry $\mathcal{A} = \mathcal{PT}$ that ensures real or complex-conjugate eigenvalues means the eigenvalues occur symmetrically about the $x$-axis in the complex plane (Fig. 1a). Here, we demonstrate that the results translate seamlessly from Schrodinger-like equations ($\mathcal{PT}$-symmetry) to diffusion equations (anti-$\mathcal{PT}$ symmetry) and beyond.

To remind the reader about these ideas, we start with a diffusive system governed by equation $\partial_t|\psi(t)\rangle = \mathcal{D}|\psi(t)\rangle$. This can be written in the Schrodinger-like form with an $H_\mathcal{D} = i\mathcal{D}$. If the matrix $\mathcal{D}$ has an antilinear symmetry $\mathcal{A}$, it follows that the Hamiltonian $H_\mathcal{D}$ anticommutes with $\mathcal{A}$, i.e. $\mathcal{A}H_\mathcal{D} = -H_\mathcal{D}\mathcal{A}$. Such systems are called anti-$\mathcal{PT}$-symmetric systems, and their eigenvalues occur in pairs that are symmetric about the $y$-axis, i.e. they occur in pairs

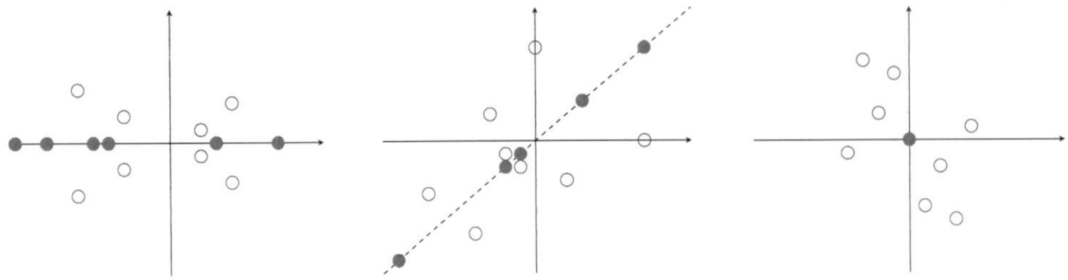

(a) $\eta H = H^\dagger \eta$ (b) $\eta H = e^{-i\pi/2} H^\dagger \eta$ (c) $\eta H = -H\eta$

Figure 1: Schematic illustration of symmetries of eigenvalues for non-Hermitian Hamiltonians with different symmetries. (a) In a $\mathcal{PT}$-symmetric case, eigenvalues are reflection-symmetric about the horizontal axis, (b) For Hamiltonians with anyonic $\mathcal{PT}$-symmetry, they are reflection-symmetric about the dotted line with slope $-\tan(-\pi/4) = 1$. (c) For Hamiltonians with a chiral symmetry, eigenvalues are reflection-symmetric through the origin, i.e. occur in pairs $(\epsilon_k, -\epsilon_k)$. Note that eigenvalues of anti-$\mathcal{PT}$-symmetric Hamiltonians (not shown) are reflection-symmetric about the vertical axis.

$(\epsilon_k, -\epsilon_k^*)$ [59, 60, 61, 62, 63]. Note that a Hamiltonian can be both $\mathcal{PT}$-symmetric and anti-$\mathcal{PT}$-symmetric with respect to different parity operations. The simplest example is a prototypical dimer with balanced gain and loss described by the $2 \times 2$ Hamiltonian $H_2 = (J\sigma_x + i\gamma\sigma_z)/2$. We leave it to the reader to verify that $H_2$ is $\mathcal{PT}$ symmetric with respect to $\mathcal{PT} = \sigma_x *$, but *it is also anti-$\mathcal{PT}$-symmetric with respect to the operator $\mathcal{PT} = \sigma_z *$.*

Following the procedure outlined in Sec. 2, Hermitian conserved operators for a diffusive system can thus be generated recursively starting from the initial observable $\eta_1$ that satisfies a modified intertwining relation,

$$\eta_1 H_{\mathcal{D}} = -H_{\mathcal{D}}^\dagger \eta_1, \tag{22}$$

$$\eta_{k+1} = i\eta_k H_{\mathcal{D}} = \eta_{k+1}^\dagger. \tag{23}$$

This recursive procedure generalizes to other systems where the equations of motion are linear, but do not fall neatly into either Schrodinger- or diffusion-equation categories. The conserved operators for such systems are defined by the intertwining relation

$$\eta H = e^{i\varphi} H^\dagger \eta. \tag{24}$$

Physically, such Hamiltonians have eigenvalues that are symmetric about a line with slope $-\tan(\varphi/2)$ in the complex eigenvalue plane, i.e they occur in pairs $(\epsilon_k, e^{-i\varphi}\epsilon_k^*)$ (Fig. 1b). Equivalently, such a Hamiltonian satisfies the equation $\mathcal{A}H = e^{i\varphi} H\mathcal{A}$ for some antilinear operator $\mathcal{A}$. Note that $\varphi = 0$ gives us the $\mathcal{PT}$-symmetric case whereas the anti-$\mathcal{PT}$-symmetric case corresponds to $\varphi = \pi$. Due to its similarity with quasiparticle-statistics definitions in two dimensions, Hamiltonians that require Eq.(24) for their conserved observables are said to have "anyonic $\mathcal{PT}$ symmetry" [64, 65].

For an anyonic-$\mathcal{PT}$-symmetric Hamiltonian, we can reiterate the process of solving Eq.(24) by spectrally decomposing all the operators in question. The recursive procedure for obtaining the constants of motion is also modified a bit,

$$\eta_{k+1} = e^{i\varphi/2} \eta_k H = \eta_{k+1}^\dagger. \tag{25}$$

Finally, note that virtually the same behavior can be deducted by combination of equation-of-motion and other symmetries that the Hamiltonian may have. For example, if a Hamiltonian has a chiral symmetry, there exists an $\eta = \Pi$ that anticommutes with the Hamiltonian, i.e. $\eta H = -H\eta$. This, combined with the intertwining relation, implies that the spectrum of $H$ occurs in reflection-through-the-origin symmetric pairs $(\epsilon_k, -\epsilon_k)$ (Fig. 1c). Consequently, it gives rise to conserved pseudochirality in a simple, non-Hermitian model [66].

## 5. Discussion and Summary

Now, we present the consequences of conserved quantities for lossy, non-Hermitian systems, briefly discuss generalizations of this approach to time-periodic or discrete-time systems, and then conclude with a summary.

### *5.1. Consequences for passive, $\mathcal{PT}$-symmetric systems*

$\mathcal{PT}$-symmetric Hamiltonians are realized in classical wave-systems where balanced gain and loss can be implemented by ignoring the thermal and quantum fluctuation effects [67]. Conserved quantities in such systems lead to phase locking at the EP or in the $\mathcal{PT}$-broken region because the norm of an arbitrary state $\langle\psi(t)|\psi(t)\rangle$ grows with time whereas the expectation values $\langle\psi(t)|\eta_k|\psi(t)\rangle$ remain constant [68]. But what about the quantum domain?

The non-Hermitian Hamiltonians that arise for truly quantum systems through post-selection [42, 49] are always lossy. In such cases, the passive $\mathcal{PT}$-transition across the EP is marked by the emergence of a slowly decaying mode *whose decay-rate decreases when the local loss strength increases* [28, 43]. Since the lossy Hamiltonian does not have complex-conjugate eigenvalues, in principle, the intertwining relation cannot be satisfied. However, the state norm $\langle\psi(t)|\psi(t)\rangle$ and the expectation values $\langle\psi(t)|\eta_k|\psi(t)\rangle$ decay at different rates in the passive $\mathcal{PT}$-symmetry broken region. Due to the exponential-in-time separation between these two quantities, the phase locking occurs for passive models [14]. These considerations show that all such effects continue be valid for *any non-Hermitian Hamiltonian that results from a $\mathcal{PT}$-symmetric Hamiltonian by a complex identity shift.*

### *5.2. Conserved quantities in time-periodic models*

Our analysis for conserved observables is applicable to a static Hamiltonian $H \neq H^\dagger$. In general, when the Hamiltonian $H(t)$ is time dependent, time-translational invariance is broken and so are the conservation laws. If the Hamiltonian is time periodic with period $T$, the time-translational symmetry is restored for discrete shifts $t \to t+pT$ where $p$ is an integer. The long-term dynamics of the system are then governed by the time evolution operator for one period,

$$G_F(T) = \mathbb{T}e^{-i\int_0^T dt' H(t')} \equiv e^{-iTH_F}, \tag{26}$$

where $\mathbb{T}$ denotes the time-ordered product that accounts for the non-commuting nature of $H(t)$ at different instances of time. Note that Eq.(26) also defines the (static) Floquet Hamiltonian $H_F$ [69, 70]. $G_F(T)$ determines the stroboscopic dynamics, namely dynamics only at integer times $pT$, while the dynamics during the intermediate times $pT \leq t \leq (p+1)T$, called the micromotion, is determined by the family of operators

$$K(t) = \mathbb{T}e^{-i\int_0^t dt' H(t')}. \tag{27}$$

It is easy to show that if the Hamiltonian $H(t)$ is Hermitian, then so is the Floquet Hamiltonian $H_F$. Similarly, if the Hamiltonian $H(t)$ is $\mathcal{PT}$-symmetric, the resultant Floquet Hamiltonian $H_F$ also has an antilinear symmetry, albeit with possibly different $\mathcal{P}$ and $\mathcal{T}$ operators [71]. Under the stroboscopic evolution, an observable $\eta$ is conserved provided it satisfies

$$G_F^\dagger \eta G_F = \eta. \tag{28}$$

Equation (28) defines the "intertwining relation" for systems that are described by a time-evolution operator $G$ instead of a Hamiltonian. In particular, it defines stroboscopically conserved quantities for non-Hermitian, discrete time quantum walks. The recursive procedure for generating other intertwining operators follows through with

$$\eta_{k+1} = \eta_k G. \tag{29}$$

Thus, a complete characterization of conserved quantities for non-Hermitian, time-periodic Hamiltonians as well as non-unitary discrete time quantum walks can be carried out with the explicit, analytical, recursive procedure we have reviewed in this article.

### *5.3. Summary*

In this article, we have reviewed conserved quantities that arise in the dynamics of systems that are governed by non-Hermitian Hamiltonians with different antilinear symmetries. For isolated quantum systems governed by Hermitian Hamiltonians, the conserved observables are linearly independent operators that commute with the Hamiltonian. In contrast, for a quantum channel, generically, there are no conserved quantities except the trace of the density matrix. Classical or quantum systems governed by effective, non-Hermitian Hamiltonians lie between these two limits. Non-Hermitian Hamiltonians with different antilinear symmetries ($\mathcal{PT}$, anti-$\mathcal{PT}$, anyonic-$\mathcal{PT}$) give rise to different intertwining relations that characterize conserved observables.

The fact that eigenvector-projections of Hermitian operators remain conserved during time evolution is quite obvious, although often somewhat neglected when discussing conservation laws in quantum mechanics. With that insight, we have reviewed the spectral decomposition method for constructing conserved quantities for the non-Hermitian cases, and then outlined a simple recursive, analytical solution for a complete set of linearly independent intertwining operators. Note that each conserved operator $\eta_k$ generates a one-parameter unitary group $U_k(\theta) = \exp(i\theta\eta_k)$ by exponentiation. However, the unitary transformation does not leave the Hamiltonian invariant, i.e. $U_k(\theta)H \neq HU_k(\theta)$, and therefore it does not have a direct physical meaning.

In this article, we have further quantified the number of conserved observables for non-Hermitian Hamiltonians with either diabolic degeneracies (level crossings) or EP degeneracies, the latter being most commonplace for non-Hermitian matrices. While in the first case we discovered new conserved observables, we showed that that crossing through an EP degeneracy does not yield any new ones. Thus, in the complex eigenvalue plane, the number of conserved observables remains equal to the maximum rank of the Hamiltonian matrix.

## References

[1] Goldstein H 2002 *Classical mechanics* (San Francisco: Addison Wesley) ISBN 0201657023
[2] Landau L D 1976 *Mechanics* (Oxford New York: Pergamon Press) ISBN 9780750628969
[3] Feynman R 2013 *The Feynman lectures on physics* (Uttar Pradesh, India Dorling Kindersley: Pearson India Education Services Pvt Ltd) ISBN 8131792110
[4] Close F E 2010 *Neutrino* (Oxford New York: Oxford University Press) ISBN 0199695997
[5] Baym G and Kadanoff L P 1961 *Phys. Rev.* **124**(2) 287–299 URL https://link.aps.org/doi/10.1103/PhysRev.124.287
[6] Mahan G 2000 *Many-Particle Physics* (Boston, MA: Springer US) ISBN 978-1-4757-5714-9
[7] Sakurai J J 2011 *Modern quantum mechanics* (Boston: Addison-Wesley) ISBN 0805382917
[8] Sudarshan E C G, Mathews P M and Rau J 1961 *Phys. Rev.* **121**(3) 920–924 URL https://link.aps.org/doi/10.1103/PhysRev.121.920
[9] Gorini V, Kossakowski A and Sudarshan E 1976 *Journal of Mathematical Physics* **17** 821 URL https://doi.org/10.1063/1.522979
[10] Lindblad G 1976 *Comm. Math. Phys.* **48** 119–130 URL https://projecteuclid.org:443/euclid.cmp/1103899849
[11] Manzano D 2020 *AIP Advances* **10** 025106 URL https://doi.org/10.1063/1.5115323
[12] Buča B and Prosen T 2012 *New Journal of Physics* **14** 073007 URL https://doi.org/10.1088/1367-2630/14/7/073007
[13] Albert V V and Jiang L 2014 *Phys. Rev. A* **89**(2) 022118 URL https://link.aps.org/doi/10.1103/PhysRevA.89.022118
[14] Bian Z, Xiao L, Wang K, Zhan X, Onanga F A, Ruzicka F, Yi W, Joglekar Y N and Xue P 2020 *Physical Review Research* **2** URL https://doi.org/10.1103/physrevresearch.2.022039
[15] Bender C M and Boettcher S 1998 *Phys. Rev. Lett.* **80**(24) 5243–5246 URL https://link.aps.org/doi/10.1103/PhysRevLett.80.5243
[16] Bender C M, Brody D C and Jones H F 2002 *Phys. Rev. Lett.* **89**(27) 270401 URL https://link.aps.org/doi/10.1103/PhysRevLett.89.270401
[17] Bender C M 2007 *Reports on Progress in Physics* **70** 947–1018 URL https://doi.org/10.1088%2F0034-4885%2F70%2F6%2Fr03
[18] Mostafazadeh A 2002 *Journal of Mathematical Physics* **43** 205–214 URL https://doi.org/10.1063/1.1418246
[19] Mostafazadeh A 2010 *International Journal of Geometric Methods in Modern Physics* **07** 1191–1306 URL https://doi.org/10.1142/s0219887810004816

[20] Joglekar Y N, Thompson C, Scott D D and Vemuri G 2013 *The European Physical Journal Applied Physics* **63** 30001 URL https://doi.org/10.1051/epjap/2013130240
[21] El-Ganainy R, Makris K G, Khajavikhan M, Musslimani Z H, Rotter S and Christodoulides D N 2018 *Nature Physics* **14** 11–19 URL https://doi.org/10.1038/nphys4323
[22] Kato T 1995 *Perturbation Theory for Linear Operators* (Springer Berlin Heidelberg) URL https://doi.org/10.1007/978-3-642-66282-9
[23] Özdemir Ş K, Rotter S, Nori F and Yang L 2019 *Nature Materials* **18** 783–798 ISSN 1476-4660 URL https://doi.org/10.1038/s41563-019-0304-9
[24] Bender C M, Berntson B K, Parker D and Samuel E 2013 *American Journal of Physics* **81** 173–179 URL https://doi.org/10.1119/1.4789549
[25] Schindler J, Li A, Zheng M C, Ellis F M and Kottos T 2011 *Physical Review A - Atomic, Molecular, and Optical Physics* **84** 1–5 ISSN 10502947
[26] Schindler J, Lin Z, Lee J M, Ramezani H, Ellis F M and Kottos T 2012 *Journal of Physics A: Mathematical and Theoretical* **45** 444029 URL https://doi.org/10.1088/1751-8113/45/44/444029
[27] Chitsazi M, Li H, Ellis F and Kottos T 2017 *Physical Review Letters* **119** 093901 URL https://doi.org/10.1103/physrevlett.119.093901
[28] de J León-Montiel R, Quiroz-Juárez M A, Domínguez-Juárez J L, Quintero-Torres R, Aragón J L, Harter A K and Joglekar Y N 2018 *Communications Physics* **1** URL https://doi.org/10.1038/s42005-018-0087-3
[29] Wang H, Lai Y H, Yuan Z, Suh M G and Vahala K 2020 *Nature Communications* **11** 1610 URL https://doi.org/10.1038/s41467-020-15341-6
[30] Duchesne D, Aimez V, Morandotti R, Christodoulides D N, Salamo G J, Volatier-Ravat M, Siviloglou G A and Guo A 2009 *Physical Review Letters* **103** 1–4 ISSN 0031-9007
[31] Rüter C E, Makris K G, El-Ganainy R, Christodoulides D N, Segev M and Kip D 2010 *Nature Physics* **6** 192–195 ISSN 17452473 (*Preprint* 1003.4968) URL http://dx.doi.org/10.1038/nphys1515
[32] Regensburger A, Bersch C, Miri M A, Onishchukov G, Christodoulides D N and Peschel U 2012 *Nature* **488** 167–171 URL https://doi.org/10.1038/nature11298
[33] Weidemann S, Kremer M, Helbig T, Hofmann T, Stegmaier A, Greiter M, Thomale R and Szameit A 2020 *Science* **368** 311–314 URL https://doi.org/10.1126/science.aaz8727
[34] Peng B, Özdemir Ş K, Lei F, Monifi F, Gianfreda M, Long G L, Fan S, Nori F, Bender C M and Yang L 2014 *Nature Physics* **10** 394–398 URL https://doi.org/10.1038/nphys2927
[35] Zhu X, Ramezani H, Shi C, Zhu J and Zhang X 2014 *Phys. Rev. X* **4**(3) 031042 URL https://link.aps.org/doi/10.1103/PhysRevX.4.031042
[36] Lü X Y, Jing H, Ma J Y and Wu Y 2015 *Phys. Rev. Lett.* **114**(25) 253601 URL https://link.aps.org/doi/10.1103/PhysRevLett.114.253601
[37] Li J, Harter A K, Liu J, de Melo L, Joglekar Y N and Luo L 2019 *Nature Communications* **10** 855 ISSN 2041-1723 URL https://doi.org/10.1038/s41467-019-08596-1
[38] Chen T, Gou W, Xie D, Xiao T, Yi W, Jing J and Yan B 2020 Quantum zeno effects across a parity-time symmetry breaking transition in atomic momentum space (*Preprint* 2009.01419)
[39] Wu Y, Liu W, Geng J, Song X, Ye X, Duan C K, Rong X and Du J 2019 *Science* **364** 878–880 URL https://doi.org/10.1126/science.aaw8205
[40] Xiao L, Wang K, Zhan X, Bian Z, Kawabata K, Ueda M, Yi W and Xue P 2019 *Phys. Rev. Lett.* **123**(23) 230401 URL https://link.aps.org/doi/10.1103/PhysRevLett.123.230401
[41] Klauck F, Teuber L, Ornigotti M, Heinrich M, Scheel S and Szameit A 2019 *Nature Photonics* **13** 883–887 URL https://doi.org/10.1038/s41566-019-0517-0
[42] Naghiloo M, Abbasi M, Joglekar Y N and Murch K W 2019 *Nature Physics* **15** 1232–1236 URL https://doi.org/10.1038/s41567-019-0652-z
[43] Joglekar Y N and Harter A K 2018 *Photonics Research* **6** A51 URL https://doi.org/10.1364/prj.6.000a51
[44] Znojil M 2009 *Journal of Mathematical Physics* **50** 122105 URL https://doi.org/10.1063/1.3272002
[45] Znojil M 2015 *International Journal of Theoretical Physics* **54** 3867–3870 URL https://doi.org/10.1007/s10773-014-2501-2
[46] Mostafazadeh A 2020 *Entropy* **22** 471 URL https://doi.org/10.3390/e22040471
[47] Graefe E M, Günther U, Korsch H J and Niederle A E 2008 *Journal of Physics A: Mathematical and Theoretical* **41** 255206 URL https://doi.org/10.1088/1751-8113/41/25/255206
[48] Hodaei H, Hassan A U, Wittek S, Garcia-Gracia H, El-Ganainy R, Christodoulides D N and Khajavikhan M 2017 *Nature* **548** 187–191 URL https://doi.org/10.1038/nature23280
[49] Quiroz-Juárez M A, Perez-Leija A, Tschernig K, Rodríguez-Lara B M, Magaña-Loaiza O S, Busch K, Joglekar Y N and de J León-Montiel R 2019 *Photonics Research* **7** 862 URL https://doi.org/10.1364/prj.7.000862
[50] Znojil M and Geyer H B 2006 *Physics Letters B* **640** 52–56 URL https://doi.org/10.1016/j.physletb.2006.07.028
[51] Hatano N and Nelson D R 1996 *Phys. Rev. Lett.* **77**(3) 570–573 URL https://link.aps.org/doi/10.1103/PhysRevLett.77.570

[52] Hatano N 1998 *Physica A: Statistical Mechanics and its Applications* **254** 317–331 URL https://doi.org/10.1016/s0378-4371(98)00010-7

[53] Lee C H and Thomale R 2019 *Phys. Rev. B* **99**(20) 201103 URL https://link.aps.org/doi/10.1103/PhysRevB.99.201103

[54] Helbig T, Hofmann T, Imhof S, Abdelghany M, Kiessling T, Molenkamp L W, Lee C H, Szameit A, Greiter M and Thomale R 2020 *Nature Physics* **16** 747–750 URL https://doi.org/10.1038/s41567-020-0922-9

[55] Bender C M, Cooper F, Meisinger P N and Savage V M 1999 *Physics Letters A* **259** 224–231 URL https://doi.org/10.1016/s0375-9601(99)00468-5

[56] Bender C M, Brody D C and Jones H F 2003 *American Journal of Physics* **71** 1095–1102 URL https://doi.org/10.1119/1.1574043

[57] Simón M, Buendía Á and Muga J 2018 *Mathematics* **6** 111 URL https://doi.org/10.3390/math6070111

[58] Alaña A, Martínez-Garaot S, Simón M A and Muga J G 2020 *Journal of Physics A: Mathematical and Theoretical* **53** 135304 URL https://doi.org/10.1088/1751-8121/ab7781

[59] Peng P, Cao W, Shen C, Qu W, Wen J, Jiang L and Xiao Y 2016 *Nature Physics* **12** 1139–1145 URL https://doi.org/10.1038/nphys3842

[60] Choi Y, Hahn C, Yoon J W and Song S H 2018 *Nature Communications* **9** URL https://doi.org/10.1038/s41467-018-04690-y

[61] Li Y, Peng Y G, Han L, Miri M A, Li W, Xiao M, Zhu X F, Zhao J, Alù A, Fan S and Qiu C W 2019 *Science* **364** 170–173 ISSN 0036-8075 (*Preprint* https://science.sciencemag.org/content/364/6436/170.full.pdf) URL https://science.sciencemag.org/content/364/6436/170

[62] Zhang F, Feng Y, Chen X, Ge L and Wan W 2020 *Phys. Rev. Lett.* **124**(5) 053901 URL https://link.aps.org/doi/10.1103/PhysRevLett.124.053901

[63] Wilkey A, Vemuri G and Joglekar Y N 2020 *Active Photonic Platforms XII* ed Subramania G S and Foteinopoulou S (SPIE) URL https://doi.org/10.1117/12.2566926

[64] Longhi S and Pinotti E 2019 *EPL (Europhysics Letters)* **125** 10006 URL https://doi.org/10.1209/0295-5075/125/10006

[65] Arwas G, Gadasi S, Gershenzon I, Friesem A, Davidson N and Raz O 2021 Anyonic parity-time symmetric laser (*Preprint* 2103.15359)

[66] Rivero J D H and Ge L 2020 *Phys. Rev. Lett.* **125**(8) 083902 URL https://link.aps.org/doi/10.1103/PhysRevLett.125.083902

[67] Purkayastha A, Kulkarni M and Joglekar Y N 2020 *Phys. Rev. Research* **2**(4) 043075 URL https://link.aps.org/doi/10.1103/PhysRevResearch.2.043075

[68] Joglekar Y N, Onanga F A and Harter A K 2018 *Phys. Rev. A* **97**(1) 012128 URL https://link.aps.org/doi/10.1103/PhysRevA.97.012128

[69] Joglekar Y N, Marathe R, Durganandini P and Pathak R K 2014 *Physical Review A* **90** 040101 URL https://doi.org/10.1103/physreva.90.040101

[70] Lee T E and Joglekar Y N 2015 *Phys. Rev. A* **92**(4) 042103 URL https://link.aps.org/doi/10.1103/PhysRevA.92.042103

[71] Harter A K and Joglekar Y N 2020 *Progress of Theoretical and Experimental Physics* **2020** URL https://doi.org/10.1093/ptep/ptaa181

# $\mathcal{PT}$ symmetry, pattern formation, and finite-density QCD

**Moses A. Schindler[1], Stella T. Schindler[2], and Michael C. Ogilvie[1]**

[1] Department of Physics, Washington University, St. Louis, MO 63130
[2] Center for Theoretical Physics, Massachusetts Institute of Technology, Cambridge, MA 02139

E-mail: stellas@mit.edu

**Abstract.** A longstanding issue in the study of quantum chromodynamics (QCD) is its behavior at nonzero baryon density, which has implications for many areas of physics. The path integral has a complex integrand when the quark chemical potential is nonzero and therefore has a sign problem, but it also has a generalized $\mathcal{PT}$ symmetry. We review some new approaches to $\mathcal{PT}$-symmetric field theories, including both analytical techniques and methods for lattice simulation. We show that $\mathcal{PT}$-symmetric field theories with more than one field generally have a much richer phase structure than their Hermitian counterparts, including stable phases with patterning behavior. The case of a $\mathcal{PT}$-symmetric extension of a $\phi^4$ model is explained in detail. The relevance of these results to finite density QCD is explained, and we show that a simple model of finite density QCD exhibits a patterned phase in its critical region.

## 1. Introduction

Determining the phase diagram of quantum chromodynamics (QCD), the theory of the strong force, is an important goal with broad implications for nuclear and particle physics, astrophysics, and cosmology. The QCD phase diagram is often described as a function of the quark chemical potential $\mu$ and the temperature $T$. Decades of work have produced a sophisticated picture of the behavior of QCD at nonzero temperature and zero chemical potential. When $\mu \neq 0$, we are much less certain of the behavior of QCD. Lattice simulations with $\mu \neq 0$ run into the sign problem: the Euclidean path integral has a complex integrand and does not have a probabilistic interpretation. Such sign problems may be non-deterministic non-polynomial (NP) hard in the general case [1]. Although much effort has been directed towards the development of algorithms to circumvent the sign problem [2, 3], we do not yet have a clear picture of QCD phase structure at finite density from lattice simulations. Astrophysical evidence relevant to QCD phase structure is equivocal at best, and exploration of the most important regions of the $\mu$-$T$ plane by direct experiments is still in its early stages.

Figure 1 shows a possible phase diagram for QCD in the $\mu$-$T$ plane. In the low-$T$, low-$\mu$ region, quarks and gluons are bound in hadrons. At very high $\mu$ and low $T$, there are strong arguments based on asymptotic freedom that QCD must be a color superconductor [4]. However, we do not know where in the $\mu$-$T$ plane that behavior sets in. At high $T$ quarks and gluons are predicted to appear as a hot, dense plasma. There is good evidence from a variety of models that there is a first-order phase transition line emerging from the $T = 0$ axis and terminating in a second-order critical point [5, 6]. A number of experimental programs, including BES at RHIC [7] and CBM at FAIR [8], are specifically targeting this critical behavior.

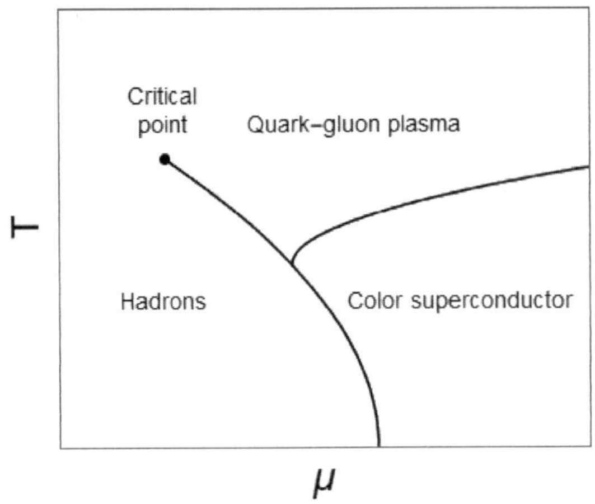


**Figure 1.** A possible phase diagram of QCD in the $\mu$-$T$ plane. In the bottom left corner of the phase diagram, quarks and gluons are bound inside hadrons. In extreme conditions like compact stars and the early universe, it is hypothesized that quarks and gluons can become unbound and form exotic phases of matter. This paper focuses on the transition between hadrons and quark-gluon plasma, which is hypothesized to take the form of a first-order critical line that terminates in a second-order critical endpoint.

In this paper we look at finite-density QCD through the lens of non-Hermitian and $\mathcal{PT}$ symmetric quantum theory. Although QCD at nonzero density and temperature has a path integral formulation with a complex integrand, the path integral is invariant under the combined action $\mathcal{CK}$ of charge conjugation $\mathcal{C}$ and complex conjugation $\mathcal{K}$, a form of $\mathcal{PT}$ symmetry [9]. We have developed new algorithmic and analytic techniques for studying $\mathcal{PT}$-symmetric quantum field theories in order to better understand QCD. A key development has been the design of a new algorithm that allows for lattice simulation of $\mathcal{PT}$-symmetric field theories [10, 11]. This algorithm recasts some models with complex actions into equivalent models with real actions. This has allowed us to compare lattice simulations with analytical results for models in the Z(2) symmetry class. In the process of this work, we have also learned about the relation of $\mathcal{PT}$-symmetric field theories with complex actions to equivalent real representations, some of which are local and some of which are nonlocal. The latter are closely connected to the relationship between $\mathcal{PT}$-symmetric and higher-derivative models.

The biggest surprise of our work has been the connection between $\mathcal{PT}$-symmetric field theories and pattern formation. Pattern formation is known to occur in many systems at a wide range of length and energy scales in physics [12]. Patterns often arise from competition between attractive and repulsive forces. Many examples are found in condensed matter physics; an exotic example is nuclear pasta, which has been predicted to arise in the inner crust of a neutron star. Here, competition between the Coulomb and nuclear forces leads protons and neutrons to cluster together in patterns which have been whimsically associated with pasta (lasagna, spaghetti, gnocchi, etc.) [13, 14, 15].

$\mathcal{PT}$-symmetric field theories provide natural realizations for some experimentally observed phenomena which are impossible in normal field theories. Conventional field theories obey spectral positivity. For a Hermitian scalar field $\phi(x)$, this implies that its two-point function can be written as a sum over decaying exponentials

$$\langle\phi(x)\phi(y)\rangle = \int_0^\infty d\rho(m^2) \int \frac{d^dk}{(2\pi)^d} \frac{1}{k^2+m^2} e^{ik\cdot x} \tag{1}$$

where $\rho(m^2) \geq 0$. This immediately rules out pattern formation, where order parameters exhibit periodic behavior. Spectral positivity also rules out propagators with sinusoidally modulated exponential decay. It is easy to see that such behavior is natural in $\mathcal{PT}$-symmetric field theories when there are complex-conjugate eigenvalues of the Hamiltonian. As we will show in Sec. 3, there is a close relation between modulated exponential decay, pattern formation and second-order critical points in some $\mathcal{PT}$-symmetric field theories. Furthermore, there is a close connection between the balance of attractive and repulsive forces associated with pattern formation and the imaginary couplings often found in $\mathcal{PT}$-symmetric models. Recall that in conventional field theories the force due to scalar particle exchange is always attractive; in contrast, an imaginary coupling in a $\mathcal{PT}$-symmetric field theory leads to a repulsive coupling. In the model we study in Sec. 3, an imaginary coupling produces a Lifshitz instabilty which can be understood as the cause of pattern formation. This also strengthens the general connection between $\mathcal{PT}$-symmetric models and higher-derivative theories.

In Sec. 2, we review some general results for $\mathcal{PT}$-symmetric quantum field theories, including a new algorithm for simulating a large class of $\mathcal{PT}$-symmetric scalar field theories. Section 3 explores the phase structure of $\mathcal{PT}$-symmetric scalar field theories. A $\mathcal{PT}$-symmetric extension of the conventional $\phi^4$ model is studied using both analytical methods and lattice simulation. Section 4 discusses the role of $\mathcal{PT}$ symmetry in QCD at nonzero density. We summarize this work and offer concluding remarks in Sec. 5.

## 2. $\mathcal{PT}$ symmetry and quantum field theory

In 1998, a groundbreaking paper by Bender and Boettcher showed that a complex extension of the harmonic oscillator, $\mathcal{H} = -\partial_x^2 + x^2(ix)^\epsilon$, has a countably infinite spectrum of real and positive eigenvalues for $\epsilon \geq 0$ [16]. Bender and Boettcher observed that the invariance of this class of models under $\mathcal{PT}$ symmetry, where $\mathcal{P}$ is parity and $\mathcal{T}$ is time reversal, is an important symmetry of $\mathcal{H}$. We emphasize an important notational convention: in the context of non-Hermitian physics the letters $\mathcal{P}$ and $\mathcal{T}$ refer generically to any given pair of linear and antilinear operators; they need not represent parity inversion and time reversal operators specifically. Every eigenvalue of a $\mathcal{PT}$-symmetric operator must be either real or part of a complex-conjugate eigenvalue pair. It has become conventional to say that an operator with entirely real eigenvalues has an unbroken $\mathcal{PT}$ symmetry; otherwise the symmetry is said to be broken. This terminology is statement about the reality of the full spectrum of excitations in a quantum-mechanical system, and is not directly related to spontaneous symmetry breaking in field theories. If one or more tuples of eigenvalues are coalesced at the same value, the operator is said to be at an exceptional point [17]. $\mathcal{PT}$-unbroken Hamiltonians exhibit many of the properties of Hermitian systems: unitarity, a positive norm under a $(\mathcal{CPT})$ inner product [17], eigenfunctions that possess analogs of interlacing zeros [18, 19], orthogonality, and completeness [20, 21]. These behaviors break down in a characteristic manner when moving a parameter like $\epsilon$ in $\mathcal{H}$ past an exceptional point and into the broken symmetry regime. Non-Hermitian systems without $\mathcal{PT}$ symmetry do not in general exhibit any of these properties. For further details, see Ref. [22].

The first experimental realizations of systems with $\mathcal{PT}$ symmetry were in optical waveguides with gain and loss [23, 24, 25]. The paraxial wave equation of optics is isomorphic to the Schrödinger equation, albeit with the electric field $E$ taking the place of the quantum-mechanical wavefunction $\psi$, refractive index $n(x)$ instead of potential $V(x)$, and propagation direction $z$ replacing time $t$. We can understand the appearance of an imaginary, or non-Hermitian, term in a Hamiltonian as representing the exchange of energy between a system and its environment, with the direction of energy flow indicated by the term's sign. In the $\mathcal{PT}$-symmetric Schrödinger equation, the potential term must satisfy Im $V(x) = -$ Im $V^*(-x)$, which means the system loses precisely as much energy to its environment as it gains back. Researchers have studied $\mathcal{PT}$ symmetry in many areas of physics; see, e.g., the reviews [26, 27, 28].

$\mathcal{PT}$-symmetric quantum field theories can have many analogous properties to conventional quantum field theories [22]. For example, the spectrum of a $\mathcal{PT}$-QFT may be real and bounded below. It is possible to construct an inner product under which a $\mathcal{PT}$-quantum field theory exhibits unitary time evolution [29]. This has implications for important physics models; for example, when one analyzes the Lee Model in a $\mathcal{PT}$-symmetric framework, there is no ghost; under a correctly defined inner product, all states have positive norm [30].

The path integral approach to quantum mechanics and quantum field theory is a powerful tool, but the development of $\mathcal{PT}$ symmetric quantum theories has largely focused on the canonical approach emphasizing Hilbert space structure. We have developed a Euclidean path integral technique which recasts a broad class of $\mathcal{PT}$-symmetric scalar field theories with complex actions into real forms. If this real form of the functional integral has a positive integrand, then the sign problem is removed and the path integral can be simulated with standard lattice methods.

As an example, let us consider the general case of a single $\mathcal{PT}$-symmetric scalar field $\chi(x)$ with action

$$S(\chi) = \sum_x \left[\frac{1}{2}(\partial_\mu \chi(x))^2 + V(\chi(x)) - ih(x)\chi(x)\right], \tag{2}$$

where $V(\chi) = V^*(-\chi)$. In the lattice field theory, we take the spatial derivative term to be the standard finite-difference expression. The external field $h(x)$ is dependent on Euclidean spacetime location, and is used to deduce correlation functions for $\chi$ in the new representation. We now rewrite the kinetic and potential terms as Fourier transforms. We write the single-site kinetic term as:

$$\exp\left[\frac{1}{2}\left(\partial_\mu \chi\right)^2\right] = \int d\pi_\mu(x) \exp\left[\frac{1}{2}\pi_\mu(x)^2 + i\pi_\mu(x)\partial_\mu \chi_x\right]. \tag{3}$$

The weight term $w[\chi(x)] \equiv \exp\left[-V(\chi(x))\right]$ associated with the potential $V$ can be written as the transform of a real dual weight $\tilde{w}[\tilde{\chi}(x)]$. If $\tilde{w} > 0$, then we can define the real dual potential $\tilde{V}(\tilde{\chi}) \equiv -\log[\tilde{w}(\tilde{\chi})]$. This condition of dual weight positivity is equivalent by Bochner's theorem to the condition that the weight $w[\chi(x)]$ is positive definite. After a lattice integration by parts, it is straightforward to integrate out $\chi$ to obtain the action

$$Z = \int \prod_x d\pi_\mu(x) \exp\left\{-\sum_x \left[\frac{1}{2}\pi_\mu^2(x) + \tilde{V}(\partial \cdot \pi(x) - h(x))\right]\right\}, \tag{4}$$

which is positive everywhere. In such a representation, standard lattice simulation methods may be used. This representation is local and may be used in any dimension; algorithmic implementation is easy. The extension to multiple scalar fields is trivial. This transformation can be understood as a kind of duality transform appropriate for real-valued fields. The dual positive weight condition puts restrictions on potentials $V$ that can be simulated using this method. However, $\tilde{V}$ need only be real for $V$ to be $\mathcal{PT}$ symmetric, so this method may be used to create an infinite number of $\mathcal{PT}$-symmetric field theories in any space-time dimensionality.

This technique is quite similar to the method used in the analysis of the wrong-sign $x^4$ model, where the functional integration contour is complex, lying in a nontrivial Stokes wedge [31]. The method we have presented applies to scalar field theories when the functional integration contour is along the real axis. It seems likely that similar techniques can be used to reduce some models defined on complex contours to a real form.

## 3. Phase structure in $\mathcal{PT}$-symmetric field theories

In this section we review recently developed techniques to determine the phase diagram of a $\mathcal{PT}$-symmetric scalar quantum field theory and the application of those techniques to a $\mathcal{PT}$-symmetric extension of Hermitian $\phi^4$ theory, providing some additional details on the phase structure [11]. For standard scalar field theories, there is a well-defined procedure for studying phase structure within the path integral formalism using perturbation theory. Writing the Lagrangian density $\mathcal{L}$ in terms of a set of real fields $\phi^a(x)$ where $a$ runs from 1 to $N$, we have:

$$\mathcal{L}(\phi^a) = \frac{1}{2}(\partial_\mu \phi^a)^2 + V(\phi^a). \tag{5}$$

If the potential $V$ is bounded from below, there is always at least one spacetime-independent solution $\phi_0^a(x)$ which achieves the global minimum of the potential, and is thus a candidate for building a perturbative solution around. The mass matrix is real and symmetric, and has non-negative eigenvalues, with zero eigenvalues associated with Goldstone bosons.

This can change profoundly in a $\mathcal{PT}$-symmetric scalar field theory with a complex action. Consider such a theory in $d$ dimensions. We suppose there is a set of fields $\phi^a$ that transform trivially under $\mathcal{P}$ and $\mathcal{T}$ transformations and a set of fields $\chi^b$ that transform nontrivially, such that the action is invariant under the combined action of the operators $\mathcal{PT}$. We find a perturbative solution $(\phi_0^a, \chi_0^b)$ by minimizing $V$ over the class of spacetime-independent solutions. We assume that $\mathcal{PT}$ symmetry is maintained, which implies that $\phi_0^a$ is real, $\chi_0^b$ is imaginary and $V(\phi_0, \chi_0)$ is real. The mass matrix $\mathcal{M}$ associated with this solution is given in block form by

$$\mathcal{M} = \begin{pmatrix} \dfrac{\partial^2 V}{(\partial \phi^a)^2} & \dfrac{\partial^2 V}{\partial \phi^a \partial \chi^b} \\ \dfrac{\partial^2 V}{\partial \phi^a \partial \chi^b} & \dfrac{\partial^2 V}{(\partial \chi^b)^2} \end{pmatrix} \tag{6}$$

evaluated at $(\phi_0^a, \chi_0^b)$. This mass matrix is not necessarily Hermitian but is $\mathcal{PT}$-symmetric:

$$\mathcal{M} = \Sigma \mathcal{M}^* \Sigma, \tag{7}$$

where $\Sigma$ is a diagonal matrix with entries $+1$ associated with the $\phi^a$ fields and $-1$ associated with the $\chi^b$ fields. The characteristic equations for $\mathcal{M}$ and $\mathcal{M}^*$ are the same, so they have the same eigenvalues. As a consequence, every eigenvalue of $\mathcal{M}$ must be either real or the complex-conjugate of another eigenvalue.

The zeros of $\det\left(q^2 + \mathcal{M}\right)$ are the poles in momentum space of the matrix propagator for the $\phi$ and $\chi$ fields. For standard quantum field theories, stable perturbative vacua have propagator poles that are real and negative as a function of $q^2$, so $q^2 > 0$ implies $\det\left(q^2 + \mathcal{M}\right) > 0$ in this case. Poles with $q^2 < 0$ lead to exponentially decaying propagators, consistent with spectral positivity. In $\mathcal{PT}$-symmetric models, complex-conjugate poles lead to sinusoidally-modulated exponential decay, which is inconsistent with spectral positivity. This behavior is familiar from $\mathcal{PT}$ symmetry breaking in quantum mechanics, where by definition energy eigenvalues become complex. Such behavior has been known for some time in condensed matter physics, where the boundary between two parameter regions where the propagators change behaviors is known as a disorder line [32]. We use the term disorder line to avoid confusion in models that have first-order critical lines and second-order critical endpoints.

The one-loop effective potential of our generic $\mathcal{PT}$-symmetric scalar field theory at a classical homogeneous solution is given by

$$V_{\text{eff}}(\phi_0, \chi_0) = V(\phi_0, \chi_0) + \frac{1}{2}\int \frac{d^d q}{(2\pi)^d} \log\det\left(q^2 + \mathcal{M}\right). \tag{8}$$

| Region | $\det(\mathcal{M})$ | Zeros of $\det(q^2+\mathcal{M})$ | Behavior of the propagator |
|---|---|---|---|
| Normal | Positive | All zeros are negative | Exponential decay |
| $\mathcal{PT}$ broken | Positive | One or more pairs of zeros are complex conjugates | Modulated exponential decay |
| Patterned | Positive | Even # of positive zeros | Stable to $q^2=0$ fluctuations, but unstable to $q^2>0$ fluctuations |
| Unstable | Positive | Odd # of positive zeros | Instability to $q^2=0$ fluctuations |
| Unstable | Negative | — | Instability to $q^2=0$ fluctuations |

**Table 1.** The phase diagram of a non-Hermitian $\mathcal{PT}$-symmetric quantum field theory, described in terms of the behavior of its matrix propagator. Note that two new behaviors can occur in $\mathcal{PT}$-QFTs that are generally not present in conventional Hermitian QFTs. First, the propagator may possess complex zeros, and due to $\mathcal{PT}$ symmetry these zeros must come in complex-conjugate pairs. This leads to a propagator with exponential decay that is modulated by a sinusoidal term, and the propagator may even be partly negative. Nonetheless this complex region exhibits stable behavior and field configurations with a typical appearance. There is also a region with tachyonic modes that can support stable ground-state field configurations, although these field configurations are not homogeneous.

If $\det\left(q^2+\mathcal{M}\right)$ becomes negative for real $q^2$, the one-loop term develops an imaginary part, indicating an instability of that particular solution. If the eigenvalues of the mass matrix are all either positive or in a complex-conjugate pair, we have $\det\left(q^2+\mathcal{M}\right)>0$, and the classical solution $(\phi_0,\chi_0)$ is stable against small fluctuations. Such solutions may be stable or metastable, but they are not unstable. If a homogeneous solution has a mass matrix with negative roots, then $V_{\text{eff}}(\phi_0,\chi_0)$ will develop an imaginary part, indicating instability. The most interesting case occurs when $\det\mathcal{M}>0$ but $\det\left(q^2+\mathcal{M}\right)$ has an even number of positive zeros. The condition $\det\mathcal{M}>0$ tells us that the solution is stable against $q^2=0$ fluctuations, but the propagator is unstable to fluctuations for some values of $q^2>0$. When a homogeneous solution is not stable, we might consequently assume the full theory is unstable. But there is a second possibility: simply, that the stable solution is not homogeneous. This is what happens when there are an even number of positive zeros: the ground state field configurations are patterned. In normal field theories, such phases are not associated with stable vacua because they do not represent global minima of the vacuum energy density. In $\mathcal{PT}$-symmetric theories, a homogeneous solution may be the global minimum across all homogeneous solutions but be unstable against $q^2\neq 0$ fluctuations. We summarize our conjectured phase structure for $\mathcal{PT}$-symmetric field theories in Table 1.

### *3.1. A $\mathcal{PT}$-symmetric model with Z(2) symmetry*

We now consider a $\mathcal{PT}$-symmetric extension of a conventional $\phi^4$ model which is amenable to both analysis and lattice simulation:

$$S(\phi,\chi)=\sum_x \frac{1}{2}(\nabla_\mu\phi)^2+\frac{1}{2}(\nabla_\mu\chi)^2+V(\phi,\chi), \tag{9}$$

where we set

$$V(\phi,\chi)=\frac{1}{2}m_\chi^2\chi^2-ig\phi\chi+U(\phi)+h\phi. \tag{10}$$

The potential $U$ for $\phi$ is the double-well potential $U(\phi) = \lambda(\phi^2 - v^2)^2$. Equation (9) represents a Hermitian scalar field $\phi(x)$ coupled to a $\mathcal{PT}$-symmetric scalar field $\chi(x)$ by the imaginary strength $ig$. The model is invariant not only under the Z(2) symmetry $(\phi, \chi) \to (-\phi, -\chi)$, but also under the generalized $\mathcal{PT}$ symmetry of complex conjugation combined with $(\phi, \chi) \to (\phi, -\chi)$. $\mathcal{PT}$-symmetric models similar to this one can easily be constructed for other universality classes besides Z(2). For example, an O($N$)-symmetric model with an $N$-dimensional Hermitian field $\vec{\phi}$ can be extended to a $\mathcal{PT}$-symmetric model O($N$) model with an additional field $\vec{\chi}$ and an imaginary term in the Lagrangian of the form $-ig\vec{\phi} \cdot \vec{\chi}$.

Because $\chi$ enters quadratically in the action $S$, it can easily be integrated out of the functional integral, giving a nonlocal effective action of the form

$$S_{\text{eff}} = \sum_x \left[\frac{1}{2}(\partial_\mu \phi(x))^2 + \lambda(\phi^2 - v^2)^2 + h\phi\right] + \frac{g^2}{2} \sum_{x,y} \phi(x)\Delta(x-y)\phi(y),$$

where

$$\Delta(x-y) = \int \frac{d^d q}{(2\pi)^d} \frac{1}{q^2 + m_\chi^2} e^{iq\cdot(x-y)}. \tag{11}$$

This model has been extensively studied in the case $m_\chi = 0$, where it is known to give rise to pattern-forming regions; see, *e.g.*, [33] and references therein. The $m_\chi = 0$ limit is sometimes described in the condensed matter literature as Coulomb frustrated because the extra interaction acts against the symmetry-breaking behavior of the $\phi^4$ model [33, 34, 35].

The value of the order parameter at tree level, $\phi_0$, is determined by minimizing the potential $V$ or equivalently by minimizing the effective potential associated with $S_{\text{eff}}$:

$$V_{\text{eff}}(\phi_0) = \lambda(\phi_0^2 - v^2)^2 + g^2\phi_0^2/2m_\chi^2 - h\phi_0. \tag{12}$$

The effect of $\chi$ on $\phi_0$ for $h = 0$ is to restore the symmetric value $\phi_0 = 0$ at sufficiently large values of $g$. However, $\phi_0$ tells us the value of the zero-momentum component of $\phi$. Pattern formation is associated with Fourier modes $\tilde{\phi}(q)$ with nonzero $q$, which requires additional calculation. One approach to understanding pattern formation is to expand $S_{\text{eff}}$ in a derivative expansion. The last, nonlocal term in $S_{\text{eff}}$ generates an infinite series of local higher-order derivative terms

$$\int d^4x \frac{g^2}{2m_\chi^2} \phi(x) \sum_{n=0}^{\infty} \left(\frac{-\nabla^2}{m_\chi^2}\right)^n \phi(x). \tag{13}$$

The $n = 0$ term in this expansion is the last term in $V_{\text{eff}}$. The $n = 1$ represents a correction to the kinetic term, and higher $n$ terms give higher-derivative terms in $S_{\text{eff}}$. Crucially, the $n = 1$ term of the series is negative, indicating that the quadratic derivative term of $S_{\text{eff}}$ becomes negative for sufficiently large $g$. When the quadratic kinetic term is sufficiently negative, the homogeneous phase is unstable to perturbations with nonzero wave number. Thus the occurrence of a pattern-forming region is a manifestation of a Lifshitz instability [36].

A global picture of the phase structure follows from the inverse $\phi$ propagator obtained from $S_{\text{eff}}$ at tree level:

$$G^{-1}(q^2) = q^2 + U''(\phi_0) + \frac{g^2}{q^2 + m_\chi^2} \tag{14}$$

where $U''(\phi_0) = -4\lambda v^2 + 12\lambda\phi_0^2$. The allowed phases of the model are determined using $\phi_0$ and $G(q)$. The poles of the propagator $G$ are zeros of $(q^2 + m_\chi^2)(q^2 + U) + g^2 = 0$. This quadratic in $q^2$ has real coefficients, so its roots are either both real or form a complex-conjugate pair. Specifically, we can write the poles of $G(q^2)$ as $q^2 = r_\pm$ with

$$r_\pm = -\frac{1}{2}\left(m_\chi^2 - 4\lambda v^2 + 12\lambda\phi_0^2\right) \pm \frac{1}{2}\sqrt{(m_\chi^2 + 4\lambda v^2 - 12\lambda\phi_0^2)^2 - 4g^2}. \tag{15}$$

If $r_- < 0$ and $r_+ > 0$, the position-space propagator grows exponentially and the system is unstable. If $r_+, r_- < 0$ then the position-space propagator decays exponentially and our system behaves like a normal, stable quantum field theory. If $r_+ = r_-^*$ then the position-space propagator decays exponentially with sinusoidal modulation but the homogeneous vacuum is also stable in this region of the phase diagram. The condition $r_+, r_- > 0$ leads to pattern formation.

An equivalent approach to determining the phase structure is to start from $S$ rather than $S_{\text{eff}}$ and find the static solution $(\phi_0, \chi_0)$ which minimizes $V(\phi, \chi)$. Unless the underlying $\mathcal{PT}$ symmetry of $S$ is broken, $\phi_0$ will be real and $\chi_0$ will be purely imaginary. Linearizing the propagator around the static solution, we find the inverse propagator for the $(\phi, \chi)$ set of fields is $q^2 + \mathcal{M}$, where $\mathcal{M}$ is the $2 \times 2$ mass matrix

$$\mathcal{M} = \begin{pmatrix} \dfrac{\partial^2 V}{\partial \phi^2} & \dfrac{\partial^2 V}{\partial \phi \partial \chi} \\ \dfrac{\partial^2 V}{\partial \phi \partial \chi} & \dfrac{\partial^2 V}{\partial \chi^2} \end{pmatrix}, \tag{16}$$

which is

$$\mathcal{M} = \begin{pmatrix} U''(\phi_0) & ig \\ ig & m_\chi^2 \end{pmatrix}. \tag{17}$$

The mass matrix $\mathcal{M}$ is not Hermitian, but it satisfies a $\mathcal{PT}$ symmetry condition

$$M = \sigma_3 M^* \sigma_3. \tag{18}$$

As in our general analysis, this condition implies that the eigenvalues of $\mathcal{M}^*$ must be the same as those of $\mathcal{M}$, and thus they are either both real or form a complex pair. The zeros of the inverse matrix propagator can be obtained as the zeros of the characteristic equation

$$\det(q^2 + \mathcal{M}) = (q^2)^2 + q^2 \operatorname{Tr}(\mathcal{M}) + \det(\mathcal{M}) = 0 \tag{19}$$

The coefficients $\operatorname{Tr}(\mathcal{M})$ and $\det(\mathcal{M})$ are real, implying that the roots are either both real or form a complex-conjugate pair. The zeros of the characteristic equation are the propagator poles and so the two methods give the same results.

In Figure 2, we plot the regions for the four distinct phases of the model as a function of $\langle \phi \rangle$ for the parameter set $m^2 = 1/2$, $\lambda = 1/10$ and $v = 3$. The graph on the left shows the phase diagram in the $\phi$-$g$ plane, while the graph on the right shows the phase structure in the $h$-$g$ plane. The different regions are classified by the nature of the poles of the $\phi$ propagator in the $q^2$ complex plane. We denote the region of parameter space where both poles are real and negative as "Normal" (orange in Fig. 2). This leads to exponential decay of the $\phi$ propagator, as it does in conventional field theories. In the region labeled Complex (blue), the poles as a function of $q^2$ in the $\phi$ propagator are complex conjugates. This region is similar to the so-called broken $\mathcal{PT}$ region of $\mathcal{PT}$-symmetric quantum mechanical models. The $\phi$ propagator in this region also decays exponentially, but with sinusoidal modulation. This behavior violates spectral positivity and does not occur in Hermitian models. This behavior in $\mathcal{PT}$-symmetric field theories is the field theory analog of complex-conjugate energy eigenvalues in $\mathcal{PT}$-symmetric quantum mechanics. The boundary between the Normal and Complex regions is called a disorder line. The region labeled Patterns (green) is the region where both poles are real and positive; it is in this region where persistent patterns occur. In the Unstable region (red), both poles are real with one positive and one negative. This region is not thermodynamically stable. It is inaccessible as an equilibrium state in the canonical ensemble where $h$ is a free parameter.

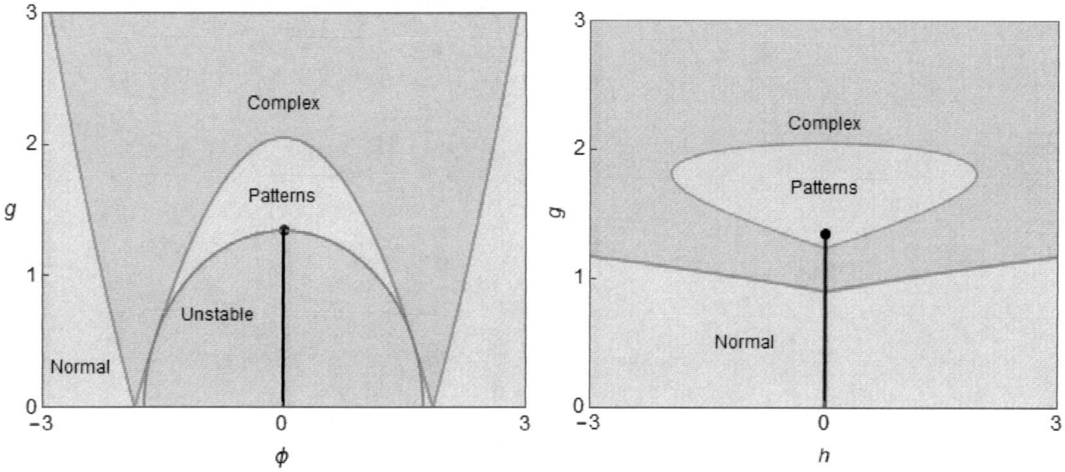


**Figure 2.** The phase diagram of eq. (9), plotted in two different parameter spaces. In the $\langle\phi\rangle - g$ plane, the critical line runs up the $g$ axis until it terminates in a critical endpoint where it intersects the disorder line that marks the boundary between the patterned and the complex regions. When we make a change of variables to the $g - h$ plane there is no unstable region, and the critical point lies within the patterned region.

The phase diagram in the $h$-$g$ plane shows a cut at $h = 0$, as in the familiar case of a ferromagnetic Ising model. This is a first-order phase transition, across which $\phi_0$ jumps. The two sides of the cut may be in the Normal, Complex or Pattern regions, as shown in the figure. The unstable phase does not appear in the $h$-$g$ phase diagram, and there are also metastable states from the other three phases which do not appear. These unstable and metastable states have moved through the cut into an analytic continuation from the stable regions. The critical line terminates at a second-order critical point where the $\phi$ field becomes massless. In the $\phi$-$g$ plane, this occurs at the boundary between the unstable and patterned regions, but in the $h$-$g$ plane, the metastable pattterned states are lost, and the critical end point appears in the middle of the patterning phase.

The location of the boundary between the patterned and unstable regions in the $\phi$-$g$ plane is controlled by the parameter $m_\chi^2$. If we take $m_\chi^2 \to \infty$, then the effects of the $\chi$ field on $\phi$ disappear, and both the complex and patterned phases disappear. In this limit, the phase diagram is that of the usual $\phi^4$ model. On the other hand, the limit $m_\chi^2 \to 0$ turns the Yukawa (screened Coulomb) potential induced by $\chi$ into the long-range Coulomb potential. This is a smooth limit in the $\phi$-$g$ plane and in the phase diagram in Fig. 3. The unstable region disappears in this limit. Similarly, the pattern-forming region disappears in the limit $m_\chi^2 \to \infty$.

The effective action $S_{\text{eff}}$ is a function of $g^2$, and can be continued to $g^2 < 0$. This corresponds to the continuation $g \to ig$ in $S$, in which case the action $S$ becomes real, and neither pattern formation nor complex $q^2$ poles can occur. The small areas of normal behavior seen in Fig. 2 in between the complex and unstable phases are connected to the larger normal region when the phase diagram is plotted as a function of $g^2$ and $g^2 < 0$ values are included as in Fig. 4. In the limit $m_\chi^2 \to 0$, the unstable region occurs only for $g^2 < 0$.

The behavior we have found is based on the analysis of $S_{\text{eff}}$ at tree-level, and is thus independent of dimensionality. As we know, fluctuations at one loop and beyond can strongly effect critical behavior in a dimension-dependent way. We know that in the simple $\phi^4$ model, fluctuations destroy spontaneous symmetry breaking below $d = 2$, the lower critical dimension

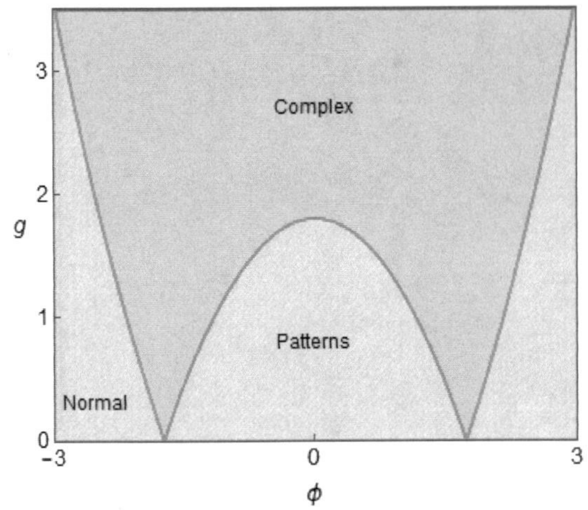


**Figure 3.** The phase diagram of eq. (9) in the $m_\chi^2 = 0$ limit, where all other parameters are the same as in Figure 2. Note that the unstable region has completely disappered for $g > 0$ in this limit.

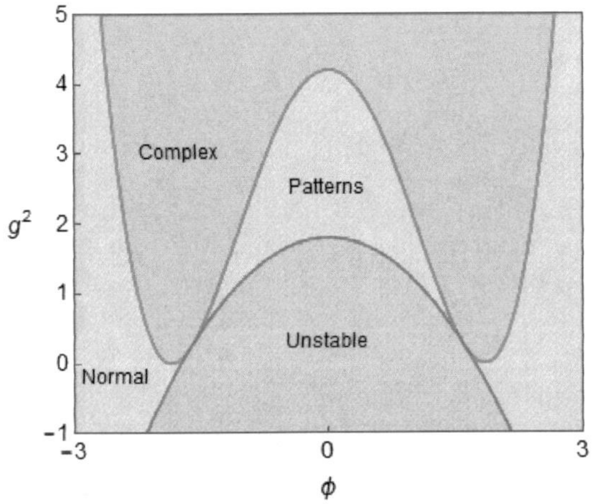


**Figure 4.** The phase diagram of eq. (9) in the $\phi$-$g^2$ plane, where all parameters are the same as in Figure 2, but the phase diagram has been extended to $g^2 < 0$. For $g^2 < 0$ there are only normal and unstable phases, as expected for a Hermitian theory. Note that the small region of normal behavior sandwiched between the complex and unstable regions in Fig. 2 is connected to the larger normal region, when viewing the system in the extended $g^2$ plane.

for the model. Examination of the infrared behavior of the $\phi$ propagator in our model shows that it generally has infrared behavior no worse than $1/q^2$, as in the simple $\phi^4$ model. The only exception is the point where both masses are zero, which occurs at tree level when $\mathrm{Tr}(\mathcal{M}) = 0$ and $\det(\mathcal{M}) = 0$. This does not change the overall phase diagram for $d \geq 2$. A similar analysis applies in other models.

*3.2. Simulations of the Z(2)-symmetric model*

We can complement our analysis with lattice simulations. Using the method described in Sec. 2, we cast the action for $\mathcal{PT}$-extended $\phi^4$ theory in eq. (9) into an entirely real, positive form

$$\tilde{S} = \sum_x \left[ \frac{1}{2}(\nabla_\mu \phi)^2 + \frac{1}{2}\pi_\mu^2 + \frac{1}{2m_\chi^2}(\nabla \cdot \pi - g\phi)^2 + \lambda(\phi^2 - v^2)^2 + h\phi \right], \tag{20}$$

which can be simulated using the Metropolis algorithm.

We have performed an extensive set of simulations of the model in two dimensions using the Metropolis algorithm applied to the real action $\tilde{S}$. The parameters $m_\chi^2$, $\lambda$, and $v$ were fixed to the same values shown in the phase diagrams of the previous section: $m_\chi^2 = 0.5$, $\lambda = 0.1$ and $v = 3$. The simulations shown here were performed using a $64 \times 64$ lattice, typically with a hot start followed by 20,000 sweeps after a period of equilibration. Pattern formation has also been observed in three dimensions [10, 11], but the two-dimensional case is much easier to visualize, and our analytical results are independent of dimension.

In Fig. 5 we show configurational snapshots of the order parameter $\langle\phi\rangle$ taken at the end of these long runs. The color scale consists of four equally-spaced bins running from approximately -3 to 3, ranging from dark to light. Fifteen of these figures exhibit patterning behavior, and one (top right) does not. This patterned region is in reasonable agreement with the analytic predictions of Sec. 3.1, given the limitations of finite lattice size, finite lattice spacing, and tree-level perturbation theory. In particular, the simulation results confirm the overall phase diagram determined analytically.

The patterned field configurations take the form of curved stripes and dots of large positive $\langle\phi\rangle$ value floating in a sea of large negative $\langle\phi\rangle$ value, as in other pattern-forming systems. Typically, there is a small transition region between the two most extreme values in each configuration. A single configuration may contain both dots and stripes of various lengths, with irregular orientation and placement relative to one another. As we increase $h$ and move rightward through Fig. 5, we notice that patterns tend to decrease in characteristic length scale. Twisted and connected stripes gradually shorten into individual strands, which shorten until becoming balls, and eventually these balls too disappear. On the other hand, as we increase $g$ and move downward in the figure, the characteristic width of patterns tends to become smaller until passing the limitations of what we can observe with our lattice size and spacing. As we vary $g$ and $h$, the observed pattern morphologies change smoothly, as does the average action $\langle S\rangle$.

In Fig. 6, we take the position-space Fourier transform of Fig. 5 and show its absolute value on a $64 \times 64$ momentum lattice with $q = 0$ at the center. We clip the maximum value of the Fourier transform at each site at 10 and have colors running from 0 (black) to 10 (white). We immediately notice that the ring-shaped configurations in Fig. 6 are associated with the patterned configurations in Fig. 5. The presence of a momentum-space ring confirms the explanation in Sec. 3.1 that $q^2 > 0$ modes are the common source of pattern formation here. In the sole homogeneous field configuration at the top right, the Fourier transform has a single large value at the origin, representing a nonzero expected value for $\phi$.

Finally, in Fig. 7 we plot a histogram of the expectation value of $\phi(x)$ alongside the configurations which produced the histograms. These configurations all have $h = 0$, with $g$ increasing downward. We recall that in the $g$-$h$ plane the critical line lies along $h = 0$, running through the normal and complex regions of parameter space before terminating in a critical endpoint inside the patterned region (see Fig. 2). At $g = 0.70$, the field configuration is homogeneous and the histogram shows a single peak centered around a large negative number. This is indicative of the broken symmetry we expect in a non-patterned region. When we move to $g = 0.90$, the field configuration is patterned, the expectation value of the field is zero, and its

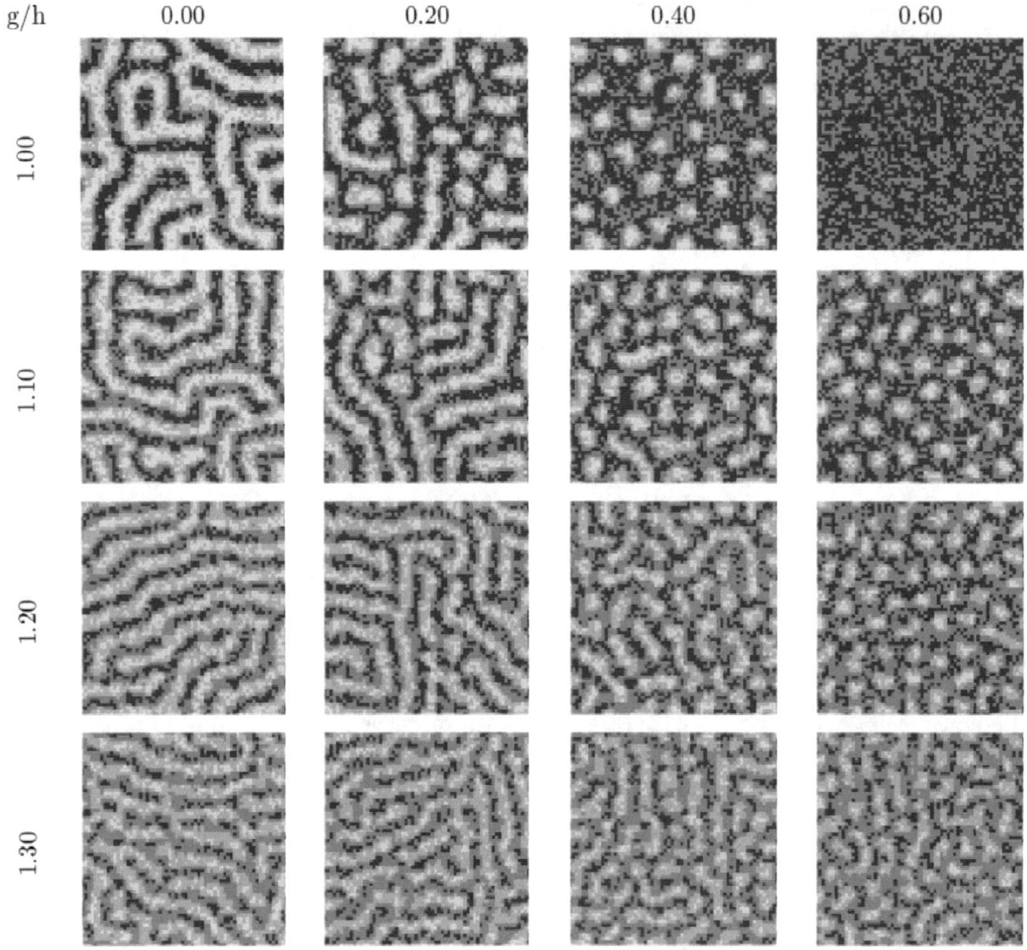


**Figure 5.** Snapshots of equilibrium field configurations $\langle\phi(x)\rangle$ from eq. (20) for a range of parameter values $(g,\, h)$ simulated on the lattice. Fifteen of these configurations exhibit patterned behavior. The remaining configuration, in the top left corner, does not exhibit patterns; any variation in color reflects expected small fluctuations. Each patterned configuration exhibits a distribution of shapes. Patterned configurations vary smoothly into one another as we adjust $g$ and $h$, as does the order parameter $\langle\phi_0\rangle$. As we increase $g$ patterns tend to become thinner and as we increase $h$ the shapes tend to become shorter, until we exit the patterned region. The Fourier transforms of these configurations are shown in Fig. 6. The patterned region is in reasonable agreement with the analytic calculations of Sec. 3.1, given the limitations of tree-level analytics, coupled with finite lattice size and spacing.

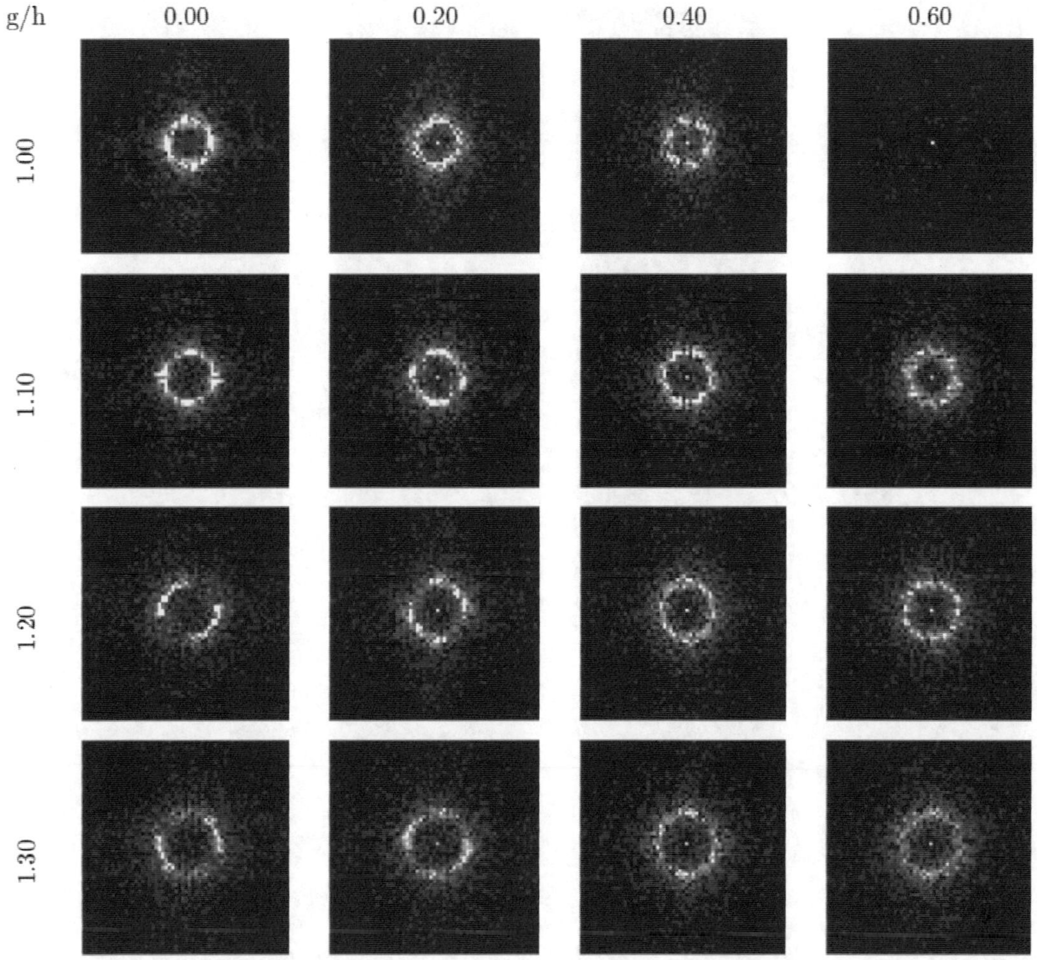


**Figure 6.** Absolute value $|\tilde{\phi}(k)|$ of the Fourier transforms of the position-space field configurations $\langle\phi(x)\rangle$ in Fig. 5. Patterned configurations in position space correspond to a ring-shaped configuration in momentum space. The sole non-patterned position-space configuration in Fig. 5 corresponds to a sharp peak at the origin in momentum space (top right). The presence of rings suggests the common underlying origin of patterning: propagator poles with $q^2 > 0$. In some cases a small number of modes on the ring are excited, which may occur due to finite size effects, lattice pinning, or locking into an atypical region of configuration space.

histogram is bimodal, with the field splitting up into high- and low-valued regions. As we move up the critical line, the two peaks slowly move towards the origin until coalescing into a single hump at the critical endpoint. At $g = 1.70$ the curve of the histogram is still fairly broad and we observe patterned field configurations even though the system no longer sits on the critical line. This is consistent with our picture of a critical endpoint in the middle of the patterned region in the $h$-$g$ plane.

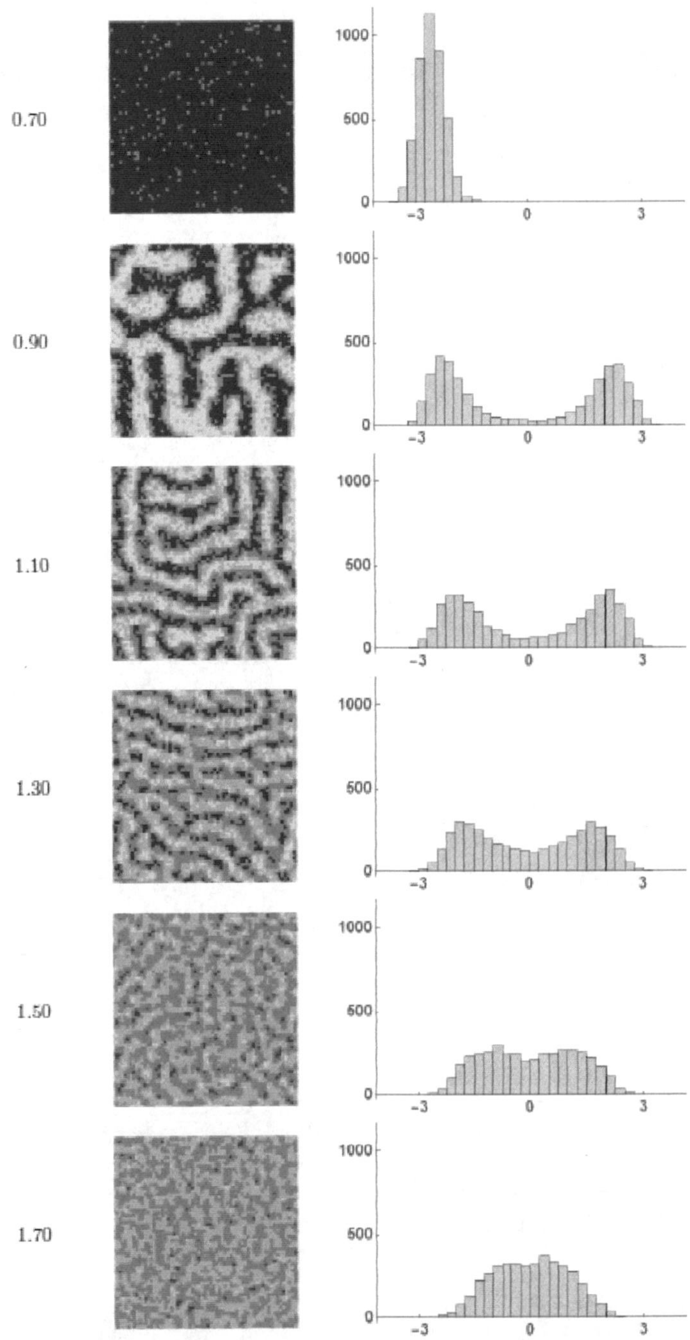


**Figure 7.** A characteristic set of $h = 0$ configurations for eq. (20) and histograms of the expectation value $|\langle\phi\rangle|$ at each of the $64^2$ values of $x$. The critical line of this system starts at $g = 0$ and runs up the $h = 0$ axis. The $g = 0.70$ configuration exhibits a homogeneous field configuration and asymmetric histogram, and thus lies outside the patterned region. As we move up the $g$ axis to 0.90, we pass into the patterned region and the histogram takes on a symmetric bimodal form which persists for a large range of $g$ values. Around $g = 1.50$ we pass the critical endpoint and into a region where the histogram is a symmetric single-peaked distribution.

### *3.3. Phase transition dynamics*

Because pattern formation is often seen in phase transition dynamics, it is interesting to consider the dynamics of our model as compared with similar behavior seen in phase transition dynamics of conventional $\phi^4$ models [36, 37, 38]. The order parameter $\phi$ is not conserved, so we can model its dynamics using $S_{\text{eff}}$ and model A dynamics, *i.e.*, with a Langevin equation

$$\frac{\partial\phi(x,t)}{\partial t} = -\Gamma\frac{\delta S_{\text{eff}}}{\delta\phi(x,t)} + \eta(x,t). \tag{21}$$

where $\Gamma$ is a decay constant and $\eta$ is a white noise term. If the white noise is normalized to $\langle\eta(x,t)\eta(y',t')\rangle = 2\Gamma\delta(x-x')\delta(t-t')$, the distribution of $\phi(x,t)$ will converge for long times to the distribution given by the path integral using $S_{\text{eff}}$. The difference between our model and the standard model A dynamics of a $\phi^4$ field theory, obtained by taking $g = 0$, is the nonlocal term in $S_{\text{eff}}$ induced by $\chi$. Linearizing the Langevin equation around a homogeneous solution $\phi_0$ and transforming to momentum space, the Langevin equation becomes

$$\frac{\partial\tilde{\phi}(q,t)}{\partial t} = -\Gamma\left(q^2 + U''(\phi_0) + \frac{g^2}{q^2+m_\chi^2}\right)\tilde{\phi}(q,t) + \tilde{\eta}(q,t) = -\Gamma G^{-1}(q)\tilde{\phi}(q,t) + \tilde{\eta}(q,t) \tag{22}$$

If the nonlocal term were not present, $\phi_0$ would be unstable when $U''(\phi) < 0$, and spinodal decomposition would occur. Comparing the dynamics when $g \neq 0$ to the purely local model when $g = 0$, we see that the large-$q$ behavior is identical, but $g \neq 0$ changes the small-$q$ behavior. In the region where pattern formation occurs in equilibrium, model A dynamics is that of spinodal decomposition for large $q$, but relaxational for small $q$.

We thus interpret the equilibrium patterning behavior of this model as a form of arrested spinodal decomposition. Starting from a homogeneous solution $\phi_0$ with $U''(\phi) < 0$ , the early-time Langevin evolution will produce the exponentially growing modes of spinodal decomposition for large $q$, but for small $q$ fluctuations are damped. Spinodal decomposition is arrested at a characteristic scale in momentum space, with $\phi_0$ stabilized by the nonlocal term. From this point of view, the chief dynamical difference between the patterned region and the unstable region is that in the patterned region, low $q$ fluctuations are suppressed but grow exponentially with $t$ in the unstable region.

The presence of a characteristic scale, obvious in the Fourier transforms of configuration snapshots, is reminiscent of model B dynamics for a conserved order parameter [36, 37, 38]. Model B dynamics is appropriate for simulations of Ising models when the total magnetization is kept constant, which is natural when the Ising model is interpreted as a binary allow lattice gas. The analog for a standard $\phi^4$ models is to hold the average value of $\phi$ fixed. Linearized model B dynamics of a standard $\phi^4$ model is described by

$$\frac{\partial\tilde{\phi}(q,t)}{\partial t} = -\Gamma q^2\left(q^2 + U''(\phi_0)\right)\tilde{\phi}(q,t) + \tilde{\eta}(q,t). \tag{23}$$

The extra factor of $q^2$ suppresses time evolution of the zero-momentum mode of $\tilde{\phi}(q,t)$, which is essentially the average value of $\phi$. In the early stages of spinodal decomposition with model A dynamics, it is the $q = 0$ mode which increases most quickly. On the other hand, in model B dynamics the most rapidly growing modes have $q^2 > 0$, because of the extra factor of $q^2$ in the linearized Langevin equation. This leads to a characteristic scale in momentum space for spinodal decompositions with model B dynamics. In both cases, the system will eventually leave the spinodal region $U''(\phi) < 0$ as phase separation completes. This does not happen in our $\mathcal{PT}$-symmetric model: when $g \neq 0$, large-scale (small-$q$) fluctuations relax quickly, but large homogeneous regions remain unstable to fluctuations of nonzero $q$. In particular, coarsening at arbitrarily large scales does not occur.

## 4. $\mathcal{PT}$ symmetry and QCD at nonzero density

Quantum Chromodynamics (QCD) at nonzero temperature $T$ and chemical potential $\mu$ is an important example of a quantum field theory with a generalized $\mathcal{PT}$ symmetry. Much of what we know about the phase structure of QCD and other gauge theories comes from Euclidean lattice simulations combined with a few key theoretical ideas. The symmetries and order parameters associated with quark confinement and chiral symmetry breaking are of particular interest as principal determinants of gauge theory phase structure. None of this changes when the chemical potential is nonzero, but there are additional complications. The presence of non-positive weights in the functional integral has proved to be a very difficult problem. We lack effective algorithms for lattice simulation in this case, and analytical methods must also be modified. The $\mathcal{CK}$ symmetry of finite-density QCD is a powerful tool. The sign problem is closely associated with the Euclidean formalism, and we will focus in this section on quark confinement and its relation to $\mathcal{PT}$ symmetry.

### *4.1. The sign problem and the origin of $\mathcal{CK}$ symmetry in finite density QCD*

The Euclidean Lagrangian for an SU($N$) gauge field with fermions in the fundamental representation is

$$\mathcal{L}_{\text{QCD}} = \bar{\psi}(i\not{D} - m)\psi + \frac{1}{2g^2}\text{Tr}(F_{\mu\nu}F^{\mu\nu}). \tag{24}$$

The gauge field $A_\mu(x)$ is an SU($N$) Lie algebra-valued field which may be decomposed as $A_\mu(x) = A_\mu^a(x)T^a$, where the $T^a$'s are generators of the group and the index $a$ runs from 1 to $N^2 - 1$. The physical case of QCD is three colors of quarks, or $N = 3$. The fermion field $\psi$ in general carries flavor. In the case of QCD, the lightest quark flavors are the $u$ and $d$. The gluon field strength tensor $F_{\mu\nu}$ and the covariant derivative $D$ are matrices with color indices, given by

$$F_{\mu\nu} = \left(\partial_\mu A_\nu^a - \partial_\mu A_\nu^a + f^{abc} A_\mu^b A_\nu^c\right) T^a \tag{25}$$

$$D_\nu = \partial_\nu + iA_\nu + \mu\delta_{\nu,t}. \tag{26}$$

The coefficients $f^{abc}$ are the structure constants of SU($N$), and the parameter $\mu$ in the covariant derivative is the chemical potential. Note that the introduction of a nonzero chemical potential not only breaks Euclidean spacetime invariance by establishing a preferred frame, but leads to the non-Hermitian behavior at finite density: the operator $iD_\nu$ is Hermitian when $\mu = 0$, but not when $\mu \neq 0$.

The free energy $F(T, \mu)$ of a gauge theory may be calculated from the partition function $Z$ via $F = -T \log Z$, with $Z$ given by the Euclidean path integral

$$\begin{aligned} Z &= \int [dA_\mu]\,[d\psi][d\bar{\psi}]\, e^{-S} \\ &= \int [dA_\mu]\, \det M(\mu)\, e^{S_{YM}}, \end{aligned} \tag{27}$$

where the Fadeev-Popov determinant necessary in continuum perturbation theory is included in the measure $[dA_\mu]$. In the second line, we have performed the functional integral over the fermion fields, giving rise to the fermion functional determinant $\det M(\mu)$, with $M = i\not{D} - m$. We impose nonzero temperature by giving Euclidean spacetime the topology $R^3 \times S^1$, with the circumference of $S^1$ given by $\beta = T^{-1}$. The fermion determinant has an expansion in terms of closed fermion paths through spacetime. Each closed path $p$ carries with it a non-Abelian version of the Aharanov-Bohm phase

$$W(p) = \text{Tr}\left\{\mathcal{P}\exp\left[i \oint_p dx^\mu A_\mu(x)\right]\right\}, \tag{28}$$

the Wilson loop. Here, $\mathcal{P}$ indicates path ordering along $p$ of $A_\mu(x)$. For $N \geq 3$ these loops are complex. In the absence of a chemical potential, complex contributions to the fermion determinant from a given Wilson loop are canceled becaue they occur in pairs, e.g. $W(p)+W^*(p)$, tracing the same path but in opposite directions. At nonzero temperatures, Wilson loops may have a nonzero winding number $n$ around the timelike, $S^1$ direction. A winding number $n > 0$ represents a net current of fermions in the timelike direction, while $n < 0$ represents a net current of antifermions. If $\mu \neq 0$, each Wilson loop in the expansion of the fermion determinant picks up an additional factor of $e^{\beta n \mu}$ so that $W(p) + W^*(p)$ becomes $e^{n\beta\mu} W(p) + e^{-n\beta\mu} W^*(p)$, and the imaginary parts do not cancel. Because of this, the fermion determinant gives rise to complex weights in the path integral. These complex weights give rise to the sign problem in finite-density QCD. This sign problem appears to be an artifact of the Euclidean formalism, but efforts to study finite density QCD using the Hamiltonian formalism have not provided a compelling alternative. The key role of paths which wind around the $S^1$ direction in Euclidean space indicates the importance of topology within the Euclidean formalism.

Our understanding of the $\mathcal{CK}$ symmetry of SU($N$) gauge theories at finite-density is closely related to the sign problem. Within the Euclidean formalism, it is most natural to use complex conjugation $\mathcal{K}$ rather than time reversal $\mathcal{T}$ as the fundamental antilinear operation. Systems at finite density, with chemical potential $\mu$ nonzero, explicitly break invariance under the linear charge conjugation operation $\mathcal{C}$. Finite density QCD is invariant under the combined action of $\mathcal{C}$ and $\mathcal{K}$; this $\mathcal{CK}$ symmetry is the generalized $\mathcal{PT}$ symmetry of finite density QCD. Charge conjugation interchanges $W(p)$ and $W^*(p)$, as does complex conjugation $\mathcal{K}$; in addition, $\mathcal{K}$ conjugates any complex coefficients in the expansion. Because the fermion determinant is real when $\mu = 0$, we know that the coefficients in the expansion are real. Thus, the combined action $\mathcal{CK}$ leaves the fermion determinant invariant, and finite-density gauge theories are $\mathcal{CK}$ symmetric. If we allow the chemical potential to be complex, an extension of this argument allows us to understand the relation

$$\det M(\mu) = \det M^*(-\mu), \tag{29}$$

as a consequence of $\mathcal{CK}$ symmetry, although this identity also may be proven using gamma-matrix identities.

### *4.2. Quark confinement and center symmetry*

The most striking feature of QCD is quark confinement: at low temperature and density, quarks and gluons are confined inside baryons and mesons. In this section and the next, we review some well-known basic features of confinement; see, *e.g.*, Ref. [39]. for more details. In lattice simulation of gauge fields without other particles, the Wilson loop has an area law behavior

$$\langle \mathrm{Tr_F} W(p) \rangle \sim \exp\left[-\sigma A\right], \tag{30}$$

where $\sigma$ is the string tension, which measures the confining force. At nonzero temperature, the string tension may also be measured using the Polyakov loop $P$, also known as the Wilson line. The Polyakov loop is essentially a Wilson loop that uses a compact direction in space-time to close the curve using a topologically nontrivial path in space time. The Polyakov loop is given by

$$P(\vec{x}) = \mathcal{P} \exp\left[i \int_0^\beta dx_4\, A_4(x)\right]. \tag{31}$$

The one-point function of the Polyakov loop measures the free energy $F_Q$ needed to add a heavy quark to the system

$$\langle \mathrm{Tr_F} P(\vec{x}) \rangle \sim e^{-\beta F_Q}. \tag{32}$$

In a confined phase, $\langle \mathrm{Tr_F}\, P(\vec{x}) \rangle = 0$, which we interpret as $F_Q = \infty$.

Quark confinement is associated with the breaking of a global symmetry of the gauge action, center symmetry. The center of a Lie group is the set of all elements that commute with every other element; the center of $\mathrm{SU}(N)$ is $\mathrm{Z}(N)$. The action of the gauge field is invariant under global center symmetry representations because the gauge fields are in the adjoint representation of $\mathrm{SU}(N)$, on which center symmetry transformations are trivial. This is particularly clear in lattice gauge theories, where a link variable $U_\mu(x)$ in the gauge group is associated with each lattice site $x$ and direction $\mu$. The link variable is identified with continuum path-ordered exponential of the gauge field from $x$ to $x + \hat{\mu}$: $U_\mu(x) = \exp\left[iaA_\mu(x)\right]$. Consider a center symmetry transformation on all the links in the timelike direction at a fixed time. In $\mathrm{SU}(N)$ gauge theories at finite temperature this is $U_4(\vec{x}, t) \to zU_4(\vec{x}, t)$ for all $\vec{x}$ and fixed $t$, with $z \in \mathrm{Z}(N)$. Lattice gauge actions such as the Wilson action are composed of sums of small Wilson loops called plaquettes, which are invariant under this global symmetry. However, the Polyakov loop transforms as $P(\vec{x}) \to zP(\vec{x})$. Unbroken global $\mathrm{Z}(N)$ symmetry thus implies $\langle \mathrm{Tr}_F P(\vec{x}) \rangle = 0$. Global $\mathrm{Z}(N)$ symmetry defines the confining phase of a pure gauge theory, and the Polyakov loop is the order parameter for the deconfinement transition at nonzero temperature. Above a critical temperature, $\langle \mathrm{Tr_F}\, P(\vec{x}) \rangle \neq 0$, indicating a spontaneous breaking of $\mathrm{Z}(N)$ symmetry.

Fermions in the fundamental representation, such as quarks, explicitly break the $\mathrm{Z}(N)$ symmetry of a confining gauge system, and $F_Q$ is always finite in this case. If $\mu = 0$, charge conjugation symmetry implies that $F_Q = F_{\bar{Q}}$. At nonzero density, charge conjugation is explicitly broken, and in general we expect $F_Q \neq F_{\bar{Q}}$. This is also the statement $\langle \mathrm{Tr_F}\, P(\vec{x}) \rangle \neq \langle \mathrm{Tr_F}\, P^\dagger(\vec{x}) \rangle$ where both are real. In a Hermitian theory it would be surprising that the expectation values of two Hermitian conjugate operators have two different real values: we would expect that the expectation values would be complex conjugates, perhaps with vanishing imaginary part. This behavior is readily explained in terms of $\mathcal{PT}$ symmetry: the operator $P - P^\dagger$ is odd under $\mathcal{C}$ and $\mathcal{K}$ and has a real expectation value.

### *4.3. Deconfinement and Z(N) symmetry*

At very high temperatures, perturbation theory is reliable because of asymptotic freedom: the running gauge coupling becomes arbitrarily small as $T \to \infty$. Perturbation theory indicates that $\mathrm{Z}(N)$ symmetry is broken in this region, even when only gauge bosons are considered. The one-loop effective potential for a gauge boson in the background of a static Polyakov loop $P$ can be easily evaluated in a gauge where $A_4$ is time-independent and diagonal [40, 41]. The contribution added when $T > 0$ can be written as

$$V_{\mathrm{eff}}^{1l}(P) = -\frac{2}{\pi^2} \sum_{n=1}^{\infty} \frac{1}{n^4} \mathrm{Tr_{Adj}} P^n. \tag{33}$$

where the factor of two represents the two helicity states of each mode. This can be interpreted as a sum of contributions from gauge boson worldlines wrapping around the compact direction an arbitrary number of times.

From this form, it is easy to see that $V_{\mathrm{eff}}^{1l}(P)$ is minimized when all the moments $\mathrm{Tr_{Adj}} P^n$ are maximized. This occurs when $P \in \mathrm{Z}(N)$, which gives $\mathrm{Tr_{Adj}} P^n = N^2 - 1$. This indicates that the one-loop gluon effective potential favors the deconfined phase. The pressure $p$ is the negative of the free energy density at the minimum,

$$p(T) = 2(N^2 - 1)\frac{\pi^2 T^4}{90}, \tag{34}$$

which is exactly $p$ for a blackbody with $2(N^2 - 1)$ degrees of freedom. From this we see directly the spontaneous breaking of $\mathrm{Z}(N)$ symmetry by gauge bosons at high temperature.

In many systems, broken symmetry phases occur at low temperatures and symmetry is restored at high temperatures. The phase structure of gauge theories as a function of temperature is unusual because the broken-symmetry phase is the high-temperature phase. A lattice construction of the effective action for Polyakov loops, valid for strong-coupling, is instructive [42, 43, 44, 45]. The spatial link variables may be integrated out exactly if spatial plaquette interactions are neglected. Each spatial link variable then appears only in two adjacent temporal plaquettes, and may be integrated out exactly using the same techniques that are used in the Migdal-Kadanoff real-space renormalization group [43, 46]. The resulting effective action has the form

$$S_{\text{eff}} = -\sum_{\langle jk\rangle} K \left[ \mathrm{Tr_F} P_j \mathrm{Tr_F} P_k^\dagger + \mathrm{Tr_F} P_k \mathrm{Tr_F} P_j^\dagger \right] \tag{35}$$

where $K$ is a function of the lattice gauge coupling $g^2$ and the temperature in lattice units: $n_t = 1/Ta$. A strong-coupling calculation gives a explicit form for $K$ as $K \simeq \left(1/g^2N\right)^{n_t}$ to leading order. In the weak-coupling limit, a Migdal-Kadanoff bond-moving argument gives $K \simeq 2N/g^2 n_t$. This effective action represents a $Z(N)$-invariant nearest-neighbor interaction of a spin system where the Polyakov loops are the spins. It depends only on gauge-invariant quantities. Standard expansion techniques show that the $Z(N)$ symmetry is unbroken for small $K$, and broken for large $K$. This model explains why the high-temperature phase of gauge theories is the symmetry-breaking phase: the relation between $K$ and the underlying gauge theory parameters is such that $K$ is small at low temperatures and large at high temperatures, exactly the reverse of a classical spin system where the coupling is proportional to $T^{-1}$.

The lattice construction of the Polyakov loop effective action is a concrete realization of Svetitsky-Yaffe universality [47], which states that a second-order deconfinement transition in a $(d+1)$-dimensional gauge theory is in the universality class of classical spin systems in $d$ dimensions with the same global symmetry. Lattice simulations indicate that all pure $SU(N)$ gauge theories have a deconfining phase transition at some temperature $T_d$, above which center symmetry is broken. In accordance with predictions based on universality, the deconfinement transition for an $SU(2)$ gauge theory in $3+1$ dimensions has been well established as being in the universality class of the three-dimensional Ising model, exhibiting a second-order transition at $T_d$. The deconfinement transition for $SU(3)$ in $3+1$ dimensions is first order. This is consistent with Landau-Ginsburg predictions for a system with a $Z(3)$ symmetry. The transitions for $N > 3$ appear to be first-order in $3+1$ dimensions as well, with a smooth limit as $N$ goes to infinity [48, 49].

### *4.4. Quarks and effective actions*

The addition of quarks to QCD directly affects deconfinement because of the loss of center symmetry. The effects can be seen directly in the effective potential for quarks moving in a nontrivial Polyakov loop background. The one-loop contribution from nonzero temperature and density can be written as [50]

$$V_F(\theta) = -\frac{2N_f}{\beta} \mathrm{Tr_F} \int \frac{d^d k}{(2\pi)^d} \left[ \ln\left(1 + P e^{\beta\mu - \beta\omega_k}\right) + \ln\left(1 + P^\dagger e^{-\beta\mu - \beta\omega_k}\right)\right], \tag{36}$$

where the factor of two accounts for spin degeneracy, $N_f$ denotes the number of flavors, and $\omega_k$ is the fermion energy. The only change from free fermions at nonzero temperature and density are the factors of $P$ and $P^\dagger$.

From a field theory point of view, $V_F$ is a contribution to the effective potential from quarks at nonzero temperature and density. This has a simple physical interpretation. The Polyakov loop is the non-Abelian analog of the Bohm-Aharanov phase factor of electromagnetism for a particle winding around the Euclidean time direction at finite temperature. If we were working in Minkowski space, the Boltzmann weight for a classical particle at $\vec{r}$ with momentum $\vec{k}$ and charge $Q$ would be

$$e^{\beta\mu-\beta\omega_k-\beta Q A_0(\vec{r})} \tag{37}$$

where $A_0$ is the static potential. After rotation to Euclidean space this factor becomes

$$e^{\beta\mu-\beta\omega_k+i\beta Q A_4(\vec{r})}. \tag{38}$$

This coupling of field to particle gives rise to the familiar Coulomb interaction between particles at finite temperature. The non-Abelian analog of this Boltzmann factor for particles is

$$Pe^{\beta\mu-\beta\omega_k} \tag{39}$$

while for antiparticles it is

$$P^\dagger e^{-\beta\mu-\beta\omega_k}. \tag{40}$$

Thus $V_F$ is the free energy density of quarks in a non-Abelian background field, and is associated with the quark contribution to the the local pressure via $V_F = -\beta p_F$.

More physically, fermion trajectories are weighted by factors of $e^{\beta\mu}$ and $P$ for each time they wind around the lattice in the Euclidean time direction, while antifermion trajectories are weighted by factors $e^{-\beta\mu}$ and $P^\dagger$. In the low-temperature and low-density region, the Polyakov loop factors suppress the contributions of colored excitations. For $\mu > 0$, quark contributions are increased, eventually overcoming the effects of the Polyakov loops and leading to a change from a dense gas of hadrons to a dense gas of quarks and gluons. The complex nature of $V_F$ is manifest in the differences between the terms representing fermions and antifermions: at nonzero density $P$ appears with a factor $e^{\beta\mu}$, while $P^\dagger$ has a factor $e^{-\beta\mu}$. While complex, $V_F$ is $\mathcal{CK}$-symmetric because both $\mathcal{C}$ and $\mathcal{K}$ interchange $P$ and $P^\dagger$.

A variety of models have been used to study the properties of QCD related to $\mathcal{PT}$ symmetry. There is strong evidence from various calculations that sinusoidally-modulated correlation functions, a hallmark of $\mathcal{PT}$-symmetric field theories, appear in models of finite-density QCD. On the basis of a flux tube model equivalent to a Z(3) spin model with a sign problem, Patel suggested that there might be an oscillatory signal in two-point baryon number density correlation functions [51, 52]. Signals of oscillatory behavior were subsequently found in Polyakov loop correlation functions using a variety of methods. In [53], strong-coupling expansions were used for heavy quarks in SU(3) lattice gauge theory to demonstrate the existence of parameter regions where Polyakov loops have sinusoidally-modulated exponential decay. Akerland and de Forcrand demonstrated similar behavior in a Z(3) model with a complex action, using both mean field theory and lattice simulation [54].

In [55, 56], a class of phenomenological continuum models were studied that combine one-loop thermodynamics with the effects of confinement for the case of SU(3) gauge bosons and two flavors of quarks at finite temperature and density. These models are all described by an effective potential which is the sum of three terms:

$$V_{\text{eff}}(P) = V_g(P) + V_f(P) + V_d(P). \tag{41}$$

The potential term $V_g(P)$ is the one-loop effective potential for gluons, and the potential term $V_f(P)$ contains all quark effects, including the one-loop expression for quark thermodynamics. The potential term $V_d(P)$ represents confinement effects. Three different forms for $V_f(P)$ were considered: heavy quarks, massless quarks, and a Polyakov-Nambu-Jona Lasinio (PNJL) model [57]. These models generalize Nambu-Jona Lasinio models to include confinement-deconfinement effects as well as chiral symmetry effects. The effective potentials of such models will be a function of both the chiral order parameter $\bar{\psi}\psi$ and $P$. Two different forms for the confining potential $V_d(P)$ were used as well, so a total of six different models were studied. The PNJL models were the most sophisticated models studied, because they include the effects of chiral symmetry breaking as well as the deconfinement transition. In all cases, regions of parameter space at low $T$ and intemediate values of $\mu$ were found where complex-conjugate eigenvalues appear in the Polyakov loop mass matrix. This is an indicator for $\mathcal{PT}$ symmetry breaking, and leads to sinusoidally-modulated exponential decay of Polyakov loop correlation functions. Such regions are separated from the region where normal exponential decay occurs by disorder lines, where the eigenvalues of the mass matrix move continuously from real to complex values. For the PNJL models, the region where complex eigenvalues occurs is associated with the critical endpoint of a line of first-order phase transitions, similar to the line in Fig. 1.

Subsequent work [58] showed that the methods used for QCD could be used to describe the statistical mechanics of a large class of models with liquid-gas transitions using a $\mathcal{PT}$-symmetric action. From a field theory perspective, three-dimensional $\mathcal{CK}$-symmetric models are obtained from dimensional reduction of a four-dimensional field theory at finite temperature and density. The simplest cases of interest are models with a single type of particles, interacting via a scalar field $\sigma$ and a vector field $A_\mu$. Both $\sigma$ and $A_\mu$ will be taken to have masses. The potential induced by $\sigma$ will be attractive, while that caused by the static vector potential $A_4$ will be repulsive between particles. The particles of the underlying theory are integrated out, and after dimensional reduction and redefinition of fields, a Lagrangian of the general form

$$L_{3d} = \frac{1}{2}(\nabla\phi_1)^2 + \frac{1}{2}m_1^2\phi_1^2 + \frac{1}{2}(\nabla\phi_2)^2 + \frac{1}{2}m_2^2\phi_2^2 - F(\phi_1, \phi_2) \tag{42}$$

is obtained. Here $\phi_1$ is associated with the attractive force and $\phi_2$ with the repulsive force. The field $\phi_1$ is naturally a four-dimensional scalar, but $\phi_2$ can be obtained from the fourth component of a vector interaction. The function $F$ can be interpreted as $\beta p\,(\phi_1, \phi_2)$, where $\beta$ is the inverse of the tempurature $T$ and $p$ is a local pressure. In particular, $p\,(\phi_1, \phi_2)$ is the local pressure of the gas of particles in the grand canonical ensemble in the presence of the background fields $\phi_1$ and $\phi_2$. This is the simplest class of three-dimensional field theories with liquid-gas phase transitions that may be derived from renormalizable four-dimensional field theories. For a classical fluid, both attractive and a repulsive potential terms are required, as shown by the early example of the van der Waals equation. In relativistic theories, attractive potentials are associated wth scalar exchange, but repulsive potential effects require in addition vector exchange. The use of a massive vector field is natural at nonzero temperature and density, resulting in a potential between particles that is the difference of two Yukawa potentials.

The key feature of $L_{3d}$ is that it is not real, but instead satisfies the $\mathcal{PT}$ symmetry condition

$$L_{3d}(\phi_1, \phi_2) = L_{3d}^*(\phi_1, -\phi_2). \tag{43}$$

The charge conjugation transformation $\mathcal{C}$ is the linear operator which takes charged particles into their antiparticles. This naturally takes $A_\mu \to -A_\mu$ as in the case of QED, which implies $\phi_2 \to -\phi_2$; the $\phi_1$ field is left invariant. The antilinear operator used is complex conjugation. A nonzero chemical potential $\mu$ explicitly breaks $\mathcal{C}$ symmetry and leads to a complex $L_{3d}$, but the antilinear symmetry $\mathcal{CK}$ remains [9, 55, 56, 53]. Because $L_{3d}$ is complex, this class of models has a sign problem. $\mathcal{PT}$ symmetry implies that the saddle points of $L_{3d}$ have $\phi_2$ purely imaginary; at these saddle points, $L_{3d}$ is real. Analytic continuation of the fields into the complex plane leads to a resolution of the sign problem at tree level, in the sense that the action is real at the saddle points. More generally, unbroken $\mathcal{PT}$ symmetry implies that the expected value of $\langle\phi_2\rangle$ must be zero or purely imaginary because $\langle i\phi_2\rangle^* = \langle i\phi_2\rangle$. Different models are obtained with different forms for $F\,(\phi_1, \phi_2)$. For relativistic fermions of mass $m$, $F$ is taken to be

$$F = \int \frac{d^3k}{(2\pi)^3} \log\left[1 + \exp\left(-\beta\sqrt{k^2 + \left(m - g\beta^{-1/2}\phi_1\right)^2} + \beta\mu + i\beta^{1/2}e\phi_2\right)\right]. \tag{44}$$

*4.5. $\mathcal{PT}$ symmetry, pattern formation and lattice models of finite-density QCD*

In this section, we apply for the first time the methods of Sec. 3 to an effective model of QCD: a static fermion model, in order to search a pattern-forming region in parameter space. For lattice gauge theories at finite temperature, the static approximation represents very well the effect of heavy fermions. In the case where heavy fermions satisfy $\beta M \gg 1$, their effects may be represented in the action as

$$S_F = -\sum_x 2N_f \log\left\{\det\left[1 + e^{\beta\mu-\beta M}P(x)\right]\det\left[1 + e^{-\beta\mu-\beta M}P^\dagger(x)\right]\right\}, \tag{45}$$

where we take the sum over space, not spacetime. Note that for $\mu \neq 0$, $S_F$ is complex but $\mathcal{CK}$ symmetric. This action is often approximated by

$$S_F \approx -h_F \sum_x \left[e^{\beta\mu}\mathrm{Tr_F}P(x) + e^{-\beta\mu}\mathrm{Tr_F}P^\dagger(x)\right] \tag{46}$$

where $h_F = N_f e^{-\beta M}$. Recalling the spin model interpretation, we see that a heavy particle behaves like an external field $h_F$ coupled to the Polyakov loop.

We examine a highly simplified model obtained from lattice models and determine its phase structure. In the fermion determinant, we make a mean field approximation of the matrix $P(x)$ by $(\mathrm{Tr_F}P(x)/N_c)I$, where $N_c$ is the number of colors. Because $\mathrm{Tr_F}P(x)$ is complex for $N_c \geq 3$, we write $\mathrm{Tr_F}P(x)/N_c$ as $e^{\chi(x)+i\phi(x)}$ and $\mathrm{Tr_F}P^\dagger(x)/N_c = e^{\chi(x)-i\phi(x)}$. We make the further simplifying assumption that antifermion contributions are heavily suppressed by a factor of $e^{-\beta(\mu+M)}$ and ignore them. As in the liquid-gas models discussed in the previous section, we assume that the field $\chi$ gives rise to an attractive interaction $V_A$ between static fermions, while the field $\phi$ induces a repulsive interaction $V_R$. As before, the simplest choice for $V_A$ and $V_R$ is two Yukawa potentials, and the complete lattice action is

$$S = \frac{1}{2\beta}\sum_{x,y}\left\{\frac{1}{g_\chi^2}\left[(\nabla\chi)^2 + m_\chi^2\chi^2\right] + \frac{1}{g_\phi^2}\left[(\nabla\phi)^2 + m_\phi^2\phi^2\right]\right\} - 2N_cN_f\sum_x \log\left[1 + ze^{\chi(x)+i\phi(x)}\right], \tag{47}$$

where $z = e^{\beta(\mu-M)}$. This model has a clear relation to the sine-Gordon model. This is perhaps unsurprising because the sine-Gordon model is itself equivalent to a classical Coulomb gas. It was first used to describe a lattice gas with attractive and repulsive interactions by Fisher and Park [59], who showed that this model can describe a standard liquid gas transition for $z > 0$. However, they were interested in the critical singularity associated with hard-core repulsion which occurs when $z < 0$ and showed explicitly that this critical behavior is in the $i\phi^3$ universality class. We are interested in the normal liquid-gas transition for $z > 0$, which is in the Z(2) universality class, and possible patterned phases. We can apply the methods of Sec. 3 directly to find the behavior of eq. (47). We set $2N_fN_c = 1$ to match the work of Fisher and Park and take the parameter values $m_\chi/m_\phi = 0.2$, $g_\chi^2/m_\chi^2 = 0.3$, and $g_\phi^2/m_\phi^2 = 0.5$. The phase diagram in $n$-$T$ space is shown in Fig. 8. We note that the phase diagram of eq. (47) looks much like that of our $\mathcal{PT}$-symmetric extension of the $\phi^4$ model, eq. (9), consisting of all four expected regions of behavior: normal, unstable, complex, and patterned. We take this as an indication of the ubiquity of pattern formation in $\mathcal{PT}$-symmetric quantum field theories and the possibility of such patterns in finite-density QCD.

## 5. Concluding remarks and future directions

The inherent non-Hermiticity of finite-density QCD has caused major obstacles to understanding its phase structure, manifested by a complex action and a sign problem. The invariance of

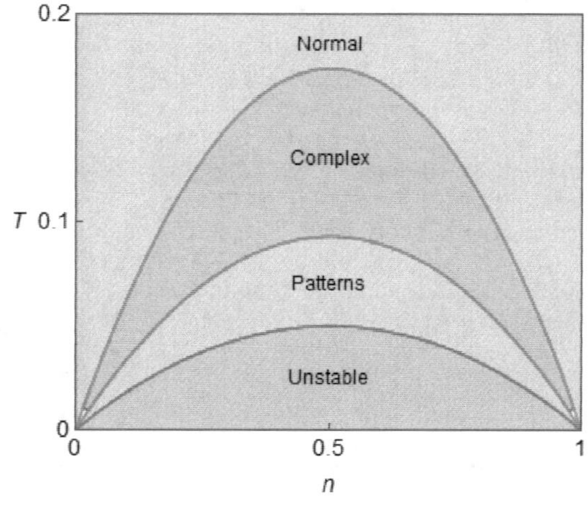


**Figure 8.** Phase diagram of the lattice gas with action eq. (47) in the $n$-$T$ plane, calculated using the techniques of Sec. 3. Note the similarity of this phase diagram to Fig. 2.

finite-density QCD under the combined operations of $\mathcal{CK}$ is a powerful tool. We have developed techniques to determine the phase diagram of scalar $\mathcal{PT}$-symmetric quantum field theories. These techniques may be applied to effective models of the deconfinement transition and hold promise for deepening our understanding of QCD phase structure.

In Sec. 2, we reviewed a recently developed technique to cast a large class of $\mathcal{PT}$-symmetric quantum field theories with complex actions into real positive forms that may be simulated on the lattice. We can now construct arbitrarily many new $\mathcal{PT}$-symmetric quantum field theories both as extensions of Hermitian universality classes, as well as an indirect construction from the real form. This method may also shed new light on the relationship between $\mathcal{PT}$-symmetric models, higher derivative theories, and other real models with local and nonlocal interactions.

In Sec. 3, we examined the phase diagram of $\mathcal{PT}$-symmetric quantum field theories. We described the standard procedure for finding the stable and unstable regions of a conventional quantum field theory and discussed how this technique breaks down in the non-Hermitian case. We derived the generalization of this technique to the $\mathcal{PT}$-symmetric case, and confirmed its validity for a $\mathcal{PT}$-symmetric extension of $\phi^4$ theory using lattice simulations. $\mathcal{PT}$-symmetric scalar field theories may exhibit pattern formation in the vicinity of critical points, a feature impossible in conventional field theories.

Section 4 described the origin of $\mathcal{CK}$ symmetry in QCD at nonzero chemical potential in terms of the Polyakov loop, and the strong evidence for $\mathcal{PT}$-breaking behavior, manifesting as sinusoidal modulation of the Polyakov loop two-point function. We also exhibited a simplified model of finite-density QCD with a patterning phase. It is possible that finite-density QCD exhibits patterned fields in its critical region, which manifest as agglomerations of deconfined matter floating in a sea of confined phase, and vice versa. These possibilities pose new challenges and opportunities for theory, experiment, and astrophysical observation.

## Acknowledgments

STS was supported by a Graduate Research Fellowship from the U.S. National Science Foundation under Grant No. 1745302; the U.S. Department of Energy, Office of Science, Office of Nuclear Physics, from DE-SC0011090; and fellowship funding from the MIT Department of Physics. MCO would like to acknowledge helpful discussion with Hiromichi Nishimura.

## References

[1] Aarts G 2016 *J. Phys. Conf. Ser.* **706** 022004 (*Preprint* 1512.05145)
[2] Aarts G and Stamatescu I O 2008 *JHEP* **09** 018 (*Preprint* 0807.1597)
[3] Cristoforetti M, Di Renzo F and Scorzato L (AuroraScience) 2012 *Phys. Rev. D* **86** 074506 (*Preprint* 1205.3996)
[4] Alford M G, Schmitt A, Rajagopal K and Schäfer T 2008 *Rev. Mod. Phys.* **80** 1455–1515 (*Preprint* 0709.4635)
[5] Fukushima K and Hatsuda T 2011 *Rept. Prog. Phys.* **74** 014001 (*Preprint* 1005.4814)
[6] Fu W j, Pawlowski J M and Rennecke F 2020 *Phys. Rev. D* **101** 054032 (*Preprint* 1909.02991)
[7] Adamczyk L *et al.* (STAR) 2017 *Phys. Rev. C* **96** 044904 (*Preprint* 1701.07065)
[8] Friman B, Hohne C, Knoll J, Leupold S, Randrup J, Rapp R and Senger P (eds) 2011 *The CBM physics book: Compressed baryonic matter in laboratory experiments* vol 814
[9] Meisinger P N and Ogilvie M C 2013 *Phil. Trans. Roy. Soc. Lond. A* **371** 20120058 (*Preprint* 1208.5077)
[10] Ogilvie M C and Medina L 2018 *PoS* **LATTICE2018** 157 (*Preprint* 1811.11112)
[11] Schindler M A, Schindler S T, Medina L and Ogilvie M C 2020 *Phys. Rev. D* **102** 114510 (*Preprint* 1906.07288)
[12] Seul M and Andelman D 1995 *Science* **267** 476–483 ISSN 0036-8075 (*Preprint* https://science.sciencemag.org/content/267/5197/476.full.pdf) URL https://science.sciencemag.org/content/267/5197/476
[13] Ravenhall D G, Pethick C J and Wilson J R 1983 *Phys. Rev. Lett.* **50** 2066–2069
[14] Hashimoto M a, Seki H and Yamada M 1984 *Progress of Theoretical Physics* **71** 320–326 ISSN 0033-068X (*Preprint* https://academic.oup.com/ptp/article-pdf/71/2/320/5459325/71-2-320.pdf) URL https://doi.org/10.1143/PTP.71.320
[15] Caplan M E and Horowitz C J 2017 *Rev. Mod. Phys.* **89** 041002 (*Preprint* 1606.03646)
[16] Bender C M and Boettcher S 1998 *Phys. Rev. Lett.* **80** 5243–5246 (*Preprint* physics/9712001)
[17] Bender C M 2007 *Rept. Prog. Phys.* **70** 947 (*Preprint* hep-th/0703096)
[18] Bender C M, Boettcher S and Savage V M 2000 *J. Math. Phys.* **41** 6381 (*Preprint* math-ph/0005012)
[19] Schindler S T and Bender C M 2018 *J. Phys. A* **51** 055201 (*Preprint* 1704.02028)
[20] Mezincescu G A 2000 *J. Phys. A* **33** 4911–4916 (*Preprint* quant-ph/0002056)
[21] Weigert S 2003 *Phys. Rev. A* **68** 062111 (*Preprint* quant-ph/0306040)
[22] Bender C M, Dorey P E, Dunning C, Fring A, Hook D W, Jones H F, Kuzhel S, Lévai G and Tateo R 2019 *PT Symmetry* (WSP)
[23] El-Ganainy R, Makris K G, Christodoulides D N and Musslimani Z H 2007 *Opt. Lett.* **32** 2632–2634 URL http://ol.osa.org/abstract.cfm?URI=ol-32-17-2632
[24] Guo A, Salamo G J, Duchesne D, Morandotti R, Volatier-Ravat M, Aimez V, Siviloglou G A and Christodoulides D N 2009 *Phys. Rev. Lett.* **103**(9) 093902 URL https://link.aps.org/doi/10.1103/PhysRevLett.103.093902
[25] Rüter C, Makris K G, El-Ganainy R, Christodoulides D N, Segev M and Kip D 2010 *Nat. Phys.* **6**(3) 192 URL https://doi.org/10.1038/nphys1515
[26] Feng L, El-Ganiany R and Ge L 2017 *Nature Photonics* **11** 752–762
[27] El-Ganiany R, Makris K G, Khajavikhan M, Musslimani Z H, Rotter S and Christodoulides D N 2018 *Nature Physics* **14** 11–19
[28] Miri M A and Alu A 2019 *Science* **363** eaar7709
[29] Bender C M, Brody D C and Jones H F 2004 *Phys. Rev. Lett.* **93**(25) 251601 URL https://link.aps.org/doi/10.1103/PhysRevLett.93.251601
[30] Bender C M, Brandt S F, Chen J H and Wang Q h 2005 *Phys. Rev. D* **71** 025014 (*Preprint* hep-th/0411064)
[31] Bender C M, Brody D C, Chen J H, Jones H F, Milton K A and Ogilvie M C 2006 *Phys. Rev. D* **74** 025016 (*Preprint* hep-th/0605066)
[32] Stephenson J 1970 *Phys. Rev. B* **1**(11) 4405–4409
[33] Muratov C B 2002 *Phys. Rev. E* **66**(6) 066108 URL https://link.aps.org/doi/10.1103/PhysRevE.66.066108
[34] Ortix C, Lorenzana J and Di Castro C 2008 *Phys. Rev. Lett.* **100**(24) 246402 URL https://link.aps.org/doi/10.1103/PhysRevLett.100.246402
[35] Ortix C, Lorenzana J and Di Castro C 2009 *Physica B: Condensed Matter* **404** 499–502 ISSN 0921-4526 URL https://www.sciencedirect.com/science/article/pii/S092145260800570X
[36] Chaikin P M and Lubensky T C 1995 *Principles of Condensed Matter Physics* (Cambridge: Cambridge University Press)
[37] Cross M C and Hohenberg P C 1993 *Rev. Mod. Phys.* **65**(3) 851–1112 URL https://link.aps.org/doi/10.1103/RevModPhys.65.851
[38] Bray A 1994 *Advances in Physics* **43** 357–459 URL https://doi.org/10.1080/00018739400101505

[39] Ogilvie M C 2012 *J. Phys. A* **45** 483001 (*Preprint* 1211.2843)
[40] Gross D J, Pisarski R D and Yaffe L G 1981 *Rev. Mod. Phys.* **53** 43
[41] Weiss N 1981 *Phys. Rev.* **D24** 475
[42] Polonyi J and Szlachanyi K 1982 *Phys.Lett.* **B110** 395–398
[43] Ogilvie M 1984 *Phys.Rev.Lett.* **52** 1369
[44] Green F and Karsch F 1984 *Nucl.Phys.* **B238** 297
[45] Gross M and Wheater J 1984 *Nucl.Phys.* **B240** 253
[46] Billo M, Caselle M, D'Adda A and Panzeri S 1997 *Int.J.Mod.Phys.* **A12** 1783–1846 (*Preprint* hep-th/9610144)
[47] Svetitsky B and Yaffe L G 1982 *Nucl. Phys.* **B210** 423
[48] Lucini B, Teper M and Wenger U 2002 *Phys.Lett.* **B545** 197–206 (*Preprint* hep-lat/0206029)
[49] Lucini B, Teper M and Wenger U 2004 *JHEP* **0401** 061 (*Preprint* hep-lat/0307017)
[50] Meisinger P N and Ogilvie M C 2002 *Phys.Rev.* **D65** 056013 (*Preprint* hep-ph/0108026)
[51] Patel A 2012 *PoS* **LATTICE2012** 096 (*Preprint* 1210.5907)
[52] Patel A 2012 *Phys. Rev. D* **85** 114019 (*Preprint* 1111.0177)
[53] Nishimura H, Ogilvie M C and Pangeni K 2016 *Phys. Rev. D* **93** 094501 (*Preprint* 1512.09131)
[54] Akerlund O, de Forcrand P and Rindlisbacher T 2016 *JHEP* **10** 055 (*Preprint* 1602.02925)
[55] Nishimura H, Ogilvie M C and Pangeni K 2014 *Phys. Rev. D* **90** 045039 (*Preprint* 1401.7982)
[56] Nishimura H, Ogilvie M C and Pangeni K 2015 *Phys. Rev. D* **91** 054004 (*Preprint* 1411.4959)
[57] Fukushima K 2004 *Phys.Lett.* **B591** 277–284 (*Preprint* hep-ph/0310121)
[58] Nishimura H, Ogilvie M C and Pangeni K 2017 *Phys. Rev. D* **95** 076003 (*Preprint* 1612.09575)
[59] Park Y and Fisher M E 1999 *Phys. Rev. E* **60**(6) 6323–6328 URL https://link.aps.org/doi/10.1103/PhysRevE.60.6323

Journal of Physics: Conference Series **2038** (2021) 012023 doi:10.1088/1742-6596/2038/1/012023

# Benign ghosts in higher-derivative systems

**Andrei Smilga**
SUBATECH, Université de Nantes, 4 rue Alfred Kastler, BP 20722, Nantes 44307, France
E-mail: smilga@subatech.in2p3.fr

**Abstract.** A brief review of the physics of systems including higher derivatives in the Lagrangian is given. All such systems involve *ghosts*, i.e. the spectrum of the Hamiltonian is not bounded from below and the vacuum ground state is absent. Usually this leads to collapse and loss of unitarity. In certain special cases, this does not happen, however: ghosts are *benign*. This happens, in particular, in exactly solvable higher-derivative theories, but exact solvability seems to be a sufficient but not a necessary condition for the benign nature of the ghosts.

We speculate that the Theory of Everything is a higher-derivative field theory, characterized by the presence of such benign ghosts and defined in a higher-dimensional bulk. Our Universe represents then a classical solution in this theory, having the form of a 3-brane embedded in the bulk.

## 1. Motivation

In the vast majority of problems in theoretical mechanics and theoretical physics, the Lagrangians depend on generalized coordinates and generalized velocities, but not on generalized accelerations or still higher derivatives of dynamic variables. In particular, this is true for the Lagrangian of the Standard Model.

However, having unravelled by ∼1975 the underlying structure of matter at the scale $\gtrsim 10^{-17}$ cm, the theorists met a major impasse: the attempts to construct quantum theory of gravity have not been successful so far. There are two main reasons for that.

(i) A direct attempt to quantize Eisnstein's gravity gives a nonrenormalizable quantum theory. Such a theory does not make sense not only because of the impossibility of perturbative calculations, but also due to impossibility to define path integrals: in a nonrenormalizable theory, the lattice path integral has no continuum limit.

(ii) In any gravity theory, time and spatial coordinates are intertwined so that the classical equations of motion do not represent a Cauchy problem — the evolution in independent flat time. As a result, causality is lost. Einstein's equations admit strange Gödel solutions with closed time loops [1]. Thus, even the classical general relativity has fundamental difficulties whose resolution is not in sight. And things do not become more assuring when one tries to quantize it.

Today the prevailing opinion of the experts is that the Theory of Everything that includes quantum gravity is a superstring theory of some kind. However, the latter has its own serious difficulties.

(i) Its building is impressive and beautiful, but the lowest stories of this building are hid in mist. String theory is simply not formulated at the fundamental nonperturbative level.

(ii) More that 30 years have now passed since the superstring revolution of 1985, but string theory still cannot boast of phenomenological successes. It has not contributed much in our understanding of why the world we see is this and not that.

Bearing all this in mind, it is interesting to pursue as far as we can an alternative, more conservative line of reasoning, assuming that the TOE is not a string theory, but a conventional field theory living in flat space with universal flat time.

- The first question that one should be able to answer in this approach is how to explain the observed fact that the space-time we live in is curved. And the only way to do so is to assume that our (3+1)–dimensional Universe represents a thin curved film embedded in a flat higher-dimensional bulk, like a 2-dimensional soap bubble is embedded in flat 3-dimensional space.
  The fundamental TOE should be formulated in this bulk, and our Universe should represent a classical solution in this theory, a kind of Abrikosov string, but extended not in one, but in three spatial directions, and very thin in the transverse directions so that the latter cannot be perceived.
- Then the question is how this higher-dimensional theory can look like. The problem is that we cannot write any Lagrangian, familiar from 4-dimensional physics, like

$$\mathcal{L} \;=\; -\frac{1}{2g^2}\mathrm{Tr}\{F_{\mu\nu}F_{\mu\nu}\}\,. \tag{1}$$

  We cannot do that because at $D > 4$ the coupling constant $g$ acquires dimension, and the theory is no longer renormalizable.
  What one can try to do is to add extra derivatives. For example, the 6-dimensional Lagrangian

$$\mathcal{L}^{D=6} \;=\; \alpha\mathrm{Tr}\{F_{\mu\nu}\Box F_{\mu\nu}\} + \beta\mathrm{Tr}\{F_{\mu\nu}F_{\nu\alpha}F_{\alpha\mu}\} \tag{2}$$

  involves dimensionless coupling constants $\alpha, \beta$ and is renormalizable.
- We thus arrive at a higher-derivative (HD) theory, and that was my own main motivation to study such theories — I tend to believe that the Holy Grail TOE is a HD theory living in a higher-dimensional bulk, with our Universe representing a curved 3-brane embedded there [2,3]. [1]
  But we should say right now that all such theories have a common striking feature — they are *ghost-ridden.* This means that the quantum Hamiltonian of any such theory does not have a ground state and its spectrum involves the levels with arbitrarily low energies. Neither has this Hamiltonian a "sky state" — the spectrum also involves the levels with arbitrarily high energies.

For many years people thought that such theories are inherently sick and did not study them. Indeed, the absence of the vacuum ground state is a very strange and unusual feature. And in many cases such theories are sick, indeed. They involve a *collapse* leading to violation of unitarity. But it has become clear recently that the collapse and violation of unitarity is a feature of *many*, but not of *all* ghost-ridden theories. There are quantum mechanical and field theory systems with "benign ghosts" — their spectrum is bottomless and there is no vacuum state, but still the quantum problem is well posed, there is no collapse and the evolution operator is unitary.

Unfortunately, there has been much confusion in the literature devoted to this problem. But correct understanding of this issue is now being gradually achieved. We mostly follow in our talk the review [4].

[1] Unfortunately, we have now absolutely no idea *what* this theory might be.

## 2. Pais-Uhlenbeck oscillator

The dynamics of HD systems was first studied by Ostrogradsky who developed back in 1848 the Hamiltonian method in classical mechanics independently of Hamilton and applied it to the ordinary and HD systems [5]. The inherent feature of the Ostrogradsky Hamiltonian is a term involving the linear dependence of one of the canonical momenta, so that the classical energy can acquire both arbitrary large positive and arbitrary large negative values [6].

The dynamics of a simplest quantum HD Hamiltonian was first studied in Ref. [7]. The classical Pais-Uhlenbeck Hamiltonian is derived from the Lagrangian

$$L = \frac{1}{2}\left[\ddot{x}^2 - (\omega_1^2+\omega_2^2)\dot{x}^2 + \omega_1^2\omega_2^2 x^2\right] \tag{3}$$

and involves two pairs of canonical variables: $x, p_x$ and $v, p_v$. It reads

$$H = p_x v + \frac{p_v^2}{2} + \frac{(\omega_1^2+\omega_2^2)v^2}{2} - \frac{\omega_1^2\omega_2^2 x^2}{2} \tag{4}$$

One of the Hamilton equations of motion gives $\partial H/\partial p_x = v = \dot{x}$.

Pais and Uhlenbeck noticed that the spectrum of the quantum version of (4) (with $p_x \to \hat{p}_x = -i\partial/\partial x$, $p_v \to \hat{p}_v = -i\partial/\partial v$) is not bounded neither from above, nor from below — the ground state of $\hat{H}$ is absent. It was later understood that the absence of the ground state is an inherent feature of quantum HD systems [8].[2]

To determine the spectrum of the PU Hamiltonian, it is convenient to perform the following quantum canonical transformation [3] [10]:

$$\begin{aligned} X_1 &= \frac{1}{\omega_1}\frac{\hat{p}_x+\omega_1^2 v}{\sqrt{\omega_1^2-\omega_2^2}}, & \hat{P}_1 \equiv -i\frac{\partial}{\partial X_1} &= \omega_1\frac{\hat{p}_v+\omega_2^2 x}{\sqrt{\omega_1^2-\omega_2^2}}, \\ X_2 &= \frac{\hat{p}_v+\omega_1^2 x}{\sqrt{\omega_1^2-\omega_2^2}}, & \hat{P}_2 \equiv -i\frac{\partial}{\partial X_2} &= \frac{\hat{p}_x+\omega_2^2 v}{\sqrt{\omega_1^2-\omega_2^2}}. \end{aligned} \tag{5}$$

(We assumed that $\omega_1 > \omega_2$. The case of equal frequencies will be treated a little later.) In these terms, the Hamiltonian reduces to a difference of two oscillator Hamiltonians:

$$H = \frac{\hat{P}_1^2+\omega_1^2X_1^2}{2} - \frac{\hat{P}_2^2+\omega_2^2X_2^2}{2}. \tag{6}$$

Its spectrum is

$$E_{nm} = \left(n+\frac{1}{2}\right)\omega_1 - \left(m+\frac{1}{2}\right)\omega_2 \tag{7}$$

with positive integer $n, m$.

Consider the case $\omega_1 = 2\omega_2$. Then

$$E_{nm} = \left(2n - m + \frac{1}{2}\right)\omega_2\,. \tag{8}$$

[2] One can in principle get rid of the linear term in the Hamiltonian and thereby "exorcise the ghost" by imposing certain constraints that effectively reduce the dimensionality of the phase space [9]. But this just means that, if these contraints are explicitly resolved and one goes back to the Lagrangian formalism, higher derivatives disappear.

[3] For a general theory of quantum canonical transformations see [11] and references therein. One can also first perform the canonical transformation at the classical level and quantize afterwards.

We see that the spectrum is discrete, not bounded and that each level is infinitely degenerate. The same properties hold for any rational ratio $\omega_1/\omega_2$.

An interesting situation arises for incommensurable frequencies. In that case, the spectrum is *everywhere dense*: for any energy $E$ one can find an arbitrary close eigenvalue $E_{nm}$. However, the spectrum is not continuous, but rather *pure point*: the wave functions of all eigenstates are normalizable.

In the degenerate case, $\omega_1 = \omega_2 \equiv \omega$, the situation is somewhat more complicated. We cannot perform the canonical transformation (5) anymore and the Hamiltonian is not reduced to a difference of two oscillator Hamiltonians. The solution of the problem can be obtained by the following trick [7, 12].

At the first step, we get rid of the potential terms $\propto x^2$ and $\propto v^2$ in the Hamiltonian

$$\hat{H} = \hat{p}_x v + \frac{\hat{p}_v^2}{2} + \omega^2 v^2 - \frac{\omega^4 x^2}{2} \tag{9}$$

by representing its eigenfunctions as

$$\Psi(x,v) = e^{-i\omega^2 xv}\phi(x,v)\,. \tag{10}$$

The Hamiltonian acting on $\phi(x,v)$ has the form

$$\hat{H}_\phi = \frac{1}{2}\hat{p}_v^2 + v\hat{p}_x - \omega^2 x\hat{p}_v\,. \tag{11}$$

At the second step, we perform the quantum canonical transformation

$$x \to \frac{1}{\omega}x + \frac{1}{4\omega^2}\hat{p}_v, \quad \hat{p}_x \to \omega\hat{p}_x, \quad v \to v + \frac{1}{4\omega}\hat{p}_x, \quad \hat{p}_v \to \hat{p}_v\,. \tag{12}$$

The transformation (12) is the superposition of the scale transformation $x \to x/\omega, \hat{p}_x \to \omega\hat{p}_x$ and the unitary transformation

$$\hat{O} \to \exp\left\{\frac{i\hat{p}_x\hat{p}_v}{4\omega}\right\}\hat{O}\exp\left\{-i\frac{\hat{p}_x\hat{p}_v}{4\omega}\right\}.$$

It gives the new Hamiltonian

$$\hat{H}'_\phi = \frac{\hat{p}_x^2 + \hat{p}_v^2}{4} + \omega(v\hat{p}_x - x\hat{p}_v) \tag{13}$$

acting on the wave function $\phi'(x,v)$ related to $\phi(x,v)$ according to

$$\phi(x,v) = \exp\left\{\frac{i}{4\omega^2}\frac{\partial^2}{\partial x\partial v}\right\}\phi'(\omega x, v)\,. \tag{14}$$

The first term in the Hamiltonian (13) describes free motion in the transformed $(x,v)$ plane, and the second term is proportional to the "angular momentum" operator

$$\hat{l} = v\hat{p}_x - x\hat{p}_v\,. \tag{15}$$

Expressed in the original variables, this operator reads

$$\hat{l} = \frac{v\hat{p}_x}{2\omega} - \frac{\omega}{2}x\hat{p}_v + \frac{1}{4\omega}\left(\hat{p}_v^2 - \frac{\hat{p}_x^2}{\omega^2}\right) + \frac{3\omega}{4}(v^2 - \omega^2 x^2)\,. \tag{16}$$

A dedicated reader may check that it commutes with the original Hamiltonian (9).

The eigenfunctions of (13) are the same as for the free Laplacian, but the energy is shifted by $l\omega$, where $l$ is the integer eigenvalue of (15). They are

$$\phi'_{lk}(\omega x, v; t) \ \propto \ J_l(kr)e^{-il\theta}e^{-it(l\omega+k^2/4)}\,, \tag{17}$$

where $(r,\theta)$ are the polar coordinates in the plane $(\omega x, v)$.

The eigenfunctions of the original Hamiltonian are

$$\Psi_{lk}(x,v;t) \ \propto \ e^{-i\omega^2 xv}\exp\left\{\frac{i}{4\omega^2}\frac{\partial^2}{\partial x \partial v}\right\}\left[J_l\left(k\sqrt{v^2+\omega^2x^2}\right)\left(\frac{\omega x - iv}{\omega x + iv}\right)^{l/2}\right]e^{-it(l\omega+k^2/4)} \tag{18}$$

They are not normalizable and describe a continuum spectrum rather than a pure point spectrum characteristic for a nondegenerate system with $\omega_1 \neq \omega_2$. Any real energy is admissible, and each energy level is infinitely degenerate: the eigenfunctions $\Psi_{l+1,k}$ and $\Psi_{l,\sqrt{k^2+4\omega}}$ have the same energy.

The difference in the quantum behaviour of the degenerate ($\omega_1 = \omega_2$) and non-degenerate ($\omega_1 \neq \omega_2$) systems matches the difference in their classical behaviour. The motion described by the Hamiltonian (6) is finite and that is why the spectrum is pure point. And in the degenerate case the situation is different. The equation of motion

$$\left[\frac{d^4}{dt^4} + 2\omega^2\frac{d^2}{dt^2} + \omega^4\right]x \ = \ 0\,. \tag{19}$$

admits not only ordinary oscillatory solutions $x(t) \propto e^{i\omega t}$, but also the solutions

$$x(t) \propto te^{i\omega t}\,, \tag{20}$$

where the amplitude of oscillations grows with time. An infinite classical motion produces a continuum spectrum in the quantum problem.

One can observe that both in the non-degenerate and in the degenerate case both the classical and the quantum dynamics of the system are quite benign. The wave functions are known explicitly. The Hamiltonian is Hermitian and the quantum evolution is unitary. For example, the evolution operator for the system (6) is simply a product of the evolution operators for the individual oscillators:

$$\mathcal{K}(X_1', X_2'; X_1, X_2; t) \ = \ \sum_{n=0}^{\infty}\psi_n^*(X_1')\psi_n(X_1)e^{-i\omega_1 t(n+1/2)}\sum_{m=0}^{\infty}\psi_m^*(X_2')\psi_m(X_2)e^{i\omega_2 t(m+1/2)}\,, \tag{21}$$

where $\psi_n(X)$ are the standard oscillator eigenfunctions.

One can meet in the literature a statement that the Pais-Uhlenbeck oscillator with degenerate frequencies is not unitary due to the presence of Jordan blocks [13]. But this is not correct. Infinite-dimensional Jordan blocks appear in this problem if one tries to describe the motion in terms of the oscillator wave functions, as if the motion were finite. And their appearance is simply a manifestation of the fact that the spectrum in this case is continuous, and not of the violation of unitarity. We refer the reader to Refs. [14], [15], [4], where we clarify this point.

The *Minkowskian* path integral for the evolution operator (21) can be expressed either in the Lagrangian form

$$\sim \int\prod_t dx(t)\exp\left\{i\int dt\, L(\ddot{x},\dot{x},x)\right\} \tag{22}$$

or in the Hamiltonian form

$$\sim \int \prod_t dx(t)dv(t)dp_x(t)dp_v(t) \exp\left\{ i \int dt \, [p_v \dot{v} + p_x \dot{x} - H(p_v, p_x; v, x)] \right\} . \tag{23}$$

Indeed, substituting in (23) the Hamiltonian (4) and integrating over $\prod_t dp_x(t)$, we obtain the factor

$$\prod_t \delta[v(t) - \dot{x}(t)] . \tag{24}$$

Performing the integral over $\prod_t dv(t)$ with this factor and doing the Gaussian integral over $\prod_t dp_v(t)$, we reproduce (22) with the Lagrangian (3).

In ordinary quantum theories including a well-defined vacuum state we can perform the Wick rotation $t \to -i\tau$ and define an Euclidean path integral. This does not work, however, for the integral (23). The contribution to this integral coming from the first term in the Hamiltonian (4)

$$\prod_\tau \int_{-\infty}^{\infty} dp_x(\tau) \exp\left\{ \int d\tau \, p_x(\tau) \left[ i\frac{dx(\tau)}{d\tau} - v(\tau) \right] \right\} \tag{25}$$

diverges and does not give anything like (24). (The latter circumstance means that the variable $v(\tau)$ can no longer be interpreted as the velocity.) The fact that the Euclidean path integral diverges agrees well with that the evolution operator (21) is not defined at imaginary time — at $t = -i\tau$ the second series diverges with a vengeance.

Another way of going into Euclidean space was suggested in [16]. Instead of performing an analytic continuation of the Euclidean path integral (23), one could try to do it at the Lagrangian level, capitalizing on the fact that the Euclidean counterpart of the Lagrangian (3) is positive definite. Trading $t$ for $i\tau$ in (22) [4], one obtains a converging path integral.

However, if one tries to take the expression for the Euclidean evolution operator thus derived and goes back to real Minkowsky time, one obtains something which does not describe a unitary evolution with a Hermitian Hamiltonian [16–18]. It is thus better not to go to Euclidean space whatsoever, but keep the time real all the time.

One can sometimes meet a statement that the only way to define a Minkowski path integral is by an analytic continuation from Euclidean space. If true, that would mean that, given the fact that the Euclidean path integral is not defined, the Minkowski path integral is not defined either. But it is not so. The fact that Minkowski path integrals can be defined independently of Euclidean ones is especially clearly seen in the semi-classical limit where the Minkowski evolution operator can be evaluated as[5]

$$\langle x't'|xt\rangle \approx \Delta \exp\{iS_{\rm cl}\} , \tag{26}$$

where $S_{\rm cl}$ is the action on the corresponding classical trajectory (which is assumed to be large in the units of $\hbar$) and $\Delta$ is a well defined Gaussian integral. The corrections to $\Delta$ in any perturbative order can also be evaluated. Finally, Minkowski path integrals can be calculated numerically using the technique of Ref. [20].

[4] The standard Wick rotation $t \to -i\tau$ works if one changes the sign in the defnition of $L$.

[5] See also Eq. (3.62) in the book [19] that expresses the Minkowskian path integral for the evolution operator in the problem with a linearly rising potential $V(x) = -fx$. (Obviosly, the spectrum of this problem does not have a bottom.)

Journal of Physics: Conference Series **2038** (2021) 012023 doi:10.1088/1742-6596/2038/1/012023

## 3. Philological digression

We have proven that the nondegenerate quantum higher-derivative systems have no ground state. How should one *call* this phenomenon? This question is actually not so irrelevant. Experience shows that an inexact, not carefully chosen name may excite false associations and lead eventually to wrong claims.

Traditionally, one says that such bottomless systems involve *ghosts*. Let us explain where this name came from and what it means. By "ghosts" one usually has in mind the states with negative norm. Negative norm means negative probability and production of such states means violation of unitarity. Such ghosts appear e.g. in gauge theories. These are scalar photons in QED, and scalar gluons and Faddeev-Popov ghosts in non-Abelian theories.

In the framework of the Gupta-Bleuler quantization procedure for QED, one introduces the creation and annihilation operators $a^\dagger_\mu(\boldsymbol{k})$ and $a_\mu(\boldsymbol{k})$ for all four photon polarizations, which satisfy the following commutation relations:

$$[a_\mu(\boldsymbol{k}), a^\dagger_\nu(\boldsymbol{q})] \;=\; -\eta_{\mu\nu}\delta(\boldsymbol{k}-\boldsymbol{q})\,. \tag{27}$$

The commutator $[a_0, a^\dagger_0]$ is then negative. That means that the scalar photon state has a negative norm. Indeed, introduce the Fock vacuum $|\Phi\rangle$ whose norm is positive. We then have

$$\Big|\,|a^\dagger_0|\Phi\rangle\Big|^2 \;=\; \langle\Phi|a_0a^\dagger_0|\Phi\rangle \;=\; \langle\Phi|[a_0, a^\dagger_0]|\Phi\rangle \;=\; -\langle\Phi|\Phi\rangle\,. \tag{28}$$

In gauge theories, such ghosts are harmless: one can define a reduced physical space involving only the physical states with positive norm (transverse photons or transverse gluons) and show that the ghosts states are not created in collisions of physical particles. [6]

Let us go back to the Pais-Uhlenbeck oscillator. Consider the non-degenerate case, which is more simple. Introducing the creation and annihilation operators in the usual way, we may write the quantum Hamiltonian as

$$H \;=\; \omega_1 a_1 a^\dagger_1 - \omega_2 a_2 a^\dagger_2 + C\,. \tag{29}$$

The excitations of the second term give a tower of states with decreasing energies all the way down to $-\infty$. Suppose, however, that we do not like these negative energies so much that we want to get rid of them by any cost. Then one can formally define a state $|\Phi\rangle$ which is annihilated by $a^\dagger_2$ rather than by $a_2$. This state is strange and unhandy: its wave function is [22]

$$\Phi(X_2) \;\propto\; \exp\left\{\frac{\omega_2}{2}X_2^2\right\} \tag{30}$$

so that the state is not normalizable. Still, formally, the state $|\Phi\rangle$ has the lowest energy. We can even bring it to zero or to any other finite value by choosing an infinite positive constant $C$ in (29). Then the operator $a_2$ acting on $|\Phi\rangle$ would increase its energy, and we can *call* it a creation operator $b^\dagger$. And $a^\dagger_2$ is now interpreted as the annihilation operator $b$. The price one has to pay for that is that the commutator $[b, b^\dagger]$ is now negative and the states describing excitations above the "vacuum" $|\Phi\rangle$ are ghost states, they have negative norm.

Personally, I find this construction rather awkward. It is much better to talk about negative energies, keeping the norm positive. But, historically, people thought in these terms and that is why they (wrongly) believed for a long time that higher derivatives necessarily entail the violation of unitarity.

R. Woodard suggested to call this phenomenon "Ostrogradskian instability" without the reference to ghosts. This name has, however, its own drawbacks. First of all, Ostrogradsky

[6] This is explained in many textbooks including my own book [21].

knew nothing about quantum Hamiltonians and their spectra. The observation of the absence of the ground states in HD quantum systems belongs not to him, but to Pais and Uhlenbeck.

In addition, the word "instability" invokes wrong images. Having heard it, a physicist imagines a ball on the top of the hill and thinks of the exponential growth of deviations from the equilibrium. But for certain HD systems, there is no such growth. The spectrum has no bottom, but there is no instability — neither at the classical nor at the quantum level.

Thus, we prefer not to use this word and not invent anything new, but rather to stick to the traditional and more familiar to most people name "ghost", not invoking, however, the negative metric description. For us, a *ghost system* is by definition a system where the Hamiltonian does not have a ground state but involves oscillatory-type excitations with arbitrarily negative energies. However, all the states of this Hamiltonian have positive norm. We will call the excitations with negative energies ghosts, but will distinguish the systems with *benign* ghosts, where the quantum problem is well defined, as it is for the Pais-Uhlenbeck oscillator, and the systems with *malignant* ghosts involving a collapse and loss of unitarity. The latter systems are more common. They will be discussed in Sect. 5. And some examples of the benign ghost systems will be given in Sects. 6,7.

The last comment concerns the paper of C. Bender and P. Mannheim [23] who suggested to bust ghosts by replacing the original Hilbert space spanned over the eigenfunctions $\Psi(x, v)$ of the Hamiltonian (4) or, which is equivalent for different frequencies, the Hilbert space spanned by the oscillator wave functions of the Hamiltonian (6) by *another* Hilbert space involving the functions depending on real $X_1$ and *imaginary* $X_2 = iY_2$ and normalized in *that* region. Then the wave function (30) becomes a good normalizable vacuum state, $\int_{-\infty}^{\infty} |\Phi|^2 dY_2 < \infty$, and all other eigenstates have positive energies (and positive norm).

Personally, I do not quite see a point of doing so. One should understand that this complexification brings us to *another* quantum problem having little to do with the original one. No ghosts in the former does not mean no ghosts in the latter.

## 4. Including interactions

We discussed so far only mechanical systems, but it is easy to write down a field-theory generalization of (3). We can write

$$\mathcal{L} = \frac{1}{2}\left[(\Box\phi)^2 - (M_1^2 + M_2^2)(\partial_\mu\phi)^2 + M_1^2 M_2^2 \phi^2\right] . \tag{31}$$

When $M_1 \neq M_2$, one can go over to the Hamiltonian, perform an approriate canonical transformation and bring it again in the Lagrangian form to obtain

$$\mathcal{L} = \frac{1}{2}\left[(\partial_\mu\Phi_1)^2 - M_1^2\Phi_1^2\right] - \frac{1}{2}\left[(\partial_\mu\Phi_2)^2 - M_2^2\Phi_2^2\right] . \tag{32}$$

This Lagrangian includes an ordinary scalar field $\Phi_1$ and a ghost field (i.e., in our terminology, a field whose quanta carry negative energies) $\Phi_2$. We can expand each field in Fourier modes, in which case the Lagrangian is split into an infinite number of noninteracting sectors with a definitite momentum $\boldsymbol{p}$. In each such sector, the dynamics is described by the Hamiltonian (6) with $\omega_{1,2}^2 = \boldsymbol{p}^2 + M_{1,2}^2$. In free theory, the negative sign of the second term in (6) or (32) is irrelevant. The classical equations of motion and their solutions do not depend on this sign. The spectra of the quantum Hamiltonians are different for different signs, but in the absence of interactions, energy does not really play a dynamical role and serves only for bookkeeping.

A trouble may set it, however, if interactions are included. Heuristically, one could expect in this case an instability due to copious production of ghosts. Let us see whether an interactive system has or not this instability.

We modify the Lagrangian (3) by adding there nonlinear terms. Consider for example the system [24]

$$L = \frac{1}{2}\left[\ddot{x}^2 - 2\omega^2\dot{x}^2 + \omega^4 x^2\right] - \frac{1}{4}\alpha x^4 . \tag{33}$$

The classical equations of motion read

$$\left(\frac{d^2}{dt^2} + \omega^2\right)^2 x - \alpha x^3 = 0 . \tag{34}$$

The classical trajectories depend on four initial conditions. There is an obvious stationary point

$$x(0) = \dot{x}(0) = \ddot{x}(0) = x^{(3)}(0) = 0 . \tag{35}$$

The behaviour of the system at the vicinity of this point depend on the sign of $\alpha$. If $\alpha < 0$, the Ostrogradsky Hamiltonian acquires an extra negative contribution to the energy and all the trajectories other than $x(t) = 0$ are unstable — they run away to infinity in finite time. This instability is even worse that Lyapunov's exponential growth of perturbations. We are dealing here with a *collapse*: the classical dynamical problem is not well posed beyond a certain time horizon.

The situation is better for positive $\alpha$. The stationary point (35) lies in the center of an "island of stability" — the trajectories with initial conditions at its vicinity do not go astray, but exhibit an oscillatory behaviour. However, this island has a shore. When the deviations of intial conditions from (35) are large enough, the trajectory collapses. To chart this shore, one should perform a numerical study. Assume for simplicity that $\dot{x}(0), \ddot{x}(0)$ and $x^{(3)}(0)$ stay zero, and $x(0) = c > 0$. Then a critical value $c_{\rm crit}$ can be found, above which the trajectories go astray and run to infitiny, but the system only exhibits benign oscillations around zero when $c < c_{\rm crit}$. This value is roughly $c_{\rm crit} \approx 0.3\,\omega^2/\sqrt{\alpha}$ (the dependence on $\omega$ and $\alpha$ follows, of course, from simple scaling arguments). For illustration, we plotted in Fig.1 the solution to the equations of motion (34) for $\omega = \alpha = 1$ and $x(0)$ just above $c_{\rm crit}$. After some quasiharmonic oscillations, the trajectory finally goes astray and runs to infinity. And for $x(0) < c_{\rm crit}$, it keeps oscillating forever.

The same behaviour is seen if three other initial conditions are shifted from zero. The island has a finite area (or rather a finite phase space volume).

A similar island of stability was observed in [25] for the model Hamiltonian [7]

$$H = \frac{1}{2}\left[p_x^2 - (p_y^2 + y^2) + x^2 y^2\right] , \tag{36}$$

where a free particle is coupled to the ghost oscillator. If we pose the initial conditions

$$x(0) = y(0) = a, \qquad p_x(0) = p_y(0) = 0 , \tag{37}$$

the system undergoes benign oscillations at the vicinity of the origin if $a \lesssim 0.59$ and goes astray, running to the infinity, for larger deviations.

What happens in the quantum case? One cannot give a quite definite answer to this question without a special numerical study, but certain heuristic arguments that the quantum problem is also malignant and collapsing in this case will be given below.

Classical and quantum collapse are well known for certain special ordinary (not HD) systems. The simplest such system describes the 3-dimensional motion of a particle with the Hamiltonian

$$H = \frac{\vec{p}^2}{2m} - \frac{\kappa}{r^2} . \tag{38}$$

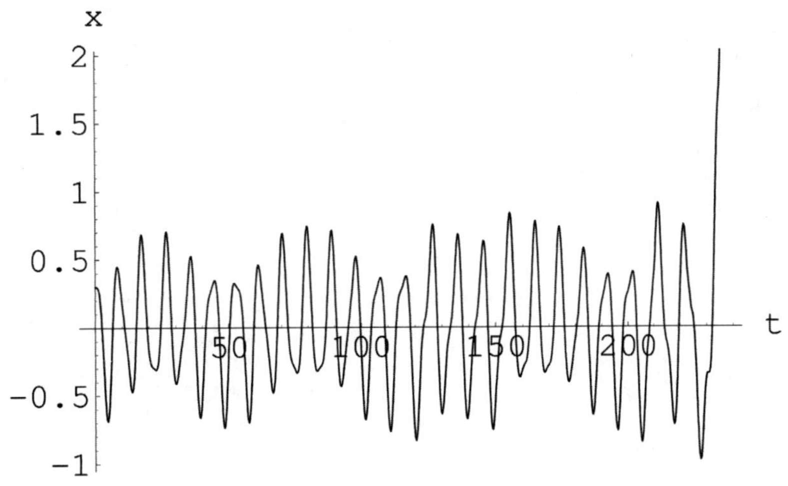


**Figure 1.** Oscillating and collapsing

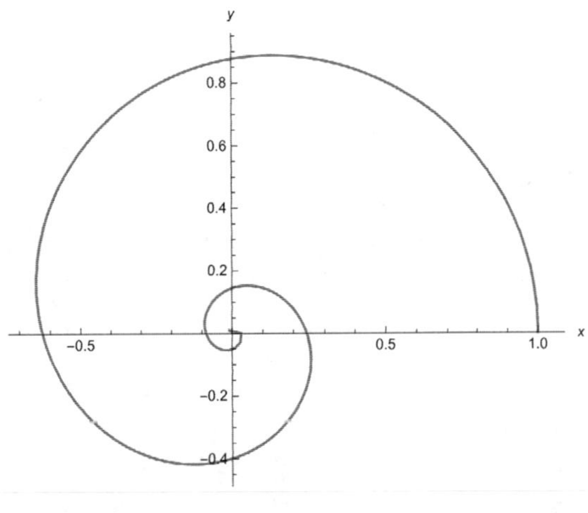


**Figure 2.** Falling on the center for the Hamiltonian (38) with $m = 1$ and $\kappa = .05$. The energy is slightly negative. The particles with positive energies escape to infinity.

Classically, for certain initial conditions, the particle falls to the center in a finite time. An example of such a trajectory is given in Fig. 2. The quantum dynamics of this system depends on the value of $\kappa$. If $m\kappa < 1/8$ (where $m$ is the mass of the particle), the ground state exists and unitarity is preserved. If $m\kappa > 1/8$, the spectrum is not bounded from below and, which is worse, the quantum problem cannot be well posed until the singularity at the origin is removed. For example, one can pose $V(r) = -\kappa/r^2$ for $r > a$ and $V(r) = -\kappa/a^2$ for $r \le a$. The spectrum then depends on $a$ [26–28]. Without such a cutoff, the probability "leaks" into the singularity

[7] The authors of this paper were interested in the toy Hamiltonian (36) in association with a certain cosmological problem.

and unitarity is violated. And for $m\kappa < 1/8$ quantum fluctuations cope successfully with the attractive force of the potential and prevent the system from collapsing.

Going back to the system (33), a variational analysis shows that, in constrast to the system (38) with small $\kappa$, the ground state in the spectrum is absent [14]. We see that for *some* HD systems involving collapsing classical trajectories, their ghosts are malignant and unitarity is violated by the same mechanism as for the strong attractive potential (38). On the other hand, one can also conjecture that

*If the classical dynamics of the system is benign, its quantum dynamics is also benign, irrespectively of whether the spectrum has or does not have a bottom.*

## 5. Benign ghosts: classical and quantum mechanics

In some special cases, a nontrivial interacting ghost system may be completely benign: all the classical trajectories (and not only the trajectories restricted to a limited region of phase space) are stable and behave well at all times. Capitalizing on the conjectured property formulated above, the corresponding quantum problems are also well posed — a special study of this complicated question is not required!

### *5.1. Exactly solvable models*

*5.1.1. A simplest nontrivial model* The first example of such a system was found in [29] [8]. The Hamiltonian involves two pairs of the dynamic variables, $(x, p)$ and $(D, P)$, and reads

$$H = pP + DV'(x)\,, \tag{39}$$

where $V(x)$ is an arbitrary smooth even function. The simplest nontrivial case is

$$V(x) = \frac{\omega^2 x^2}{2} + \frac{\lambda x^4}{4}\,, \qquad \lambda > 0\,. \tag{40}$$

The kinetic part of the Hamiltonian (39) is not positive definite, and, obviously, the spectrum of its quantum counterpart has no bottom. If one introduces the variables

$$X_{1,2} = \sqrt{\frac{\omega}{2}}x \pm \frac{1}{\sqrt{2\omega}}D\,, \qquad P_{1,2} = \frac{1}{\sqrt{2\omega}}p \pm \sqrt{\frac{\omega}{2}}P\,, \tag{41}$$

the Hamiltonian (39) acquires the form

$$H = \frac{P_1^2 + \omega^2 X_1^2}{2} - \frac{P_2^2 + \omega^2 X_2^2}{2} + \frac{\lambda}{4\omega}(X_1 - X_2)(X_1 + X_2)^3\,. \tag{42}$$

In other words, this is the Hamiltonian (6) with degenerate frequencies, where an extra quartic interaction of a special form is added.

A nice distinguishing feature of the system (39) is its integrability. Indeed, it involves besides $H$ another integral of motion:

$$N = \frac{P^2}{2} + V(x)\,. \tag{43}$$

The Poisson bracket $\{H, N\}$ vanishes. This allows one to find the solution analytically.

The classical equations of motion are

$$\ddot{x} + V'(x) = 0\,, \qquad \ddot{D} + V''(x)D = 0\,. \tag{44}$$

[8] It was found by studying a certain higher-derivative supersymmetric system, but we will not explore here the supersymmetric aspects of this problem.

The first equation is especially simple. It describes oscillations in the quartic potential (40). The solutions are the elliptic functions whose parameters depend on the integral of motion $N$ :

$$x(t) \;=\; x_0 \mathrm{cn}[\Omega(t-t_0), k] \tag{45}$$

where

$$\alpha = \frac{\omega^4}{\lambda N}, \quad \Omega = [\lambda N(4+\alpha)]^{1/4}, \quad k^2 \equiv m = \frac{1}{2}\left[1 - \sqrt{\frac{\alpha}{4+\alpha}}\right],$$
$$x_0 = \left(\frac{N}{\lambda}\right)^{1/4} \sqrt{\sqrt{4+\alpha} - \sqrt{\alpha}}\,. \tag{46}$$

Here $k$ is the parameter of the Jacobi elliptic functions [30] and $\mathrm{cn}(\tau)$ is a periodic function with the period

$$4K \;=\; 4\int_0^{\pi/2} \frac{d\theta}{\sqrt{1-k^2\sin^2\theta}}\,. \tag{47}$$

The equation for $D$ represents an elliptic variety of the Mathieu equation. Generically, it is not the simplest kind of equation, but in our case the solutions can be found in an explicit form. One of the solutions is

$$D_1(t) \;\propto \dot{x}(t) \;\propto \mathrm{sn}[\Omega t, k]\mathrm{dn}[\Omega t, k] \tag{48}$$

(we have chosen $t_0 = 0$). The second solution can be found from the condition that the time derivative of the Wronskian $W = \dot{D}_1 D_2 - \dot{D}_2 D_1$ vanishes. We find

$$D_2(t) \propto \dot{x}(t) \int^t \frac{dt'}{\dot{x}^2(t')} \;\;\propto\;\; \mathrm{sn}[\Omega t, k]\,\mathrm{dn}[\Omega t, k] \int^t \frac{dt'}{\mathrm{sn}^2[\Omega t', k]\,\mathrm{dn}^2[\Omega t', k]}\,. \tag{49}$$

Two independent solutions (48) and (49) exhibit oscillatory behaviour with constant and linearly rising amplitude, correspondingly (see Fig. 3). This linear growth has the same nature as in (20) and does not lead to any inconsistency.

The eigenvalues and eigenfunctions of the quantum Hamiltonian can also be found explicitly in this case [29]. Most of the states belong to the continuum spectrum (note again the similarity with the situation for the free PU oscillator with degenerate frequencies discussed in Sect.3). There are, however, not one, but many bands. The bands with positive energies start from $E = \omega$, $E = 2\omega$, etc, and extend upwards, while the bands with negative energies start from $E = -\omega$, $E = -2\omega$, etc, and extend downwards. The energies depend on $N$:

$$E_n(N) \;=\; \frac{\pi n}{2K(k)}\left[\lambda N\left(4 + \frac{\omega^4}{\lambda N}\right)\right]^{1/4} \tag{50}$$

with $n = \pm 1, \pm 2, \ldots$ and $K(k)$ defined in (47). This means that the levels with $E \in (\omega, 2\omega)$ and $E \in (-2\omega, -\omega)$ are not degenerate, the levels with $E \in (2\omega, 3\omega)$ and $E \in (-3\omega, -2\omega)$ are doubly degenerate, the levels with $E \in (3\omega, 4\omega)$ and $E \in (-4\omega, -3\omega)$ are 3-fold degenerate, etc. The wave functions are

$$\Psi_{nN}(x, D) \propto \frac{1}{\sqrt{N - V(x)}} \exp\left\{ iD\sqrt{2[N - V(x)]} + \frac{iE_n(N)}{\sqrt{2}} \int^x \frac{dy}{\sqrt{N - V(y)}} \right\}. \tag{51}$$

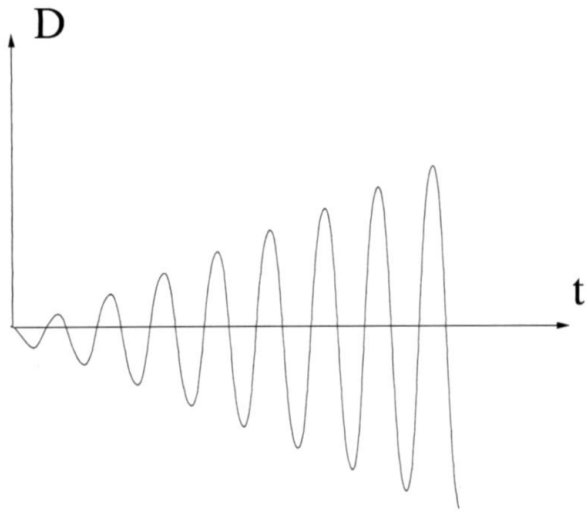


**Figure 3.** A typical behaviour of $D(t)$, as follows from the solution of (44).

If setting $n = 0$ in Eq.(50) and substituting this in (51), we obtain an infinity of zero-energy states with the wave functions

$$\Psi_{0N}(x, D) \propto \frac{1}{\sqrt{N - V(x)}} \exp\left\{iD\sqrt{2[N - V(x)]}\right\} . \tag{52}$$

The functions (52) are not normalizable, but we can in fact choose the basis with normalizable eigenfunctions. Indeed, one can show that any function

$$\Psi_0^{(g)}(x, D) = \int_{-\infty}^{\infty} g\left(\frac{P^2}{2} + V(x)\right) e^{iPD} dP \tag{53}$$

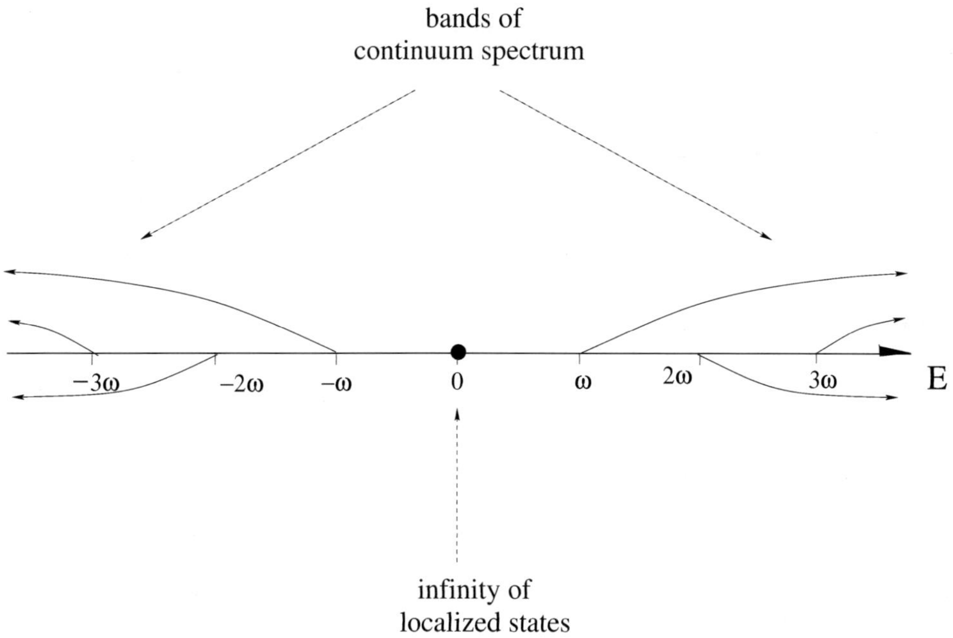


**Figure 4.** Spectrum of the Hamiltonian (39).

is a solution to the Schrödinger equation $\hat{H}\Psi = 0$. The solutions (52) are obtained from (53) by setting $g(N) = \delta(N - N_0)$. But we can also choose $g_k(N) = N^k e^{-N}$ giving the normalizable functions $\Psi_{0k}$. The full spectrum of the quantum Hamiltonian is represented in Fig. 4. It is Hermitian and the evolution operator is unitary.

*5.1.2. Ghosts in the Toda chain* There are many other exactly solvable systems with benign ghosts [31]. A large class of such systems is related to ordinary ghost free nonlinear exactly solvable systems.

Consider as an example the closed Toda chain with three particles. The Hamiltonian is

$$H = \frac{1}{2}(p_1^2 + p_2^2 + p_3^2) + V_{12} + V_{23} + V_{31}\,, \tag{54}$$

where $V_{12} = e^{q_1 - q_2}$, etc. Besides the energy, the system involves an obvious integral of motion $P = p_1 + p_2 + p_3$, as well as the less obvious cubic invariant

$$I = \frac{1}{3}(p_1^3 + p_2^3 + p_3^3) + p_1(V_{12} + V_{31}) + p_2(V_{12} + V_{23}) + p_3(V_{23} + V_{31})\,. \tag{55}$$

The Poisson brackets $\{H, I\}$, $\{H, P\}$ and $\{I, P\}$ vanish. For the system with three degrees of freedom, we have three integrals of motion that are in involution. The system is exaclty solvable.

In contrast to the open Toda chain, where the potential represents the sum of only two terms $V_{12} + V_{23}$, the potential in (54) keeps the three particles together. If we impose the requirement $q_1 + q_2 + q_3 = 0$ (the center of mass does not move), the motion is finite, with the classical trajectories representing complicated nonlinear oscillations. In the quantum problem [32], we impose the constraint $\hat{P}\Psi(q_1, q_2, q_3) = 0$, in which case the spectrum of $\hat{H}$ is discrete. The eigenfunctions of (54) are simultaneously the eigenfunctions of $\hat{I}$. [9]

Now note that $\hat{I}$ is cubic in momenta and its eigenvalues can be both positive and negative. The following simple property holds:

**Proposition 1.** *For each state with energy $E$ and positive eigenvalue of $\hat{I}$, there exists a state with the same energy and the negative eigenvalue of $\hat{I}$.*

*Proof.* The operator $\hat{H}$ is invariant under the following discrete symmetry:[10]

$$\hat{P}: \quad q_1 \leftrightarrow -q_2, \; q_3 \to -q_3 \,. \tag{56}$$

The Hamiltonian commutes with $\hat{P}$, so that, if $\Psi$ is an eigenfunction of $\hat{H}$, $\hat{P}\Psi$ is also an eigenfunction of $\hat{H}$ with the same energy. But $\hat{I}$ *anticommutes* with $\hat{P}$. It follows that, if $\hat{I}\Psi = \lambda\Psi$, then

$$\hat{I}\hat{P}\Psi \; = \; -\hat{P}\hat{I}\Psi \; = \; -\lambda\hat{P}\Psi \,.$$

□

In other words, the spectrum of $\hat{I}$ is symmetric under reflection $\lambda \to -\lambda$. To prove that this spectrum extends indefinitely down to $-\infty$ (and extends indefinitely up to $\infty$), it is sufficient to consider the *classical* trajectories with large values of $|I|$. The highly excited eigenstates of $\hat{H}$ and $\hat{I}$ are related to these trajectories by WKB correspondence. But it is obvious that both the energy and $|I|$ can be arbitrary large. Simply take the initial conditions with $q_j(0) = 0$ and $p_2(0) = p_3(0) = -p_3(0)/2 = M \to \infty$.

There is only one step to go. We *call* $\hat{I}$ rather that $\hat{H}$ the Hamiltonian. The new Hamiltonian does not have a ground state and is hence ghost-ridden. On the other hand, the spectral problem for $\hat{I}$ is quite well defined. There is no collapse and no loss of unitarity.

A nice feature of the Hamiltonian $\hat{I}$, compared to the quantum counterpart of (39) is its discrete spectrum.

For the quantum Toda problem, it does not matter which operator, $\hat{H}$ or $\hat{I}$, is called the Hamiltonian: they have the same spectrum. But the classical dynamics associates with $H$ and $I$ are different. For some reason, the scholars have not been so much interested in the latter. I only was able to find in the literature the Hamilton equations of motion associated with the cubic invariant for the open Toda chain [33]. In our case, the Hamilton equations of motion for $I$ read

$$\begin{aligned}
\dot{q}_1 &= p_1^2 + V_{12} + V_{31}, \qquad & \dot{p}_1 &= (p_1 + p_3)V_{31} - (p_1 + p_2)V_{12}\,, \\
\dot{q}_2 &= p_2^2 + V_{23} + V_{12}, \qquad & \dot{p}_2 &= (p_2 + p_1)V_{12} - (p_2 + p_3)V_{23}\,, \\
\dot{q}_3 &= p_3^2 + V_{31} + V_{23}, \qquad & \dot{p}_3 &= (p_3 + p_2)V_{23} - (p_3 + p_1)V_{31}\,,
\end{aligned} \tag{57}$$

to be compared with the standard equations of motion

$$\ddot{q}_1 = \dot{p}_1 = V_{31} - V_{12}, \qquad \ddot{q}_2 = \dot{p}_2 = V_{12} - V_{23}, \qquad \ddot{q}_3 = \dot{p}_3 = V_{23} - V_{31}\,. \tag{58}$$

To feel the difference, compare the classical trajectories associated with $H$ and $I$ (see Fig. 2).

The initial conditions for both trajectories are identical and the conserved values of $H$ and $I$ are the same. The patterns of the two trajectories are similar—oscillations with beats. But for the equations of motion based on $I$, the period of beats is smaller. [11]

[9] The operator $\hat{I}$ is defined by Eq. (55) where all the momenta $\hat{p}_j$ stay consistently on the left or consistently on the right: the two orderings give the same result.

[10] It is also invariant under cyclic permutations of $q_{1,2,3}$, but this is irrelevant for our purposes.

[11] There is no wonder that the classical trajectories associated with $H$ and $I$ are different. The phase space has four essential coordinates: $p_{1,2}, q_{1,2}$. The fixed values of two integrals of motion define a two-parametric surface in this space. Both trajectories lie on this surface, but they need not coincide.

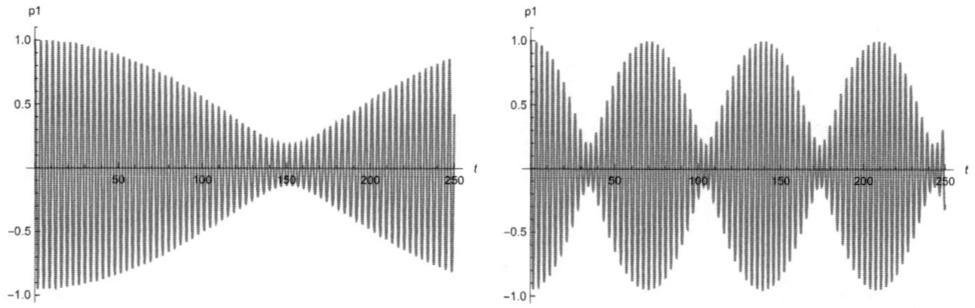


(a) $H$ (b) $I$

**Figure 5.** The dependence $p_1(t)$ for the equations of motion based on $H$ and on $I$. The inital conditions are $q_1(0) = q_2(0) = q_3(0) = 0, p_1(0) = 1, p_2(0) = p_3(0) = -.5$.

The solutions to (58) for $q_j(t)$ also oscillate, while these solutions to (57) grow quasilinearly. This difference is due to the fact that, for the system (57), the velocities do not coincide with the momenta.

The presence of two integrals of motion imposes essential constraints on the trajectories. We *conjecture* that they do not exhibit collapse for *any* choice of initial conditions, leaving the task to prove it rigourously to a trained mathematician.

### *5.2. Not integrable models*

For the ghosts to be benign, it is not necessary that the system is exactly solvable. Many systems with benign ghosts that are *not* exactly soluble are known.

Well, a reservation is of order here. In contrast to regular periodic trajectories of exactly solvable integrable models, the trajectories of not integrable models exhibit stochastic behaviour. We can solve in this case the equations of motion only numerically and can never be sure that a surprise does not await us beyond the horizon of our calculations. Still, these numerical studies, some samples of which will be presented below, give us strong indications that there are no such surprises.

All models considered in this subsection represent certain *deformations* of the exacty solvable models presented above. In Ref. [29] a model with the Hamiltonian

$$H = pP + DV'(x) - \frac{\gamma}{2}(D^2 + P^2) \tag{59}$$

was considered. This model is not integrable and can only be studied numerically. This numerical study exhibits a benign behaviour of classical trajectories. The latter are even more handy than for the model (39) — if $\gamma$ is not too large, the linear growth of the amplitudes of the oscillations is absent. [12] One observes instead a finite motion with beats (Fig. 6).

Alternatively, one can add a nonlinear term in the Hamiltonian (6). One has found out that for many such deformations, the trajectories still behave in a benign way, not exhibiting collapse. This happens for the Hamiltonian [14]

$$H = \frac{P_1^2 + \omega_1^2 X_1^2}{2} - \frac{P_2^2 + \omega_2^2 X_2^2}{2} + \kappa(X_1 - X_2)(X_1 + X_2)^3\,; \tag{60}$$

[12] A numerical analysis shows that, if $\gamma$ exceeds some limit that depends on the values of $\omega$, $\lambda$ and of the initial conditions, the amplitude starts to grow again, and this growth is exponential. Still, we did not observe a collapse and the solution exists at all the times.

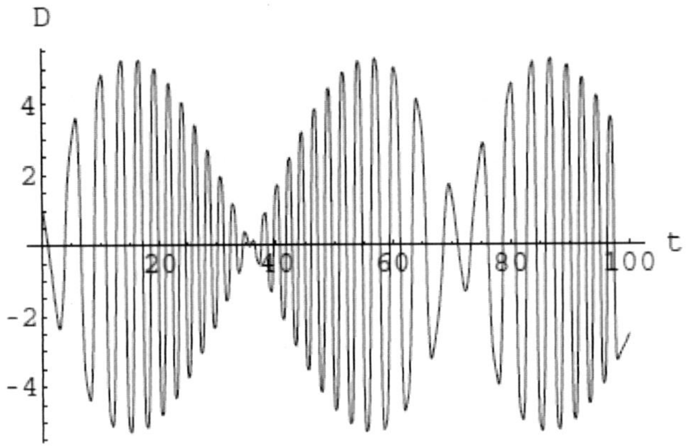


**Figure 6.** A typical trajectory for the Hamiltonian (59).

for the Hamiltonian [34, 35]

$$H = \frac{P_1^2 + \omega_1^2 X_1^2}{2} - \frac{P_2^2 + \omega_2^2 X_2^2}{2} + \lambda \sin^4(X_1 + X_2) \tag{61}$$

and for the Hamiltonian [22]

$$H = \frac{P_1^2 + \omega_1^2 X_1^2}{2} - \frac{P_2^2 + \omega_2^2 X_2^2}{2} + \lambda(X_1^4 - X_2^4) + \mu X_1^2 X_2^2, \qquad \lambda \gg \mu. \tag{62}$$

One can also perturb the cubic Toda invariant (55) by adding to it an oscillator potential:

$$I \to \tilde{I} = I + \frac{\alpha}{2}[(q_1 - q_2)^2 + (q_1 - q_3)^2 + (q_2 - q_3)^2]. \tag{63}$$

The Poisson bracket $\{\tilde{I}, p_1 + p_2 + p_3\}$ still vanish, but the bracket $\{\tilde{I}, H\}$ does not. Thus, $H$ is not an integral of motion anymore and the system is not integrable.

The trajectories can be found numerically. The trajectory for the Hamiltonian (63) with $\alpha = 1$ and the same initial conditions as in Fig. 5 is shown in Fig. 7. We observe quasiperiodic oscillations (the regularity of trajectories seen in Fig. 5 is now lost, which confirms the fact that the system is no longer integrable). There is no collapse.

## 6. Benign ghosts: field theory

### *6.1. A nonintegrable model*

The first example of a field theory enjoying benign ghosts was constructed in [36]. It is a straighforward generalization of (39). The Lagrangian of the model is

$$\mathcal{L} = \partial_\mu \phi \partial_\mu D - DV'(\phi), \tag{64}$$

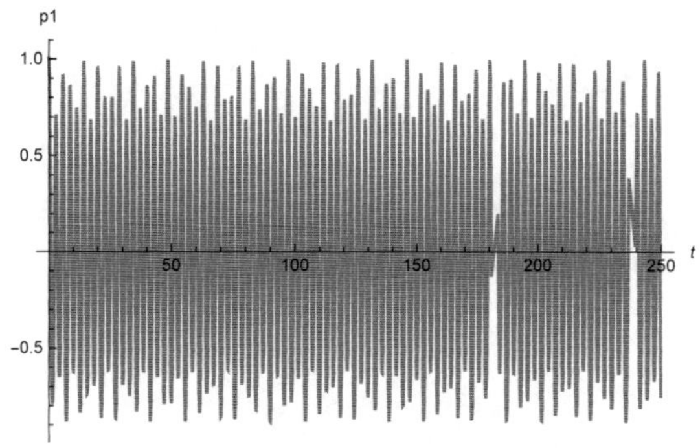


**Figure 7.** A typical trajectory for the Hamiltonian (63).

where $\mu = 0, 1$ (the model is 2-dimensional) and

$$V(\phi) = \frac{\omega^2\phi^2}{2} + \frac{\lambda\phi^4}{4}, \qquad \lambda > 0\,. \tag{65}$$

The model has two integrals of motion: the energy

$$E = \int dx \left[\dot{\phi}\dot{D} + \partial_x\phi\,\partial_x D + D\phi(\omega^2 + \lambda\phi^2)\right], \tag{66}$$

which can be both positive and negative, and the positive definite

$$N = \int dx \left\{\frac{1}{2}\left[\dot{\phi}^2 + (\partial_x\phi)^2\right] + \frac{\omega^2\phi^2}{2} + \frac{\lambda\phi^4}{4}\right\}. \tag{67}$$

With only two integrals of motion for the infinite number of degrees of freedom, the model is not integrable and can only be solved numerically. We did so in the case of only one spatial dimension. The equations of motion are

$$\begin{aligned} \Box\phi + \omega^2\phi + \lambda\phi^3 &= 0 \\ \Box D + D(\omega^2 + 3\lambda\phi^2) &= 0\,. \end{aligned} \tag{68}$$

We see that $\phi(x,t)$ satisfies a nonlinear wave equation. The solutions to this equation cannot grow — a growth is incompatible with the conservation of $N$. The amplitude of the oscillations of the field $D(x,t)$ can grow with time, however.

We played with different values of the parameters $\omega, \lambda$ and with different initial conditions and never found a collapse, only at worst a linear growth of $D(x,t)$ with time. A typical behaviour is shown in Fig. 8, where the dispersion $d(t) = \sqrt{\langle D^2\rangle_x}$ is plotted as a function of time. This particular graph corresponds to the choice of the parameters $\omega = \lambda = 1$, to the length of the spatial box $L = 20$ (periodic boundary conditions in the spatial direction were imposed), while the initial conditions at $t = 0$ were chosen as follows:

$$\begin{aligned} \phi(x,0) = 5e^{-x^2}, \qquad & D(x,0) = \cos(\pi x/20)\,, \\ \dot{\phi}(x,0) \;=\; & \dot{D}(x,0) \;=\; 0\,. \end{aligned} \tag{69}$$

The dispersion undergoes stochastic fluctuations, as is natural for a non-integrable nonlinear system. However, on the average, $d(t)$ grows linearly with time, similar to Fig. 3, and there is no trace of collapse.

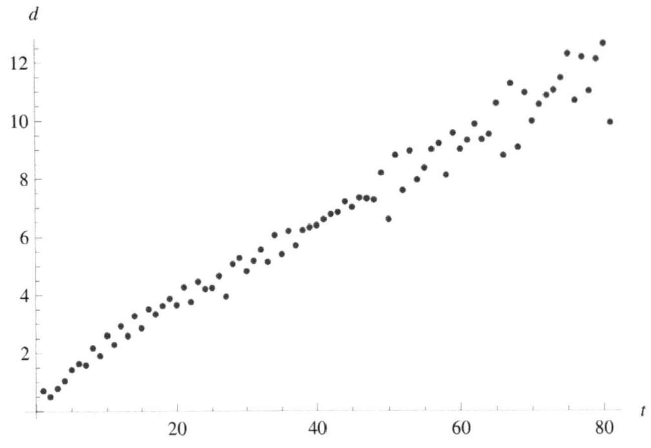


**Figure 8.** Dispersion $d = \sqrt{\langle D^2 \rangle}$ as a function of time with the initial conditions (69).

*6.2. Integrable models*
There are many exactly solvable $(1+1)$ systems, which are characterized by an infinite number of integrals of motion. *Each* of the latter can be chosen as a Hamiltonian, and we obtain thereby a set of higher-derivative exactly solvable $(1+1)$-dimensional field theories. Take as an example the Sine-Gordon model. Its equation of motion reads

$$\phi_{tt} - \phi_{xx} = -\sin\phi \,. \tag{70}$$

The conventional energy of the system,

$$E = \int_{-\infty}^{\infty} dx \left[ \frac{1}{2}(\phi_t^2 + \phi_x^2) + (1 - \cos\phi) \right] \tag{71}$$

is conserved, and this gives the standard Hamiltonian.

But there is an infinity of other integrals of motion. The next in complexity after (71) is [37]

$$E_4 = \int dx \left[ \frac{1}{4}(\phi_t - \phi_x)^4 - (\phi_{tt} - 2\phi_{tx} + \phi_{xx})^2 - (\phi_t - \phi_x)^2 \cos\phi \right] . \tag{72}$$

This expression involves higher derivatives, is not bounded from below, and the Hamiltonian $H_4$ is ghost-ridden. To the best of my knowledge, the corresponding dynamical field equations were never studied (not speaking of quantum dynamics!), but such a study would be very interesting.

## 7. Discussion

Our main message is that the presence of ghosts (unboundness of the spectrum of the Hamiltonian) does not necessarily mean a collapse and disaster — there are nontrivial interacting ghost-ridden systems where this does not happen. In spite of the presence of ghosts, HD theories should be treated seriously, one should not throw them immediately to a garbage busket, as it was habitually done for many years. Instead, they should be interrogated one by one: the question whether a HD theory is benign or not (involves or does not involve the collapse) is a nontrivial dynamical question.

Unfortunately, we do not know today a lot of such benign ghost theories, and our dream to find a HD theory which might play the role of the TOE is far from being fulfilled. The only known benign field-theory systems that we analyzed live in one spatial and one temporal dimension, while the TOE that includes gravity should be multidimensional, $D > 4$.

In Sect. 6, we discussed only the classical dynamics of the HD field theory models, but a hard question is *what is their quantum dynamics*? Ordinary field theories may be studied perturbatively. We single out the free Hamiltonian $H_0$, find its spectrum and treat nonlinear terms as perturbation. If the spectrum of $H_0$ includes a vacuum state and particle excitations above it, one may define the scattering matrix whose elements represent transition amplitudes between the particle asymptotic states. These transition amplitudes can in many cases be found perturbatively.

Here this approach does not work. Look at the system (64), (65). If the interaction is switched off, each level involves an infinite degeneracy: an "ordinary particle" (an excitation of the positive-energy oscillator) has the same energy as the state with 2 ordinary particles and one ghost, 3 particles and 2 ghosts, etc. And when the interaction is present ($\lambda \neq 0$), the spectrum is radically reshuffled. If we single out a particular Fourier mode and suppress artificially its interaction with the other modes, the Hamiltonian in this sector looks similar to (42) and its spectrum includes continuum bands as in Fig. 4. It is rather clear that even if $\lambda$ is small, the spectrum of the full Hamiltonian for the system (64) has little to do with the spectrum of the free system. [13] If trying to describe it in terms of particles, these particles (both the ordinary particles and the ghosts) must have a continuum spectrum of *masses*. [14]

Without any doubt, all this looks very strange and unusual. But strange and unusual does not mean self-contradictory and inconsistent. And one can be almost sure that the TOE, whatever it is, must be strange and unusual. After all, people tried many less strange and more usual candidates for the role of the TOE, including strings, but these efforts have not been successful so far.

We have just recalled confining theories, where the spectra of the free Hamiltonian and of the full Hamiltonian are rather different. Bearing this in mind, the scenario of *confinement of ghosts* was discussed in the literature. The main idea is that even if the ghosts are present in the bare Hamiltonian, they may disappear from the physical spectrum due to confinement [41, 42]. This conjecture was inspired by the hypothesis [41, 43, 44] that the fundamental gravity action is not the Einstein-Hilbert action, but the conformally invariant Weyl HD action

$$S^{\text{Weyl}} \sim \frac{1}{f^2} \int \sqrt{-g}\, d^4x\, C_{\mu\nu\lambda\sigma} C^{\mu\nu\lambda\sigma}\,, \tag{73}$$

where

$$C_{\mu\nu\lambda\sigma} = R_{\mu\nu\lambda\sigma} - \frac{1}{2}(g_{\mu\lambda}R_{\nu\sigma} - g_{\mu\sigma}R_{\nu\lambda} - g_{\nu\lambda}R_{\mu\sigma} + g_{\nu\sigma}R_{\mu\lambda}) + \frac{1}{6}(g_{\mu\lambda}g_{\nu\sigma} - g_{\mu\sigma}g_{\nu\lambda}) \tag{74}$$

is the Weyl tensor, or a supersymmetric generalization of this action. Conformal gravity and conformal supergravities are asymptotically free [45], so that classical (super)conformal symmetry is broken there by quantum effects. Like in QCD, this brings about a dimensional parameter — the scale where the coupling constant $f$ becomes large. It is natural to assume that this scale is of order of the Planck mass $M_{Pl}$. The EH term in the effective low-energy action along with the cosmological constant term are *induced* due to quantum effects by Sakharov's mechanism [46, 47]. [15]

[13] Also in ordinary QCD-like theories without ghosts but with confinement the spectrum of the full theory has nothing to do with the spectrum of $H_0$. Still, in those cases *(i)* there is a kinematical region where the perturbative spectrum and perturbative scattering amplitudes make sense; *(ii)* even though the spectra of excitations for the full Hamiltonian $H$ and for the free Hamiltonian $H_0$ are very different, the ground vacuum state is well defined in both cases and the physical spectrum involves only ordinary particles carrying positive energies. It is difficult to *calculate* the hadron scattering amplitudes, but they can be *defined* and *measured*.

[14] One can recall at this point supermembranes, where the mass spectrum is also continuous [38, 39].

[15] Nobody has an idea today how to get rid of the huge cosmological constant, which appears in all approaches. This is a separate very hard question.

The propagator of the metric following from (73) behaves as

$$D(k) \sim \frac{f^2}{k^4}, \tag{75}$$

which brings about the confining static potential

$$V(r) \propto f^2 \int \frac{d^3k}{k^4} e^{i\boldsymbol{kr}} \propto f^2 r. \tag{76}$$

One may thus suppose that this potential confines the ghost degrees of freedom and only ordinary excitations are left in the physical spectrum.

However,

- We now do not think that *any* gravity theory, regardless of whether it is the conventional Einstein's gravity or a HD gravity, is a part of the TOE. We argued in the Introduction that the TOE should rather be a HD theory living in a flat higher-dimensional bulk, while the gravity is an effective theory living on the 3-brane, which is our Universe.
- As was mentioned above, the physical spectrum can be very different from the spectrum of the free Hamiltonian, and many degrees of freedom that are seen in the latter may indeed disappear from the spectrum — be confined. But a general theorem proven in [4,8] and valid for *any* nondegenerate HD system dictates that the physical spectrum of its Hamiltonian does not have a ground state and must involve ghosts. Thus, confinement or any other mechanism can drastically affect the spectrum of a HD Hamiltonian, but it does not allow one to get rid of all the ghosts.

**Acknowledgements**
It is a pleasure for me to thank Andreas Fring and Francisco Correa for their invitation to give a talk at the PT seminar and all the participants of the seminar for their interest and questions.

## References

[1] Gödel K 1949 An example of a new type of cosmological solutions of Einstein's field equations of gravitation *Rev. Mod. Phys.* **21** 447
[2] Smilga A V 2003 Quantum gravity as Escher's dragon *Phys. Atom. Nuclei* **66** (2003) 2092 (*Preprint hep-th/02120133*)
[3] Smilga A V 2006 *6D superconformal theory as the theory of everything* (Gribov Memorial volume ) ed Yu L Dokshitzer, P Levai and Y Nyiri (Singapore: World Scientific) p 443 (*Preprint* hep-th/0509022)
[4] Smilga A V 2017 Classical and quantum dynamics of higher-derivative systems *Int. J. Mod. Phys.* A **32** 1730025 (*Preprint* arXiv:1710.11538 [hep-th])
[5] Ostrogradsky M 1850 Mémoire sur les équations différentielles relatives au problème des isopérimètres *Mem. Ac. St. Petersbourg* **VI 4** 385
[6] Woodard R P 2015 The theorem of Ostrogradsky *Scholarpedia* **10** 32243 (*Preprint* arXiv:1506.02210 [hep-th])
[7] Pais A and Uhlenbeck G E 1950 On field theories with non-localized action *Phys. Rev.* **79** 145
[8] Raidal M and Veermäe H 2017 On the quantisation of complex higher derivative theories and avoiding the Ostrogradsky ghost *Nucl. Phys.* B **916** 607 (*Preprint* arXiv:1611.03498 [hep-th])
[9] Chen T, Fasiello M, Lim E A and Tolley A J 2013 Higher derivative theories with constraints: exorcising Ostrogradsky's ghost *J. Cosmol. Astropart. Phys.* JCAP02(2013)042 (*Preprint* arXiv:1209.0583 [hep-th])
[10] Mannheim P D and Davidson A 2005 Dirac quantization of Pais-Uhlenbeck fourth order oscillator *Phys. Rev.* A **71** 042110 (*Preprint* hep-th/0408104)
[11] Anderson A 1994 Canonical transformations in quantum mechanics *Ann. Phys.* **232** 292
[12] Bolonek K and Kosinski P 2006 Comments on "Dirac quantization of the Pais-Uhlenbeck fourth order oscillator" *Preprint* quant-ph/0612009
[13] Bender C M and Mannheim P D 2008 Exactly solvable *PT*-symmetric Hamiltonian having no Hermitian counterpart *Phys. Rev.* D **78** 025022 (*Preprint* arXiv:0804.4190 [hep-th])
[14] A.V. Smilga 2009 Comments on the dynamics of the Pais-Uhlenbeck oscillator *SIGMA* **5** 017 (*Preprint* arXiv:0808.0139 [quant-ph])

[15] Smilga A V 2015 Exceptional points of infinite order give a continuous spectrum *Int. J. Theor. Phys.* **54** 3900 (*Preprint* arXiv:1409.8450 [math-ph])
[16] Hawking S W and Hertog T 2002 Living with ghosts *Phys. Rev.* D **65** 103515 (*Preprint* arXiv:hep-th/0107088)
[17] Andrzejewski K , Gonera J and Maslanka P 2011 Euclidean path integral and higher-derivative theories *Prog. Theor. Phys.* **125** 247 (*Preprint* arXiv:0904.3055 [hep-th])
[18] Ketov S V, Michiaki G and Yumibayashi T 2012 *Quantizing with a higher time derivative* ( Advances in quantum field theory) ed Ketov S (London: InTech) (*Preprint* arXiv:1110.1155 [hep-th])
[19] Feynman R P and Hibbs A R 1965 *Quantum mechanics and path integrals* (New York: McGraw-Hill)
[20] Blinnikov S I and Nikitin N V 1982 *Approximations to path integrals and spectra of quantum systems* Preprint physics/0309060
[21] Smilga A V 2001 *Lectures on quantum chromodynamics* (Singapore: World Scientific)
[22] Ilhan I B and Kovner A 2013 Some comments on ghosts and unitarity: the Pais-Uhlenbeck oscillator revisited *Phys. Rev.* D **88** 044045 (*Preprint* arXiv:1301.4879 [hep-th])
[23] Bender C M and Mannheim P D 2008 No ghost theorem for the fourth-order derivative Pais-Uhlenbeck oscillator *Phys, Rev. Lett.* **100** 110402 (*Preprint* arXiv:0706.0207 [hep-th])
[24] Smilga A V 2005 Benign vs. malicious ghosts in higher-derivative theories *Nucl. Phys.* B **706** 598 (*Preprint* hep-th/0407231)
[25] Carroll S N, Hoffman M and Trodden M 2003 Can the dark energy equation-of-state parameter $w$ be less than $-1$? *Phys. Rev.* D **68** 023509 (*Preprint* astro-ph/0301273)
[26] Case K M 1950 Singular potentials *Phys. Rev.* **80** 797
[27] Meetz K 1964 Singular potentials in non-relativistic quantum mechanics *Nuov. Cim.* **34** 690
[28] Perelomov A M and Popov V S 1970 Collapse onto scattering center in quantum mechanics *Teor. Mat, Fiz.* **4** 48
[29] Robert D and Smilga A V 2008 Supersymmetry vs. ghosts *J. Math. Phys.* **49** 042104 (*Preprint* math-ph/0611023)
[30] Gradstein I S and Ryzhik I M 1965 *Tables of integrals, series and products* (New York: Academic Press)
[31] Smilga A V 2021 On exactly solvable ghost-ridden systems *Phys.Lett.* A **389** (2021) 127104 (*Preprint* arXiv: 2008.12966 [hep-th])
[32] Gutzwiller M G 1981 The quantum mechanical Toda lattice II *Ann. Phys.* **133** 304
[33] Perelomov A M 1990 *Integrable systems of classical mechanics and Lie algebras* (Basel: Birkhauser) p.208
[34] Pavšič M 2013 Stable self-Interacting Pais-Uhlenbeck oscillator *Mod. Phys. Lett.* A **28** 1350165 (*Preprint* arXiv:1302.5257)
[35] Pavšič M 2016 Pais-Uhlenbeck oscillator and negative energies *Int. J. Geom. Meth. Mod. Phys.* **13** 1630015 (*Preprint* arXiv:1607.06589[gr-qc])
[36] Smilga A V 2014 Supersymmetric field theory with benign ghosts *J. Phys.* A **47** 152001 (*Preprint* arXiv:1306.6066 [hep-th])
[37] Lamb G L Jr. 1970 Higher conservation laws in ultrashort optical pulse propagation *Phys. Lett.* **32A** 251
[38] Smilga A V 1989 *Super-Yang-Mills quantum mechanics and supermembrane spectrum* (Proc. Conf. Supermembranes and physics in 2+1 dimensions Trieste 1982) ed M J Duff, C N Pope and E. Sezgin p. 182 (Singapore: World Scientific) (*Preprint* arXiv:1406.5987 [hep-th]
[39] de Witt B, Lüscher M and Nicolai H 1989 The supermembranes is unstable *Nucl. Phys.* B **320** 135
[40] Ivanov E A, Smilga A V and Zupnik B M 2005 Renormalizable supersymmetric gauge theory in six dimensions *Nucl. Phys.* B **726** 131
[41] Smilga A V 1982 *Spontaneous generation of the Newton constant in the renormalizable gravity theory* (Proc. Conf. Group theoretical methods in physics Zvenigorod 1982) ed M A Markov, V I Man'ko and A E Shabad vol 2 p 73 (London: Harwood Academ. Publishers) (*Preprint* arXiv:1406.5613 [hep-th])
[42] Donoghue J F 2017 Quartic propagators, negative norms and the physical spectrum *Phys. Rev.* D **96** 044007 (*Preprint* arXiv:1704.01533 [hep-th])
[43] Zee A 1982 A theory of gravity based on the Weyl–Eddington action *Phys. Lett.* B **109** 183
[44] Zee A 1983 Einstein gravity emerging from quantum Weyl gravity *Ann. Phys.* **151** 431
[45] Fradkin E S and Tseytlin A A 1982 Renormalizable asymptotically free quantum theory of gravity *Nucl. Phys.* B **201** 469
[46] Sakharov A D 1967 Vacuum quantum fluctuations in curved space and the theory of gravitation *Dokl. Akad. Nauk Ser. Fiz.* **177** *(1967) 70* (English translation: *Sov. Phys. Dokl.* **12** 1040)
[47] Adler S L 1982 Einstein gravity as a symmetry-breaking effect in quantum field theory *Rev. Mod. Phys.* **54** 729

PTSeminar2020 IOP Publishing
Journal of Physics: Conference Series **2038** (2021) 012024 doi:10.1088/1742-6596/2038/1/012024

# Fall-to-the-centre as a $\mathcal{PT}$ symmetry breaking transition

**Sriram Sundaram**[1], **C P Burgess**[1,2] **and Duncan H J O'Dell**[1]

[1] Department of Physics and Astronomy, McMaster University, 1280 Main St. W., Hamilton, Ontario, Canada, L8S 4M1

[2] Perimeter Institute for Theoretical Physics, 31 Caroline St. N., Waterloo, Ontario, Canada, N2L 2Y5

E-mail: `dodell@mcmaster.ca`

**Abstract.** The attractive inverse square potential arises in a number of physical problems such as a dipole interacting with a charged wire, the Efimov effect, the Calgero-Sutherland model, near-horizon black hole physics and the optics of Maxwell fisheye lenses. Proper formulation of the inverse-square problem requires specification of a boundary condition (regulator) at the origin representing short-range physics not included in the inverse square potential and this generically breaks the Hamiltonian's continuous scale invariance in an elementary example of a quantum anomaly. The system's spectrum qualitatively changes at a critical value of the inverse-square coupling, and we here point out that the transition at this critical potential strength can be regarded as an example of a $\mathcal{PT}$ symmetry breaking transition. In particular, we use point particle effective field theory (PPEFT), as developed by Burgess *et al* [1], to characterize the renormalization group (RG) evolution of the boundary coupling under rescalings. While many studies choose boundary conditions to ensure the system is unitary, these RG methods allow us to systematically handle the richer case of nonunitary physics describing a source or sink at the origin (such as is appropriate for the charged wire or black hole applications). From this point of view the RG flow changes character at the critical inverse-square coupling, transitioning from a sub-critical regime with evolution between two real, unitary fixed points ($\mathcal{PT}$ symmetric phase) to a super-critical regime with imaginary, dissipative fixed points ($\mathcal{PT}$ symmetry broken phase) that represent perfect-sink and perfect-source boundary conditions, around which the flow executes limit-cycle evolution.

## 1. Introduction

In the presence of an attractive $1/r$ potential a classical particle will follow either an elliptic, hyperbolic or parabolic trajectory. However, in more singular potentials the particle can exhibit a new type of behaviour where it spirals down onto the origin, a phenomenon called "fall to the centre" [2, 3, 4]. The least singular potential where this occurs is the attractive inverse square potential $-g/r^2$ which has precisely the same radial dependence as the centrifugal barrier and hence can overcome it for large enough $g$.

The behaviour of quantum particles in an attractive inverse square potential has been studied in an experiment by Denschlag, Umshaus and Schmiedmayer [5] who scattered cold lithium atoms from a thin (radius $\sim 1\mu$m) charged wire and observed fall to the centre. The existence of the inverse square potential in this case is easily understood: consider a neutral but polarizable atom interacting with a charged wire; the radial electric field $\mathcal{E}$ emanating from the wire falls

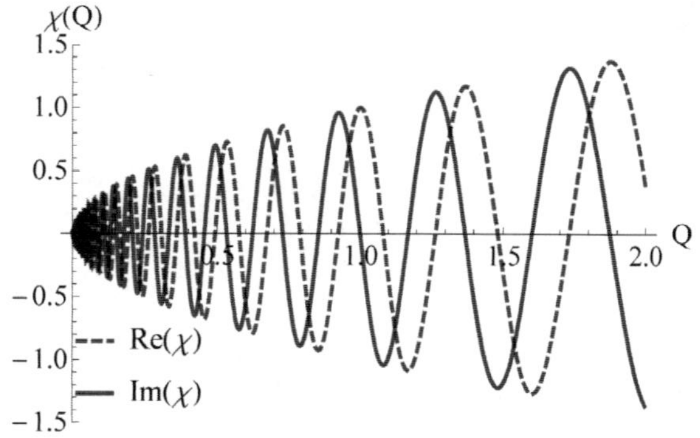


**Figure 1.** The wavefunction $Q^{-i|\sigma|+1/2} = Q^{1/2}e^{-i|\sigma|\log Q}$ in the supercritical regime with $|\sigma| = 20$. Note the logarithmic phase singularity at Q = 0 where the phase is undefined because it oscillates so fast it takes all values at once. The purple dashed line shows the real part and the black solid line shows the imaginary part.

off as $1/r$ and induces a dipole moment $d$ in the particle that in the linear response regime is proportional to the strength of the field. The interaction energy $-(1/2)d\mathcal{E}$ between the atom and the wire must then go as the inverse square $-1/r^2$ of the distance between the atom and the wire. The Schrödinger equation describing this situation takes the form

$$-\frac{\hbar^2}{2m}\frac{\partial^2\psi}{\partial Q^2} - \frac{g}{Q^2}\psi = \mathrm{i}\hbar\frac{\partial\psi}{\partial t} \tag{1}$$

(since in this paper we will only consider one-dimensional problems we have replaced the radial coordinate $r$ with the coordinate $Q$ which lies in the range $-\infty \leq Q \leq \infty$). Other physical situations where the inverse square potential appears include the Efimov effect (a counter-intuitive family of bound states of three particles with an infinite bound state spectrum given by a geometric series) [6, 7, 8, 9, 10, 11, 12, 13, 14, 15, 16, 17], Calogero-Sutherland models [18], and near the event horizon of black holes [19, 20, 21]. Analogue inverse square potentials also appear in various ways in optics, such as in the optical coherence of sunlight [22], and in Maxwell fisheye lenses where the refractive index takes an inverse square form [23, 24, 25].

The peculiar properties of inverse square potentials in quantum mechanics have been widely discussed in the literature, and good introductions can be found in Refs. [26, 27]. The key feature of Eq. (1) is that it is scale invariant under joint continuous scaling of space and time, $Q \to sQ$ and $t \to s^2t$. This means that, unlike the Coulomb potential where the Bohr radius provides a length scale, there is no natural length scale in the inverse square problem. Indeed, even if we consider the one dimensional case (as we do here) such that there is no centrifugal barrier, the system is still saved from collapse in the Coulomb case by the zero-point kinetic energy $\sim \frac{\hbar^2}{2ma^2}$ associated with a state of size $a$ beating the potential energy $-\frac{q_1q_2}{4\pi\epsilon_0 a}$ at small enough $a$. The same is *not* true in the inverse square case where both the zero-point energy and the potential scale in the same way. In fact, if we can find one solution of Eq. (1) then we have found an infinite family of them related by $s$. Thus, if we can find one bound state with (negative) energy $E$ then there is a continuum of bound states $s^2E$ with every possible negative energy and the spectrum is therefore unbounded from below.

Despite this pathology, Eq. (1) has exact stationary solutions $\psi(Q,t) = \chi(Q)\exp(-\mathrm{i}Et/\hbar)$, where $\chi(Q)$ are Hankel functions in the case of scattering states ($E > 0$) and modified Bessel functions in the case of bound states ($E < 0$). The problem lies, however, in finding appropriate

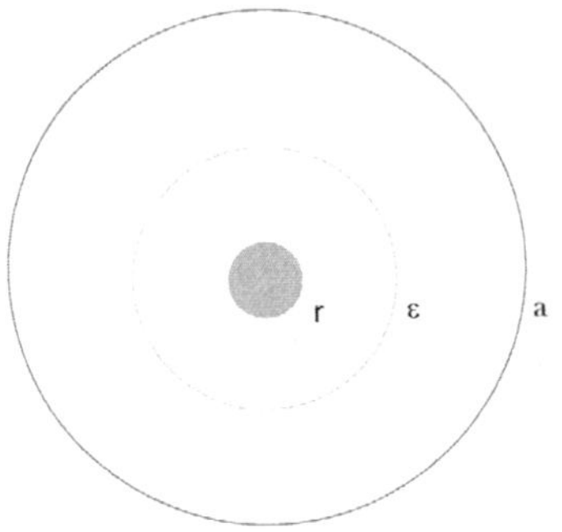


**Figure 2.** A figurative illustration of the hierarchy of length scales in the regularization of the fall-to-the-centre problem. The length $a$ is the scale we have access to in experiments which is very large compared to the regulator scale $\epsilon$ which is in turn very large compared to $r$, the size of the source located at the origin, i.e. $r \ll \epsilon \ll a$. The boundary condition is derived using the PPEFT action that describes the properties of the source at the regulator scale $\epsilon$ (which is arbitrary). Renormalizing the source-bulk coupling ensures that physical predictions at scale $a$ are independent of the regulator $\epsilon$.

boundary conditions, i.e. the linear combination of these solutions that describes a particular physical situation. In the Coulomb problem we are able to choose one of the two linearly independent solutions to the radial equation simply by its asymptotic behaviour: at small distances where the centrifugal barrier dominates, the solution to the hydrogenic radial equation with angular momentum quantum number $l$ (an integer) takes the form

$$u(r) = A\, r^{l+1} + B\, r^{-l} \tag{2}$$

and since $r^{-l}$ blows up as $r \to 0$ (assuming $l \neq 0$) we can set $B = 0$ without further thought. The same logic cannot in general be applied to the inverse square potential. At small distances where we can ignore the eigenvalue $E$ in comparison to the other terms in the eigenvalue equation, the solutions are of the form

$$\chi(Q) = C_+\, Q^{1/2+\sigma} + C_-\, Q^{1/2-\sigma} \tag{3}$$

where $\sigma = \sqrt{1/4 - \alpha}$ and $\alpha \equiv 2mg/\hbar^2$. Defining $\alpha_c \equiv 1/4$, which is the critical value at which fall to the centre first takes place, we see that when $\alpha < \alpha_c$ (subcritical regime) the two solutions are distinguishable in terms of their behaviour as $Q \to 0$, but when $\alpha > \alpha_c$ (supercritical regime) the modulus of both solutions is identical and they only differ by a phase (see Figure 1), and there is no simple criterion like boundedness at the origin for choosing one solution over the other. Clearly we need to supply a boundary condition to fix the ratio $C_+/C_-$; the fact we did not need to do this in the Coulomb case is a reflection of the latter's rather special properties (superintegrability). Furthermore, the logarithmic phase divergence present in both solutions in Eq. (3) indicates that our problem is missing some physics near the origin.

A resolution to these issues is suggested by the atom-wire problem. The inverse square potential is merely the long range behaviour of the atom-wire interaction and when they approach closely the fact that the wire has a finite width becomes important. This introduces the microscopic length scale $r$ (the radius of the wire), see Figure 2, that provides a short-distance cut-off that regulates the singularity in the inverse square potential and breaks the continuous scaling symmetry. The breaking of classical continuous scale invariance in a quantum theory is termed a *scale anomaly* [26]. Although anomalies were originally conceived in the context of particle physics, they have recently been realized in ultracold atomic systems through the Efimov effect [9, 10, 11, 12, 13, 14, 15, 16, 17], and frequency shifts of breathing modes in trapped 2D Fermi gases [28, 29, 30, 31, 32, 33]. A scale anomaly has also been observed in graphene where the relativistic nature of the dispersion relation means that it is the Coulomb potential that leads to a scale invariant system in that case [34].

The 'high-energy' physics at the microscopic length scale $r$ might be complicated, and we might even be ignorant of its detailed form, but if all we can observe is the low-energy behaviour at large distances then we can treat the problem using the techniques of effective field theory which provide systematic methods for regulating the theory. The version that we apply in this paper is known as point particle effective field theory (PPEFT) and allows us to specify the boundary condition in an intuitive and physical way through the presence of an imagined point 'particle' that sits the origin [1, 35]. The theory is conveniently formulated in terms of the action $S_{\text{total}} = S_B + S_p$, where $S_B$ is the action for the 'bulk' field $\chi$ that has Eq. (1) as its equation of motion, and $S_p$ is the point particle effective action that describes the microscopic physics localized at the origin and how it couples to the bulk field. In this paper we will review the application of PPEFT to the inverse square problem and confirm that at lowest order the effect of the microscopic physics on the inverse square problem is simply to add a *compulsory* Dirac $\delta$-function to Eq. (1) so that the time-independent Schrödinger equation reads

$$-\frac{d^2\chi}{dQ^2} - \frac{\alpha}{Q^2}\chi + \lambda\delta(Q)\chi = k^2\chi \tag{4}$$

where $k^2 = 2mE/\hbar^2$. By integrating this equation over an infinitesimal range $-\epsilon < Q < \epsilon$ we can see that the value of the coupling constant $\lambda$ determines the jump in gradient of $\chi$ at the origin and provides the necessary boundary condition at the origin. This boundary condition is unitary (with hermitian Hamiltonian) when $\lambda$ is real, but is generically not unitary if $\lambda$ is complex.

The boundary condition fixes the ratio of integration constants, $C_+/C_-$, and this ratio determines the problem's intrinsic length scale: the ratio of the solutions in Eq. (3) gives the length scale $L = (C_+/C_-)^{-\frac{1}{2\sigma}}$ which breaks the scale invariance as long as both $C_+$ and $C_-$ are finite. In this picture the coupling $\lambda$ is to be regarded as depending on the position $\epsilon$ at which the boundary condition is imposed in such a way as to ensure that the physical scale $C_+/C_-$ remains fixed. This defines a renormalization-group (RG) flow for $\lambda$ under scale transformations. The interpretation of this flow is clearest when $\sigma$ is real, i.e. $\alpha < \alpha_c$. Through the boundary condition $\lambda$ controls the relative weight of the two solutions $Q^{1/2\pm\sigma}$ at $Q = \pm\epsilon$. This relative weight changes as $Q$ changes, with the $Q^{1/2+\sigma}$ solution eventually dominating at large length scales, corresponding to an infrared (IR) fixed point for which the wavefunction will become scale invariant. In other words, we flow towards a scale invariant IR fixed point associated with $C_- = 0$. Conversely, in the opposite ultraviolet (UV) limit we have a scale invariant fixed point for which $Q^{1/2-\sigma}$ dominates, associated with $C_+ = 0$. These simple arguments will be made more concrete in the rest of this paper, but the upshot is that $\lambda$ depends on the energy/length scale of our observations, i.e. undergoes renormalization [36, 26, 21, 37, 38, 39, 1].

The full renormalization group (RG) flow of $\lambda$ with scale is nonlinear. In the subcritical regime there are two fixed points where the solution is scale invariant, as argued above. The UV fixed point is unstable, the IR one is stable, and the flow between them can in general pass through complex values of $\lambda$ (see Figures 3 and 4) so that the Hamiltonian in Eq. (4) is generically non-hermitian, although in the special case where the flow starts on the real axis $\lambda$ will always remain real. Either way, all trajectories in the subcritical regime eventually tend to the real value of $\lambda$ at the IR fixed point. However, as $\alpha$ increases the two fixed points approach one another and merge exactly at $\alpha = \alpha_c$; thereafter they proceed to evolve as complex conjugates in the complex plane [38]. This heralds a topological change in the RG flow such that the trajectories become limit cycles that can never reach the fixed points, see Figure 4. The evolution of $\lambda$ along a limit cycle is periodic on a logarithmic scale and this implies that in the supercritical regime continuous scale invariance is broken in favour of a discrete version (quantum anomaly). This is the explanation for the geometric series bound state spectrum found in the Efimov problem [8], although the Efimov case is exceptional because $\lambda$ is real and this corresponds to a self-adjoint extension of the original inverse square problem. The general case again corresponds to complex values of $\lambda$ that give a complex extension of the supercritical inverse square problem and have been used to describe inelastic scattering [39] and the absorption of particles on a charged wire [35]. Close examination of the trajectories in Figure 4 shows that the flow near the real $\lambda$ axis in the supercritical regime is unstable because it forms a separatrix between trajectories in the upper and lower half planes such that a small deviation can lead to very different flow (around one fixed point or the other). In fact, the closer the trajectory approaches the real axis during that part of its motion, the further it gets swept away during the rest. Of the two complex stationary points, one corresponds to a perfect absorber of probability, and the other a perfect emitter of probability.

One of the articles of faith of quantum mechanics used to be that the Hamiltonian should be hermitian: this leads to a real energy spectrum and guarantees unitary time evolution where probability is preserved. However, starting in 1998 it was pointed out by Bender and coworkers [40, 41, 42] that non-hermitian Hamiltonians possessing $\mathcal{PT}$ symmetry can have exclusively real eigenvalues. $\mathcal{P}$ is the parity operator that effects the transformation $Q \to -Q$, and $\mathcal{T}$ is the time-reversal operator that effects the transformations $t \to -t$ and $\mathrm{i} \to -\mathrm{i}$. Furthermore, there exists a phase transition as a function of a parameter where some of the eigenstates spontaneously break $\mathcal{PT}$ symmetry and the corresponding eigenvalues become complex. This observation has led to an explosion of interest in non-hermitian Hamiltonians in quantum mechanics [43, 44, 45, 46, 47, 48, 49, 50, 51], and optics [52, 53, 54, 55, 56], including experimental confirmation [57, 58, 59, 60].

In a far-sighted paper on RG fixed point mergers, Kaplan *et al.* [38] point out that the Berezinskii-Kosterlitz-Thouless (BKT) phase transition (vortex-antivortex pair unbinding transition) is also an example of a phase transition where conformal scaling is lost when two real fixed points merge and enter the complex plane, exactly as happens in the inverse square potential problem. In this paper we also focus on the behaviour of the fixed points of the RG flow and suggest that it is analogous to Bender's $\mathcal{PT}$ symmetry breaking transition. Indeed, the Hamiltonian in Eq. (4) is typically non-hermitian and undergoes a transition as a function of the parameter $\alpha$ from the subcritical regime, where the flow is organized by real fixed points, to the supercritical regime where it is organized by complex ones. Whereas the real fixed points display $\mathcal{T}$ symmetry, this is broken when they become complex (one representing a source, the other a sink of probability). The eigenfunctions $Q^{1/2\pm\sigma}$ also change their nature when $\alpha > \alpha_c$, going from being real to complex and developing a phase singularity at the origin. While not all aspects of the connection are clear to us at the time of writing (such as the role of $\mathcal{P}$) the basic scenarios are similar enough to warrant investigation.

The rest of this paper is organized as follows : in Section 2 we recapitulate the inverse square Hamiltonian and mention a few points not already covered in the Introduction. In Section 3 we introduce PPEFT and use it to derive a boundary condition due to a microscopic 'source' at the origin, and apply this boundary condition in Section 4 to analyze the continuity equation governing probability conservation, and in Section 5 to the 'bulk' wavefunction in both the sub- and supercritical regimes. Since the wavefunction is singular at the source, we implement renormalization of the effective source-bulk coupling in Section 6 that leads to the emergence of a scale anomaly. We also calculate the reflection and transmission probabilities; as these are observable physical quantities they should remain invariant under the RG flow. Our expressions for the reflection and transmission probabilities are therefore expressed in terms of RG invariant parameters. We finish with some perspectives and conclusions in Section 7. Note that in the

Journal of Physics: Conference Series **2038** (2021) 012024 doi:10.1088/1742-6596/2038/1/012024

rest of this paper we specialize to the wavefunction in the scattering regime ($E > 0$). However, this leads to the same basic critical behaviour and RG flow as the bound state case ($E < 0$).

## 2. Hamiltonian for the inverse square problem

The Hamiltonian we will work with is given by

$$H = \frac{P^2}{2m} - \frac{g}{Q^2}. \tag{5}$$

As noted above, both the potential and the kinetic energy scale in the same way, and this holds true in any dimension, unlike, say, $\delta$-function interactions which only scale in this way in two dimensions (in this paper we consider the 1D case). Classically, the conserved quantity corresponding to the continuous scale invariance of the inverse square Hamiltonian is the generator of the scale transformations $D = QP$,

$$\frac{dD}{dt} = 2H \tag{6}$$

which is conserved if energy is zero, and is sometimes termed an *almost* conservation law [61]. In terms of the dimensionless parameter $\alpha = 2mg/\hbar^2$, fall to the centre occurs when $\alpha > \alpha_c = 1/4$ (supercritical regime). In this regime neither the boundedness of the Hamiltonian nor normalizability turn out to be good criteria for selecting either one of the two eigenfunctions over the other, as demonstrated by Burgess *et al.* in [1]. Furthermore, the energy in the supercritical regime is unbounded from below. This is demonstrated explicitly in Ref. [36] by using a cleverly chosen trial wavefunction to show that

$$\alpha \int \frac{|\chi(Q)|^2}{Q^2} dQ > \int |\chi'(Q)|^2 \, dQ \tag{7}$$

when $\alpha > \alpha_c$. In 3D, the classical dynamics in the supercritical case gives rise to an unstable trajectory which spirals to the centre.

According to von Neumann's theorem, when a Hamiltonian is essentially self-adjoint the eigenvalue problem is mathematically well-posed. When the Hamiltonian is not self-adjoint one can attempt to prune the unsavoury parts by applying a self-adjoint extension. It turns out that the inverse square problem is only self-adjoint for the strongly repulsive case and is not self-adjoint for any attractive potential, i.e. when $\alpha > 0$. While self-adjoint extensions can be applied in both the sub- and supercritical cases (like in the Efimov problem which is in the supercritical regime), they are not unique and so there is some arbitrariness involved [62]. Furthermore, they do not necessarily remove the problem of unboundedness from below [36, 63, 64]. In the next section we introduce the PPEFT method which uses a source particle at the origin as a transparent way of choosing an appropriate boundary condition [1]. The source could be unitary (self-adjoint extension) or nonunitary (complex extension), either of which could be physically acceptable, depending on the situation. When combined with renormalization the results are independent of the regulator. PPEFT gives the same results as other methods (in particular, it also does not remove unboundedness from below), but has the advantage of being intuitive and systematic.

## 3. Point particle effective field theory and boundary condition

PPEFT starts from an action and this allows us to specify the physics and symmetries of the problem rather than imposing an arbitrary cut-off. The total action is written as $S_{\text{total}} = S_B + S_p$, where $S_B$ is the action for the Schrödinger bulk field $\chi(Q)$ describing the long distance, low energy physics, and is given by

$$S_B = \int dt\, dQ \left[\frac{\mathrm{i}\hbar}{2}\left(\chi^*\partial_t\chi - \chi\partial_t\chi^*\right) - \left(\frac{\hbar^2}{2m}\left|\nabla\chi\right|^2 + V(Q)\left|\chi\right|^2\right)\right] \tag{8}$$

with $V(Q) = -\frac{g}{Q^2}$, and $S_p$ is the action describing the coupling between the bulk field and the microscopic (short distance, high energy) source localized around $Q = 0$

$$S_p = \int dt\, dQ \mathcal{L}_p(\chi^*, \chi)\delta(Q). \tag{9}$$

The key to PPEFT is a series expansion of the lagrangian density $\mathcal{L}_p(\chi^*, \chi) = -h\chi^*\chi + \cdots$ where higher terms contain higher powers of $\chi$ and $\chi^*$ and/or their derivatives. This can be viewed as analogous to a multipole expansion where successive terms build in more information about the source but are less important at large distances. Thus, the coupling $h$ can be considered to be the 'monopole' moment, and since each term must have the same total dimension, the multipole moments of higher terms will have correspondingly higher dimensions. Referring to Figure 2, the relevant scale for the source particle is $r$ and therefore the higher order moments will generically be proportional to $r/a$ raised to some power and hence smaller, allowing us to build in finer details about source-bulk coupling in a controlled fashion. In this paper we will not use the full power of PPEFT and will only retain the leading term meaning that we have just one coupling parameter $h$.

The field equations can be obtained by extremizing the action $\delta S = 0$, to obtain the time dependent Schrödinger equation

$$\left(-\frac{\partial^2}{\partial Q^2} + U(Q)\right)\chi = \mathrm{i}\frac{2m}{\hbar}\frac{\partial\chi}{\partial t} \tag{10}$$

with

$$U(Q) = -\frac{\alpha}{Q^2} + \lambda\delta(Q), \tag{11}$$

where $\lambda = 2mh/\hbar^2$ is the redefined source-bulk coupling constant. We see that the point particle effective action has modified the potential $V(Q)$ by adding a Dirac $\delta$-function, as previously claimed in Eq. (4).

The boundary condition is obtained by integrating the Schrödinger equation over the infinitesimal region $-\epsilon \le Q \le \epsilon$, which gives

$$\lambda = \left[\frac{\partial \ln(\chi)}{\partial Q}\right]_{Q=-\epsilon}^{Q=\epsilon}. \tag{12}$$

The length scale $\epsilon$ is a cut-off or regulator which should be much shorter than the observable length scale $a$, but much larger than the size of the source $r$ located at the origin because the PPEFT action does not converge at distances where $\epsilon \approx r$. In other words, we need the hierarchy $a \gg \epsilon \gg r$ to be obeyed [see Fig. 2]. Furthermore, physical observables like reflection and transmission probabilities should both be finite and independent of the regularization scale $\epsilon$, and this is where renormalization of the source-bulk coupling $\lambda$ comes in. However, before coming to the renormalization there are two other tasks we can accomplish using the boundary condition given in (12). The first is to use it in combination with the continuity equation to analyze probability conservation and the second is to apply it to the wavefunction.

## 4. Continuity equation

The continuity equation is a basic tool for analyzing probability conservation. In a hermitian system we expect probability to be conserved, but in a nonhermitian system such basic properties can be violated. In this section we show how probability conservation is determined by the source-bulk coupling constant $\lambda$ by evaluating the probability current at the boundaries $Q = \pm\epsilon$.

Proceeding from the Schrödinger equation and its complex conjugate in the usual way we find the continuity equation

$$\partial_t \rho + \nabla . J = \frac{2}{\hbar}\rho \Im \left[ -\frac{g}{Q^2} + \frac{\hbar^2 \lambda}{2m}\delta(Q)\right] \tag{13}$$

where $\rho$ is the probability density, and $J$ is the standard probability current

$$J = \frac{\mathrm{i}\hbar}{2m}\left(\chi \partial_Q \chi^* - \chi^* \partial_Q \chi\right). \tag{14}$$

The term in the square brackets in Eq. (13) is the total potential term in the Schrödinger equation. Since $g$ is assumed to be real, the first term will make no contribution, but the second term will if $\lambda$ is complex.

The boundary condition Eq. (12) can be used to calculate the net probability current out of the origin

$$J(\epsilon) - J(-\epsilon) = \frac{\mathrm{i}\hbar}{2m}(\lambda^* - \lambda)\chi^*(\epsilon)\chi(\epsilon). \tag{15}$$

We can therefore refine our condition for probability conservation violation to saying that $\lambda$ must be complex (nonhermitian Hamiltonian) and the probability for finding the particle at the source must be nonzero. The origin will be a sink if $\Im(\lambda) < 0$ or a source if $\Im(\lambda) > 0$, as pointed out in [35].

## 5. Wavefunction of the inverse square Hamiltonian in 1D and the boundary condition

The time-independent Schrödinger equation with energy eigenvalue $E = \hbar^2 k^2/2m > 0$ and purely inverse square potential is given by

$$\left(-\frac{d^2}{dQ^2} - \frac{\alpha}{Q^2}\right)\chi(Q) = k^2\chi(Q). \tag{16}$$

Putting $z = kQ$ and $\chi(Q) = \sqrt{z}u(z)$, this is transformed to the Hankel differential equation

$$z^2 u'' + zu' + (z^2 - \sigma^2)u = 0 \tag{17}$$

where $\sigma^2 = 1/4 - \alpha$. The parameter $\sigma$ is real in the subcritical regime, but in the supercritical regime it is an imaginary number which we write as $\sigma = \pm \mathrm{i}\zeta$, where $\zeta \in \mathbb{R}$. It is notable that, unlike $\sigma^2$, the eigenvalue $k^2$ does not appear as an explicit parameter in Eq. (17) but instead occurs only as a scaling factor in the argument of the solutions. We also note in passing that the Hankel differential equation is invariant under $z \to -z$, i.e. under parity $\mathcal{P}$, and also under $\sigma \to -\sigma$.

The independent solutions to the Hankel equation are the two Hankel functions $H_\sigma^{(1)}(z)$ and $H_\sigma^{(2)}(z)$ whose properties, including asymptotics, are summarized in the Appendix. From these it can be inferred that $H_\sigma^{(1)}(z)$ asymptotes at large $z$ to a right moving wave and $H_\sigma^{(2)}(z)$ asymptotes to a left moving wave. Translating back to the original wavefunction $\chi(Q)$, and assuming an initial wave coming in from the right, the general solution to the scattering problem can be written as

$$\chi_{\text{in+ref}}(Q) = \sqrt{kQ}\left(H_\sigma^{(2)}(kQ) + RH_\sigma^{(1)}(kQ)\right), Q \geq \epsilon \tag{18}$$

$$\chi_{\text{trans}}(Q) = T\sqrt{kQ}H_\sigma^{(2)}(kQ)\ ,\ Q \leq -\epsilon \tag{19}$$

where, $R$ and $T$ are the reflection and transmission amplitudes to be determined by the boundary condition at $Q = \pm\epsilon$.

### *5.1. Boundary condition for the subcritical case*

In the subcritical case $\sigma \in \mathbb{R}$. We can find a relation between R and T by demanding continuity of the wavefunction near the origin $\chi_{\text{in+ref}}(\epsilon) = \chi_{\text{trans}}(-\epsilon)$, yielding

$$R + \mathrm{i}T\exp(\mathrm{i}\pi\sigma) = -\frac{H_\sigma^{(2)}(k\epsilon)}{H_\sigma^{(1)}(k\epsilon)} \tag{20}$$

where we have used the reflection identities of the Hankel functions given in the Appendix. For $k\epsilon \ll 1$, this relation becomes

$$R + \mathrm{i}T\exp(\mathrm{i}\pi\sigma) \approx \frac{1 - X\exp(\mathrm{i}\pi\sigma)}{1 - X\exp(-\mathrm{i}\pi\sigma)} \tag{21}$$

where

$$X = \frac{\Gamma(1-\sigma)}{\Gamma(1+\sigma)}\left(\frac{k\epsilon}{2}\right)^{2\sigma}. \tag{22}$$

We now consider the boundary condition given in Eq. (12). When expressed in terms of the scattering solution it reads

$$\lambda = \frac{\partial\ln[\chi_{\text{in+ref}}(\epsilon)]}{\partial Q} - \frac{\partial\ln[\chi_{\text{trans}}(-\epsilon)]}{\partial Q} \tag{23}$$

and we find the following expression for the coupling constant

$$\lambda = \frac{1}{\epsilon}\left(1 - \sigma\left[\frac{1 + X\exp(\mathrm{i}\pi\sigma) - R(1 + X\exp(-\mathrm{i}\pi\sigma))}{1 - X\exp(\mathrm{i}\pi\sigma) - R(1 - X\exp(-\mathrm{i}\pi\sigma))} + \frac{1 + X\exp(-\mathrm{i}\pi\sigma)}{1 - X\exp(-\mathrm{i}\pi\sigma)}\right]\right). \tag{24}$$

In general $\lambda$ is complex and hence breaks $\mathcal{PT}$ symmetry. However, in this paper we focus on the fixed points of the RG flow and we will show in Section 6 that in the subcritical case the UV and IR fixed points are real and hence preserve $\mathcal{PT}$ symmetry. Expanding Eq. (24) in powers of $X$ in the small $k\epsilon$ regime we obtain

$$\Lambda \approx -2\sigma\left[1 + X\exp(\mathrm{i}\pi\sigma)\left(\frac{1 - R\exp(-2\mathrm{i}\pi\sigma)}{1 - R}\right) + X\exp(-\mathrm{i}\pi\sigma) + \mathcal{O}(X^2)\right] \tag{25}$$

where in order to simplify the expression we have defined

$$\Lambda \equiv 2(\lambda\epsilon - 1)\ . \tag{26}$$

*5.2. Boundary condition for the supercritical case*

A very similar calculation to that given above can be performed to find the coupling constant $\lambda$ in the supercritical case, the only difference being that $\sigma$ is purely imaginary. Of the two possibles choices for the square root, we choose $\sigma = -\mathrm{i}\zeta$. We find

$$\chi_{\text{in+ref}}(Q) = \frac{\sqrt{kQ}\left(\frac{-\left(\frac{kQ}{2}\right)^{\mathrm{i}\zeta}}{\Gamma(1+\mathrm{i}\zeta)} + \frac{\exp(\pi\zeta)\left(\frac{kQ}{2}\right)^{-\mathrm{i}\zeta}}{\Gamma(1-\mathrm{i}\zeta)}\right) + R\left(\frac{\left(\frac{kQ}{2}\right)^{\mathrm{i}\zeta}}{\Gamma(1+\mathrm{i}\zeta)} - \frac{\exp(-\pi\zeta)\left(\frac{kQ}{2}\right)^{-\mathrm{i}\zeta}}{\Gamma(1-\mathrm{i}\zeta)}\right)}{\sinh(\pi\zeta)}, \quad Q \geq \epsilon \tag{27}$$

$$\chi_{\text{trans}}(Q) = \frac{T\sqrt{kQ}}{\sinh(\pi\zeta)}\left(\frac{-1}{\Gamma(1+\mathrm{i}\zeta)}\left(\frac{kQ}{2}\right)^{\mathrm{i}\zeta} + \frac{\exp(\pi\zeta)}{\Gamma(1-\mathrm{i}\zeta)}\left(\frac{kQ}{2}\right)^{-\mathrm{i}\zeta}\right), \quad Q \leq -\epsilon\,. \tag{28}$$

Demanding continuity near the origin yields the relation

$$R + \mathrm{i}T\exp(\pi\zeta) = -\frac{H^{(2)}_{-\mathrm{i}\zeta}(k\epsilon)}{H^{(1)}_{-\mathrm{i}\zeta}(k\epsilon)}\,. \tag{29}$$

For small $k\epsilon \ll 1$, this relation can be written

$$R + \mathrm{i}T\exp(\pi\zeta) \approx \frac{1 - X\exp(\pi\zeta)}{1 - X\exp(-\pi\zeta)}\,. \tag{30}$$

The coupling constant in the super critical case is then given by :

$$\lambda = \frac{1}{\epsilon}\left(1 + \mathrm{i}\zeta\left[\frac{1 + X\exp(\pi\zeta) - R(1 + X\exp(-\pi\zeta))}{1 - X\exp(\pi\zeta) - R(1 - X\exp(-\pi\zeta))} + \frac{1 + X\exp(-\pi\zeta)}{1 - X\exp(-\pi\zeta)}\right]\right) \tag{31}$$

where

$$X = \frac{\Gamma(1-\sigma)}{\Gamma(1+\sigma)}\left(\frac{k\epsilon}{2}\right)^{2\sigma}\,. \tag{32}$$

Expanding Eq. (31) in powers of $X$ in the small $k\epsilon$ regime we obtain

$$\Lambda \approx 2\mathrm{i}\zeta\left[1 + X\exp(\pi\zeta)\left(\frac{1 - R\exp(-2\pi\zeta)}{1 - R}\right) + X\exp(-\pi\zeta) + \mathcal{O}(X^2)\right]\,. \tag{33}$$

We now proceed to renormalize the source-bulk coupling constant so that the reflection and transmission probabilities do not depend upon the regulator scale $\epsilon$ and are finite. We also discuss the fixed point merger of the renormalization group as a $\mathcal{PT}$ symmetry breaking transition as the strength of the inverse square potential $\alpha$ is tuned from the subcritical to supercritical regime.

## 6. Renormalization group

### *6.1. Renormalization group and fixed point merger exhibiting a $\mathcal{PT}$ transition*

If we blindly let $\epsilon \to 0$ we run into trouble because the wavefunction develops singularities at small $Q$ [see Eq. (3)]. PPEFT then breaks down because the bulk wavefunction is singular at the source located at $Q = 0$. To deal with this we must renormalize the source-bulk coupling constant $\lambda$. Following Ref. [1], we start by writing down the RG flow equation for the coupling constant $\lambda$, or more conveniently its close relative $\Lambda = 2(\lambda\epsilon - 1)$. This can be obtained by taking the derivative with respect to $\epsilon$ of the above equation for $\Lambda$, keeping all observables (such as $k$) fixed. After some algebra this gives

$$\epsilon\frac{d}{d\epsilon}\left(\frac{\Lambda}{2\sigma}\right) = \sigma\left(1 - \left(\frac{\Lambda}{2\sigma}\right)^2\right)\,. \tag{34}$$

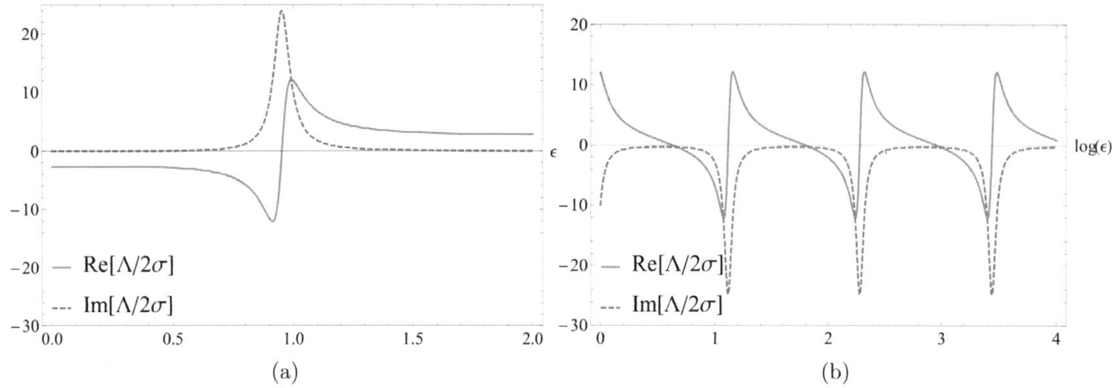


**Figure 3.** Panel (a) shows the real and imaginary parts of the RG flow as a function of $\epsilon$ for the subcritical case which means that $\sigma$ is real. Panel (b) shows the real and imaginary parts of the RG flow in the supercritical case which means that $\sigma$ is imaginary. In the supercritical case the flow exhibits a log periodic behaviour.

This equation determines how the coupling constant $\lambda$ must depend on the regulator $\epsilon$ to renormalize any divergences such that physical quantities are independent of $\epsilon$. The non-trivial zeroes of the right hand side at $\Lambda = \pm 2\sigma$ correspond to the fixed points of the flow. At fixed points the theory is scale invariant. An important consequence of the RG running of the coupling of the Dirac $\delta$-function is as follows: the vanishing of the coefficient of the $\delta$-function ($\lambda = 0$) only happens at $\Lambda = -2$, which is not a fixed point unless $\sigma = \pm 1$, but this value of $\sigma$ is impossible to realize with an attractive inverse square potential because $\sigma = \sqrt{1/4 - \alpha}$. Thus, a $\delta$-function term is inevitable: the flow will always produce one.

The above RG evolution equation, considered as a first order differential equation, has a relatively simply quadratic right hand side and can be integrated analytically. In terms of the initial condition $\lambda(\epsilon_0) \equiv \lambda_0$, one finds

$$\frac{\Lambda}{2\sigma} = \frac{\frac{\lambda_0}{2\sigma} + \tanh(\sigma \ln(\epsilon/\epsilon_0))}{1 + \frac{\lambda_0}{2\sigma} \tanh(\sigma \ln(\epsilon/\epsilon_0))}. \tag{35}$$

*6.1.1. Supercritical case* In the supercritical case when $\alpha > 1/4$, the fixed points for Eq. (34) are given by $\Lambda = \pm 2\sigma = \pm 2i\zeta$, $\zeta \in \mathbb{R}$. Thus, there are two fixed points and they are purely imaginary complex conjugates of each other. The RG evolution when $\sigma$ is imaginary is dealt with in detail in [35], but the resulting flow of the coupling $\Lambda$ as $\epsilon$ is increased is shown in Fig. 3(b). The limit cycle behaviour of the trajectories is illustrated in Fig. 4(b), and combined with the logarithmic derivative in Eq. (34) means that the flow exhibits a log periodic behaviour as a function of $\epsilon$. Note that each trajectory can be uniquely labelled by, for example, its value where it crosses the imaginary axis ($\Re[\Lambda] = 0$) and this label is then an RG invariant. In fact, the flow picks a scale $\epsilon_*$ at this point and hence breaks continuous scale invariance, exhibiting discrete scale invariance instead. The limit cycle behaviour means that trajectories that start away from the fixed points can never reach them. Physically, these fixed points correspond to the (scale invariant) scenarios of a perfect sink when $\Im(\lambda) < 0$ and a perfect source when $\Im(\lambda) > 0$ and because they are complex they break $\mathcal{PT}$ symmetry.

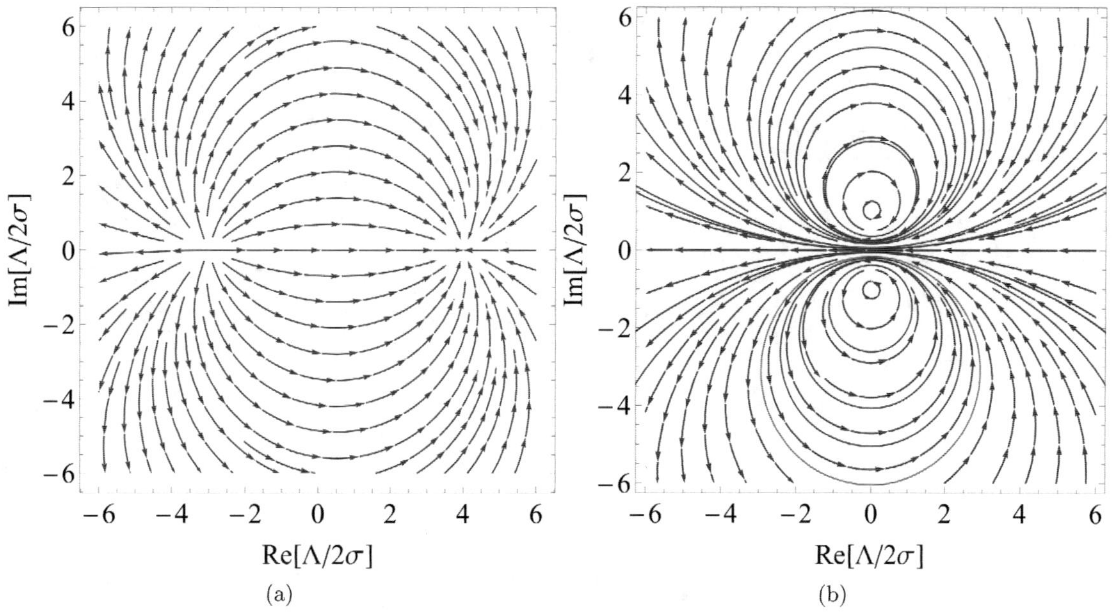


**Figure 4.** Panel (a) shows the phase portrait of the RG flow in the subcritical case (real $\sigma$) which has two real fixed points, one stable and the other unstable. Panel (b) shows a limit cycle behaviour in the supercritical case (imaginary $\sigma$). Each RG trajectory picks a length scale $\epsilon_*$ when $\Re[\hat{\lambda}] = 0$. Arrows indicate the direction of flow. Each trajectory is characterized by the values of $\epsilon_*$ and $y_*$, where $y_* = \Im[\lambda(\epsilon_*)]$.

To calculate the reflection coefficient we can use the small $\epsilon$ expansion of Eq. (35)

$$\frac{\Lambda}{2\sigma} \approx -1 - 2\left(\frac{\epsilon}{\epsilon_*}\right)^{2\sigma}, \qquad (\epsilon \ll \epsilon_*) \tag{36}$$

and substitute it into Eq. (33) to obtain

$$R = \frac{X_* \cosh(\pi\zeta) - 1}{X_* \exp(-\pi\zeta) - 1} \tag{37}$$

where,

$$X_* = \frac{\Gamma(1-\sigma)}{\Gamma(1+\sigma)}\left(\frac{k\epsilon_*}{2}\right)^{2\sigma} \tag{38}$$

and using the small $k\epsilon$ limit in Eq. (30) finally yields

$$T = -\mathrm{i}\exp(-\pi\zeta)(1-R) = \frac{\mathrm{i}X_* \exp(-\pi\zeta)\sinh(\pi\zeta)}{X_* \exp(-\pi\zeta) - 1}. \tag{39}$$

Note that all observables are expressed entirely in terms of RG invariant quantities.

*6.1.2. Subcritical case* In the subcritical case $\sigma \in \mathbb{R}$. The solution of the RG flow equation as a function of $\epsilon$ for this case is shown in Fig. 3(a). If we chose to start with real $\Lambda$ as an initial condition, the coupling will stay real as the flow evolves: hermiticity is itself an RG invariant. The RG phase portraits in sub- and supercritical cases are strikingly different: every RG trajectory in the phase portrait for the subcritical case flows from the UV fixed point $\Lambda = -2\sigma$ to the IR fixed point $\Lambda = +2\sigma$ as shown in Fig. 4(a). The two real fixed points correspond to scale invariant phases and are also $\mathcal{PT}$ symmetric. If we tune the strength of the inverse square potential $\alpha$ the critical points merge at $\alpha_c = 1/4$. Hence, we view the fixed point merger as a form of $\mathcal{PT}$ symmetry breaking transition.

To calculate the reflection coefficient, we follow a similar procedure to the supercritical case. We use the small $\epsilon$ expansion of Eq. (35)

$$\frac{\Lambda}{2\sigma} \approx -1 - 2\left(\frac{\epsilon}{\epsilon_*}\right)^{2\sigma}, \quad (\epsilon << \epsilon_*) \tag{40}$$

to obtain [1]

$$R = \frac{X_* \cos(\pi\sigma) - 1}{X_* \exp(-i\pi\sigma) - 1} \tag{41}$$

where,

$$X_* = \frac{\Gamma(1-\sigma)}{\Gamma(1+\sigma)}\left(\frac{k\epsilon_*}{2}\right)^{2\sigma}. \tag{42}$$

Finally, the transmission coefficient can be expressed as

$$T = -\mathrm{i}\exp(-i\pi\sigma)(1-R) = \frac{\mathrm{i}X_* \exp(-i\pi\sigma)\sin(\pi\sigma)}{X_* \exp(-i\pi\sigma) - 1}. \tag{43}$$

## 7. Conclusion

In this work we have argued that fall to the centre is a form of $\mathcal{PT}$ symmetry breaking transition if one focuses on the fixed point structure of the RG flow. The seemingly simple inverse square Hamiltonian presents some subtle difficulties and by itself is not a fully defined problem: one must impose a boundary condition representing additional microscopic 'source' physics at the origin to make the eigenvalue problem well posed. We use the PPEFT tools of ref. [1] to derive the boundary condition for the 1D case in both the sub- and supercritical regimes. This amounts to adding an inevitable Dirac $\delta$-function at the origin. Depending upon the nature of the physics at the origin, one can choose a unitary or a non-unitary boundary condition and the Hamiltonian becomes non-hermitian when a non-unitary boundary condition is implemented. The source-bulk coupling $\lambda$ is evaluated by a boundary condition on the wavefunction at the length scale $\epsilon$ (which is a regulator scale that is arbitrary). The RG flow of the source-bulk coupling shows several interesting properties, and in this paper we show that the fixed point merger that occurs as the strength of the inverse square potential $\alpha$ is tuned from subcritical to supercritical [39] is also a $\mathcal{PT}$ symmetry breaking transition. In particular, the two real RG fixed points of the source-bulk coupling $\lambda$ in the subcritical case ($\alpha < 1/4$), which are attractive/repulsive and preserve $\mathcal{PT}$ symmetry, merge as we tune $\alpha$ to the critical value $\alpha_c = 1/4$. As the strength of the potential is further increased above the critical value, i.e. when $\alpha > 1/4$, we enter the supercritical regime and the fixed points disappear into the complex plane, breaking $\mathcal{PT}$ symmetry. Thus, while the system can be both scale invariant (at a fixed point) *and* unitary for $\alpha < 1/4$, one must choose *either* scale invariance *or* unitarity for $\alpha > 1/4$. We do not claim that fall to the centre is an exact realization of the standard $\mathcal{PT}$ breaking transition [40, 41, 42], but it has many similar elements. One of the differences is that, although the Hamiltonian is generically nonhermitian in both phases, it can also be chosen to be hermitian ($\lambda$ real) in both phases, rare though these trajectories are. It is, however, nonselfadjoint in both phases if one does not regulate it. The precise role of $\mathcal{P}$ is also elusive (although the Hamiltonian and the boundary condition both always appear to be $\mathcal{P}$ symmetric), and this remains a topic for further investigation.

## Acknowledgments
We acknowledge the support of the Natural Sciences and Engineering Research Council of Canada (NSERC) [Ref. No. RGPIN-2017-06605 and SAPIN-2020-00045]. Research at the Perimeter Institute is supported in part by the Government of Canada through NSERC and by the Province of Ontario through MRI.

## Appendix A. Properties and asymptotics of Hankel functions
In this appendix we state some of the properties of the Hankel function that are used in the paper. The Hankel differential equation is given by:

$$z^2u'' + zu' + (z^2 - \sigma^2)u = 0 \ . \tag{A.1}$$

Hankel functions of first and second kind are defined by:

$$H_\sigma^{(1)}(z) = J_\sigma(z) + \mathrm{i}N_\sigma(z) \tag{A.2}$$

and

$$H_\sigma^{(2)}(z) = J_\sigma(z) - \mathrm{i}N_\sigma(z) \tag{A.3}$$

where $J_\sigma$ is the Bessel function and $N_\sigma$ is the Neumann function.

The asymptotic large $z$ behaviour of Hankel functions is given by:

$$H_\sigma^{(1)}(z) \sim \sqrt{\frac{2}{\pi z}} \exp\left[\mathrm{i}\left(z - \frac{\pi\sigma}{2} - \frac{\pi}{4}\right)\right] \tag{A.4}$$

$$H_\sigma^{(2)}(z) \sim \sqrt{\frac{2}{\pi z}} \exp\left[-\mathrm{i}\left(z - \frac{\pi\sigma}{2} - \frac{\pi}{4}\right)\right] \ . \tag{A.5}$$

and they have the following reflection properties:

$$H_\sigma^{(1)}(\exp(\mathrm{i}\pi)z) = -\exp(-\mathrm{i}\pi\sigma)H_\sigma^{(2)}(z) \tag{A.6}$$

$$H_\sigma^{(2)}(\exp(-\mathrm{i}\pi)z) = -\exp(\mathrm{i}\pi\sigma)H_\sigma^{(1)}(z) \ . \tag{A.7}$$

For small $z$, Hankel functions reduce to monomials:

$$H_\sigma^{(1)}(z) \approx \frac{1}{\mathrm{i}\sin(\pi\sigma)}\left(\frac{1}{\Gamma(1-\sigma)}\left(\frac{z}{2}\right)^{-\sigma} - \frac{\exp(-\mathrm{i}\pi\sigma)}{\Gamma(1+\sigma)}\left(\frac{z}{2}\right)^{\sigma}\right) \tag{A.8}$$

$$H_\sigma^{(2)}(z) \approx \frac{1}{\mathrm{i}\sin(\pi\sigma)}\left(-\frac{1}{\Gamma(1-\sigma)}\left(\frac{z}{2}\right)^{-\sigma} + \frac{\exp(\mathrm{i}\pi\sigma)}{\Gamma(1+\sigma)}\left(\frac{z}{2}\right)^{\sigma}\right) \tag{A.9}$$

## References
[1] Burgess C P, Hayman P, Williams M and Zalavári L 2017 *Journal of High Energy Physics* **2017** 106
[2] Lifshitz E M and Landau L D 1981 *Quantum Mechanics: Non-Relativistic Theory* 3rd ed (Butterworth-Heinemann)
[3] Case K M 1950 *Phys. Rev.* **80** 797
[4] Perelomov A M and Popov V S 1970 *Theoretical and Mathematical Physics* **4** 664–677
[5] Denschlag J, Umshaus G and Schmiedmayer J 1998 *Phys. Rev. Lett.* **81** 737
[6] Efimov V 1970 *Phys. Lett.* **33B** 563

[7] Efimov V 1973 *Nucl. Phys.* **A210** 157
[8] Braaten E and Hammer H W 2007 *Ann. Phys.* **322** 120
[9] Kraemer T, Mark M, Waldburger P, Danzl J G, Chin C, Engeser B, Lange A D, Pilch K, Jaakkola A, Nägerl H C and Grimm R 2006 *Nature (London)* **440** 315
[10] Pollack S E, Dries D and Hulet R G 2009 *Science* **326** 1683
[11] Ferlaino F and Grimm R 2010 *Physics* **3** 9
[12] Bhaduri R K, Chatterjee A and van Zyl B P 2011 *Am. J. Phys.* **79** 274–281
[13] Huang B, Sidorenkov L A, Grimm R and Hutson J M 2014 *Phys. Rev. Lett.* **112** 190401
[14] Pires R, Ulmanis J, Häfner S, Repp M, Arias A, Kuhnle E D and Weidemüller M 2014 *Phys. Rev. Lett.* **112** 250404
[15] Tung S K, Jiménez-García K, Johansen J, Parker C V and Chin C 2014 *Phys. Rev. Lett.* **113** 240402
[16] Moroz S, D'Incao J P and Petrov D S 2015 *Phys. Rev. Lett.* **115** 180406
[17] Kunitski M, Zeller S, Voigtsberger J, Kalinin A, Schmidt L P H, Schöffler M, Czasch A, Schöllkopf W, Grisenti R E, Jahnke T, Blume D and Dörner R 2015 *Science* **348** 551
[18] Gurappa N and Panigrahi P K 1999 *Phys. Rev. B* **59** R2490
[19] Srinivasan K and Padmanabhan T 1999 *Phys. Rev. D* **60** 024007
[20] Birmingham D, Gupta K S and Sen S 2001 *Physics Letters B* **505** 191–196
[21] Camblong H and Ordonez C 2003 *Phys. Rev. D* **68** 125013
[22] Sundaram S and Panigrahi P K 2016 *Optics letters* **41** 4222–4224
[23] Alonso M 2015 *New J. Phys.* **17** 073013
[24] Frank A, Leyvraz F and Wolf K B 1990 *Journal of Mathematical Physics* **31** 2757–2768
[25] Leonhardt U 2009 *New Journal of Physics* **11** 093040
[26] Coon S A and Holstein B R 2002 *Am. J. Phys.* **70** 513–519
[27] Essin A M and Griffiths D J 2006 *Am. J. Phys.* **74** 109–117
[28] Pitaevskii L P and Rosch A 1997 *Phys. Rev. A* **55** R853
[29] Olshanii M, Perrin H and Lorent V 2010 *Phys. Rev. Lett.* **105** 095302
[30] Taylor E and Randeria M 2012 *Phys. Rev. Lett.* **109** 135301
[31] Holten M, Bayha L, Klein A C, Murthy P A, Preiss P M and Jochim S 2018 *Phys. Rev. Lett.* **121** 120401
[32] Peppler T, Dyke P, Zamorano M, Herrera I, Hoinka S and Vale C J 2018 *Phys. Rev. Lett.* **121** 120402
[33] Murthy P A, Defenu N, Bayha L, Holten M, Preiss P M, Enss T and Jochim S 2019 *Science* **365** 268
[34] Ovdat O, Mao J, Jiang Y, Andrei E and Akkermans E 2017 *Nature communications* **8** 1–6
[35] Plestid R, Burgess C P and O'Dell D H J 2018 *Journal of High Energy Physics* **2018** 59
[36] Gupta K S and Rajeev S G 1993 *Phys. Rev. D* **48** 5940
[37] Mueller E J and Ho T L 2004 *arXiv preprint cond-mat/0403283*
[38] Kaplan D B, Lee J W, Son D T and Stephanov M A 2009 *Phys. Rev. D* **80** 125005
[39] Moroz S and Schmidt R 2010 *Annals of Physics* **325** 491–513
[40] Bender C M and Boettcher S 1998 *Phys. Rev. Lett.* **80** 5243
[41] Bender C M, Boettcher S and Meisinger P N 1999 *Journal of Mathematical Physics* **40** 2201–2229
[42] Bender C M, Brody D C and Jones H F 2002 *Phys. Rev. Lett.* **89** 270401
[43] Bender C M, Brody D C and Jones H F 2003 *Am. J. Phys.* **71** 1095–1102
[44] Bender C M and Mannheim P D 2010 *Phys. Lett. A* **374** 1616–1620
[45] Das A and Greenwood L 2009 *Phys. Lett. B* **678** 504–507
[46] Bazeia D, Das A, Greenwood L and Losano L 2009 *Phys. Lett. B* **673** 283–287
[47] Dorey P, Dunning C and Tateo R 2001 *J. Phys. A: Math. Gen.* **34** L391
[48] Abhinav K and Panigrahi P K 2010 *Annals of Physics* **325** 1198–1206
[49] Abhinav K, Jayannavar A and Panigrahi P K 2013 *Annals of Physics* **331** 110–119
[50] Mandal B P, Mourya B K and Yadav R K 2013 *Phys. Lett. A* **377** 1043–1046
[51] Mandal B P, Mourya B K, Ali K and Ghatak A 2015 *Annals of Physics* **363** 185–193
[52] Lumer Y, Plotnik Y, Rechtsman M C and Segev M 2013 *Phys. Rev. Lett.* **111** 263901
[53] Zyablovsky A A, Vinogradov A P, Pukhov A A, Dorofeenko A V and Lisyansky A A 2014 *Physics-Uspekhi* **57** 1063
[54] Raja S V, Govindarajan A, Mahalingam A and Lakshmanan M 2019 *Phys. Rev. A* **100** 033838
[55] Raja S V, Govindarajan A, Mahalingam A and Lakshmanan M 2020 *Phys. Rev. A* **101** 033814
[56] Longhi S 2018 *EPL (Europhys. Lett.)* **120** 64001
[57] Bender C M, Berntson B K, Parker D and Samuel E 2013 *Am. J. Phys.* **81** 173–179
[58] Schindler J, Li A, Zheng M C, Ellis F M and Kottos T 2011 *Phys. Rev. A* **84** 040101
[59] Guo A, Salamo G J, Duchesne D, Morandotti R, Volatier-Ravat M, Aimez V, Siviloglou G A and Christodoulides D N 2009 *Phys. Rev. Lett.* **103** 093902
[60] Rüter C E, Makris K G, El-Ganainy R, Christodoulides D N, Segev M and Kip D 2010 *Nat. Phys.* **6** 192–195

[61] Rajeev S G 2013 *Advanced mechanics: from Euler's determinism to Arnold's chaos* (OUP Oxford)
[62] Capri A Z 1977 *Am. J. Phys.* **45** 823–825
[63] Basu-Mallick B and Gupta K S 2001 *Phys. Lett. A* **292** 36
[64] Gopalakrishnan S 2006 *Self-Adjointness and the Renormalization of Singular Potentials* Ph.D. thesis Department of Physics of Amherst College

# Exactly solvable nonlinear eigenvalue problems

**Qing-hai Wang**
Department of Physics, National University of Singapore, 117551, Singapore
E-mail: qhwang@nus.edu.sg

**Abstract.** The nonlinear eigenvalue problem of a class of second order semi-transcendental differential equations is studied. A nonlinear eigenvalue is defined as the initial condition which gives rise a separatrix solution. A semi-transcendental equation can be integrated once to a first order nonlinear equation, e.g., the Ricatti equation. It is shown that the nonlinear eigenvalue problems of these semi-transcendental equations are equivalent to linear eigenvalue problems. They share the exactly same eigenvalues. The eigensolutions in the two problems are closely related. The nonlinear eigenvalue problem equivalent to the (half) harmonic oscillator in quantum mechanics is solved exactly. This is the first solvable nonlinear eigenvalue problem. The nonlinear eigenvalue problems of some extended equations are also studied.

## 1. Introduction

The nonlinear eigenvalue problem was first proposed in 2014 by Bender *et al* as an initial-value problem for the first order nonlinear differential equation [1]

$$y'(x) = \cos[\pi x y(x)]. \tag{1}$$

For large $x$, the leading asymptotic behaviour for the solution of the above equation is

$$y(x) \sim \left(m + \tfrac{1}{2}\right)/x, \quad x \to \infty, \tag{2}$$

where $m$ is an integer. When $m = 2n$ is even, the asymptotic behaviour is stable. That is, if the initial condition $y(0) = a$ gives a solution approaching to $y(x) \sim \left(2n + \frac{1}{2}\right)/x$, then a slightly different initial condition $y(0) = a + \delta$ will give a solution with the same asymptotic behaviour when $\delta$ is sufficiently small. Thus, solutions with slightly different initial conditions bundle together for large $x$. On the contrary, when $m = 2n + 1$ is odd, the asymptotic behaviour is unstable. Only a precise initial condition, say $y(0) = b_n$, can give rise a solution with $y_n(x) \sim \left(2n + 1 + \frac{1}{2}\right)/x$. Any small deviation in the initial condition, $y(0) = b_n + \delta$, will lead the solution to either $y(x) \sim \left(2n + \frac{1}{2}\right)/x$ or $y(x) \sim \left[2(n+1) + \frac{1}{2}\right]/x$. This is true regardless how small $\delta$ is, as long as $\delta \neq 0$. Therefore, the solution $y_n(x)$ is *a separatrix solution.* This unstable feature is somewhat similar to the solutions in linear eigenvalue problems near an eigensolution. In a typical linear eigenvalue problem, e.g., the harmonic oscillator, an exact eigenvalue gives rise an eigensolution which decays as $x \to \infty$ and $x \to -\infty$. A slight deviation from the exact eigenvalue will not lead to a solution which is slightly different from the eigensolution. Rather, its magnitude blows up as $x \to \infty$ or $x \to -\infty$. In this sense, we call $y_n(x)$ as *an eigensolution*, and the corresponding initial conditions $b_n$ as *the nonlinear eigenvalue.*

Later, the nonlinear eigenvalue problem was extended to the first two Painlevé equations [1, 2]:

$$\begin{aligned} y''(x) &= 6y^2(x) + x, && (3) \\ y''(x) &= 2y^3(x) + y(x). && (4) \end{aligned}$$

For these second order differential equations, two types of nonlinear eigenvalue problems were studied. One is holding the initial function $y(0)$ fixed and varying the initial slope $y'_n(0)$ as the eigenvalues. Another one is varying $y_n(0)$ with a fixed $y'(0)$. For large $n$, the nonlinear eigenvalues follow power laws in the leading asymptotic behaviour. These laws were first found numerically [1, 2]. They were then proved rigorously [3, 4]. Curiously, these power laws are deeply connected to the linear eigenvalues of some non-Hermitian Hamiltonians in the WKB approximation [5].

Recently, the first two Painlevé equations in Eqs. (3) and (4) were generalised to an infinite series equations with the movable singularities being algebraic [6, 7]

$$y''(x) = \tfrac{2(M+1)}{(M-1)^2}\left[y^M(x) + xp(y) + q(y)\right], \qquad (5)$$

where $p(y)$ and $q(y)$ are polynomials in $y$ of degree at most $M-2$:

$$\begin{aligned} p(y) &= P_{M-2}\,y^{M-2} + P_{M-3}\,y^{M-3} + \cdots + P_1\,y + P_0, \\ q(y) &= Q_{M-2}\,y^{M-2} + Q_{M-3}\,y^{M-3} + \cdots + Q_1\,y + Q_0. \end{aligned}$$

Clearly that the first Painlevé equation in Eq. (3) is the special case with $M = 2$ and the second Painlevé equation in Eq. (4) is the special case with $M = 3$.

Numerical nonlinear eigenvalues and eigensolutions were obtained for some generalised Painlevé equations. The asymptotic behaviours of large nonlinear eigenvalues exhibit similar power laws as in the first two Painlevé equations [6, 7]. A new type of nonlinear eigenvalue problem was discovered, which resembles to some characteristic features of the hyperfine structure in nuclear physics [7].

In this paper, we study the nonlinear eigenvalue problem of another second order differential equation with the form

$$\frac{\mathrm{d}^2W(x)}{\mathrm{d}x^2} = -2W(x)\frac{\mathrm{d}W(x)}{\mathrm{d}x} + q(x)\frac{\mathrm{d}W(x)}{\mathrm{d}x} + q'(x)W(x), \qquad (6)$$

where $q(x)$ is an arbitrary analytic function and $q'(x)$ is its derivative. This equation is called *semi-transcendental* because it can be integrated once to a first order nonlinear equation. In this particular example, the first order equation is equivalent to the Ricatti equation. In Sec. 2, we show an equivalent relation between the nonlinear eigenvalue problem of Eq. (6) and a linear eigenvalue problem based on this semi-transcendental feature. Unlike all the *asymptotic* results on the nonlinear eigenvalue problems studied so far, this equivalent relation is *exact*. This is the primary result of this paper. It puts the nonlinear eigenvalue problem on a much solid ground.

For some special choices of $q(x)$ in Eq. (6), the resulting Riccati equation is exactly solvable. To illustrate the equivalent relation, we present two examples in Sec. 3. The first one is an exactly solvable nonlinear eigenvalue problem which is equivalent to a half simple harmonic oscillator. The nonlinear eigenvalues and eigensolutions are solved exactly. To our best knowledge, this is the first exactly solvable nonlinear eigenvalue problem. The second example is a numerical study on the nonlinear eigenvalue problem which is equivalent to a half quartic anharmonic oscillator.

All the movable singularities of first two Painlevé equations in Eqs. (3) and (4) are poles. They can be generalised into equations in (5) with movable singularities being algebraic. Similarly, we generalise Eq. (6) into a series of equations with algebraic movable singularities in Sec. 4. The nonlinear eigenvalues of two such equations are obtained numerically. The asymptotic behaviour of large eigenvalues are found to follow simple power laws, just like the first two Painlevé equations and the generalised Painlevé equations.

## 2. Equivalence of two types of eigenvalue problems

In this section we present an equivalent relation between a nonlinear eigenvalue problem and a linear eigenvalue problem. By exploiting the semi-transcendental feature of Equation (6), we show that the two eigenvalue problems possess essentially same eigenvalues, up to a minus sign and a constant shift. The eigensolutions in the two problems are closely related.

Equation (6) is the number V equation of the Canonical Equations of Type I in the classic textbook on ordinary differential equations by Ince [8]. It can obviously be integrated once to a first-order nonlinear equation, which is equivalent to the Ricatti equation. To see this, let us first change the dependent variable as

$$y(x) = W(x) - \tfrac{1}{2}q(x). \tag{7}$$

The equation satisfied by $y(x)$ is

$$y''(x) = -2y'(x)y(x) + V'(x), \tag{8}$$

where we have defined the derivative of an arbitrary function $V(x)$ as

$$V'(x) \equiv \tfrac{1}{2}q'(x)q(x) - \tfrac{1}{2}q''(x). \tag{9}$$

For any analytic function $V(x)$, the movable singularities of $y(x)$ are all simple poles.

Equation (8) can be directly integrated to a Riccati equation,

$$y'(x) = -y^2(x) + V(x) - E, \tag{10}$$

where $E$ is a constant of integration.

If $V(x)$ is unbounded for large $x$, then $y(x)$ may either develop infinite movable singularities, or, after finite singular points, approach to one of the two asymptotes

$$y(x) \sim \pm\sqrt{V(x) - E}, \quad x \to \infty. \tag{11}$$

It can be shown that for positive $x$, the upper branch is a stable fixed point in the functional space and the lower branch is unstable. To see this, let

$$y(x) = Y(x) + \phi(x), \tag{12}$$

where $Y(x)$ is a solution of Eq. (10). Plugging into Eq. (10), we get

$$\phi'(x) = -2Y(x)\phi(x) - \phi^2(x). \tag{13}$$

The solution to this nonlinear equation is

$$\phi(x) = \frac{\mathrm{e}^{-2\int_a^x \mathrm{d}s\, Y(s)}}{\displaystyle\int_b^x \mathrm{d}t\, \mathrm{e}^{-2\int_a^t \mathrm{d}s\, Y(s)}}, \tag{14}$$

where $a$ and $b$ are two constants. Note that $\phi(x)$ depends only on $b$, but not on $a$. Depending on the sign of $Y(x)$, the denominator of Eq. (14) has two distinct asymptotic behaviours for large $x$ [9]:

$$\int_b^x \mathrm{d}t\, \mathrm{e}^{-2\int_a^t \mathrm{d}s\, Y(s)} \sim \begin{cases} \displaystyle\int_b^\infty \mathrm{d}t\, \mathrm{e}^{-2\int_a^t \mathrm{d}s\, Y(s)} \equiv C, & Y(x) \to +\infty, \quad x \to \infty; \\ \displaystyle -\frac{1}{2Y(x)} \mathrm{e}^{-2\int_a^x \mathrm{d}s\, Y(s)}, & Y(x) \to -\infty, \quad x \to \infty. \end{cases} \tag{15}$$

This in turn leads two asymptotic behaviours for $\phi(x)$,

$$\phi(x) \sim \begin{cases} \displaystyle\frac{1}{C}\mathrm{e}^{-2\int_a^x \mathrm{d}s\, Y(s)} \to 0, & Y(x) \to +\infty, \quad x \to \infty; \\ -2Y(x) \to \infty, & Y(x) \to -\infty, \quad x \to \infty. \end{cases} \tag{16}$$

Thus, the positive branch $Y(x) \sim \sqrt{V(x) - E}$ is stable and the negative branch is unstable. Note that $\phi(x)$ does not change signs for a real solution $Y(x)$. This means that, near the stable fixed point (asymptote), the solutions approach it without oscillation. This is different from the first two Painlevé equations and their generalisations studied in Refs. [1, 2, 6, 7].

A Riccati equation is equivalent to a second order linear differential equation. To see this, let us rewrite the solution of Eq. (10) as

$$y(x) = \frac{\psi'(x)}{\psi(x)}, \tag{17}$$

then $\psi(x)$ satisfies a Schrödinger equation

$$-\psi''(x) + V(x)\psi(x) = E\psi(x). \tag{18}$$

The solution to the Schrödinger equation (18) has two asymptotic behaviors for large $x$:

$$\psi(x) \sim \frac{N_\pm}{[V(x) - E]^{1/4}} \mathrm{e}^{\pm\int_0^x \mathrm{d}s \sqrt{V(s)-E}}, \quad x \to \infty. \tag{19}$$

Here we have chosen the lower integration limit to be 0 without loss of generality. Only the diminishing solutions are corresponding to the separatrix solutions in $y(x) \sim -\sqrt{V(x) - E}$. That is, $N_+ = 0$.

For the initial slope nonlinear eigenvalue problem of Eq. (8), the initial value $y(0)$ is fixed and the initial slope $y'(0)$ is the eigenvalue when it yields a separatrix solution. If we choose a homogeneous initial value as

$$y(0) = 0, \tag{20}$$

then the initial slope of $\psi$ vanishes,

$$\psi'(0) = 0. \tag{21}$$

That is, to find a separatrix solution of $y$ is equivalent to solve the Schrödinger equation with the boundary conditions $\psi(+\infty) = 0$ and $\psi'(0) = 0$. The eigenvalue of the *linear* eigenvalue problem is simply $E$. The eigenvalue of the initial slope problem of the *nonlinear* eigenvalue problem is

$$y'(0) = V(0) - E. \tag{22}$$

We emphasise that this equivalence between the two eigenvalue problems is exact.

## 3. An exactly solvable model

The nonlinear differential equation (8) can be solved exactly with a quadratic function $V$,

$$V(x) = x^2. \tag{23}$$

The corresponding Schrödinger equation (18) describe the quantum harmonic oscillator.

A separatrix solution has the form

$$y(x) = \frac{D'_\nu(x)}{D_\nu(x)}, \tag{24}$$

where $D_\nu(x)$ is the parabolic cylinder function and $\nu$ is related to the integration constant in Eq. (10) by

$$E = 2\nu + 1. \tag{25}$$

At the origin, we have the initial conditions

$$y(0) = -2\frac{\Gamma\left(-\frac{\nu}{2} - \frac{1}{2}\right)}{\Gamma\left(-\frac{\nu}{2}\right)}, \qquad y'(0) = -2\nu - 1 - 4\frac{\Gamma^2\left(-\frac{\nu}{2} - \frac{1}{2}\right)}{\Gamma^2\left(-\frac{\nu}{2}\right)}. \tag{26}$$

For real parameter $\nu$, the above $y'(0)$ never vanishes. Therefore, there is no nonlinear eigenvalues for the initial function problem with $y'(0) = 0$.

For the initial slope problem with $y(0) = 0$, we get

$$\nu = 2n, \qquad n = 0, 1, 2, \cdots. \tag{27}$$

The eigenfunctions are

$$y_n(x) = -x + 4n\frac{H_{2n-1}(x)}{H_{2n}(x)}, \qquad n = 0, 1, 2, \cdots. \tag{28}$$

The nonlinear eigenvalues are

$$y'_n(0) = -4n - 1. \tag{29}$$

Similarly, if one chooses

$$V(x) = x^4, \tag{30}$$

half of the eigenvalues of the anharmonic oscillator will be the eigenvalues of the nonlinear initial slope problem (see Table 1). There is no real eigenvalues for the initial function problem for this equation with $y'(0) = 0$.

## 4. Extended Ricatti equations

In this section, we generalise the (derivative) of the Riccati equation in Eq. (6) to a class of semi-integrable equations with movable singularities to be algebraic. Namely, there is no logarithmic or other essential singularities. This is a similar work which has been done to the first two Painlevé equations in [6, 7]. Since the term "generalised Riccati equations" was used for other purpose, we call these new equations as "extended Ricatti equations." The generic form for these equations are

$$y''(x) = -\frac{n+1}{n}y'(x)y^n(x) + \sum_{i=0}^{m}\sum_{j=0}^{k} a_{mk}x^m y^k(x). \tag{31}$$

**Table 1.** Initial slopes which give rise to separatrix solutions to the equation $y''(x) = -2y'(x)y(x) + 4x^3$ with $y(0) = 0$.

| $n$ | 0 | 1 | 2 | 3 |
|---|---|---|---|---|
| $y_n'(0)$ | −1.06036209 | −7.45569793 | −16.26182601 | −26.52847118 |
| $n$ | 4 | 5 | 6 | 7 |
| $y_n'(0)$ | −37.92300102 | −50.25625451 | −63.40304698 | −77.27320048 |
| $n$ | 8 | 9 | 10 | |
| $y_n'(0)$ | −91.79806681 | −106.92330738 | −122.60463900 | |

For example, for the nonlinear equation

$$y''(x) = -\frac{4}{3}y'(x)y^3(x) + x, \tag{32}$$

the movable singularities are cubic roots:

$$\begin{aligned} y(x) \sim \; & \frac{1}{(x-x_0)^{1/3}} + a_3(x-x_0)^1 + \frac{3x_0}{10}(x-x_0)^2 \\ & -\frac{6a_3^2}{11}(x-x_0)^{7/3} + \frac{3}{26}(x-x_0)^3 + \sum_{i=10}^{\infty} a_i(x-x_0)^{i/3}, \qquad x \to x_0, \end{aligned} \tag{33}$$

where $a_3$ is a constant to be determined by the initial conditions.

For $y(0) = 0$, the nonlinear eigenvalues for the initial slope problem are listed in Table 2. For large $n$, they follow a simple power law

$$y_n'(0) \sim -2.32805n^{4/5}, \qquad n \to \infty. \tag{34}$$

Similarly, for the equation

$$y''(x) = -\frac{4}{3}y'(x)y^3(x) + xy(x), \tag{35}$$

the movable singularities are also cubic roots:

$$\begin{aligned} y(x) \sim \; & \frac{1}{(x-x_0)^{1/3}} + a_3(x-x_0)^1 + \frac{x_0}{2}(x-x_0)^{5/3} - \frac{6a_3^2}{11}(x-x_0)^{7/3} \\ & +\frac{3}{20}(x-x_0)^{8/3} - \frac{9a_3x_0}{26}(x-x_0)^3 + \sum_{i=10}^{\infty} a_i(x-x_0)^{i/3}, \qquad x \to x_0. \end{aligned} \tag{36}$$

For $y(0) = 0$, the nonlinear eigenvalues for the initial slope problem are listed in Table 3. For large $n$, they follow a simple power law

$$y_n'(0) \sim -0.96951n^{4/9}, \qquad n \to \infty. \tag{37}$$

**Table 2.** Initial slopes which give rise to separatrix solutions to the equation $y''(x) = -\frac{4}{3}y'(x)y^3(x) + x$ with $y(0) = 0$.

| | | | | |
|---|---|---|---|---|
| $n$ | 0 | 1 | 2 | 3 |
| $y'_n(0)$ | −1.00243383 | −2.94953545 | −4.60069679 | −6.11303401 |
| $n$ | 4 | 5 | 6 | 7 |
| $y'_n(0)$ | −7.53640737 | −8.89523799 | −10.20390084 | −11.47180488 |
| $n$ | 8 | 9 | 10 | 11 |
| $y'_n(0)$ | −12.70555477 | −13.91002352 | −15.08894332 | −16.24525712 |
| $n$ | 12 | 13 | 14 | 15 |
| $y'_n(0)$ | −17.38134027 | −18.49914706 | −19.60031125 | −20.68621727 |
| $n$ | 16 | 17 | 18 | 19 |
| $y'_n(0)$ | −21.75805200 | −22.81684331 | −23.86348925 | −24.89878065 |
| $n$ | 20 | 21 | 22 | 23 |
| $y'_n(0)$ | −25.92341874 | −26.93802924 | −27.94317361 | −28.93935824 |
| $n$ | 24 | 25 | | |
| $y'_n(0)$ | −29.92704197 | −30.90664231 | | |

**Table 3.** Initial slopes which give rise to separatrix solutions to the equation $y''(x) = -\frac{4}{3}y'(x)y^3(x) + xy(x)$ with $y(0) = 0$.

| | | | | |
|---|---|---|---|---|
| $n$ | 0 | 1 | 2 | 3 |
| $y'_n(0)$ | −0.89134081 | −1.21797331 | −1.48616058 | −1.71029522 |
| $n$ | 4 | 5 | 6 | 7 |
| $y'_n(0)$ | −1.90473552 | −2.07799763 | −2.23533407 | −2.38019349 |
| $n$ | 8 | 9 | 10 | 11 |
| $y'_n(0)$ | −2.51496483 | −2.64137602 | −2.76072118 | −2.87399742 |
| $n$ | 12 | 13 | 14 | 15 |
| $y'_n(0)$ | −2.98199128 | −3.08533557 | −3.18454800 | −3.28005827 |

## 5. Conclusions

In this paper, we present an exact equivalence between a class of nonlinear eigenvalue problems and conventional linear eigenvalue problems. By exploiting the semi-integrable features of these nonlinear differential equations, we have linked the nonlinear eigenvalues to the well-known linear eigenvalues. We found the first exactly solvable nonlinear eigenvalue problem. This equivalence also provides new insights about the nature of nonlinear eigenvalue problems.

In addition, we generalised the Riccatti equations to an infinite series of nonlinear equations whose movable singularities are algebraic. Two examples are explicitly presented. The nonlinear eigenvalues exhibit a similar power law as those in the Painlevé equations. There are totally 50 second order differential equations satisfying the Painlevé criterion [8]. Many of them may be generalised to have algebraic movable singularities. We leave this to a future study.

**Acknowledgments**
The author thanks for the useful discussions with Prof Carl M. Bender. The author is grateful for the comments and suggestions on the writing from Prof Jiangbin Gong.

## References

[1] Bender C M, Fring A, and Komijani J 2014 *J. Phys. A: Math. Theor.* **47** 235204
[2] Bender C M and Komijani J 2015 *J. Phys. A: Math. Theor.* **48** 475202
[3] Long W-G, Li Y-T, Liu S-Y, Zhao Y-Q 2017 *Stud. App. Math.* **139** 505-532
[4] Long W-G and Zeng Z-Y 2020 *Preprint* arXiv:2005.03440
[5] Bender C M and Boettcher S 1998 *Phys. Rev. Lett.* **80** 5243-5246
[6] Bender C M, Komijani J, and Wang Q-h 2017 *Resurgence, Physics and Numbers (Centro di Ricerca Matematica Ennio De Giorgi, Pisa)* CRM Series vol 20 ed F Fauvet, D Manchon et al pp 67-90
[7] Bender C M, Komijani J, and Wang Q-h 2019 *J. Phys. A: Math. Theor.* **52** 315202
[8] Ince E L 1956 *Ordinary Differential Equations* (Dover, New York)
[9] Bender C M and Orszag S A 1978 *Advanced Mathematical Methods for Scientists and Engineers* (McGraw Hill, New York)

Journal of Physics: Conference Series **2038** (2021) 012026 doi:10.1088/1742-6596/2038/1/012026

# Paths of unitary access to exceptional points

**Miloslav Znojil**

The Czech Academy of Sciences, Nuclear Physics Institute, Hlavní 130, 250 68 Řež, Czech Republic

Department of Physics, Faculty of Science, University of Hradec Králové, Rokitanského 62, 50003 Hradec Králové, Czech Republic

E-mail: `znojil@ujf.cas.cz`

**Abstract.** With an innovative idea of acceptability and usefulness of the non-Hermitian representations of Hamiltonians for the description of unitary quantum systems (dating back to the Dyson's papers), the community of quantum physicists was offered a new and powerful tool for the building of models of quantum phase transitions. In this paper the mechanism of such transitions is discussed from the point of view of mathematics. The emergence of the direct access to the instant of transition (i.e., to the Kato's exceptional point) is attributed to the underlying split of several roles played by the traditional single Hilbert space of states $\mathcal{L}$ into a *triplet* (viz., in our notation, spaces $\mathcal{K}$ and $\mathcal{H}$ besides the conventional $\mathcal{L}$). Although this explains the abrupt, quantum-catastrophic nature of the change of phase (i.e., the loss of observability) caused by an infinitesimal change of parameters, the explicit description of the unitarity-preserving corridors of access to the phenomenologically relevant exceptional points remained unclear. In the paper some of the recent results in this direction are summarized and critically reviewed.

## 1. Introduction.

In the context of quantum physics the first signs of appreciation of the phenomenological relevance of exceptional points [1] appeared during the studies of the so called open quantum systems [2]. In these studies the effective Hamiltonians act in a model subspace of the full Hilbert space and are non-Hermitian. Thus, it was not too surprising to reveal that "the positions of the exceptional points" vary "in the same way as the transition point of the corresponding phase transition" [3]. The possibility of a generic connection between exceptional points and phase transitions has been born.

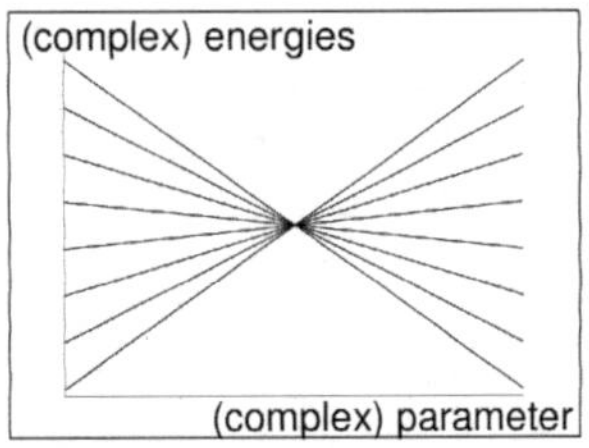


**Figure 1.** A schematic sample of the degeneracy of an $N-$plet of energies at an exceptional point of order $N$ (EPN) with $N = 8$ (EP8).

In our present paper we will summarize several aspects of this connection. In order to narrow the subject we will only consider the exceptional-point-related phenomena emerging in the theory of the closed, stable, unitary quantum systems.

*1.1. Mathematical concept of exceptional point.*
In mathematics the exceptional point (EP) can be defined as the value of an (in general, complex) parameter $g$ at which a linear operator (which is, say, non-Hermitian but analytic in $g$) loses its diagonalizability. For Hamiltonians, one of the possible consequences is schematically depicted in Fig. 1. This picture indicates that near an EP singularity there may exist an $N-$plet of eigenvalues of the operator (i.e., typically, bound state energies specified by a Hamiltonian $H(g)$) which merge in the limit of $g \to g^{(EP)}$. Simultaneously, in contrast to the non-EP dynamical scenarios, the EP or EPN degeneracy also involves the eigenvectors.

It is probably worth adding that the term "exceptional point" can be found slightly differently (and also more rigorously) defined in the Kato's book [1]. In the Kato's definition, for example, the notion of an eigenvector is not used at all. Interested readers should consult the book: in what follows the attention is shifted from the Kato's emphasis on analyticity assumptions to the role played by the eigenvectors.

*1.2. Early applications of exceptional points in quantum physics of closed systems.*
The confluence of eigenvalues as studied by mathematicians and sampled by Fig. 1 did not initially find any immediate applications in quantum physics of closed systems. Among the reasons one can find, first of all, the widespread habit of keeping all of the realistic phenomenological bound-state Hamiltonians self-adjoint. This also required, for pragmatic reasons, a replacement of the general complex parameter (say, $g \in \mathbb{C}$ in $H(g)$) by a real variable (i.e., by $\lambda \in \mathbb{R}$ in $H(\lambda)$). A combination of the two constraints rendered the mergers impossible. Only after an abstract mathematical operation of analytic continuation of Hamiltonian $H(\lambda)$ it was possible to reveal, in several models [4, 5, 6, 7], the existence of the EPs. Naturally, all of them were manifestly non-real, Im $\lambda^{(EP)} \neq 0$. Only an indirect indication of their presence near a real line of $\lambda$ could have been provided by the avoided level crossings [8], a spectral feature sampled in Fig. 2.

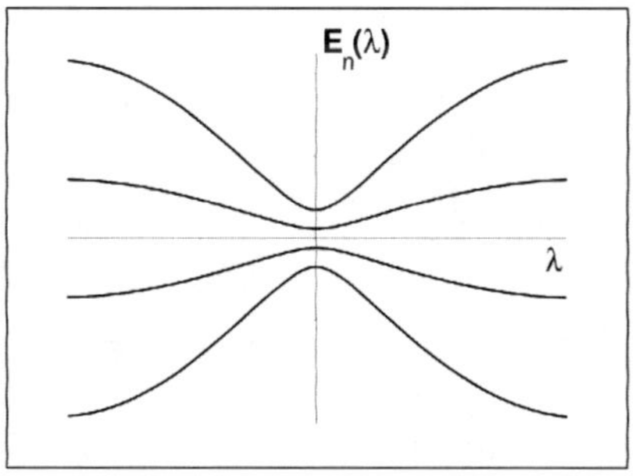


**Figure 2.** Avoided crossing of four real (i.e., observable) energy levels (arbitrary units).

In the quantum unitary-evolution setting a dramatic change of the situation only came with the Bender's and Boettcher's pioneering letter [9]. The authors revealed that a suitable weakening of the property of the self-adjointness of $H(\lambda)$ could make the EP singularities "visible" and real. What followed the discovery (cf. also the later review paper [10]) was an enormous increase of interest of the physics community in a broad variety of Hamiltonians $H(\lambda)$ possessing the *real* (i.e., in principle, experimentally accessible) EP singularities with

$\text{Im}\,\lambda^{(EP)} = 0$. Some features of the EP-related non-diagonalizable Hamiltonians were discussed in [11], and a few examples with explicit parametrizations of the vicinity of EP2 and of EP3 were considered in [12] and [13]. In 2010, the conference organized by W. D. Heiss in Stellenbosch was even *exclusively* dedicated to the role of the EPs in multiple branches of physics [14].

In our present paper we are going to interpret the EP and EPN degeneracies as sampled by Fig. 1 in a strict unitary-evolution sense. This means that we will only consider the real parameters $\lambda$ lying in a small vicinity of $\lambda^{(EPN)}$. Under this assumption we will require that the whole spectrum of energies remains real and non-degenerate either on both sides of $\lambda^{(EPN)}$ (i.e., at any not too remote $\lambda \neq \lambda^{(EPN)}$) or on one side at least (i.e., for $\lambda < \lambda^{(EPN)}$ or for $\lambda > \lambda^{(EPN)}$). We will, naturally, also admit that the value of $\lambda$ parametrizes a smooth curve passing through a larger, $d-$dimensional space $\mathbb{R}^d$ of the real parameters determining a $d-$parametric Hamiltonian $H(\lambda) = H[a(\lambda), b(\lambda), \ldots, z(\lambda)]$.

*1.3. Two-parametric example.*
For illustration let us recall the two-parametric real-matrix Hamiltonian of Ref. [15],

$$H(a,b) = \left( \begin{array}{cccc} -3 & b & 0 & 0 \\ -b & -1 & a & 0 \\ 0 & -a & 1 & b \\ 0 & 0 & -b & 3 \end{array} \right) . \tag{1}$$

Its eigenvalues

$$E_{\pm,\pm}(a,b) = \pm\frac{1}{2}\sqrt{20 - 4\,b^2 - 2\,a^2 \pm 2\,\sqrt{64 - 64\,b^2 + 16\,a^2 + 4\,b^2a^2 + a^4}} \tag{2}$$

remain real and non-degenerate inside a two-dimensional unitarity-supporting domain $\mathcal{D}^{(physical)}$ of parameters $a = a(\lambda)$ and $b = b(\lambda)$ which is displayed in Fig. 3.

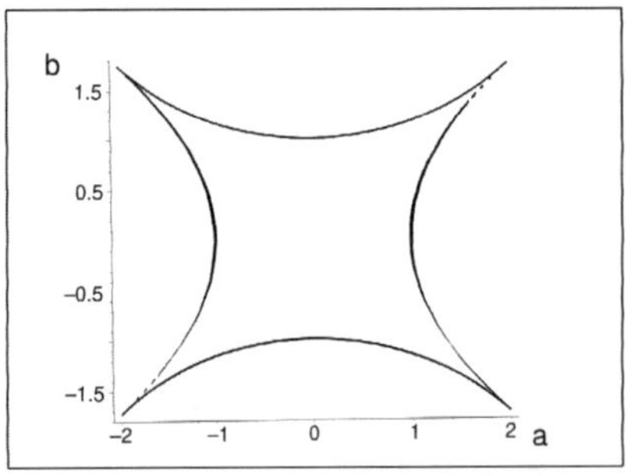


**Figure 3.** The boundary of domain $\mathcal{D}^{(physical)}$ for toy-model (1) with $d = 2$.

It is worth adding that once one moves to the EP-supporting models with more parameters, $d > 2$, the illustrative shape of the $d = 2$ domain in Fig. 3 (viz., a deformed square with protruded vertices) appears to be, in some sense, generic. For a family of solvable models with $N > 4$ such an intuition-based conjecture has been confirmed in [16, 17]. A more recent, abstract theoretical explanation of the hypothesis may be found in [18, 19]. On this background one can expect that the most interesting smooth curves parametrized by $\lambda$ would be those which end at one of the EPN vertices with maximal $N$ (i.e., with $N = 4$ in Fig. 3).

### *1.4. Paradox of stability near exceptional points.*

In a way inspired by the above example one can expect that the behavior of the closed quantum system with parameters lying deeply inside $\mathcal{D}$ would not be too surprising. In such a dynamical regime, small changes of the parameters leave the spectrum real. The formulation of predictions can be based on a conventional perturbation theory.

Close to the boundary $\partial\mathcal{D}$ the situation is different and much more interesting. Indeed, in a small vicinity of this boundary, a small change of the parameter seems to be able to cause an abrupt loss of the reality of the spectrum, i.e., in the language of the closed quantum systems, of the observability of its characteristics (represented by operators). In contrast to the models of the open quantum systems (which are not of our immediate interest now, and in which the eigenvalues having a physical meaning may be complex), a spontaneous collapse *alias* quantum phase transition caused by a small fluctuation of the interaction seems unavoidable.

Our present paper will be fully devoted to its study. In fact, a major part of the paper will provide a concise explanation that in the specific context of the closed, unitary quantum systems the latter, intuitive expectation of instabilities is unreliable (see section 2 for introduction). We will clarify why such an interpretation of dynamics is incorrect (see section 3), why a deeper clarification of the point is important (cf. section 4), and, finally, what would be a valid conclusion (section 5).

Keeping this purpose in mind, our text will start by a sketchy presentation of a (very non-representative) sample of the current state of applications of the stationary version of the formalism represented, schematically, by Fig. 4. This will be followed by a (partially critical) review of some open questions connected, first of all, with the role of the Kato's exceptional points in phase transitions. We will clarify the role of parameters in the vicinity of EPs. In this dynamical regime, a few comments will be also added on the correct analysis of stability of the non-Hermitian but unitary quantum systems with respect to small perturbations.

A concise summary of our present message will finally be formulated in section 6.

## 2. Unitary evolution in Schrödinger picture using non-Hermitian Hamiltonians.

### *2.1. Theoretical background*

The above-mentioned paradox of stability near EPs is reminiscent of the old puzzle of the stability of atoms in classical physics. In fact, the resolution of the latter puzzle belongs among the most remarkable successes which accompanied the birth of quantum mechanics. The innovation was based on Schrödinger equation representing bound states by ket-vector elements of a suitable Hilbert space $\mathcal{K}$,

$$H\,|\psi_n\rangle = E_n\,|\psi_n\rangle\,, \qquad |\psi_n\rangle \in \mathcal{K}\,, \qquad n=0,1,\ldots\,. \tag{3}$$

Subsequently, the incessant growth of the number of successful phenomenological applications of quantum theory was accompanied by the emergence of various innovative mathematical subtleties. One of the ideas of the latter type (and of a decisive relevance for the present paper) can be traced back to the papers by Dyson [20] and Dieudonné [21]. Independently, they introduced the concept of the Θ-pseudo-Hermitian Hamiltonians. These operators (with real spectra) are assumed to remain non-Hermitian in $\mathcal{K}$ but restricted by the quasi-Hermiticity relation

$$H = \Theta^{-1}H^\dagger\Theta \neq H^\dagger\,, \qquad \Theta = \Theta^\dagger > 0\,. \tag{4}$$

For details see the text below, or the older review paper [22], or its more recent upgrades [10, 23, 24, 25, 26, 27].

Briefly, the Θ-pseudo-Hermiticity innovation can be characterized as a reclassification of the status of the Hilbert space of states (cf. Fig. 4). Indeed, in conventional textbooks the choice of $\mathcal{K}$ in Schrödinger Eq. (3) is usually presented as unique. In the textbook cases of stable,

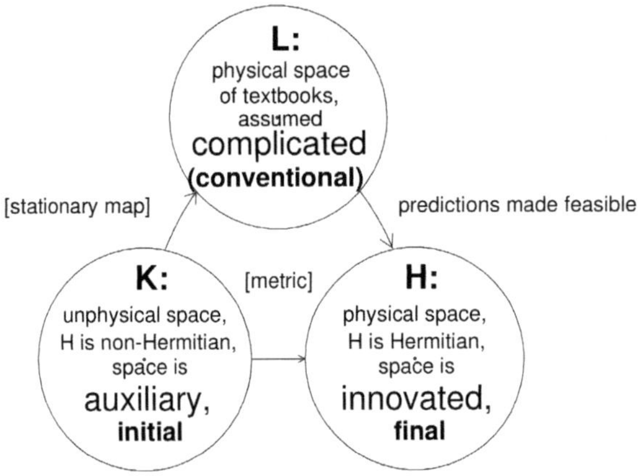


**Figure 4.** Triplet of Hilbert spaces representing a bound state $\psi$ and connected by the Dyson map $\Omega$ of Eq. (6) and by the innovative *ad hoc* amendment $I \to \Theta = \Omega^\dagger \Omega \neq I$ of the inner-product metric.

unitarily evolving quantum systems, in a way observing Stone theorem [28], also the Hamiltonian itself would necessarily be required self-adjoint in $\mathcal{K}$. After the reclassification, in contrast, the meaning of symbols $\mathcal{K}$ and $H$ is being changed. Firstly, in the Dyson's spirit one decides to admit that $H$ can be non-Hermitian in $\mathcal{K}$. In the light of the Stone theorem this means that the status of $\mathcal{K}$ must be changed from "physical" to "unphysical". Secondly, in the Dieudonné's spirit, the postulate of validity of quasi-Hermiticity relation (4) enables us to interpret operator $\Theta$ as a metric [22]. Thus, we may amend the inner product in order to convert the unphysical Hilbert space $\mathcal{K}$ into a new physical Hilbert space $\mathcal{H}$,

$$\langle \psi_1 | \psi_2 \rangle_{\mathcal{H}} = \langle \psi_1 | \Theta | \psi_2 \rangle_{\mathcal{K}} \,. \tag{5}$$

Thirdly, the factorization $\Theta = \Omega^\dagger \Omega \neq I$ of the metric enables us to introduce an operator

$$\mathfrak{h} = \Omega^{-1} \, H \, \Omega \tag{6}$$

and to interpret is as a hypothetical alternative isospectral Hamiltonian in another, alternative physical Hilbert space $\mathcal{L}$. By the assumption which dates back to Dyson [20] and which motivated also multiple applications of his approach [22], operators of the form (6) happened to be self-adjoint but technically less user-friendly.

### *2.2. Notation conventions*

Further details characterizing such an apparently redundant representation of a single state $\psi$ will be recalled and summarized below. Now, let us only point out that the Dyson's and Dieudonné's reformulation of the postulates of quantum theory was deeply motivated. It is only necessary to accept and appreciate *both* the Dyson's activity *and* the Diedonné's scepticism. Indeed, Dyson discovered and used, constructively, several positive and truly innovative aspects of the use of quasi-Hermiticity in physics and phenomenology. At the same time, the Dieudonné's well founded critical analysis of the "hidden dangers" behind the quasi-Hermiticity is still a nontrivial and exciting subject for mathematicians [29, 30]).

In the context of physics the latter "hidden dangers" were, fortunately, cleverly circumvented (cf. review [22]). Some of the corresponding technical and mathematical recommendations will be recollected below. Immediately, let us only mention that the amendment of mathematics led to an ultimate compact and explicit three-Hilbert-space (3HS) formulation of the most general non-stationary version of quasi-Hermitian quantum mechanics as first proposed in [31] and as subsequently reviewed in [23].

In what follows, we will employ the most compact notation as introduced in our latter two papers. The reason is twofold. Firstly, along the lines indicated in [23], the choice of such a notation will simplify the separation of our present perception of physics from its alternatives which often share the mathematical terminology while not sharing the phenomenological scope. Secondly, the emphasis put on notation will enable us to review the field of our present interest in a sufficiently compact and concise manner, avoiding potential misunderstandings caused by the variabiity of the notation used in the literature (cf. Table 1).

**Table 1.** Sample of confusing differences in notation conventions.

| concept | symbol | | | |
|---|---|---|---|---|
| Hilbert space metric | $\eta_+$ | $\rho$ | $\widetilde{T}$ | $\Theta$ |
| Dyson's map | $\rho$ | $\eta$ | $S$ | $\Omega$ |
| state vector | $\lvert\psi\rangle$ | $\Psi$ | $\lvert\Psi\rangle$ | $\lvert\psi\rangle$ |
| dual state vector | $\lvert\phi\rangle$ | $\rho\Psi$ | $\widetilde{T}\lvert\Psi\rangle$ | $\lvert\psi\rangle\rangle$ |
| reference | [24] | [32] | [22] | here |

*2.3. The concept of hidden Hermiticity.*

*2.3.1. Motivation.* In an incomplete sample of ambitions of the 3HS reformulation of quantum theory let us mention, first of all, the Dyson's description of correlations in many-body systems [20] inspired by numerical mathematics (where one would speak simply about a "preconditioning" of the Hamiltonian). Secondly, in combination with the assumption of $\mathcal{PT}-$symmetry [27] the 3HS approach (complemented by the mathematical Krein-space methods [33, 34]) opened new horizons in our understanding of the first-quantized relativistic Klein-Gordon equation [35, 36]. Thirdly, a transfer of the underlying "hidden Hermiticity" ideas to relativistic quantum field theory [37] and/or to the studies of supersymmetry [38] inspired a number of methodical studies of various elementary toy models [9, 39, 40]. Last but not least it is worth mentioning that the applications of the 3HS formalism even touched the field of canonical quantum gravity based on the use of Wheeler-DeWitt equation [41].

*2.3.2. Disambiguation.* The solution $\Theta = \Theta(H)$ of Eq. (4) does not exist whenever the spectrum of $H$ ceases to be real. This means that only certain parameters in non-Hermitian $H(\lambda)$ remain unitarity-compatible and "admissible", $\lambda \in \mathcal{D}$. In the admissible cases, in a way explained in [23], there exists a mapping $\Omega$ which realizes an equivalence of predictions made in $\mathcal{H}$ with those made in a third, hypothetical and, by our assumption, practically useless Hilbert space $\mathcal{L}$. The latter space is precisely the space of states used in conventional textbooks. In the present 3HS context (depicted in Figure 4), its role is purely formal because in this space, operator (6) representing the Hamiltonian and formally self-adjoint in $\mathcal{L}$ is, by assumption, too complicated to be useful or tractable (for example, it may happen to be a highly non-local pseudo-differential operator [24]).

In the literature devoted to applications of unitary quantum theory the authors working in the 3HS version of Schrödinger picture do not always sufficiently clearly emphasize the Hermiticity of the physical Hamiltonian in the physical Hilbert space $\mathcal{H} = \mathcal{H}^{(Standard)}$ (say, by writing $H = H^{\ddagger}$ [23]). Another potential source of confusion lies in the widespread habit (or rather in the abuse of language) of using shorthand phrases (like "non-Hermitian Hamiltonians") or shorthand formulae (like $H \neq H^{\dagger}$) without adding that one just temporarily dwells in an irrelevant, auxiliary, unphysical Hilbert space $\mathcal{K}$. The resulting, fairly high probability of misunderstandings is further enhanced by the diversity of conventions as sampled in Table 1.

## 3. Constructive aspects of the triple Hilbert space formalism.

### *3.1. Metric and its ambiguity.*

Two alternative model-building strategies based on the "generalized Hermiticity" (4) have been used in applications. In the first one one chooses $\mathfrak{h} = \mathfrak{h}^{\dagger}$ and $\Omega$ and reconstructs $H$ and $\Theta$. In fact, the use of such a strategy remained restricted just to nuclear physics of heavy nuclei in practice [22]. At present, almost exclusively [24], one picks up the Hamiltonian (i.e., a "trial and error" operator $H$ which is non-Hermitian in $\mathcal{K}$) and reconstructs, via Eq. (4), the (necessarily, nontrivial) metric $\Theta > 0$ (i.e., the correct physical Hilbert space of states denoted, here, by dedicated symbol $\mathcal{H}$). The approach based on the reconstruction of metric now forms the mainstream in research. The false but friendly space $\mathcal{K}$ and a non-Hermitian Hamiltonian $H$ are both assumed to be given in advance while a suitable Hermitizing inner-product metric must be reconstructed, in principle at least. The Hermiticity of any other observable $\Lambda$ in $\mathcal{H}$ must also be guaranteed. In the auxiliary space $\mathcal{K}$ this requirement has the form $\Lambda^{\dagger}\,\Theta = \Theta\,\Lambda$.

In an elementary illustration the Wheeler-DeWitt-like equation

$$H = H^{(WDW)}(\tau) = \left[ \begin{array}{cc} 0 & \exp 2\tau \\ 1 & 0 \end{array} \right] \neq H^{\dagger} \quad \text{in} \quad \mathcal{K} = \mathbb{R}^2 \tag{7}$$

yields the two real closed-form eigenvalues $E = E_{\pm} = \pm \exp \tau$ so that it can serve as a sample of the Dyson-Dieudonné definition of quasi-Hermiticity (4). A decisive advantage of the use of such a highly schematic one-parametric two-by-two real-matrix example is that one can easily solve Eq. (4) and construct *all* of the eligible physical inner-product-metric operators

$$\Theta = \Theta^{(WDW)}(\tau, \beta) = \left[ \begin{array}{cc} \exp(-\tau) & \beta \\ \beta & \exp \tau \end{array} \right] = \Theta^{\dagger}\,, \qquad |\beta| < 1\,. \tag{8}$$

These solutions form a complete set of candidates for the (Hermitian and positive definite) eligible metric [22]. In this example one notices that parameter $\beta$ is an independent variable. This observation is, indeed, compatible with the well known fact that the assignment $\Theta = \Theta(H)$ of the metric to a preselected Hamiltonian is not unique [22, 24, 42].

### *3.2. False instabilities and open systems in disguise*

In the literature devoted to applications the authors interested in non-Hermiticity often do not sufficiently clearly separate the quantum and non-quantum theories. Here, we are not going to deal with the latter branch of physics. Nevertheless, even within the range of quantum mechanics the authors often intermingle the results concerning the open and closed quantum systems. Here, almost no attention will be paid to the former family of models, either. An exception should be made in connection with the papers dealing with certain non-Hermitian but $\mathcal{PT}$−symmetric quantum systems where, typically, the authors claim that "complex eigenvalues may appear very far from the unperturbed real ones despite the norm of the perturbation is arbitrarily small" [43].

As long as the latter claims (of a top *mathematical* quality) are accompanied by certain fairly vague quantum-theoretical considerations (which could certainly prove misleading), we feel forced to point out that recently, the study of the parametric domains of unitarity near EPs [44] clarified the point (cf. also the less formal explanation in [15]). The essence of the misunderstanding can be traced back to the fact that the loss of stability was deduced, in [43], from the properties of the pseudospectrum [29]. Unfortunately, the construction was only performed using the trivial form of the inner-product metric defining just the manifestly unphysical Hilbert space $\mathcal{K}$ where $\Theta = I$. For this reason the mathematical results about pseudospectra in $\mathcal{K}$ make sense in, and only in, the open quantum systems. In these systems the predicted instabilities really do occur because the space $\mathcal{K}$ itself still keeps there the status of the physical space.

We may summarize that in the 3HS models of closed systems the Hamiltonians are in fact self-adjoint in $\mathcal{H}$. This means that the evaluation of their pseudospectra would *necessarily* require the work with norms which would be expressed in terms of the physical metric $\Theta$. Thus, once the existence of such a metric is guaranteed (which is, naturally, a nontrivial task!), the proofs of stability based on the pseudospectra will apply.

We should also add that the smallness of perturbations is a concept which crucially depends on the metric $\Theta$ defining the physical Hilbert space $\mathcal{H}$. From this point of view it is obvious that as long as the metric itself becomes necessarily strongly anisotropic in the vicinity of EPs [42], also some of the perturbations which might look small in $\mathcal{K}$ become, in such a regime, large in $\mathcal{H}$, and *vice versa* [45].

### *3.3. EP (hyper)surfaces and their geometry.*

For the lovers of closed formulae the existence as well as the geometry of access to EPs was made very explicit in paper [19]. An advertisement of the contents of this paper can be brief: a list of transmutations is given there between various versions of a special Bose-Hubbard (BH) system (represented by certain complex finite matrices) and of a discrete and truncated anharmonic oscillator (AO). It is sufficient to recall here just the ultimate message of the paper: at an EP singularity of order $N$ it is possible to match, via a phase transition, many entirely different quantum systems. Represented in their respective Hilbert spaces $\mathcal{K}$ and sharing just their dimension $N < \infty$.

In [19] the idea is illustrated via its several closed-form realizations. Incidentally, all of these models happened to be unitary in a domain $\mathcal{D}$ of a shape resembling a (hyper)cube with protruded vertices. In a broader perspective one can say that by definition [1], the latter vertices are precisely the EP extremes of our present interest. In this light, our present paper could be briefly characterized as a study of the geometry of the generic unitarity-supporting domains of parameters, with particular emphasis on understanding of the sharply spiked shapes of their surfaces $\partial\mathcal{D}$ in a small vicinity of their EP vertices and edges. Indeed, we found such phenomenologically relevant features of the geometry mathematically remarkable and worth a dedicated study.

## 4. Real-world models and predictions.

### *4.1. Mathematics: Amended inner products and exceptional points.*

The main purpose of the introductory recollection of the 3HS formalism was to prepare a turn of attention to the key role played, in the 3HS applications, by the concept of exceptional points (EPs). Although their original rigorous definition may be already found in the old Kato's monograph on perturbation theory [1], their usefulness for quantum physics of unitary systems only started emerging after Bender with Boettcher pointed out, in their pioneering letter [9], that the EPs (also known as Bender-Wu singularities [4, 7]) could also acquire an immediate phenomenological interpretation of the points of quantum phase transition. Alternatively, their properties appeared relevant in the more speculative contexts of Calogero models and/or of supersymmetry [46].

From all of the similar 3HS-applicability points of view it is necessary to start the model-building processes from a preselected candidate for the Hamiltonian which is parameter-dependent, $H = H(\lambda)$. Moreover, it must be non-Hermitian in the auxiliary Hilbert space $\mathcal{K}$ and,, at the same time, properly Hermitian and self-adjoint in an "amended" Hilbert space of states $\mathcal{H}$. Now, the key point is that in the light of assumption (4), the latter space can be represented via a mere amendment (5) of the inner product in $\mathcal{K}$. In other words, *any* solution $\Theta = \Theta(H)$ of Eq. (4) *defines* the necessary physical space $\mathcal{H} = \mathcal{H}(H)$.

In opposite direction, *many* of the eligible and Hamiltonian-dependent metrics and spaces may and will *cease* to exist *before* the variable, path-specifying parameter $\lambda$ reaches the ultimate EP value $\lambda^{(EP)} \in \partial\mathcal{D}$. For the parameters lying inside the physical domain $\mathcal{D}$, the Hamiltonian must still be assigned such a specific metric $\Theta$ and space $\mathcal{H}$ which would exist *up to the required limit* of $\lambda \to \lambda^{(EPN)}$. In this sense, for any preselected quasi-Hermitian quantum system, our knowledge and specification of the boundary $\partial\mathcal{D}$ near EPNs are of an uttermost importance.

*4.2. Realistic many-body systems.*

In the latter setting we should return, once more, to Fig. 3 illustrating the sharply spiked, fragile, parameter-fine-tuning nature of the shape of the sample domain near its EPN extremes. Due to their potential phase-transition interpretation, these extremes seem to be the best targets of a realistic experimental search.

*4.2.1. Realistic systems inclined to support an approximate decomposition into clusters.* The invertible mapping $\Omega$ as mentioned in Fig. 4 connects the ket-vector elements of two Hilbert spaces: In the notation of Ref. [23] we have

$$|\psi\succ = \Omega\,|\psi\rangle\,, \qquad |\psi\succ \in \mathcal{L}\,, \qquad |\psi\rangle \in \mathcal{K}\,. \tag{9}$$

Recently it has been revealed that precisely the same mapping (attributed to Dyson [22]) also forms a mathematical background of the so called coupled cluster method (CCM, [47]). In fact, the implementation aspects of the latter, CCM interpretation of formula (9) were already used in calculations and tested, say, in quantum chemistry. What was particularly successful are the variational (or, more precisely, bi-variational) realizations of the CCM philosophy, with emphasis put upon the construction of ground states, and with a well-founded preference of mappings (9) in the exponential form $\Omega = \exp S$ where $S$ is represented in a suitable operator basis.

The latter, apparently purely technical restriction seems to be responsible for the success of the method which is currently "one of the most versatile and most accurate of all available formulations of quantum many-body theory" [48]. In paper [48], extensive 3HS-CCM parallels have been found. The respective strengths and weaknesses of the two approaches look mutually complementary. Currently [49], their further analysis is being concentrated upon the strengths. One may expect that the consequent, mathematically consistent 3HS quantum theory might enhance the range of applicability of the more pragmatic but very precise CCM ground-state constructions. Along these lines, in particular, the new theoretical predictions may be expected to concern the EP-related many-body quantum phase transitions which could be also, in parallel, experimentally detected.

*4.2.2. Bose-Hubbard model and its open- and closed-system interpretations.* The Bose-Hubbard Hamiltonian

$$H = \varepsilon\left(a_1^\dagger a_1 - a_2^\dagger a_2\right) + v\left(a_1^\dagger a_2 + a_2^\dagger a_1\right) + \frac{c}{2}\left(a_1^\dagger a_1 - a_2^\dagger a_2\right)^2 \tag{10}$$

of Graefe et al [50] has been developed to describe an $(N-1)$-particle Bose-Einstein condensate in a double well potential containing a sink and a source of equal strengths. Besides the usual annihilation and creation operators the definition contains the purely imaginary on-site energy difference $2\varepsilon = 2i\gamma$. In the fixed$-N$ representation the Hamiltonian is a matrix: At $N = 6$ we have, for example,

$$H^{(6)}(\gamma, c, v) = \left( \begin{array}{cccccc} -5i\gamma + \frac{25}{2}c & \sqrt{5}v & 0 & 0 & 0 & 0 \\ \sqrt{5}v & -3i\gamma + \frac{9}{2}c & 2\sqrt{2}v & 0 & 0 & 0 \\ 0 & 2\sqrt{2}v & -i\gamma + \frac{1}{2}c & 3v & 0 & 0 \\ 0 & 0 & 3v & i\gamma + \frac{1}{2}c & 2\sqrt{2}v & 0 \\ 0 & 0 & 0 & 2\sqrt{2}v & 3i\gamma + \frac{9}{2}c & \sqrt{5}v \\ 0 & 0 & 0 & 0 & \sqrt{5}v & 5i\gamma + \frac{25}{2}c \end{array} \right) . \quad (11)$$

Once we fix the inessential single particle tunneling constant $v = 1$ and once we localize the EPN singularity at $\gamma = 1$ and at the vanishing strength of the interaction between particles $c = 0$, we reveal, at any $N$, that after an *arbitrarily small* $c \neq 0$ perturbation, the spectrum *abruptly* ceases to be real (see *loc. cit.*). This means that the metric $\Theta$ and space $\mathcal{H}$ cease to exist, either. The perturbed system admits, exclusively, the non-unitary, open-system interpretation in $\mathcal{K}$.

In our present framework restricted to closed systems, only the parameters contained inside the suitable physical domain $\mathcal{D} = \{\gamma, c \,|\, \gamma \in (-1, 1)\,, c \in (c_{\min}(\gamma), c_{\max}(\gamma))\}$ (with the shape resembling, locally, Fig. 3 near its spikes) would be compatible with the reality of the energies. Interested readers may find an extensive study and detailed constructive description of the shape of such a unitarity compatible domain in our rather lengthy recent paper [51].

### *4.3. Generalized Bose Hubbard models*

Up to now we paid attention to the models (sampled by the Bose Hubbard Hamiltonian (10)) with the EPN singularities possessing the trivial geometric multiplicity $K = 1$ [1]. Interested readers may find, in paper [52], an introduction into a more general category of the EPs characterized by a clustered, $K-$centered degeneracy of the wave functions with $K > 1$. In these cases the EP-related quantum catastrophes (i.e., the generalized $\mathcal{PT}-$symmetry breakdowns) appeared to be of the form of confluence of several independent EPs with $K = 1$. The paper illustrated the advanced mathematics of the degeneracy of degeneracies via low-dimensional matrix models. The emergence of unusual horizons found its mathematical formulation in the language of geometry of Riemann surfaces, accompanied by the phenomenological predictions of certain anomalous phase transitions.

A model-independent analysis of these anomalies in the dynamical EP-unfolding scenarios was based, in subsequent paper [53], on their parametrization by the matrix elements of admissible (i.e., properly scaled and unitarity-compatible) perturbations. A consistency of algebra with the EP-related deformations of the Hilbert-space geometry has been confirmed. The new degenerate perturbation techniques were developed and their implementation has been found feasible. Via a class of schematic models, a constructive analysis of the vicinity of the simplest nontrivial EPN with $K = 2$ was performed.

An implementation of the schematic recipe to the Bose-Hubbard-type generalized models may finally be found described in [51]. It was shown that there always exists a non-empty unitarity domain $\mathcal{D}$ comprising a multiplet of variable matrix elements of the admissible perturbations for which the spectrum is all real and non-degenerate. The intuitive expectations were confirmed: the physical parametric domains near EPs were found sharply spiked.

A richer structure was revealed to characterize the admissibility of the perturbations. Two categories of the models were considered. In the first one the number of bosons was assumed conserved (leading to the matrix Hamiltonians of the form (11)). The alternative assumption of the particle-number non-conservation led to the realistic $K > 1$ scenarios in which the spectra also remain real and non-degenerate. The quantum evolution controlled by the Hamiltonians of larger (or even infinite) dimensions still remains unitary.

In all of these cases, in spite of a rapid increase of the complexity of the formulae with the number of particles, the existence as well as a sharply spiked structure of $\mathcal{D}$ near EPN has again been reconfirmed. The first steps of the explicit constructive analysis of the structure of $\mathcal{D}$ were performed in the simplest case with $N = 5$ where the access to EP5 appeared mediated by eight independently variable parameters.

*4.4. Further phenomenological challenges.*

The early abstract words of warning against the deceptive nature of the concept of quasi-Hermiticity [21, 29] were recently reconfirmed by the authors of paper [54]. After a detailed analysis of the popular non-Hermitian but $\mathcal{PT}-$symmetric imaginary cubic anharmonic oscillator these authors came to the conclusion that such a *"fons et origo"* of the theory can be characterized by the singular behavior attributed to an "intrinsic" EP. Subsequently, the latter concept has been very explicitly described and discussed in [43] and [55] (cf. also [56, 57, 58]). All of these studies and discoveries contributed to the motivation of our present study since they enhanced the importance of the knowledge of the behavior of the 3HS closed-system models at parameters lying close to their EP limits.

Another, independent source of interest in the study and explicit description of the domains $\mathcal{D}$ of the unitarity-compatible "admissible" parameters in the close vicinity of EPs may be seen in the frequently experimentally observed phenomenon of the avoided level crossings. In a way sampled by Figure 3 this phenomenon occurs even in the spectra of finite-dimensional Hermitian matrices.

The related, highly desirable analytic continuation of the spectra towards their EP degeneracies is by far not an easy task. The task is intimately connected with the 3HS-inspired turn of attention to the description of quantum dynamics using non-Hermitian Hamiltonians. This opens multiple technical questions. One of them is that after one perturbs a quasi-Hermitian Hamiltonian or even only its parameter, $H(\lambda) \to H(\lambda')$, one immediately encounters the re-emergence of the well known ambiguity of the Hilbert-space inner product in Eq. (5) [24, 42, 59]. As a byproduct of this observation there appeared a need of a deep and thorough reformulation of perturbation theory itself [45], with nontrivial consequences concerning, in particular, the systems lying close to the boundary $\partial\mathcal{D}$.

Besides the technical open questions there also exists a number of the strong parallel challenges emerging in the context of quantum phenomenology. Their truly prominent samples emerged in the context of quantum cosmology and, in particular, in the attempted descriptions

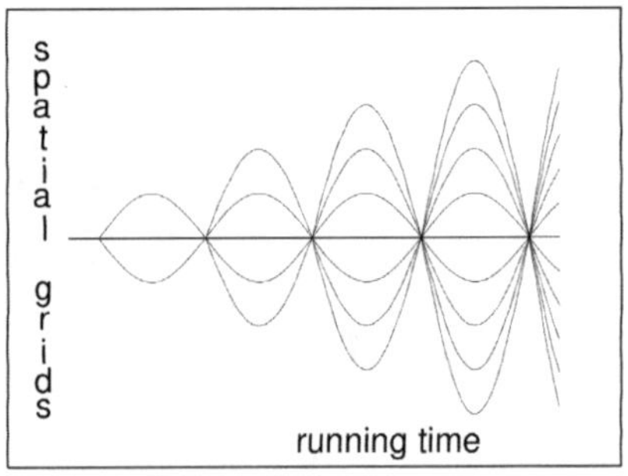


**Figure 5.** Schematic 3HS picture of the Universe evolving through a sequence of Eons separated by EPs. Sampled by "breathing" one-dimensional $N-$point grids with $N_1 = 2$, $N_2 = 4$, etc.

of the evolution of the Universe shortly after its initial Big Bang singularity. The key point is that the classical-theory-supported existence of Big Bang seems to contradict the conventional quantum-theoretical paradigm of Hermitian theory. By the latter theory the Big-Bang-type phase transitions cannot exist, being "smeared" and reduced to the mere avoided crossing behavior of the spatial coordinates called Big Bounce of the Universe [60].

A disentanglement of the puzzle could be, in principle, offered by the 3HS models in which the Big Bang would correspond to a real EP-related spectral singularity of a suitable non-Hermitian operator (cf., e.g., [61]). Such a hypothesis would admit even a highly speculative "evolutionary cosmology" pattern of Fig. 5 in which a sequence of penrosian Eons separated by the Big-Crunch/Big-Bang singularities would render the structure of the "younger" Universes richer and more sophisticated.

## 5. Exceptional-point-mediated quantum phase transitions.

### *5.1. EPs as quantum crossroads.*

In paper [19] we emphasized that a classification of passages of closed quantum systems through their EP singularities could be perceived as a quantum analogue of the classical catastrophe theory [62]. In this context let us only add that the EP-mediated phase transitions could acquire the form of a quantum process of bifurcation,

initial phase, $t < 0$,
Hamiltonian $H^{(-)}(t)$

$\Downarrow$

**process of degeneracy**

$\Downarrow$

| $t > 0$ branch A, Hamiltonian $H^{(+)}_{(A)}(t)$ | **option A** $\Longleftarrow$ | the EP "crossroad", indeterminacy at $t = 0$ | **option B** $\Longrightarrow$ | $t > 0$ branch B, Hamiltonian $H^{(+)}_{(B)}(t)$ |
|---|---|---|---|---|

Thus, in principle, the future extensions of our present models might even incorporate a multiverse-resembling branching of evolutions at $t = 0$.

Marginally, let us add that in such a branched-evolution setting one could find applications even for some results on non-unitary, spectral-reality-violating evolutions. An illustration may be found in papers (sampled by [63]) where just the search for the EP degeneracies has been performed without any efforts of guaranteeing the reality of the spectrum.

### *5.2. Perturbation theory near EPs using nonstandard unperturbed Hamiltonians.*

At the above-mentioned "cross-road" EP instant $t = 0$ the Hamiltonian ceases to be diagonalizable. This means that such an instant can be perceived as a genuine quantum analogue of the classical Thom's bifurcation singularity *alias* catastrophe [62]. The distinguishing feature of the phenomenon in its quantum form is that it is "instantaneously" incompatible with the postulates of quantum theory. Fortunately, the theory returns in full force at any, arbitrarily small time before or after the catastrophe.

In [19], several explicit and strictly algebraic, solvable-model illustrations of such a passage through the EPN singularity may be found described in full detail. Alternatively, the phenomenon can be also described in a model-independent manner. Indeed, a return to the diagonalizability can be characterized as a perturbation of the non-diagonalizable $t = 0$ Hamiltonian $H_{(EP)}$. Thus, any multiplet of states $|\vec{\psi}(t)\rangle$ can be constructed, before or after $t = 0$, as the solution of a properly perturbed Schrödinger equation

$$(H_{(EP)} + \lambda\, W)|\vec{\psi}\rangle = \epsilon\, |\vec{\psi}\rangle\,. \tag{12}$$

One has to keep in mind that the unperturbed Hamiltonian itself is an anomalous operator, the conventional diagonalization (or, more generally, spectral representation) of which does not exist. Once we have to consider here just its finite-dimensional matrix forms, the constructive approach to Eq. (12) can be based on the evaluation of the so called transition matrices $Q_{(EP)}$, defined as solutions of the Schrödinger-like linear algebraic equation

$$H_{(EP)}Q_{(EP)} = Q_{(EP)}J_{(JB)}(E_{(EP)})\,.$$

The symbol $J_{(JB)}(E_{(EP)})$ denotes here the canonical representation of $H_{(EP)}$. Once we decide to choose it in the most common Jordan-matrix form, the related transition matrices $Q_{(EP)}$ can be reinterpreted as an analogue of the unperturbed basis. In this basis, the perturbed Schrödinger equation (12) acquires the canonical form

$$[J_{(JB)}(E_{(EP)}) + \lambda\, V]|\vec{\phi}\rangle = \epsilon\,|\vec{\phi}\rangle\,. \tag{13}$$

Interested readers are recommended to consult Refs. [44] and [51] for the further details of the solution of such an equation.

For our present purposes the essence of the latter technicalities may be explained using the elementary unperturbed real-matrix Hamiltonians of Ref. [19],

$$H^{(2)}_{(EP)} = \left[\begin{array}{cc} 1 & 1 \\ -1 & -1 \end{array}\right], \qquad H^{(3)}_{(EP)} = \left[\begin{array}{ccc} 2 & \sqrt{2} & 0 \\ -\sqrt{2} & 0 & \sqrt{2} \\ 0 & -\sqrt{2} & -2 \end{array}\right], \ \ldots\,.$$

For this series of examples all of the transition matrices are non-unitary but known in closed form. At $N = 3$ one gets

$$Q^{(3)}_{(EP)} = \left[\begin{array}{ccc} 2 & 2 & 1 \\ -2\sqrt{2} & -\sqrt{2} & 0 \\ 2 & 0 & 0 \end{array}\right]$$

etc. As long as the lack of space does not allow us to reproduce here the further details, let us redirect the readers to paper [45] (in which some overall conceptual features of the EP-related perturbation approximation construction are described) and to paper [53] (in which the more complicated EPs with geometric multiplicity greater than one are taken into consideration).

Out of the most essential conclusions of the latter two studies let us pick up the single and apparently obvious fact (still not observed, say, in Refs. [29, 43]) that the class of admissible, operationally meaningful perturbations must not violate the self-adjointness of the Hamiltonian in the correctly reconstructed physical Hilbert space $\mathcal{H}$.

### *5.3. Constructions based on the differential Schrödinger equations.*

In the year 1998 Bender with Boettcher discovered the existence of the *real* (i.e., in principle, experimentally *accessible*) EPs generated by certain *local* and non-Hermitian but parity-time symmetric ($\mathcal{PT}-$symmetric) potentials [9]. The EPs were interpreted as instants of the spontaneous breakdown of $\mathcal{PT}-$symmetry. Their reality was unexpected because for the conventional local potentials the EPs are never real [4].

Among the specific studies of the non-Hermitian but $\mathcal{PT}-$symmetric differential Schrödinger equations $H\,\psi = E\,\psi$ a distinguished position belongs to paper [64] by Dorey et al who considered the angular-momentum-spiked oscillator Hamiltonians

$$H(M,L,A) = -\frac{d^2}{dx^2} + \frac{L(L+1)}{x^2} - (ix)^{2M} - A\,(ix)^{M-1}\,, \quad M = 1,2,\ldots\,, \quad L, A \in \mathbb{R} \tag{14}$$

in which the "coordinate" $x$ lied on a suitable *ad hoc* complex contour. They showed that inside a suitable domain $\mathcal{D}$ of parameters these Hamiltonians generate the strictly real bound-state-like spectra. These authors were the first to describe the shape and role of the boundaries $\partial\mathcal{D}$ formed by the EPs. Unfortunately, they did not make the picture complete because they did not construct the corresponding physical inner products.

*5.4. Harmonic oscillator.*

Once one restricts attention to the most elementary choice of $M = 1$ yielding the one-parametric harmonic-oscillator Hamiltonian $H(1, L, A) = H^{(HO)}(L)$, the model becomes exactly solvable at all real $L \in \mathbb{R}$ [65]. For this reason the HO domain of unitarity $\mathcal{D}^{[HO]}$ has an elementary, multiply connected form of a "punched" interval with EPs (i.e., with elements $L^{(EP)}$ of boundary $\partial\mathcal{D}^{[HO]}$) excluded,

$$\mathcal{D}^{[HO]} = \left(-\frac{1}{2}, \frac{1}{2}\right) \bigcup \left(\frac{1}{2}, \frac{3}{2}\right) \bigcup \left(\frac{3}{2}, \frac{5}{2}\right) \bigcup \ldots .$$

This property (cf. Fig. 6) enabled us to pay more attention, in paper [66], to one of the

♡

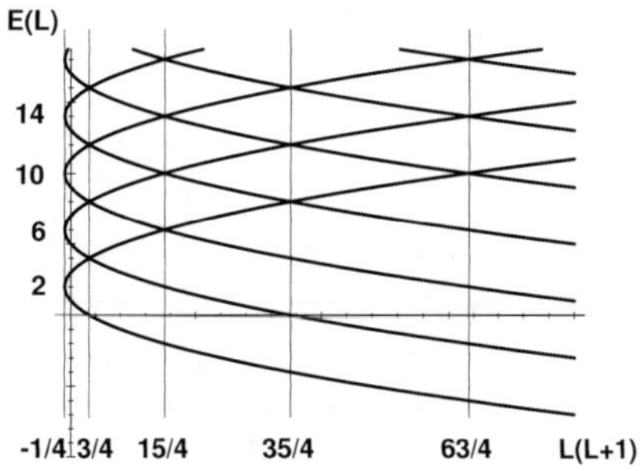


**Figure 6.** Spectrum of Hamiltonian (14) at $M = 1$.

key challenges connected with the theory, viz., to the constructive analysis of the practical consequences of the nontriviality and of the ambiguity of the related angular-momentum-dependent metrics $\Theta = \Theta(L)$. Our main result was the construction of a complete menu of the infinite-parametric assignments $H \to \Theta(H)$ of an eligible metric to the Hamiltonian.

The very possibility of doing so makes the HO model truly unique. For technical as well as phenomenological reasons we restricted our attention just to the parameters $L$ which lied close to the points of the boundary of the domain of the unitarity, i.e., not far from the set of EPs

$$\partial\mathcal{D} = \left\{-\frac{1}{2}, \frac{1}{2}, \frac{3}{2}, \ldots\right\} . \quad (15)$$

The basic technical ingredient in the construction of the metrics (see its details as well as the rather long explicit formulae in [66]) was twofold. Firstly, the availability of the closed-form diagonalization of $H^{(HO)}(L)$ enabled us to replace the Hamiltonian, at any one of its EP limits, by an equivalent matrix called canonical or Jordan-block representation. Thus, at $L^{(EP)} = -1/2$, for example, such a representation has the elementary block-diagonal form

$$J^{(-1/2)}_{(EP)} = \left( \begin{array}{cc|cc|cc} 2 & 1 & 0 & 0 & 0 & \ldots \\ 0 & 2 & 0 & 0 & 0 & \ldots \\ \hline 0 & 0 & 6 & 1 & 0 & \ldots \\ 0 & 0 & 0 & 6 & 0 & \ldots \\ \hline 0 & 0 & 0 & 0 & 10 & \ldots \\ \vdots & \vdots & \vdots & \vdots & \ddots & \ddots \end{array} \right) + \text{corrections}\,.$$

Secondly, the highly nontrivial fact that all of the unavoided energy-level crossings occurred pairwise and simultaneously led to the decomposition of the metric-determining relation $H^\dagger(L)\Theta(L) = \Theta(L)\,H(L)$ (cf. Eq. (4)) to a set of its finite-dimensional (in fact, two-by-two) matrix components numbered by the separate degenerate energies $E^{(EP)} = 2, 6, 10, \ldots$ .

In such a setup, every value $L^{(EP)} = -1/2, 1/2, 3/2, \ldots$ may be perceived as an instant of a quantum phase transition which involves all levels at once. In a way accounting, in an exhaustive manner, for the non-uniqueness, the one-parametric ambiguity of every two-by-two submatrix of $\Theta(L)$ (once more, recall Eq. (8) for illustration) contributes, independently, to the ultimate infinite-parametric ambiguity of selection of the physics-determining inner product in the infinite-dimensional physical Hilbert space $\mathcal{H}^{(HO)}$.

## 6. Summary.

The Dyson's traditional 3HS recipe of Fig. 4 was based on a "de-Hermitization" interpretation $\mathfrak{h} \to H$ of Eq. (6). This means that in the realistic (i.e., typically, many-body) quantum systems eligible for such a treatment, the failure of a conventional numerical evaluation of predictions based on the use of conventional $\mathfrak{h}$ (as encountered in papers [20, 22]) found its not quite expected remedy in an apparently counterintuitive transition to a non-selfadjoint Hamiltonian $H$.

At present the flowchart of the whole theoretical pattern is usually inverted. In the manner reported in [22] the construction of innovative models starts from a non-selfadjoint candidate $H$ for the Hamiltonian [9]. Then, one has to guarantee the theoretical consistency of the model in a way prescribed by the diagram

input:
non–Hermitian $H$ with real spectrum,
unphysical but friendly Hilbert space $\mathcal{K}$

$\longrightarrow$

output:
metric $\Theta = \Omega^\dagger\Omega$ (s. t. $H^\dagger\Theta = \Theta H$),
optimal physical Hilbert space $\mathcal{H}$

$\nearrow\!\!\swarrow$ **equivalent predictions**

reference:
textbook Hamiltonian $\mathfrak{h} = \Omega H \Omega^{-1} = \mathfrak{h}^\dagger$ .
unfriendly physical Hilbert space $\mathcal{L}$

The model-building process is initiated by the choice of a *bona fide* Hamiltonian $H$ which is defined and non-Hermitian in auxiliary space $\mathcal{K}$. The theory is then based on an exact or approximate re-Hermitization of $H$ via $\Theta$. In multiple applications, the subsequent (and, often, difficult) reference to the lower-case Hamiltonian $\mathfrak{h}$ or to the map of Eq. (6) is usually made just in an approximate manner. Fortunately, the task can often be facilitated via the variability of parameters in $H = H(\lambda)$ so that the necessary knowledge of the physical domain $\mathcal{D}$ of the admissible values of these parameters is often successfully realized.

In applications, such a form of the 3HS formalism is to be kept user-friendly, offering reasonably calculable predictions. Besides the expected technical friendliness, an equally important merit of the 3HS formalism may be seen in an emerging access to new and unusual phenomena.

In our present selection, all of the phenomena under consideration were characterized by the proximity of EPs, treated as forming the boundary $\partial\mathcal{D}$ of the domains of "acceptable" *alias* "physical" (i.e., unitarity-compatible) parameters of the model in question. We reviewed and slightly extended several recent related results. From the abstract methodical point of view we put emphasis upon the suitability and amendments of the necessary (although, sometimes, less usual) perturbation-type construction techniques. This enabled us to clarify several counterintuitive facts characterizing the behavior of the closed quantum systems in the small vicinities of EPs of higher orders.

As one of our main conclusions the readers should remember the fact that in the vicinity of EPs, the technique of perturbations offered one of the most efficient tools of the parametrization and classification of the "admissible" (i.e., unitarity-preserving) multiparametric closed-quantum-system Hamiltonians $H(\lambda) = H[a(\lambda), b(\lambda), \ldots, z(\lambda)]$.

**Keywords**
exceptional points; quasi-Hermitian quantum theory; perturbations; quantum catastrophes;

**Acknowledgments**
Work supported by the Excellence project PřF UHK 2020 of the University of Hradec Králové.

## References


[1] Kato T 1966 *Perturbation Theory for Linear Operators* (Berlin: Springer-Verlag)
[2] Moiseyev N 2011 *Non-Hermitian Quantum Mechanics* (Cambridge: CUP)
[3] Heiss W D, Mller M and Rotter I 1998 *Phys. Rev.* E **58** 2894
[4] Bender C M and Wu T T 1969 *Phys. Rev.* **184** 1231
[5] Moiseyev N and Friedland S 1980 *Phys. Rev.* **A 22** 618
[6] Heiss W D and Sannino A L 1991 *Phys. Rev.* **A 43** 4159
[7] Alvarez G 1995 *J. Phys. A: Math. Gen.* **28** 4589–4598
[8] Znojil, M 2020 *Ann. Phys. (NY)* **413** 168050
[9] Bender C M and Boettcher S 1998 *Phys. Rev. Lett.* **80** 5243–5246
[10] Bender C M 2007 *Rep. Prog. Phys.* **70** 947–1018
[11] Mostafazadeh A 2002 *J. Math. Phys.* **41** 6343
[12] Guenther U and Stefani F 2005 *Czech. J. Phys.* **55** 1099
[13] Guenther U, Rotter I and Samsonov B F 2007 *J. Phys. A: Math. Theor.* **40** 8815
[14] http://www.nithep.ac.za/2g6.htm (accessed 2018 Jan 28)
[15] Znojil M and Růžička F 2019 *J. Phys. Conf. Ser.* **1194** 012120
[16] Znojil M 2007 *J. Phys. A: Math. Theor.* **40** 4863–4875
[17] Znojil M 2007 *J. Phys. A: Math. Theor.* **40** 13131–13148
[18] Znojil M 2019 *Phys. Rev.* A **100** 032124
[19] Znojil M 2020 *Proc. Roy. Soc.* A **476** 20190831.
[20] Dyson F J 1956 *Phys. Rev.* **102** 1217
[21] Dieudonné J 1961 *Proc. Int. Symp. Lin. Spaces* (Oxford: Pergamon) pp 115–122
[22] Scholtz F G, Geyer H B and Hahne F J W 1992 *Ann. Phys. (NY)* **213** 74–101
[23] Znojil M 2009 *Symm. Integ. Geom. Methods and Applications* **5** 001
[24] Mostafazadeh A 2010 *Int. J. Geom. Meth. Mod. Phys.* **7** 1191–1306
[25] Bagarello F, Gazeau J P, Szafraniec F H and Znojil M 2015 *Non-Selfadjoint Operators in Quantum Physics: Mathematical Aspects* ed F Bagarello, J-P Gazeau et al (Hoboken, NJ: John Wiley & Sons)
[26] Christodoulides D and Yang J K 2018 *Parity-time Symmetry and Its Applications* (Singapore: Springer)
[27] Bender C M 2018 *PT Symmetry in Quantum and Classical Physics* (Singapore: World Scientific).
[28] Stone M H 1932 *Ann. Math.* **33** 643–648
[29] Trefethen L N and Embree M 2005 *Spectra and Pseudospectra* (Princeton, NJ: PUP)
[30] Antoine J P and Trapani C 2015 *Non-Selfadjoint Operators in Quantum Physics: Mathematical Aspects*, ed F Bagarello, J-P Gazeau et al (Hoboken, NJ: John Wiley & Sons) pp 345–402
[31] Znojil M 2008 *Phys. Rev. D* **78** 085003
[32] Fring A and Moussa M H Y 2016 *Phys. Rev. A* **93** 042114
[33] Langer H and Tretter Ch 2004 *Czech. J. Phys.* **54** 1113–1120
[34] Albeverio S and Kuzhel S 2015 *Non-Selfadjoint Operators in Quantum Physics: Mathematical Aspects*, ed F Bagarello, J-P Gazeau et al (Hoboken, NJ: John Wiley & Sons) pp 293–344
[35] Mostafazadeh A 2003 *Class. Quantum Grav.* **20** 155
[36] Znojil M 2017 *Ann. Phys.* (NY) **385** 162–179
[37] Bender C M and Milton K A 1997 *Phys. Rev.* D **55** R3255
[38] Andrianov A A, Ioffe M V, Cannata F and Dedonder J P 1999 *Int. J. Mod. Phys.* A **14** 2675
[39] Buslaev V and Grecchi V 1993 *J. Phys. A: Math. Gen.* **26** 5541
[40] Jones H F and Mateo J 2006 *Phys. Rev.* D **73** 085002.
[41] Mostafazadeh A 2004 *Ann. Phys.* (NY) **309** 1
[42] Krejčiřík D, Lotoreichik V and Znojil M 2018 *Proc. Roy. Soc. A: Math. Phys. Eng. Sci.* **474** 20180264
[43] Krejčiřík D, Siegl P, Tater M and Viola J 2015 *J. Math. Phys.* **56** 103513
[44] Znojil M 2018 *Phys. Rev.* A **97** 032114
[45] Znojil M 2020 *Entropy* **22** 000080
[46] Znojil M 2020 *Symmetry* **12** 892
[47] Čížek J 1966 *J. Chem. Phys.* **45** 4256
[48] Bishop R F and Znojil M 2020 *Eur. Phys. J. Plus* **135** 374
[49] Znojil M, Bishop R F and Veselý P 2021 in preparation
[50] Graefe E M, Günther U, Korsch H J and Niederle A E 2008 *J. Phys. A: Math. Theor.* **41** 255206
[51] Znojil M 2020 *Proc. Roy. Soc. A: Math., Phys. & Eng. Sci.* A **476** 20200292
[52] Znojil M and Borisov D I 2020 *Nucl. Phys.* B **957** 115064
[53] Znojil M 2020 *Symmetry* **12** 1309
[54] Siegl P and Krejčiřík D 2012 *Phys. Rev.* D **86** 121702(R)
[55] Guenther U and Stefani F 2019 "IR-truncated PT-symmetric $ix^3$ model and its asymptotic spectral scaling graph" (arXiv:1901.08526)
[56] International conference "Mathematical Aspects of Physics with Non-Self-Adjoint Op-

erators: 10 Years After", CIRM, Marseille, 1-5 February, 2021, https://www.cirm-math.fr/Schedule/screen_display.php?id_renc=2153
[57] Semorádová I 2021 "Spectral approximation of Schrdinger operators with complex potentials: diverging eigenvalues", conference [56], videorecorded talk: https://bbb1.cirm-math.fr/playback/presentation/2.0/playback.html?meetingId=b81c5b6372740ae284eca117b09dbdc9cc6e84c7-1612163540943
[58] Guenther U 2021 "Asymptotic spectral scaling graphs for IR-truncated $mathcalPT-$symmetric model Hamiltonians with $V = -(ix)^{2n+1}$ potentials and beyond", conference [56]
[59] Znojil M, Semorádová I, Růžička F, Moulla H and Leghrib I 2017 *Phys. Rev. A* **95** 042122
[60] Ashtekar A, Pawlowski T and Singh P 2006 *Phys. Rev.* D **74** 084003
[61] Znojil M 2016 *Non-Hermitian Hamiltonians in Quantum Physics*, ed F Bagarello, R Passante and C Trapani (Cham: Springer) pp 383–399
[62] Zeeman E C 1977 *Catastrophe Theory-Selected Papers 1972-1977* (Reading: Addison-Wesley)
[63] Znojil M 2019 *Ann. Phys.* (NY) **405** 325–339
[64] Dorey P, Dunning C and Tateo R 2001 *J. Phys. A: Math. Gen.* **34** 5679–5703
[65] Znojil M 1999 *Phys. Lett.* A **259** 220–223
[66] Znojil M 2020 *Sci. Reports* **10** 18523

# AUTHOR INDEX